VERTEBRATE DISSECTION

ninth edition

VERTEBRATE DISSECTION

ninth edition

Dominique G. Homberger
Professor of Biological Sciences
Louisiana State University and A&M College
Baton Rouge, Louisiana

Warren F. Walker, Jr.
Emeritus Professor of Biology
Oberlin College
Oberlin, Ohio

THOMSON
BROOKS/COLE
Australia • Canada • Mexico • Singapore • Spain
United Kingdom • United States

THOMSON

BROOKS/COLE

Publisher: Peter Marshall
Acquisitions Editor: Nedah Rose
Development Editor: Laura Kenney
Assistant Editor: Christopher Delgado
Editorial Assistant: Jennifer Keever
Technology Project Manager: Travis Metz
Marketing Manager: Ann Caven
Marketing Assistant: Sandra Perin
Advertising Project Manager: Linda Yip
Project Manager, Editorial Production: Belinda Krohmer
Print/Media Buyer: Judy Inouye
Permissions Editor: Kiely Sexton
Production Service: Graphic World Publishing Services
Text Designer: Kim Menning
Copy Editor: Betty Litt
Illustrator: Karen Westphal
Cover Designer: Gary Palmatier
Cover Printer: CTPS
Compositor: Graphic World, Inc.
Printer: CTPS

Printed in China
2 3 4 5 6 7 07 06

For more information about our products, contact us at:
Thomson Learning Academic Resource Center
1-800-423-0563
For permission to use material from this text, contact us by:
Phone: 1-800-730-2214
Fax: 1-800-730-2215
Web: http://www.thomsonrights.com

Library of Congress Control Number: 2003102213

ISBN 0-03-022522-1

Brooks/Cole—Thomson Learning
10 Davis Drive
Belmont, CA 94002
USA

Asia
Thomson Learning
5 Shenton Way #01-01
UIC Building
Singapore 068808

Australia/New Zealand
Thomson Learning
102 Dodds Street
Southbank, Victoria 3006
Australia

Canada
Nelson
1120 Birchmount Road
Toronto, Ontario M1K 5G4
Canada

Europe/Middle East/Africa
Thomson Learning
High Holborn House
50/51 Bedford Row
London WC1R 4LR
United Kingdom

Latin America
Thomson Learning
Seneca, 53
Colonia Polanco
11560 Mexico D.F.
Mexico

Spain/Portugal
Paraninfo
Calle/Magallanes, 25
28015 Madrid, Spain

To our many students, past and present,
some of whom are now professors in their own right.

PREFACE TO THE NINTH EDITION

New to the Ninth Edition

During the past decade since the publication of the 8th edition of this book, functional and evolutionary morphology as a biological discipline has continued to grow vigorously and has generated new insights into comparative vertebrate anatomy. At the same time, the experiences in the comparative anatomy course of one of us (DGH) have enabled us to undertake what is perhaps the most extensive revision of this book since its publication in 1954. No part has been unaffected by this revision, and most parts have been tested extensively by students involved in the laboratory sessions and special research projects.

New dissection approaches that combine minimal destruction with optimal preservation of the three-dimensional relationships of organs and structures have been introduced. Numerous passages have been completely rewritten to facilitate comprehension. Several sections have benefitted from original research undertaken by one of us (DGH).

The quality of many figures was improved by making extensive modifications or by completely redrawing them. Numerous new figures were created by a professional scientific illustrator, Karen Westphal, in close collaboration with one of us (DGH) to ensure anatomical accuracy. Other figures were drafted by us or modified from the literature and subsequently professionally executed by C.H. Wooley. We also used figures from the companion textbook, *Functional Anatomy of the Vertebrates* by Liem, Bemis, Walker, and Grande (2001), to help illustrate context and background information. Finally, we used color in all parts of the book to clarify structures and concepts.

In the sections on mammals, the description and dissection instructions for the rabbit were greatly expanded. The number of figures representing the skeletal elements of the rabbit was significantly increased. Whenever the anatomical configurations differ between cats and rabbits, the descriptive sentences were identified by introductory phrases, such as "In rabbits . . ." or "In cats . . . ." The more comprehensive treatment of both the cat and the rabbit provides instructors with the choice of dissecting either mammal species, or both species at the same time to compare representatives of carnivorous and herbivorous mammals.

A major innovation in this new edition is the inclusion of materials that provide a comparative context and background information to provide a rationale for the dissections and to add interest to the dissection exercises. These materials have been placed in boxes called "Anatomy in Action" that are clearly set off from the dissection instructions so as not to distract from them. Whenever appropriate, we have referred to recent studies and novel ideas to demonstrate the vibrancy and excitement of contemporary comparative anatomy and its relevance to ecology, evolution, and the health sciences.

The novel design of this new edition further facilitates a "pick-and-choose," modular approach for the treatment of comparative anatomy. Each chapter is color-coded to facilitate cross-checking of materials. As in earlier editions, we introduce each organ system, and often each major group of animals (fishes, amphibians, mammals), with relevant information "in a nutshell" on the embryonic development of the system, the division of the system into its parts, and the evolutionary history of the system.

The new edition also introduces a two column format, which will make the material easier to read, especially in a laboratory setting in which students necessarily shift their attention between the text and the specimens.

Philosophical and Didactic Approach

The philosophical and didactic approach to morphology in the 9th edition remains the same as in previous editions. The central theme of this book is the study of the body constructions, or *bauplans,* of representatives of major chordate groups, such as ancestral chordates, jawless vertebrates, jawed fishes, amphibians, and mammals. Some reptilian representatives are included in the chapters on the skeleton. In the case of the eyeball, brain, and heart, the instructions relate to organs of the sheep, because they are difficult to study in smaller specimens, such as cats and rabbits. An understanding of the morphology of these chordates and vertebrates will provide students with a firm basis to enable them to infer and mentally visualize the morphological and functional transformations that were responsible for the evolution of the various vertebrate classes. It will also provide students with an appreciation for the diversity of chordates and vertebrates.

Instructions for dissections are included for a hemichordate and two ancestral chordates to provide a broader evolutionary context for the study of vertebrates. In a second chapter, the anatomy of the lamprey *(Petromyzon)* is presented as a model of a jawless fish and to introduce students to the basic organization of vertebrates. The next ten chapters are organized by organ systems, and each chapter traces the evolutionary transformations of a particular organ system by dissecting representative vertebrates from three levels of organization. The spiny dogfish *(Squalus acanthias)* is used as a model for a jawed, but lungless fish; the mudpuppy *(Necturus maculosus)* as a model for a tetrapod at the threshold between the bauplan of a fish and that of a fully terrestrial tetrapod; and the cat *(Felis catus)* or rabbit *(Oryctolagus cuniculus)* as models of carnivorous or herbivorous mammals, respectively.

Flexibility and modularity has been a major goal throughout this book, because instructors of comparative anatomy courses differ in their approaches, interests, and time constraints. Many instructors use a systems approach by following the evolution of each organ system separately across the various vertebrate classes, and our book has been organized with this approach in mind. But other instructors follow an organismal approach by emphasizing the structural, functional, and mechanical coherence of each organism and by studying particular species one after the other. Such an approach is easily accommodated by our book because of its clear organization and color-coded chapters. Furthermore the detailed Table of Contents and Index will allow instructors to select and rearrange the materials, as well as create a syllabus as a guide for the students.

Our book includes enough material for an intensive course, but the instructions are written in such a manner that individual dissections can be omitted or replaced by demonstrations. For example, the dissections of the ancestral chordates and lamprey can be omitted or replaced by demonstrations of previously dissected specimens, so that the course can concentrate on the shark, mudpuppy, and one of the mammal species. In some courses, only the shark and one of the mammal species are dissected and studied in detail. If only a limited amount of time is available, selected parts of particular organ systems can be omitted, such as the muscles of the hind limb, certain sense organs, the lumbar plexus, or the lymphatic system. Regardless of what may have to be omitted or dissected in a particular course, all parts are described thoroughly and coherently because, in our opinion, it is a better educational experience for a student to dissect a limited amount of material well than a lot of material superficially.

The background and contextual information in the "Anatomy in Action" boxes is not meant to replace a textbook. These materials provide "in a nutshell" the needed theoretical background for understanding an organism and its parts, or places the anatomical observations within a broader ecological and evolutionary context. Some of the materials explore constructional principles of anatomy; others discuss how the configuration of structures is molded by physical and other constraints imposed on them by the organism and its environment. Several "Anatomy in Action" boxes trace the embryonic development of structures to provide the necessary background to understand the anatomical organization of adult organisms. If students study these materials in advance of a laboratory session, their dissections will be more meaningful to them, and instructors will have more time at their disposal to help with the dissection and to provide additional information.

In describing mammals, we continue to favor the use of standardized terms agreed upon in anatomical codes, such as the *Nomina Anatomica Veterinaria* and the more recently published *Terminologia Anatomica* for human anatomy. The latter also provides Anglicized terms, and we have adopted these whenever appropriate and, if necessary, modified them according to the former as a guide for quadrupedal mammals. We have applied these anatomical terms to non-mammalian vertebrates when appropriate, but fishes and amphibians possess some unique structures that require special terms. We present a list of the more common word roots in Appendix A. Studying this will help students understand the derivation of anatomical terms and will help fix them in mind, see also "A Note on the Handling of Specimens" in Appendix B.

Acknowledgments

As always, we are very much indebted to the outstanding staff of our publishers for all of the assistance and encouragement that they have given us. We have worked particularly closely with our Development Editors, first with Melanie Cann at Harcourt, Brace & Jovanovich and subsequently with Laura Kenney, a freelance editor. Belinda Krohmer, our Project Editor, and Michael McConnell of Graphic World Publishing Services, have skillfully guided us through the complexities of the production phase of the book. We are especially grateful to Nedah Rose, our Acquisitions Editor, for her support and encouragement throughout the process. A full list of contributors to the publishing process appears on the copyright page.

Individual parts of this book have been reviewed by William E. Bemis, University of Massachusetts at Amherst; Kurt Schwenk, University of Connecticut; Cheryl Wilga, University of Rhode Island; and J. H. Youson, University of Toronto at Scarborough. We are grateful to them for having taken time from their busy schedules to provide us with constructive feedback. Their suggestions have been very helpful, but we take responsibility for any errors that remain. Students and other colleagues have given us additional suggestions, and we hope that they will continue to call our attention to any errors or deficiencies. Suggestions may be sent to us by mail or via the web page of the publisher.

As always, Warren Walker is grateful to his wife, Tensy Walker, for her support and help in checking many parts of this book.

Dominique G. Homberger
Warren F. Walker, Jr.

CONTENTS

CHAPTER 12

A NOTE TO THE STUDENT: ANATOMICAL TERMINOLOGY

When studying anatomy for the first time, you will encounter many unfamiliar terms that must be mastered because effective communication is facilitated by their use. For example, the term "biceps brachii muscle" is much simpler to use to refer to the muscle in question than to circumscribe it as "the muscle that originates by two heads on the proximal end of the bone of the upper arm and the raven beak-shaped process of the shoulder blade in human beings and inserts on the proximal end of the medial bone of the lower arm." Some understanding of the derivation of these anatomical terms will help to fix their meaning and spelling in your mind. Most are based on Latin or Greek roots, and, as you become familiar with the more common roots, you will recognize the terminology for the shorthand that it is. The root *chondro—,* for example, always means "cartilage," and it is used in many combinations: Chondrichthyes (cartilaginous fish), chondrocranium (cartilaginous braincase), perichondrium (connective tissue surrounding cartilage), or chondrocyte (cell within cartilage). The more frequently and important roots used in anatomy and for the names of organisms are presented in Appendix A. Referring to this list from time to time will help you master anatomical terminology. There are also certain conventions, described below, that make the terminology more rational.

Terms for Structures

Anatomists first described structures by their appearance in human beings because human anatomy was their primary concern. Over the centuries a plethora of names have been proposed, and many structures have a long list of synonyms. To bring some order out of this chaos, human anatomists have periodically agreed upon codes of terminology, the most recent being the *Terminologia Anatomica (International Anatomical Terminology).* Anatomical terms are in Latin or Greek, but they usually are translated into the vernacular for each language. Veterinary anatomists have agreed upon a *Nomina Anatomica Veterinaria* in which they have brought most of the terms for domesticated mammals into close agreement with the terminology for human beings. The major differences between veterinary and human anatomical terms concern those for direction and orientation, which differ between a biped and quadruped. Avian anatomists have agreed upon a *Nomina Anatomica Avium,* and many of these terms are applicable to reptiles.

In preparing this edition we have used terms from the *Terminologia Anatomica* and *Nomina Anatomica Veterinaria* for most mammalian structures. We have applied these terms or terms from the *Nomina Anatomica Avium* to other vertebrates where this appears to be reasonable, but many fishes posses unique structures that require special terms. Favored terms are placed in boldface when first used for each animal. In a few cases in mammals where our favored term is not the same as the scientific term, the latter is provided in parentheses and italics, e.g., **kidney** *(ren).* "Ren" is not used in English as a noun, but it is helpful to know the term because it forms the basis for the adjective "renal." We have avoided introducing synonyms unless they are in common use, in which case the synonym is given in regular type, e.g., **cleidobranchialis muscle** (clavodeltoid). Eponyms, or terms derived from a person's name, are usually avoided in formal anatomical terminology, but we have included some of them that are in common use as synonyms, e.g., **auditory tube** (Eustachian tube).

Terms for Directions, Planes, and Sections

Terms for Direction

Many terms for orientation and direction are the same in human and comparative anatomy, but there are certain differences due to the bipedal posture of human beings. A structure located toward the head of a quadruped in relation to a point of reference is described as **cranial** (e.g., cranial vena cava); one toward the tail, as **caudal** (Fig. N-1). The term **rostral** replaces "cranial" within the head and indicates a location toward the tip of the snout in relation to the point of reference. Comparable directions of structures within human beings are described as **superior** (e.g., superior vena cava) and **inferior.**

A structure toward the back of a quadruped in relation to a point of reference is described as **dorsal**; one toward the belly, as **ventral.** The terms "dorsal" and "ven-

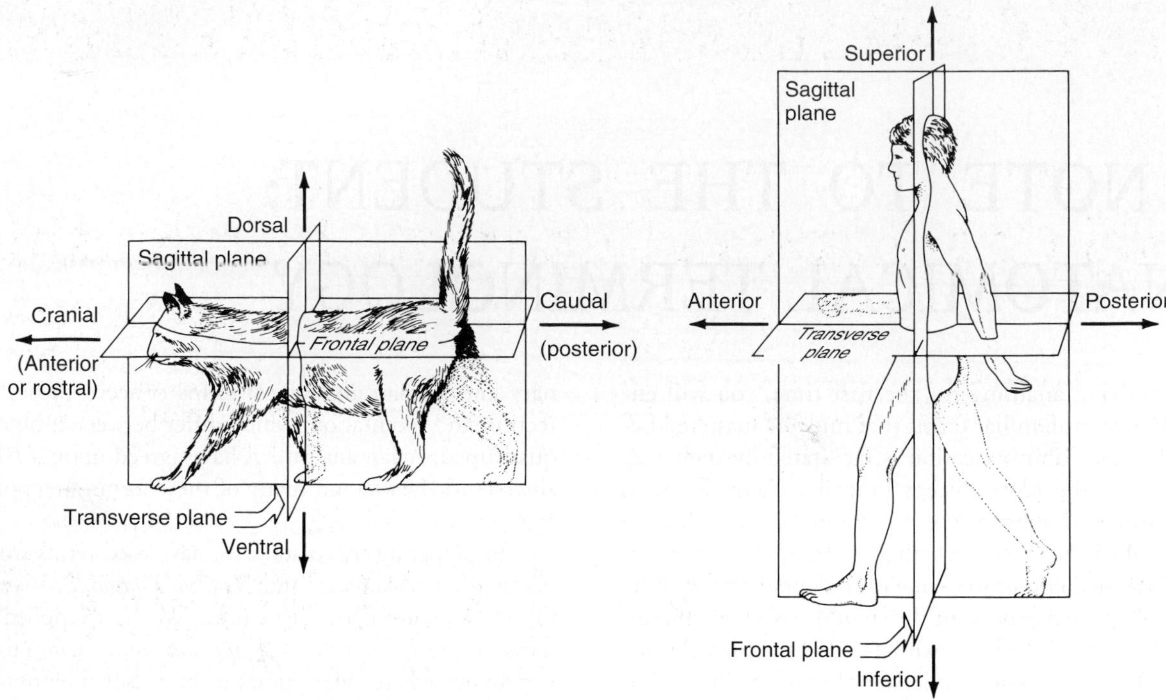

Figure N-1

Diagrams showing the planes of the body and the differences in the terms for direction in a quadruped and in a human being.

tral" are not used in human anatomy; comparable directions are described as **posterior** and **anterior.** The codes recommended that the terms "anterior" and "posterior" not be used in a quadruped for the cranial and caudal locations within the body, and usually we have followed these recommendations except for well-established terms for certain organs (e.g., posterior cardinal veins in a fish).

Other terms for locations and directions of structures are used in the same way in all animals. **Lateral** refers to the side of the body in relation to the point of reference; **medial,** to a position toward the midline. **Median** is used for a structure in the midline. **Distal** refers to a part of some organ, such as an appendage or blood vessel, that is farthest removed from the center of the body or the origin of a blood vessel; **proximal,** to the opposite end of the structure, i.e., the part nearest the point of reference.

Left and **right** are self-evident terms, but it should be emphasized that, in anatomy, these terms refer to the *specimen's* right or left, regardless of the way the specimen is viewed by the observer.

Adverbs may be formed from the above adjectives by adding the suffix **-ly** or **-ad** to the root, in which case the term implies motion in a given direction. To say that a structure extends caudally, or caudad, means that it is moving toward the caudal end of the body.

Planes and Sections of the Body

The body of a specimen is sectioned in various planes to reveal the location and spatial relationships of internal organs. In a quadruped, a longitudinal, vertical section which passes through the median, longitudinal axis of the body is a **median** or **sagittal section** (Fig. N-1). Such a section lies in the **median** or **sagittal plane.** Sections or planes parallel with, but lateral to, the medial plane are described as **parasagittal.**

A section cut across the body from dorsal to ventral, and at right angles to the longitudinal axis, is a **transverse** or **cross section,** and it lies in the **transverse plane.**

A **frontal** (coronal) **section** or **plane** is one lying in the longitudinal axis, and passing horizontally from side to side.

Basal Chordates and Chordate Relationships

Chapter 1

Although the focus of this book is on the anatomy of representative vertebrates, we will understand them better if we review briefly their relationships to other animals. First of all, vertebrates belong to a group or clade[1] known as the **Craniata** (Fig. 1-1). Craniates are animals with a well-formed head, including a brain surrounded by cartilages and/or bones that form a braincase or **cranium.** Nearly all craniates also are **vertebrates** because they have at least an incipient backbone or vertebral column composed of a series of **vertebrae.** Among craniates, only the marine jawless hagfishes lack any trace of vertebrae, so technically they are not vertebrates.

Distinctive Features of Chordates

Craniates are a subphylum of a larger clade, the phylum **Chordata** (Fig. 1-1), which includes about 45,000 species ranging from small, soft-bodied sea squirts through the large assemblage of vertebrates. Ancestral chordates evolved five distinctive morphological characters related to their unique methods of feeding and locomotion. These features are retained in modified form in later derived species, or at least in their larvae or embryos. We call such an assemblage of features **synapomorphies** (Gr. *syn-* "together" + *morph-* "shape," "structure") because they are morphological structures that define or hold a clade together. Differences in the degree of expression of these five synapomorphies are important criteria used to characterize the three subphyla of Chordates: **Tunicata, Cephalochordata,** and **Craniata** (Fig. 1-1).

Basal chordates, like many other invertebrates, are filter feeders that trap food particles from the surrounding seawater. Their unique adaptations for collecting plankton and detritus form the basis for two of their synapomorphies: the endostyle and pharyngeal slits. Food particles in a stream of water passing through the pharynx are trapped in a sheet of mucus secreted by a distinctive **endostyle,** a longitudinal groove in the floor of the pharynx that is lined with glandular and ciliated cells. Some endostyle cells also bind iodine to protein and are homologous to the secretory cells of the vertebrate thyroid gland. Excess water that enters the pharynx with the food escapes through a series of **pharyngeal slits** (Fig. 1-2). In craniates, gills develop in the pharynx between the pharyngeal slits, so the pharynx also acquires a respiratory function, but this is not the case in the basal chordates. The pharyngeal slits may open directly to the outside, as they do in craniates, or first into a surrounding atrial chamber, as they do in basal chordates. Adult terrestrial vertebrates lack pharyngeal slits, but their embryos have pharyngeal pouches. An endostyle and pharyngeal slits evolved in ancestral chordates and continue in some form in their descendants.

Three additional synapomorphies are related to the pattern of locomotion of basal chordates: a postanal tail, a notochord, and an unpaired, dorsal, tubular nerve cord (Fig. 1-2). A **postanal tail** is an extension of the body caudal to the anus, which increases the length of the body that participates in locomotion. It is not found in earthworms or other nonchordates because their anus is at the very caudal end of the body. The **notochord** is a hydroskeleton of liquid-filled cells that is encased in a firm sheath of connective tissue. It forms an incompressible yet bendable rod of cells that prevents the body from being compressed and shortening when longitudinal muscle fibers in the trunk and tail contract. Contraction of these fibers results in side-to-side bending of the trunk and tail that brings about swimming movements rather than a telescoping of the trunk as occurs in an earthworm. The **nerve cord** of chordates is quite different from the paired, ventral, solid nerve cords of annelids and arthropods. It integrates the activities of longitudinal muscle fibers in the trunk and tail.

Chordate Relationships

Chordates share with echinoderms (sea stars, sea urchins, and related species), hemichordates (acorn worms), and several minor phyla a few features that place them into a larger

[1]A *clade* is an assemblage of related animals that includes their common ancestor and all of the descendants of this ancestor. Clades range in size from a group of related species (a genus or subgenus) to a phylum or group of related phyla.

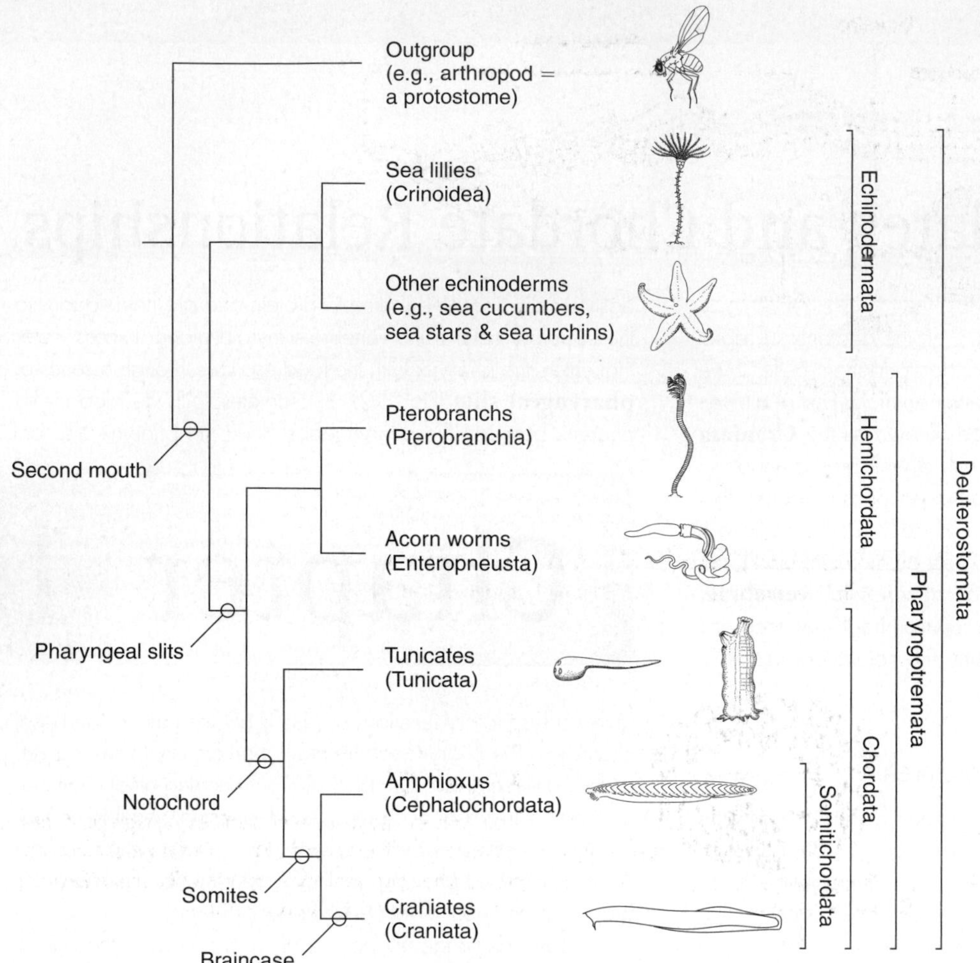

Figure 1-1
Cladogram of the evolution of craniates. Cladograms depict relationships by a series of branching points, or nodes, where groups diverge. Each node is characterized by the evolution of one or more synapomorphies, one of which usually forms the basis for the new clade name. Notice that clades form a series of nested groups. One distinctive synapomorphy is named for each of the five nodes in the evolution of craniates from a remote ancestor. Clades are an outgrowth of the Linnaean system of grouping organisms into phyla, classes, orders, families, genera, and species. These terms are used, but our increasing knowledge of evolution has generated more levels of relationships than can be described in a handful of taxonomic categories. *(From Liem, Bemis, Walker, and Grande, 2001.)*

clade, the **Deuterostomata** (Fig. 1-1). A brief excursion into embryology is needed to understand this relationship. The gastrula stage of the embryo of nearly all animals has an opening, the **blastopore,** which leads into the embryonic gut cavity or **archenteron** (Fig. 1-3). In mollusks, annelid worms, arthropods, and related coelomates, the blastopore becomes the adult mouth or contributes to it. These phyla are placed in the clade **Protostomata** because the first opening into the archenteron becomes the adult mouth. In deuterostomes, a new or second mouth invaginates into the archenteron, and the blastopore becomes the anus or closes and lies near the anus. Protostomes and deuterostomes also differ in patterns of embryonic cleavage, mesoderm and coelom formation, and larval morphologies. All of these appear to be fundamental differences.

PHYLUM HEMICHORDATA

Before studying the chordates, we will examine briefly the hemichordates because they appear to be closely related to chordates. The approximately 90 species of contemporary

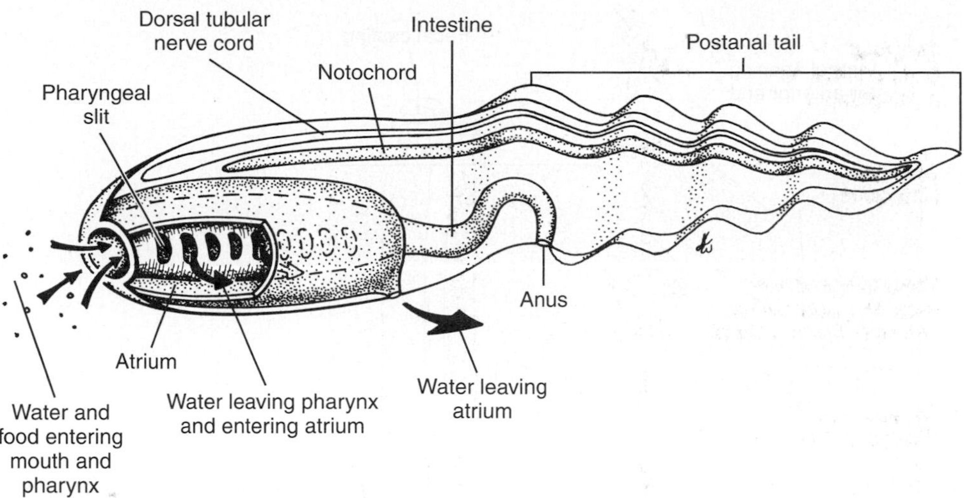

Figure 1-2

Diagram of a hypothetical generalized chordate based loosely on a larval tunicate. Four of the five distinctive features of chordates are shown. The fifth, the endostyle, is a longitudinal groove within the floor of the pharynx.

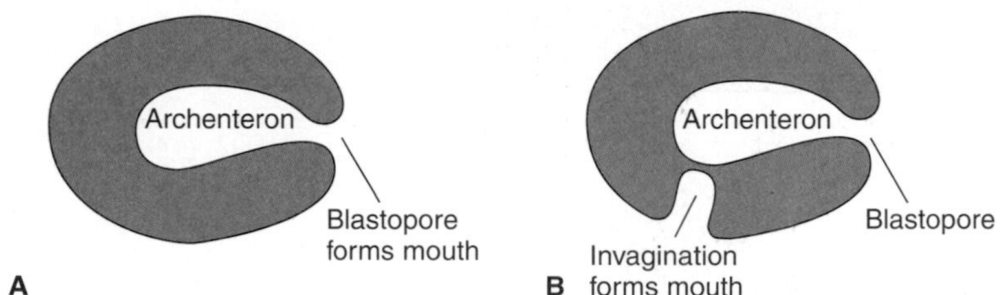

Figure 1-3

Diagrams in lateral view of sections through gastrulae showing the differences in the way the mouth forms in protostomes (**A**) and deuterostomes (**B**). *(From Liem, Bemis, Walker, and Grande, 2001.)*

hemichordates include a few very small, colonial, deep-sea species known as pterobranchs, as well as wormlike enteropneusts, or acorn worms, which are more abundant and range in length from a few centimeters to 2 meters. We will look at a representative of the class Enteropneusta. **Enteropneusts** burrow in the sand and mud of tidal flats and shallow coastal waters. *Balanoglossus* and *Saccoglossus* are common genera along North American coasts.

Examine the external features of one of the enteropneusts (Fig. 1-4**A**). You may have to use low magnification to see certain structures. The body is divided into three distinct regions: an anterior proboscis, a collar, and a long trunk. The **proboscis** attaches by a narrow stalk to the encircling collar just posterior to it. In some species, the proboscis nesting in the collar resembles an acorn in its cup, hence the common name acorn worm. The proboscis, which is covered with cilia, and the collar assist the enteropneust in burrowing. The collar coelom (Fig. 1-4**B**) fills with seawater, thus inflating the collar and anchoring the worm in its burrow. The deflated proboscis is lengthened and pushed forward into the substratum by the action of circular muscle fibers within it; its coelom is filled and anchors the worm. Finally, the collar coelom is emptied, and the worm is pulled forward toward the proboscis.

The proboscis is extended out of the burrow when the animal is feeding, and its cilia carry minute food particles that fall on it into the **mouth.** The mouth is located inside the collar just ventral to the proboscis stalk (Fig. 1-4**B** and **C**). Food is carried into the ventral part of the pharynx, and excess water enters the dorsal part of the pharynx to escape through the numerous pharyngeal slits. The two parts of the pharynx are partially separated by a pair of longitudinal folds. Look for the paired **external pharyngeal slits** dorsolaterally on the anterior part of the trunk. As many as 150 pairs have been counted on a 40-cm specimen. Each external slit connects with the pharynx through a pharyngeal pouch and an unusual U-shaped **internal pharyngeal slit** that is formed by the downward growth of a **tongue bar.** At one stage of development the pharyngeal slits of amphioxus (a cephalochordate) have an identical appearance. Food continues from the ventral part of the pharynx into a simple straight **intestine.** Materials that are not digested and absorbed are eliminated

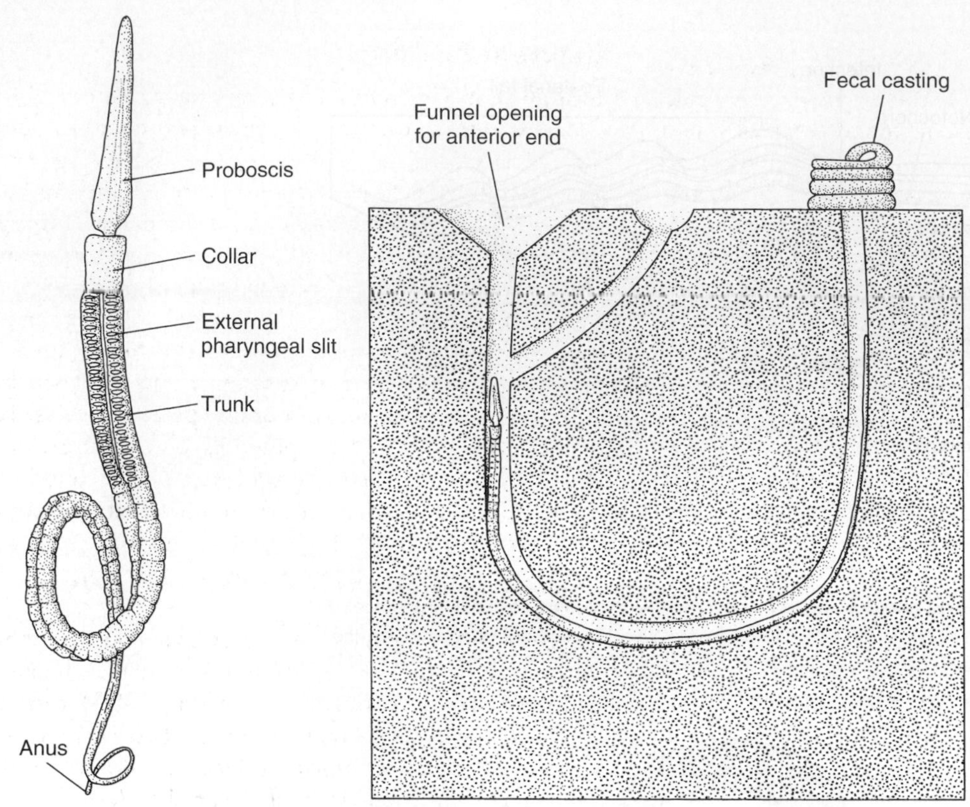

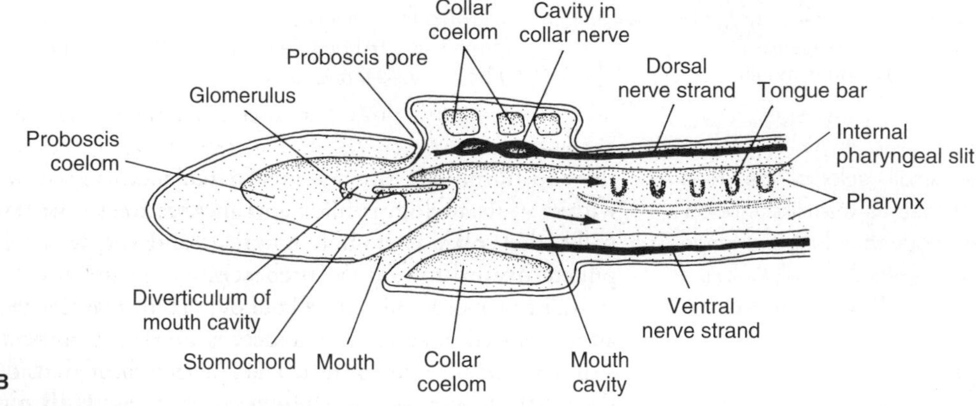

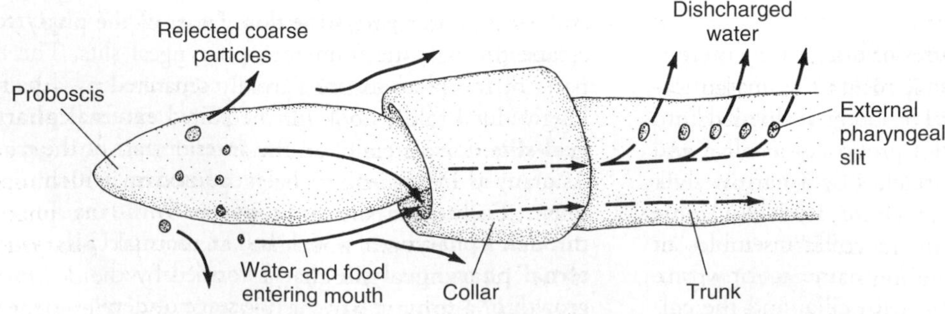

Figure 1-4

A, External view of *Saccoglossus* and a drawing showing how it lives in a burrow in the sea floor. B, Sagittal section of the anterior end. C, Cilia on the proboscis sort fine from coarse particles and direct the flow of water carrying fine particles to the mouth. *(From Dorit, Walker, and Barnes, 1991.)*

through a terminal **anus** and form distinctive fecal castings on the sea floor (Fig. 1-4**A**). There is no postanal tail.

Prominent **genital folds** can be found in sexually mature specimens just ventral to the caudal external pharyngeal slits and extending a short distance caudad. The sexes are separate but cannot be distinguished externally. In some species conspicuous **hepatic ridges** will be seen caudal to the genital ridges. The hepatic ridges are the external manifestation of digestive glands, the **hepatic caeca** that bud off the intestine. Notice a longitudinal middorsal ridge and a similar midventral ridge on the trunk. Each contains a superficial and solid **nerve strand** that extends forward into the collar (Fig. 1-4**B**). An echinoderm-like subepidermal neuron plexus connects with the nerve strands. Gas exchange and excretion occur through the body wall, but a knot of capillaries, known as the **glomerulus** (Fig. 1-4**B**), protrudes into the collar coelom and may have an excretory function.

Hemichordates share pharyngeal slits with chordates, but not the other chordate synapomorphies. Food is collected by mucus secreted on the surface of the proboscis and not by an endostyle within the pharynx. Hemichordates have a stiffening rod called the **stomochord** in the proboscis (Fig. 1-4**B**), but it is doubtful that this is homologous to a notochord. Hemichordates do get their name, however, from this *"half notochord."* Dorsal and ventral solid nerve strands are quite different from the chordate nerve cord. There is no postanal tail. Because they share only pharyngeal slits with chordates, hemichordates are not considered chordates; however, the presence of pharyngeal slits suggests an evolutionary affinity, so hemichordates and chordates can be placed together in the clade **Pharyngotremata** (Fig. 1-1).

PHYLUM CHORDATA, SUBPHYLUM TUNICATA

We begin our study of chordates by examining a representative of the subphylum **Tunicata** (also called Urochordata). All tunicates are marine, and most are encased in a leathery membrane called the tunic, which gives them their common name. There are approximately 2,000 species. About 100 species are pelagic, but the most familiar tunicates are the sessile sea squirts belonging to the class **Ascidiacea.** They attach to submerged objects in coastal waters or occasionally lie partly buried in the sand or mud. Many sea squirts are colonial, but some are solitary. It is easy to study the anatomy of solitary sea squirts, such as *Molgula,* which are abundant along the Atlantic coast.

External Features

Examine a specimen of *Molgula* in a pan of water and notice its saclike appearance. Notice the two spoutlike openings, or **siphons,** near the top of the animal (Fig. 1-5**A** and **B**). Water enters the organism through the topmost aperture **(incurrent siphon)** and is discharged through the opening that is set off on one edge **(excurrent siphon).** The margin of the incurrent siphon bears six small tentacles. Also notice the external covering, or **tunic.** In addition to some protein, the tunic contains many fibers composed of **tunicin,** a mucopolysaccharide similar to the cellulose found in plant cell walls. Tunicin is secreted by the epidermis of the underlying body wall. Minute hairlike processes extend out from the tunic and help to anchor the animal to its substrate. The lower part of the tunic may have sand grains adhering to the hairs, because *Molgula* often lies partly buried in the sand.

Although the shape of the animal appears rather asymmetrical, the sea squirt has a modified bilateral symmetry. If you compare a larva (Fig. 1-5**C**) with an adult, you will notice that the region between the two siphons represents the dorsal surface, and the rest of the edge of the adult represents the ventral surface. The incurrent siphon lies anteriorly and the excurrent siphon posteriorly.

Dissection

To expose the inside of the animal, cut through the tunic beneath the excurrent siphon and extend the cut around the base of the sac to a point near the incurrent siphon. Reflect the tunic, observing that it is attached to the rest of the body only at the siphons. Detach the tunic at these points. A number of bundles of longitudinal and circular muscle fibers lie within the thin body wall, or **mantle,** and aid in expelling water. The mantle is nearly transparent, and many of the internal organs can be seen through it. To see them more clearly, carefully peel off the mantle without unduly injuring organs that may adhere to it. This part of the dissection should be done with the specimen submerged in water.

Study the dissection and compare it with Figures 1-5**A** and **B.** You may need low magnification to see certain structures. The incurrent siphon leads into a large, thin-walled, vascular **pharynx,** which occupies most of the inside of the body. Many longitudinal bars, between which lie microscopic **pharyngeal slits,** characterize the pharynx wall of *Molgula.* To see the slits clearly, remove a piece of the pharynx wall, prepare a wet mount of it, and view it through a microscope. The slits appear as arclike clefts arranged in spirals. Although they cannot be seen in this type of preparation, the blood vessels that enable gas exchange

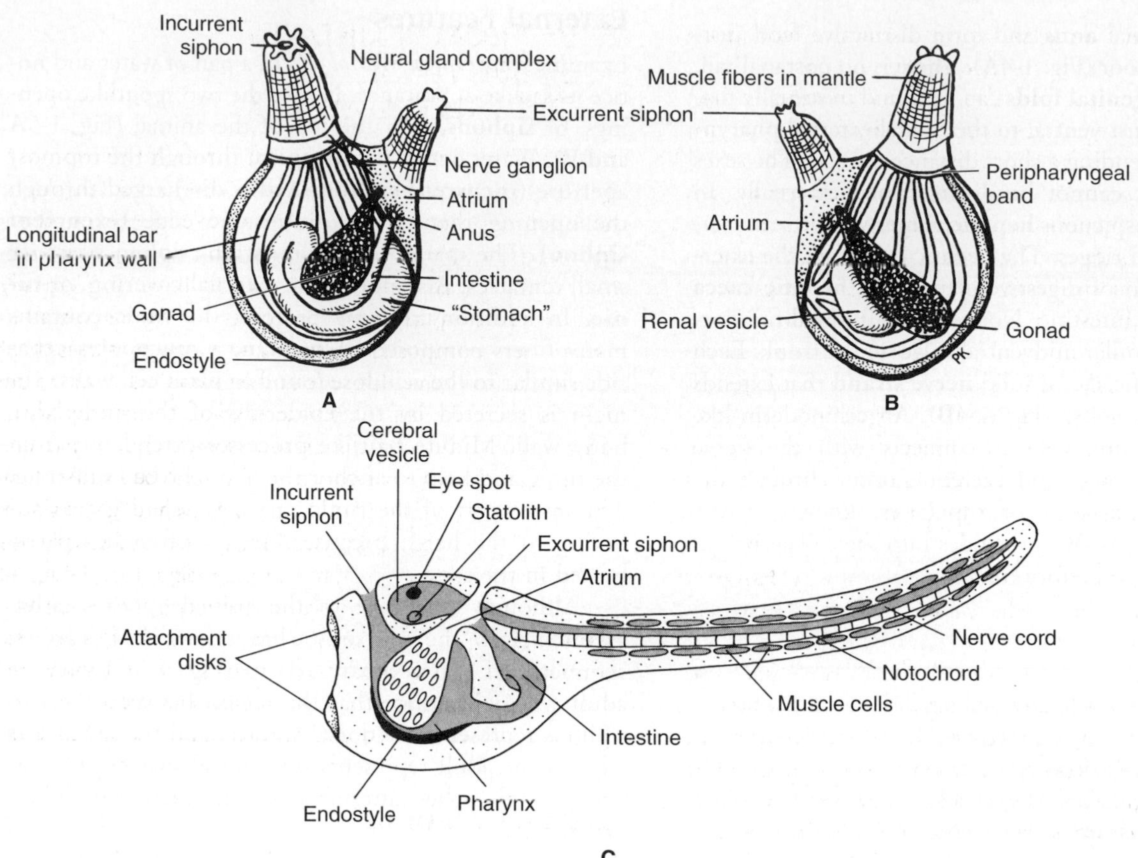

Figure 1-5

A and B, Left and right sides, respectively, of a dissection of *Molgula*. The entire tunic and most of the mantle have been removed. C, Diagrammatic lateral view of a larval ascidian.

with the environment are contained within the bars between the slits; gills are not present. The bars are covered with cilia, which create the current of water that passes through the animal. The pharyngeal slits do not lead directly to the outside but into a delicate chamber, the **atrium,** located on each side of the pharynx. The lateral portions of the atrium may not be seen, but they converge posteriorly to form a more conspicuous median atrial chamber, which opens to the surface through the excurrent siphon.

The fold along the ventral surface of the pharynx is the **endostyle.** Some of its cells are ciliated; some are glandular and secrete mucus that entraps minute food particles in the water; and some produce iodinated proteins, as do the cells of the vertebrate thyroid gland. The food-containing mucus band is moved toward the incurrent siphon by the cilia. Near the anterior end of the pharynx it is carried to the dorsal side by lateral **peripharyngeal bands,** which may be difficult to see. The mucus band moves posteriorly to the esophagus along a middorsal fold called the **dorsal lamina.** The pharynx of the sea squirt is primarily a food-gathering device, although some gases are exchanged between the blood and environment through its walls.

The rest of the digestive tract of *Molgula* lies on the left side of the pharynx. A short **esophagus** leads from the posterior end of the pharynx to a slightly expanded portion of the gut, sometimes called the **"stomach,"** and this is followed by the **intestine** proper. The esophagus, stomach, and first part of the intestine form a C-shaped loop. The intestine then doubles on itself, goes back beside the stomach and esophagus, and opens at the **anus** into the median portion of the atrial chamber. There is no stomach in the vertebrate sense. Little food is stored in this "stomach." The stomach, together with minute glandular folds evaginated from it, secretes enzymes that act upon carbohydrates, fats, and proteins. The intestine appears to be primarily absorptive.

Sea squirts reproduce asexually by budding; they also reproduce sexually. In sexually mature individuals, large **gonads** will be seen. *Molgula* has one gonad on each side of the pharynx. Inconspicuous **genital ducts** lead from the gonads to the median portion of the atrium. Tunicates are hermaphroditic but generally not self-fertilizing.

An oval **renal vesicle** lies ventral to the right gonad. Although some concretions of uric acid accumulate in the vesicle and stay there until the animal dies, most nitrogen is excreted as ammonia and is lost by diffusion through the pharynx wall. The significance of the vesicle is not understood.

A small oval-shaped structure, the **neural gland complex,** is seen on the dorsal edge of the pharynx between the two siphons. It consists of a group of dense cells that are connected by a duct and coelomic funnel (probably not visible) to the pharynx cavity. Early investigators compared the group of dense cells to part of the vertebrate pituitary gland, but studies by Ruppert (1990) show that these cells are phagocytic and not secretory. Ruppert further demonstrated that sea water is drawn into the ciliated funnel, down the duct, and across the group of dense phagocytic cells of the neural gland complex, and then enters the blood in adjacent pharyngeal blood vessels. The phagocytic cells clear the sea water of minute particles. Ruppert postulates that the neural gland complex restores blood volume, which may be reduced when the body wall contracts vigorously and squirts water from the siphons. Carefully pull off the neural gland complex, and you will see beneath it an elongated **nerve ganglion.** Other internal organs are not usually seen in this type of dissection.

Apart from the chordate feeding apparatus, which includes the endostyle and abundant pharyngeal slits, little in the structure of an adult sea squirt suggests a chordate. The locomotor chordate synapomorphies are present only in the free-swimming, tadpole-shaped larva (Fig. 1-5**C**), which does not feed but moves about for a short time in order to disperse the population. The larval tail, which lies posterior to the anus, is its locomotive organ. It contains muscle fibers, a notochord, and the distinctive unpaired, dorsal, tubular nerve cord. [The alternate term for the subphylum, Urochordata, derives from the restriction of the notochord to the tail (Gr. *oura*, "tail").] The nerve cord expands anteriorly into a cerebral vesicle containing receptors for equilibrium and light. These enable the larva to find a suitable substratum on which to settle and attach by its anterior end. The postanal tail is lost at metamorphosis and the cerebral vesicle is reduced to the nerve ganglion.

PHYLUM CHORDATA, SUBPHYLUM CEPHALOCHORDATA

The subphylum **Cephalochordata** includes the lancelet or amphioxus (*Branchiostoma*[2]) and two related genera. About 45 species of superficially fishlike animals belong to the subphylum Cephalochordata. These animals have an extremely long notochord that extends forward beyond the nerve cord to the very front on the animal. The extreme forward extension of the notochord is correlated with the burrowing habits of the animal and gives the name Cephalochordata to the subphylum. These animals are found in shallow waters in tropical and temperate oceans throughout the world. They can be locally abundant; one fishery in China had an annual catch of 35 tons! Amphioxus, like the sea squirt, is a filter feeder and usually lies partly buried with only its front end protruding (Fig. 1-6**A**). At times it swims to new feeding sites, but its locomotion is not very efficient because stabilizing fins are poorly developed. Because amphioxus shows all the chordate synapomorphies very clearly, we will examine this organism in more detail.

External Features

Examine a preserved specimen of amphioxus submerged in a pan of water. Mature specimens range in length from 4 to 6 cm. You will need low magnification to see certain structures. The shape of amphioxus is streamlined, or fusiform, being elongate, flattened from side to side (compressed), and pointed at each end (Fig. 1-6**B**). Segmental, V-shaped muscle blocks, the **myomeres,**[3] can be seen through the transparent epidermis. The apex of each V points cranially. Note that the myomeres extend nearly the entire length of the body. The lines of separation between the myomeres are connective tissue partitions called **myosepta.**

A **dorsal fin** extends along the top of the body, and a **ventral fin** can be seen beneath the caudal quarter of the animal. The dorsal and ventral fins are continuous around the tail and expand slightly in this region to form a **caudal fin.** A pair of ventrolateral fins, or **metapleural folds,** continues forward from the rostral end of the ventral fins (see Fig. 1-8).

[2]Although *Branchiostoma* (Costa, 1834) has priority over *Amphioxus* (Yarrel, 1896) as the technical name, the term amphioxus is so familiar that it is customary to retain it, at least as the common name.

[3]*Myotome* and *myomere* are terms used for muscle segments. Although often used synonymously, *myotome* is an embryonic muscle segment that has differentiated from a somite, and *myomere* is an adult muscle segment.

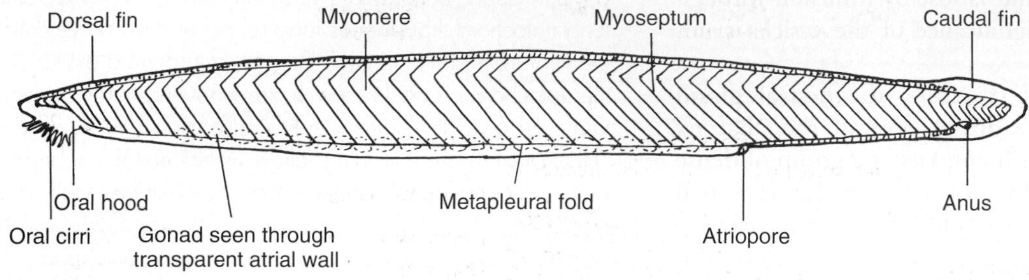

Figure 1-6
A, Amphioxus in its habitat. B, Lateral view of an adult amphioxus.

The metapleural folds end a short distance from the front of the body. In front of the metapleural folds and ventral to the cranial few myomeres, you will see a transparent chamber called the **oral hood.** The mouth is located deep within this chamber and will not be seen at this time, but the opening of the oral hood on the ventral surface can be seen. It is fringed with small tentacles called **oral cirri** (singular: **cirrus**), which often are folded across its opening. The oral cirri contain chemoreceptive cells, and they aid in excluding large material, permitting only water and small food particles to enter.

Water that enters the pharynx passes through pharyngeal slits into an atrial chamber whose opening **(atriopore)** you will see between the caudal ends of the paired metapleural folds. The intestine opens by an **anus** located on the left side of the caudal fin, so there is a short **postanal tail.**

If the specimen is mature, you will see on each side a row of whitish oval **gonads.** They lie just ventral to the myomeres in the branchial region of the body. Their number ranges from about 20 to 35, varying slightly with the species. The gametes are discharged directly into the atrium. The sexes are separate.

Whole Mount Slide

Study a stained microscope slide of a small specimen of amphioxus under the low power of the microscope (Fig. 1-7). Note its fusiform shape, and find the structures described earlier: myomeres; dorsal, ventral, and caudal fins; metapleural folds; oral hood; oral cirri; atriopore; and anus. The myomeres have been cleared to render them somewhat transparent, so they will not be seen as plainly as in the preserved specimen, but you should at least see indications of them, just ventral to the dorsal fin.

Observe that small transparent blocks called **fin ray boxes** support the dorsal and ventral fins. In order to see the **metapleural folds,** you will have to focus sharply on the surface of the specimen. Each will appear as a horizontal line, parallel and slightly dorsal to the ventral edge of the body.

Because the small specimens used to make slides are not sexually mature, the **gonads** will not be fully developed or may even be absent. If present, they will appear as a row of lightly staining, oval structures close to the ventral edge of the body.

Notice the **notochord** located in the back, dorsal to the dark-staining **digestive tract.** It extends nearly the entire length of the animal in the general position of the vertebral column of vertebrates (Anatomy in Action 1-1).

The single, dorsal, tubular **nerve cord** appears as a dark-staining band lying dorsal to the notochord. The dark pigment granules along its ventral edge may help you recognize its position. Each granule represents part of a simple **photoreceptor.** Notice that the granules are particularly numerous near the front of the animal. The nerve cord ends in a blunt point rostrally. A distinct expanded brain is not present, but cell groups have been identified in the cranial end of the neural tube that appear to be homologous to groups of cells found in the brain of craniates (Butler, 2000; Holland and Holland, 1999). A prominent pigment spot can be seen in front of the nerve cord. It is

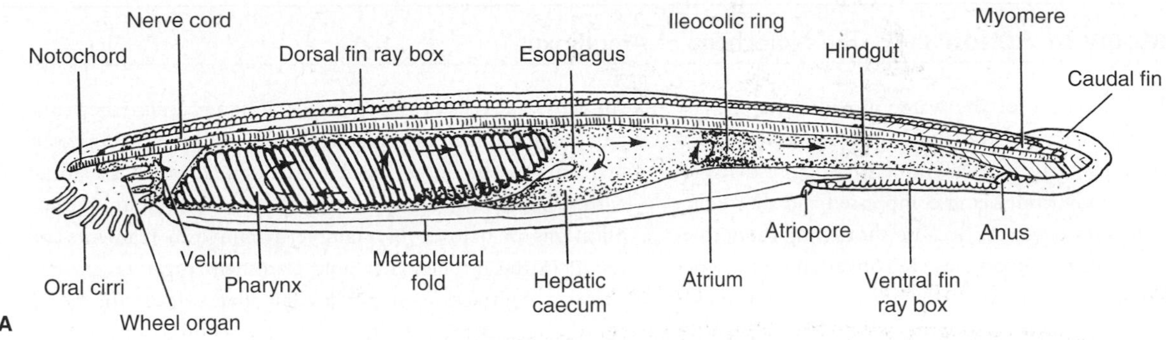

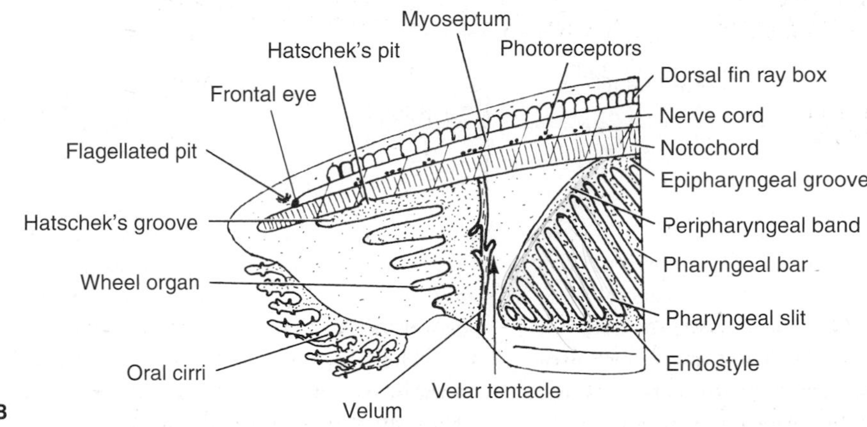

Figure 1-7

Anatomy of amphioxus. **A**, Lateral view of a whole mount slide of a juvenile specimen. Arrows show the directions of mucus and food movement. **B**, Enlargement of the anterior end.

part of a median, light-sensitive **frontal eye.** Sharp focusing will reveal a clear saclike structure just dorsal to the front of the nerve cord. It is called the **flagellated pit** and is believed to be a chemoreceptor. It occurs only on the left side of the snout. In the embryo it connects with the tubular nerve cord.

Examine the region of the **oral hood** in detail. The oral cirri have small processes along their edges, and a skeletal rod of cartilage-like material supports each cirrus. All the rods connect with a common basal rod. Ciliated grooves, or bands, are located on the inside of the lateral walls of the oral hood. In a lateral view they appear as large, dark-staining, fingerlike bands extending forward from a common basal band. Coordinated action of the cilia on these bands rotates and draws a current of water into the organism. For this reason the group of bands is called the **wheel organ.** The most dorsal band, called **Hatschek's groove,** is longer than the others. Hatschek's groove is partly lined with endocrine cells, which secrete hormones (similar to vertebrate pituitary hormones) into the blood. Slightly anterior to the middle of Hatschek's groove you will see a region where the groove is deeper and forms a pit that extends dorsally to overlap the right side of the notochord. This is **Hatschek's pit.** It is the entrance into **Hatschek's nephridium,** a tubule that extends caudad to enter the front of the pharynx.

Caudal to the wheel organ you will notice a dark-staining line that is approximately in the transverse plane. This is the **velum**—a transverse partition that forms the posterior wall of the oral hood. The mouth, which cannot be seen in this view, is located in its center. The mouth is fringed with **velar tentacles,** which can be seen. They too act as strainers and probably contain chemoreceptive cells.

The mouth leads into a large **pharynx.** Most of the lateral walls of the pharynx are perforated by numerous elongated **pharyngeal slits** with ciliated **pharyngeal bars** between them. In favorable specimens, supporting rods may be seen within the pharyngeal bars. Mature specimens have more than 200 bars. The pharyngeal bars provide a very large ciliated surface that plays a major role in moving water and food particles through the pharynx. Food is entrapped in mucus, and the water escapes through the pharyngeal slits. The pharynx is primarily a

Anatomy in Action 1-1 The Notochord of Amphioxus

The buoyancy of the water provides the primary support for aquatic animals, so the notochord is not so much a supporting rod as a structure that prevents the body from being compressed and shortening when the myomeres contract. Because shortening cannot occur, contractions of the myomeres are converted into a series of lateral undulations that sweep down the animal and power swimming. The notochord also stiffens the body enough for the animal to push forward or backward into the sand.

As in other chordates, the notochord of amphioxus consists of vacuolated turgid cells encased in a firm collagenous sheath. The cells are flattened discs stacked like a column of checkers. The cells of amphioxus are unique because they contain paramyosin muscle filaments. The ability of paramyosin filaments to remain contracted for extended periods of time with relatively little energy expenditure has several unique applications in nature; for example, paramyosin filaments are found in the muscles that hold clamshells together. In amphioxus, contraction of paramyosin filaments in the notochord increases the turgidity of the notochord; thus, the notochord acts as an adjustable hydroskeleton. Whether the animal is swimming or burrowing forward or backward, the leading end of the body is stiffer than the trailing end, which oscillates to a greater extent.

food-concentrating mechanism. Although blood vessels pass through the pharyngeal bars, no gills are present. The major site of gas exchange appears to be the general body surface. As in the sea squirt, water in amphioxus does not pass through the pharyngeal slits directly to the outside of the animal but into an **atrium.** The only part of the atrium seen in this view is the clear space ventral to the pharynx and continuing beneath the gut to the **atriopore.** The atrium is formed by the ventral outgrowth of folds of the body wall around the pharyngeal slits. The encasement and protection of the delicate pharyngeal apparatus are essential in a burrowing species.

The longitudinal band that extends along the entire floor of the pharynx is the **endostyle.** Its function is the same as in tunicates. It secretes mucus that entraps minute food particles, and its cilia carry the mucus anteriorly. The string of mucus passes to the dorsal side of the pharynx along the pharyngeal bars and **peripharyngeal bands.** A peripharyngeal band is located on each side of the front of the pharynx and appears as a dark-staining line extending from the ventral edge of the velum diagonally dorsad and caudad just above the anterior pharyngeal slits. The mucus sheet is carried caudally along a middorsal **epipharyngeal groove.** Experiments have shown that certain endostylar cells concentrate radioactive iodine in a manner similar to cells of the vertebrate thyroid gland. In amphioxus, the iodinated proteins are discharged into the gut (rather than into the blood, as in craniates) and absorbed further caudad.

The caudal end of the pharynx floor extends diagonally dorsad and, just behind the last pharyngeal slit, the pharynx leads into a short narrow **esophagus.** The outlines of the esophagus often are obscured by a large hepatic caecum but can be seen by looking carefully. The dorsal edge of the esophagus lies just ventral to the notochord, and its floor will appear as a longitudinal line extending caudally a short distance from the bottom of the last gill slit.

The diameter of the alimentary tract increases twofold or threefold just caudal to the esophagus, because the large **hepatic caecum** has evaginated at this point. The hepatic caecum extends toward the front of the animal, lying along the ventral side of the esophagus and the right side of the pharynx. It is located within the atrium.

Some smaller food particles are carried into the hepatic caecum, where the particles are digested by enzymes secreted there and absorbed directly. Absorbed food is stored as glycogen and lipid in certain caecal cells; thus, the caecum has some functions (e.g., storage) in common with the vertebrate liver and others (e.g., enzyme secretion) not associated with the vertebrate liver. The **midgut** becomes quite narrow caudal to the hepatic caecum. Food particles that are not digested in the hepatic caecum are carried down the midgut by ciliary action because there are no muscles in the wall of the digestive tract. The deeply stained segment of the digestive tract is called the **ileocolic ring.** At this point, beating cilia cause the cord of mucus and food along the length of the entire midgut to rotate. This rotary action aids in the discharge of enzymes by the hepatic caecum and the mixing of enzymes with the food mass. A still narrower **hindgut** follows the ileocolic ring and opens on the body surface at the **anus.** Remains of microorganisms, including the shells of diatoms, often are seen in the hindgut.

Other internal structures are not seen in this type of preparation.

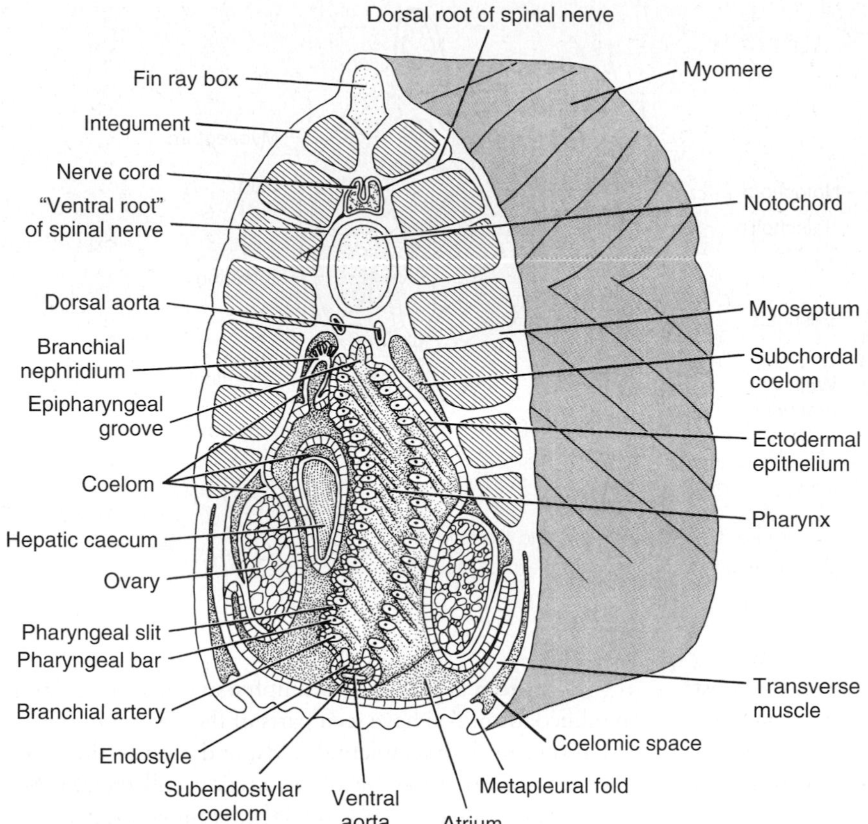

Figure 1-8
Diagrammatic cross section through amphioxus at the level of the posterior part of the pharynx. The section is shown in anterior view; thus, the left side of the animal is on the right side of the drawing.

Cross Sections

Common Features in the Sections

You will understand better the anatomy of amphioxus if you examine slides of representative cross sections. While studying such sections, compare them with the diagrams shown in Figures 1-8 and 1-9, and correlate the appearance of organs in this view with their appearance in the whole mount.

Many features will look much the same in any section. The body surface is covered by a simple **integument** consisting of an epidermis composed of a single layer of columnar epidermal cells underlain by a dermis composed of a gelatinous subcutis containing a few connective tissue fibers. There are no gland or pigment cells. The **dorsal fin** will be recognized along with the hollow-appearing, supporting **fin ray box.** The cavity within the fin ray box is an extension of the coelom. Cross sections of **myomeres** will appear as several oval chunks of muscle tissue deep to the integument. You may see parts of the coelom surrounding them. The myomeres are separated from each other by the **myosepta.** The **nerve cord** is the large tubular structure slightly ventral to the dorsal fin. Its cavity, the **neurocoel,** is very narrow. In favorable sections you will see lateral extensions of the nerve cord that superficially resemble the dorsal and ventral roots of vertebrate spinal nerves. The dorsal extensions pass between the myomeres and are nerves carrying sensory fibers into the nerve cord and motor fibers away from the nerve cord to the ventral nonmyomeral muscles. The apparent ventral roots are not nerves; rather they are long thin extensions of muscle cells from both the myomeres and the notochord. These muscle extensions act as nerves as they synapse with nerve cells within the nerve cord. A similar pattern of muscle innervation is found in round worms (nematodes) and some other invertebrates.

The **notochord** lies just beneath the nerve cord. It consists of vacuolated cells that contain paramyosin muscle fibers and are distended with fluid. The cells of the notochord are held tightly together by a firm connective tissue sheath. You will see the sheath, but the details of the cells will not be apparent.

Section Through the Oral Hood

In a section taken near the front of the animal (Fig. 1-9**A**), you will see a space, the **vestibule,** lying ventral to the notochord and myomeres and flanked laterally by the thin walls of the oral hood. The ciliated bands of the **wheel organ** appear as thicker patches of epithelium on the in-

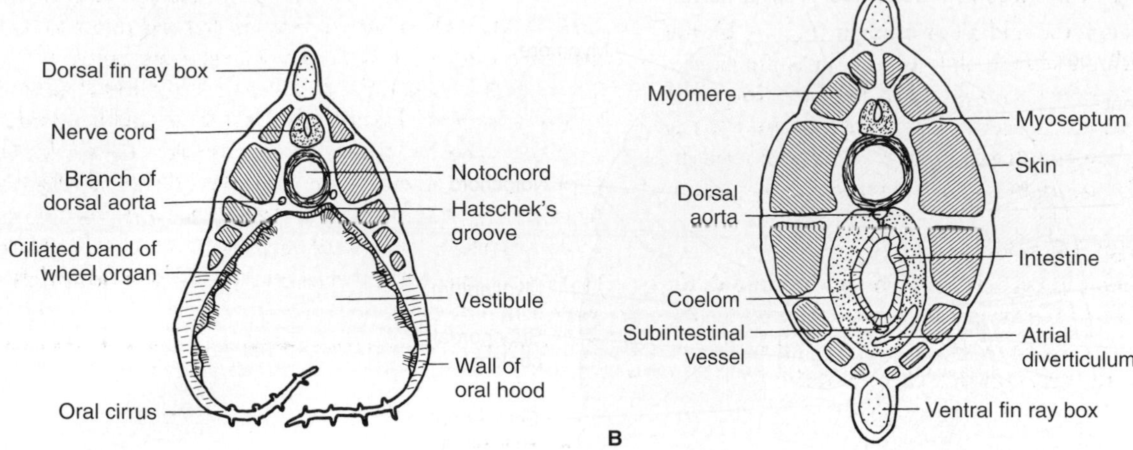

Figure 1-9

Diagrammatic cross sections through amphioxus at the level of the oral hood (**A**) and the hindgut (**B**). This is a posterior view; thus, the specimen's right side is on the right side of the drawing.

side of the wall of the hood. **Hatschek's groove,** the most dorsal of these, is located a bit to the right of the median plane. The oral hood opens ventrally, but some sections may be taken just caudal to opening. Pieces of **oral cirri** will be seen in the section.

Section Through the Pharynx

In a section through the pharynx (Fig. 1-8), the paired **metapleural folds** will be seen projecting from the ventrolateral portion of the body. There is a prominent **coelomic space** in each. The wrinkled body wall between them contains a **transverse muscle sheet** that extends from the myomeres on one side to those on the other. This layer serves to compress the atrium dorsal to it and thus aids in expelling water. The **pharynx** occupies most of the center of the section and is surrounded laterally and ventrally by the **atrium.** Note the numerous **pharyngeal bars** that form the wall of the pharynx and the **pharyngeal slits** between them. The deeply grooved **endostyle** can be seen in the floor of the pharynx, and a similar **epipharyngeal groove** is visible in its roof.

In certain sections, pieces of the **gonads** protrude into the atrium from the body wall, carrying the lining of the atrium before them. An ovary consists of many large nucleated cells; testis tissue appears as small dark dots or fine tubules.

If the section is taken near the caudal end of the pharynx, the hollow, oval **hepatic caecum** will be observed lying on the right side of the pharynx. It first appears to lie completely within the atrium but actually is covered with a layer of atrial epithelium because it has pushed into the atrium from behind (Fig. 1-7**A**).

A **coelom** is present in amphioxus but in a highly modified form. You have seen parts of the coelom in the fin ray boxes, in the metapleural folds, and around the myomeres. Close examination of the section will reveal other subdivisions. A pair of **subchordal coeloms** is located ventrolateral to the notochord and slightly lateral to several of the most dorsal pharyngeal bars. The atrium in this region is a narrow space between the pharyngeal bars and the subchordal coelom. Another coelomic canal, the **subendostylar coelom,** will be found ventral to the epithelium of the endostyle. It connects with the subchordal coelom by small coelomic canals within every other pharyngeal bar; portions of these canals may be found. The bars containing the coelomic canals are the primary pharyngeal bars; secondary bars (tongue bars) grow down between them from the roof of the pharynx during embryonic development as they do in hemichordates. Another portion of the coelom will be seen lateral to the gonads. Finally, a very narrow coelomic space may be seen between the hepatic caecum and the surrounding cells of the atrial epithelium.

Excretion occurs via clusters of small cells that may be seen protruding into the subchordal coelom in the branchial region. These cells have been called **solenocytes** because of their resemblance to the excretory cells of some worms, but many investigators now call them **branchial nephridia.** Excretory products are discharged into the subchordal coelom, where they are picked up by the branchial nephridia through slotted tubelike extensions into the atrium (Ruppert and Smith, 1988). Ruppert (1994) finds that the branchial nephridia share many features with vertebrate kidney tubules and may be homologous to them.

Section Through the Intestine

In a section through the midgut or hindgut (Fig. 1-9**B**), the metapleural folds are absent; the **ventral fin** is present instead. Such a section also will show the **intestine** lying within a large coelomic space, as the separate coelomic canals of the pharyngeal region have coalesced. A caudal atrial diverticulum lies on the right side of the intestine and coelom.

Section Through the Anus

A section through the anus passes through the **caudal fin.** Note that this fin is narrower and higher than either the dorsal or ventral fin, with which it is continuous. The intestine opens at the **anus** on the left side of the fin.

All chordate synapomorphies show very clearly in the cephalochordates. Cephalochordates have features that also occur in craniates, notably the myomeres and a ventral glandular diverticulum from the gut. In addition, cephalochordates have evolved special features that are adaptations for their burrowing mode of life. Among these are the atrial chamber, a contractile notochord, and an extreme rostral extension of the notochord. These features preclude cephalochordates from being the direct ancestors of craniates, but the features cephalochordates share with craniates indicate that they are more closely related to craniates than to tunicates. Because they share myomeres, which develop from embryonic somites, cephalochordates and craniates form the clade **Somitichordata** (Fig. 1-1).[4]

[4]The term *"Protochordata"* has been widely used for a grouping of the Tunicata and Cephalochordata. This is a useful term to describe noncraniate chordates, but, as shown in Figure 1-1, phylogenetic analysis shows that cephalochordates are more closely related to craniates that to tunicates.

Anatomy of the Lamprey

Chapter 2

We introduce the craniates and vertebrates by studying the sea lamprey, *Petromyzon marinus*. Lampreys and the distantly related hagfishes are the only surviving jawless craniates. Lampreys are the easier of the two to obtain and, although specialized in many ways, give us a glimpse of ancestral craniate and vertebrate structure.

Distinctive Features of Craniates

As we have seen in Figure 1-1, the chordate subphylum **Craniata** is closely related to the basal chordates and particularly to the cephalochordates. Craniates became more active than other chordates, and most of their distinctive derived features are correlated with this more active lifestyle. Active animals need a concentration of sense organs on the front end of the body, which first encounters new environments. Craniate **sense organs** include the nose, eyes, ears, and lateral line organs. Lateral line organs are an aquatic sensory system that detects movements in the water. A distinct **brain** processes this sensory information and initiates appropriate motor responses. The brain and some of the sense organs are housed in a braincase or **cranium,** so craniates have a well-developed head and are referred to as **cephalic.**

Craniates share muscle segments **(myomeres)** with the cephalochordates. Hagfishes (order **Myxiniformes**) lack any trace of a vertebral column, whereas all other craniates have at least traces of a **vertebral column** and therefore are also **vertebrates.** The vertebral column develops around the spinal cord and the notochord and replaces the notochord in most species; however, the lamprey and a few groups of basal jawed fishes retain the notochord into the adult stage.

Increased activity is not possible without increased levels of food processing, gas exchange, and distribution of food molecules, gases, and other materials throughout the body, as well as effective mechanisms for eliminating waste products. A **muscularized** gut tube, as opposed to reliance on ciliary action, enabled craniates to obtain more food and move it through the digestive tract more efficiently. The earliest craniates were jawless[1] and probably continued to be filter feeders or scavengers of soft food. Skeletal **visceral arches** evolved between the pharyngeal slits, and muscles acting on them enabled the pharynx to expand, contract, and pump in a greater amount of water and food. Regions of the digestive tract became specialized for specific purposes, and a **liver** and **pancreas** evolved. Gills attached to the visceral arches increased gas exchange. A circulatory system with a multichambered **heart** allowed for efficient distribution of materials throughout the body, and excretory organs became well developed.

Basal Vertebrates

Nearly 43,000 species of extant craniates have been described, making the craniates by far the largest subphylum of the Chordata. The marine hagfishes left no fossil record, but many groups of jawless fishes are known from the first half of the Paleozoic Era (300 to 500 million years ago). Most of these had an extensive armor of bony plates that developed within their skin, so sometimes they are collectively called **ostracoderms,** but this is not a natural evolutionary grouping. Seven distinct orders of extinct ostracoderms are known. Of this large array of agnathous vertebrates,[1] only about 40 species of lampreys (order **Petromyzontiformes**) have survived to recent times.

We will study the evolution of vertebrate organ systems by examining each system in turn in a representative series of species, but it is helpful to begin by briefly examining all organ systems in an adult lamprey. This chapter, which focuses on *P. marinus* (the sea lamprey), will acquaint you with the basic organization of vertebrates. Lampreys have survived by adapting to a very specialized mode of life. *P. marinus* feeds by attaching to other fishes by means of a suctorial buccal funnel, rasping away the flesh of the prey, and sucking its blood and other body fluids. The lamprey's eel-like shape, the absence of an extensive dermal armor, and many specializations of the digestive and respiratory tracts correlate with this unique type of

[1]Jawless craniates have been grouped into the class **Agnatha,** but this is not a natural grouping because jawless craniates include several distinct lines of evolution. The term *agnatha* is often used in a descriptive sense.

predatory lifestyle, but the lamprey also retains the distinctive characteristics of craniates and vertebrates in a relatively unspecialized and ancestral form.

THE ADULT LAMPREY

External Features

Examine the external features of a lamprey, noting its eel-like shape and scaleless skin. The body can be divided into **head, trunk,** and **caudal regions.** The head extends through the gill or branchial area; the caudal region, or tail, extends from the cloacal aperture to the tip of the tail. The body is rounded in cross section in the head and cranial part of the trunk. More caudally, the trunk and tail become laterally compressed, or flattened from side to side. This increases the surface area that can be thrust against the water as the body undulates during locomotion.

Notice that the only fins present are in the median plane—two **dorsal fins** and a symmetrical **caudal fin** (Fig. 2-1). Lateral or paired fins, which usually are found in jawed fishes, are absent. The median fins are supported by slender, cartilaginous **fin rays,** which you can see clearly by cutting across the fin in the frontal plane.

Notice the **buccal funnel** at the front of the head. It is fringed with **papillae,** which are sensory and also enable the lamprey to form a tight seal when it attaches to another fish. The area within the buccal funnel is called the **oral disc** (Fig. 2-3). It is lined with **horny teeth,** which are composed of cornified cells and hence differ from the true teeth of jawed vertebrates. A protrusible **tongue** (often called the piston) is situated near the center of the funnel. It is outlined by a ring of dark tissue and bears small, rasplike, horny teeth. The **mouth opening** lies just dorsal to the tongue. No jaws are present. Absence of jaws is a feature that characterized the basal vertebrates, but the horny teeth and tongue are specializations for the animal's blood-sucking mode of feeding. A single **median naris** (also called the nasohypophyseal opening) is located far back on the top of the head. A median naris is an unusual but ancient feature; it also is found in the fossils of many ostracoderms. The manner in which the naris is displaced during embryonic development from the more usual ventral position is shown in Figure 2-2. This condition derives from the tremendous enlargement of the upper lip of the larva to form the buccal funnel.

You will see an oval area just caudal to the naris that often is slightly depressed and generally a lighter color than the rest of the skin (Fig. 2-1). This area marks the location of the **pineal eye complex,** an ancient feature also found in the ostracoderms. Experiments have shown that the pineal eye complex detects changes in light and initiates diurnal color changes in larval lampreys, probably by its effect on the hypothalamus and hypophysis, or pituitary gland. Other physiological activities are most likely adjusted to the diurnal cycle in a similar way. A pair of conventional but lidless **lateral eyes** lies on the sides of the head. Seven pairs of oval **external gill slits** are located behind the lateral eyes.

If you let the head dry slightly and examine it with low magnification, you may detect groups of pores, or little bumps, arranged in short lines. One group is found caudal to the lateral eye; a second group begins ventral to the eye and extends rostrally and dorsally; and a third

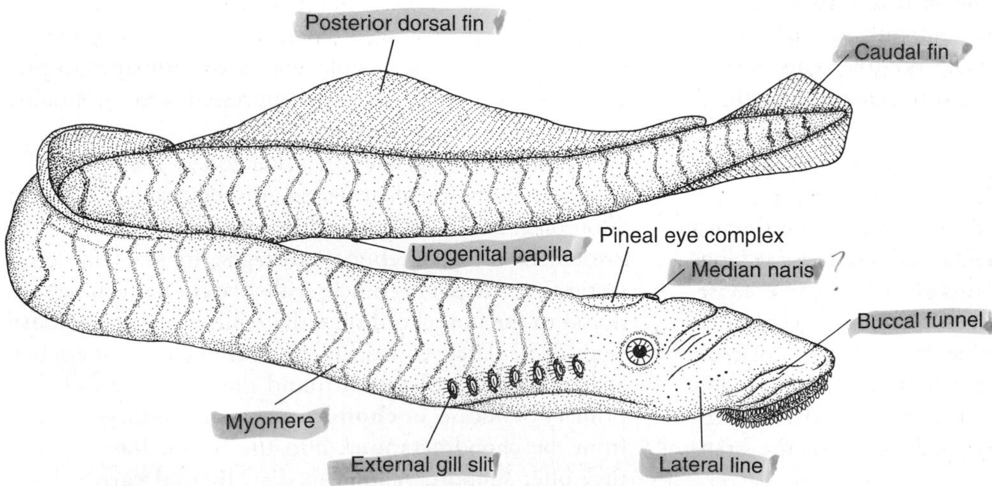

Figure 2-1
Lateral view of the sea lamprey, *Petromyzon marinus.*

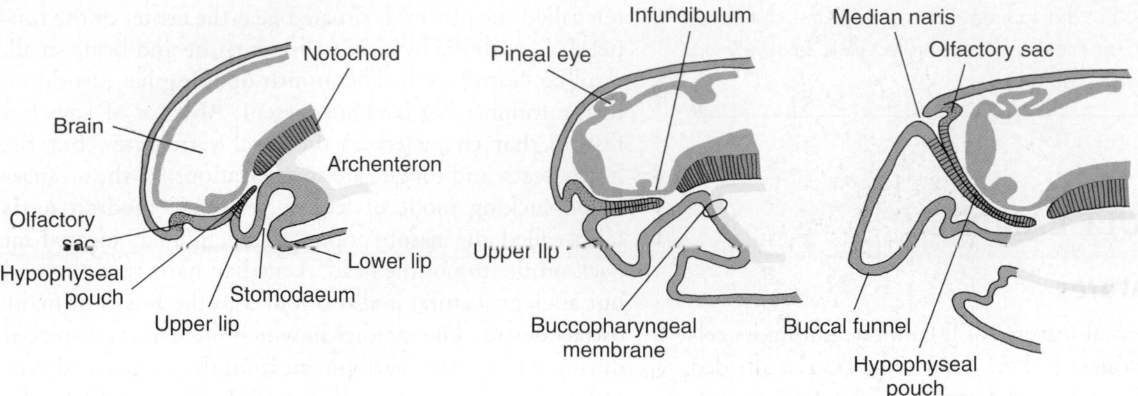

Figure 2-2

Diagrams in the sagittal plane of three stages in the development of the head of *Petromyzon.* Notice in particular how the originally independent hypophyseal pouch and olfactory sac are crowded together and pushed onto the top of the head by enlargement of the upper lip. The hypophysis of the lamprey invaginates rostral to the stomodaeum rather than from within the stomodaeum as it does in jawed vertebrates. *(From Parker and Haswell, 1972, after Dohrn.)*

group is located on the ventral side of the head, caudal to the buccal funnel. These pores, together with other less conspicuous ones, are parts of the **lateral line system,** a group of sense organs associated with detecting vibrations and movements in the water (see Chapter 8).

The body musculature consists chiefly of segmented **myomeres**[2] whose outlines can be seen through the skin of the trunk and tail, especially in smaller specimens. Each myomere is roughly W shaped, with the top of the "W" pointing cranially. Each myomere is continuous from its dorsal to ventral end, with no interruption near its middle, as occurs in jawed fishes.

On the underside of the caudal end of the trunk you will see a shallow pit called the **cloaca.** It receives the excretory and genital products, which leave through the tip of a small **urogenital papilla** and, just cranial to the papilla, the **anus,** or opening of the intestine. The opening of the cloaca to the surface is called the **cloacal aperture.**

Sagittal and Cross Sections

Study the internal structure of the lamprey by examining a midsagittal section (Fig. 2-3) and a series of cross sections (Fig. 2-4). Work in groups for this portion of the exercise. Some students should prepare cross sections, and the other students should prepare sagittal sections. A series of cross sections that will match those shown in Figure 2-4 is as follows: **(A)** through the pineal and lateral eyes; **(B)** through a pair of external gill slits near the middle of the branchial region; **(C)** just caudal to the last external gill slit; **(D)** about 3 cm behind the branchial region; **(E)** near the middle of the trunk (not illustrated), about 3 cm cranial to the cloaca; and **(F)** through the tail. In preparing the sagittal section, use a large knife and be particularly careful to cut the head and branchial region as close as possible to the sagittal plane. If the halves are unequal in size, take the larger one and dissect it down to the sagittal plane.

The Skeletal and Muscular Systems

Study the most complete sagittal section, correlating the appearance of structures in this view with their appearance in the cross sections. Notice that the main skeletal axis is a long **notochord** extending from the caudal end of the body to a point beneath the middle of the brain. It has a gelatinous texture but is enclosed in a strong fibrous sheath. It is firm yet flexible and serves primarily to prevent the body from being compressed and shortening when the myomeres contract.

The rest of the lamprey's skeleton is composed of a unique type of cartilage based on the protein **lamprin** (Robson et al., 1993) rather than on collagen, which is the protein in most vertebrate cartilages. In favorable cross sections, cartilaginous blocks, called **arcualia,** will be seen dorsal to the notochord on either side of the **spinal cord** (Fig. 2-5). They constitute the rudiments of vertebral arches. Other cartilages surround the brain and form the primary braincase, or **chondrocranium.** Cartilages extend from the chondrocranium into the buccal funnel where they offer support. A long median **lingual cartilage** extends into the tongue and supports its movements during feeding. Still other cartilages are found beneath the skin

[2]See Footnote 3, Chapter 1.

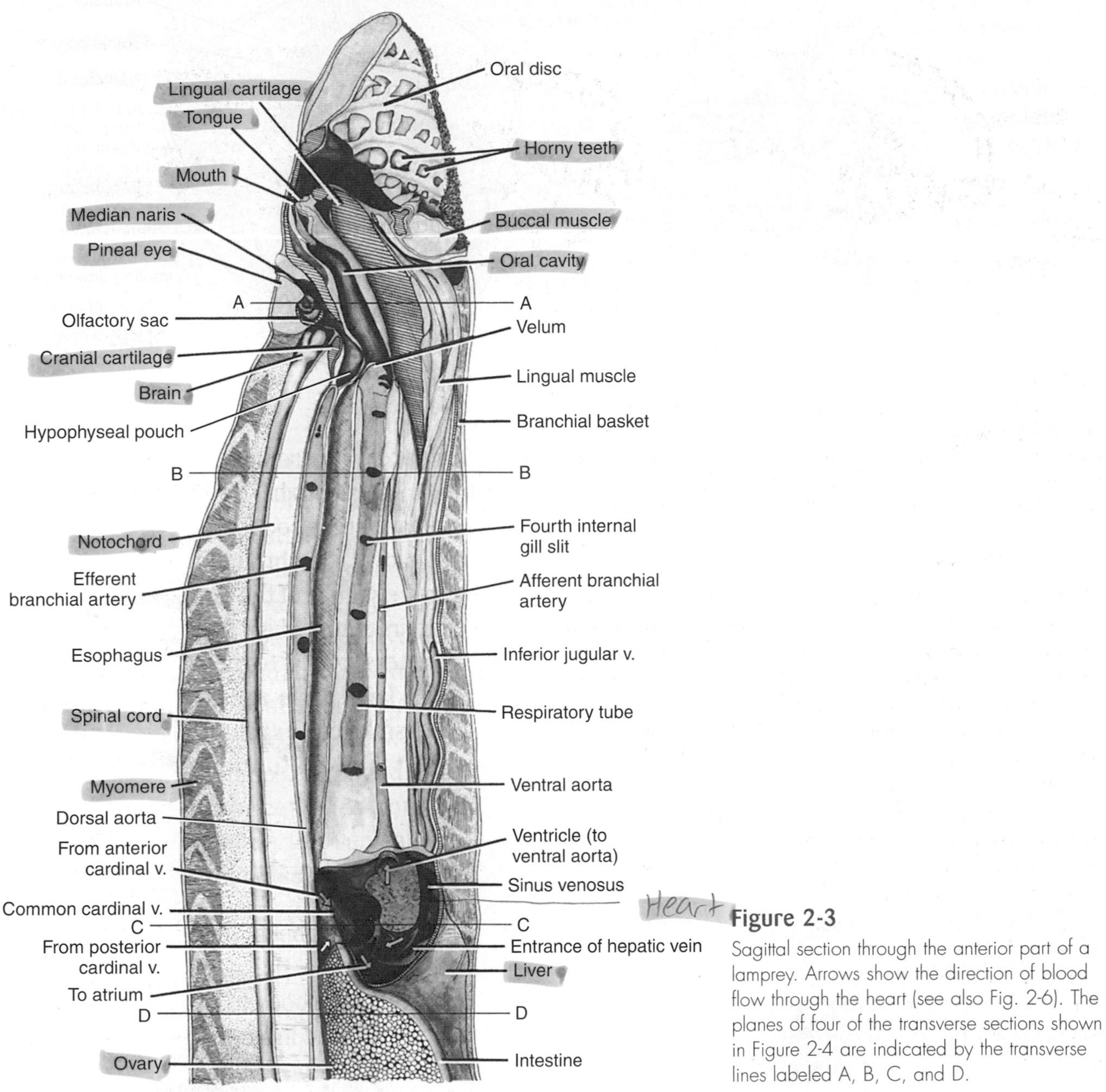

Figure 2-3
Sagittal section through the anterior part of a lamprey. Arrows show the direction of blood flow through the heart (see also Fig. 2-6). The planes of four of the transverse sections shown in Figure 2-4 are indicated by the transverse lines labeled A, B, C, and D.

lateral to the gill region and posterior to the heart. They form a **branchial basket** that supports the gill region. Examine a special preparation of the skeletal system, if available, to see this system better. In lampreys and other agnathans, the branchial basket lies lateral to the gill pouches rather than medial to the gills and next to the cavity of the pharynx, where the visceral arches of all other vertebrates are located. Despite this difference in position, the branchial basket of lampreys and the visceral arches of jawed fishes probably are homologous. Both develop from ectomesenchyme cells derived from the embryonic neural crest. Kimmel, Miller, and Keynes (2001) have shown that in both lampreys and jawed fishes, this ectomesenchyme in the branchial region first lies laterally, just beneath the surface ectoderm and lateral to the mesoderm and developing branchial pouches. It remains in this position in lampreys so the branchial basket develops superficially. In jawed fishes this ectomesenchyme continues to migrate and moves medial to the mesoderm so the visceral arches develop deeply next to the pharyngeal lumen.

The muscular system consists primarily of the segmented myomeres, which were already observed during

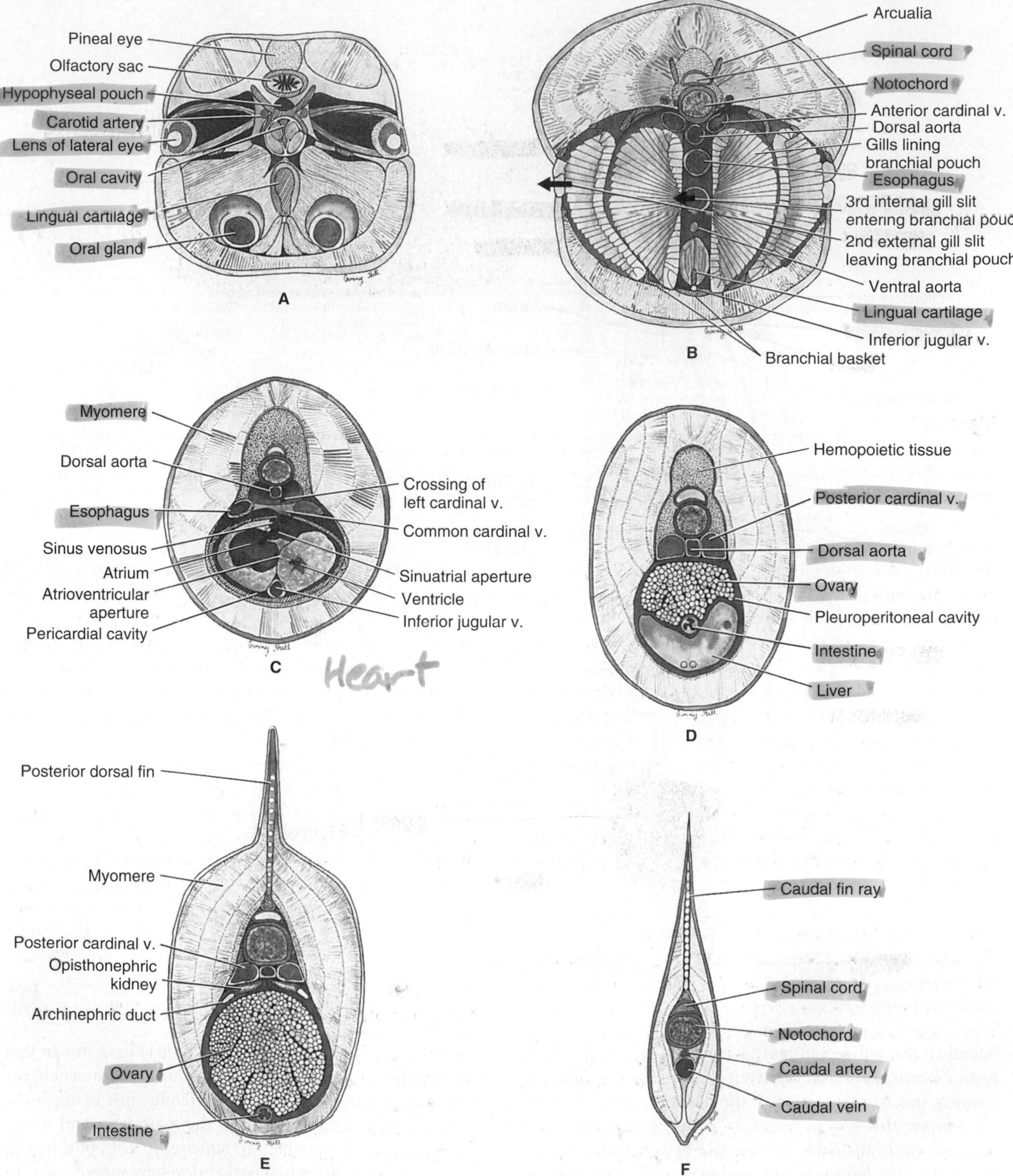

Figure 2-4

Representative transverse sections through a lamprey: **A,** through the pineal eye complex and lateral eyes (plane A in Fig. 2-3); **B,** through an external gill slit (plane B); **C,** through the heart (plane C); **D,** through the cranial end of the liver (plane D); **E,** through the caudal part of the trunk; **F,** through the tail. The black arrows in **B** show the flow of the respiratory current of water when the lamprey is not feeding, that is, when the water is entering a branchial pouch through an internal gill slit from the respiratory tube and leaving through the external gill slit.

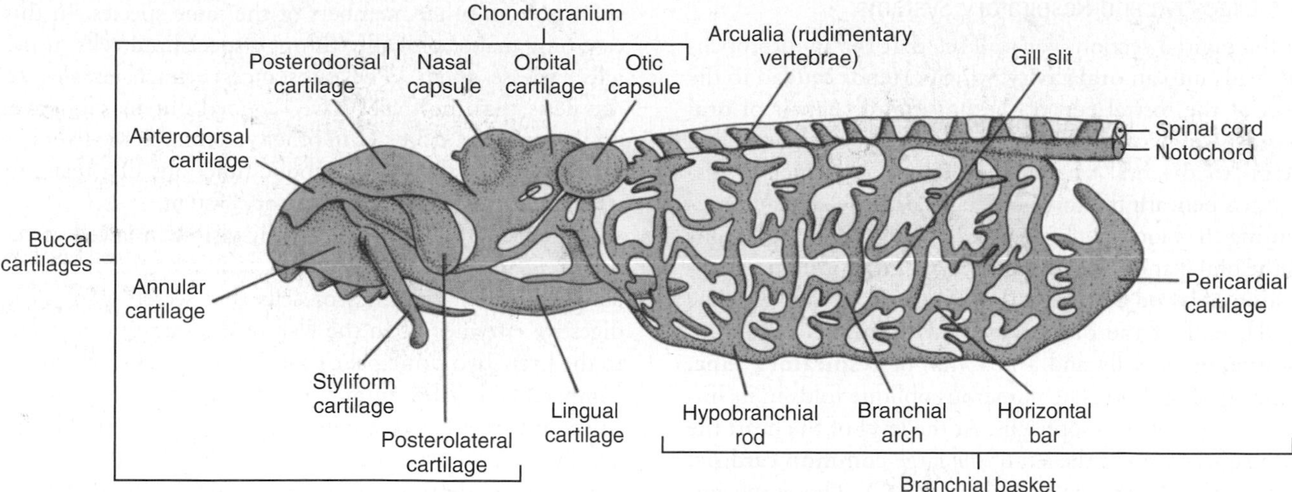

Figure 2-5

Lateral view of the chondrocranium, buccal and lingual cartilages, and branchial basket of a lamprey. The nasal capsule, orbital cartilage, and otic capsule are parts of the chondrocranium associated with the nose, eye, and inner ear, respectively. *(From Young, 1981, after Parker.)*

the survey of external features. Note their appearance in the sections. Each myomere consists of bundles of longitudinal skeletal muscle fibers that attach to connective tissue **myosepta** between the myomeres. The apices of the folds of the myomeres, which you observed in a surface view, extend forward and backward in overlapping cones so a cross section cuts through several myomeres. Waves of contraction pass from the front of the body to the end of the tail, alternating between the two sides of the body. This causes the lateral undulations of the trunk and tail by which the animal swims. The movements of the buccal funnel and tongue during feeding are caused by the intricate musculature you see associated with these structures.

The Nervous System and Sense Organs

The **brain** and the **spinal cord** can be seen lying dorsal to the notochord. The lamprey has the tripartite brain (forebrain, midbrain, and hindbrain) characteristic of all vertebrates, but you probably cannot distinguish these parts in this type of dissection. The lamprey's brain is similar to that of the dogfish (Chapter 9), except that the cerebellum is barely developed in the lamprey.

Notice the connections of the median naris in the sagittal section. It first leads into a dark **olfactory sac** located rostral to the brain. The internal surface of the sac is greatly increased by numerous folds. A **hypophyseal pouch** continues from the entrance of the olfactory sac and passes ventral to the brain and the rostral end of the notochord. Respiratory movements of the pharynx squeeze the end of the hypophyseal pouch much as we squeeze the bulb of a medicine dropper, thereby moving water in and out of the olfactory sac. The adenohypophysis, a part of the pituitary gland that lies within the cranial cavity ventral to the brain, is derived from an embryonic hypophyseal pouch. In jawed vertebrates the hypophysis invaginates from the roof of the mouth, or stomodaeum; Fig. 2-2. The pituitary gland is difficult to see in this type of dissection.

The clear area dorsal to the olfactory sac represents the site of the **pineal eye complex.** Actually, there are two median eyes here; the pineal eye lies dorsal to a smaller parietal (parapineal) eye, but details cannot be seen in a dissection of this type. In addition to being light sensitive, both eyes have glandular activity. The well-developed **lateral eyes** show in one of the cross sections (Fig. 2-4**A**). The eye of fishes will be studied in more detail later (Chapter 8), but in the lamprey you will recognize the large spherical **lens** characteristic of fishes, a dark **pigmented layer,** and possibly the whitish **retina** between the pigmented layer and the lens.

If you make another cross section (not illustrated) just caudal to the lateral eye, you will see part of the ear. It will appear as a bit of tissue imbedded in a cavity of the chondrocranium, the **otic capsule** (Fig. 2-5), lying lateral to the brain. The ear of lampreys and all fishes is an **internal ear.** Fishes lack the middle and external ears that enable terrestrial vertebrates to detect airborne vibrations. As you will learn in Chapter 8, the inner ear normally has three semicircular ducts, but the lamprey has only two. This condition was also found in some ostracoderms.

The Digestive and Respiratory Systems

In the sagittal section, you will see that the mouth opening leads into an **oral cavity,** which extends caudad to the level of the rostral end of the notochord. A pair of **oral glands,** which can be seen in Figure 2-4**A,** lies ventral and lateral to the oral cavity. Their secretion, which is discharged beneath the tongue, acts as an anticoagulant, preventing the blood of the prey from clotting. The secretions of the oral glands also contain enzymes that initiate the digestion of blood cells and other body cells of the prey. The caudal end of the oral cavity leads into two tubes—an **esophagus** dorsally and a pharynx, or **respiratory tube,** ventrally (Fig. 2-3). The numerous oblique folds in its lining distinguish the esophagus. At the level of the heart the esophagus passes to the left of the large **common cardinal vein,** which enters the heart (Fig. 2-4**C**). The esophagus leads to the **intestine,** the first part of which lies between the liver and gonad (Figs. 2-3 and 2-4**D**). This part of the intestine can expand greatly during feeding. The intestine continues as a narrow straight tube to the cloaca. Longitudinal folds increase the internal surface area of the intestine. One fold is particularly prominent and has a somewhat spiral course; hence, it is called the **spiral valve.** As in other chordates that feed on minute food particles, an enlarged stomach that stores material is not present. Much of the intestine atrophies after spawning.

A large and, in the sagittal section, triangular **liver** lies just caudal to the heart. It develops embryonically as a ventral outgrowth of the intestine. During larval life a gallbladder (Fig. 2-7) and bile duct are present, but these structures are lost at metamorphosis. The adult liver produces bile pigments, and bile salts and acids. Because adult lampreys do not have a gallbladder and bile duct, the digestive functions, if any, of the bile salts and acids are uncertain. However, Li et al. (2002) discovered that these substances act as pheromones, which are substances released by individuals into the environment that initiate responses from other members of the same species. In this case only males, probably through specialized cells in the gills, release them. The substances attract females to the nests that the males have scooped out in the gravel on the stream bottom. As in other vertebrates, the liver has many food storage and metabolic functions that relate to its connections with the circulatory system.

The pancreas, which is grossly visible in jawed vertebrates, is not present as a grossly visible organ in the sea lamprey; instead, patches of cells that secrete pancreatic digestive enzymes lie in the wall of the intestine adjacent to the liver. Two principal clumps of other cells, which are comparable to the endocrine portion of the pancreas (islets of Langerhans), are imbedded in the intestinal wall and liver near the junction of the esophagus and intestine. Experimental destruction of these clumps causes a rise in blood sugar.

The intestine and liver lie within a division of the coelom known as the **pleuroperitoneal cavity.** The intestine of vertebrates usually is attached to the dorsal wall of the body cavity by a long dorsal mesentery, but this mesentery is reduced in the lamprey to a few strands of tissue surrounding blood vessels that pass to the intestine and liver.

Returning to the respiratory tube, you will see that its entrance is guarded by a series of tentacles that constitute the **velum;** its wall is perforated by seven **internal gill slits.** By looking at a cross section and using a probe, you will note that each internal gill slit leads into an enlarged gill pouch, or **branchial pouch** (Fig. 2-4**B**), lined with gill lamellae. The pouches open to the surface through the **external gill slits** (Anatomy in Action 2-1).

The Circulatory System

Study the **heart** in sagittal and transverse sections (Figs. 2-3 and 2-4**C**). Its structure is shown diagrammatically in Figure 2-6. The heart lies in another division of the coelom, the

Anatomy in Action 2-1 Feeding and Respiration in Lampreys

Some species of fresh water lampreys, called brook lampreys, do not feed after metamorphosing from the larval stage. Their gut soon atrophies, so they die soon after reproducing. The large sea lamprey feeds for several years as an adult, and the dorsal esophagus and ventral, blind-ending respiratory tube is correlated with its method of feeding. When not attached to its prey, a respiratory current of water enters the mouth and respiratory tube, crosses the gills, and is discharged through the external gill slits. A one-way flow of water across the gills is the normal pattern in fishes. When the sea lamprey is attached to its prey, this course of water flow is no longer possible, and the lamprey uses a tidal ventilation of its gills by pumping water in and out of the external gill slits (Fig. 2-4**B**). Some water may seep into the respiratory tube, but this would not interfere with feeding. The most active phase of respiration is expiration; the branchial muscles constrict the pouches, and water is forcibly expelled. Inspiration results primarily from the elastic recoil of the branchial basket.

pericardial cavity, and consists of three chambers: the sinus venosus, the atrium, and the ventricle. The thin-walled, tubular **sinus venosus** is located between the atrium and ventricle. The sinus venosus receives blood low in oxygen content from the body and leads into the large **atrium** through a **sinuatrial aperture.** The atrium of preserved specimens generally is filled with dried and hardened blood and, hence, is dark in color. It is located lateral to the sinus venosus and fills most of the left side of the pericardial cavity. It leads by an **atrioventricular aperture** into a more muscular **ventricle** located in the right ventral portion of the pericardial cavity. Valves present in the apertures keep blood moving in the correct direction. A distinct conus arteriosus, which is a fourth heart chamber often found in jawed fishes, is not present, so the ventral aorta leaves directly from the ventricle.

The pericardial cavity is separated from the pleuroperitoneal cavity by a **transverse septum.** In the lamprey a large **pericardial cartilage,** which is attached to the branchial basket, stiffens this septum (Fig. 2-5). Because the pericardial wall is stiff and will not collapse, contraction of the ventricle probably results in an increase of the "open" space within the pericardial cavity and, hence, a reduction of pressure around the atrium and sinus venosus. This reduced pressure would help "suck" blood in from the veins. Blood pressure in the veins is very low in fishes.

The **ventral aorta** extends forward ventral to the respiratory tube (Fig. 2-3), giving off seven paired **afferent branchial arteries** that lead to capillaries in the gills. The branchial arteries may not be seen. In cross sections, the **dorsal aorta** can be found just ventral to the notochord. It receives blood high in oxygen content from the gills by way of **efferent branchial arteries** and continues caudally, supplying the body musculature and viscera. In the tail, the dorsal aorta is called the **caudal artery** (Fig. 2-4**F**). The head is supplied chiefly by a pair of **carotid arteries** (Fig. 2-4**A**) that leave from the front of the dorsal aorta.

A pair of **anterior cardinal veins** (Fig. 2-4**B**) and an **inferior jugular vein** drain the head and the branchial region (except for the gills) and their numerous venous sinuses. The anterior cardinal veins are located lateroventral to the notochord; the inferior jugular vein lies ventral to the prominent tongue musculature in the floor of the branchial region. The anterior cardinal veins enter the right common cardinal vein, and the inferior jugular vein enters the sinus venosus.

A **caudal vein** is located ventral to the caudal artery. The characteristic renal portal system of fishes is absent in the lamprey; the caudal vein does not go to the kidneys but instead goes directly to the paired **posterior cardinal veins.** These veins can be seen in cross sections of the trunk on either side of the dorsal aorta (Fig. 2-4**D** and **E**). At the level of the heart, the right posterior and anterior cardinal veins unite to form a **right common cardinal vein** (duct of Cuvier) that passes ventrally on the right side of the esophagus to enter the sinus venosus (Figs. 2-3 and 2-6). Most jawed fishes have a comparable left common cardinal vein, but in the lamprey the left posterior and anterior cardinal veins curve medially to join the right common cardinal vein (Fig. 2-4**C**). An inconspicuous but important hepatic portal system of veins carries blood from the alimentary canal to the liver. Blood is carried from minute spaces in the liver through a **hepatic vein** to the sinus venosus of the heart (Fig. 2-3).

Lampreys have no spleen. Blood-forming (hemopoietic) tissue is represented by the dark, specialized connective tissue dorsal to the spinal cord.

The Urogenital System

The excretory organs consist of paired **opisthonephric kidneys** that appear as flaps suspended from the dorsal wall of the caudal half of the pleuroperitoneal cavity (Fig. 2-4**E**). You may have to push the gonad aside to see them. Each kidney is drained by an **archinephric duct,** which runs along its free border. Cut away the lateral body wall in the cross section that contains the cloaca and observe that the two archinephric ducts unite at the

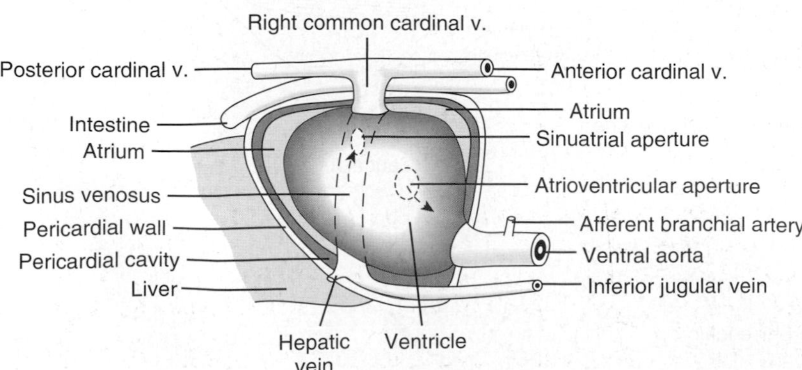

Figure 2-6
Lateral view of the right side of the heart of the lamprey. The position of the sinus venosus, which lies between the atrium and ventricle, is shown by the dotted lines.

caudal end of the pleuroperitoneal cavity to form a **urogenital sinus,** which opens at the tip of the **urogenital papilla.** You may be able to pass a bristle into an archinephric duct and out the urogenital papilla.

Although the **ovary** or **testis** develops from paired primordia, the gonad of the adult is a large median organ that fills most of the pleuroperitoneal cavity. It is suspended by a mesentery. Genital ducts are absent in both sexes. The gametes are discharged directly from the gonad into the coelom. They leave the coelom through paired **genital pores** located on either side of the urogenital sinus at the extreme posterior end of the pleuroperitoneal cavity. Search for the pores by carefully probing this area.

THE AMMOCOETE

In the spring, some adult sea lampreys leave the ocean or large lakes where they live and ascend streams to spawn and die. In about 3 weeks, the eggs hatch into larvae called ammocoetes because, when first discovered, the larvae were thought to be a distinct species, named *Ammocoetes*. The ammocoetes burrow into the sand, much like amphioxus, and live as filter feeders for 5 to 7 years. After attaining a length of about 12 cm and storing much fat, the larvae metamorphose over a period of 3 to 4 months. Their sense organs become better developed, their oral region changes, and the pharynx, now called the *respiratory tube,* separates from the newly developed esophagus. These changes adapt the animal to its new mode of life.

Whole Mount Slide

Examine a whole mount slide of a small ammocoete with the low power of a microscope (Fig. 2-7). Note its fusiform body shape and the **dorsal** and **caudal fins,** which are continuous with each other. The fine dark lines, or specks, on the body surface are pigment cells, or **chromatophores.** If the pigment is dispersed, you can see that each chromatophore consists of a central area from which branching processes radiate. Although the **myotomes** have been rendered transparent in preparing the slide, their outlines may show as faint lines on the surface.

The **spinal cord** appears as a dorsal, dark-staining band that is enlarged rostrally to form the **brain.** In favorable specimens you can see that the brain is composed of several divisions or lobes separated by constrictions. The largest and most caudal lobe is the hindbrain **(rhombencephalon);** the next, the midbrain **(mesencephalon);** and the most rostral (sometimes subdivided), the forebrain **(prosencephalon).** The **notochord** appears as a light-staining, longitudinal band ventral to the spinal cord and the caudal two divisions of the brain.

The small surface protuberance rostral to the brain is the **median naris,** which leads into the **hypophyseal pouch** lying ventral to the brain. If apparent, each incompletely developed **lateral eye** is represented by a round dark spot lying between the mesencephalon and prosencephalon. An evagination from the caudal portion of the roof of the prosencephalon is the primordium of the **pineal eye complex;** the large clear vesicle overlapping the front of the rhombencephalon is the primordium of the **inner ear.**

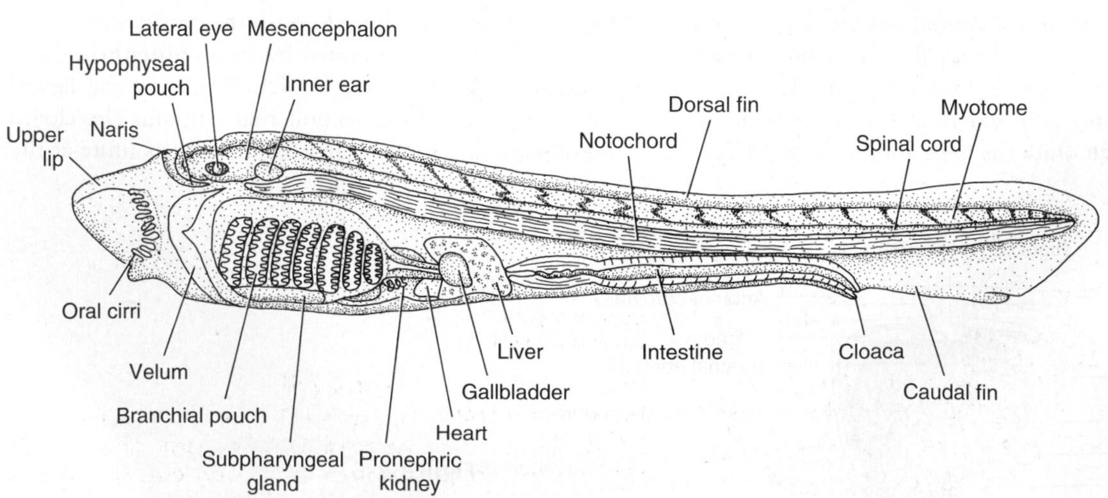

Figure 2-7
Lateral view of a whole mount slide of the ammocoete of the lamprey.

Many of the specializations of the adult digestive and respiratory systems are absent in the ammocoete because it feeds by sifting minute food particles from the water. Observe that the **upper lip** has already enlarged to form the primordium of the buccal funnel. The **lower lip** appears as a transverse shelf. The **mouth opening** is surrounded by a series of **oral cirri** that function as strainers and sensory organs. Behind the mouth is a clear chamber, the **oral cavity,** which is bounded caudally by a pair of large muscular flaps, the **velum.** Movements of the velum bring a current of water and food into the pharynx. It is important to note that the feeding current is caused by muscular action of the velum and pharynx rather than by ciliary action as in basal chordates. This is more efficient and is probably a factor that permits the ammocoete to attain a considerably larger size than amphioxus. Observe the seven large **branchial pouches** in the pharyngeal region. They are lined with **gill lamellae** and open through small, round, **external gill slits,** which may be seen by sharp focusing on the surface. Ventral to the pharynx you will find a large, dark-staining, elongated body called the **subpharyngeal gland.** In young larvae the subpharyngeal gland secretes mucus, which is discharged through a duct to the pharynx. The lining of the pharynx and gill pouches produces additional mucus. The mucus forms a longitudinal cord in the pharynx that traps food particles, and the cord is carried back into the **esophagus.** The subpharyngeal gland has many of the characteristics of the basal chordate endostyle and may be its homologue. Certain of its cells also produce iodinated proteins, including thyroglobulin. During metamorphosis these cells are transformed into the thyroid gland of the adult lamprey. The alimentary canal widens posterior to the esophagus to form the **intestine,** which continues to the **cloaca.**

The **liver** is located adjacent to the caudal portion of the larval esophagus and contains a large clear vesicle, the **gallbladder.** Notice the **heart** lying ventral to the esophagus in front of the liver. Between the heart and esophagus you will see a few bell-shaped or fingerlike processes, or tubules, which are parts of the larval **pronephric kidney.**

Cross Section Through the Pharynx

Examine a cross section through the pharynx with the low power of a microscope and correlate the appearance of structures in this view with their appearance in the whole mount slide. A somewhat flattened **spinal cord** will be seen dorsally lying above the large **notochord** (Fig. 2-8). A column of fat cells, which will become the hemopoietic tissue in the adult, lies above the spinal cord. Notice that the spinal cord contains the characteristic central canal of chordates and that the notochord is composed of vacuolated cells surrounded by a dense connective tissue sheath. The **dorsal aorta** lies ventral to the notochord, and the paired **anterior cardinal veins** flank the notochord ventrolaterally.

A large central **pharynx** and lateral **branchial pouches** occupy most of the body ventral to the dorsal

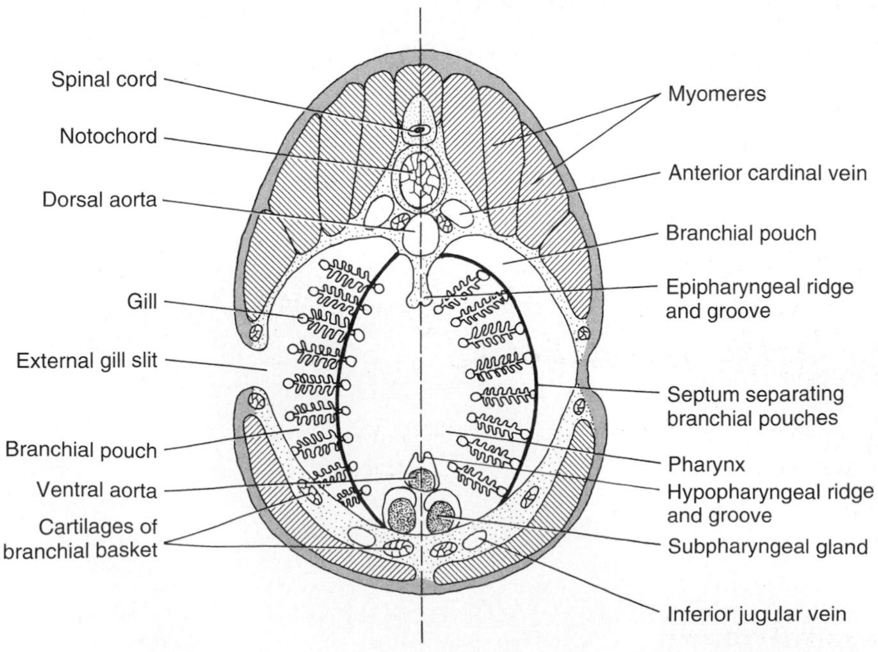

Figure 2-8

Composite diagrammatic transverse section through the pharynx of an ammocoete. A section taken at the level of an external gill slit is shown on the left side, and a section between the external gill slits is shown on the right side.

aorta. The branchial pouches and the septa separating them slant posteriorly as they extend from the pharynx toward the body surface; hence, a transverse section may cut through parts of two pouches. The relationship of the feathery gill lamellae to the septum to which they attach is complex. Lamellae on the more anterior surface of a septum point toward the body surface (left side of Fig. 2-8), whereas those on the posterior surface are directed toward the pharynx (right side of Fig. 2-8). Many blood vessels can be seen in the gills. Each branchial pouch discharges on the body surface through a single **external gill slit.** The large **subpharyngeal gland** forms a ridge in the floor of the pharynx. It may appear as two or three glandular masses, depending on the level of the section. A **ventral aorta** and **hypopharyngeal ridge** overlie the subpharyngeal gland. An **epipharyngeal ridge** extends into the pharynx from its roof. Cartilages of the **branchial basket** and associated muscles can be seen in the body wall lateral to the branchial pouches.

The ammocoete is of great phylogenetic interest and may give us some idea of the nature of ancestral craniates. It resembles amphioxus in that it has all the chordate characteristics but lacks such specializations as the atrium surrounding the pharynx and the extreme rostral extension of the notochord. It also lacks many of the specializations of the adult sea lamprey for blood sucking, yet it has all of the essential craniate features. Its well-developed sense organs, brain, myomeres, and caudal fin enable it to be more active than any of the basal chordates. It uses efficient muscular movements of the velum and pharynx in filter feeding, and muscular contractions of the gut carry food posteriorly. A larger and more active animal than amphioxus, the ammocoete no longer depends on cutaneous gas exchange but has evolved gills at a site where there is a good flow of water (i.e., in the branchial pouches). These gill pouches, therefore, are now part of both the feeding and the respiratory mechanisms.

The External Anatomy and Integument of Vertebrates

Chapter 3

In this chapter, you will see for the first time representatives of the jawed vertebrates, called **Gnathostomata,** the anatomy and evolution of which is the focus of this laboratory manual. A survey of the evolution of craniates and vertebrates is covered in all textbooks of comparative and functional anatomy, so we will not consider this topic in detail here. Figure 3-1 is a diagram of the major aspects of craniate evolution, and includes a summary of the groups we will mention or examine in detail.

A few words of explanation are appropriate. The lampreys, extinct jawless vertebrates ("ostracoderms"), and the hagfishes often are placed in a class "Agnatha." This is a convenient descriptive term, but these disparate groups are unrelated in an evolutionary sense. The terms "Ostracoderms" and "Reptilia" are placed in quotes to indicate that they are not natural monophyletic clades but rather are paraphyletic groupings.[1] "Ostracoderms" is a catchall grouping for many different types of extinct jawless fishes. "Reptilia" is not a natural clade because it does not include all of their descendents, for example, the birds. However, reptile is a convenient term to describe the nonavian sauropsids (turtles, tuatara, squamates, and crocodiles) and is widely used in this sense. The osteichthyan fishes (bony fishes) are often considered to be a class (Osteichthyes), but evolutionary phylogeneticists use the term to include all vertebrates (including tetrapods) that have a bony internal skeleton. This is why we qualify the term by adding "fishes." The situation is analogous to the use of the term gnathostomata for all jawed vertebrates. How craniate groups are interpreted doubtless will be based on the textbook you are using.

Species To Be Studied

Reviewing the evolutionary history of the craniates and vertebrates in Figure 3-1 allows us to appreciate the opportunities and limitations of a basic course in comparative anatomy for which this manual is intended. We understand now, for example, that contemporary reptiles did not evolve from the amphibians we know today as living animals. To grasp fully how today's variety of reptiles evolved from early tetrapods, we would need to study mostly fossil material. Fossils, however, are incomplete remains of organisms. Therefore, to learn about the evolutionary transformation of all the anatomical structures, including the soft tissues and organs, we resort to the comparative study of the anatomy of extant vertebrates.

We recognize also that all vertebrates living today have a long evolutionary past, although some vertebrates have modified their anatomy in more fundamental ways than others. Thus, no living vertebrate is "more primitive," "lower," "more advanced," or "higher" than any other extant species. When we compare the anatomy of an amphibian, such as *Necturus,* with the anatomy of a mammal, we do not do so because *Necturus* is "more primitive" than a mammal, but because *Necturus* has retained a greater number of ancestral tetrapod features than have mammals and, therefore, is a good model for a basal tetrapod.

We also realize that no particular evolutionary lineage is inherently more interesting than any other. Traditionally, however, the evolutionary line leading from the earliest fishlike vertebrates to the mammals has held special fascination for us. Therefore, this manual considers mainly the anatomy of animals that illustrate distinctive stages in that evolutionary line, namely, the anatomy of a lamprey as a model of an ancestral jawless vertebrate, that of a shark as a model of an ancestral jawed vertebrate, that of a salamander as a model of an ancestral tetrapod, and that of a cat or rabbit as models of a mammal. The choice of the particular species for each evolutionary stage has been dictated by purely practical considerations, namely, abundance and ease of procurement.

Sharks belong to the subclass **Elasmobranchii** of the class **Chondrichthyes** (Fig. 3-1) because their gills are borne on platelike transverse sheets of connective tissue and each external

[1]A paraphyletic group is not a natural clade because it may contain groups that are not closely related ("Ostracoderms"), or it may not contain all of the descendent groups of an ancestral group ("Reptilia").

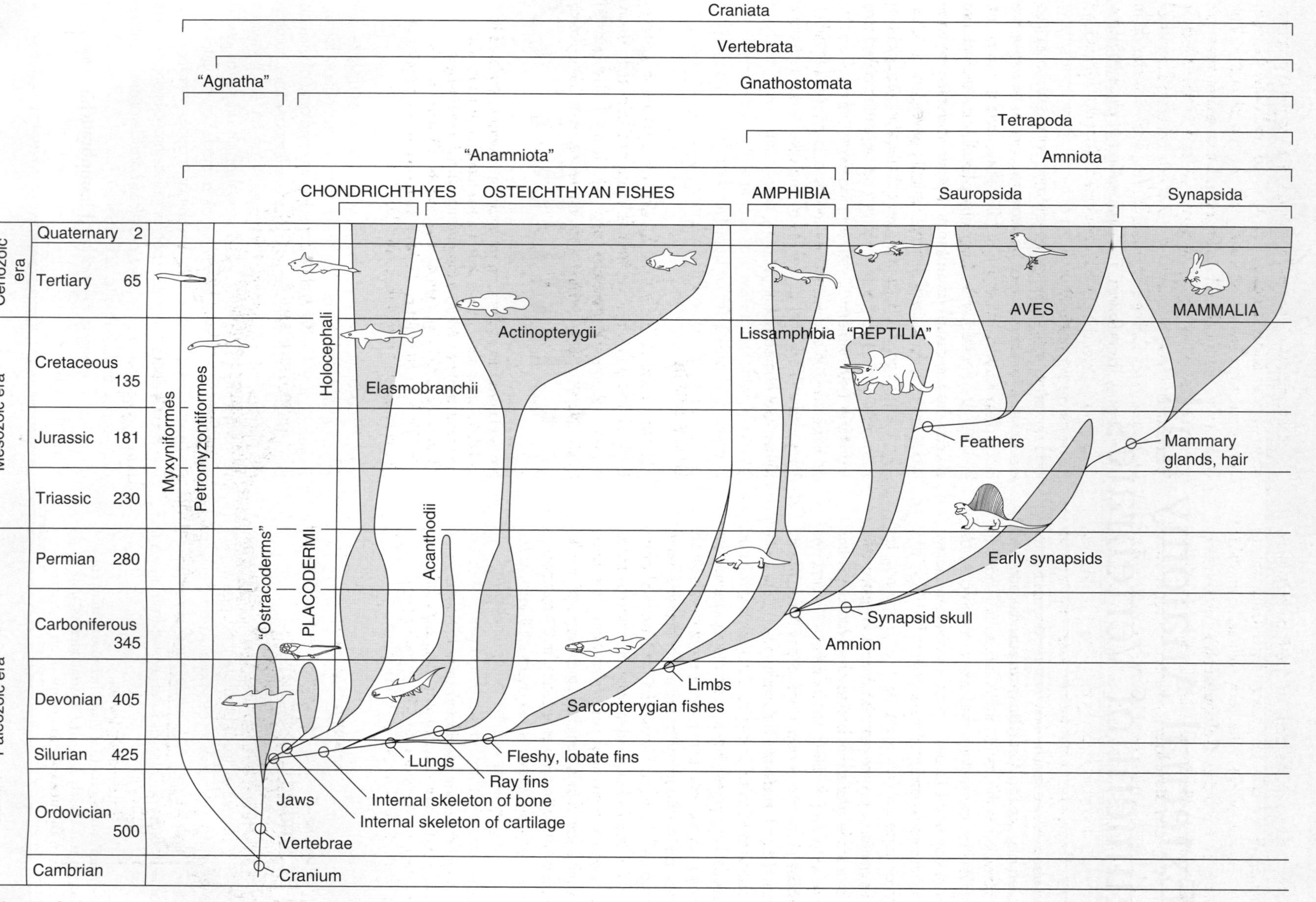

Figure 3-1

Diagram of the major groups of craniates and vertebrates showing their hypothesized relationships, their distribution in geologic time, and an estimate of their relative abundance. The ages of the geologic periods are given in millions of years since the start of the period. Groups in capital letters are traditional classes; those in quotation marks are paraphyletic. Circles indicate points where major evolutionary changes occurred.

gill slit opens independently on the body surface (Fig. 3-2). In the subclass **Holocephali,** the gills are covered by a fleshy operculum and have a common gill slit to the body surface. Extant elasmobranchs belong to the orders **Galeomorphii** (large typical sharks), **Squaliformes** (smaller sharks, including the dogfish—the subject of our study), and **Batoidea** (flattened, largely bottom-dwelling skates and rays). A good example of a shark is the spiny dogfish of the North Atlantic and North Pacific Oceans, *Squalus acanthias.* Adult males range in length from 0.7 to 1 m; females are slightly larger. They prefer water temperatures ranging from 6°C to 15°C; hence, they migrate north in the spring and south in the fall. A migratory school may include thousands of individuals. They are voracious and prey on most species of fishes and crustaceans smaller than themselves. They are considered edible in Europe and are beginning to be used as food in this country.

Ancestral tetrapods belong to the class Amphibia (Fig. 3-1). Many were rather large animals (some up to 1.5 m long), but the group is represented today by an infraclass[2] **Lissamphibia** of small, usually smooth- and slimy-skinned animals that fall into three orders: the burrowing, limbless caecilians of the tropics (order **Gymnophiona**); the frogs and toads (order **Anura**); and the salamanders (order **Urodela**). The mudpuppy *(Necturus)* is a satisfactory example of a salamander. Unfortunately, it is unusual in one respect: it is paedomorphic and retains certain larval features into sexual maturity. Larval features retained by *Necturus* should not be confused with ancestral characteristics. *Necturus* is distributed throughout most of the eastern half of the United States. The most widespread species is *Necturus maculosus.* It is most abundant in clear waters of lakes and larger streams, but it is also found in weed-choked, turbid, and smaller bodies of water. It is most active at night, when it forages for small fish, crayfish, aquatic insect larvae, and mollusks.

[2]Whether the lissamphibians are considered a subclass or an infraclass depends on the way a particular investigator subdivides the amphibians, the type of evolutionary interrelationships recognized, and the degree of resolution being sought.

Contemporary mammals include the egg-laying monotremes (order **Monotremata**) of Australia and New Guinea, which are such unusual animals that they are placed in their own group, the subclass **Prototheria.** Other extant mammals give birth to living young and are placed in the subclass **Theria.** The several orders of pouched marsupials, most of which are also confined to Australia and New Guinea, form the infraclass **Metatheria.** The many orders of placental mammals, characterized by a well-developed placenta, are placed in the infraclass **Eutheria.**

Directions in this manual are written in such a way that they can apply to the domestic cat *(Felis catus),* which belongs to the order **Carnivora,** or to the rabbit *(Oryctolagus cuniculus),* which belongs to the order **Lagomorpha.** If some students are provided with cats and others with rabbits, interesting comparisons can be made between ecologically divergent species. This comparison illustrates the basic "blueprint" (bauplan) of mammals and, at the same time, reveals important differences that have been superimposed on a common structural pattern as these species evolved independently and adapted to different patterns of locomotion and to a carnivorous (cat) or herbivorous (rabbit) mode of life.

Body Shape

For each species that we study, we will examine its general body shape and major external features and consider certain aspects of the **integument** and skin *(cutis).*[3] The general body shape of a vertebrate is mainly reflective of the animal's

[3]As explained earlier, the term we favor for structures is placed in boldface when first used. If this is a mammalian structure and the term is not an Anglicization of the official *Terminologia Anatomica* (TA) term, then we also give the TA term in italics. The TA term often forms the basis for related terms, in this case for cutaneous = pertaining to the skin.

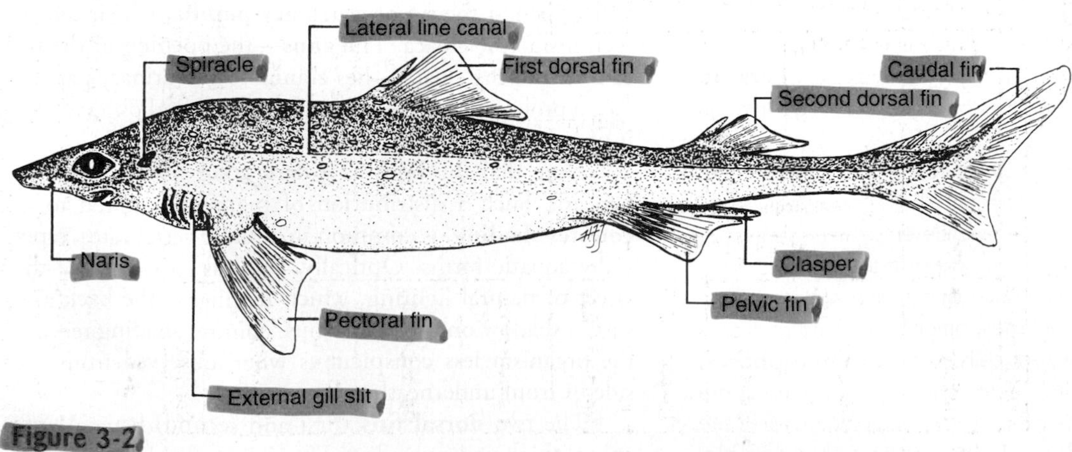

Figure 3-2
Lateral view of the spiny dogfish, *Squalus acanthias.*

particular mode of locomotion. At one end of the spectrum of vertebrate shape, we find the purely aquatic fishes with their streamlined bodies, which are adapted to reduce the resistance of the surrounding water during swimming. Lateral undulations of the fish's trunk and tail provide the major propulsive forces. The paired appendages, or fins, may provide additional thrust, but they are used in sharks primarily in vertical maneuvering. They have no role in generating lift during horizontal movement, as once believed (Wilga and Lauder, 2002). At the other end of the spectrum are the terrestrial mammals, in which the paired appendages have become powerful limbs that not only support the body above the ground but also propel the body forward. Most mammals do not need a streamlined body, and a pronounced neck allows the head to be moved independently from the trunk. Between these two extremes in body shape we find the terrestrial urodele amphibians, which have limbs, but a poorly defined neck region. Their limbs are used in locomotion, but lateral undulations of the trunk and tail help advance and retract the limbs. In this manual we will consider the three vertebrate body shapes just described, but keep in mind that many other body shapes and modes of locomotion have evolved in all classes of vertebrates.

The Integument

The surface of vertebrates is covered by the **integument,** which protects the body in many ways. It protects against abrasion, undue exchanges of water and salts with the external environment, and ultraviolet radiation; however, the skin does not isolate the body from the outside world. Because it contains many sensory receptors, it also establishes a connection between the organism and its environment.

Briefly, the skin consists of two layers of tissue: an outer **epidermis** of stratified epithelial cells derived from the embryonic ectoderm, and an inner **dermis** *(corium)* of dense connective tissue derived from the mesoderm. In fishes and amphibians, the epidermis is thin and permeable to water and gases. As vertebrates adapted to a terrestrial environment, the epidermis became thicker, and its outer cells became impregnated with lipid materials and the horny protein **keratin,** which together reduce water loss and gas exchange through the skin.

Accessory integumentary structures (e.g., pigment cells, glands) and a variety of bony and horny structures (e.g., scales, feathers, hair) develop from the epidermis, dermis, or both. Pigments play an important role in many animals, serving for species recognition, concealment, and sometimes advertisement. The pigment of fishes, amphibians, and reptiles is contained within specialized cells, the **chromatophores,** which, although derived from embryonic neural crest cells, are located in the dermis. The most common chromatophores are stellate (star-shaped) **melanophores** containing the dark pigment **melanin.** The pigment moves within the melanophores of fishes, amphibians, and some reptiles under the influence of nerves, hormones, or both. When the animal is dark, the pigment is dispersed throughout the cell; when the animal is light, the pigment is withdrawn to the center. The melanophores of birds and mammals transfer their pigment to epithelial cells.

All skin glands are derivatives of epidermal cells, but the larger glands invaginate into the dermis. The glands of fishes and amphibians primarily produce a protective coat of mucus. Reptiles and birds have few cutaneous glands, but mammals have a variety of glands.

Feathers and hair also are derivatives of epidermal cells, but scales (of which there are many types) may develop from either or both layers of the skin. Bony formations in the skin have sometimes been called an "exoskeleton," but the term **integumentary skeleton** is more appropriate. These formations are cellular structures that develop within the skin and are not a secretion on the skin surface, as is the exoskeleton of invertebrates.

THE SPINY DOGFISH

External Features

Examine a specimen of the spiny dogfish, noting the streamlined, or **fusiform,** shape that enables the animal to move easily through the water (Fig. 3-2). The body regions blend into one another and are not as well demarcated as they are in tetrapods. Nevertheless, the body can be divided into a **head,** which includes the gill region and extends caudad to the pectoral girdle; a **trunk,** which continues to the **cloacal aperture;** and a **tail** caudal to the cloacal aperture. The **cloaca** is a chamber on the ventral side that receives the intestine and the urinary and genital ducts. The urinary ducts, and the genital ducts in the male, open at the tip of a **urinary papilla,** which can be seen inside the cloaca. The **anus**—the opening of the intestine into the cloaca—lies cranial to the urinary papilla. Pass a probe through the cloacal aperture and observe that it passes into the intestine via the anus.

Note that the body is a dark color above and light beneath. Such a distribution of pigment, referred to as **counter shading,** is common in aquatic vertebrates, especially aquatic forms. Optically, it tends to neutralize the effect of natural lighting, which highlights the back and casts a shadow on the belly; thus, counter shading renders the organism less conspicuous when observed from the side or from underneath.

The two **dorsal fins** (first and second) have a large **spine** on their front edge. The spines are defensive and are associated with modified skin glands that secrete a slightly irritating substance. Ringlike markings on the spines give an indication of the fish's age; the life span of dogfishes ranges from 25 to 30 years. Large, paired **pectoral fins** are located just behind the gill region. Paired **pelvic fins** are located at the caudal end of the trunk. Males have stout, longitudinally grooved copulatory organs, called **claspers,** on the medial side of their pelvic fins. The tail ends in a large **caudal fin** of the **heterocercal** type, that is, the fin is asymmetrical; the body axis turns up into the larger dorsal lobe of the fin, and most of the fin rays are ventral to the axis. Fibrous fin rays, called **ceratotrichia,** support all of the fins. Cartilages, which will be seen later, lie within the base of each fin and provide further support.

A heterocercal tail occurs in many early fishes and probably was the ancestral type. Wilga and Lauder (2002) have studied the vortices shed into the wake by the action of the heterocercal tail in the leopard *(Triakis semifasciata)* and bamboo *(Chiloscyllium punctatum)* sharks. They have analyzed quantitatively the orientation of the vortices and the forces of the jets through the core of the vortices relative to body orientation. They conclude that, according to the classic model of tail action, the heterocercal tail does generate a significant backwards and downwards force that helps to propel these species and lifts the caudal end of the body. This lift force rotates the body about its center of mass and drives the head downward, thereby compensating for lift forces generated at the front of the body.

The **mouth,** which is supported by jaws, is located on the underside of the head and is bounded laterally by deep **labial pockets.** A **labial fold,** containing a cartilage, lies between the mouth and pocket. Large **eyes** are set in sockets on each side of the head. The rim of each socket forms immovable **eyelids.** Paired external nostrils, called **nares,** are on the underside of the pointed snout (see Fig. 8-1**A**). The opening of each naris is partially subdivided by a little flap of skin so that the stream of water flows into the **olfactory sac** on one side and out the other. Pass a probe into a naris and notice that the olfactory sac does not communicate with the oropharyngeal cavity, as it does in some sarcopterygian fishes and all tetrapods.

A row of five **external gill slits** is located cranial to the pectoral fin. Caudal to the eye you will see a large opening called the **spiracle.** Small parallel ridges, representing a reduced gill called the **pseudobranch,** can be seen on a fold of tissue that is separated from the rostral wall of the spiracular passage by a deep recess. Probe to determine the extent of this recess. This fold of tissue is a **spiracular valve,** which can be closed to prevent water from leaving the pharynx through the spiracle during respiration. This ensures that all water must cross the gills as it exits through the external gill slits.

If you look with low magnification at the top of the head between the spiracles, you will see a pair of tiny **endolymphatic pores,** one pore on each side of the midline. These pores communicate with the inner ear, which in most fishes is an organ used both for equilibrium and for detection of higher-frequency vibrations or sound waves. Vibrations of low frequency and movements in the water are detected by the lateral line system. The position of one canal of this system, **lateral line canal,** is indicated by a fine, light horizontal stripe extending along the side of the body. It is nearer to the dorsal than to the ventral surface. Also notice the patches of pores on the head through which a jellylike substance exudes if the area is squeezed. They are the openings of the **ampullae of Lorenzini.** These ampullae, which function as electroreceptors, are a modified part of the lateral line system.

Accessory Structures of the Integument

The skin of sharks contains chromatophores; glands, usually in the form of simple, scattered mucous cells; and hardened dermal structures. In the spiny dogfish, the hardened structures are minute dermal denticles, or **placoid scales,** which cover the animal and which you can feel by moving your hand cranially over the surface. (At one time elasmobranch skin, sold as shagreen, was used as a fine abrasive for polishing wood.) To see the placoid scales, you will have to use low magnification or, better still, observe a special microscopic preparation (Fig. 3-3**A**). (**Melanophores** may be seen at the same time.) Each scale consists of a basal plate of acellular bone imbedded in the dermis from which a spine perforates the epidermis and projects caudad (Fig. 3-3**B**). These scales resemble teeth in many ways because both develop embryonically in the same way. However, scales and teeth differ in size, location in the body, and function. The spiny scales reduce turbulence in the flow of water next to the skin and, hence, reduce drag during swimming. Placoid scales appear to be remnants of the extensive dermal armor of early jawless fishes, the "ostracoderms" (Anatomy in Action 3-1).

AMPHIBIANS AND REPTILES

External Features

Examine a specimen of *Necturus* and compare it with *Squalus.* The body is elongate, with a modest-sized, flattened **head;** an incipient **neck** region; a long **trunk;** and

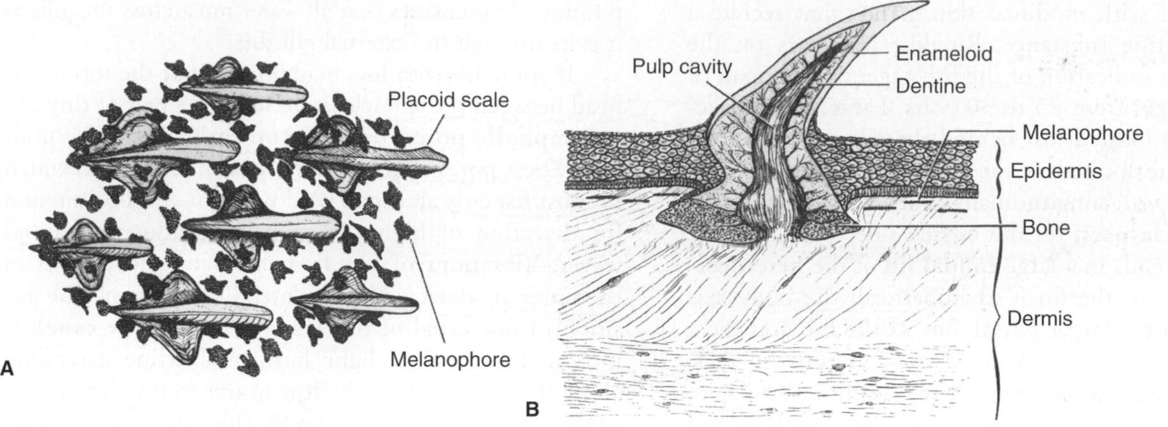

Figure 3-3
Skin and placoid scales of a shark (the head end of the shark points to the left). **A,** Magnified surface view, with melanophores. **B,** Magnified vertical section through a placoid scale and the skin. *(B, Redrawn from Dean, 1895.)*

Anatomy in Action 3-1 The Scales of Fishes

Many ostracoderms, such as heterostrachans, had thick bony scales or plates composed of several layers (Fig. **A**). A basal layer consisted of layered, avascular bone and was overlain by a thicker layer of spongy vascular bone containing many cavities for blood vessels. This bone was capped by denticles composed of a dentinelike material often called **cosmine.** Dentine is very similar to bone; however, it lacks bone cells and contains long microscopic tubules that were left behind by the retracting bone cells. In cosmine, these tubules are clustered into tufts. A very hard, acellular material lacking tubules and resembling enamel tops the cosmine. True enamel develops from the ectoderm, but this material, often called **enameloid,** may be hardened dentine or cosmine, which, like bone, is derived from the dermis.

In the evolution of chondrichthyan fishes, the bony layers of the ancestral scales were lost, leaving only the surface denticles as placoid scales (Fig. 3-3**B**). Gars *(Lepisosteus),* a surviving group of early actinopterygians, retains thick scales capped with many layers of enamel or enameloid, called **ganoine.** These are **ganoid scales.** Early sarcopterygian fishes also had thick scales, but the cosmine layer was far thicker than the enamel or enameloid. These are called **cosmoid scales.** Most recent fishes have retained only a thin layer of laminated bone (Fig. **B**). These scales are known as **cycloid scales** if their surface is smooth and **ctenoid scales** if the caudal edge that reaches the surface of the skin bears minute processes called cteni.

The spines of fishes are usually enlarged bony scales. As we noted, the fin rays of cartilaginous fishes are horny ceratotrichia, but the bony fin rays of actinopterygian fishes, which are called lepidotrichia, evolve from rows of small bony scales on each surface of the fins.

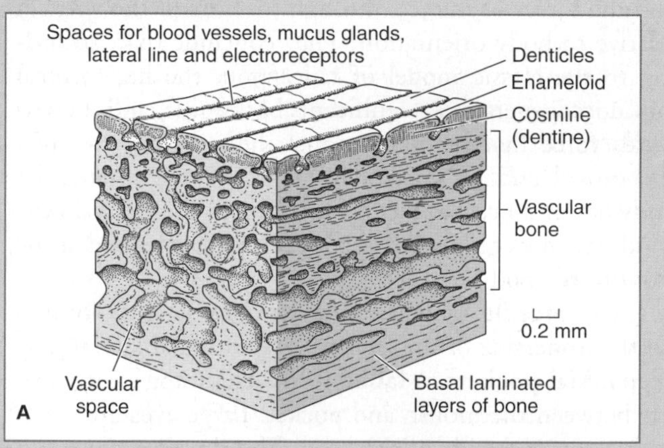

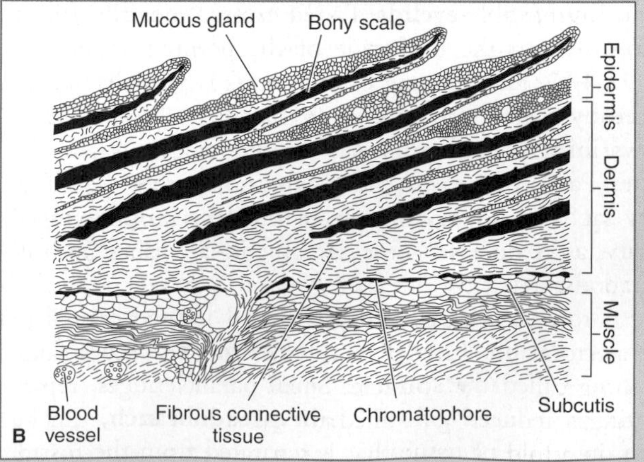

A, Three-dimensional block sectioned from a bony scale of a heterostracan, one of the extinct basal jawless vertebrates. **B,** Vertical section through the skin and cycloid scales of a teleost. *(A and B from Liem, Bemis, Walker, and Grande, 2001.)*

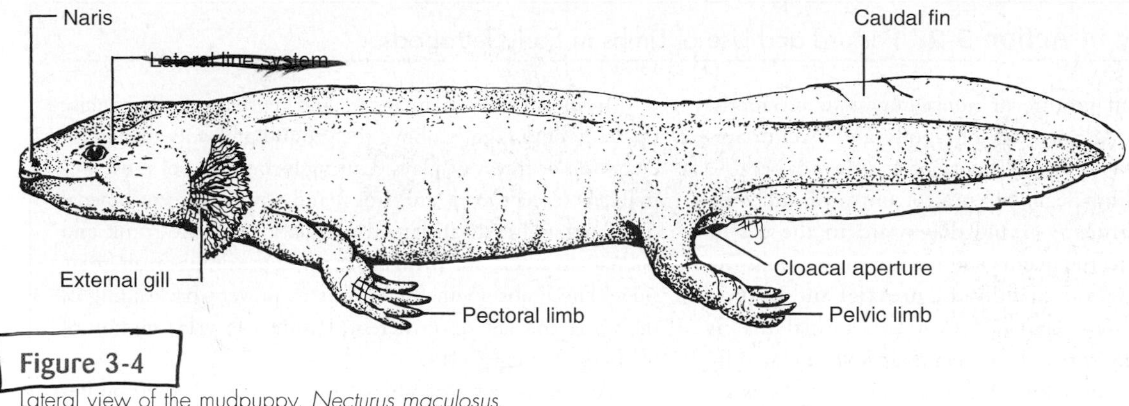

Figure 3-4
Lateral view of the mudpuppy, *Necturus maculosus*.

a powerful, laterally compressed **tail** (Fig. 3-4). All salamanders resemble ancestral tetrapods in that they have retained a well-developed tail, but the tail is especially powerful in *Necturus* because it is an aquatic species that swims by lateral undulations of the trunk and tail. There are no median fins, except for traces on the tail, and these lack fin rays.

The paired fins of fishes have become transformed into pectoral and pelvic limbs in tetrapods. Limbs are not large in *Necturus* because this species does not venture onto land, but they show the three segments characteristic of tetrapod limbs. In the pectoral appendage these are the upper arm **(brachium),** forearm **(antebrachium),** and hand **(manus).** The elbow joint **(cubitus)** is located between the brachium and antebrachium; the wrist **(carpus)** is located in the proximal part of the manus. Corresponding parts of the pelvic appendage are the thigh **(femur),**[4] shank **(crus),** and foot **(pes);** corresponding joints are the knee **(genu)** and ankle **(tarsus).** Only four digits are present. We think of five digits as being the normal number, but the earliest tetrapods of the Late Devonian period had seven or more. Amphibian ancestors of the amniotes did have five digits, but lissamphibians never have more than four digits on the manus, although most have five digits in the pes. The first digit of the manus of *Necturus* is considered homologous to the ancestral second digit. If the entire limb is pulled out to the side at right angles to the body with the palm of the hand (or sole of the foot) facing ventrally, the cranial side is said to be **preaxial;** the caudal side, **postaxial.** The posture and use of limbs in early tetrapods are described in Anatomy in Action 3-2.

The **mouth** is terminal, and a pair of widely spaced external nostrils, called **nares,** lies dorsal to the front of the mouth. The nares communicate with the rostral end of the oral cavity by way of internal nostrils, called **choanae,** thus permitting air to be taken into the mouth cavity. Small **eyes** are present but lack eyelids. The absence of eyelids is a larval feature found only in aquatic amphibians. Movable eyelids are not necessary in aquatic animals because the external surface of the eye (the cornea) is not in danger of desiccation. Metamorphosed terrestrial amphibians have movable eyelids, which help to protect, cleanse, and moisten the eyeball. Unlike frogs and most reptiles, salamanders lack an external eardrum, or tympanic membrane. There is an internal ear, however, and vibrations reach it primarily by way of skull bones. The **lateral line system** of fishes is retained in *Necturus* and appears, when the specimen has dried a bit, as rows of depressed dashes above and below the eyes, on the cheek, and on the ventral surface of the head. A less obvious row of dashes extends caudad along the side of the trunk. The lateral line system, too, is a larval feature that is found only in aquatic amphibians; metamorphosed terrestrial amphibians lose it.

Three pairs of prominent **external gills** can be seen at the caudal end of the head. Although some gas exchange takes place through the highly vascularized skin and *Necturus* comes to the surface to gulp air (see Anatomy in Action 10-2), these gills are the major respiratory organs. External gills are larval structures and are lost by salamander species that metamorphose into terrestrial adults. The gills of larval amphibians are supported by elements of the visceral skeleton, as are the gills of fishes, but amphibian gills project outward from the body surface, and fish gills lie within gill pouches or are covered by an operculum. *Necturus* also has two **gill slits,** which can be seen between the bases of the external gills. The fold of skin extending across the ventral surface of the head between the gills is called the **gular fold.**

Finally, observe the **cloacal aperture** at the caudal end of the trunk. Lips bearing tiny papillae in the male and small folds in the female bound it.

[4]The term *femur* can be used for the thigh or the bone within it. The *Terminologia Anatomica* term for the bone is *os femoris*.

Anatomy in Action 3-2 Posture and Use of Limbs in Early Tetrapods

The limb posture of nonmammalian tetrapods, including ancestral tetrapods and extant amphibians and reptiles, is described as **splayed,** because the upper limb segments project outward and only the lower limb segments extend downward to the pes or manus. In Figure **A,** the manus and pes are shown pointing laterally, so that you can identify the **preaxial** and **postaxial sides** of the limb segments. In a live animal that is standing or moving forward, the pes points craniolaterally and the manus cranially (Fig. **B**). In the hindlimb, this reorientation is accomplished by a rotation of the pes at the tarsus. In the forelimb, the reorientation is more extensive and takes place in the elbow and distal end of the antebrachium, with the result that the two bones of the antebrachium cross partially at their distal end and the preaxial sides of the antebrachium and manus now faces medially.

In most extant salamanders and reptiles, the limbs raise the body only slightly above the ground and only during locomotion or threat displays. During locomotion of most amphibians (except toads and frogs) and reptiles (except bipedal dinosaurs and turtles), lateral undulations of the trunk and tail remain crucial for swinging forward the limbs on alternate sides. The limbs' main function is to prevent backsliding of the trunk and tail, and, thereby, ensure a forward motion of the body (see Fig. 7-10).

Alligators and crocodiles, however, can lift their bodies significantly above the ground so that they are capable of running and even galloping. Their limb posture and locomotion are models for a possible intermediary stage between those of ancestral tetrapods and of quadrupedal mammals.

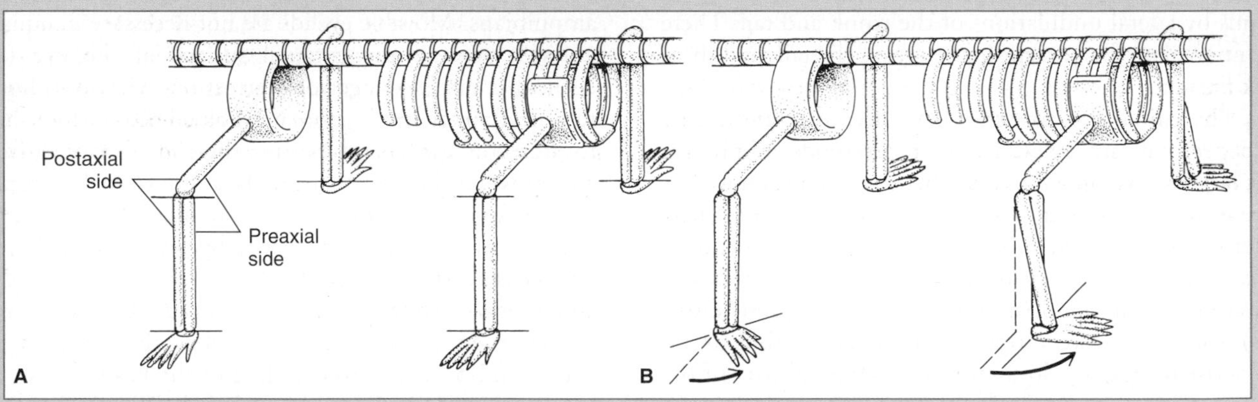

Diagrams of limb posture in an ancestral tetrapod. The hands and feet are pointing laterally better to indicate the preaxial and postaxial sides of the limbs **(A)**. When the animal walks, the hand and foot rotate forward **(B)**. *(Modified from Starck, 1979.)*

Accessory Structures of the Integument

The skin of the earliest tetrapods of the Late Devonian period is not known, but small bony scales may have been retained from their piscine ancestor. Extant lissamphibians lack scales, with the exception of bony nodules in the dermis of some caecilians. The skin of *Necturus* and most contemporary amphibians is relatively thin and not cornified. Many alveolar mucous glands secrete mucus that protects the skin and keeps it moist. Exchanges of gases, water, and some ions between the environment and the body occur through the skin. Many species have cutaneous glands that secrete toxic substances. The parotid glands caudal to the eyes of toads secrete a mild irritant that discourages unwary predators. Some South American frogs have poison glands that produce a very toxic material, which indigenous people use to prepare poison darts.

We have no direct knowledge of the integument of synapsids ancestral to mammals. Presumably, like the integument of extant reptiles, the integument of early synapsids was adapted to the terrestrial habitats in which these animals lived. Reptilian integument can be seen on demonstration specimens of a lizard, alligator, or turtle. Reptile skin is much thicker than amphibian skin, and the epidermis is heavily cornified, forming conspicuous **horny scales** in squamates. Between the individual horny scales, the epidermis is thinner, less cornified, and flexible. In certain species and certain regions of the body, bony plates that form in the dermis underlay these horny scales. On the head and on the external rim of the mouth, the horny scales are enlarged to form **horny plates.** In turtles, the shell is formed of horny scales that have been modified to form large horny plates that are supported by bony plates. The bony plates of the dorsal part of the shell, called **carapace,** are composed primarily of dermal bone, but they also incorporate the endoskeletal vertebrae and ribs. Some of the dermal plates in the ventral part of the shell, **plastron,** include remnants of the original bony armor of ancestral fishes. The tips of the digits of reptiles bear **claws,** which are heavily cornified epidermal structures covering a core of bone. Cutaneous glands are largely absent in the dry horny skin of reptiles. Only a few glands remain in limited areas, such as along the ventral surface of the thigh of some lizards. These may secrete pheromones for sexual communication.

MAMMALS

External Features

Examine either a cat or a rabbit and compare it with *Necturus.* The diagnostic **hair** of mammals is at once evident. Notice that the evolutionary trend for increased differentiation of body regions, which began in ancestral tetrapods, has continued in mammals. The **head** *(caput)* is large and separated from the **trunk** *(corpus)* by a distinct and movable **neck** *(collum)* (Fig. 3-5). A mammal moves its head in many directions as it explores its environment and obtains food. In the trunk we can recognize a **back** *(dorsum),* **thorax, abdomen,** and **pelvis.** A **tail** *(cauda)* is typically present in mammals but, except in the whales and porpoises, is greatly reduced in size compared with that of an ancestral tetrapod. In some terrestrial mammals it is used as a balancing organ or prehensile organ during locomotion, or to brush away insects. In a few others, such as the rabbit and deer, it is rudimentary and used mainly as a social signal. The tail is lost as an external structure in apes and humans.

Locomotion is accomplished by thrusts of the paired appendages, which consist of the usual parts: **brachium, antebrachium,** and **manus** in the pectoral appendage; and thigh or **femur,**[5] **crus,** and **pes** in the pelvic appendage. In both the cat and the rabbit, the most medial (or first) digit of the manus is vestigial, and the corresponding digit of the pes has been completely lost as an externally visible structure. Observe either on a mounted specimen or on a skeleton that mammalian limbs are carried under the body and are not splayed as in amphibians and reptiles (Anatomy in Action 3-3). A carnivore walks on its toes with the carpus and tarsus raised off the ground. This method of locomotion is referred to as **digitigrade,** in contrast to the **plantigrade**

[5]See Footnote 4.

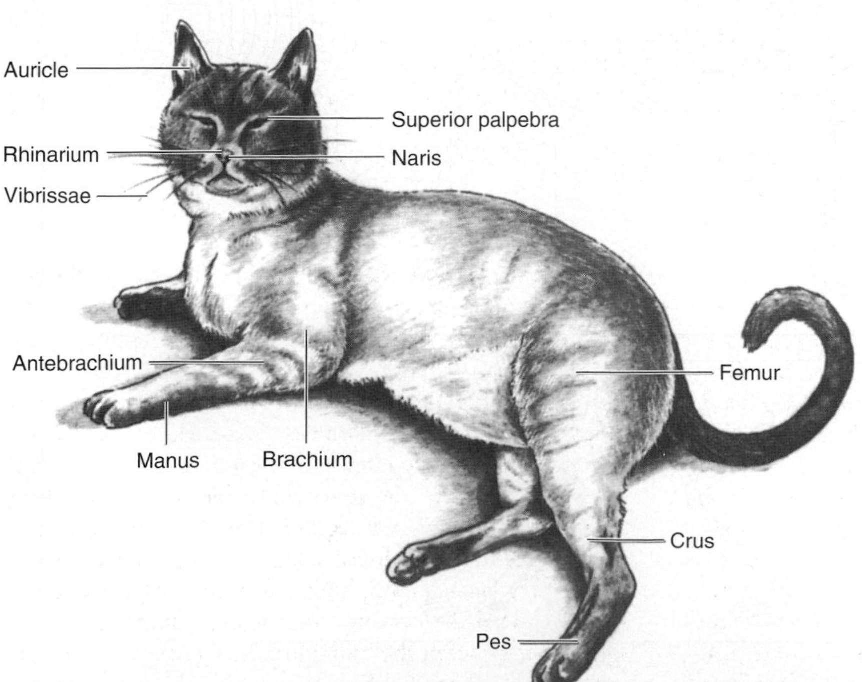

Figure 3-5
External features of a cat, *Felis catus.*

Anatomy in Action 3-3 The Evolution of the Posture and Use of Limbs in Mammals

As mammals evolved from early synapsids, their elbows and knees were brought closer to the sides of the trunk, so that their legs were carried nearer the body axis and could be moved back and forth in a vertical plane. In fast-moving mammals, such as carnivores and lagomorphs, all parts of the limbs move in the same vertical plane that is oriented in the direction of locomotion. This reorientation of the limbs provided more efficient support of the body. It also made longer steps and strides possible[6] and lateral undulations of the vertebral column unnecessary to move the limbs forward (see Anatomy in Action 3-2 and Fig. 7-10).

The limbs of ancestral synapsids were in a splayed position and projected laterally from the trunk, so that the preaxial and postaxial sides of the limbs faced cranially and caudally, respectively (Fig. **A**). The evolutionary transformation into the limb posture of mammals (Fig. **B**) involved two changes that must have occurred gradually and *in tandem,* but are shown as separate events to clarify this complex evolutionary process. One change involved the rotation of the brachium of the forelimb so that the elbow points caudally and the rotation of the femur of the hindlimb so that the knee points cranially (Fig. **B**). As a result of these reorientations of the brachium and femur, the elbow and knees were moved near the sides of the trunk and closer to the body axis (see Fig. 7-22). The original hinge joint at the knee was retained. A cranially pointing knee also results in a cranially pointing hind foot so the angle joint is also primarily a hinge joint. A second change was a rotation of the radius about the ulna at the elbow and distal radioulnar joints. When the front foot is prone, these two bones now cross and the front foot points cranially (Fig. **C**).

As a consequence of the complex reorientation of the limbs in mammals, the ancestral preaxial and postaxial sides of the limbs have changed their orientation. In the hindlimb, the originally preaxial side of the entire limb now faces medially and the postaxial side faces laterally. In the forelimb, however, the situation is more complex. The originally preaxial sides of the brachium and antebrachium face laterally and the postaxial sides face medially. In contrast, the originally preaxial side of the manus faces medially and the originally postaxial side faces laterally, which is the same orientation as that of the pes.

In order to understand the complex reorientation of limbs in mammals, place your own appendages into the ancestral tetrapod position and then rotate them into the cursorial mammal position.

[6]A *step* is the distance an animal is carried forward by one foot on the ground (the left hind foot, for example). A *stride* is the distance an animal travels between the placement of one foot on the ground and the next placement of the same foot. A stride is equal to four steps in a quadruped plus any distance traveled when all of the feet are briefly off of the ground, as in a gallop.

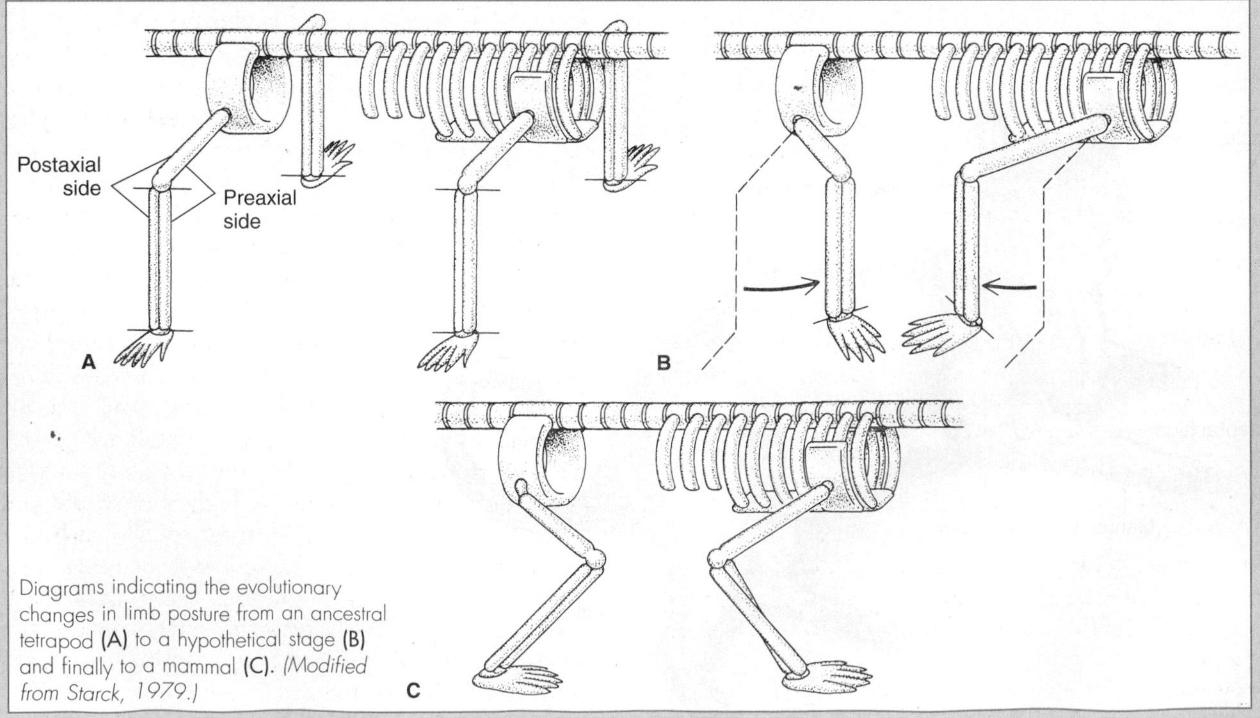

Diagrams indicating the evolutionary changes in limb posture from an ancestral tetrapod **(A)** to a hypothetical stage **(B)** and finally to a mammal **(C)**. *(Modified from Starck, 1979.)*

locomotion in humans and bears, in which the entire sole of the foot is flat on the ground, or **unguligrade** locomotion in ungulates such as the horse and cow, in which the animal walks on the tips of its toes. The rabbit's mode of locomotion is digitigrade with respect to the front feet, but the hind legs are modified for bounding. Just before a bound, the pes is in the plantigrade position. The terminal segment of each toe bears a **claw;** in the cat this segment is hinged in such a way that the claw can be retracted or extended (see Fig. 6-15**C** and **D**).

Examine the head. Notice that the **mouth** *(os)* is bounded by fleshy **lips** *(labia),* the upper one being deeply cleft in the rabbit, hence the term harelip. The paired external nostrils, called **nares,** are close together on the nose and surrounded by moist bare skin known as the **rhinarium.** The **eyes** *(oculi)* are large and are protected by movable upper and lower eyelids, called **palpebrae.** Spread the palpebrae apart and observe a third lid, called the **nictitating membrane,** in the medial corner of the eye. The nictitating membrane can be drawn across most of the eye, thus helping to moisten and cleanse this organ. Mammals have a prominent external ear consisting of a conspicuous external flap, called the **auricle,** or *pinna,* and an external ear canal that extends into the head from the base of the auricle. The eardrum, or **tympanic membrane,** is located at the end of the external ear canal. It will not be seen at this time. The part of the head that includes the jaws, mouth, nose, and eyes is referred to as the **facial region;** the rest, containing the brain and ears, is referred to as the **cranial region.** .

The cloaca of nonmammalian vertebrates has become divided in therian mammals (marsupials and placentals) so that the intestine and urogenital ducts open independently at the surface. The opening of the intestine, called the **anus,** is found just ventral to the base of the tail. In females, the combined opening of the urinary and reproductive ducts appears as a second passage, the **vaginal vestibule,** which is bounded by small folds ventral to the anus. In males, the urogenital duct opens at the tip of a **penis.** Look for the sac-shaped **scrotum** containing the testes near the penis. In rabbits, the testes may be retracted into the abdominal cavity. The entire area of the anus and external genitals is called the **perineum** in both sexes. We will examine the external genital organs in more detail in Chapter 12.

Carefully feel along the ventral surface of the thorax and abdomen on each side of the midline, and you will find two rows of nipples *(papillae mammae)* hidden in the fur. These bear the minute openings of the mammary glands. The nipples are more prominent in females, but rudiments can sometimes be found in males. There are usually four or five pairs in cats and six in rabbits, but the number varies. The mammary glands can be palpated only in lactating females.

Accessory Structures of the Integument

If available, examine microscopic slides of mammal skin and demonstrations of some of the integumentary accessory structures. The basic structure of mammal skin is shown diagrammatically in Figure 3-6. The plane of the section of the slides does not pass neatly through the long axes of hairs and glands; therefore, you will have to examine different parts of the slides to reconstruct the entire gland or hair. Most mammals lack horny scales, but the **epidermis** is thick and keratinized. New cells are produced by mitosis in the basal layer of the epidermis, the **stratum germinativum.** As they move toward the surface, they accumulate lipids and keratin and die. A thick layer of dead cells forms the horny, waterproof **stratum corneum** on the surface. Various intermediate layers of the epidermis can be recognized depending on the skin region studied.

The **dermis** is much thicker than the epidermis. You may see a few pigment-producing melanophores, also called **melanocytes,** in the dermis just beneath the epidermis, but most of the melanin they produce is transferred to epithelial cells and hair. The dermis, which is composed of connective tissue, contains many blood vessels, some of which lie in dermal papillae that push into the epidermis and come close to the body surface. The extent of blood flow through the skin can be controlled, a very important factor in regulating body temperature. Flow rate increases when the body is too warm and decreases when body heat needs to be conserved.

Unlike reptile skin, mammal skin is rich in glands, all of which are epithelial, develop from the epidermis, and invaginate into the dermis. Alveolar **sebaceous glands** usually discharge their oily secretion, the **sebum,** into the follicles surrounding the hairs. Sebum helps lubricate, waterproof, and condition hairs. Coiled, tubular sweat glands are of two varieties. **Apocrine sweat glands,** which are abundant in the armpits and in the genital area of humans, also discharge into the hair follicles and produce secretions in which bacteria may grow. The combination of the metabolic waste products of the bacteria and the secretions is responsible for body odors. The specialized scent glands of some mammals are modified apocrine glands. **Eccrine sweat glands** produce a more watery solution that is discharged on the body surface and is important in cooling the body in some species, such as human beings and horses. The openings of some eccrine sweat glands can be seen on your fingertips using low magnification. In heavily furred mammals, in which the evaporative loss of body water by panting helps reduce body temperature, eccrine sweat glands tend to be limited to the snout, tail base, or soles of the feet. The distinctive **mammary glands,** from which mammals derive their

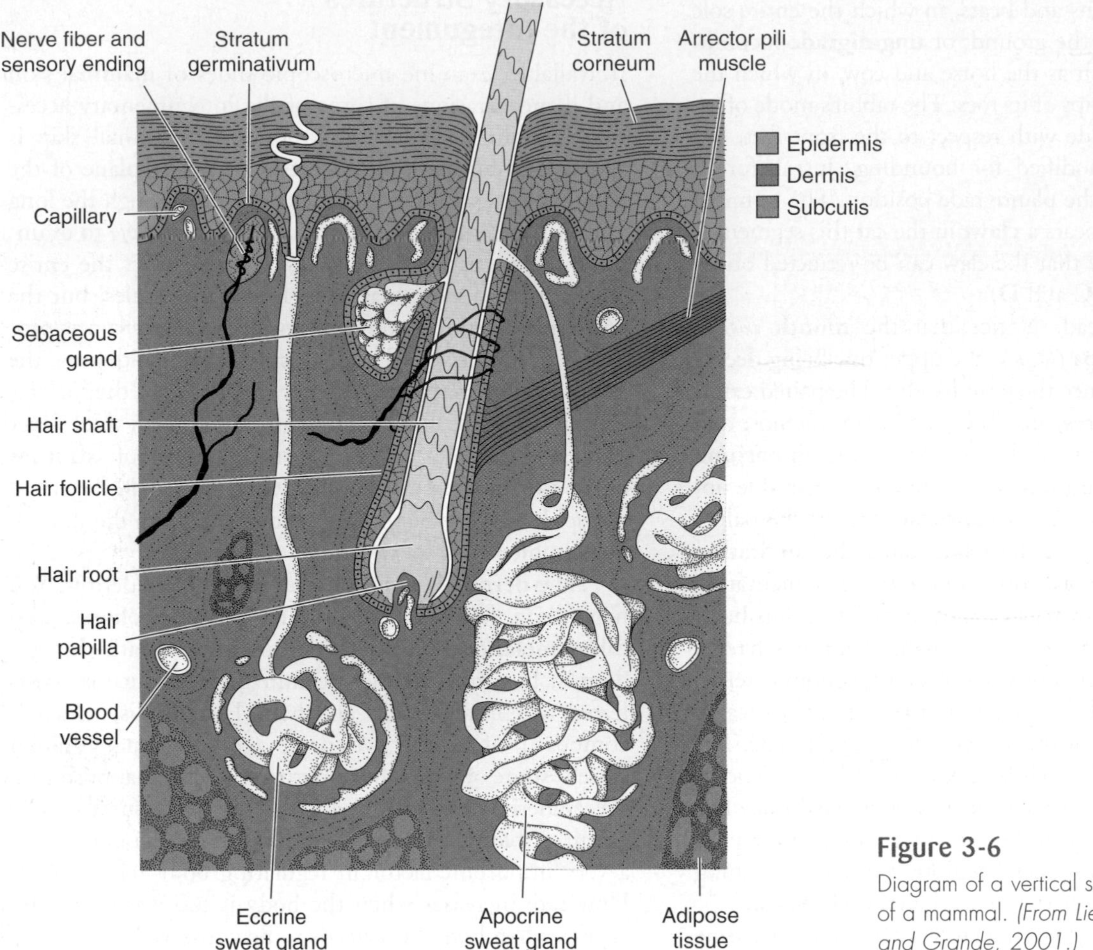

Figure 3-6

Diagram of a vertical section through the skin of a mammal. *(From Liem, Bemis, Walker, and Grande, 2001.)*

name, resemble sebaceous glands in that they are composed of branched alveolar units, and they resemble sweat glands in that they have contractile **myoepithelial cells** surrounding the secretory cells. Blackburn (1991) has proposed that mammary glands evolved from neither sebaceous nor sweat glands, but as a neomorphic mosaic combining characteristics of both, as well as new features.

The most conspicuous integumentary derivative is the protective and insulating covering of **hair** *(capillus).* Notice on the slides that a **follicle** of living epidermal cells surrounds a **hair shaft,** which is composed of dead cornified cells. The base of the shaft expands as the **hair root,** where mitosis of cells produces the shaft. A small smooth muscle, the **arrector pili,** attaches to the follicle (Fig. 3-6). Because the hairs slant toward the surface, contraction of the arrector pili muscles pulls the hairs into a more erect position, increasing the thickness of the hair layer and the thermal insulating effects of the hair.

Hair replaces the horny scales of reptiles and probably early synapsids in most mammals, but scalelike structures may still be found on the tails of certain rodents, and they have redeveloped over the bony plates of the armadillo shell. Hair is a newly evolved feature of mammals and is not considered to be homologous with either horny scales or feathers because details of the embryonic development of hair are different. Moreover, the distribution of hair, as seen, for example, on the back of one's hand, suggests that hairs evolved in small clusters between the horny scales in the ancestor of furry mammals, and that subsequently the scales were lost as hairs grew more densely. In most mammals the hair forms a dense fur over the body, being modified in certain places such as the **eyelashes** *(cilia)* and tactile **whiskers** *(vibrissae)* on the heads of most mammals. There are, however, many departures from this pattern. Hair is reduced in humans and lost in the adults of such highly aquatic mammals as whales. In some other species, hair has become adapted for very specialized purposes. The quills of a porcupine are a case in point.

Claws are retained in most mammals but have been transformed into **nails** *(ungulae)* in primates and into

hoofs in ungulates. Other common integumentary structures are the **foot pads** *(tori)* on the feet of most mammals. These are simply thickenings of the stratum corneum.

Aside from the widely distributed integumentary structures just mentioned, some mammals have other hard accessory structures of the skin. Bony plates form in the dermis of the armadillo and support the horny plates that cover most of the dorsal surface of the animal. The toothless baleen whales have large, fringed, cornified plates of **baleen** that hang down from the roofs of their mouths and entrap plankton. Horns and antlers are other types of integumentary structures (Anatomy in Action 3-4).

Anatomy in Action 3-4 Horns and Antlers in Mammals

The heads of some mammals, especially ungulates, carry processes, such as horns and antlers, which greatly vary among mammals. All of these processes are covered by the epidermis, at least during certain stages of their development, and some of them may be supported by a core of dermal bone that grows out of the dermal skull roof.

Bovids (cattle, sheep, buffaloes, and antelopes, except the pronghorn) possess **horns** that consist of a core of dermal bone, a layer of vascularized and innervated dermis, a layer of living epidermis, and a superficial sheath of cornified epidermis. The latter makes up the horn in the narrow sense (Fig. **A**). The pronghorn possesses horns whose structure is similar to that of bovid horns, except that their cornified epidermal sheath is shed and regrown annually (Fig. **B**). Horns are present in both sexes and are used mostly for defense.

Giraffes bear permanent horns of dermal bone, which remain covered by hairy skin (Fig. **C**).

Cervids (deer, reindeer, and caribous) possess **antlers** (Fig. **D**). Antlers are bony outgrowths from the dermal skull roof, and their epidermal covering does not cornify. After the antlers have reached their final size, the skin, also called **velvet,** is shed. Antlers are grown and shed annually. In deer, antlers grow only in males, which use them in male-male competition for females. In reindeer and caribous, both sexes grow antlers and are also used for defense.

Rhinoceroses sport one or two "horns" on their snout (Fig. **E**). These horns are purely epidermal structures without a core of bone. The epidermis giving rise to the solidly cornified epidermis appears to consist of cornified tubules. Contrary to earlier interpretations, these tubules are not homologous to hair, and the rhinoceros horn does not consist of amalgamated hair.

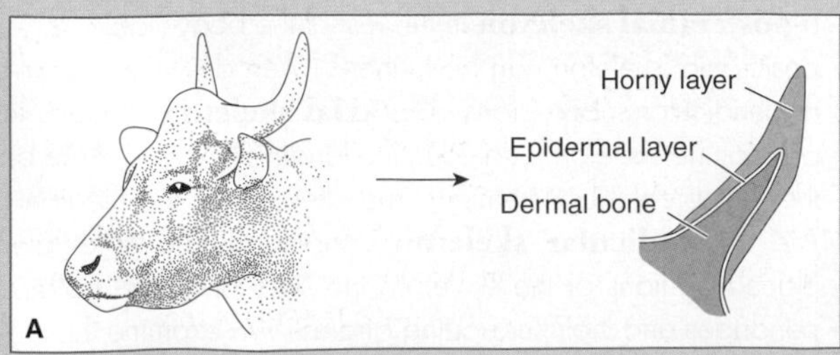

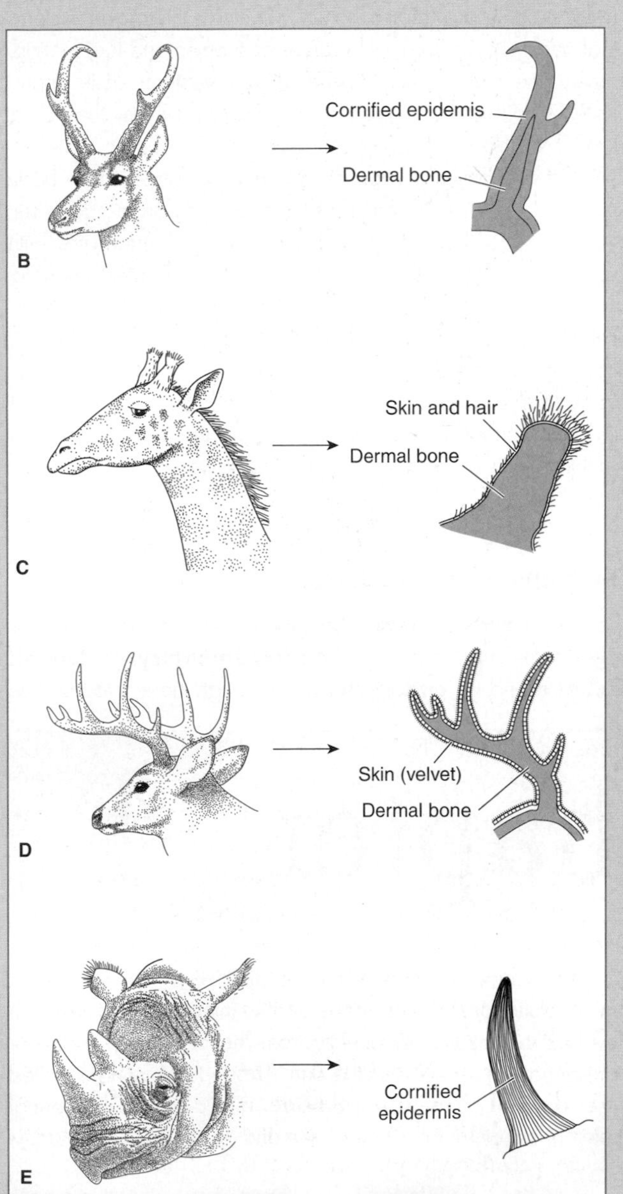

Examples of horns and antlers in mammals. **A**, Cow *(Bovis)*. **B**, Pronghorn *(Antilocapra)*. **C**, Giraffe *(Giraffa)*. **D**, White-tailed deer *(Odocoileus)*. **E**, Black rhinoceros *(Diceros)*. *(From Liem, Bemis, Walker, and Grande, 2001.)*

The Cranial Skeleton

Chapter 4

In the next several chapters, we will study the organ systems that deal with the general functions of support and locomotion, namely, the musculoskeletal and nervous systems. It is appropriate to consider the skeleton first because it forms the framework of the body.

The vertebrate skeleton is internal. It develops within body tissues, as opposed to forming from secretions on the body surface, as it does in insects and many other invertebrates with exoskeletons. Depending on the group of vertebrates, the notochord, cartilage, or bone may contribute to the adult skeleton. Skeletal elements support the body, protect internal organs, store important mineral ions such as calcium and phosphorus, occasionally house blood-forming tissues (e.g., red bone marrow), and form lever systems that transfer muscle forces to other structures such as the appendages and jaws.

Divisions of the Skeleton

The vertebrate skeleton can be subdivided in various ways. It may first be subdivided into the **integumentary,** or **dermal, skeleton** and the **endoskeleton.** Although these two become united in various degrees, they are distinct in their embryonic and evolutionary origins. Bone of the integumentary skeleton develops embryonically directly from the mesenchyme[1] in (or just beneath) the dermis of the skin. This type of bone is called either **dermal** or **membrane bone.** It follows that the dermal skeleton is superficial. Bony scales and plates, and their derivatives, are dermal in nature. In contrast, the endoskeleton arises in deeper body layers and consists of cartilage or bone that develops within and around cartilaginous centers of growth. Although such bone has the same histological structure as dermal bone, it is convenient to differentiate it as **cartilage replacement bone.**

The integumentary skeleton was particularly prominent in the early jawless fishes called "ostracoderms," which had an extensive dermal armor. The dermal armor is reduced in contemporary species, but it is still represented by the bony scales of bony fishes and by dermal bones in much of the head and parts of the shoulder girdle of most vertebrates. In addition, some dermal bone has evolved independently of bony scales in the dermis of the trunk of a few species. The bony plates of the turtle shell and armadillo are examples (Chapter 3). As the dermal skeleton has regressed during evolution, the endoskeleton has become much more prominent.

The skeleton also can be subdivided into visceral and somatic portions. The **visceral skeleton,** as the name implies, is associated with the "inner tube" (gut) of the body. It consists of skeletal arches, called visceral arches, that form in the wall of the pharynx. It contributes to the jaws and supports the gills in aquatic vertebrates. Aside from its location, the visceral skeleton differs from most of the somatic skeleton in that it develops from ectodermal mesenchyme that is derived from the neural crest. The **somatic skeleton** is associated with the "outer tube" of the body (body wall and appendages). It includes the dermal skeleton and all of the endoskeleton except for the visceral arches. Parts of the skull develop from neural crest cells, but most of the somatic skeleton develops from mesodermal mesenchyme.

It is also convenient to subdivide the skeleton according to its location in the body: a **cranial skeleton** in the head and a **postcranial skeleton** in the rest of the body (Table 4-1). The postcranial skeleton can be further broken down into axial and appendicular subdivisions. The **axial skeleton** includes those parts of the skeleton located in the longitudinal axis of the body: the vertebrae, ribs, sternum, and skeleton of the median fins. The **appendicular skeleton** consists of the more laterally placed portions of the skeleton: the skeleton of the paired appendages and their supporting girdles. We examine the cranial skeleton in this chapter and the rest of the skeleton in the next two chapters.

[1]Mesenchyme is an embryonic connective tissue composed of stellate, migratory cells that contribute to many adult tissues. Most mesenchyme is of mesodermal origin, but some (ectomesenchyme) derives from the ectodermal neural crest.

Table 4-1 Divisions of the Vertebrate Skeleton*

Cranial skeleton
Chondrocranium
Visceral skeleton (splanchnocranium)
Mandibular arch
Hyoid arch
Branchial arches
Dermal bones (dermatocranium)
Postcranial skeleton
Axial skeleton
Vertebral column (and notochord)
Ribs
Sternum
Medial fins
Appendicular skeleton
Pectoral girdle and appendages
Pelvic girdle and appendages

*All structures listed belong to the endoskeleton except for the dermatocranium and certain elements (e.g., the clavicle and interclavicle) of the pectoral girdle, which belong to the dermal skeleton. In addition, all structures listed belong to the somatic skeleton except for the splanchnocranium, which belongs to the visceral skeleton.

The cranial skeleton comprises three groups of elements that are distinct in some vertebrates but become confusingly united and mixed in most species: (1) the **chondrocranium**[2]; (2) the **visceral skeleton (splanchnocranium);** and (3) associated **dermal bones (dermatocranium).** The chondrocranium surrounds a variable amount of the brain and forms protective capsules about the olfactory sacs and inner ears. The splanchnocranium is composed of **visceral arches,** most of which originally supported gills. In many vertebrates, parts of the visceral skeleton also form or contribute to the jaws, form ear ossicles that transmit sound waves, help to encase the brain, and support the tongue. The dermal bones cover the chondrocranium and visceral arches in most fishes. Some dermal bones are lost in terrestrial vertebrates, but many persist to help form the braincase, the jaws, and the facial portion of the cranial skeleton.

[2]The term *chondrocranium* sometimes is used to include all parts of the head skeleton derived from cartilage (chondrocranium proper and visceral arches), but we use the term in its narrower sense, which excludes the visceral arches. Some authors use the term *neurocranium* for the chondrocranium in this restricted sense.

FISHES

The cranial skeleton of *Squalus* shows more clearly than that of most fishes the basic configuration of the chondrocranium and splanchnocranium, but *Squalus* is atypical in having an endoskeleton that is entirely cartilaginous. *Squalus* also lacks the dermal plates that covered parts of the cranial skeleton in early fishes. The only bony tissue present in sharks is the plate of acellular bone at the base of the placoid scales. The loss of ossification in extant sharks is thought to have evolved secondarily. A nonmineralized skeleton has a lower density and may improve the buoyancy of sharks, which do not possess a swim bladder (see Anatomy in Action 7-1). To observe the dermal bones that cover the endochondral skull bones in most fishes, including those ancestral to tetrapods, we will study the bowfin, *Amia,* whose skull has retained many characters of the ancestral actinopterygian bony fishes.

Chondrocranium

The chondrocranium is a complex box of cartilage, or cartilage replacement bone, that we can best understand by describing briefly the major features of its embryonic development (Fig. 4-1). Two pairs of longitudinal cartilages that lie beneath the brain form the basis for the chondrocranium in all vertebrates. The caudal pair, called **parachordals,** is located on each side of the cranial end of the notochord, which extends only as far forward as the pituitary gland, or hypophysis. The cranial pair, called **trabeculae,** is situated rostral to the hypophysis (Fig. 4-1A). The parachordals enlarge and unite to form the **basal plate.** The trabeculae form the **ethmoid plate,** which extends rostrally between the nasal sacs. Together the basal and ethmoid plates form the floor of the chondrocranium (Fig. 4-1B). A variable number of **occipital arches,** which are serially homologous to developing vertebrae, unite with the caudal end of the basal plate and surround the caudal end of the brain. An **otic capsule** develops around each inner ear, and a **nasal capsule** develops around each nasal sac. (An **optic capsule** begins to form in the wall of the developing eyeball and may ossify as sclerotic plates, but it does not unite with the rest of the chondrocranium because the eyeballs need to be free to rotate.) The lateral walls of the chondrocranium between the eyeballs develop from a complex set of **orbital cartilages,** which coalesce with each other and with other parts of the chondrocranium, leaving foramina for the passage of cranial nerves that emerge from the cranial cavity (Fig. 4-1C). In most vertebrates, dermal bones cover the brain dorsally, and only two rods of cartilage form the roof of the chondrocranium: the occipital arches and the **synotic**

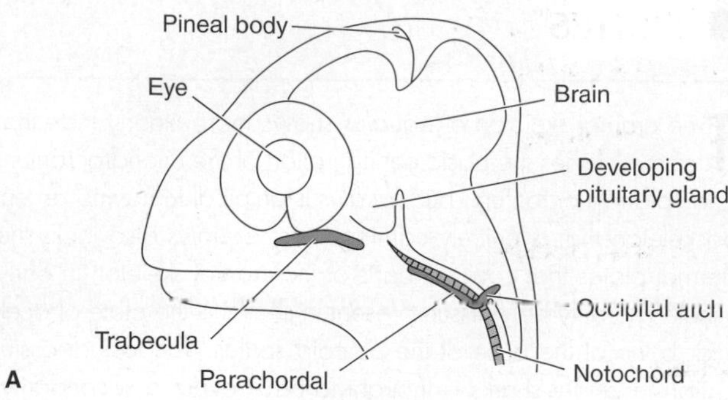

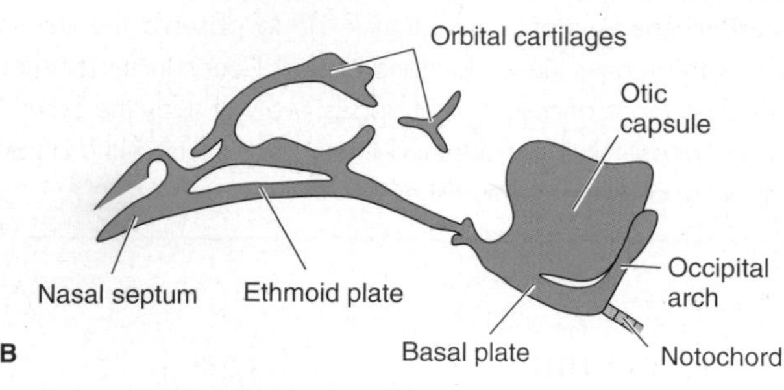

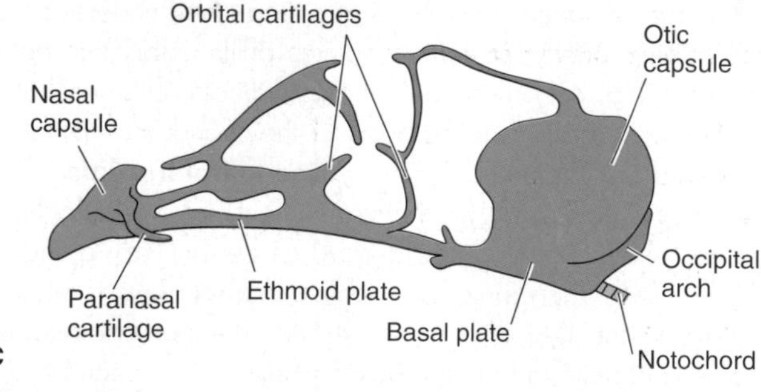

Figure 4-1

Lateral views of three stages in the embryonic development of the chondrocranium of the lizard, *Lacerta*. **A**, Embryo showing the developing chondrocranium *in situ*. **B**, Chondrocranium in an early embryo. **C**, Chondrocranium of a later embryo. *(From Liem, Bemis, Walker, and Grande, 2001; after de Beer.)*

tectum between the otic capsules. Because dermal bones are absent in cartilaginous fishes, the chondrocranium is complete dorsally in these vertebrates.

Complete Chondrocranium

Study a preparation of the chondrocranium of *Squalus* (Fig. 4-2). The embryonic chondrocranium is greatly expanded in the adult, more so than in the adults of other groups of vertebrates because it forms the entire skull. Its pointed, trough-shaped **rostrum** is at the rostral end; the caudal end has a rather square shape. A pair of large sockets for the eyes, the **orbits,** lie on the sides. Ventrally between the orbits the chondrocranium is very narrow; its dorsal surface is much wider.

It is convenient to describe the chondrocranium by the embryonic regions described in the introduction to this section, even though these are not clearly demarcated in the adult. The **occipital region,** which develops from the occipital arches, is the very caudal portion of the chondrocranium in the midline. It surrounds a large hole, the **foramen magnum,** through which the spinal cord enters the cranial cavity. A pair of bumps, the **occipital condyles,** can be seen ventral to the foramen magnum on each side of an area that resembles the body or centrum of a vertebra. They develop from parts of a vertebra that has been incorporated in the occipital region, and they articulate the chondrocranium with the vertebral column. In contrast to cartilaginous fishes, bony fishes, ancestral tetrapods, reptiles, and birds have a single rounded condyle located directly ventral to the foramen magnum.

The paired **otic capsules** are the large caudolateral corners of the chondrocranium that extend from the occipital region to the orbits. Between them, on the dorsal side, is a large depression called the **parietal fossa.** You will see two pairs of openings within the fossa that lead to the paired inner ears. The smaller cranial pair is the **endolymphatic foramina;** the larger caudal pair is the **perilymphatic foramina.** Look for ridges on each otic capsule: two obliquely oriented ridges dorsally and a horizontal ridge laterally. These cover the **semicircular canals** within which lie the **semicircular ducts** of the inner ear. Finally, there are two large pairs of foramina on the caudal edge of the chondrocranium lateral to the occipital condyles. The more medial **vagus foramen** is for the passage of the vagus nerve; the more lateral **glossopharyngeal foramen** is for the glossopharyngeal nerve.

Ventrally, the flat, broad **basal plate** connects the paired otic capsules. Some of the notochord persists in the chondrocranium of the adult spiny dogfish, and a white strand of calcified cartilage that surrounds the notochord indicates its position. This strand can be seen through the translucent hyaline cartilage along the midventral line of the basal plate. The small hole in the midline, anterior to the strand of calcified cartilage, is the **carotid foramen** for the passage of the internal carotid arteries that supply the brain.

The **optic region** is the area that lies between and includes the orbits. The **antorbital process, supraorbital crest,** and **postorbital process** form the cranial, dorsal, and caudal walls, respectively, of each orbit. Ventrally, the orbit is open. Most of the floor of the chondrocranium is narrow between the two orbits, but near the caudal part of the orbit the floor is wider and bears a prominent pair of lateral bumps called the **basitrabecular processes.** As will be seen later (Fig. 4-5**B**), the upper jaw articulates with the basitrabecular processes. Note that the roof of

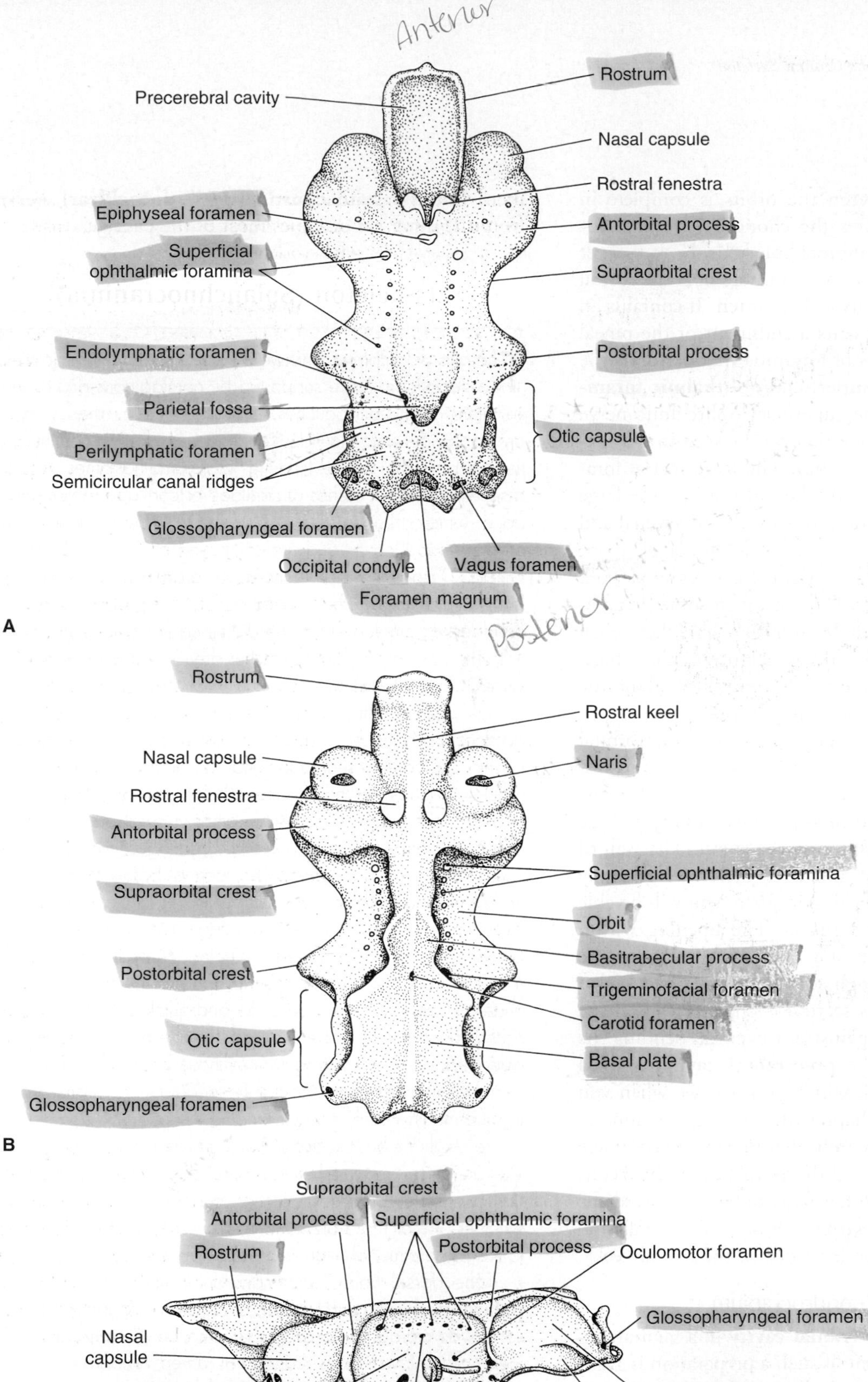

Anterior
Rostrum
Precerebral cavity
Nasal capsule
Rostral fenestra
Epiphyseal foramen
Antorbital process
Superficial
ophthalmic foramina
Supraorbital crest
Endolymphatic foramen
Postorbital process
Parietal fossa
Otic capsule
Perilymphatic foramen
Semicircular canal ridges
Glossopharyngeal foramen
Occipital condyle
Vagus foramen
Foramen magnum
Posterior
A
Rostrum
Rostral keel
Nasal capsule
Naris
Rostral fenestra
Antorbital process
Superficial ophthalmic foramina
Supraorbital crest
Orbit
Basitrabecular process
Postorbital crest
Trigeminofacial foramen
Carotid foramen
Otic capsule
Basal plate
Glossopharyngeal foramen
B
Supraorbital crest
Antorbital process
Superficial ophthalmic foramina
Rostrum
Postorbital process
Oculomotor foramen
Glossopharyngeal foramen
Nasal
capsule
Otic capsule
Facet for hyomandibula
Naris
Orbit
Optic foramen
Trigeminofacial foramen
Trochlear foramen
Abducens foramen
Optic
pedicle
Basitrabecular process
C

Figure 4-2

Chondrocranium of *Squalus*. **A**, Dorsal view. **B**, Ventral view. **C**, Lateral view.

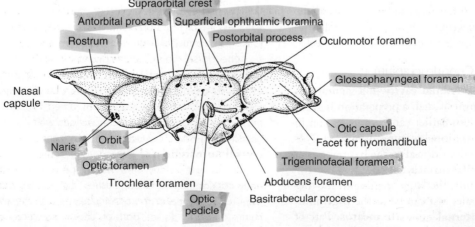

the chondrocranium between the orbits is complete in *Squalus.* In most vertebrates the chondrocranium is incomplete dorsally because dermal bones for the roof cover the brain dorsally. The small median hole near the rostral end of the roof is the **epiphyseal foramen.** It contains an epiphyseal stalk that represents a rudiment of the pineal eye (Chapter 8). The series of foramina that perforate the supraorbital crest are the **superficial ophthalmic foramina,** for the passage of the superficial ophthalmic nerve and its branches. Other foramina can be seen in the medial wall of the orbit (Fig. 4-2**C**). The large rostral foramen is the **optic foramen** for the optic nerve; the large caudal **trigeminofacial foramen** is for the trigeminal and facial nerves. The remaining foramina are for smaller cranial nerves and blood vessels. A small cartilaginous stalk resembling a golf tee is left in the orbit in some preparations. This is the **optic pedicle,** and it anchors the eyeball within the orbit. It is a distinctive feature of the Chondrichthyes whose orbits are open ventrally and of the closely related, extinct Placodermi (Fig. 3-1).

The entire chondrocranium rostral to the antorbital processes is called the **nasal region.** It consists of a pair of delicate round **nasal capsules,** which are attached to the rostral surface of each antorbital process, and of the long median **rostrum,** which supports the snout. The wall of each nasal capsule is very thin and, therefore, generally broken in preparations. If it is complete, you will be able to see its external opening, **naris.** The opening within the capsule is for passage of the olfactory tract of the brain. The rostrum is trough shaped dorsally and keeled ventrally, making its cross section T shaped. This shape strengthens the rostrum against dorsoventral bending. Its dorsal concavity, called the **precerebral cavity,** is filled with a gelatinous material, which you will see when you study the sense organs (Chapter 8). The large opening at its caudal end is the **precerebral fenestra.** In complete specimens, it is closed toward the cranial cavity by a connective tissue sheet. Two other large openings toward the cranial cavity, **rostral fenestrae,** can be seen ventrally between the nasal capsules on each side of the rostral keel.

Sagittal Section of the Chondrocranium

Examine the inside of the **cranial cavity** in a sagittal section of the chondrocranium, if such a preparation is available. Certain structures seen earlier can be noted again in this view, but there are additional features of interest. Notice the depression in the floor dorsal to the basitrabecular processes. This is the **sella turcica,** a recess for the hypophysis or pituitary gland. The large opening in the medial wall of the otic capsule, just caudal to the trigeminofacial foramen, is the **internal acoustic meatus.** Part of the glossopharyngeal nerve enters here to pass beneath the inner ear and emerge through the glossopharyngeal foramen, but the **statoacoustic (vestibulocochlear) nerve** from the inner ear occupies most of the passage.

Visceral Skeleton (Splanchnocranium)

The gills of lampreys, and we believe those of extinct jawless "ostracoderms," are supported by cartilaginous arches that lie lateral to the gill pouches, beneath the skin. These arches are not divided into individual cartilages but form a continuous, flexible branchial basket (see Fig. 2-5). Gnathostomes possess a series of visceral arches of cartilage or cartilage replacement bone, which are located medial to the gill pouches next to the oropharyngeal cavity and are composed of jointed skeletal elements. Despite these differences, the branchial basket of jawless vertebrates and the visceral skeleton of gnathostomes are considered homologous as we discussed in Chapter 2.

The jaws of cartilaginous fishes are formed by an enlarged visceral arch of cartilage in the wall of the mouth cavity. This arch is in series with the other visceral arches and usually is regarded as the first or **mandibular arch.** Kimmell, Miller, and Keynes (2001) present additional developmental and gene regulatory evidence that the mandibular arch of gnathostomes is part of the same series as the other visceral arches, and it probably is homologous to one of the cartilages that support the velum of lampreys. In other fishes, dermal bones attach to the underlying mandibular arch and contribute to the formation of the jaws. Jaws may have evolved originally as part of the respiratory mechanism and not for feeding (Mallatt, 1996). Jaws may have enabled an early fish to open its mouth quickly and widely, draw in a current of water, and quickly close the mouth so that the water could be expelled from the pharynx across the gills. Regardless of how jaws may have been first used, they soon became an integral part of the feeding mechanism. Opening the mouth widely and expanding the pharynx also sucked in prey with the water, and jaws could be used to seize prey. We are uncertain whether the mandibular arch was the very first in a series of visceral arches in ancestral vertebrates, or whether one or more premandibular visceral arches lay rostral to the present mandibular arch and were later lost.

The dorsal half of the mandibular arch, known as the **palatoquadrate cartilage,** must be braced against the skull so that its caudal end can act as a stable fulcrum for movements of the lower jaw. In ancestral jawed vertebrates, it was most likely braced by direct connections with the chondrocranium (Fig. 4-3A). This type of jaw suspension is termed **ancestral autostylic.** The second visceral arch, called the **hyoid arch,** was not involved in this type of jaw suspension. In early cartilaginous and bony fishes, the palatoquadrate cartilage retained one or more articulations with the chondrocranium, and the dorsal part of the hyoid arch, called the hyomandibular cartilage or **hyomandibula,** functioned as a prop between the otic capsule and the caudal end of the palatoquadrate cartilage (Fig. 4-3B). This is the **amphistylic** type of jaw suspension. Any gill slit that may have lain between the mandibular arch and the hyoid arch was reduced to a spiracle or lost entirely. An amphistylic jaw suspension also occurs in choanate or lobe-fined fishes, which are ancestral to terrestrial vertebrates. In derived cartilaginous and bony fishes, including *Squalus,* the primary brace for the palatoquadrate is the hyomandibula, but the palatoquadrate cartilage also articulates with the chondrocranium through its orbital process (Fig. 4-3C). This is the **hyostylic** type of jaw suspension. It permits the jaws to swing forward or downward during feeding but restricts lateral movements. Holocephalans and lungfishes have strong jaws adapted for crushing shellfish and other hard food. Their palatoquadrate cartilage has

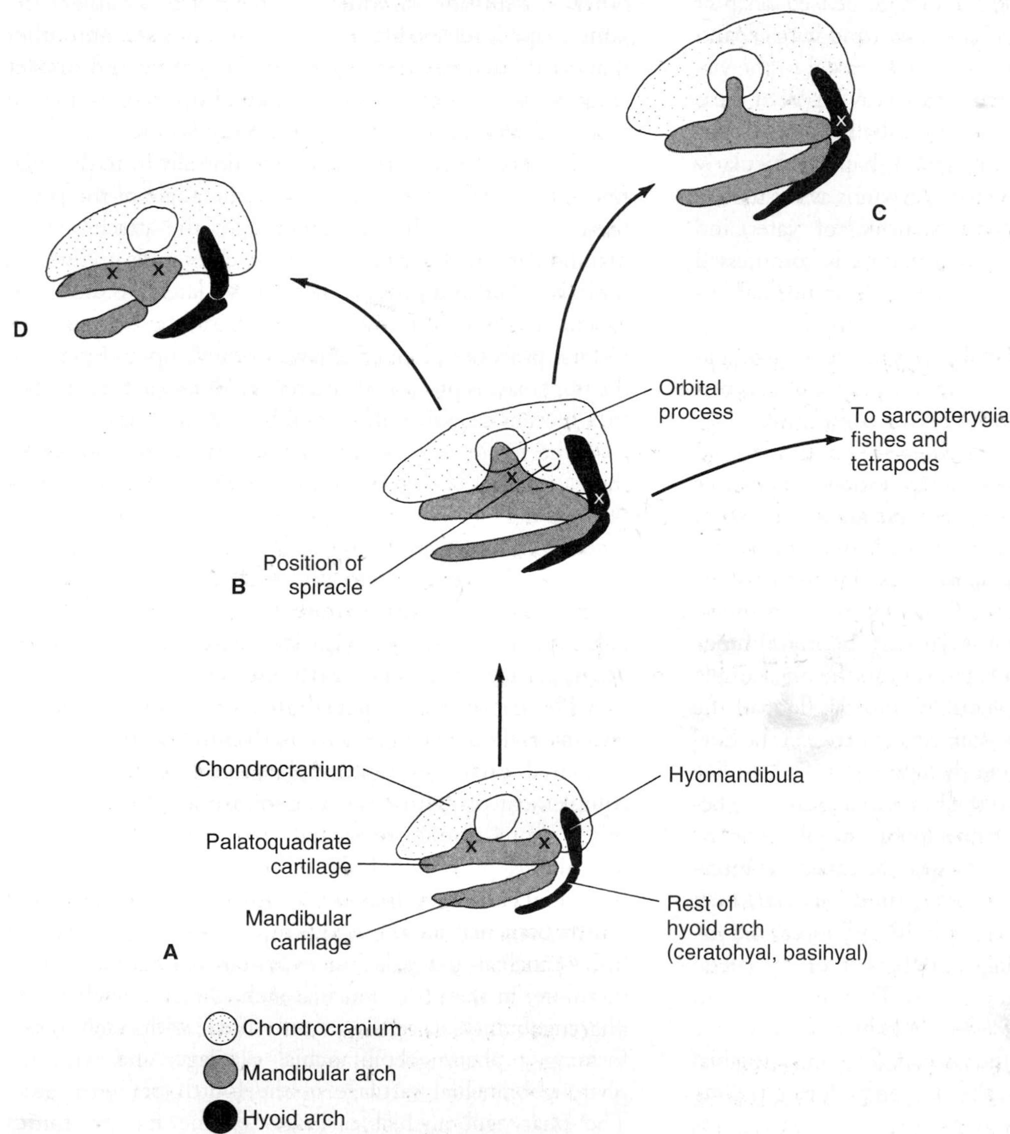

Figure 4-3
Four stages in the probable evolution of jaw suspension in fishes. The embryonic cartilaginous elements are shown, but many of these ossify in bony fishes. A, Ancestral autostylic suspension. B, Amphistylic suspension of early cartilaginous and bony fishes. C, Hyostylic suspension of most cartilaginous and bony fishes. D, Secondary autostylic suspension of holocephalans and lungfishes. X = points where the palatoquadrate cartilage attaches to other skeletal elements. Great movement of the palatoquadrate is permitted in the hyostylic type of suspension, but its excursion is limited by the basitrabecular processes of the chondrocranium and by ligaments.

united firmly with the chondrocranium (Fig. 4-3**D**). This is the **secondary autostylic** type of jaw suspension.

Examine a preparation of the visceral skeleton of *Squalus.* The visceral skeleton lies ventral to the chondrocranium and vertebral column and just rostral to the pectoral girdle. The pericardial cavity containing the heart lies directly ventral to the caudal end of the visceral skeleton, and its caudal end rests on the pectoral girdle (see Chapter 11). Muscles that pull the lower jaw down and help expand the pharynx occupy the rest of the space ventral to the visceral skeleton and around the pericardial cavity. All of these structures form a functional unit, as we shall see (see Chapter 10). The visceral skeleton changes in shape, particularly during suction feeding. The pharynx expands as the mouth opens and food is drawn in with a current of water, and then the mouth is closed and the pharynx is compressed when water is expelled across the gills. During normal respiration the mouth does not open but the pharynx expands, drawing water in through the spiracle (Wilga and Motta, 1998). Figure 4-4 shows the visceral arches when the mouth is closed and the pharynx compressed. Figure 4-5 shows the visceral arches when the mouth is open and the pharynx expanded. These figures show the possible range of movements, but the movements are not as extensive during respiration (see Anatomy in Action 10-1).

First identify the seven visceral arches that compose the visceral skeleton (Fig. 4-4**A**). The first arch, which is modified to form the upper and lower jaws, is the **mandibular arch.** The second arch, which extends from the otic capsule to the angle of the jaws and ventrally into the floor of the mouth, is the **hyoid arch.** The spiracle lies between the dorsal portion of the hyoid arch and the upper jaw. The last five visceral arches, which are called the **branchial arches,** lie between the pharyngeal pouches and support the gills. The tissue between the pharyngeal pouches can be called the **interbranchial septa.**[3] Note that the third visceral arch corresponds to the first branchial arch. In addition to the visceral arches, thin **labial cartilages** can be seen on the lateral surfaces of the jaws (Figs. 4-4 and 4-5). They are located in the labial folds previously observed (Chapter 3). They are too superficial to be a part of the visceral skeleton. The labial cartilages swing forward when the jaws open during feeding (Fig. 4-5**A**) and in doing so pull the labial folds over the lateral side of the gape to produce a tubular opening that facilitates suction feeding (Wilga and Motta, 1998).

Study the mandibular arch, or first visceral arch, in more detail. The paired **palatoquadrate cartilages** form the upper jaw; the paired **mandibular cartilages** form the lower jaw (Figs. 4-4 and 4-5). Both paired cartilages bear several rows of sharp triangular teeth, which are loosely attached to the surface of the jaws and are similar to one another. A dentition in which the teeth are essentially the same shape is referred to as **homodont.** *Squalus* and other sharks can use these teeth to tear up large prey into smaller chunks. In many fishes, the teeth are simple cones used to seize and hold prey, which is swallowed whole.

Two prominent processes extend dorsally from the palatoquadrate cartilage. The one above the angle of the jaw is the **adductor mandibulae process** for the attachment of mandibular muscles. The process, which extends up into the orbit, is the **orbital process** and it passes lateral to the chondrocranium and just rostral to the basitrabecular process. Orbital processes permit the jaws to move up and down as the upper jaw is protruded and retracted during feeding, but they prevent excessive caudal and lateral motions.

The hyoid arch, or second visceral arch, consists of three cartilages. The midventral element of the hyoid arch is called the **basihyal cartilage.** A **ceratohyal cartilage** extends from this element to the angle of the jaw, and a **hyomandibula** continues to the otic capsule of the chondrocranium. Ligaments unite the caudal end of the palatoquadrate cartilage to the hyoid arch, which connects the upper jaw to the chondrocranium.

The five paired branchial arches, or third to seventh visceral arches, and the two midventral unpaired **basibranchial cartilages** form a pharyngeal basket. The full complement of jointed elements of a branchial arch comprises, from dorsal to ventral, the paired **pharyngobranchial, epibranchial, ceratobranchial,** and **hypobranchial cartilages.** In *Squalus,* only the second, third, and fourth branchial arches follow this pattern. The first and fifth branchial arches lack a hypobranchial cartilage. Furthermore, in the fifth branchial arch, the epibranchial and pharyngobranchial cartilages are fused with each other, forming a pharyngoepibranchial cartilage, and with the pharyngobranchial cartilage of the fourth branchial arch. The pharyngobranchial cartilages of the first to fourth branchial arches are anchored to the vertebral column by individual ligaments. Look at Figure 4-5**B** to understand how the individual branchial arches connect midventrally with the two basibranchial cartilages. The unpaired caudal basibranchial cartilage is considerably larger than the small shield-shaped rostral basibranchial cartilage. It is a triangular, wide plate with a caudal spinous process that curves ventrally. It lies directly above the pericardial cavity with th[e] heart (see Fig. 11-9). It is unclear whether th[e]

[3]The tissue between successive pharyngeal pouches supports the gills and contains many associated structures: the skeletal visceral arches, muscles, nerves, and blood vessels. There is no fully agreed-upon term for this entire complex. Some authors call it the *gill arch, branchial arch,* or *visceral arch,* using the term arch in a broad sense. To avoid confusion, we limit the term *arch* to the skeletal elements. In this and previous editions we have called the entire complex the *interbranchial septum.* It also may be called a *branchiomere* (Bemis and Grande, 1992).

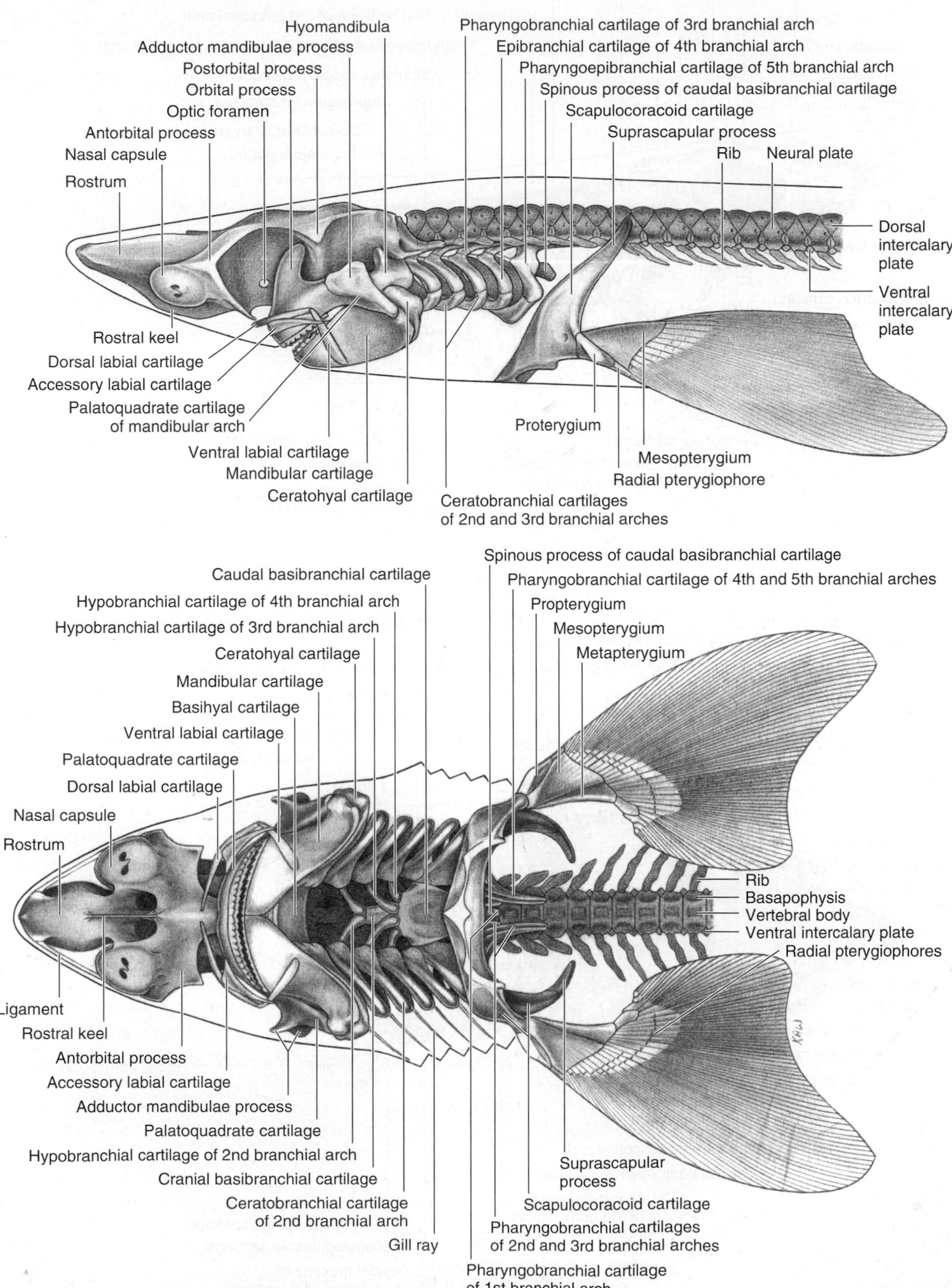

Figure 4-4

Cranial skeleton of *Squalus* and associated parts of the vertebral column and pectoral girdle when the mouth is closed and the oropharyngeal cavity is compressed. A, Lateral view. B, Ventral view.

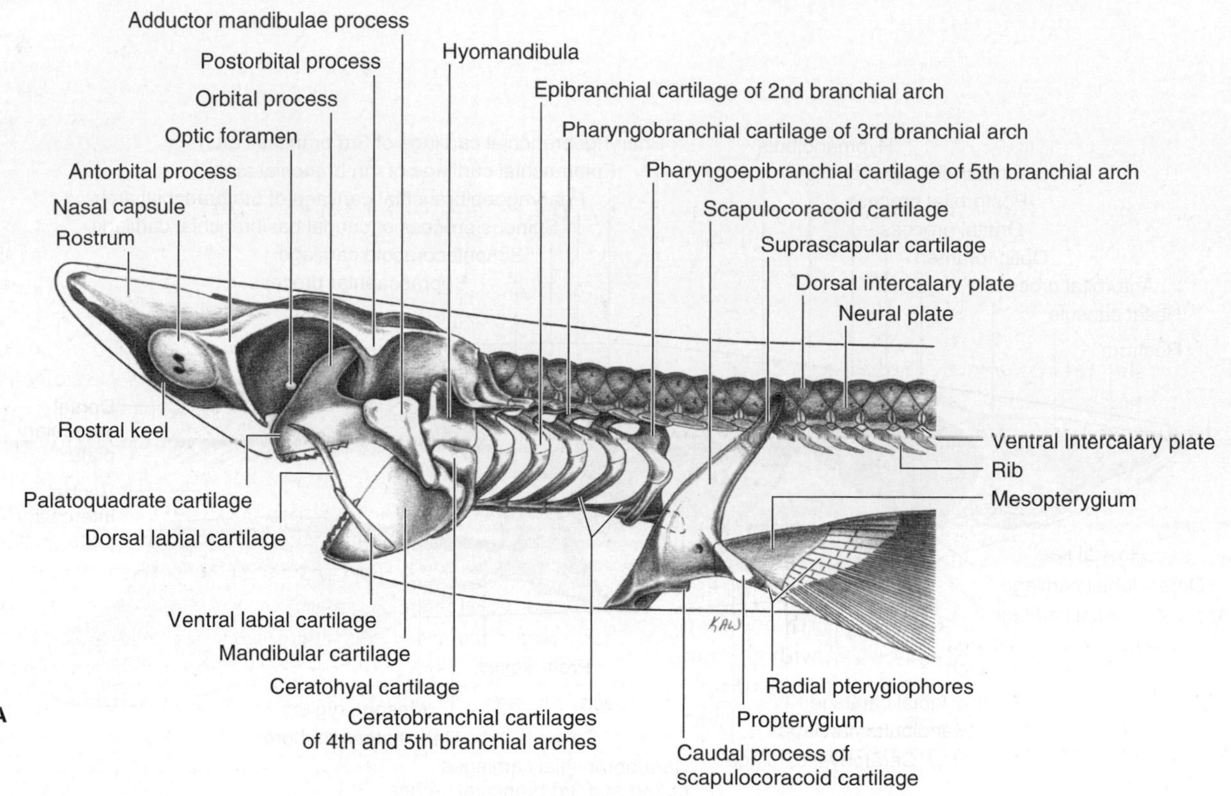

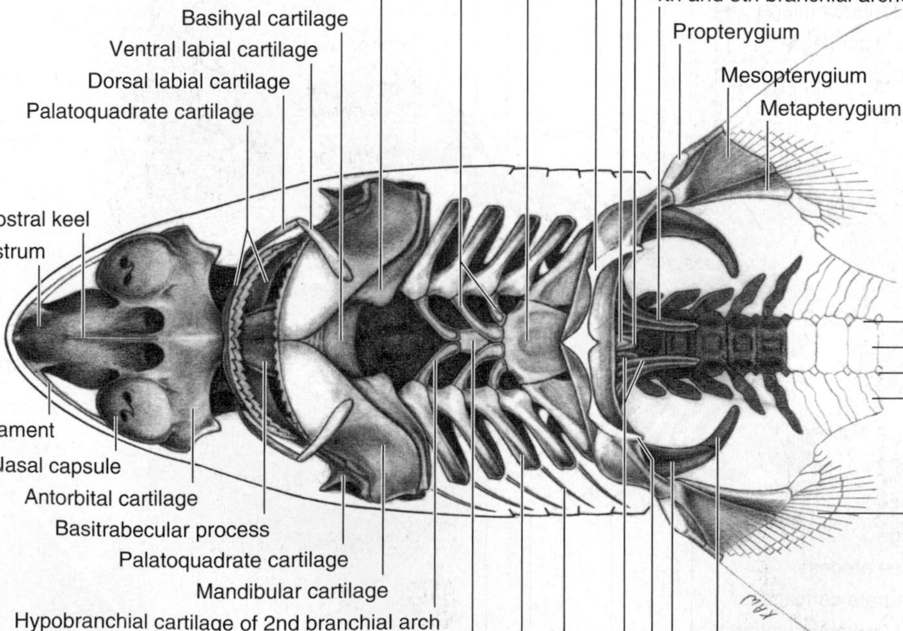

Figure 4-5

Cranial skeleton of *Squalus* and associated parts of the vertebral column and pectoral girdle when the mouth is open and the oropharyngeal cavity is expanded. Notice the change in shape of the visceral skeleton and the increase in the size of the oropharyngeal cavity. The accessory labial cartilages are not shown. **A**, Lateral view. **B**, Ventral view.

branchial cartilages evolved through the fusion of several basibranchial cartilages that used to be associated with individual branchial arches. Functionally, however, the relatively complex midventral interconnections result in a flexible midventral cartilaginous plate with two main transverse joints. Observe the varying orientation of the individual elements of the branchial arches. This complex arrangement enables the compression and expansion of the pharyngeal basket by folding and unfolding the individual branchial arches.

Gill rays, which serve as internal scaffolding of the interbranchial septa, are found on the hyomandibular and ceratohyal cartilages of the hyoid arch and on the epibranchial and ceratobranchial cartilages of the first to fourth branchial arches. The fifth branchial arch is not associated with gills and serves as attachment place for the esophagus (Figure 10-9).

Median projections from the branchial arches, called **gill rakers,** prevent food in the pharynx from entering the pharyngeal pouches.

Dermal Bones (Dermatocranium)

The third component of the head skeleton of most fishes is the dermatocranium, which covers the chondrocranium and visceral skeleton and extends as far caudad as the pectoral girdle. Although cartilaginous fishes have small dermal placoid scales in their skin (see Chapter 3), they lack dermal bones in the cranial skeleton and pectoral girdle. Evidence suggests that large dermal plates may never have been present in the ancestors of the chondrichthyans, although they were present in other fish groups, including the ancestors of terrestrial vertebrates, certain sarcopterygian fishes (Fig. 3-1). Interrelationships of the three components of the head skeleton (chondrocranium, visceral skeleton, dermal bones) are diagrammed in Figure 4-6. An appreciation of the extent of the dermal bones present in most fishes can be gained by studying them in the bowfin, *Amia calva,* a relatively basal actinopterygian.

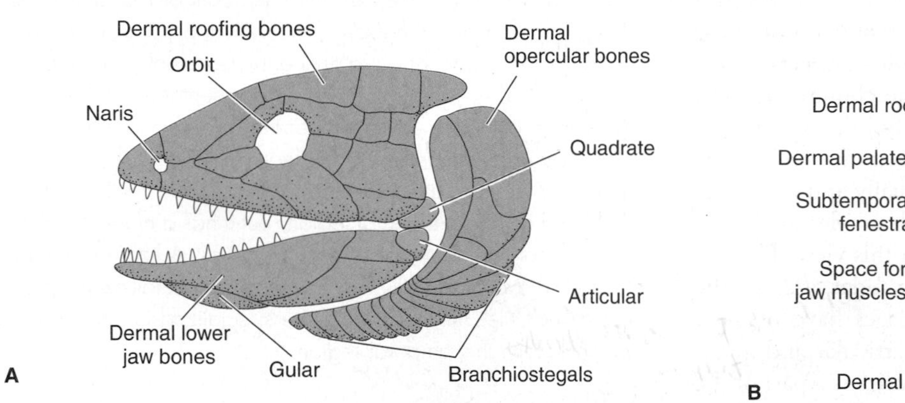

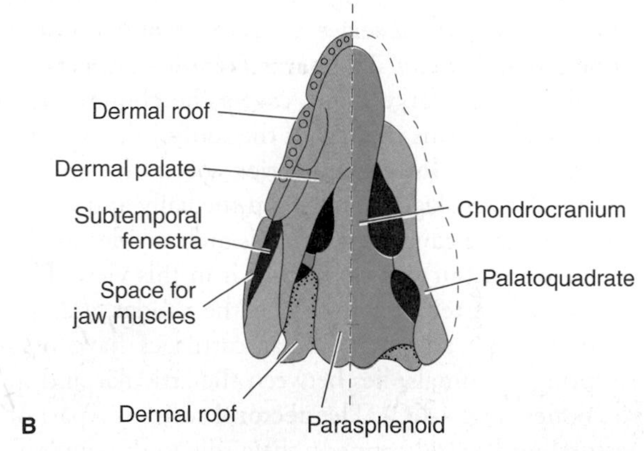

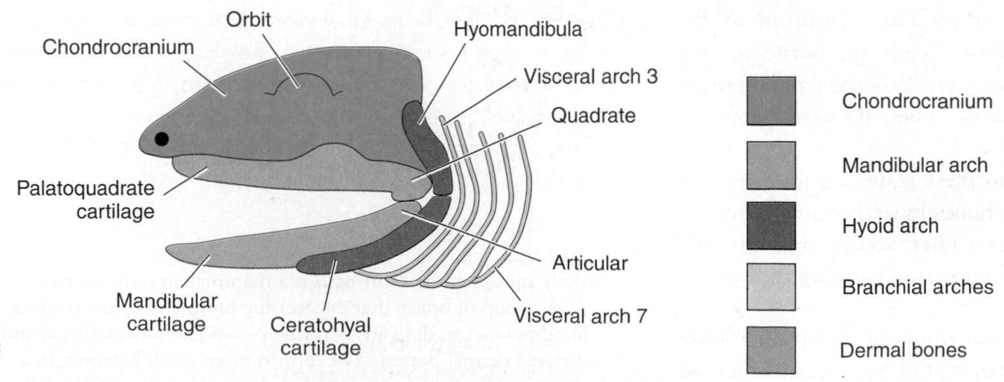

Figure 4-6
Diagrams of the components of the head skeleton of a generalized bony fish. **A,** Lateral view of the skull showing the superficial dermal bones that cover most of the other components. **B,** Ventral view of the skull with the dermal bones removed from the right side of the drawing. **C,** Lateral view of the cranial skeleton after removal of the dermal bones, revealing the chondrocranium and visceral arches.

Examine a skull of *Amia* (Fig. 4-7**A**). Nearly all of the bones you see are dermal because dermal bones sheathe all of the chondrocranium and most of the visceral arches. Only the **quadrate bone,** which ossifies in the palatoquadrate cartilage of the mandibular arch, and the **hyomandibula,** which ossifies in the hyoid arch, can be seen through a gap in the dermal roof lying between the upper jaw and the opercular bones. These two bones are cartilage replacement bones. The gap allows some mobility to the upper jaw, palate, and opercular bones during feeding and breathing movements. The parts of the skull that can move are shown in different colors. If a preparation is available in which the dermal bones have been removed from one side, you can see the chondrocranium and branchial arches.

For purposes of description, we group the dermal bones into seven series of bones, which are labeled in Figure 4-7**A:** dermal roof, opercular series, gular series, palatal series, parasphenoid series, lower jaw series, and pectoral series. The **dermal roof** covers the chondrocranium dorsally and laterally and forms the upper jaw. The **opercular** and **gular series** cover the branchial arches laterally and ventrally. The **palatal series** forms much of the roof of the oropharyngeal cavity. The **parasphenoid series,** which is actually a single large bone, covers the chondrocranium ventrally and forms the rest of the roof of the oropharyngeal cavity. The **lower jaw series** covers the embryonic mandibular cartilage laterally and medially and forms the lower jaw. The caudal end of this arch ossifies as the **articular bone,** but you do not see it in this view. The jaw joint of all vertebrates in which the caudal ends of the palatoquadrate and mandibular cartilages have ossified, excepting mammals, lies between the articular and quadrate bones (Fig. 4-6**C**). The **pectoral series** is a part of the pectoral girdle, and connects the girdle to the caudodorsal corners of the skull. There is no neck in fishes.

The teeth of *Amia* vary in size, but all are conical and essentially similar to each other. This condition of the dentition is called **homodont.** Teeth are borne on the margins of the upper and lower jaws, on the palatal bones, and on the parasphenoid bone. They are used primarily for holding prey.

The individual bones in these series are identified in Figure 4-7**B.** Many of these bones have the same names as those in terrestrial vertebrates. They occupy the same relative positions in fishes and terrestrial vertebrates, but the homologies are not entirely certain.

You may have noticed that the head skeleton of *Amia* is relatively much shorter than that of *Squalus* and that the proportions among the various parts of the head differ considerably between the two fishes. Take some time to compare the head skeletons of *Amia* and *Squalus.* In *Amia,* the pectoral fin girdle is attached directly behind the dermal roof of the skull, and the pectoral fin emerges directly from underneath the opercular series. The jaw opens terminally, that is, at the front end of the head. The openings between the gill arches are covered laterally by the dermal opercular series. Notice that connective tissue interconnects various bones and groups of bones. The connections allow movements between the multiple parts of the dermal skull. The skull of *Amia* can be considered an early prototype of the skull of bony fishes.

AMPHIBIANS AND REPTILES

The skulls[4] of contemporary amphibians and reptiles are specialized in many ways and, therefore, are not good representatives of the skulls of ancestral terrestrial vertebrates. We can better understand the components of the tetrapod skull by first describing briefly the skull of an extinct anthracosauroid, a group transitional in many ways between early amphibians and reptiles.[5] Our description is based on *Protogyrinus* and *Paleoherpeton.* When you then examine the skull of *Necturus* or of the snapping turtle, *Chelydra,* you will appreciate how their skulls compare with that of a truly early terrestrial vertebrate.

Necturus is widely used as a representative of the class Amphibia, but you must remember that *Necturus* is paedomorphic and possesses some larval features in its skull. The turtle skull also is specialized in many ways, including the loss of teeth and a very short facial region, but it has a more complete dermal roof than many other extant species. A lizard or alligator skull, if available, will show some ancestral features not present in the turtle, but these skulls are specialized in other ways, particularly in the temporal region.

The Anthracosauroid Skull

The three groups of skeletal elements present in the head region of fishes (chondrocranium, visceral skeleton, and dermal bones) were represented in the terrestrial anthracosauroids, but changes in methods of gas exchange and feeding and in the major sense organs that occurred during the transition from water to land affected skull morphology. The opercular and gular series of

[4]Strictly speaking, the term *skull* is a mammalian term for the unified group of bones that encases the brain and forms the face (chondrocranium, derivatives of the palatoquadrate cartilage, and associated dermal bones). The term is often used, however, in a looser sense for the entire cranial skeleton. Usually you can tell from the context which meaning is intended.

[5]Anthracosauroids currently are grouped with reptiles, but they also resemble ancestral amphibians and sometimes have been grouped with them.

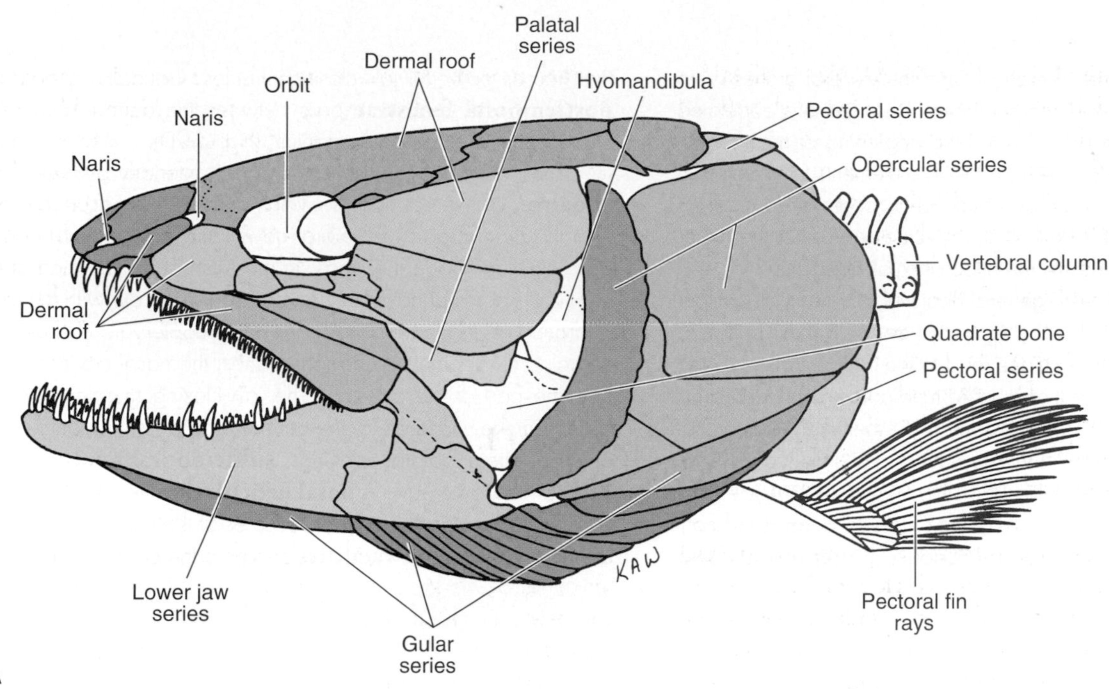

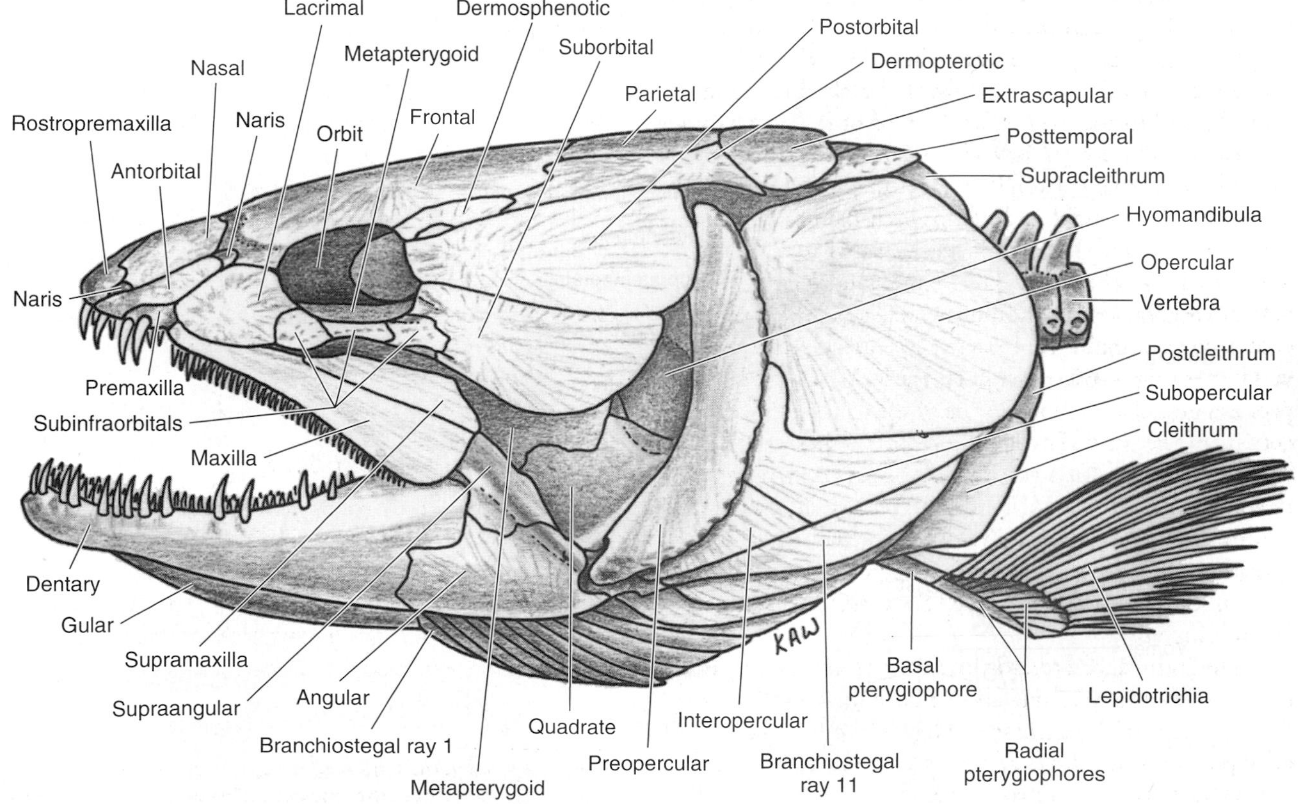

Figure 4-7

Lateral views of the skull of *Amia* with the pectoral girdle and fin attached. **A,** The groups of dermal bones are labeled. Colors indicate moveable units within the skull. Connective tissue, the orbit, vertebral column, and the visceral skeleton are not colored. **B,** Individual bones. The ceratohyal bone, which anchors the branchiostegal rays and belongs to the visceral skeleton, is not visible in this view.

dermal bones of fishes were lost, and the visceral arches were greatly reduced. These morphological changes are correlated with the reduction of the gills and the beginning of neck formation as adaptations to terrestrial life. The elements that remained in anthracosaurids can be grouped, for purposes of description, into four units: (1) the skull in its restricted sense, (2) the lower jaw, (3) the teeth, and (4) the hyobranchial apparatus.

First notice several general features of the skull proper (Figs. 4-8**A** and 4-9**A**). The external nostrils, **nares,** perforate the front of the dermal roof and lead to the nasal cavities, a pair of **orbits** lies caudal to them, and a median **parietal foramen** for a median eye lies caudal to the orbits between the parietal bones. An **otic notch** of unknown functional significance is located on the caudodorsal part of the dermal roof. Aside from this notch, the temporal portion of the roof is solid (**anapsid** condition), that is, it completely covers dorsally a **temporal fossa** that houses powerful jaw-closing muscles. Caudally, a pair of **posttemporal fenestrae** pass between the dermal roof and otic capsule of the chondrocranium (Fig. 4-9**D**).

The palate (Fig. 4-8**B**) is perforated anteriorly by a pair of **choanae,** or internal nostrils, which connect the nasal cavities with the cranial part of the oral cavity and are part of the air passages leading to the lungs. In most fishes the nasal cavities are blind sacs and have only an olfactory role, but with the appearance of choanae in the group of sarcopterygian fishes ancestral to tetrapods (the choanate fishes) the nasal cavities also become part of the airways. The jaw-closing muscles, which arise from the underside of the dermal roof, pass through a pair of large, lateral palatal openings, **subtemporal fenestrae,** to insert on the lower jaw. A **basal articulation** lies between the pterygoid and basitrabecular processes of the chondrocranium, and **interpterygoid vacuities** separate the palate and brain-

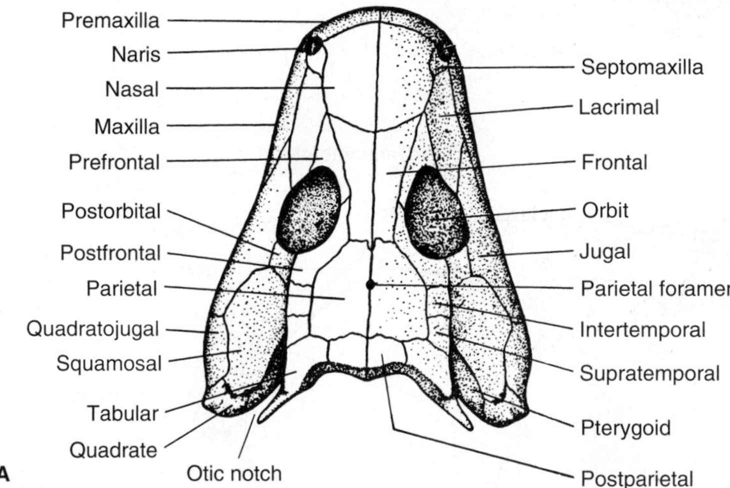

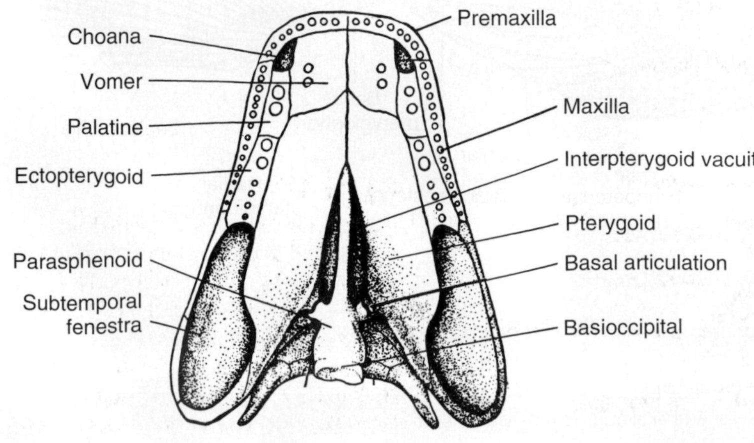

Figure 4-8

Diagrams of a basal tetrapod skull based on the Carboniferous anthracosauroid *Protogyrinus* and *Palaeoherpeton*. **A,** Dorsal view. **B,** Palatal view. *(Modified from Romer and Parsons, 1986.)*

case rostral to the basal articulation. This articulation and these vacuities permit some movement between the palate and braincase during feeding.

The many individual bones present in the dermal roof and palate of anthracosauroids are shown in Figures 4-8 and 4-9 and listed in Table 4-2. Most of them persist in later amphibians and reptiles, so you should become familiar with their names, even though the evolutionary trend in the skull since anthracosauroids has been one of bone loss and fusion.

The chondrocranium covers the back, the underside, and most of the lateral surfaces of the brain (Fig. 4-9**C** and **D**). It does not cover the top of the brain, and there is a gap in each of its lateral walls. It is ossified to a large extent, but the rostral ethmoid region and nasal capsules are unossified. As in most fishes, but not *Squalus,* the occipital condyle is a single knob ventral to the foramen magnum. The chondrocranium is perforated by foramina for nerves and blood vessels and by an **oval window** *(fenestra vestibuli)*

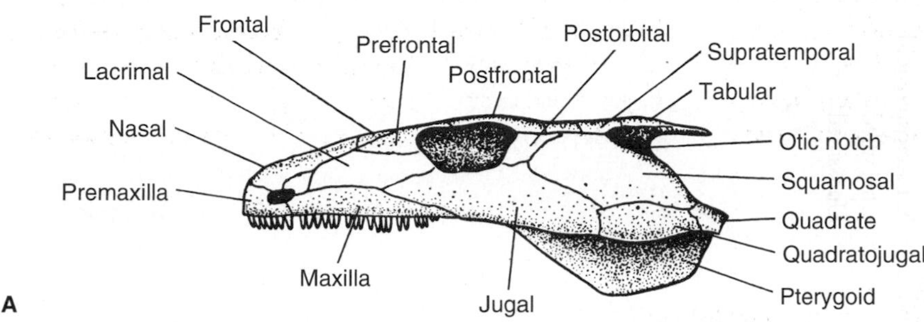

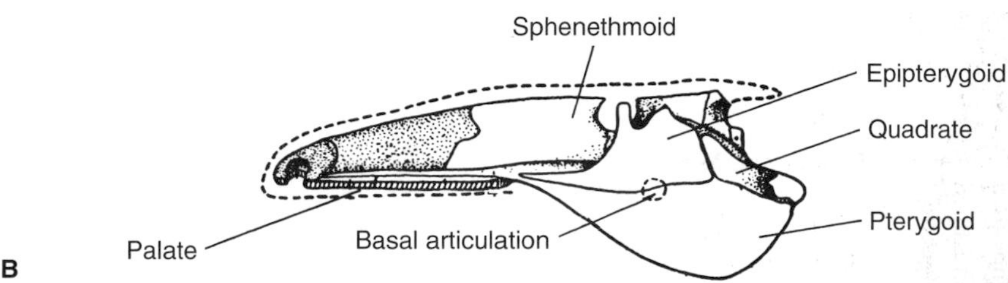

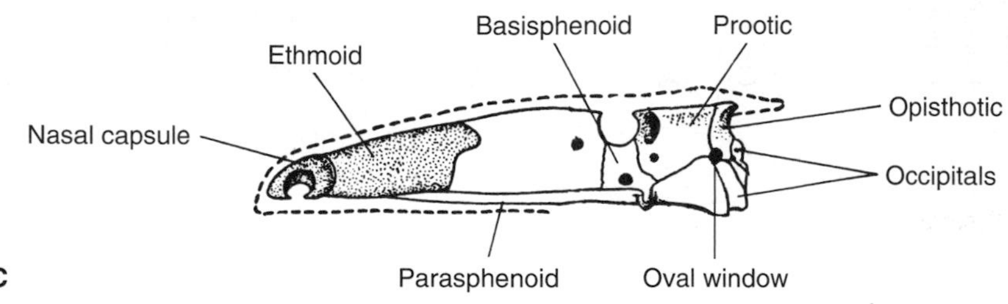

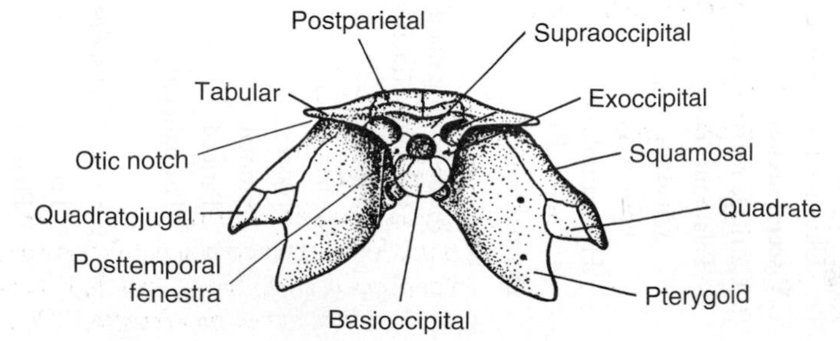

Figure 4-9

Diagrams of the anthracosauroid skull based on *Protogyrinus* and *Palaeoherpeton*. **A**, Lateral view. **B**, Lateral view after removal of the dermal skull roof, whose outline is indicated by dotted lines. **C**, Lateral view of the chondrocranium. **D**, Caudal view of an intact skull. *(Modified from Romer and Parsons, 1986.)*

Table 4-2 Components of the Tetrapod Skull and Lower Jaw*

Anthracosauroid	*Necturus*	Turtle	Lizard	Alligator	Mammal
SKULL					
Chondrocranium					
(Cartilage replacement bone)					
Basioccipital (1)	Unossified	X (1)	X (1)	X (1)	X (1) } Occipital
Exoccipital	X	X	X	X Fused with opisthotic	X } Occipital
Supraoccipital (1)	Unossified	X (1)	X (1)	X (1)	X (1) } Occipital
Opisthotic }	X	X	X	X	X } Petrous part of temporal
	Operculum				
Prootic }	X	X	X	X	X } Petrous part of temporal
Basisphenoid (1)	Unossified	X (1)	X (1)	X (1)	X (1) Body of basisphenoid
Sphenethmoid (1)	Unossified	Unossified	X Orbitosphenoid	X Laterosphenoid	X (1) Presphenoid
Unossified ethmoid region	Unossified	Unossified	Unossified	Unossified	X Ethmoid
Unossified nasal capsule	Unossified	Unossified	Unossified	Unossified	X Turbinates
Visceral arches					
(Cartilage replacement bone)					
Palatoquadrate					
Quadrate	X	X	X	X	X Incus
Epipterygoid	O	X	X	X	X Wing of basisphenoid
Hyomandibula					
Columella	X	X	X	X	X Stapes
Dermal bones					
Roof					
Tooth-bearing marginal bones					
Premaxilla (incisive)	X	X	X	X	X
Maxilla	O	X	X	X	X
Median series					
Nasal	O	O	X	X	X
Frontal	X	X	X	X	X
Parietal	X	X	X	X	X
Postparietal	O	O	O	O	X Interparietal, often a part of occipital
Circumorbital series					
Lacrimal	O	O	X	X	X
Prefrontal	O	X	X	X	O
Postfrontal	O	O	X	X	O
Postorbital	O	X	X	X	O
Jugal	O	X	X	X	X Zygomatic

Temporal series					
Intertemporal	O	O	O	O	O
Supratemporal	O	O	X	X	O
Tabular	O	O	O	O	X ? Part of occipital
Cheek bones					
Squamosal	X	X	X	X	X Squamous part of temporal
Quadratojugal	O	X	O	O	O
Palate and underside of chondrocranium					
Parasphenoid	X	X Fused with basisphenoid	X Fused with basisphenoid	X Reduced	X ? Part of basisphenoid
Vomer	X	X (1)	X	X	X (1)
Palatine	O	X	X	X	X
Ectopterygoid	O	O	X	X	X ? Part of basisphenoid
Pterygoid	X	X	X	X	X Pterygoid
LOWER JAW					
Visceral arches					
(Cartilage replacement bone)					
Mandibular cartilage					
Articular	Unossified	X	X	X	X Malleus
Dermal bones					
Lateral series					
Dentary	X	X	X	X	X
Splenials (2)	X	O	X	X	O
Surangular	O	X	X	X	O
Angular	X	X	X	X	X Tympanic part of temporal (endotympanic, a new cartilage replacement bone)
Medial series					
Coronoids (3)	O	X	X	X	O
Prearticular	O	X	X	X Fused with articular	X Anterior process of malleus

*The components of the skull and lower jaw of an anthracosauroid, together with the part of the skeleton to which they belong, are shown in the left-hand column. The homologies between these elements and those of certain other tetrapods are shown in the right-hand columns. An X indicates that the element is present; an O, that it is absent. All the elements are paired unless indicated to the contrary by a number in parentheses: (1) indicates a median element; (2) or (3) indicates that two or three of the elements are present on each side.

located on the lateral surface of the otic capsule, for the attachment of the columella (see below).

The caudal portion of the palatoquadrate cartilage has ossified as the **quadrate bone** and articulates with the lower jaw. The portion of the palatoquadrate adjacent to the basitrabecular process of the chondrocranium ossifies as the **epipterygoid bone** (Fig. 4-9**B**). This bone articulates with the braincase at a basal articulation and helps to fill in the gap in its side. The rostral part of the palatoquadrate is lost in adult anthracosaurids.

The columella (called the stapes in mammals), which evolved from the hyomandibula of fishes, can be added to these major components of the skull. In anthracosauroids, the columella is a large bone that extends between the oval window on the otic capsule of the chondrocranium to the quadrate bone and buttresses the chondrocranium against the dermal roof. In frogs and turtles the columella is a slender, rod-shaped bone that transmits high-frequency sound waves from the tympanic membrane, across the middle ear cavity, to the otic capsule. It is doubtful that anthracosauroids had a tympanic membrane. Their columella was too massive to have responded to high-frequency airborne vibrations, but it may have conducted lower-frequency vibrations travelling through the ground and jaw bones to the otic capsule.

The lower jaw of early terrestrial vertebrates consists of a mandibular cartilage surrounded by a sheath of dermal bones. The caudal end of the mandibular cartilage, which articulates with the quadrate, generally ossifies as the **articular bone.** The rest of the mandibular cartilage either remains cartilaginous or disappears. The teeth of anthracosauroids are similar in shape to those of fishes; they are small, conical, numerous, and similar to one another (homodont). They usually are arrayed in two series: a lateral series on the lateral margins of the jaws and a medial series on the palate and medial side of the lower jaw.

As stated, the mandibular arch of the visceral skeleton becomes incorporated into the skull and lower jaw, and the dorsal part of the hyoid arch (hyomandibula) becomes the columella. The rest of the hyoid arch unites with various portions of the more cranial branchial arches to form the **hyobranchial apparatus.** In contemporary tetrapods, the anterior part of this apparatus forms the skeletal base of the newly evolved mobile tongue. Ligaments extend from the hyoid to the skull base. The hyobranchial apparatus forms a sling for the support of the tongue and pharynx floor and serves as the attachment site for muscles associated with feeding and swallowing. Portions of the more caudal branchial arches form the cartilages of the larynx.

The *Necturus* Skull

General Features and Entire Skull

Study a skull of *Necturus.* It remains in the larval stage and retains fewer of the features and elements of the early tetrapod skull than the skulls of many other adult amphibians or reptiles. First examine the top of the skull (Fig. 4-10**A** and **B**). Most of the bones that you see belong to the dermal roof, although other bones have been exposed through the loss of some of the original elements of the roof. The V-shaped, tooth-bearing bone on each side of the front of the upper jaw is the **premaxilla.** The **naris** is located in the notch caudal to the premaxilla's lateral wing. Continuing caudally along the middorsal line, the next elements are a large pair of **frontal bones.** Paired **parietal bones** extend from the frontals nearly to the foramen magnum. Each parietal bone also has a narrow process that extends craniolaterally to the frontals. The **orbits** are located ventral to these processes.

Lateral to the caudal half of the parietal bone you will see two small otic bones. The **opisthotic bone** forms the very caudal angle of the skull and extends cranially to a tiny window of cartilage. The **prootic bone** lies cranial to this window. These bones form the otic capsule of the chondrocranium. The thin sliver of bone on the margin of the skull, lateral to the otic bones, is the **squamosal,** which is a component of the original dermal roof. Cranial and ventral to it, at the point where the lower jaw articulates, you will see the **quadrate,** which is the only part of the original palatoquadrate present in *Necturus.* Only the lateral portion of the quadrate, which bears the articular facet for the lower jaw, ossifies. The medial and cranial portions of the quadrate remain cartilaginous and usually are shriveled up in dry skull preparations.

The paired **exoccipitals** form the caudal end of the skull, lateral and ventral to the foramen magnum. Each bears an **occipital condyle.** Therefore, there is a pair of occipital condyles in *Necturus,* unlike the single condyle of early tetrapods.

Look at the underside of the skull (Fig. 4-10**C**). The palate consists of two pairs of dermal bones. The cranial pair, the **vomers,** lie caudal to the premaxillae and, like the premaxillae, bear a row of teeth. The caudal pair, the **pterygoids,** extends caudally from the vomers to the otic region. The pterygoids also have a row of teeth rostrally. Portions of the vomers and pterygoids can be seen from the dorsal and lateral sides. The large median bone on the ventral surface, lying between the vomers and the pterygoids and continuing to the exoccipitals, is the dermal **parasphenoid.** A small section of the unossified cartilaginous **ethmoid plate** of the chondrocranium is exposed rostral to the parasphenoid. The prootic and opisthotic bones can also be seen in the ventral view lying caudal to the pterygoids and lateral to the caudal end of the parasphenoid. A cartilaginous area, containing the **oval window,** separates them. A tiny disc-shaped bone bearing a little stem, the **stylus,** may cover the oval window. This bone represents the **columella** combined with the **operculum.** The operculum develops from the wall of the otic

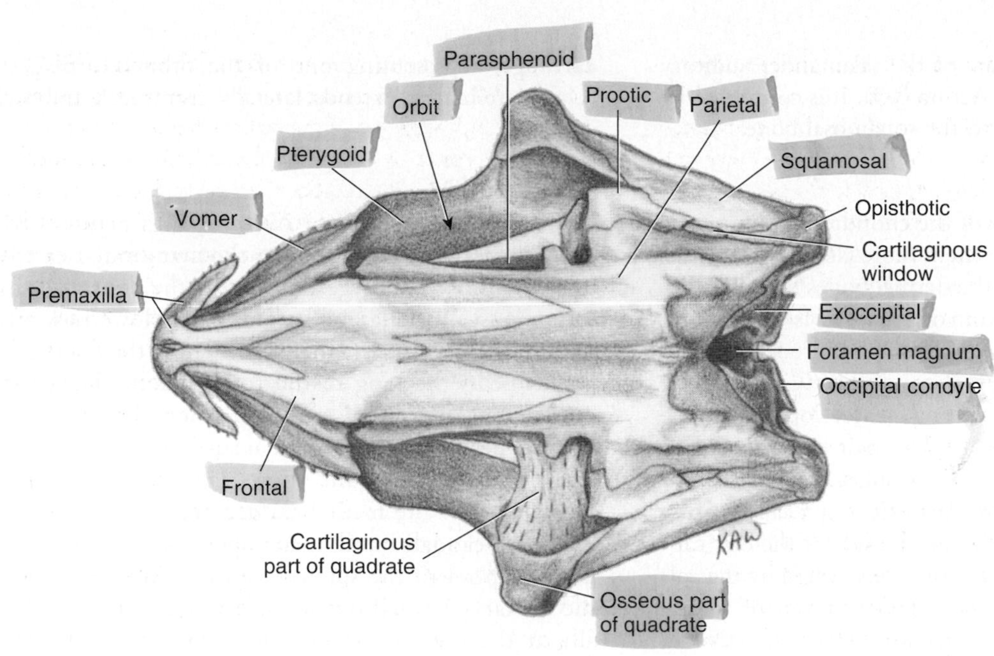

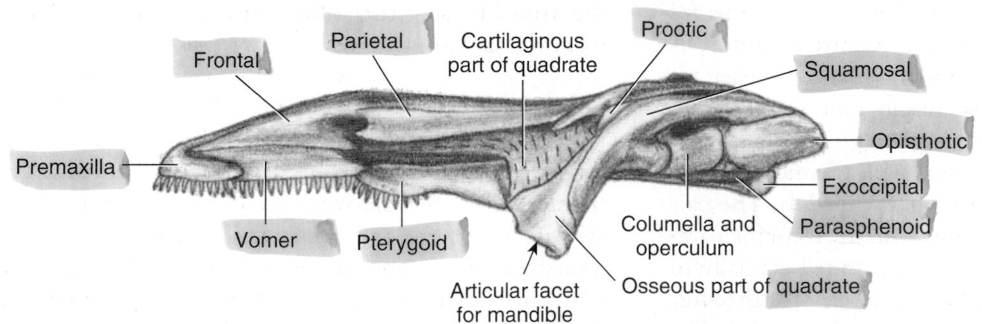

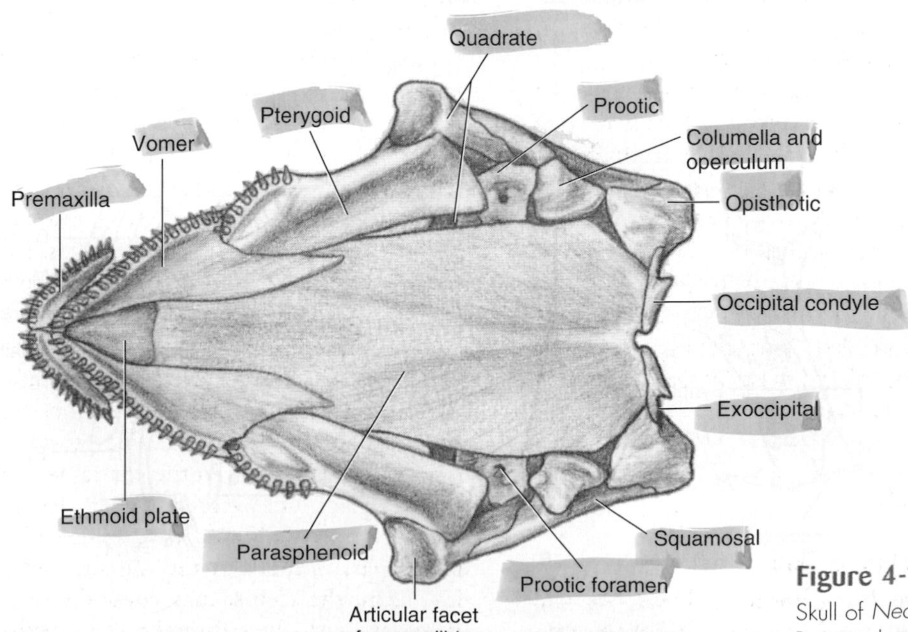

Figure 4-10
Skull of *Necturus*. **A**, Dorsal view. **B**, Lateral view. **C**, Ventral view.

capsule and is a unique part of the salamander auditory apparatus (see Anatomy in Action 8-3). It is connected by its stylus and by ligaments to the squamosal bone.

The Chondrocranium

Although you can see parts of the chondrocranium in the complete skull, you will see them more clearly by examining a preparation in which the dermal bones have been removed. The chondrocranium of *Necturus* is not a good representative of an adult amphibian chondrocranium because it is largely in the larval condition. (Compare the following description with that of the development of the chondrocranium on page 39.) The pair of large, round caudolateral swellings is the otic capsules. You will see two ossifications in each: a cranial **prootic** and a caudal **opisthotic** (Fig. 4-11**A**). The hole on the side of the otic capsule is the oval window, which may be covered by the columella. The only other ossifications are a pair of ventral **exoccipitals,** which form in the embryonic occipital arch. A delicate cartilaginous bridge, the **basioccipital arch,** connects them. Dorsally, the otic capsules are connected by another cartilaginous bridge, the **synotic tectum.** Often the **quadrate** (a part of the mandibular arch) is left on preparations of the chondrocranium and will be seen craniolaterally to the prootic bone.

The shelves of cartilage united with the medioventral edge of each otic capsule are the **parachordals.** The pair of cartilaginous rods that continue rostrally from the parachordals are the **trabeculae.** They are united cranially to form an **ethmoid plate,** from which a pair of **trabecular horns** project between very delicate **nasal capsules.** The nasal capsules usually are destroyed. A small **antorbital cartilage,** representing one of the orbital cartilages of other vertebrates, extends laterally from each trabecular cartilage.

The Lower Jaw

Compare the lower jaw of *Necturus* with Figure 4-11**B.** Only three dermal bones cover the mandibular cartilage in this animal. The largest of these is the **dentary,** which forms most of the lateral surface of the lower jaw and a small portion of the medial surface near the front of the jaw. It bears the long rostral row of teeth. The shorter caudal row of teeth lies on the **splenial.** Most of the splenial lies on the dorsomedial surface of the jaw, but a small piece of the bone shows laterally. The **angular** forms most of the medial surface of the jaw. This bone is widest caudally and then tapers to a point, which passes ventral to the splenial and continues rostrally to the dentary. A small portion of the angular shows laterally at the caudoventral corner of the jaw. The **mandibular cartilage,** which is unossified in *Necturus,* bears the articular facet that articulates with the quadrate to form the jaw joint.

The Teeth

Notice that the teeth of *Necturus* are essentially fishlike, that is, they are small, conical, and all of the same type (homodont). They are relatively numerous, and you can see parts of a lateral and medial series. The lateral series are located on the premaxillae and dentaries; the medial series, on the vomer, pterygoid, and splenial bones. The teeth attach loosely to the jaws and may be broken off in laboratory specimens.

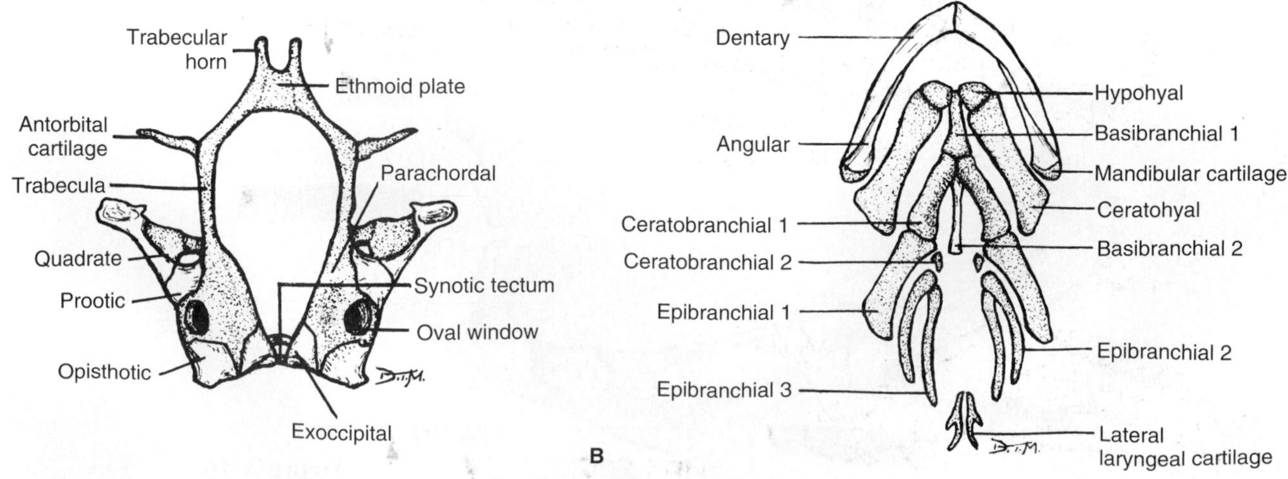

Figure 4-11

A, Ventral view of the chondrocranium and quadrate of *Necturus* after removal of the dermal bones.
B, Ventral view of the lower jaw, hyobranchial apparatus, and laryngeal cartilages. Cartilage is stippled.

The Hyobranchial Apparatus

Skeletal material that is preserved in liquid is better than dried material for studying the hyobranchial apparatus because it is nearly entirely cartilaginous. The hyobranchial apparatus is composed of parts of four visceral arches: the hyoid arch and the first three branchial arches (Fig. 4-11**B**). The hyoid arch is the most cranial and largest component. It consists on each side of a short **hypohyal** cartilage, which lies just lateral to the midventral line, and a longer **ceratohyal** cartilage, which extends toward the angle of the jaw. A median cartilage, designated **basibranchial 1,** extends caudally from the hypohyals to the first branchial arch. This arch, too, is composed of two cartilages on each side, the more medial being **ceratobranchial l;** the more lateral, **epibranchial l.** The next two arches (branchial arches 2 and 3) are greatly reduced and at first sight appear to consist only of **epibranchials 2** and **3.** On closer examination, you will see a small cartilage, **ceratobranchial 2,** which connects them with the first branchial arch. The branchial arches support the three external gills. Two gill slits are located on either side of the base of the second branchial arch (see Fig. 10-13). A small median cartilage, designated **basibranchial 2,** extends caudad from the base of the first branchial arch and is usually partly ossified. The last two branchial arches (branchial arches 4 and 5) do not contribute to the hyoid apparatus and will not be seen. There is some doubt as to their evolutionary fate, but traces of them may persist. A raphe in one of the branchial muscles, which contains a few cartilage cells, may represent a part of the fourth branchial (i.e., sixth visceral) arch. It also is probable that the rest of this arch plus the fifth branchial arch have contributed to the **lateral laryngeal cartilages** of the laryngotracheal chamber.

The Reptile Skull

In many ways the skulls of certain living reptiles are better examples of an early tetrapod skull than is that of *Necturus.* The skull of the snapping turtle, *Chelydra,* is described and illustrated in Figures 4-12 and 4-13. Illustrations of the lizard skull (Anatomy in Action 4-1) and alligator skull (Anatomy in Action 4-2) are included for comparison.

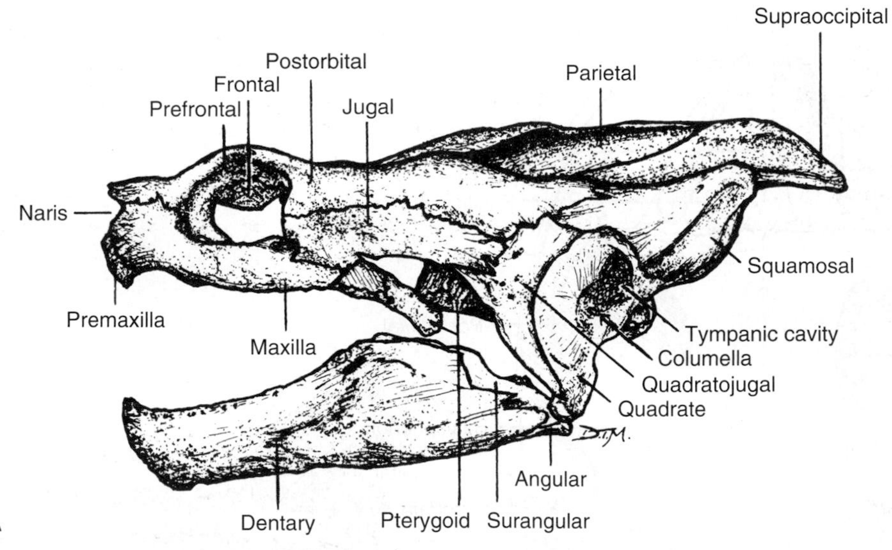

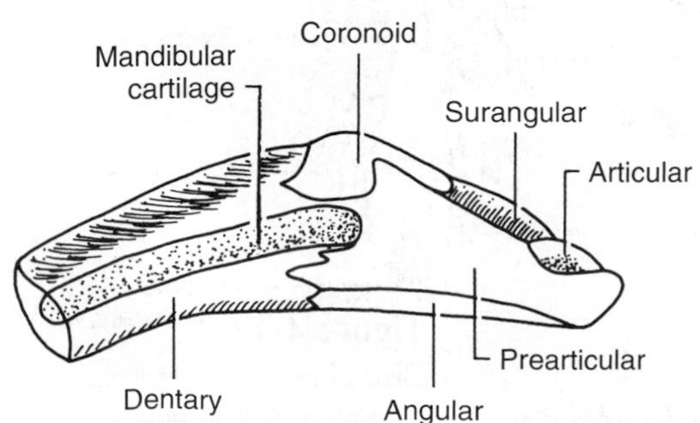

Figure 4-12

Skull of the snapping turtle, *Chelydra*. **A,** Lateral view of the skull and lower jaw. **B,** Medial view of the right half of the lower jaw.

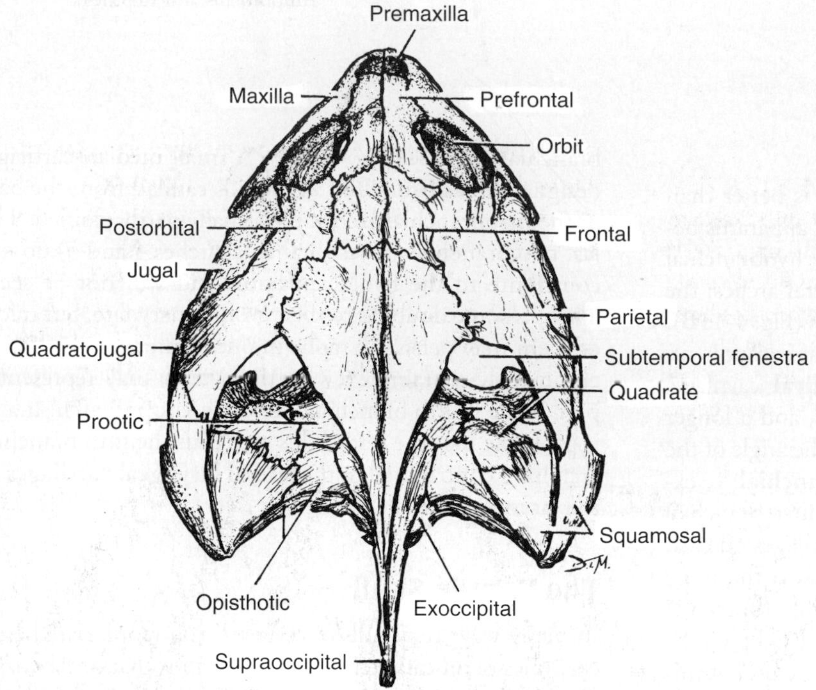

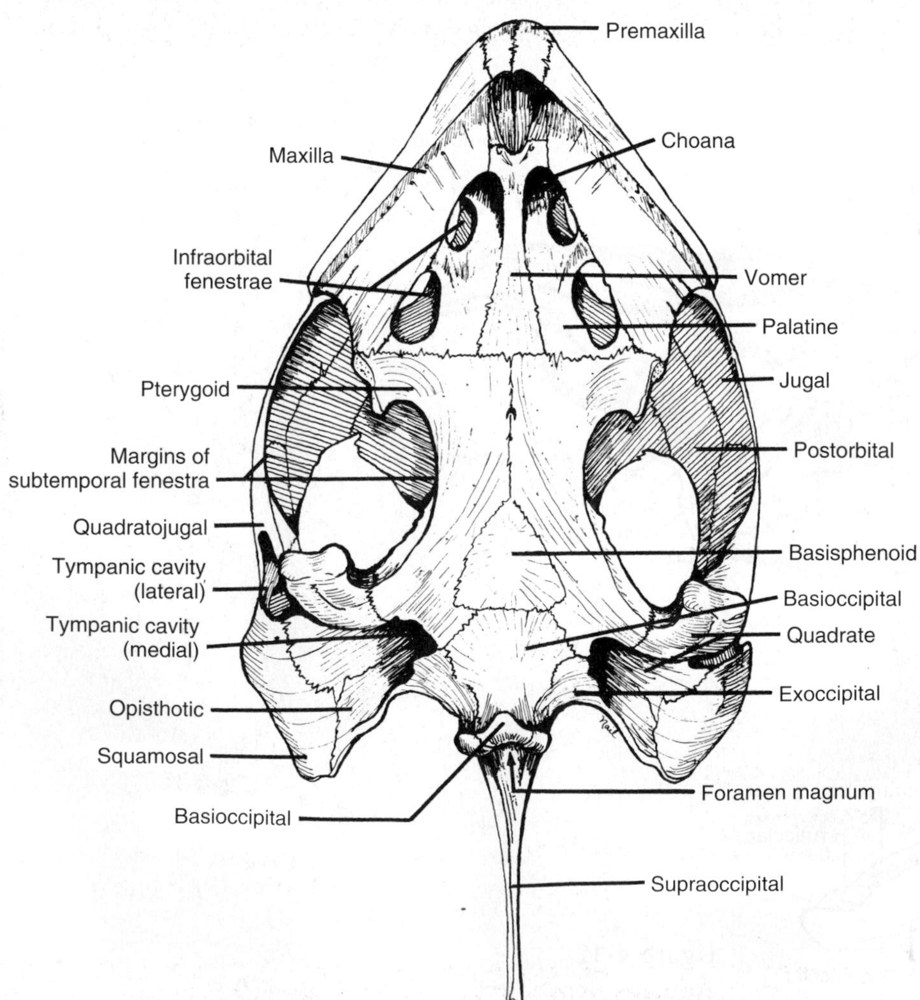

Figure 4-13
Skull of the snapping turtle, *Chelydra.* A, Dorsal view. B, Ventral view.

General Features of the Skull

Examine a skull of *Chelydra* (Figs. 4-12 and 4-13). Teeth are absent and are replaced by a horny beak sheathing the jaw margins. The horny sheathing often is missing on dried skulls. The dermal roof and palate can be seen surrounding the small, partly ossified chondrocranium. Paired external nostrils, or **nares,** are present in a whole specimen with its soft tissue, but they converge towards a single opening in the dermal roof of a skull. Paired nasal cavities remain distinct in the skull, and they open into the mouth cavity by paired internal nostrils, or **choanae,** located at the front of the palate. The **orbits** are situated far rostrally because the snout is short. The dermal roof in the temporal area has been "eaten away," or emarginated, from its caudal border in most turtles, but it is complete in sea turtles. Although this emargination is related to the attachment and bulging of the powerful jaw muscles, it is not homologous to the temporal fenestrae of other reptiles. The

Anatomy in Action 4-1 The Lizard Skull

The tegu lizard, *Tupinambis nigropunctatus,* ranges from the southern United States into South America. In some respects, the skull of the tegu lizard is more similar to the ancestral tetrapod skull than the skull of a turtle because it has a longer snout, retains simple conical teeth that are loosely attached to the jaws, and has a larger complement of bones. It retains the **nasal, postfrontal,** and **ectopterygoid bones,** as well as paired **vomers,** but the temporal region is highly specialized. Like most reptiles, lizards have a **diapsid skull** with two **temporal fenestrae** in the temporal region of the dermal skull roof. These fenestrae are associated with the origins of jaw-closing

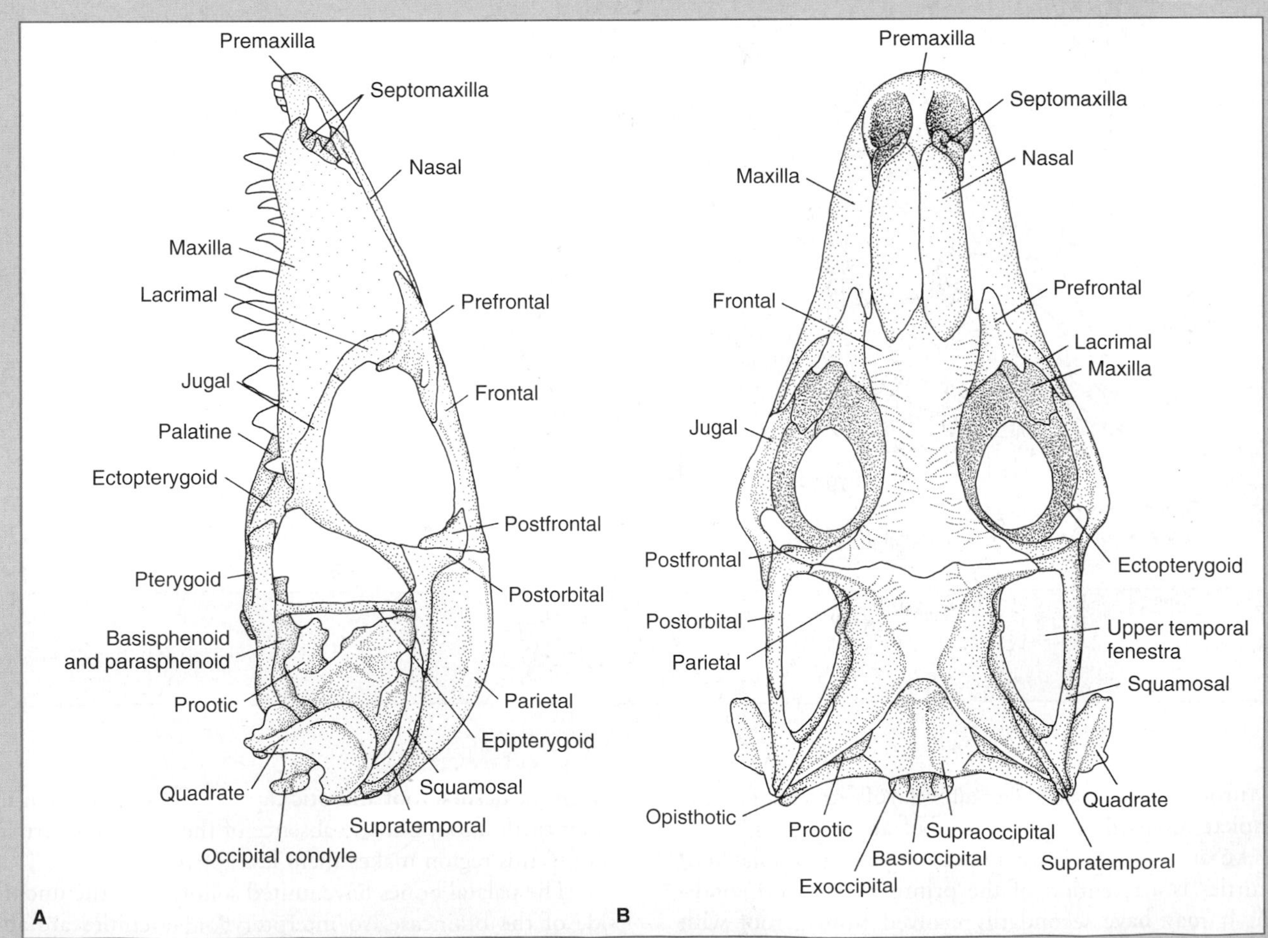

The skull and lower jaw of the tegu lizard, *Tupinambis.* **A,** Lateral view. **B,** Dorsal view. *(After Jollie, 1962.)*

continued

Anatomy in Action 4-1 The Lizard Skull *continued*

muscles. The upper temporal fenestra is easy to recognize. A bar of bone formed by the postorbital and squamosal bones bounds the lower temporal fenestra dorsally, but lizards have a modified diapsid skull because they have lost the lower bar of bone, which normally extends from the jugal bone to the quadrate bone. The lower bar ancestrally was composed of an extension of the jugal and the quadratojugal bones. Loss of this bar makes it more difficult to visualize the lower temporal fenestra. Because this lower bar no longer buttresses the quadrate, the quadrate is free to move at the quadrate/squamosal joint. Lizards have a more kinetic skull than many other reptiles.

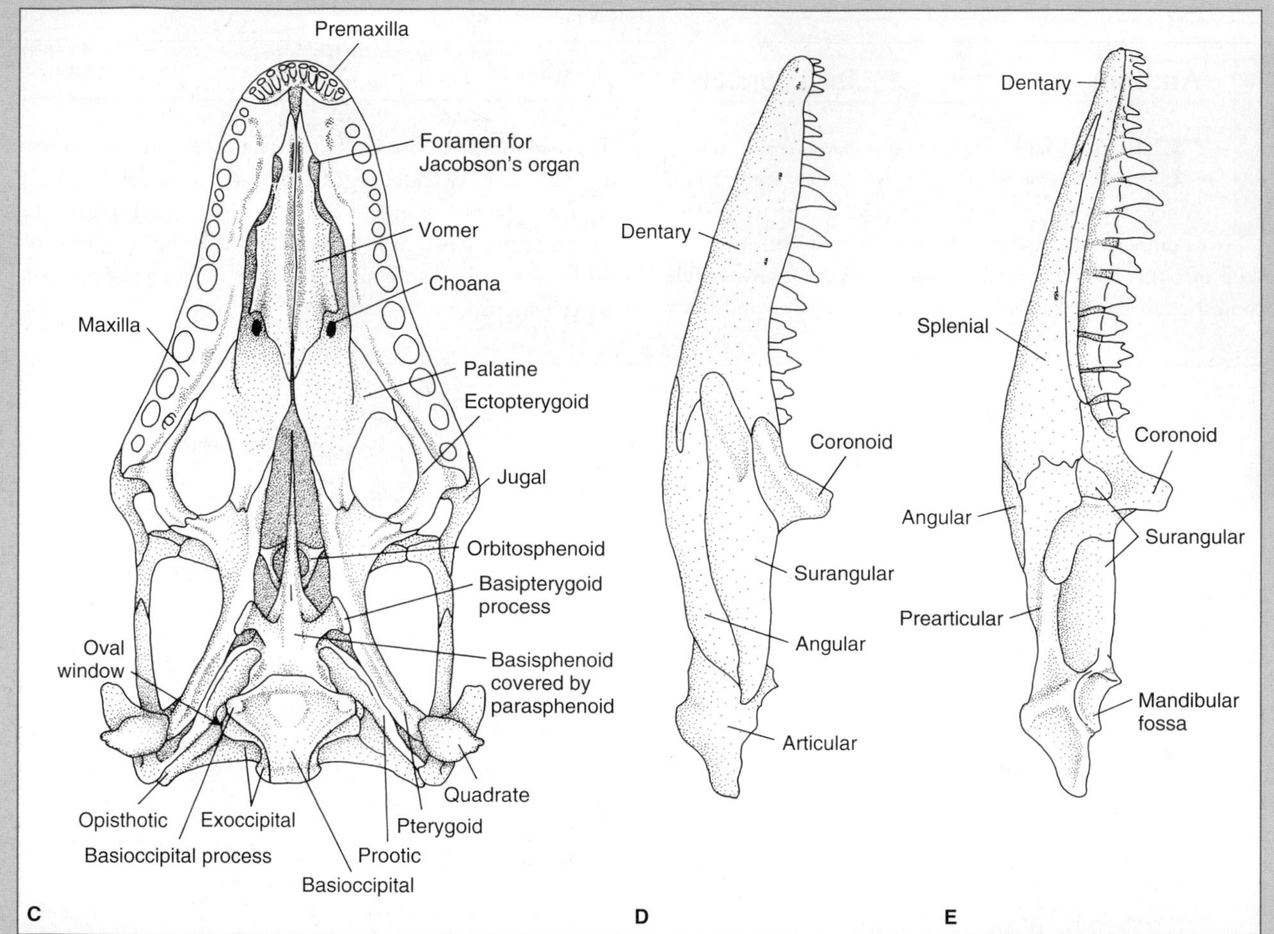

The skull and lower jaw of the tegu lizard, *Tupinambis*. **C**, Ventral view. **D**, Lateral view of the lower jaw. **E**, Medial view of the lower jaw. *(After Jollie, 1962.)*

sea turtle skull in particular allows you to visualize the complete **anapsid temporal roof** of an anthracosauroid, but we are not certain that the complete temporal roof of turtles is a retention of the primitive tetrapod condition. It may have secondarily evolved from a roof with temporal fenestrae, which most reptiles have (Anatomy in Action 4-1 and 4-2). A pair of large **posttemporal fenestrae** can be seen in a caudal view of the sea turtle skull between the dermal roof and otic capsule. They are present in other turtles, too, but the absence of the overlying dermal roof in this region makes them less apparent.

The palatal bones have united solidly with the underside of the braincase, so interpterygoid vacuities are absent. Such spaces in early tetrapods, which are retained in many contemporary reptiles, allow the palate to move relative to the braincase and dermal roof during feeding

Anatomy in Action 4-2 The Alligator Skull

The alligator skull resembles the lizard skull in having a long facial region and a fuller complement of bones than that seen in turtles. It also retains teeth, but these are firmly set in sockets, a condition called **thecodont.** The temporal region of the skull is also diapsid, and both upper and lower temporal fenestrae are clearly demarcated by complete bars of bone (Fig. **A**). Sheets of bone extend ventrally and medially from the premaxilla, maxilla, palatine, and part of the pterygoid bones and meet in the midline to form a **secondary palate,** which enables the alligator to open its mouth and feed under water without the water entering the respiratory passages (Fig. **B**). The secondary palate displaces the choanae caudally so that they open into the pharynx and not the mouth cavity. Soft tissue can close these openings in a living specimen.

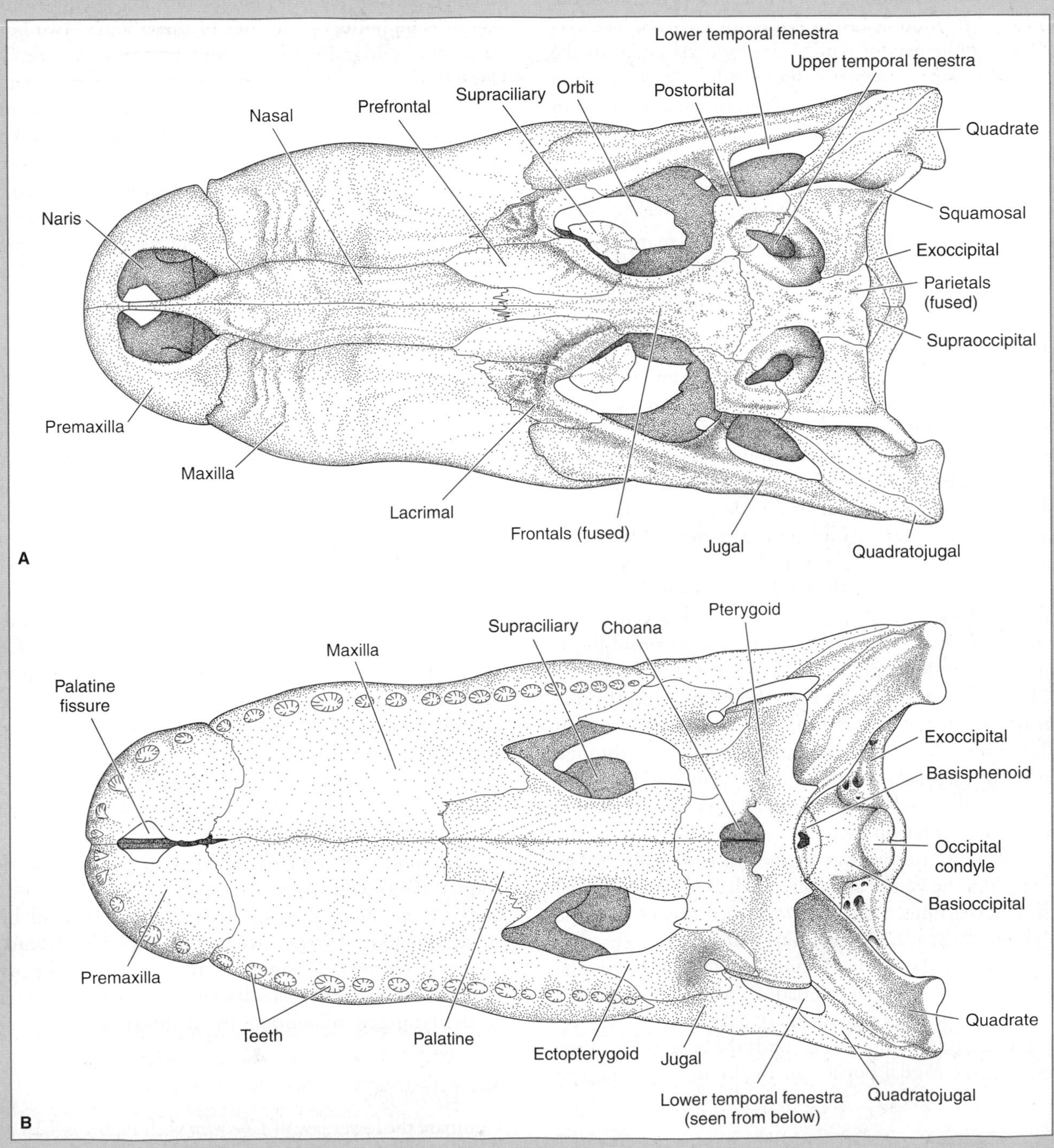

Alligator skull in dorsal **(A)** and ventral **(B)** views. *(After Liem, Bemis, Walker, and Grande, 2001, from Jollie, 1962.)*

movements. Skulls that allow for movement between skull components are called **kinetic** skulls. The turtle skull, which lacks this capacity for intracranial movement, is **akinetic.** The pair of large **subtemporal fenestrae** can be seen between the palatal bones and the lateral margins of the dermal roof. Two pairs of **infraorbital fenestrae** can be found in the palate beneath the orbits.

Examine the caudal portion of the skull and note the position of the middle ear, or **tympanic cavity.** The external and middle ears of reptiles are located caudal to the quadrate. Often, as in turtles, the quadrate partly encases and constricts the tympanic cavity so that it appears as an hourglass-shaped cavity divided into medial and lateral portions.

Composition of the Skull

The dermal bones along either lateroventral margin of the roof (Figs. 4-12 and 4-13) are a very small **premaxilla** ventral to the naris, a large **maxilla** continuing beneath the orbit, a **jugal** or zygomatic[6] caudal to the maxilla, and a **quadratojugal** just rostral to the middle ear cavity. The jugal reaches only the caudoventral corner of the orbit. The middorsal elements are a pair of large **prefrontals** caudal to the nares and dorsal to much of the orbits; a pair of small **frontals** that do not reach the orbit in *Chelydra;* and a pair of large **parietals** that extend to the prominent, middorsal, occipital crest. Each parietal also sends a wide flange ventrally that covers part of the brain not covered by chondrocranial elements. Two other dermal bones complete that portion of the roof situated between the dorsal and marginal bones. The large **postorbital** lies between the orbit and temporal emargination, and the **squamosal** forms a cap on the extreme caudolateral corner of the skull.

Dermal bones also make up the palate. Palatal processes of the premaxillae and maxillae form the very front of this region. A median **vomer,** which is formed through a fusion of originally paired elements, is located caudal to the premaxillae, and the **palatines** lie lateral to the vomer. The choanae enter on each side of the rostral end of the vomer in most turtles, but in the sea turtles and a few other turtles the vomer and palatines have sent out medial shelflike extensions that unite with each other along the midline to form a small **secondary palate** lying ventral to the **primary palate.** The secondary palate displaces the openings of the choanae caudad. A pair of large **pterygoids** forms the rest of the primary palate.

The chondrocranium surrounds the base, some of the sides, and the caudal portion of the brain, but it fails to cover the rest of the brain, which is covered instead only by the dermal bones of the roof and by the epipterygoid. Four cartilage replacement bones surround the foramen magnum: dorsally the **supraoccipital,** which forms the **occipital crest;** laterally the paired **exoccipitals;** and ventrally the **basioccipital.** The **occipital condyle** has expanded from its primitive position on the basioccipital to include parts of the exoccipitals. Distinct portions of the condyle are borne on all three of these bones. Two bones (the opisthotic and prootic) ossify in the otic capsule. The **opisthotic** can be seen in a dorsal view extending laterally from the supraoccipital and exoccipital to the squamosal. A **prootic** is located rostral to this. The **basisphenoid** can be seen ventrally lying between the pterygoids just rostral to the basioccipital. The dermal **parasphenoid** has united with the basisphenoid, so it is not seen as a distinct element. The interorbital and ethmoid regions of the chondrocranium and the nasal capsules remain unossified and usually are missing in dried skulls.

The caudal part of the embryonic palatoquadrate cartilage has ossified as the **quadrate.** This large bone occupies the caudoventral corner of the skull, extending from the jaw joint dorsally to the squamosal. Most of it passes rostral to the middle ear, but a part of it extends caudally to the middle ear. The central portion of the palatoquadrate cartilage has ossified as the **epipterygoid.** This bone helps to fill in a gap in the side of the chondrocranium. The epipterygoid is a distinct element in a sea turtle skull, but it has fused to adjacent elements in *Chelydra.* It occupies a small triangular area between the pterygoid and the ventral flange of the parietal, rostral to a large **prootic foramen,** which permits passage of the trigeminal and the abducens nerves. It is important to recognize that although the bones that ossify in the chondrocranium and palatoquadrate cartilage are cartilage replacement bones, they have different embryonic and evolutionary origins. Their separate origins may be obscure in an adult skull because the bones often are fused or sutured together.

In good preparations of the skull, a very slender, rodlike **columella** can be seen extending from the lateral part of the tympanic cavity through a small hole in the quadrate to the otic capsule. Its medial end is enlarged to form a footplate that fits into the **oval window** of the inner ear. Its lateral end would extend to the tympanic membrane in an intact specimen with soft tissues.

The Lower Jaw

Compare the lower jaw of *Chelydra* with Figure 4-12**A** and **B.** The caudal end of the mandibular cartilage has ossified as the **articular bone** and can be recognized by its smooth articular surface. The rest of the cartilage remains unossi-

[6]This bone is called the *jugal* in most tetrapods, but its mammalian homologue is the *zygomatic.* The mammalian bone was named before the homology between these bones was recognized.

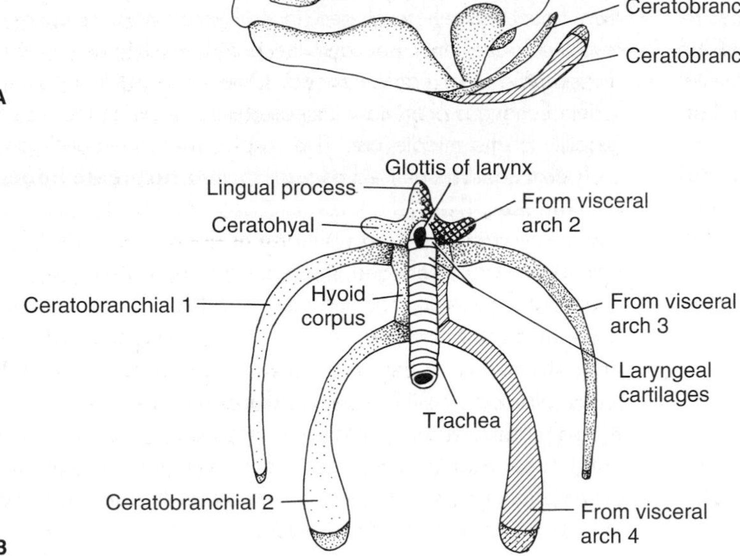

Figure 4-14

A, Lateral view of the hyobranchial apparatus of a *Chelydra in situ* ventrocaudally to the skull.
B, Dorsal view of the hyobranchial apparatus with the associated larynx and trachea. The derivation of the various parts is shown on the right side by different hatching.

fied, but in the adult it often can be found lying in a groove on the medial surface of the bony jaw. Remaining lower jaw bones are dermal. Most of the lateral surface of the jaw is formed by the **dentary,** but a small **surangular** lies on the lateral surface between the dentary and the articular bones. The rostral half of the medial surface also is formed by the dentary, but three additional elements form the medial surface caudally. These are a **coronoid** dorsally; a **prearticular** ventral to it and bridging the groove for the mandibular cartilage; and an **angular** beneath the prearticular, which forms the ventral border of the jaw. Small portions of the coronoid and angular can be seen from the lateral surface as well.

The Hyobranchial Apparatus

Study the hyobranchial apparatus of the turtle (Fig. 4-14). Its median body, the ossified **hyoid corpus,** is lodged in the base of the tongue and supports the larynx. Its rostral cartilaginous portion is derived from the **basihyal;** its caudal ossified part, from the two **basibranchials** of ancestral tetrapods. A cartilaginous **lingual process** extends rostrally from the hyoid corpus into the tongue. Cartilaginous and bony horns, the **cornua,** project laterodorsally from the hyoid corpus around the side of the neck, as visceral arches would in a fish. The first pair of cornua are short cartilaginous **ceratohyals,** which are connected by ligaments to the base of the skull. The next two pairs of ossified cornua represent the **first** and **second ceratobranchials** (i.e., visceral arches three and four). More caudal visceral arches have contributed to the cartilages of the larynx.

MAMMALS

Evolution of the Mammal Skull

The synapsid line of evolution to mammals diverged from the sauropsid line to reptiles and birds soon after the origin of amniotes in the late Carboniferous period (Fig. 3-1). Early synapsids, all of which are extinct, probably resembled reptiles in that they were ectothermic[7] animals that moved about on land relatively slowly with a splayed limb posture. Mammals became active, being endothermic animals with a constant and relatively high level of metabolism and a keen awareness of the world about them. These characteristics required gathering and processing more food, increased gas exchange, and a rapid distribution of materials throughout the body. No organ system went unchanged. The skull was affected primarily by changes in certain sense organs, by the great enlargement and

[7]The term *ectothermic* refers to animals, such as most fishes, amphibians and reptiles, whose body heat and temperature depends more on that of their environment than on metabolic heat production. Cold-blooded is a common synonym. Many ectotherms are also *poikilothermic* because their body temperature fluctuates with that of their environment, but others living in thermally stable habitats, a fish in a tropical coral reef, for example, are not. Both the terms *endothermic* and *homeothermic* refer to warm-bloodedness, and these terms are often used synonymously. Strictly speaking, endothermic refers to internal, metabolic heat production and homeothermic to the maintenance of a constant body temperature independent of environmental temperature. Both of these features characterize birds and animals.

increased complexity of the brain, and by changes in breathing and feeding mechanisms needed by an endothermic animal. This section summarizes the major skull transformations related to these changes. Further aspects of the changes will be discussed in more detail during the examination of the mammal skull, and changes in individual bones are listed in Table 4-2.

The median eye no longer reaches the skull surface through a parietal foramen as it does in many amphibians and reptiles because it is transformed into the pineal gland, an endocrine organ (Fig. 4-15). As the senses of hearing and olfaction became well developed in early mammals, a cochlea evolved in the inner ear, and the portion of the otic capsule housing the inner ear expanded. Other changes that accompanied changes of the jaw increased the sound-transmitting capability of the middle ear. The nasal cavities enlarged, and the nasal capsules ossified as scroll-shaped **turbinate bones,** or **conchae.** These changes increased the olfactory and air-conditioning surface area within the nasal cavities.

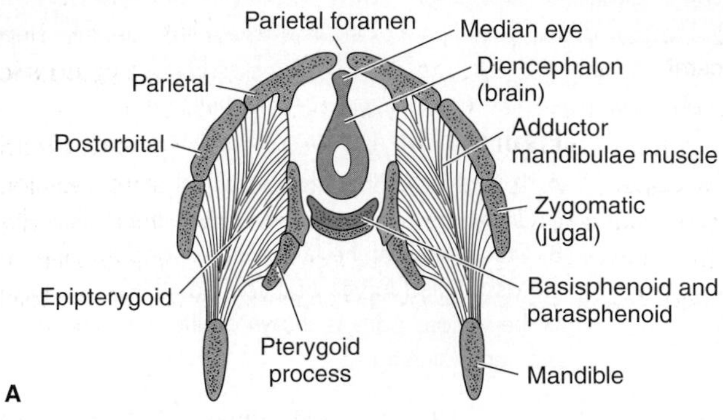

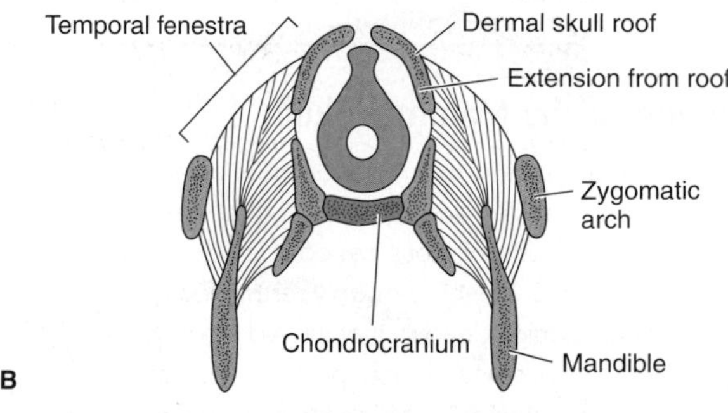

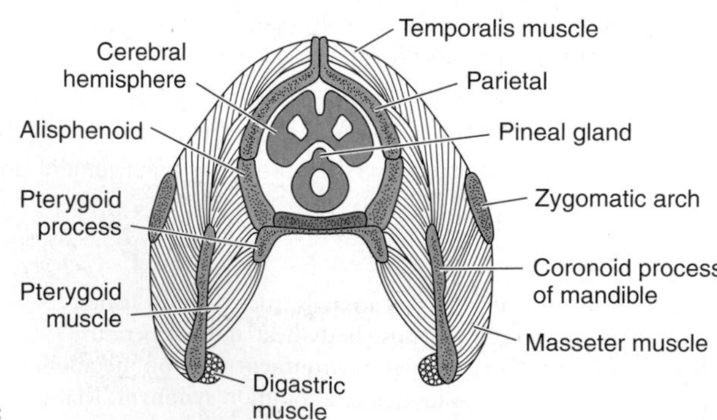

Figure 4-15
Evolution of the temporal fenestrae, the jaw musculature, and braincase of the tetrapod skull as seen in cross sections caudal to the orbit. **A,** Anthracosauroid with an anapsid temporal roof. **B,** Early synapsid with the synapsid temporal roof. **C,** Mammal. See Figure 4-6 for the color code of the derivation of the elements. *(From Liem, Bemis, Walker, and Grande, 2001.)*

As the brain enlarged, the braincase expanded. The original chondrocranium became more flattened so that it formed only the front, floor, and back of the mammalian braincase (Fig. 4-15**C**). Its rostral part, which remained cartilaginous in early tetrapods, ossified as the **ethmoid bone** in mammals. Dorsally and laterally the braincase is formed primarily by sheetlike extensions of certain dermal bones that grow ventrally and inward from the original dermal skull roof, in the same way that the parietal bones of turtles extend down from the roof. Notice in Figure 4-15**A** that the original dermal skull roof, which in this region is formed by the parietal, postorbital, and jugal, lies lateral to the jaw muscles and that these extensions from the roof lie medial to the muscles (Fig. 4-15**B** and **C**). Although derived from the dermal skull roof, these bony extensions do not constitute the original skull roof. The braincase is completed laterally by the **epipterygoid bone,** which becomes the wing, or **alisphenoid,** of the basisphenoid bone in mammals. The epipterygoid is derived from that part of the palatoquadrate cartilage that forms a movable articulation between the palatoquadrate cartilage and the chondrocranium in early tetrapods.

Their high level of metabolism requires mammals to eat a great deal of food, as well as to exchange a large volume of gases in their lungs. Head mobility increased in mammals in part by the division of the originally single occipital condyle and its shift to the sides of the foramen magnum, and in part by the modification of the joint between the first two cervical vertebrae. Mammals can move their heads up and down at the joint between the skull and first vertebra and rotate it at the joint between the first two vertebrae as they explore their environment and feed.

Powerful jaws and a strong skull are needed. Skull strength was increased partly by the reduction in size and eventual loss of many small bones, and partly by the fusion of others into large compound bony elements. The occipital, sphenoid, and temporal bones are examples of compound elements (Table 4-2). Torsional forces on the snout and upper jaw are resisted by the evolution of a **secondary,** or **hard palate,** a horizontal shelf of bone in the roof of the mouth, which is continued caudally by a fleshy shelf often called the **soft palate** (Fig. 4-16) (Thomason and Russell, 1986). In a transverse section the bony part of the secondary palate resembles an I-beam. Shelflike transverse processes of the premaxillary, maxillary, and palatine bones extended toward the midline and united with each other ventral to the ancestral primary palate to form the bony part of the secondary palate (Fig. 4-16). This displaces the openings of the internal nostrils, or choanae, caudally, thus permitting mammals to breathe and manipulate food simultaneously within their mouth cavity. (You can see an analogue of an early stage in the evolution of a secondary palate by looking at the roof of the mouth of a turtle, which has an incipient secondary palate, in Fig. 4-13**B**). The evolution of the secondary palate also allows infant mammals to suckle (see Anatomy in Action 7-2). The rostral portion of the ancestral primary palate, now situated dorsal to the secondary palate, regressed to some extent, and this area is occupied by enlarged nasal cavities.

Important changes in dentition and feeding mechanisms in mammals are correlated with increased energy needs. Mammals do not simply seize and swallow their food; they also cut it up and chew it. Mastication increases the efficiency of chemical digestion by exposing more surface area of the food to digestive enzymes. On the line of evolution through earlier synapsids toward mammals, the teeth became firmly set in sockets, a condition called **thecodont,** and differentiated into different types of teeth that crop, cut, and crush the food, a condition called **heterodont** (Fig. 4-17**C**). Precise occlusion of the teeth was needed, and this affected both tooth replacement and jaw muscles. Approximately 80% of jaw growth occurs before the first set of teeth, or milk teeth, have fully erupted. Milk teeth are lost and replaced by a single set of permanent teeth as the jaws enlarge. Subsequent tooth replacement is not continuous, as it is in amphibians and reptiles, but is limited.

Changes in the feeding mechanism in mammals were accompanied by significant changes in the jaw musculature and the parts of the skull to which these muscles attach. The relatively simple jaw musculature of early tetrapods became subdivided, larger, and stronger. The single adductor mandibulae muscle, which extended from the underside of the temporal region of the skull roof to the lateral surface of the lower jaw in ancestral tetrapods, became subdivided in mammals into a **temporalis muscle** and a **masseter muscle** (Fig. 4-15**C**). **Pterygoid muscles** extend from the palate to the medial side of the lower jaw. Contraction of these muscles generate a bite force as the lower jaw is pulled dorsally toward the upper jaw. Smaller muscle forces moved the lower jaw fore and aft and from side to side. These movements bring the teeth together very precisely, cutting and crushing the food between them.

Areas where several temporal roofing bones articulate with each other are points of mechanical weakness. Stresses from the pull of jaw muscles tend to be kept low in such regions, and where bone is not stressed it tends to regress. **Temporal fenestrae** may evolve in these areas (Figs. 4-15**B** and 4-17**A**). As temporal fenestra enlarged, the edges of the opening would have provided a firmer attachment for jaw muscles than a flat surface of bone. Also, the opening enabled jaw muscles to bulge through it when they contracted. On the line of evolution from earlier synapsids toward mammals, a single temporal fenestra evolved that originally lay ventral to the postorbital bone and part of the squamosal bone. A fenestra in this position is known as a **synapsid fenestra.** At first the synapsid fenestra was small, but it gradually enlarged. In early mammals, most of the original lateral surface of the dermal skull roof disappeared in the temporal region so that the temporal fenestra eventually merged with the orbit (Fig. 4-17**C**). The lateral roof of the braincase that you see in a mammal skull is formed by extensions from the ancestral roofing bones, which extended medial to the jaw muscles. All that is left of the original dermal skull roof in this region is a strip of bone in the middorsal line, another strip bordering the occipital region, and the handlelike **zygomatic arch** lying ventral to the temporal fenestra and orbit.

The muscle forces that pull the lower jaw upward toward the upper jaw are opposed by equal and opposite reaction forces that push down on the lower jaw. These forces are distributed in reptiles in such a way that the teeth come together in a moderately strong bite force. Forces acting on the jaw joint

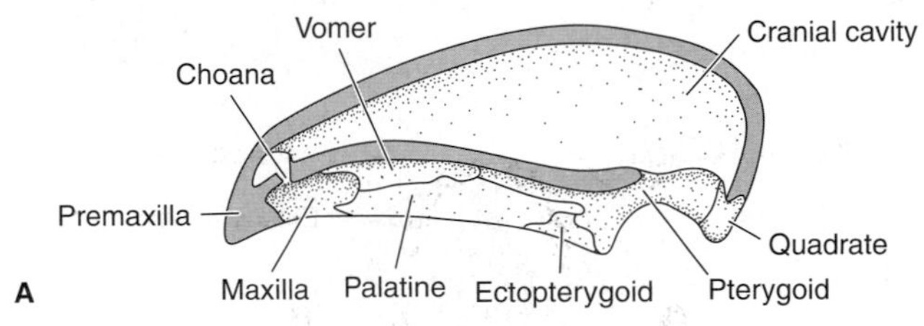

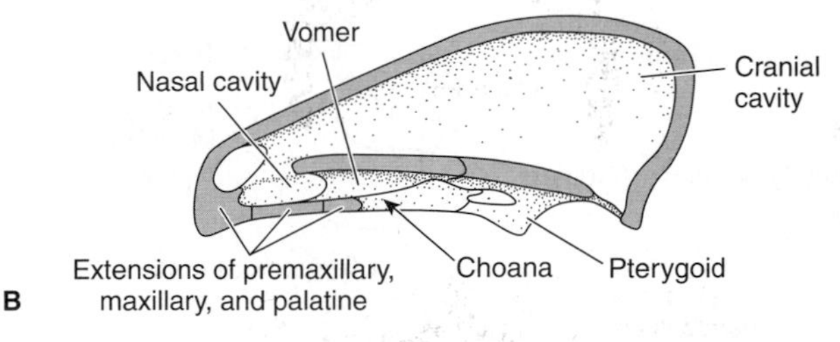

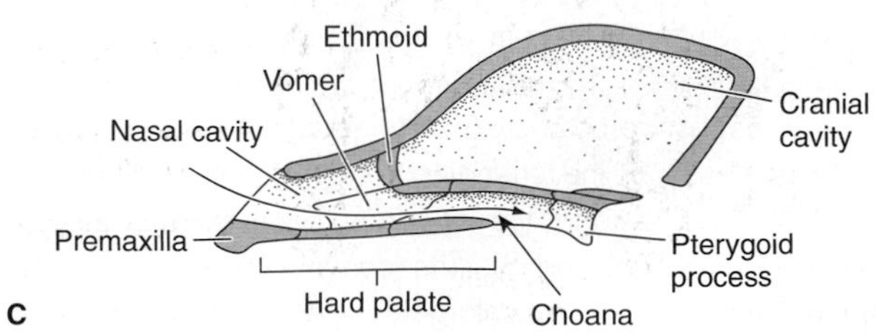

Figure 4-16

Evolution of the mammalian secondary palate. **A,** Early synapsid retains the condition seen in ancestral tetrapods. **B,** More advanced synapsid (a therapsid) demonstrates an early stage in the evolution of the secondary palate. **C,** Mammal in which the secondary palate is fully developed. *(From Liem, Bemis, Walker, and Grande, 2001; after Romer and Parsons, 1986.)*

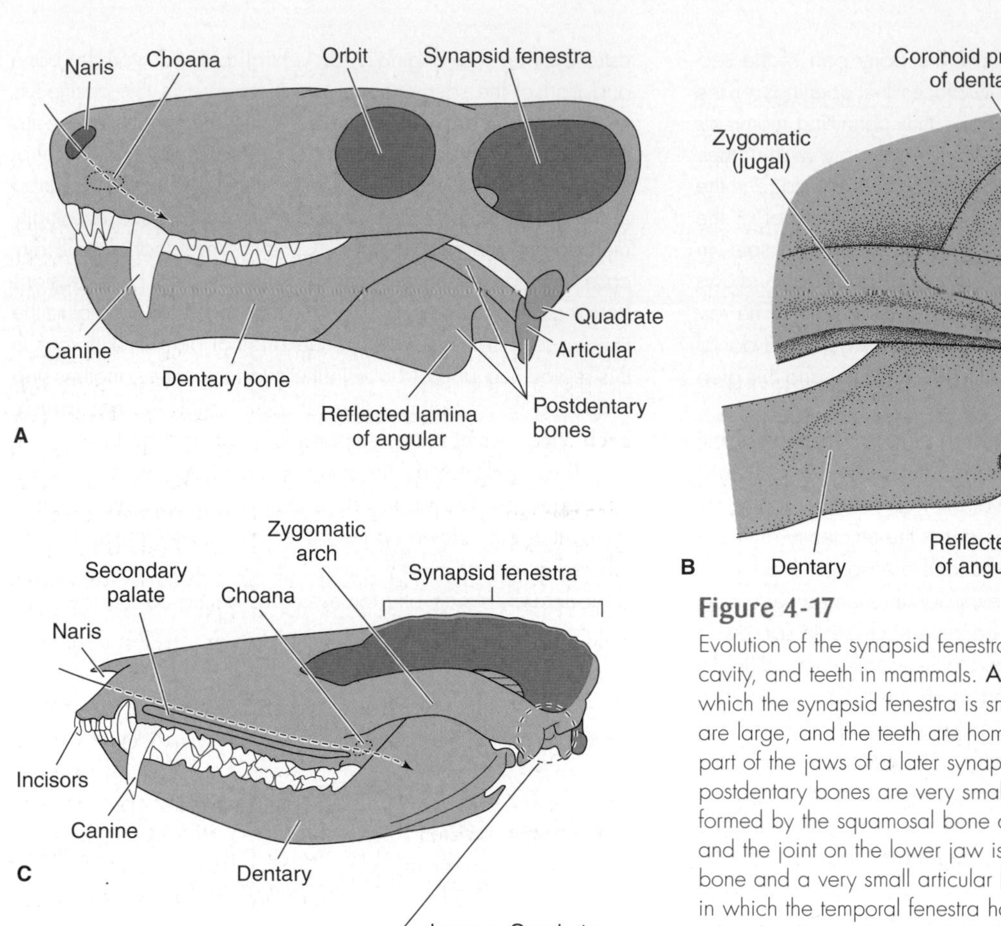

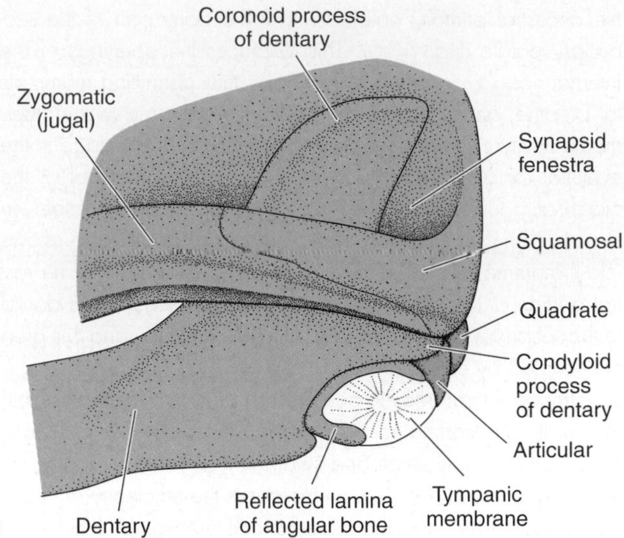

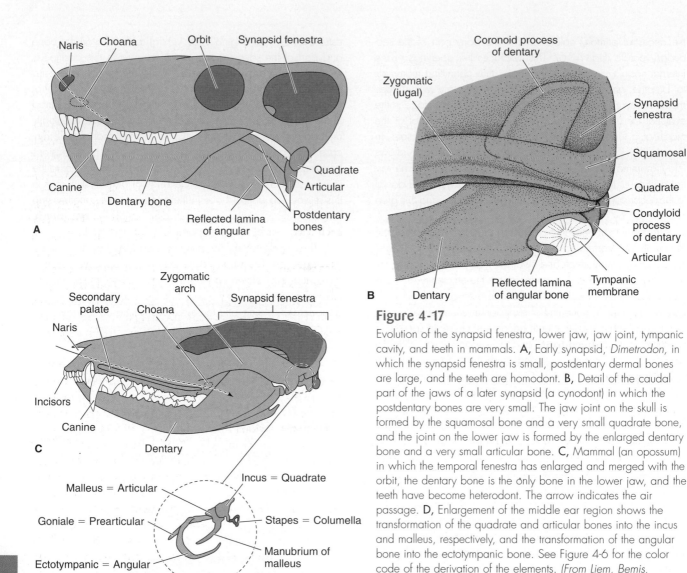

Figure 4-17

Evolution of the synapsid fenestra, lower jaw, jaw joint, tympanic cavity, and teeth in mammals. **A,** Early synapsid, *Dimetrodon,* in which the synapsid fenestra is small, postdentary dermal bones are large, and the teeth are homodont. **B,** Detail of the caudal part of the jaws of a later synapsid (a cynodont) in which the postdentary bones are very small. The jaw joint on the skull is formed by the squamosal bone and a very small quadrate bone, and the joint on the lower jaw is formed by the enlarged dentary bone and a very small articular bone. **C,** Mammal (an opossum) in which the temporal fenestra has enlarged and merged with the orbit, the dentary bone is the only bone in the lower jaw, and the teeth have become heterodont. The arrow indicates the air passage. **D,** Enlargement of the middle ear region shows the transformation of the quadrate and articular bones into the incus and malleus, respectively, and the transformation of the angular bone into the ectotympanic bone. See Figure 4-6 for the color code of the derivation of the elements. *(From Liem, Bemis, Walker, and Grande, 2001;* **B** *after Allin, 1975.)*

also are strong, and the large quadrate and articular bones resist them, as you have seen in a turtle skull (Fig. 4-12). In the course of evolution toward mammals, forces acting on the jaws were redistributed. The bite force increased considerably, whereas reaction forces acting at the jaw joint were greatly reduced. Because of the changing redistribution of forces, the postdentary bones and the quadrate became progressively smaller (Fig. 4-17A and B). The dentary, to which the jaw muscles attach, became enlarged, and eventually the only bone in the lower jaw. At the transition from earlier synapsids to mammals, the dentary reached the squamosal bone of the skull roof and contributed to the original jaw joint, which lay just craniolateral to this region (Fig. 4-17B). The functional jaw joint was now between the quadrate–squamosal in the skull proper and the articular–dentary in the lower jaw. This condition did not last long. The reduced quadrate and articular bones, which also participated in the transmission of sound waves from the lower jaw to the stapes, separated from the jaw apparatus and moved into the middle ear as additional auditory ossicles: the **incus** and **malleus,** respectively (Fig. 4-17D). (Ear evolution, which is closely related to jaw changes, is discussed in Chapter 8.) The mammalian jaw joint is now a squamosal–dentary joint.

The delicate auditory ossicles of mammals (the malleus, incus, and stapes) are protected in eutherian mammals by the formation of a plate of bone beneath the middle ear or tympanic cavity. Two elements contribute to this encasement in most species: a cartilage replacement **endotympanic bone,** which has no homologue in other vertebrates, and a dermal **tympanic bone,** which is homologous to the angular bone of the lower jaw of amphibians and reptiles.

General Features of the Skull

The basic features and composition of a mammalian skull can be studied using any mammal as a specimen. Our description is based primarily on the cat (a carnivore), but some descriptions of the rabbit (a gnawing herbivore) are included for comparison. If human and other skulls are available, you should examine them to appreciate the diverse variations of a common pattern that arise as adaptations to different environments and modes of life.

Examine the skull of a cat or rabbit (Figs. 4-18, 4-19, and 4-20). Like other vertebrate skulls, the mammal skull has a **facial region** containing the jaws, nose, and eyes, and a **cranial region** housing the brain and ears. In the facial region of the rabbit skull, the upper jaw region is highly fenestrated (Fig. 4-19**B**). Notice the paired **nares** at the rostral end of the skull and the large circular **orbits.** Although the two nares may appear contiguous in a dry skull, they are separated in life by a fleshy and cartilaginous septum.

The **foramen magnum** can be seen in the occipital region at the caudal end of the skull. A pair of **occipital condyles** flanks it. Vertical nodding movements of the head are permitted by the articulation between the paired condyles and the first cervical vertebrae, the atlas; side-to-side head-shaking movements are permitted by the articulation between the first and second cervical vertebrae, the atlas and axis. Head movements must not place excessive stress upon the spinal cord, which enters the skull through the foramen magnum.

The large round swelling on the ventral side of the skull, cranial to each occipital condyle, is the **tympanic bulla,** which forms the floor of the tympanic cavity. In life, the eardrum (tympanic membrane) is lodged in the opening of the bony **external acoustic meatus** on the lateral surface of the tympanic bulla. In rabbits, the external acoustic meatus is tubular (Fig. 4-18**B**). The tongue-shaped bony bump on the lateral surface of the tympanic bulla just caudal to the external acoustic meatus is the **mastoid process.** The similar **paracondylar process** is located on the caudal surface of the tympanic bulla (Fig. 4-19).

The handlelike bridge of bone on the side of the skull, extending from the front of the orbit to the external acoustic meatus, is the **zygomatic arch.** It is part of the facial portion of the skull formed by the zygomatic (jugal) bone lying between the zygomatic processes of the maxillary and temporal bones. The **mandibular fossa** for the articulation with the lower jaw is located at the base of the zygomatic process of the temporal. In the cat, this fossa is a transverse groove allowing mostly the opening and closing of the jaw. In the rabbit, it is smaller and flatter, and it allows the mandible to rotate sideways and move back and forth. **Postorbital processes** bound the orbit caudally. In the cat, these are large and project down from the skull roof and up from the zygomatic arch (Fig. 4-19). In the rabbit, the zygomatic arch lacks a postorbital process; a small postorbital process bounds the orbit on the skull roof (Fig. 4-18**B**). In the cat, the faint **temporal line** curves caudally from the base of the dorsal postorbital process to an indistinct middorsal ridge of bone, the **sagittal crest** (Fig. 4-18**A**). It delimits the large area of origin of the temporal muscle above the zygomatic process of the temporal bone. In the rabbit, a prominent temporal ridge delimits the much smaller area of origin of the much smaller temporal muscle. The temporal muscle passes medial to the zygomatic arch to insert onto the lower jaw. The different feeding mechanisms of the cat and rabbit will be considered later (see Anatomy in Action 7-3). Note that the lateral wall of the braincase in mammals does not correspond to the skull roof of early tetrapods (Fig. 4-15, p. 64). In the cat, a **nuchal crest** extends from the caudal end of the sagittal crest ventrally and laterally to the mastoid process. Neck muscles attach to the caudal surface of the skull and nuchal crest.

The shelf of bone that extends across the palatal region between the cheek teeth is the **secondary,** or **hard, palate** (Fig. 4-20). In a live animal, it is extended caudally by a fleshy **soft palate.** The remains of the ancestral primary palate, which forms the roof of the mouth in some reptiles (Anatomy in Action 4-1**C**), lie dorsal to the hard palate (Fig. 4-16**C**). The hard palate of mammals has evolved independently from that in alligators and sea turtles. Internal nostrils, or **choanae,** will be seen dorsal to the caudal margin of the hard palate. A nearly vertical plate of bone, the **pterygoid,** continues caudally from each side of the hard palate and ends in a **pterygoid hamulus.** A **pterygoid fossa,** for the attachment of pterygoid jaw muscles, is located caudolateral to the pterygoid hamulus. In the cat, it is much smaller than in the rabbit, reflecting the size difference of the pterygoid muscles in these two mammals.

Components of the Skull

Our description of the individual elements of the skull is based on the cat. The skulls of most mammals are composed of the same bones, but the bones differ in proportions and certain structural details. To visualize the skull bones better, you should examine a set of disarticulated bones along with the entire skull.

Dermal bones form most of the top and lateral sides of the skull (Fig. 4-19). A small, tooth-bearing **incisive bone** (premaxilla) is located ventrolateral to each external nostril. It bears the upper incisor teeth and contributes a small, shelflike process to the hard palate. The **maxilla** completes the upper jaw. It contributes a process to the

(Text continues on page 71)

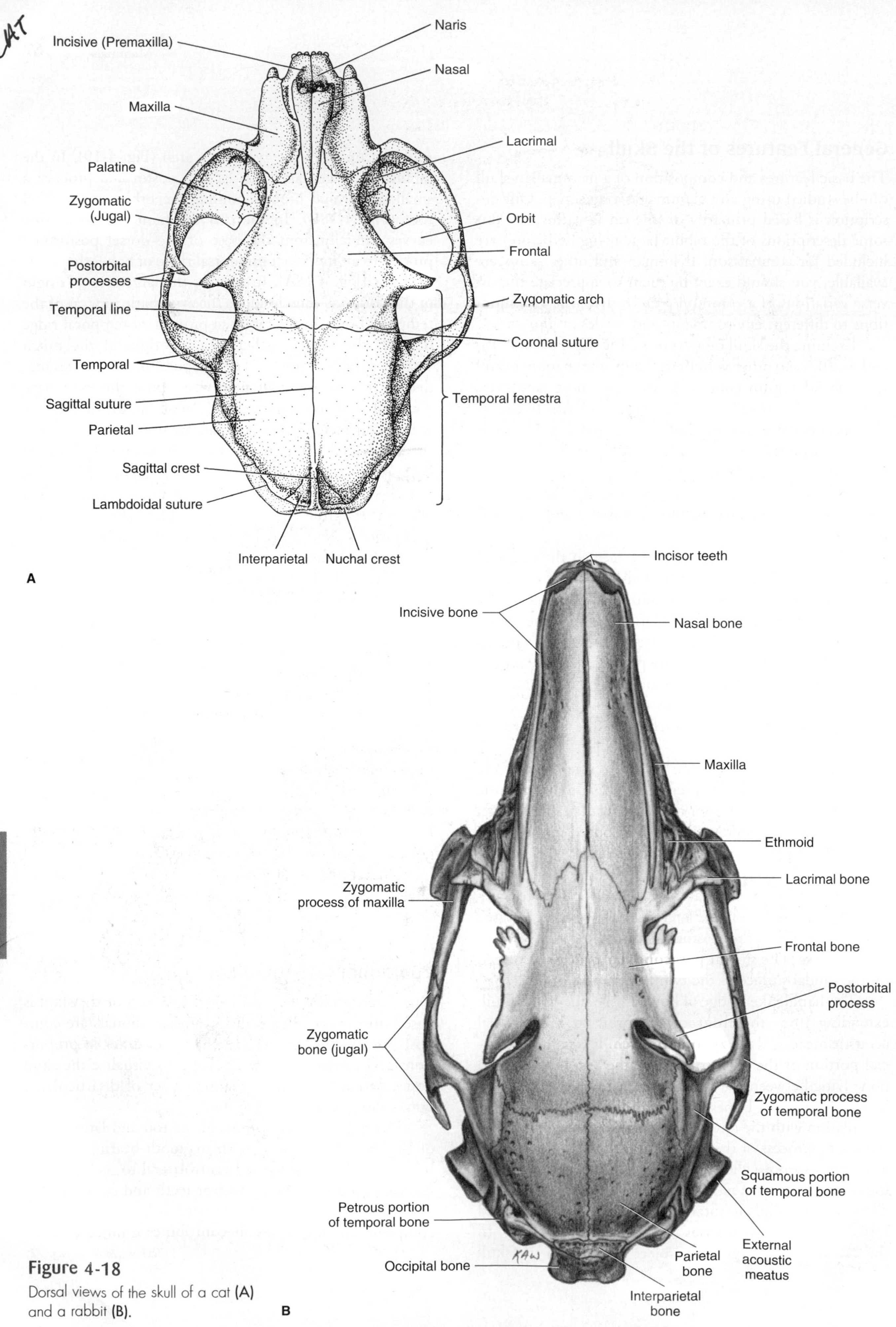

CAT
Incisive (Premaxilla)
Naris
Nasal
Maxilla
Lacrimal
Palatine
Zygomatic (Jugal)
Orbit
Frontal
Postorbital processes
Zygomatic arch
Temporal line
Coronal suture
Temporal
Sagittal suture
Temporal fenestra
Parietal
Sagittal crest
Lambdoidal suture
Interparietal
Nuchal crest
A
Incisor teeth
Incisive bone
Nasal bone
Maxilla
Ethmoid
Lacrimal bone
Zygomatic process of maxilla
Frontal bone
Postorbital process
Zygomatic bone (jugal)
Zygomatic process of temporal bone
Squamous portion of temporal bone
Petrous portion of temporal bone
External acoustic meatus
Parietal bone
Occipital bone
Interparietal bone
KAW
B

Figure 4-18
Dorsal views of the skull of a cat **(A)** and a rabbit **(B)**.

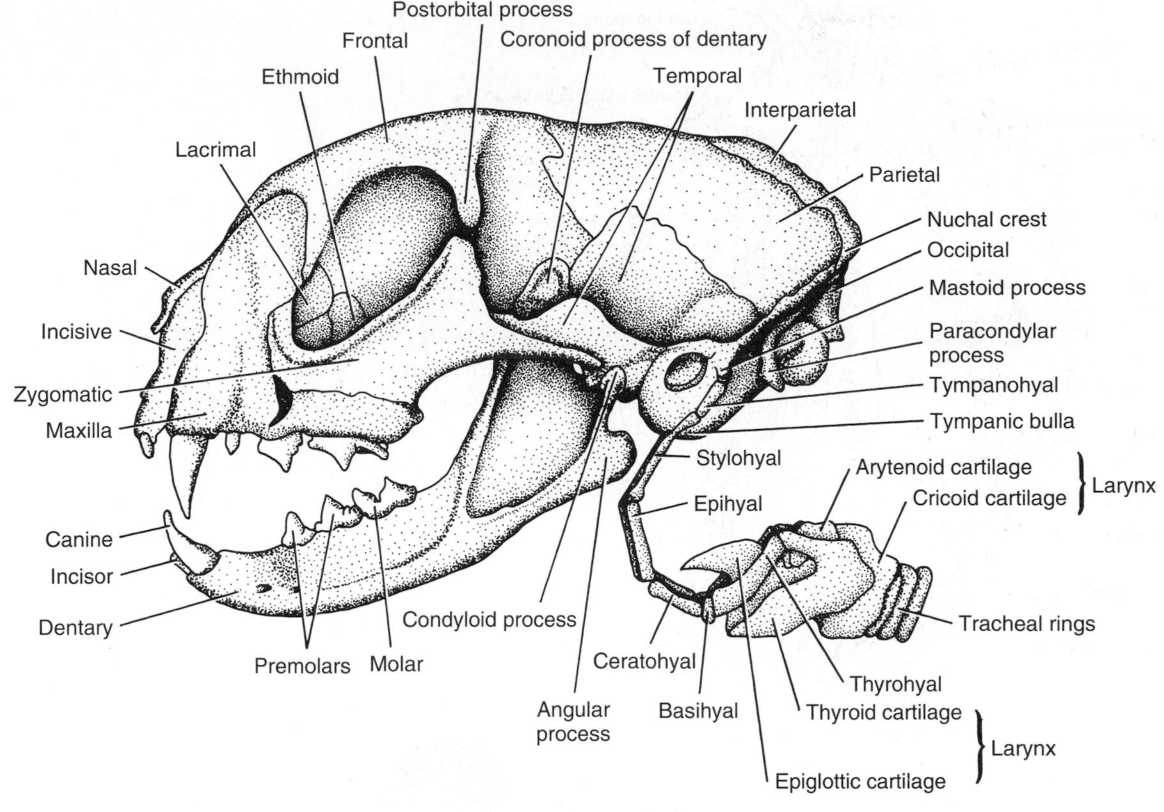

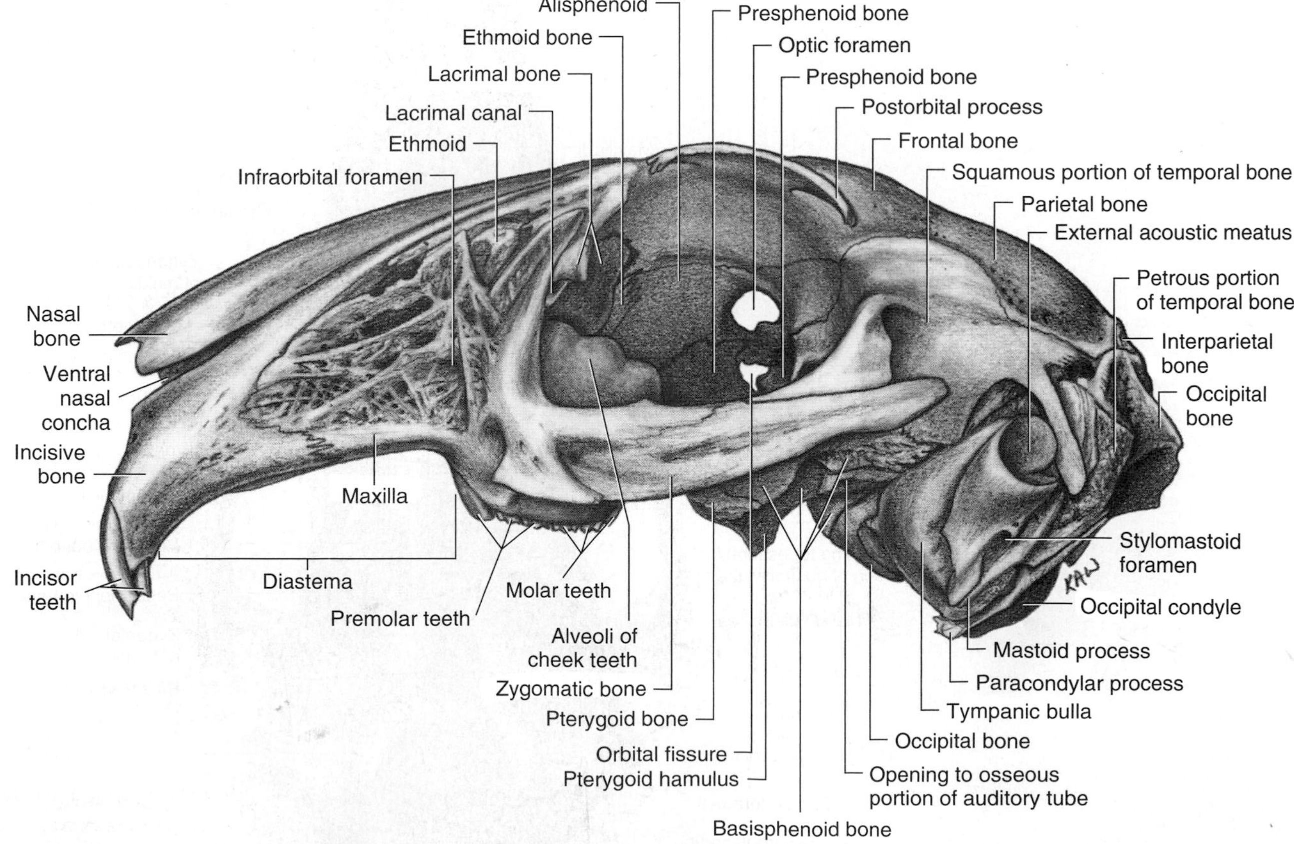

Figure 4-19

A, Lateral view of the skull, lower jaw, hyoid apparatus, and larynx of a cat. B, Lateral view of the skull of a rabbit.

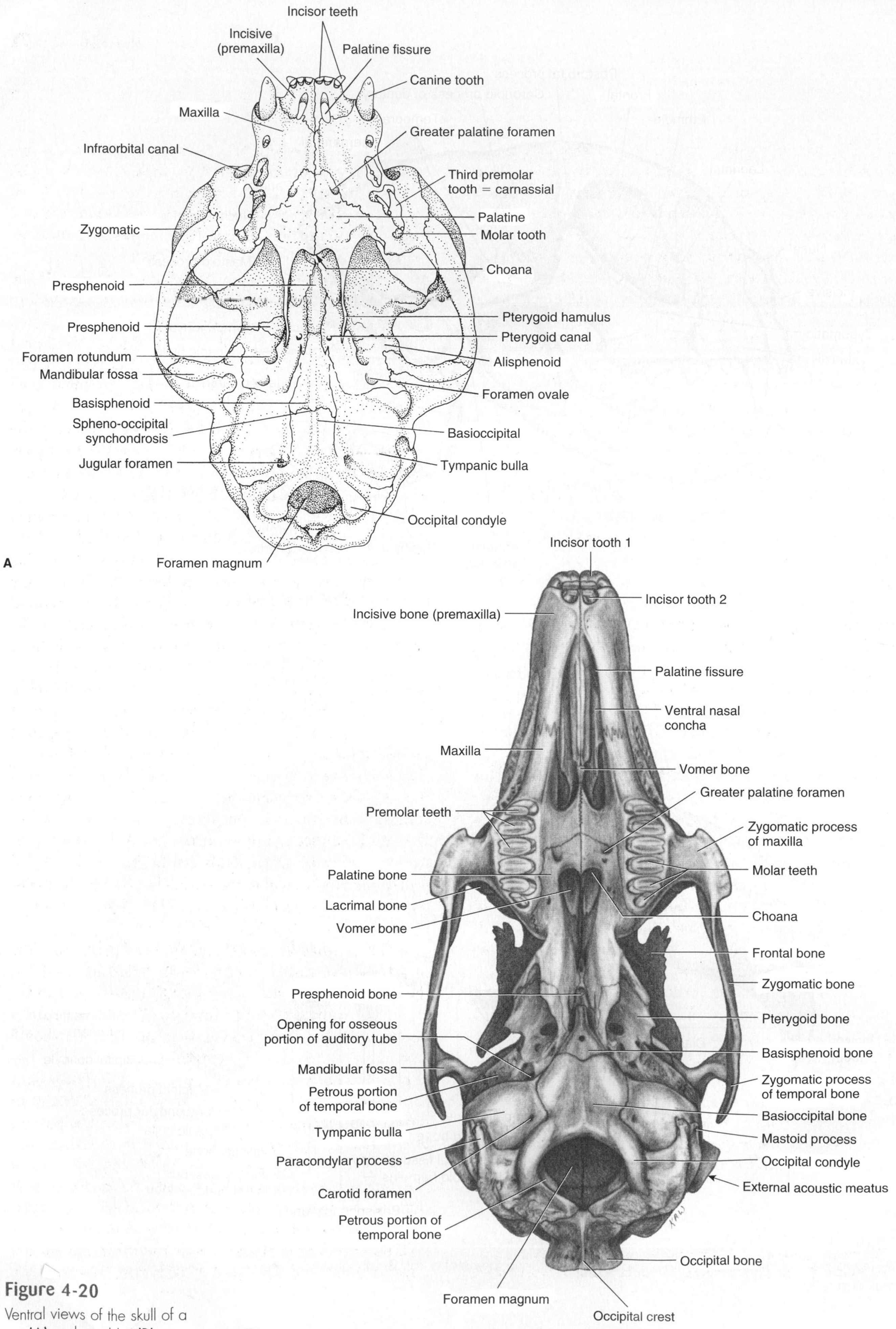
Incisor teeth
Incisive (premaxilla)
Palatine fissure
Canine tooth
Maxilla
Greater palatine foramen
Infraorbital canal
Third premolar tooth = carnassial
Palatine
Zygomatic
Molar tooth
Choana
Presphenoid
Presphenoid
Pterygoid hamulus
Pterygoid canal
Foramen rotundum
Alisphenoid
Mandibular fossa
Foramen ovale
Basisphenoid
Spheno-occipital synchondrosis
Basioccipital
Jugular foramen
Tympanic bulla
Occipital condyle
Foramen magnum
A
Incisor tooth 1
Incisor tooth 2
Incisive bone (premaxilla)
Palatine fissure
Ventral nasal concha
Maxilla
Vomer bone
Greater palatine foramen
Premolar teeth
Zygomatic process of maxilla
Molar teeth
Palatine bone
Lacrimal bone
Choana
Vomer bone
Frontal bone
Zygomatic bone
Presphenoid bone
Pterygoid bone
Opening for osseous portion of auditory tube
Basisphenoid bone
Mandibular fossa
Zygomatic process of temporal bone
Petrous portion of temporal bone
Basioccipital bone
Tympanic bulla
Mastoid process
Paracondylar process
Occipital condyle
External acoustic meatus
Carotid foramen
Petrous portion of temporal bone
Occipital bone
Foramen magnum
Occipital crest
B

Figure 4-20
Ventral views of the skull of a cat (A) and a rabbit (B).

rostral portion of the hard palate, sends a dorsal process to the orbit, and its zygomatic process forms the most rostral portion of the zygomatic arch. In the rabbit, a bulbous protrusion of the maxilla extends into the orbit; it lodges the sockets or alveoli for the long roots of the upper cheek teeth. Also in the rabbit, a long space called a **diastema,** which is devoid of teeth, lies along the ventral sides of the premaxilla and maxilla. A **lacrimal bone** is located in the craniomedial wall of the orbit just caudal to the dorsal process of the maxilla, with which it surrounds the opening to the **lacrimal canal.** The portion of the zygomatic arch ventral to the orbit is formed by the **zygomatic bone** (jugal[8]). The caudal portion of the zygomatic arch is formed by the zygomatic process of the temporal bone.

The large **temporal bone** forms part of the lateral wall of the braincase and encases the inner and middle ear cavities. The mammalian temporal has evolved through the fusion of a number of bones that are independent in amphibians and reptiles (Table 4-2). Its zygomatic process and the portion that contributes to the wall of the braincase is called its **squamous portion.** This portion evolved from the squamosal bone of early tetrapods, but the part contributing to the braincase represents the inward extension of the original squamosal of the dermal roof deep to the jaw muscles. The mastoid process and its inward extension into the cranial cavity, which will be seen in the sagittal section of the skull (Fig. 4-21, p. 73), is called the **petrous portion** of the temporal bone. This hard, rocklike portion encloses the inner ear and is homologous to the opisthotic and prootic bones of the chondrocranium of amphibians and reptiles (Figs. 4-10**C** and 4-13**A,** pp. 55 and 58). The thick part of the tympanic bulla adjacent to the external acoustic meatus is the **tympanic portion** of the temporal bone and is homologous with the dermal angular bone of the lower jaw of nonmammalian tetrapods. Finally, an endochondral **endotympanic bone,** which is a new evolutionary acquisition of mammals, forms the rest of the tympanic bulla and completes the encasement of the tympanic cavity. It has no homologue in other vertebrates.

By looking into the external acoustic meatus, you may be able to glimpse the auditory ossicles, but you should examine a special preparation to see them clearly. The **malleus**[9] is roughly mallet shaped. It has a long narrow handle, which attaches to the tympanic membrane, and a rounded head, which articulates with the incus. The **incus** is anvil shaped. It has a concave surface for the articulation with the malleus and two processes that extend from the main surface of the bone. One of these processes articulates with the head of the stirrup-shaped **stapes.** Two narrow bony columns extend from the head of the stapes to a flat oval footplate, which fits into the oval window of the otic capsule enclosing the inner ear. A stapedial artery passes between these columns of the stapes in embryonic mammals and in the adult of some species.

Dermal bones along the top of the skull include the paired **nasal bones** dorsal to the nares. They are small in the cat but very large in the rabbit. The large paired **frontals** lie dorsal and medial to the orbits, and the paired **parietals** follow caudally to the frontals (Fig. 4-18). A long suture unites the squamous portion of the temporal bone with the parietal bone on each side. Much of the parietals and frontals represent the shelflike ventral extensions of the original skull roofing bones of early tetrapods, which grew medial to the temporal muscle to provide an enlarged area of origin and a bony braincase (see Fig. 4-15, p. 64).

A large, median **occipital bone** surrounds the foramen magnum, forms the caudal surface of the skull, and includes the paracondyloid processes and most of the nuchal crest. Ventrally, the occipital bone forms the floor of the braincase between the tympanic bullae. It is a compound bone comprising the four separate occipital elements of the chondrocranium of nonmammalian tetrapods and, in most mammals, certain dermal bony elements (Table 4-2). The unpaired **interparietal bone** lies in the middorsal line in front of the occipital bone and between the caudal part of the paired parietals. It derived from the originally paired postparietal bones present in the dermal skull roof of early tetrapods. The occipital bone develops from the caudal portion of the embryonic chondrocranium. Portions of the sphenoid, the ethmoid, and the turbinate bones represent the rest of the ancestral chondrocranium. A tiny lateral portion of the **ethmoid** frequently can be seen in the medial wall of the orbit caudal to the lacrimal bone, but the major parts of this bone can be seen better in a sagittal section of the skull (Fig. 4-21, p. 73).

The sphenoid bone forms the floor and part of the sides of the braincase rostral to the tympanic bullae and caudal to the hard palate (Fig. 4-20). It is a single bone in human beings, but it is divided into a rostral presphenoid bone and a caudal basisphenoid bone in most mammals. You should view these bones also in a disarticulated skull. The **basisphenoid bone** is an extensive element, and four regions can be recognized: (1) the plate of bone on the underside of the skull just rostral to the occipital, (2) the caudal portion of the pterygoid process, (3) the caudal three foramina of a row of four at the back of the orbit, and (4) a winglike process extending dorsally between the squamous portions of the temporal and the frontal. Most of the basisphenoid bone of mammals is homologous to the basisphenoid bone of early tetrapods. The portion that includes the three foramina and the dorsal wing is homologous to the

[8]See Footnote 6, p. 62.

[9]The names of the mammalian auditory ossicles derive from their distinctive shapes. Their homologies can be found in Table 4-2.

epipterygoid bone of early tetrapods and is called the wing of the basisphenoid, or the **alisphenoid.** The pterygoid process probably corresponds to the pterygoid bone in the palate of early tetrapods, with the addition of either the ectopterygoid or parasphenoid bones, or both.

The **presphenoid bone** includes a narrow midventral strip of bone lying between the bases of the pterygoid processes and lateral extensions that pass dorsal to the pterygoid processes to form the medial wall of the orbits. The lateral extension of the presphenoid contains the large, most rostral optic foramen in the row of four foramina referred to previously, and it shares a common suture with a ventral extension of the frontal. The presphenoid is homologous to the sphenethmoid of nonmammalian tetrapods.

Some of the original dermal palatal bones of ancestral tetrapods are incorporated in the sphenoid bone; two additional dermal palatal bones can be seen in the more rostral parts of the skull. The originally paired vomers have united to form a single **vomer bone,** which appears as a midventral strip of bone rostral to the presphenoid and dorsal to the hard palate. In the cat, you will have to look into the internal nostrils to see it. It forms the ventral part of the vertical septum that separates the nasal cavities. In the rabbit, a caudal part of the vomer can be seen in front of the maxillary portion of the hard palate. Paired **palatine bones** form the caudal portion of the hard palate caudal to the maxillae, the wing rostral to the pterygoid processes, and a small part of the medial wall of the orbits caudoventrally to the lacrimals.

Sutures join most of the skull bones to one another. Sutures are slightly flexible joints due to the elasticity of the connective tissue **sutural ligaments** between them. Sometimes bones at a suture interlock by complex foldings of their edges. During growth, new bone forms in the connective tissue at the periphery of the individual bones. In live animals, the surfaces of the bones are covered by a dense connective tissue, the **periosteum,** which crosses the sutures and is anchored to the sutural ligaments. In mature individuals, the sutural ligaments may ossify so that the bones may fuse. Sutures usually carry the names of the adjacent bones (e.g., the frontomaxillary suture). A few sutures have special names. The suture between the occipital and parietal bones is the **lambdoidal suture;** that between the frontal and parietal bones, the **coronal suture;** and that between the parietal bones, the **sagittal suture** (Fig. 4-18**A**). The frontal (coronal) and sagittal planes of the body pass through the comparable sutures in humans.

The bones of the chondrocranium, which are cartilage replacement bones, are joined by cartilage. Such joints, which are also only slightly flexible due to the elasticity of the cartilage, are called **synchondroses.** An example is the sphenooccipital synchondrosis, which you can see on the ventral surface of the skull between the basisphenoid and occipital (Fig. 4-20**A**).

Interior of the Skull

Examine a pair of sagittal sections of a skull cut in such a way that the nasal septum shows on one half and the conchae on the other. Note the large cranial cavity for the brain (Fig. 4-21). It can be divided into three parts: a **caudal cranial fossa** in the occipital-otic region, which houses the cerebellum; a large **middle cranial fossa** for the cerebrum, and a small **rostral cranial fossa,** which houses the olfactory bulbs of the brain, just caudal to the nasal region. The **tentorium,** a partial transverse septum of bone, separates the middle and caudal cranial fossae in the cat. In many other mammals, the tentorium is only a sheet of connective tissue. The internal part of the petrous portion of the temporal bone, containing the inner ear, can be seen in the lateral wall of the caudal cranial fossa dorsal to the tympanic bulla. It is perforated by a large foramen, the **internal acoustic opening.** The small fossa dorsal to the internal acoustic opening lodges a lobule of the cerebellum. The **sella turcica** is a saddle-shaped notch in the basisphenoid bone on the floor of the middle cranial fossa and lodges the hypophysis. A large **sphenoidal air sinus** can be seen in the presphenoid, and a **frontal air sinus** lies in the frontal bone dorsal to the rostral cranial fossa.

A sievelike plate of bone, the **cribriform plate** of the **ethmoid bone,** whose **cribriform foramina** communicate with the nasal cavities, forms the rostral wall of the rostral cranial fossa. This plate of bone also can be seen by looking through the foramen magnum of a complete skull of a cat. The vertical **perpendicular plate** of the ethmoid, which can be seen on the larger sagittal section, extends from the cribriform plate rostrally between the two nasal cavities. It connects with the nasal bones dorsally and with the vomer ventrally and forms the bony portion of the **nasal septum.** The very front of this septum is cartilaginous. Most of the ethmoid bone is shaped like the letter T, where the top of the T is the cribriform plate and the stem is the perpendicular plate. In addition, the ethmoid bone has small lateral processes that pierce the medial wall of the orbits caudal to the lacrimal bone. These processes often are visible in the medial wall of the orbits (Figs. 4-19 and 4-22).

Each nasal cavity is largely filled with thin, complex scrolls of bone called the **nasal conchae** or **turbinates.** They show better on the section that lacks the perpendicular plate of the ethmoid (Fig. 4-21). Although the turbinates ossify from the nasal capsules of the embryonic chondrocranium, they become attached to the bones surrounding the nasal cavities. A small **dorsal nasal concha,** or nasoturbinate, lies rostral to the frontal sinus and lateral to the perpendicular septum of the nasal bone. A **ventral nasal concha,** or maxilloturbinate, lies in the rostroventral part of the nasal cavity. You can see its attachment to the maxilla by looking into the cavity through the nares. The large and very complexly folded **middle nasal concha,** or

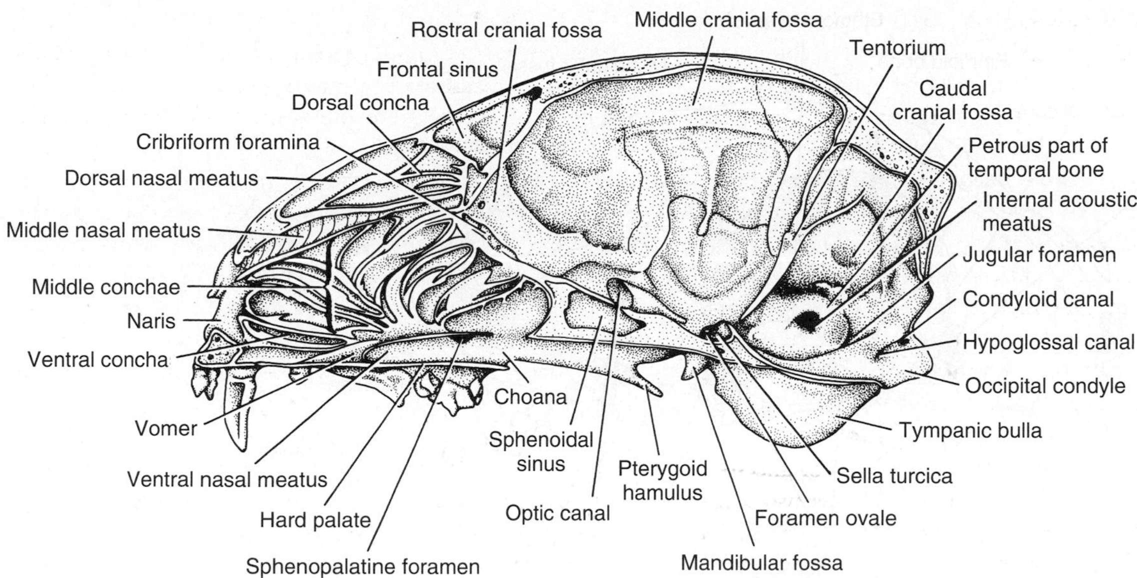

Figure 4-21

Sagittal section of the skull of a cat.

ethmoturbinate, lies between the other conchae and fills most of the nasal cavity. Some of the air entering a nasal cavity passes directly caudally through an uninterrupted passage, the **ventral nasal meatus,** which lies between the nasal conchae and the hard palate, but much of the air travels through the nasal conchae, which are covered by a moist nasal epithelium. The nasal conchae increase the surface area available for olfaction and for heating, cleansing, and moistening the inspired air. Nasal conchae, therefore, are associated with endothermy, and their presence in ancestral mammals is evidence that these animals were endothermic[10] and homeothermic. The nose is discussed in greater detail in Chapter 8.

Foramina of the Skull

Numerous foramina for nerves and blood vessels perforate the skull. These can be considered either at this time or after the nerves have been studied in Chapter 9. In any case, the foramina should be studied on a sagittal section of the skull so that both their external and internal aspects can be seen. Certain of the passages should be probed to determine their course.

First, study the foramina for the 12 cranial nerves. The first cranial nerve, the olfactory nerve, consists of many small subdivisions that enter the cranial cavity through the **cribriform foramina** in the cribriform plate of the ethmoid (Fig. 4-21). The second cranial nerve, the optic nerve, passes through the **optic canal** to emerge from the **optic foramen,** the most rostral and largest of the row of four foramina in the caudomedial wall of the orbit (Fig. 4-22). Next caudad, and largest in this row, is the **orbital fissure,** through which the third, fourth, and sixth nerves (oculomotor, trochlear, and abducens nerves, respectively) pass on their way to the extrinsic muscles of the eyeball. The orbital fissure also permits passage of the ophthalmic division of the fifth cranial nerve. (As its name implies, the fifth cranial nerve, or trigeminal nerve, has three divisions: the ophthalmic, maxillary, and mandibular divisions.) The third and smallest foramen in the row of four foramina, the **foramen rotundum,** transmits the maxillary division of the fifth nerve, and the last foramen in the row, the **foramen ovale,** transmits the mandibular division. The seventh and eighth cranial nerves (facial and vestibulocochlear nerves, respectively) travel together through the **internal acoustic meatus** (Fig. 4-21), located in the petrous portion of the temporal bone. After passing through the internal acoustic meatus, the facial and vestibulocochlear nerves separate. The vestibulocochlear nerve passes directly into the brain, and the facial nerve continues through the **facial canal** within the petrous portion of the temporal bone to emerge through the **stylomastoid foramen** on the ventral surface of the skull. The stylomastoid foramen is located beneath the tip of the mastoid process (Fig. 4-22). The ninth, tenth, and eleventh nerves (glossopharyngeal, vagus, and accessory nerves, respectively) accompany the internal jugular vein through the

[10]See footnote 7, p. 63

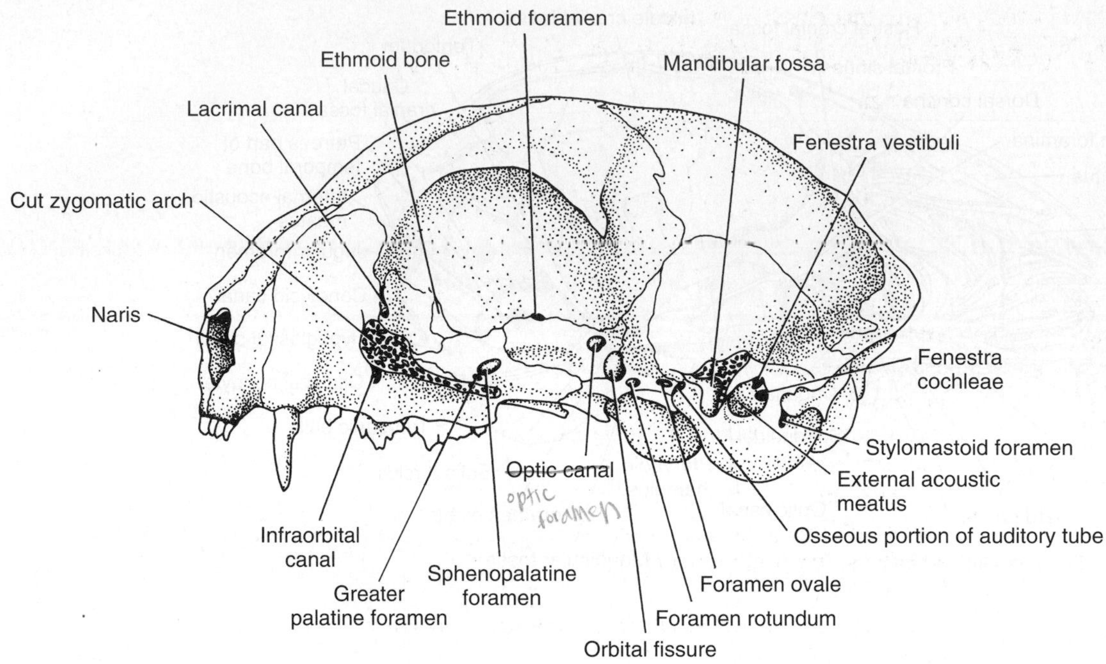

Figure 4-22

Lateral view of the skull of a cat following the removal of the zygomatic arch to reveal the skull foramina.

jugular foramen, which can be seen on the floor of the caudal cranial fossa caudal to the internal acoustic opening (Fig. 4-21). The jugular foramen opens on the ventral surface of the skull beside the posteromedial edge of the tympanic bulla (Fig. 4-20**A**). The twelfth cranial nerve (hypoglossal nerve) passes through the **hypoglossal canal,** which is visible on the floor of the caudal cranial fossa caudal to the jugular foramen (Fig. 4-21). Probe the hypoglossal canal to see that it extends cranially and emerges on the ventral surface of the skull at the jugular foramen.

As the ophthalmic and maxillary divisions of the trigeminal nerve (fifth cranial nerve) are distributed to various parts of the head, some of their branches pass through other foramina and are often in company with blood vessels. A branch of the ophthalmic division enters the nasal cavity by passing through the small **ethmoid foramen,** or series of ethmoid foramina, in the medial wall of the orbit (Fig. 4-22). The ethmoid foramen lies in the frontal bone very near its suture with the presphenoid bone. One of the branches of the maxillary division emerges through an **infraorbital canal** located in the rostral part of the zygomatic arch (Fig. 4-20**A**). However, before reaching the infraorbital canal, this nerve branch has subsidiary branches that pass through several small foramina near the front of the orbit to supply the alveoli of the molar and premolar teeth of the upper jaw. In the cat, a second branch of the maxillary division can be seen leaving the rostral portion of the orbit through the **sphenopalatine foramen** and entering the ventral part of the nasal cavity (Fig. 4-22). The sphenopalatine foramen is the larger and more medial of two foramina that lie close together in the orbital process of the palatine bone. After entering the nasal cavity, part of this nerve branch continues rostrally through the nasal passages and finally drops down to the roof of the mouth cavity through the **palatine fissure,** which is located just lateral to the midline at the rostral end of the hard palate (Fig. 4-20**A**). This fissure is much larger in the rabbit than in the cat. In certain mammals the palatine fissure also carries an incisive duct that leads from the mouth cavity to the vomeronasal organ (Jacobson's organ, Chapter 8), which is located in the nasal cavity. The caudal end of the **greater palatine foramen** lies lateral and rostral to the sphenopalatine foramen (Fig. 4-22). A third branch of the maxillary division runs through this canal to its rostral end on the hard palate in or near the suture between the palatine and maxillary bones (Fig. 4-20**A**). A fourth branch of the maxillary division backtracks to enter the orbital fissure (Fig. 4-22). It then passes into the small **pterygoid canal,** whose rostral opening may be seen in the floor of the orbital fissure. The caudal opening of the pterygoid canal appears as a tiny hole located on the ventral side of the skull near the base of the pterygoid process of the basisphenoid (Fig. 4-20**A**). After emerging from the pterygoid canal, the fourth

Table 4-3 Mammalian Cranial Nerves and Their Foramina*

Nerve	Foramen
I. Olfactory	Cribriform foramina
II. Optic	Optic canal
III. Oculomotor	Orbital fissure
IV. Trochlear	Orbital fissure
V. Trigeminal	
Ophthalmic division	Orbital fissure
One branch	Ethmoid foramina
Maxillary division	Foramen rotundum
One branch	Infraorbital canal
One branch	Sphenopalatine foramen, palatine fissure
One branch	Greater palatine foramen
One branch	Pterygoid canal
Mandibular division	Foramen ovale
One branch	Mandibular foramen, mental foramina
VI. Abducens	Orbital fissure
VII. Facial	Internal acoustic meatus, facial canal, stylomastoid foramen
VIII. Vestibulocochlear	Internal acoustic meatus
IX. Glossopharyngeal	Jugular foramen
X. Vagus	Jugular foramen
XI. Accessory	Jugular foramen
XII. Hypoglossal	Hypoglossal canal, jugular foramen

*In those cases in which a nerve goes through two or more foramina before reaching the organ it supplies, the foramina are listed in sequence from the brain. Foramina are based on a cat.

branch of the maxillary division enters the tympanic cavity through the osseous portion of the **auditory tube,** a large opening at the rostral edge of the tympanic bulla (Fig. 4-22). The ascending pharyngeal artery also enters the skull through the auditory tube. A summary of the cranial nerves and the foramina through which they pass is given in Table 4-3.

Remaining major foramina of the skull do not carry nerves. A **lacrimal canal** for the nasolacrimal (tear) duct extends from the orbit into the nasal cavity beside in the lacrimal bone (Fig. 4-22). You may also be able to find the small **carotid canal** through which the internal carotid and ascending pharyngeal arteries pass (the internal carotid artery is vestigial in the cat; Anatomy in Action 11-8). The caudal opening of the carotid canal appears as a tiny hole in the rostromedial wall of the jugular foramen. These two foramina are separate in rabbits. From here, the carotid canal extends rostrally, dorsal to the tympanic bulla, and enters the cranial cavity. Its point of entrance can be seen in the caudal cranial fossa rostral to the petrous portion of the temporal bone and ventral to the tentorium. A **condyloid canal** for a small vein lies in the caudal cranial fossa dorsal to the hypoglossal canal (Fig. 4-21).

If you have access to a specimen in which the tympanic bulla has been removed, you will be able to see two openings on the underside of the petrous portion of the temporal bone. The more dorsal opening is the **oval window** *(fenestra vestibuli)* for the stapes; the more ventral opening is the **round window** *(fenestra cochleae)* for the release of vibrations from the inner ear (Fig. 4-22).

The Lower Jaw

With the transfer of certain bones from the lower jaw of the early synapsids to the ear region of mammals and with the loss of some other bones, the enlarged paired dentary bones are left as the sole elements in the mammalian lower jaw. Examine the lower jaw, or mandible, of an adult cat (Fig. 4-23**A** and **B**) or rabbit (Fig. 4-24) and notice that the dentary bones are united anteriorly by a **mandibular symphysis,** which consists of fibrous connective tissue in juvenile specimens. The horizontal part of the mandible that bears the teeth is its **body;** the part caudal to this is its **ramus.** A large depression, the **masseteric fossa,** occupies most of the lateral surface of the ramus. Part of the masseter muscle inserts here. Part

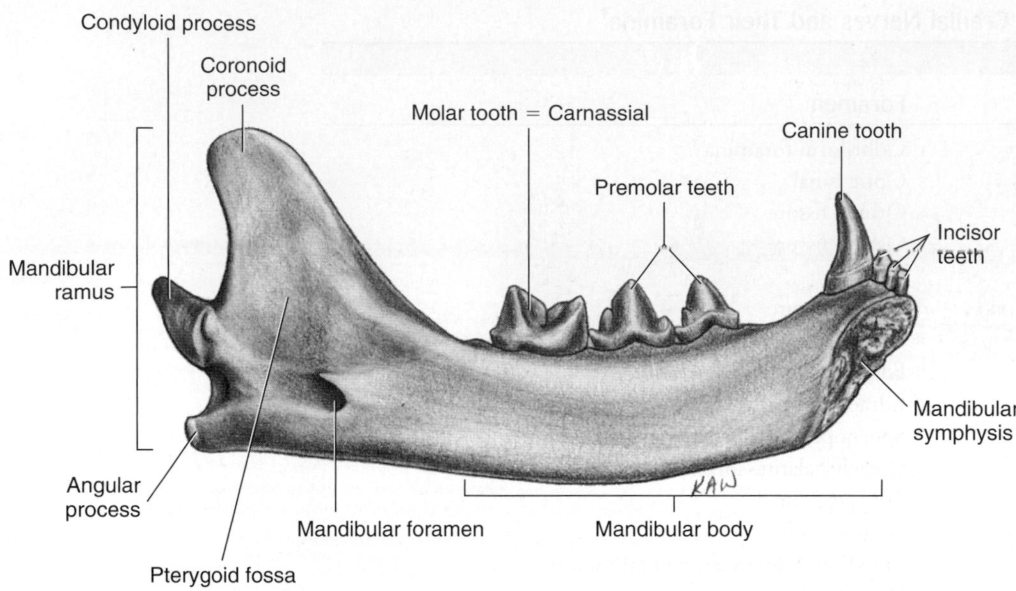

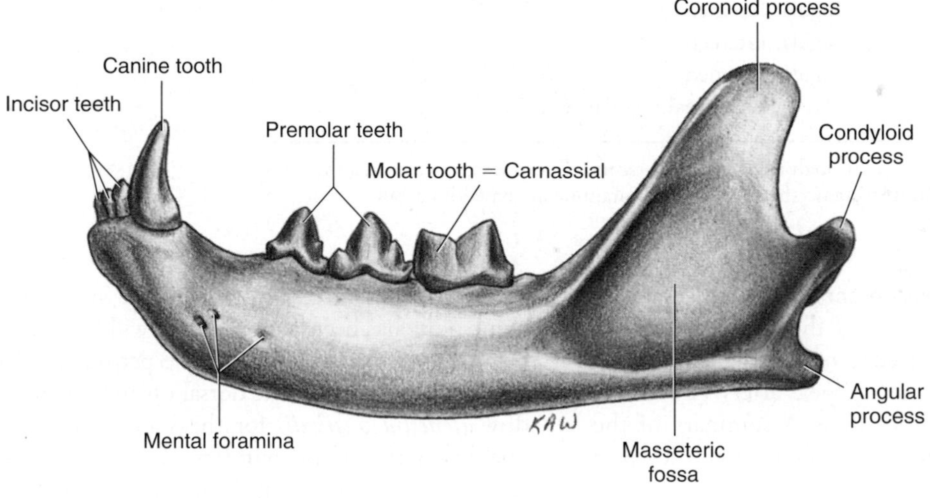

Figure 4-23

Lower jaw of an adult cat as seen in lateral (A) and medial (B) views.

continued

of the deeper pterygoid muscle inserts in a smaller **pterygoid fossa** on the medial side of the ramus.

The mandibular ramus of both the cat and rabbit bears three processes: a dorsal **coronoid process** to which the temporalis muscle attaches, a middle **condyloid process** for articulating with the mandibular fossa at the caudal end of the zygomatic arch, and a ventral **angular process** to which part of the masseter and pterygoid muscles attach. In cats, the coronoid process is much larger because its temporal muscle is much larger. The condyloid process is transversely elongated in cats and is smaller and flatter in rabbits. The angular process for the insertion of part of the pterygoid muscle and part of the masseter muscle is much larger in rabbits because the large masseter muscle is instrumental for generating grinding forces for the mastication of plant materials (Anatomy in Action 7-3).

One or two small **mental foramina** are seen on the lateral surface of the body near its rostral end. On the medial side of the ramus a large **mandibular foramen** is visible. A branch of the mandibular division of the trigeminal nerve, which supplies the teeth and the skin covering the lower jaw, enters the mandibular foramen and emerges through the mental foramina. Blood vessels accompany this nerve branch.

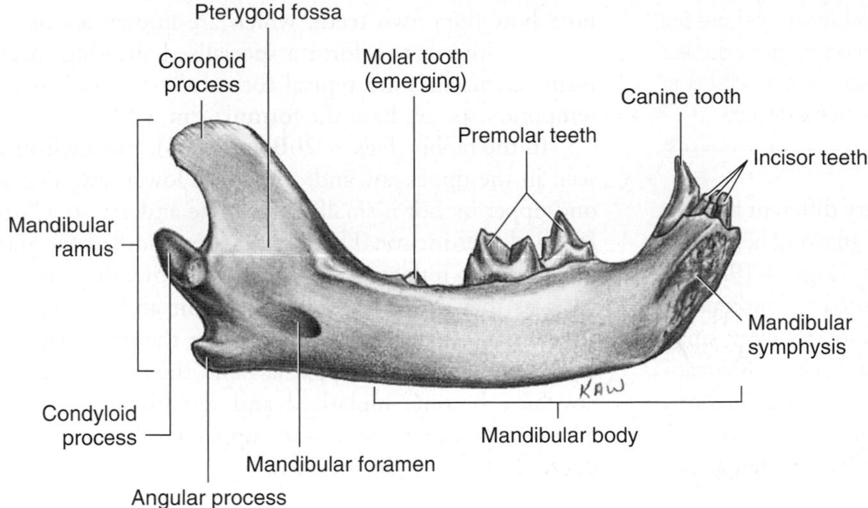

C

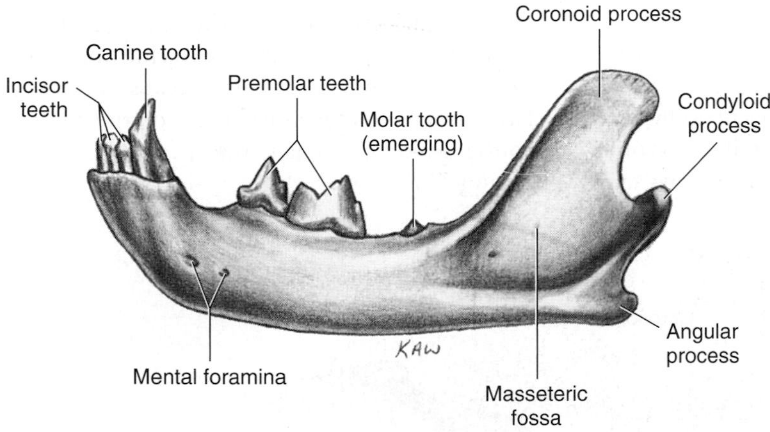

D

Figure 4-23 ***continued***

Lower jaw of a juvenile cat as seen in lateral **(C)** and medial **(D)** views.

Proportions of the various parts of the jaw are somewhat different in the jaw of a juvenile cat or other mammal (Fig. 4-23**C** and **D**). Only the milk teeth are present. As the animal and jaw grows, larger permanent teeth emerge, and the molar teeth develop and erupt.

Teeth

The teeth of mammals are quite different from those of other vertebrates. They are restricted to the jaw margins, are set in deep sockets **(thecodont),** are differentiated into various types **(heterodont),** and their replacement is limited to two sets **(diphyodont).** Most adult mammals have in each side of each jaw a series of nipping **incisors** at the front, a large piercing **canine** caudal to these, a series of cutting **premolars** caudal to the canine, and finally a series of chewing or grinding **molars.** The number of each kind of teeth present in a particular group of mammals may be expressed as a dental formula. For ancestral placental mammals this formula was:

$$3/3\ 1/1\ 4/4\ 3/3 \times 2 = 22.$$

The upper and lower jaws on each side of the body had three incisor teeth, one canine, four premolars, and three molars. Larger permanent teeth replace milk incisors, canines, and premolars as the jaws increase in size during the juvenile period. The molars appear sequentially later in life and are not replaced. The number of teeth and their structure are adapted to the diet of mammals. The molars, especially, are subject to much divergence among the various groups of mammals.

The teeth of cats and rabbits are very different because the cat is a carnivore and the rabbit is a gnawing herbivore. In the cat, each side of the upper jaw (Figs. 4-19**A** and 4-20**A**) normally has three incisors rooted to the incisive bone. One canine, three premolars, and one very small molar, all of which are rooted in the maxilla, follow caudally. In the lower jaw (Fig. 4-23**A** and **B**), three incisors, one canine, two premolars, and one large molar are normally rooted in the body of the mandible. The dental formula of the cat is therefore:

$$3/3\ 1/1\ 3/2\ 1/1 \times 2 = 16.$$

During the course of evolution, the cat has lost the first premolar in the upper jaw, the first two premolars in the lower jaw, and all molars except the first. The gap left between the canine and the premolars is called a **diastema.** The first of the remaining premolars and the molar of the upper jaw are more or less vestigial. However, the last premolar of the upper jaw (phylogenetically premolar number four) and the lower molar have become large and complex in structure. Articulate the jaws and note how these two teeth, which are known as the **carnassials,** intersect to form a specialized shearing mechanism. Carnassials are typical for carnivores, and in contemporary species have the formula Pm 4/M 1.

In the rabbit (Figs. 4-20**B** and 4-25), two incisors are seen in the upper jaw and one in the lower jaw. The second upper incisor is small and peglike and acts as a "stop" for the lower incisor. The incisors are modified for gnawing and, as in rodents, grow throughout life; they are worn away as they grow. Canines are absent and a long space **(diastema)** separates the incisors from the premolars. The three premolars in the upper jaw and the two in the lower jaw have become molarized and resemble molar teeth. Three molars are present in the upper and lower jaws. The dental formula of the rabbit is:

$$2/1\ 0/0\ 3/2\ 3/3 \times 2 = 28.$$

The molars and premolars of rabbits are specialized for grinding plant food. They are **hypsodont** (i.e., high crowned) and **lophodont** (i.e., they bear transverse enamel ridges). Notice that the distance between the left and right tooth rows in the upper jaw is greater than in the lower jaw. A rabbit can grind food only on one side of its mouth at a time because grinding requires lateral and oblique movements of the lower jaw. The configuration of the jaw joint permits this (Anatomy in Action 7-3).

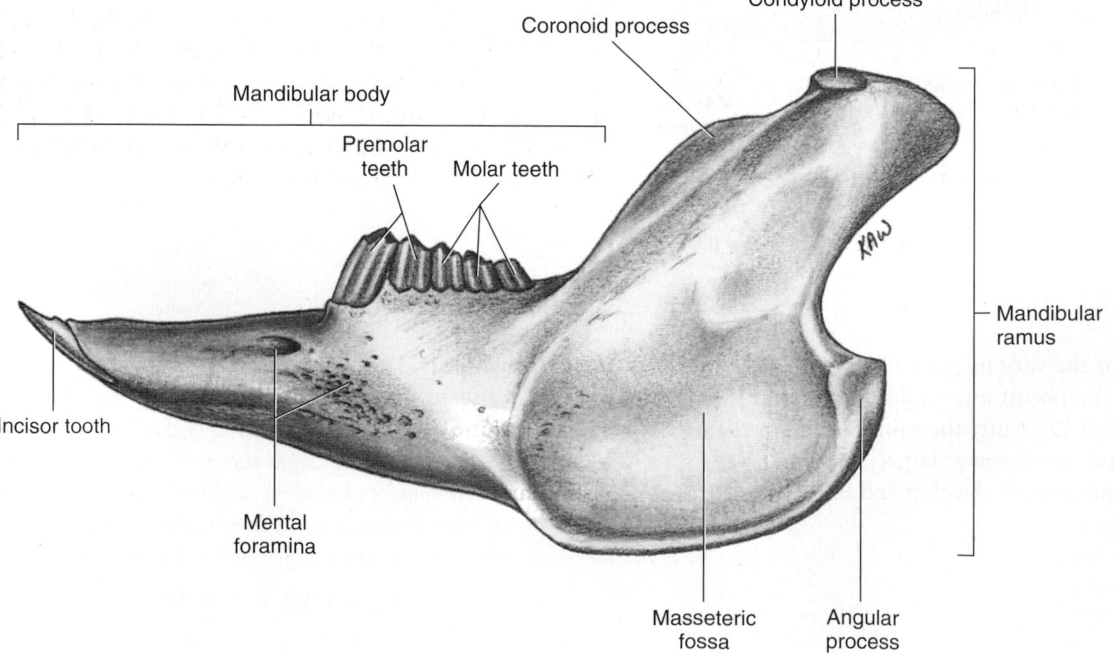

Figure 4-24
Lateral view of the lower jaw of the rabbit.

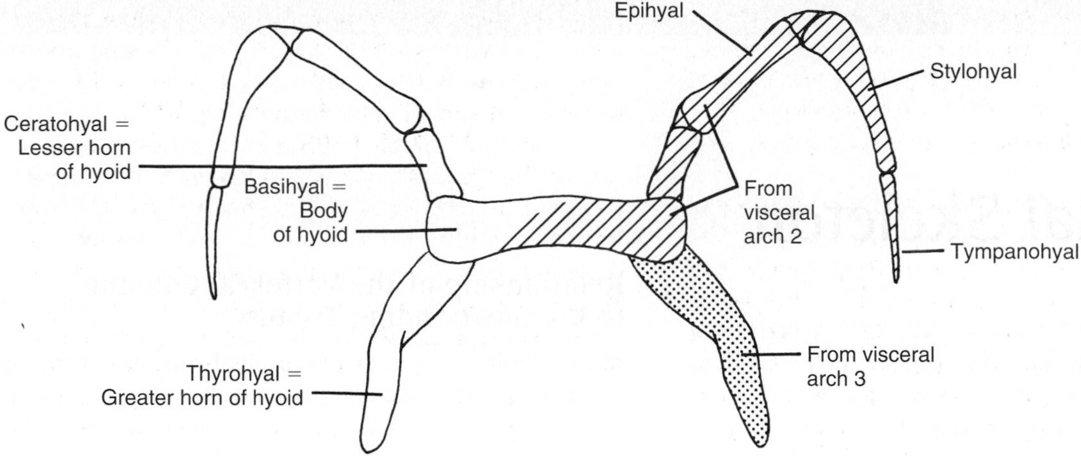

Figure 4-25
Diagram of a spread-out ventral view of the hyoid apparatus of the cat showing its components and their origins.

The Hyobranchial Apparatus

The **hyobranchial apparatus** of mammals, often simply called the hyoid apparatus, is formed from only the ventral parts of the ancestral hyoid arch and first branchial arch. More caudal ancestral branchial arches are incorporated in the laryngeal cartilages. Study the hyoid apparatus of the cat (Figs. 4-19**A** and 4-25). It may be in place on the skeleton or removed and mounted separately. It consists of a transverse bar of bone, the **basihyal,** or **hyoid body,** from which two pairs of processes, or horns, extend cranially and caudally. The caudal or **greater horns of the hyoid** are the larger horns. Each consists of but one bone, which is called the **thyrohyal,** because it extends caudally to attach to the thyroid cartilage of the larynx (Fig. 4-19**A**). Each of the cranial or **lesser horns of the hyoid** consists of a small **ceratohyal,** connected by a chain of hyoid ossicles to the skull. From ventral to dorsal these are the **epihyal, stylohyal,** and **tympanohyal.** The latter attaches to the tympanic bulla medial to the stylomastoid foramen. These three hyoid ossicles, the ceratohyal, and the basihyal are derivatives of the hyoid arch. The thyrohyal develops from the first branchial arch.

In the rabbit and human the hyoid apparatus consists of a single bone, the hyoid, which has a hyoid body, paired lesser horns, and paired greater horns. A stylohyoid ligament, which extends from the styloid process at the base of the skull to the lesser horn, replaces the chain of ossicles seen in cats as the suspension for the hyoid apparatus. The styloid process itself represents a part of the hyoid arch that has fused onto the temporal bone.

The Axial Skeleton

We begin our consideration of the postcranial skeleton by examining its axial components, which consist of the vertebral column, ribs, and sternum (if one is present). Most

Chapter 5

parts of these structures are represented by cartilaginous rudiments in the embryo and are replaced in most adult species by bone as development continues, but some parts of the vertebrae may ossify directly within the mesenchyme.

FISHES

In all vertebrates, the vertebral column develops around the notochord and spinal cord, which it surrounds and protects. The notochord enlarges and persists in the adults of lampreys, a few groups of basal actinopterygians fishes, and lungfishes, but it is replaced by the centra, or vertebral bodies, of the vertebrae in other vertebrates. Because the buoyancy of the water largely supports fishes, the vertebral column needs to provide little axial support. Rather, it acts primarily as a compression strut, enabling the fish to push through a relatively dense medium, the water. The vertebrae and ribs together participate in the undulatory movements of swimming. Because all vertebrae perform nearly the same function and are subjected to similar stresses, there is less regional differentiation of the vertebral column in fishes than in tetrapods. The first trunk vertebra articulates with the skull, and the caudal vertebrae are modified to protect the caudal artery and vein from being compressed by the powerful locomotor muscles of the tail. Trunk vertebrae are essentially the same in most species, although some regional divergence occurs in the more derived actinopterygians. In fishes, individual vertebrae are not as strong or as securely linked together as they are in terrestrial vertebrates.

Relationship of the Vertebral Column to the Surrounding Tissues

Make a fresh cross section of the tail of your specimen and observe that the vertebral column lies at the intersection of several connective tissue septa that separate the surrounding muscles (Fig. 5-1). A **dorsal skeletogenous septum** extends from the top of the vertebrae to the middorsal line of the body; a **ventral skeletogenous septum,** from the bottom of the vertebrae to the midventral line; and a **horizontal skeletogenous septum,** from each side of the vertebrae to the lateral surface of the body. The last is the most difficult to see and may be confused with portions of the **myosepta** that separate the cone-shaped muscle segments, or **myomeres.** The horizontal skeletogenous septum forms an arc that reaches the skin at the position of the lateral line canal, which appears as a small hole in the skin. It divides the myomeres horizontally into dorsal **epaxial** and ventral **hypaxial** portions. In the trunk, the vertebral column and surrounding tissues are related in much the same way. The only difference is that the ventral septum has split into two laminae, each one passing on one side of the body cavity (coelom) (Fig. 10-6).

Caudal Vertebrae

The vertebral column of fishes consists of only **trunk vertebrae** and **caudal vertebrae,** but these become somewhat modified at the attachments of the chondrocranium and the fins in the median plane of the body. Study the structure of the caudal vertebrae of *Squalus* on a special preparation or by dissecting the tail of your own specimen. If you dissect the tail, remove a piece about 5 cm long, and carefully expose the vertebrae by cleaning away all the surrounding muscle and connective tissue. Do not select a piece adjacent to either the second dorsal fin or caudal fin.

Make a cross section through the joint between two vertebrae near one end of your piece and examine it along with the lateral aspect of other vertebrae. Each vertebra consists of a cylindrical central portion, called the **centrum,** or *vertebral body,* which bears a dorsal and ventral arch of cartilage. The dorsal arch, which protects the spinal cord, is the **neural arch** or *vertebral arch;* its cavity is the **vertebral canal** (Fig. 5-2**A**). The ventral arch, which surrounds and protects the caudal artery and vein, is the

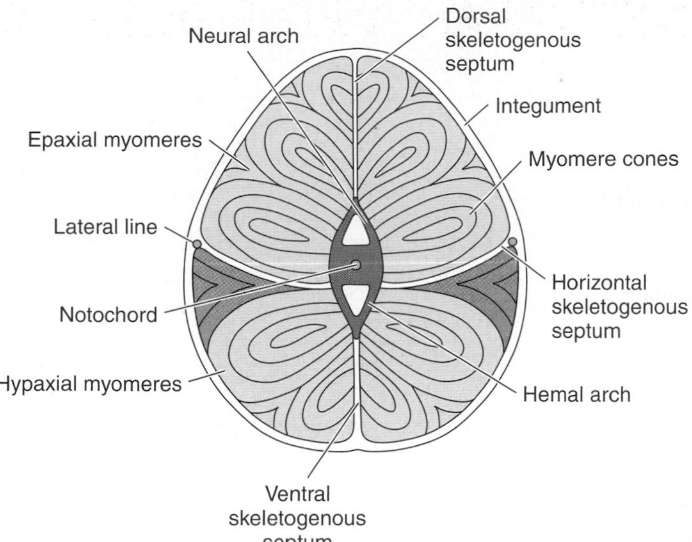

Figure 5-1
Cross section through the tail of the spiny dogfish showing the relationship of the vertebral column to surrounding tissues. The section passes through the center of a caudal vertebra where the notochord has a small diameter. The red parts of the myomeres consist of slow, oxidative muscle fibers; the white parts consist of fast glycolytic fibers (see Chapter 7). *(Modified from Liem, Bemis, Walker, and Grande, 2001.)*

hemal arch; the passage through it is the **hemal canal.** You may see the blood vessels. The artery lies dorsal to the vein. The top of each arch may extend a short distance into the dorsal and ventral skeletogenous septa as a **spinous process** and a **hemal spine,** respectively. The embryonic development of fish vertebrae is described in Anatomy in Action 5-1.

Make a sagittal section through several caudal vertebrae and study it along with the lateral aspect of other vertebrae (Fig. 5-2**B**). Notice that each side of the neural arch is composed of two roughly **triangular** blocks of cartilage. The block located directly on top of the centrum with its apex pointing dorsally is called the **neural plate.** The other block, located above the joint between the centra with its apex directed ventrally, is the **dorsal intercalary plate.** Each plate of the vertebral arch is perforated by a small hole, or **foramen,** for either the dorsal or the ventral root of a spinal nerve. A continuous vertebral arch is pe-

Anatomy in Action 5-1 Vertebral Development in Fishes

Each vertebra develops from sclerotomal mesenchymal cells that migrate from the segmented embryonic somites and accumulate around the spinal cord and notochord in an intersegmental position. A vertebra, thus, comes to lie between two segmental myomeres, which in turn have developed directly from the somites. Much of the vertebral mesenchyme forms cartilaginous **vertebral arches,** sometimes called arcualia. Some of these form the neural arch around the nerve cord; others form the hemal arch around the caudal blood vessels. The centrum develops between these arches and largely replaces the notochord. Its mode of development varies widely in vertebrates, but in fishes it appears to derive in part from the bases of the arches, called arch bases, and in part from the direct deposition of cartilage or bone in concentric rings around the notochord through perichordal ossification and even within the notochord sheath. The various components of a vertebra can be seen in a transverse section through the vertebra of some adult actinopterygian fishes (see figure).

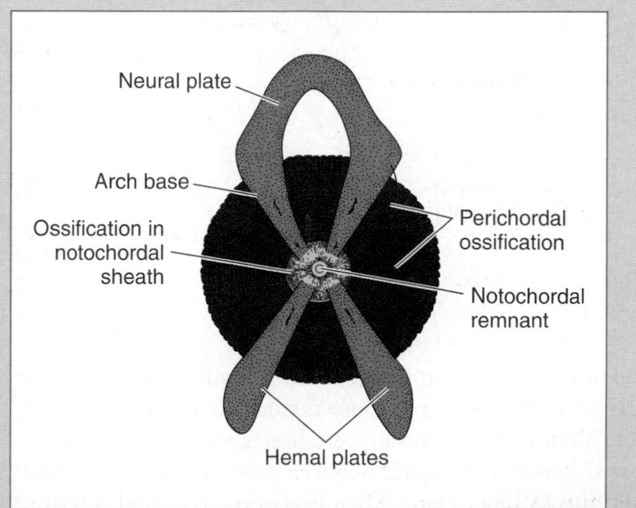

Transverse section through a trunk vertebra of an actinopterygian fish. *(From Liem, Bemis, Walker, and Grande, 2001.)*

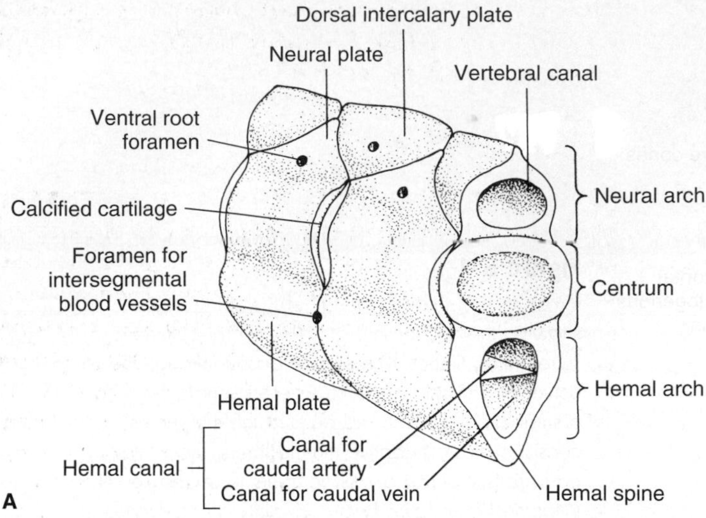

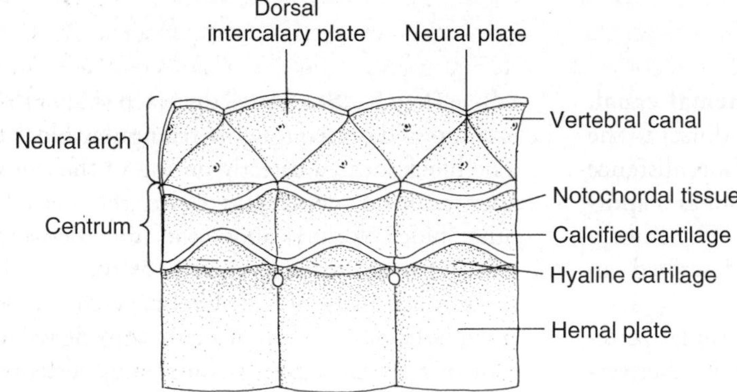

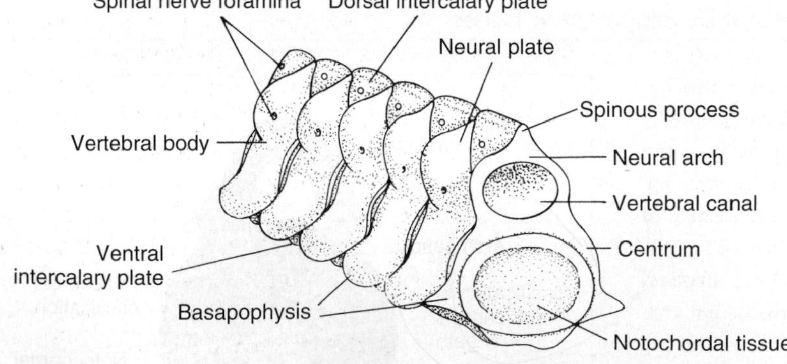

Figure 5-2
Vertebrae of *Squalus*. A, Three-dimensional craniolateral view of two caudal vertebrae. B, Sagittal section through three caudal vertebrae. C, Three-dimensional craniolateral view of several trunk vertebrae.

culiar to cartilaginous fishes; such a condition does not restrict lateral bending because cartilage is much more flexible than bone. In other vertebrates with ossified vertebrae, there is some space between successive neural arches, and this facilitates lateral bending of the vertebral column. The neural arch of other vertebrates probably is homologous to the neural plate because it develops primarily from a comparable group of cells.

The hemal arch is composed on each side of just one block, called a **hemal plate.** A small foramen for intersegmental branches of the caudal artery and vein can be found between the bases of the hemal arches.

Examine the centrum in the sagittal section. It is shaped like a biconcave spool, a shape termed **amphicoelous.** The concavity at each end contains a gelatinous substance, which is **notochordal tissue** that has persisted from the embryonic stage. A small strand of notochord extends through each centrum from one concavity to the next. The cartilage immediately adjacent to the notochord has become calcified and thus appears white. Calcified cartilage in cartilaginous fishes consists of minute mineralized blocks called **tesserae,** which form a prismatic layer in the matrix. Calcification strengthens and stiffens the cartilage, but it is not the same as ossification. The area of calcified cartilage has an hourglass shape (Fig. 5-2**B**). The rest of the cartilage is glasslike, or **hyaline.**

Trunk Vertebrae

Compare the caudal vertebrae with a special preparation of the trunk vertebrae (Figs. 5-2**C,** 4-4, and 4-5). The structure is basically the same, but the trunk vertebrae lack hemal arches. Instead, they have short ventrolateral processes, termed **basapophyses,** which project from the sides of the vertebral body. Basapophyses are serially homologous to the proximal part of the hemal arches. Small **ventral intercalary plates** lie between successive basapophyses, but they are often lost during the preparation of the specimen. They are the ventral counterparts of the dorsal intercalary plates.

Ribs

Notice on a mounted skeleton that short ribs articulate with the basapophyses and extend into the horizontal skeletogenous septum at the point where it intersects the myosepta (see Figs. 4-4 and 4-5). Ribs in this position are called dorsal or **intermuscular ribs** because they lie between the epaxial and hypaxial parts of the myomeres. Many fishes have ventral or **subperitoneal ribs,** so called because they lie in the myosepta, just beneath the peritoneum, which lines the body cavity. Some fishes have both types of ribs. Ribs strengthen the myosepta, which serve as attachment sites for the muscle fibers of the myomeres.

Median Fin Supports

Look at the dorsal fins on a mounted skeleton of *Squalus* and notice that each is supported proximally by several cartilages and distally by fibrous rods, the **ceratotrichia,** described in Chapter 3 (p. 29). The cartilages, which collectively may be called **pterygiophores,** are in two series. Those that rest on the vertebral column are **basals;** the more distal ones are **radials.** There is some question as to whether the pterygiophores develop phylogenetically from the spinous processes of the vertebrae or independently within the dorsal skeletogenous septum. The caudal fin, which is of the heterocercal type (Chapter 3), is supported proximally by enlarged spinous processes and hemal spines and distally by ceratotrichia.

AMPHIBIANS

Because the body of terrestrial vertebrates is not supported by water, their vertebral column is very important for both locomotion and support. Individual vertebrae are more completely ossified than those of most fishes, and successive vertebral arches articulate by pairs of overlapping **zygapophyses** *(articular processes)* (Fig. 5-3**A**). The zygapophyses unite the vertebrae firmly; they resist vertical bending yet permit lateral bending. Considerable regional differentiation evolved in the vertebral column and ribs of tetrapods, correlated with the different forces that act on these parts of the axial skeleton. With the loss of gills and the bony connection between the skull and pectoral girdle, a neck region appears. The number of **cervical vertebrae** varies, but there is always at least one, the **atlas. Trunk vertebrae** follow the cervical vertebrae. At least one **sacral vertebra** and **rib** articulate with the pelvic girdle and transfer forces between the trunk and hind legs. The number of **caudal vertebrae** depends on the length of the tail.

The composition of the vertebrae varies considerably, but the most primitive type in tetrapods was the **rhachitomous vertebra,** found in early amphibians (Fig. 5-3**A**). Such vertebrae consisted of a neural arch, beneath which was a notochord partly constricted by three blocks of bone, which made up the centrum. A single U-shaped **intercentrum** lay beneath the notochord at the front of each vertebra. A pair of small **pleurocentra** was located caudodorsal to the intercentrum. Hemal arches, or chevron bones, also were present in the caudal region, where they attached to the intercentra.

The structure of vertebrae diverged among early tetrapods and modern lissamphibians (Fig. 5-3**B** and **C**), but in the evolution from ancestral amphibians through both the sauropsid line of evolution to reptiles and birds and the synapsid line to mammals, the intercentrum decreased in size, and the pleurocentrum enlarged to become the definitive centrum (Fig. 5-3**D** and **E**). The notochord disappeared in the adult sauropsids and synapsids.

Most of the vertebrae bear distinct ribs to which trunk muscles attach. Although early terrestrial vertebrates had legs, fishlike lateral undulations of the trunk contributed to locomotion (see Figure 7-10). Additionally, rib movements are important in ventilating the lungs in most amniotes. The cranial trunk ribs usually unite ventrally with a **sternum** of cartilage or cartilage replacement bone. Tetrapod ribs develop somewhat differently from fish ribs. In tetrapods, the major portion of a rib occupies the same position as the subperitoneal rib of a fish, but its more dorsal attachment on the vertebra resembles that of an intermuscular rib. Most tetrapod ribs also have two processes that articulate with the vertebrae. A dorsal **tuberculum** articulates with a type of transverse

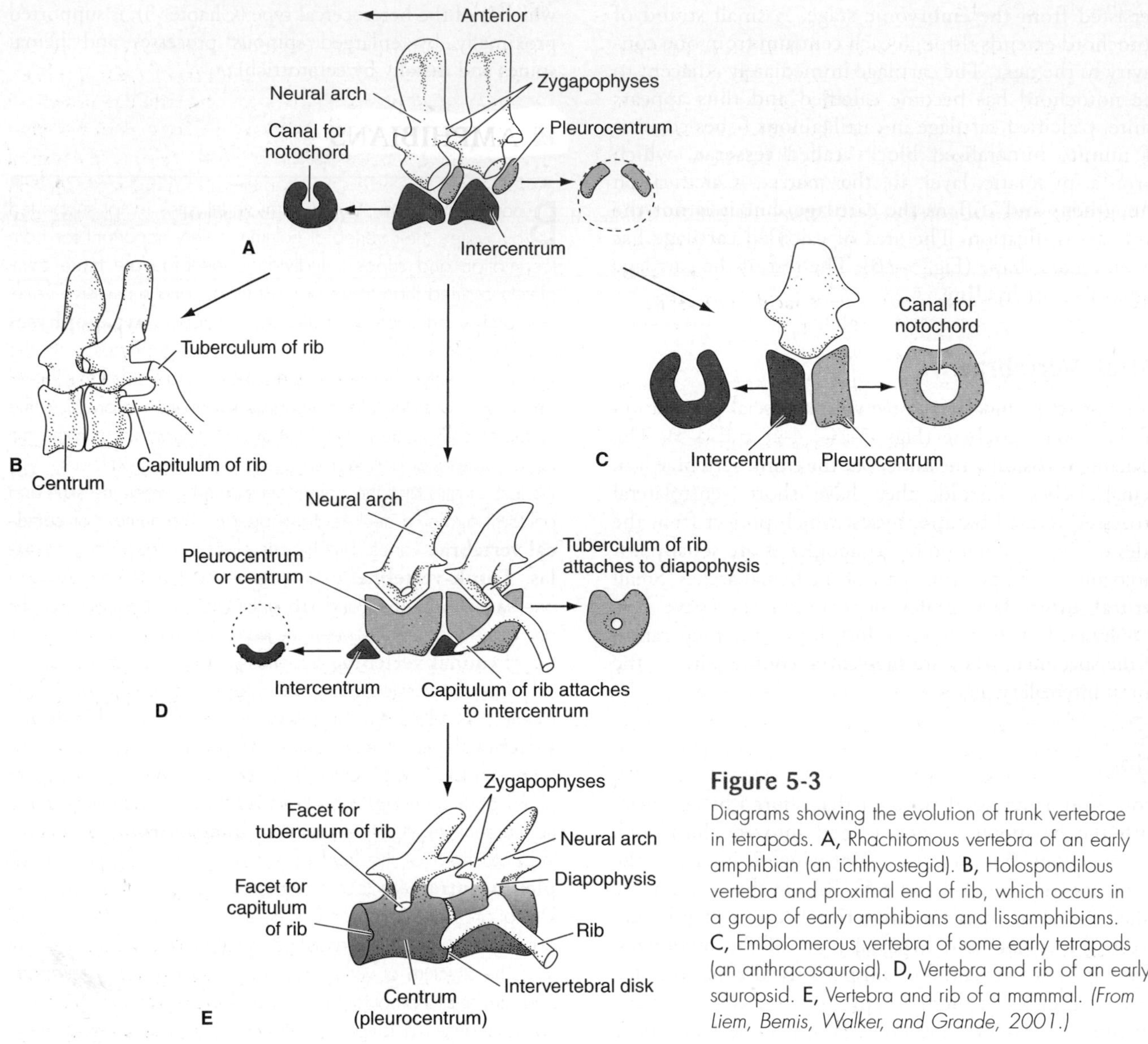

Figure 5-3

Diagrams showing the evolution of trunk vertebrae in tetrapods. **A,** Rhachitomous vertebra of an early amphibian (an ichthyostegid). **B,** Holospondilous vertebra and proximal end of rib, which occurs in a group of early amphibians and lissamphibians. **C,** Embolomerous vertebra of some early tetrapods (an anthracosauroid). **D,** Vertebra and rib of an early sauropsid. **E,** Vertebra and rib of a mammal. *(From Liem, Bemis, Walker, and Grande, 2001.)*

process on the base of the neural arch called a **diapophysis** (Fig. 5-3D). A more ventral rib **capitulum,** or head, articulated primitively with the intercentrum. With the reduction and loss of the intercentrum in mammals, the capitulum evolved an intervertebral articulation (Fig. 5-3E). Occasionally, the capitulum articulates with a small lateral process of the centrum called a **parapophysis.**

Vertebral Regions

Examine the vertebral column and ribs on a skeleton of *Necturus.* There is more regional differentiation than in the vertebral column of *Squalus.* The most cranial vertebra, called the **atlas,** lacks distinct ribs and is specialized for articulating with the paired occipital condyles of the skull. It is, of course, located in the incipient neck region and is the only vertebra that can be called a **cervical vertebra. Trunk vertebrae** extend from the atlas to the level of the pelvic girdle. All trunk vertebrae have ribs that articulate with the vertebrae. (No ribs are fused to the vertebrae.) A single **sacral vertebra** (usually the nineteenth vertebra) and its pair of ribs are modified for connecting with the pelvic girdle. The remaining vertebrae are **caudal vertebrae;** most lack ribs and bear **hemal arches.**

Trunk Vertebrae

Study a trunk vertebra on a mounted skeleton of *Necturus* and also examine an isolated specimen (Fig. 5-4). Dorsally, a vertebra consists of a **neural arch** that overlies the spinal cord, which is located in the **vertebral canal.** A small **spinous process** projects caudad from the top of the neural arch. Two pairs of small processes for the articulation of successive vertebrae project laterally from the front and back of the neural arch. The cranial pair of processes, whose smooth articular facets face dorsally, are the **cranial zygapophyses;** the caudal pair, whose facets face ventrally, are the **caudal zygapophyses.** Note how the zygapophyses of adjacent vertebrae overlap.

Ventrally, the vertebra consists of a biconcave, or amphicoelous, **centrum.** Notochordal tissue persists in the concavities. A prominent **transverse process** projects caudolaterally from each side of the neural arch and centrum. Observe that it is rather high from its dorsal edge to its ventral edge. The dorsal part of the transverse process represents a **diapophysis;** the ventral part, a **parapophysis.**

Ribs

The ribs of *Necturus* are shorter than the ribs of ancestral amphibians but have the two proximal processes characteristic of terrestrial vertebrates, a condition called bicipital. The ventral **capitulum** articulates with the parapophysis; the dorsal **tuberculum,** with the diapophysis. The distal portion of the rib is its shaft, or **body.** In extant amphibians, ribs do not participate in lung ventilation.

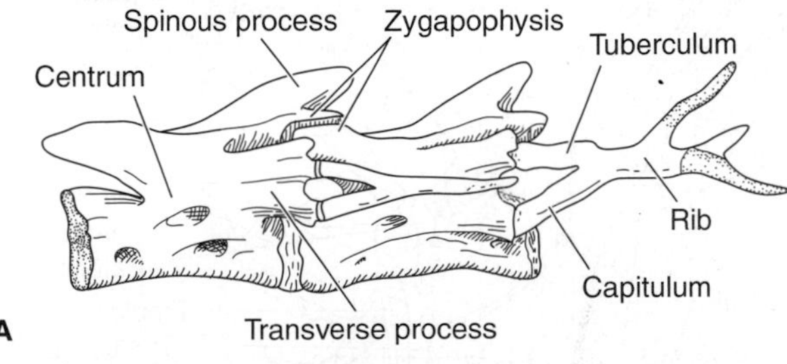

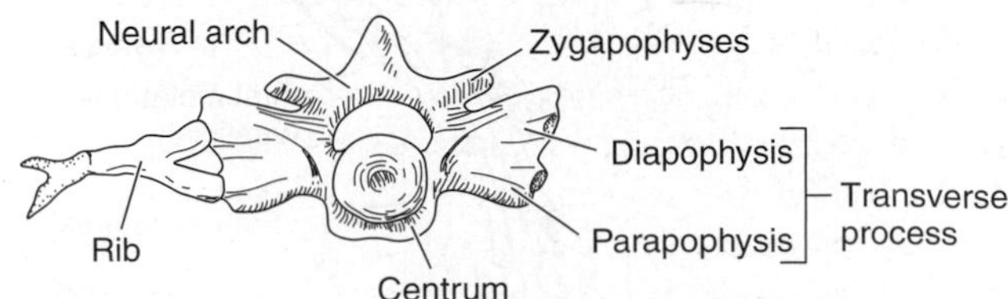

Figure 5-4

Lateral **(A)** and posterior views **(B)** of a trunk vertebra and rib of *Necturus.* Cranial is toward the left. *(From Liem, Bemis, Walker, and Grande, 2001; after Goodrich, 1930.)*

Sternum

Necturus does not have the small cartilaginous sternum found just caudal to the pectoral girdle of most urodeles. However, small pieces of cartilage, which may represent an incompletely developed sternum, are located in the ventral portions of the myosepta caudodorsal to the pectoral girdle and in the midventral connective tissue septum. You may notice them during the dissection of the muscles (Chapter 7).

MAMMALS

The axial skeleton of mammals functions as a suspension girder for the body and transfers forces between the trunk and appendages. In the evolution toward mammals, the appendages assumed a more important role in locomotion, and lateral undulations of the trunk and tail played a lesser role (Anatomy in Action 3-2). Except for cetaceans and some other aquatic mammals, the tail seldom has a role in locomotion. It does have other functions, however, such as assisting with balance in arboreal species, clearing insects from the body in larger herbivorous mammals, and facilitating social interactions as in dogs. The tail is lost as an external organ in humans and some other species such as apes.

A change in the plane of articulation on some zygapophyses in cats and some other species that run rapidly allows for considerable extension and flexion of the back, which increases stride length. Head movements are much freer in mammals than in amphibians and reptiles because of the longer cervical region and specialization of the first two cervical vertebrae.

Ribs that are moveably articulated with the vertebrae are limited to the thoracic region and, together with the newly evolved muscular diaphragm, encase the thoracic organs and help ventilate the lungs. Small embryonic ribs are present on other vertebrae, but during embryonic development they fuse onto the sides of the vertebrae, forming complex transverse processes. The rib component of a transverse process is called a **pleurapophysis.** In an adult mammal, it is not always possible to distinguish the part of the transverse process that developed from a rib from the part that developed from a diapophysis, parapophysis, or both, so the entire transverse process is often called a pleurapophysis.

Because of the many functions of the trunk skeleton and the changing stresses along its length, the vertebral column of mammals displays more regional differentiation than that seen in early tetrapods. No two vertebrae are exactly alike, but they can be sorted into five broad groups.

The individual vertebrae of mammals are similar to those of extant reptiles. The mammalian vertebral centrum evolved from the pleurocentrum, and the ancestral intercentrum was lost

(Fig. 5-3**E**). **Intervertebral discs** of fibrocartilage lie between centra. Traces of notochordal tissue, called the nucleus pulposus, may be found within the intervertebral discs.

Vertebral Groups

Examine mounted skeletons of a cat and any other mammals that are available, and also a string of disarticulated vertebrae, noting the five vertebral regions (Figs. 5-5 and 5-6). The most cranial group of vertebrae is the **cervical vertebrae.** There are more cervical vertebrae in mammals than in early tetrapods. With few exceptions, mammals have seven cervical vertebrae, all of which lack ribs. **Thoracic vertebrae,** which bear ribs, follow the cervical vertebrae, and **lumbar vertebrae** follow the thoracic vertebrae. The lumbar vertebrae lack ribs but have very large transverse processes. The number of vertebrae in these two regions varies among mammals. Cats normally have thirteen thoracic and seven lumbar vertebrae; rabbits, twelve thoracic and seven lumbar vertebrae; humans, twelve thoracic and five lumbar vertebrae. **Sacral vertebrae** follow the lumbar vertebrae and are fused together to form a solid point of attachment, the **sacrum,** for the pelvic girdle. The number of vertebrae contributing to the sacrum varies among species. Many mammals, including cats, have three; rabbits, four; humans, five. The number correlates with the stresses to which the sacrum is subjected. The sacrum of a human must support all of the body weight; that of a rabbit is subject to the stresses associated with its bounding mode of locomotion. The remaining vertebrae are **caudal vertebrae.** The number of caudal vertebrae varies with the length of the tail, but all mammals have some. Even in humans, who have no external tail, there are three to five small caudal vertebrae that are fused together to form a **coccyx** to which certain anal and gluteal muscles attach.

Thoracic Vertebrae

We will study the thoracic vertebrae first because they are the most similar to the vertebrae of *Necturus.* They show the components of a vertebra very clearly because they are less specialized than the vertebrae of other regions. Examine one from near the middle of the thoracic series. Identify the **neural arch** with its long **spinous process,** the **vertebral canal,** and the **centrum** (Fig. 5-7**D**). Each end of the centrum is flat, a shape termed **acoelous.** In life, small fibrocartilaginous **intervertebral discs** are located

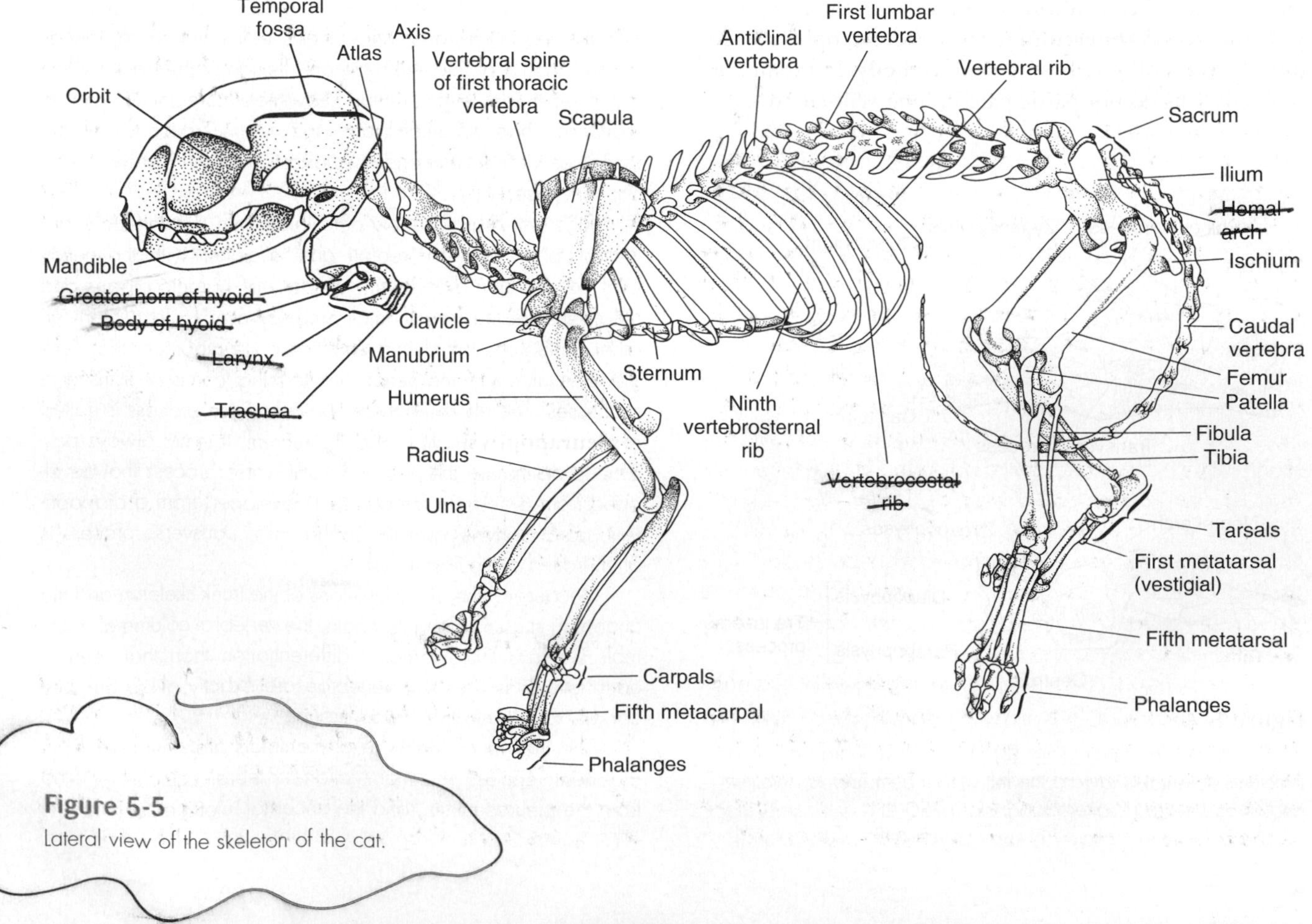

Figure 5-5
Lateral view of the skeleton of the cat.

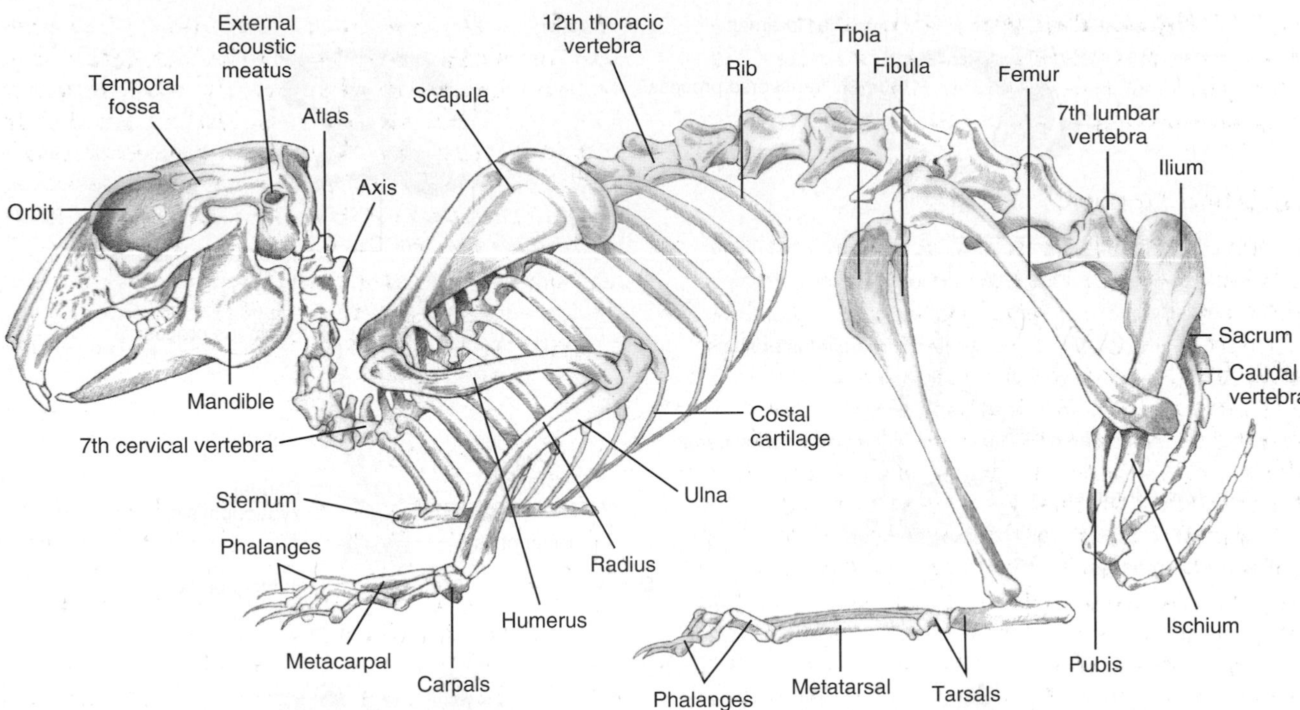

Figure 5-6
Lateral view of the skeleton of a rabbit. *(Modified from Barone, Pavaux, Blin, and Cuq, 1973.)*

between successive centra. Articulate two vertebrae and notice how the back of one neural arch overlaps the front of the one behind it. Disarticulate the vertebrae and look for smooth articular facets in the area where the vertebral arches came together. These are the zygapophyses *(articular processes).* The facets of the pair of **cranial zygapophyses** are located on the cranial surface of the neural arch and face dorsally; those of the pair of **caudal zygapophyses** are on the caudal surface of the neural arch and face ventrally. A lateral transverse process, called a **diapophysis,** projects from each side of the neural arch. Notice the smooth articular facet for the tuberculum of a rib on its proximal tip. In the cranial and middle portion of the thoracic series, the rib capitulum articulates between vertebrae, so part of the facet for a capitulum is located on the front of one centrum and part on the back of the next cranial one. The somewhat constricted portion of the neural arch between the diapophysis and centrum is called its **pedicle.** When several vertebrae are articulated, you will see openings for the spinal nerves, the **intervertebral foramina,** between successive pedicles.

Compare one of the thoracic vertebrae from the middle of the series with those near the cranial and caudal ends of the thoracic region and observe the gradual changes in the size and inclination of the spinous processes, the plane of the zygapophyseal articulations, the size and orientation of the transverse processes, and the positions of the articular facets for the ribs (Figs. 5-5 and 5-6). The changes that you observe are a consequence of changes in the forces acting on the vertebral column. For example, the plane of the articular facets of the zygapophyses changes as the major plane of vertebral movement changes. The most caudal thoracic vertebrae lack transverse processes because their ribs lack a tuberculum, and the facets for their rib heads are entirely on one centrum. Articulate the last two thoracic vertebrae and notice how a process of the pedicle, which extends caudally lateral to the zygapophyses, reinforces the articulation of the zygapophyses. This is an **accessory process.** A skilled observer can find differences among the vertebrae sufficient to identify each one precisely—as the third or eighth thoracic vertebra, for example. However, it will be sufficient for you to be able to distinguish the major group to which a vertebra belongs. All thoracic vertebrae have at least one articular facet for a rib; no other vertebrae have such facets.

Lumbar Vertebrae

Lumbar vertebrae are characterized by their large size and by prominent, bladelike transverse processes (Fig. 5-7**E**). The transverse process is a **pleurapophysis,** because a rib has united with a small diapophysis. Lumbar vertebrae also have a small bump for the attachment of ligaments and tendons dorsal to the articular facet of each cranial

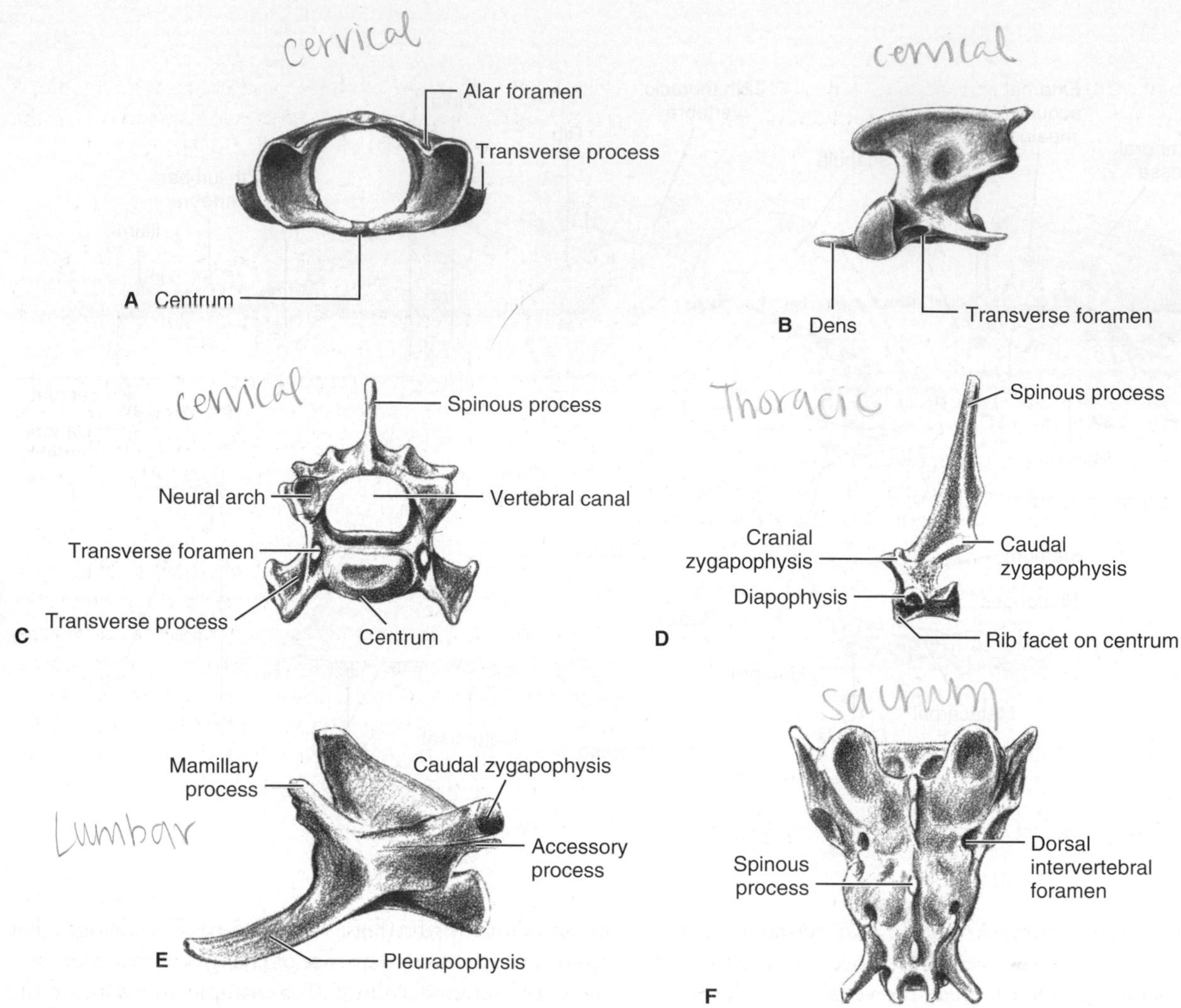

Figure 5-7

Drawings of selected vertebrae of the cat. In those shown in lateral view, cranial is toward the left. **A,** Atlas in cranial view. **B,** Axis in lateral view. **C,** Cervical vertebra in caudal view. **D,** Thoracic vertebra in lateral view. **E,** Lumbar vertebra in lateral view. **F,** Sacrum in dorsal view.

zygapophysis. This is called a **mamillary process.** You can also see traces of mamillary processes on the more caudal thoracic vertebrae. Most of the lumbar vertebrae also have accessory processes.

Sacral Vertebrae

You can easily distinguish the sacral vertebrae because they fuse into a single piece, the **sacrum,** and because they present a broad lateral surface for articulating with the pelvic girdle (Fig. 5-7**F**). Examine the sacrum closely and you will be able to detect the spinous processes, zygapophyses, pleurapophyses, and other features of the individual vertebrae that compose it. Notice how the distal ends of the pleurapophyses have fanned out and united with each other lateral to the intervertebral foramina. This condition has produced separate dorsal and ventral foramina for the respective rami of spinal nerves.

Caudal Vertebrae

Caudal vertebrae are characterized by their small size and progressive incompleteness. The more cranial vertebrae in this group have the typical vertebral parts, but more caudally there is little left but an elongated centrum. The caudal vertebrae of some mammals have V-shaped hemal arches, usually called **chevron bones,** which protect the caudal artery and vein. The cat has traces of such bones on the anterior caudal vertebrae, but they usually are lost in a mounted skeleton. The points of articulation of a chevron bone may be visible as a pair of tubercles, **hemal processes,** at the cranial end of the ventral surface of a centrum.

Cervical Vertebrae

Most of the cervical vertebrae can be recognized by their characteristic transverse processes, each of which (except for the most caudal) is perforated by a **transverse foramen** through which the vertebral blood vessels pass. The portion of the transverse process situated lateroventral to the center of the transverse foramen has developed from an embryonic rib and hence is a pleurapophysis. However, the basal part of the transverse process represents a diapophysis dorsally and a parapophysis ventrally (Cave, 1975). The last cervical vertebra normally lacks a transverse foramen and, aside from the absence of rib facets, closely resembles the first thoracic vertebra. Most of the cervical vertebrae also have low spinous processes and wide neural arches.

The first two cervical vertebrae, the **atlas** and the **axis,** respectively, are very distinctive. The atlas is ring shaped, with winglike transverse processes perforated by the transverse foramina (Fig. 5-7**A**). Its neural arch lacks a spinous process and is perforated on each side by an **alar foramen,** through which the vertebral artery enters the skull and the first spinal nerve leaves the spinal cord. Cranially, the neural arch has facets that articulate with the paired occipital condyles of the skull; caudally, it has facets that articulate with the centrum of the axis. The centrum of the atlas is reduced to a thin transverse rod.

The axis is characterized by an elongated spinous process that extends rostrally over the neural arch of the atlas; very small transverse processes; rounded articular surfaces at the cranial end of its body; and a median, tooth-shaped process, the **dens,** which projects from the front of the centrum of the axis into the atlas. In life, a strong **transverse ligament** within the atlas crosses the dorsal surface of the dens (Anatomy in Action 5-2).

Ribs

Study the **ribs** *(costae)* of a cat from disarticulated specimens and on a mounted skeleton. Most of them have both proximal processes characteristic of tetrapod ribs: a proximal **capitulum** articulating with the centrum and a more distal **tuberculum** articulating with the transverse process (Fig. 5-8). The last three ribs have only a capitulum. The portion of the rib between its two proximal processes is its **neck;** the long distal part is its **body,** or shaft. A **costal cartilage** extends from the end of the shaft.

Those ribs whose costal cartilages attach directly to the sternum are called **vertebrosternal ribs** (Fig. 5-5); those whose costal cartilages unite with other costal cartilages before reaching the sternum are **vertebrocostal ribs;** and those whose costal cartilages have no distal attachment are **vertebral,** or floating, **ribs.** Sometimes vertebrosternal ribs are called **true ribs** and all the others **false ribs.** Cats normally have nine vertebrosternal ribs, three vertebrocostal ribs, and one vertebral rib.

Sternum

The **sternum** (Figs. 5-5 and 5-6) is composed of a number of ossified segments called the **sternebrae.** The first of these constitutes the **manubrium,** the last is the **xiphisternum,** and those between make up the **body** of the sternum. A cartilaginous **xiphoid process** extends caudad from the xiphisternum. Costal cartilages unite with the sternum between the sternebrae. In many mammals, but not in the cat and most other cursorial mammals, the clavicles of the pectoral girdle articulate with the cranial end of the manubrium. All parts of the axial skeleton participate in the support and movement of the head and trunk (Anatomy in Action 5-3).

Anatomy in Action 5-2 Specializations of the Atlas and Axis

The atlas and axis are specialized to form a sort of "universal joint" that permits a wide range of movements of the head. Up-and-down movements of the head occur primarily at the occipital–atlas joint; axial rotations occur at the atlas–axis joints and angular rotations occur through bending of the cervical vertebral column. Rotational movements were made possible during the transition from earlier synapsids to mammals by the loss of the zygapophyses between the atlas and axis and by their replacement with new articular surfaces between the body of the axis and the neural arch of the atlas. Although the dens lies in the axis of rotation, Farrish Jenkins (1969) showed that it does not function primarily as a pivot. Rather, together with the transverse ligament that crosses it dorsally, the dens prevents excessive flexion at the atlas–axis joint and drooping of the head, which would otherwise occur with the loss of the zygapophyses between these vertebrae. Excessive flexion between the atlas and axis could injure the spinal cord. The dens evolved as an outgrowth of the centrum of the axis.

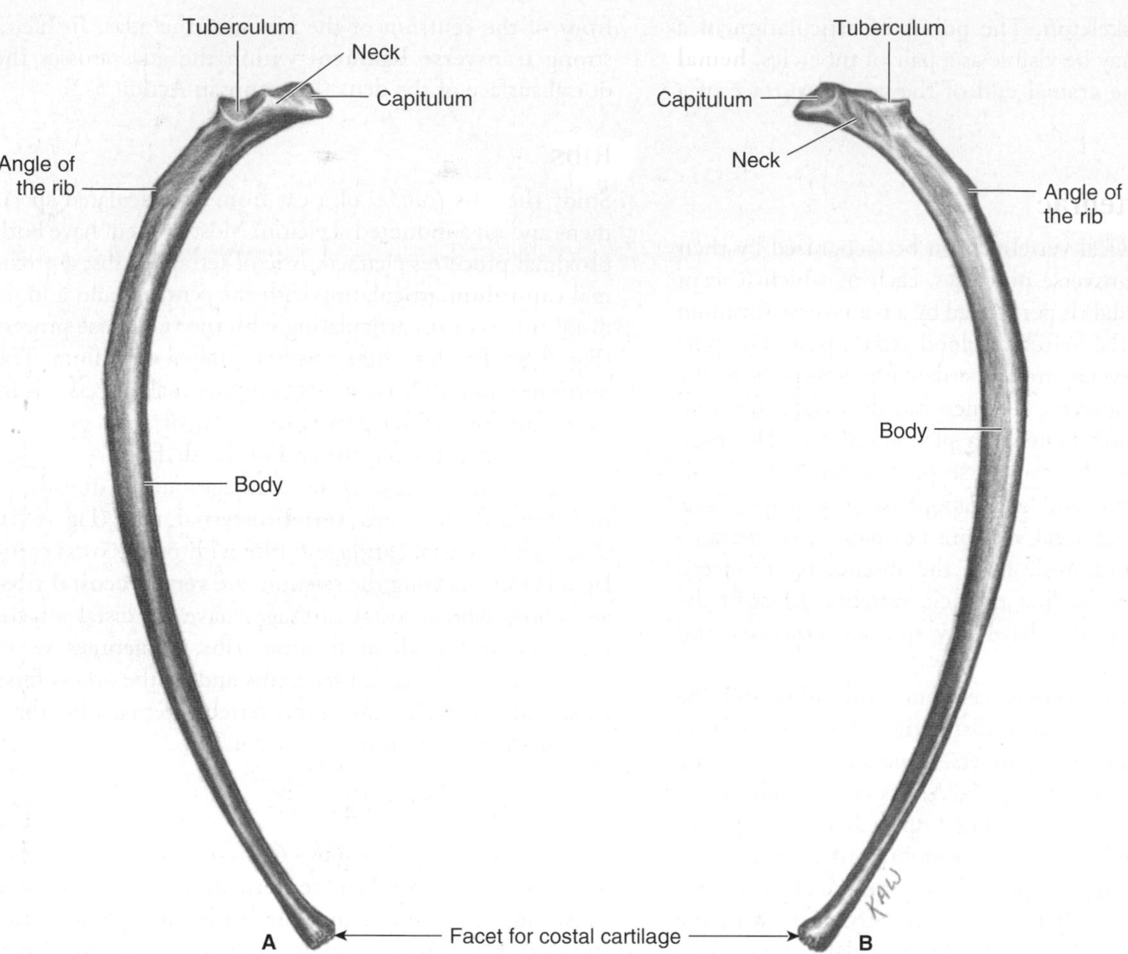

Figure 5-8
The 8th rib of a cat shown in caudal (A) and cranial (B) views.

Anatomy in Action 5-3 Biomechanics of the Cat Axial Skeleton

The vertebral column of mammals is the primary girder of the skeleton. It carries the weight of the trunk and head and transfers it to the appendages. Through its extension and flexion, it participates in the locomotor movements of the body. To facilitate interpretation of its structure, early anatomists compared the vertebral column to a cantilever bridge, but E. Slijper (1946) and others pointed out that a bridge is a static structure, whereas the vertebral column is a dynamic structure that must support the head and body in many positions and often participates in locomotion. Slijper compared the entire axial skeleton and associated muscles to an elastic bow (see figure).

The thoracic and lumbar vertebrae form one arch, or bow; the cervical vertebrae form another bow that arches in the opposite direction. The vertebral centra are the supporting elements of the bows. These bows are dynamic, and their curvature changes as the head is lowered or raised, or as the back is flexed or extended. The thoracic–lumbar bow tends to straighten out because of the elasticity of certain dorsal ligaments, the tonus of the dorsal muscles, and the weight of inner organs. This straightening is prevented by the "bowstring," which is composed of the sternum and the ventral abdominal muscles (e.g., the rectus abdominis) assisted dorsally by the subvertebral muscles (e.g., the psoas and quadratus lumborum muscles). The bow and bowstring are connected caudally by the pelvic girdle and cranially by the stout cranial ribs, which are held in place by the action of the scalenus and other muscles.

The bowstring of the cervical bow is formed by the splenius and other dorsal cervical muscles, and the **ligamentum nuchae,** which connects the back of the skull with the spinous processes of the cervical and cranial thoracic vertebrae. Thus, the vertebral column is a dynamic girder that can both support the body and participate in its movements.

The spinous processes are lever arms that transmit the pull of the muscles that are attached to them to a center of rotation, or fulcrum, between centra. The direction of their inclination tends to be perpendicular to the major muscle forces acting upon them. Because several muscles may attach onto a single spinous process, the situation becomes quite complex. In carnivores and many other mammalian quadrupeds, the tips of the spinous processes of the lumbar vertebrae point cranially, partly in response to a very powerful longissimus dorsi muscle (see Figs. 7-39 and 7-40), whereas the spinous processes of the thoracic vertebrae slope caudally, partly in response to a powerful splenius muscle. The vertebra near the middle of the trunk, where the angle of inclination of the spinous processes reverses, is called the **anticlinal vertebra** (Fig. 5-5).

Because the spinous processes are lever arms, an increase in their height increases the mechanical advantage of muscles acting upon them. The great length of the cranial thoracic spinous processes is related to the fact that the splenius and other muscles support the relatively heavy head.

The spinous processes are blade shaped because muscles pull upon them primarily in the sagittal plane of the body. These muscle forces are resisted by an increase in the anteroposterior dimension of the spines. The spines need not be thick because few forces pull them sideways.

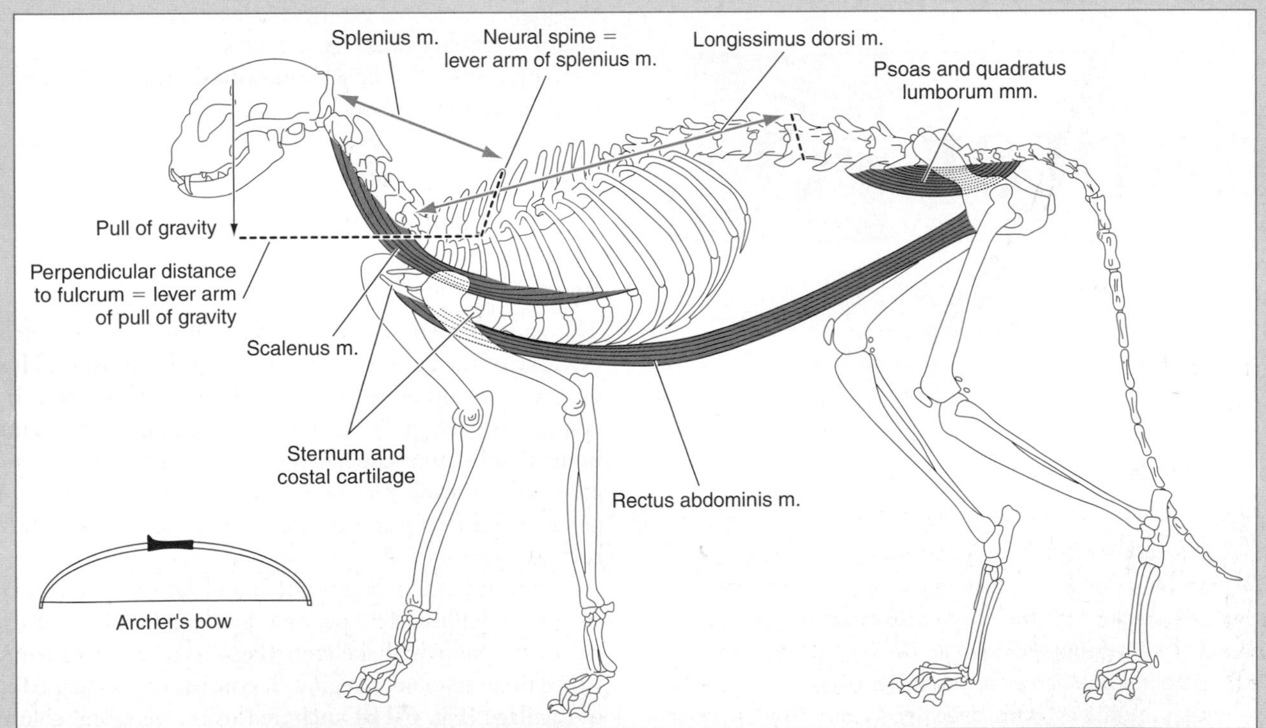

Biomechanics of the axial skeleton of the cat. The scapula has been removed. *(From Liem, Bemis, Walker, and Grande, 2001; after Slijper, 1946.)*

The Appendicular Skeleton

The appendicular portion of the postcranial skeleton consists of the skeletal supports of the paired fins of fishes and the limbs of terrestrial vertebrates, and the girdles in the body wall to which the appendages attach. Most vertebrates have both a **pectoral,** or shoulder, **girdle** and a **pelvic,** or hip, **girdle** and **appendages.** Nearly the entire appendicular skeleton is part of the endoskeleton because it is composed of cartilage or cartilage replacement bone. Several dermal bones contribute to the pectoral girdle of most species so that this girdle consists of both endoskeletal and dermal components.

Chapter 6

FISHES

A few extinct "ostracoderms" had finlike pectoral spines or flaps of uncertain structure and function, but most agnathous vertebrates lack an appendicular skeleton. Recall the absence of paired appendages in the lamprey (Chapter 2). Both pectoral and pelvic appendages are present in all jawed vertebrates, unless they have been secondarily lost. Although the movements of paired fins are important in the locomotion of some species of fishes, their primitive function was to aid in maneuvering and in stabilizing against roll, pitch, and yaw as the animal moved through the water.

Just how paired fins first evolved is not known, although similar genetic modules underlie the development of appendages in a wide range of animals from arthropods to vertebrates (Capdevila and Belmonte, 2000). Considerable independent evolutionary origin must have occurred among fishes because fin structure is quite different in the various groups of early jawed fishes. The earliest cartilaginous fishes had fan-shaped, **broad-based paired fins** that were supported by many parallel rods and attached to the trunk by a broad base (Fig. 6-1**A**). In derived species of cartilaginous fishes, the fin base has narrowed to three basal elements, or **pterygiophores** (Figs. 4-4**B**, 4-5**B**, and 6-1**B**). Such a **tribasic fin** continued to provide stability and had the advantage that it could twist and turn at its narrow base and thus increase maneuverability. Among bony fishes, the earliest actinopterygians also had a fan-shaped tribasic fin (Fig. 6-2**A**), but the sarcopterygian fishes that gave rise to tetrapods had a more elongate and lobate-shaped fin supported by a single basal element (Fig. 6-4**A**). It is from a **monobasic fin** of this type that the tetrapod limb evolved.

The Pectoral Girdle and Fin

Squalus

The appendicular skeleton of *Squalus* will be used as an example of the condition in chondrichthyan fishes. However, as with other parts of the skeleton of *Squalus,* it is atypical in that it is entirely cartilaginous. The dermal bones that became associated with the pectoral girdle at an early point during vertebrate evolution are not present in any cartilaginous fish, but they can be seen in bony fishes such as *Amia.*

Examine the pectoral girdle and fin on a skeleton of the spiny dogfish. The pectoral girdle is located just caudal to the branchial skeleton (Figs. 4-4, 4-5, and 6-1**A**), hence there is no neck region. It consists of a U-shaped bar of cartilage (Fig. 6-1**B**) anchored to the vertebral column through connective tissue. At the top of each limb of the U is a separate **suprascapular cartilage,** 1 to 2 cm long. The rest of the girdle is formed of a single piece, the **scapulocoracoid cartilage.** In some sharks, the scapulocoracoid has clearly formed from paired elements that have fused in the midventral line. The part of the scapulocoracoid that is located ventral to the **glenoid surface,** the point where the fin attaches, is called the **coracoid bar.** It is located in the same position as the coracoid bone in the girdle of tetrapods. The rest of the scapulocoracoid cartilage is called the **scapular process.** A small **coracoid**

foramen for blood vessels and nerves opens just cranial to the glenoid surface.

The pectoral girdle serves as a point of origin not only for fin muscles but also for certain hypobranchial and branchiomeric muscles that act on the lower jaw, hyoid arch, and branchial arches; thus, it plays a role in feeding and breathing movements as well as locomotion. The pectoral girdle functionally is an integral part of the head (Gudo and Homberger, 2002). We return to this topic in Chapters 7 and 10.

The pectoral fin is narrow at its base but widens distally. It is supported proximally by a series of cartilages, collectively called **pterygiophores,** and distally by fibrous fin rays, called **ceratotrichia.** The ceratotrichia develop in the dermis of the skin on each surface of the fin. The three large pterygiophores that articulate with the girdle at the glenoid surface are the **basal pterygiophores;** the smaller, more distal elements are **radial pterygiophores.** The three basal pterygiophores are, from cranial to caudal, the **propterygium, mesopterygium,** and **metapterygium.** Note that the metapterygium is the longest.

Amia

Most of the pectoral skeleton of the actinopterygian, *Polypterus* (Fig. 6-2**A**), is ossified, with a distinct scapula and coracoid, and ossified basal and radial pterygiophores. Examine a skeleton of the bowfin or choupique, *Amia* (Fig. 6-2**B**). Notice that the endoskeletal part of the pectoral girdle consists, on each side, of a small area of unossified cartilage, which lies between the pectoral fin proper and a conspicuous arch of bone located caudomedial to the dermal gill covering (operculum, Fig. 4-7**B**). This arch of bone, which is of dermal origin, forms most of the pectoral girdle. It consists of four elements. A large ventral **cleithrum** begins ventral to the gill arches and continues dorsally a bit beyond the fin and endoskeletal girdle. The rostroventral ends of the two cleithra are united. A **supracleithrum** extends from the cleithrum nearly to the roof of the skull; a **posttemporal,** which is a part of the pectoral girdle even though it has surface sculpturings similar to those of the skull bones, attaches the supracleithrum to the back of the skull. Finally, a small **postcleithrum** is located caudal to the junction of the cleithrum and supracleithrum.

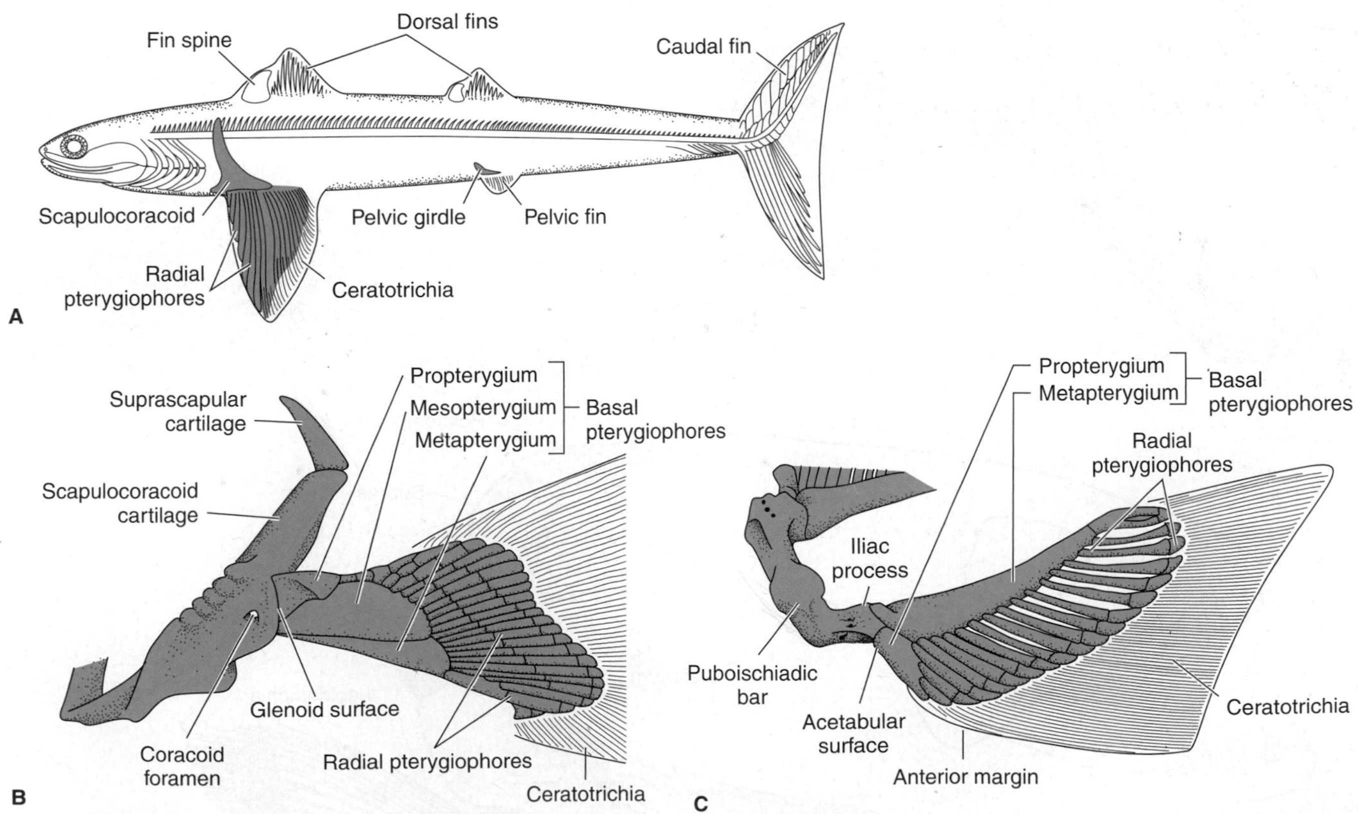

Figure 6-1

Girdles and paired fins of cartilaginous fishes in which only endoskeletal elements are present. **A,** Girdles and broad-based fins of an ancestral shark, *Cladoselache*. **B,** Pectoral girdle and tribasic fin of a recent shark, *Squalus*. **C,** Pelvic girdle and fin of *Squalus*. Endoskeletal elements of the appendicular skeleton are shown in blue and dermal elements in red in this and many other figures in this chapter. *(From Liem, Bemis, Walker, and Grande, 2001;* ***A,*** *after Zangerl, 1981;* ***B,*** *after Jollie, 1962.)*

The **clavicle** and **interclavicle,** which are ventral dermal girdle elements characteristic of the pectoral girdle of many fishes and terrestrial vertebrates (Fig. 6-2**A**), may be represented in *Amia* by slender, rather superficial slivers of bone that overlap the rostral end of the ventral part of the cleithrum (Grande and Bemis, 1998). They are not usually retained in commercial preparations of the skull and pectoral girdle of *Amia.*

The Pelvic Girdle and Fin

The pelvic girdle of *Squalus* and other fishes consists of a simple transverse rod of cartilage or bone, the **puboischiadic bar,** which is located in the ventral abdominal wall just cranial to the cloaca (Fig. 6-1**A** and **C**). Each end extends dorsally as a short **iliac process,** but these processes do not reach the vertebral column. There are often several small nerve foramina near the base of each iliac process.

Like the pectoral fin, the pelvic fin consists of a series of proximal cartilaginous pterygiophores and distal ceratotrichia (Fig. 6-1**C**). There are only two **basal pterygiophores** in the pelvic fin of *Squalus*—a long **metapterygium** that extends caudally from the pelvic girdle and clearly forms the main support of the fin, and a short **propterygium** that projects laterally from the girdle. Many **radial pterygiophores** extend into the fin from the two basal pterygiophores. In males, the skeleton of the

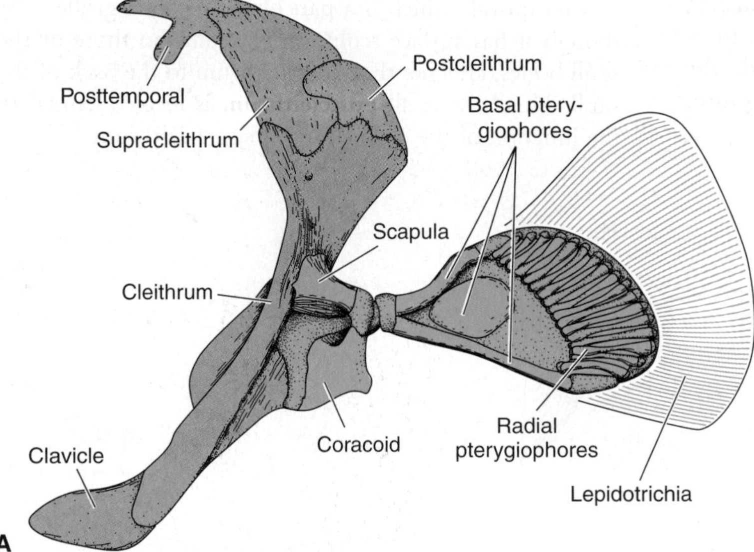

Figure 6-2
Pectoral girdle and fins of actinopterygian fishes in lateral views. **A,** *Polypterus.* Note that the endoskeletal girdle and the fin are partly ossified. **B,** Endoskeletal girdle and appendicular elements are largely unossified in *Amia.* (*A, From Liem, Bemis, Walker, and Grande, 2001.*)

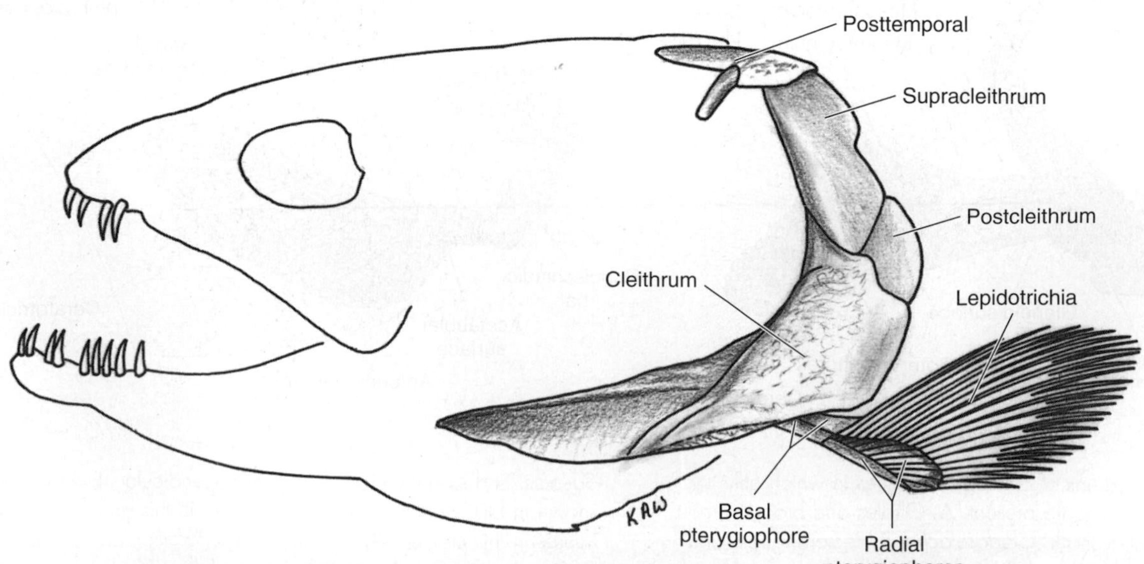

clasper, which transfers sperm to the female, is formed by enlarged and modified radials that extend caudad from the end of the metapterygium. The fin attaches to the **acetabular surface** of the pelvic girdle.

AMPHIBIANS AND REPTILES

In the evolution from fishes to tetrapods, the monobasic fin was transformed into a limb consisting of three segments (Fig. 6-3). In the pectoral appendage, these are the **brachium,** which contains the **humerus;** the **antebrachium,** which contains the **radius** and **ulna;** and the **manus,** which contains the **carpals, metacarpals,** and **phalanges.** Corresponding segments and bones in the pelvic appendage are the thigh or **femur,** with the femur[1]; the **shank** or *crus,* with its **tibia** and **fibula;** and the **pes,** with its **tarsals, metatarsals,** and **phalanges.**

It is easy to visualize the derivation of the tetrapod limb from the uniserial monobasic fin seen in *Eusthenopteron,* a sarcopterygian fish close to the ancestry of tetrapods (Fig. 6-4A and B). In the uniserial monobasic fin, the main axis of the fin extends down its postaxial edge. In tetrapods (Fig. 6-3), the humerus or femur in the proximal segment, the ulna or fibula in the next segment, and the ulnare or fibulare lie along this axis. All elements have recognizable homologues in the uniserial monobasic fin. All radial pterygiophores extend cranially from this axis in the uniserial monobasic fin.[2] The radius and tibia of tetrapods are derived from the first radial. Embryonic and genetic evidence suggests that the manus and pes, with the possible exception of its proximal elements, are neomorphs, or newly evolved structures with no homologues in fishes (Sordino et al., 1995; Wagner and Chui, 2001). Other fossil discoveries indicate that the earliest tetrapods had a variable number of digits, often as many as eight (Coates and Clark, 1990). The idealized five-fingered limb, or **pentadactyl limb** is ancestral for amniotes, but not for all tetrapods.

In early amphibians (basal tetrapods) and amniotes, the limbs were not carried under the body as they are in cats or dogs; instead they were splayed, that is, the humerus and fe-

[1]See Footnote 4, Chapter 3.

[2]The lungfishes (Dipnoi) have a biserial monobasic fin in which the axis extends through the center of the fin, and radial pterygiophores extend both caudally and cranially from it.

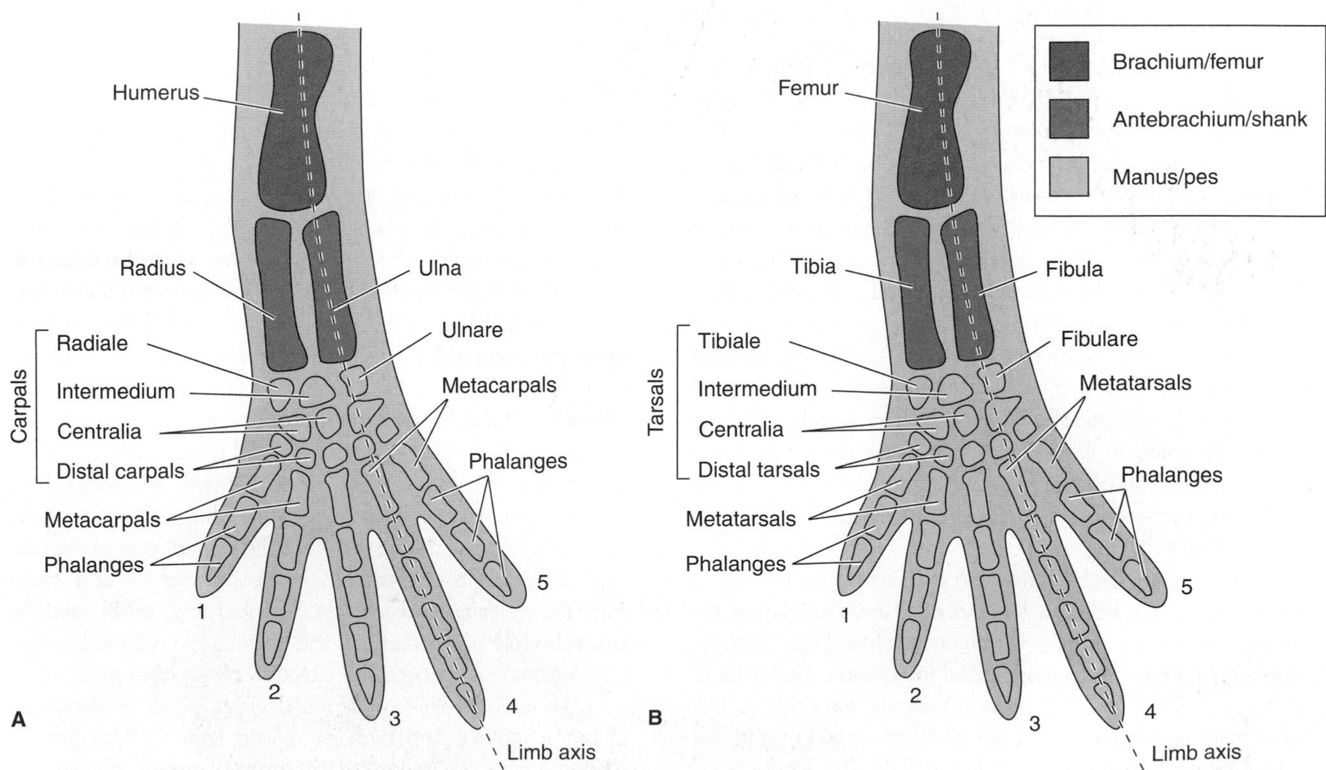

Figure 6-3
Limb segments and the bones that they contain in early tetrapods: **A,** Pectoral appendage. **B,** Pelvic appendage. The arrangement and numbering of the digits are based on a basal amniote. *(From Liem, Bemis, Walker, and Grande, 2001.)*

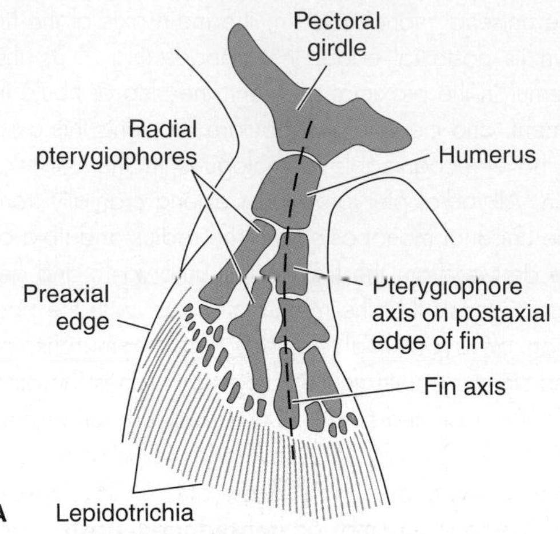

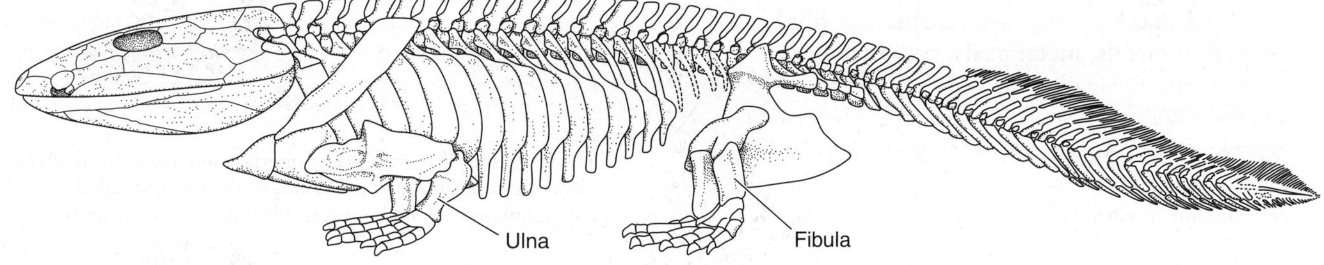

Figure 6-4
Fins and limbs at the transition from water to land. **A,** Uniserial monobasic pectoral fin of *Eustenopteron,* a fish close to the ancestry of tetrapods. **B,** Skeleton of *Ichthyostega,* an early tetrapod. Note the large girdles and the splayed position of the limbs. *(From Liem, Bemis, Walker, and Grande, 2001;* **B,** *after Jarvik, 1980.)*

mur projected laterally so that they were held horizontally when the animal moved (Fig. 6-4**B,** and Anatomy in Action 3-3). The antebrachium and shank extended vertically downward to the ground. The front foot pointed craniolaterally, but maintaining this placement as the humerus was retracted required that the radius rotate at the elbow (see Fig. 7-10). The hind foot pointed cranially when it was first placed on the ground, but it rotated at the ankle joint as the femur was retracted and pointed caudally when the foot was removed from the ground. There was no rotation of lower leg bones. An increase in toe length from medial to lateral helped to maintain toe contact with the ground as the leg was retracted. Early amniotes had two phalanges in the first toe, three in the second, four in the third, five in the fourth, and then an abrupt decrease to three or four in the fifth toe. This is expressed as a **phalangeal formula** of 2-3-4-5-3 or 2-3-4-5-4.

Originally, the limbs were used as a supplement to lateral undulations of the body in locomotion. However, during the evolution from early synapsids to mammals, the limbs became increasingly important in support and locomotion. The limbs of mammals rotated closer to, or even beneath, the body so that the humerus and femur move back and forth close to, or in, the vertical plane (see Anatomy in Action 3-4). The elbow is directed caudally, and the manus points cranially so that the radius and ulna cross when the manus is in its prone position with the palm on the ground. Both knee and hind foot point cranially, so there is no need for the lower leg bones to cross. Similar changes occurred in the sauropsid line of evolution, particularly in the hind limb of dinosaurs and birds.

Correlated with the increased importance of the appendages for locomotion and body support, the girdles became more massive and stronger. In the endoskeletal part of the pectoral girdle of ancestral tetrapods (Fig. 6-5**A**), there was only one ossification on each side, the **scapulocoracoid.** However, in extant amphibians and in the sauropsid line of evolution, a **scapula** ossified dorsal to the glenoid cavity, and an **anterior coracoid** ossified ventral to this cavity (Fig. 6-5**B**). Although these elements were large and platelike, the endoskeletal parts of the pectoral girdles of opposite sides neither united with each other ventrally nor connected directly with the vertebral column. Muscles transfer forces between the pectoral girdle and trunk skeleton in tetrapods (see Chapter 7). Correlated with the evolution of a distinct neck the dermal part of the pectoral girdle lost its connection with the back of the skull. However, the cleithrum and clavicle persisted (Fig. 6-5**B**), and the **interclavicle** became larger. The interclavicle is a median ventral element that connects the clavicles of opposite sides.

The endoskeletal part of the pectoral girdle of extant amphibians, reptiles, and birds is derived from the type just described and usually is close to it in essential pattern. However, in the synapsid line of evolution leading toward mammals, a third ossification, the **posterior coracoid** (or, simply, coracoid), appeared in the endoskeletal part of the girdle caudal to the anterior coracoid (Fig. 6-5**C**). In mammals (Fig. 6-5**D** and **E**), the

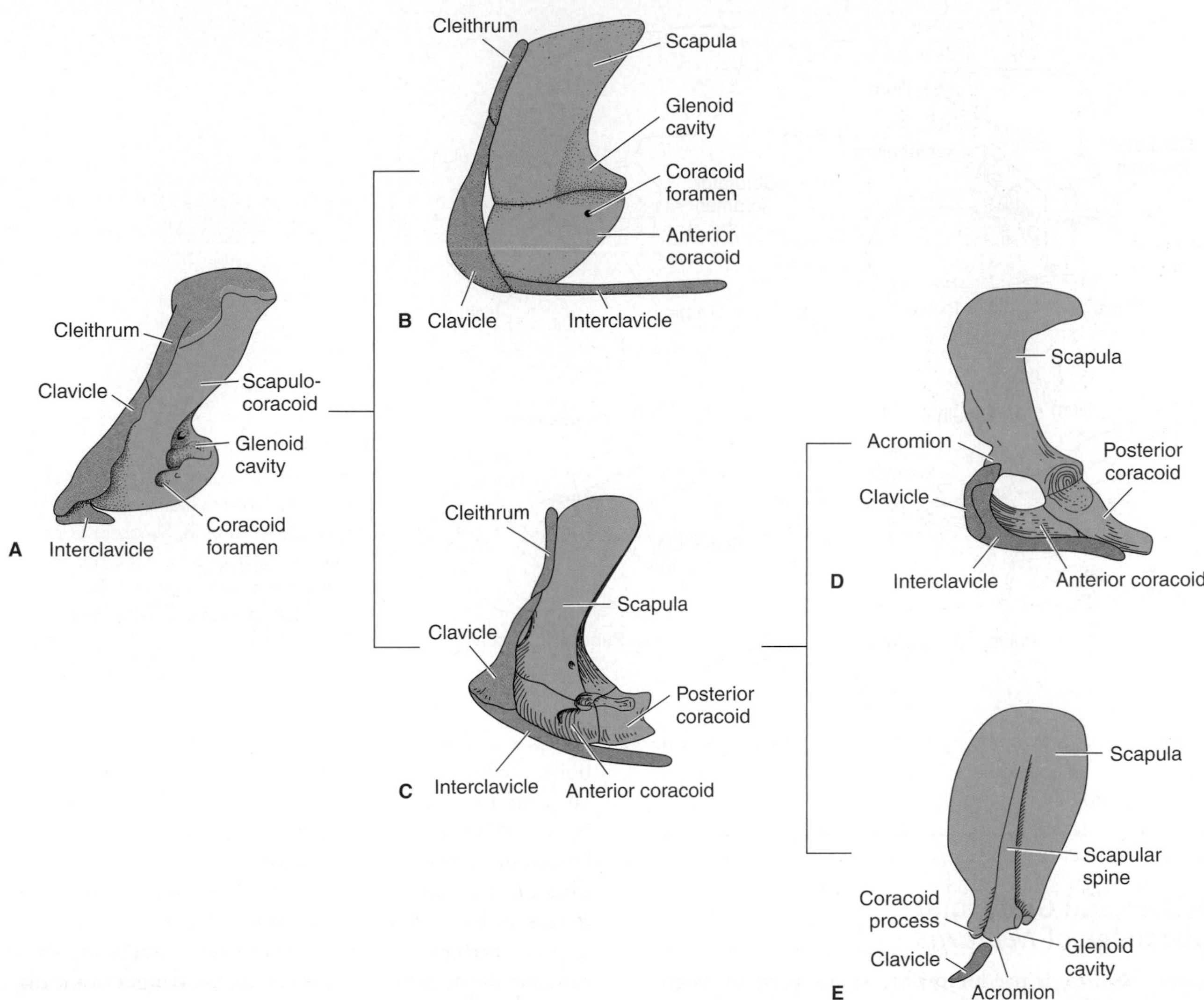

Figure 6-5
Cladogram of the evolution of the pectoral girdle in tetrapods. Note the decreasing importance of the dermal bones of the girdle and the increasing importance of the endoskeletal girdle. A, Early tetrapod, *Eryops*. B, Early sauropsid. C, Early synapsid, *Dimetrodon*. D, Monotreme mammal, the platypus, *Ornithorhynchus*. E, Therian mammal, an opossum, *Didelphis*. *(Modified from Liem, Bemis, Walker, and Grande, 2001.)*

scapular area is greatly expanded and the coracoid region is reduced. The anterior coracoid is completely lost in therian mammals, and the posterior coracoid is represented by only a small **coracoid process** of the scapula. Expansion of the dorsal part of the pectoral girdle and regression of the ventral part are correlated with the shift in limb position. Because much of the weight of the body is transferred to the ground through limb bones, ventral limb muscles become less important in raising the body from the ground and supporting it. Dorsal muscles, which are well situated to move the limb fore and aft, become more important. In all extant tetrapods, the dermal portions of the pectoral girdle have become reduced. The cleithrum is retained only in early sauropsids and synapsids; an interclavicle persists in early mammals; but only the clavicle is retained in therians, and it often is reduced in size.

The pelvic girdle (Fig. 6-6) also enlarged during the transition from water to land. In tetrapods it typically consists of three cartilage replacement bones: a **pubis** and an **ischium** ventral to the acetabulum, the articular socket for the femur, and an **ilium** that extends from the acetabulum to the sacral rib and vertebrae. The ventral elements of opposite sides unite with each other at a midventral **pelvic symphysis.** In ancestral tetrapods, the ventral elements formed a broad plate, and the ilium connected with only a single sacral vertebra (Fig. 6-6A). In the line of evolution toward mammals, the ventral elements became relatively smaller, and the ilium expanded, turned cranially, and

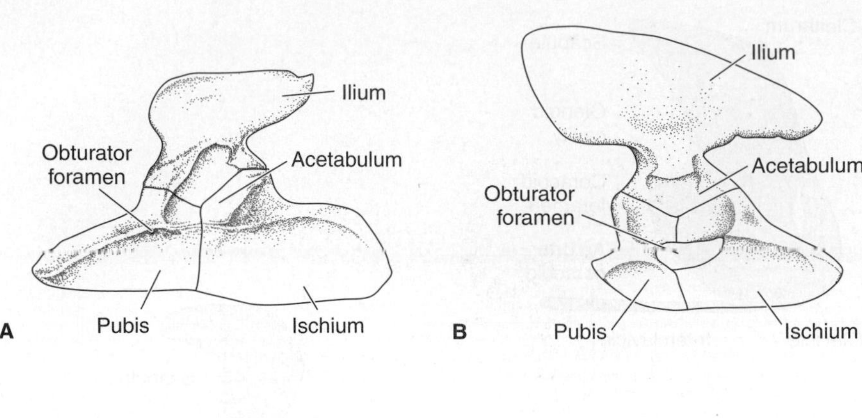

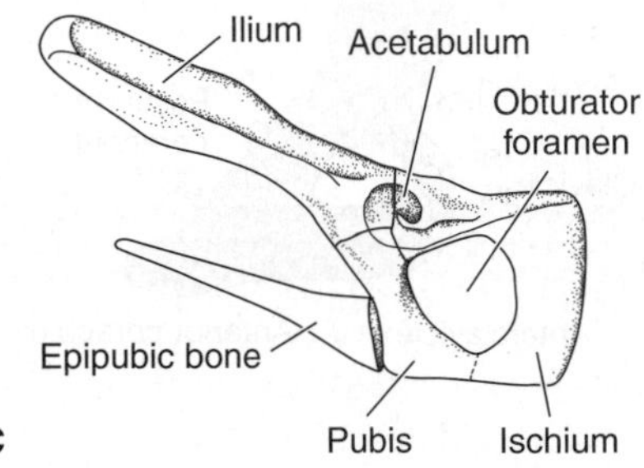

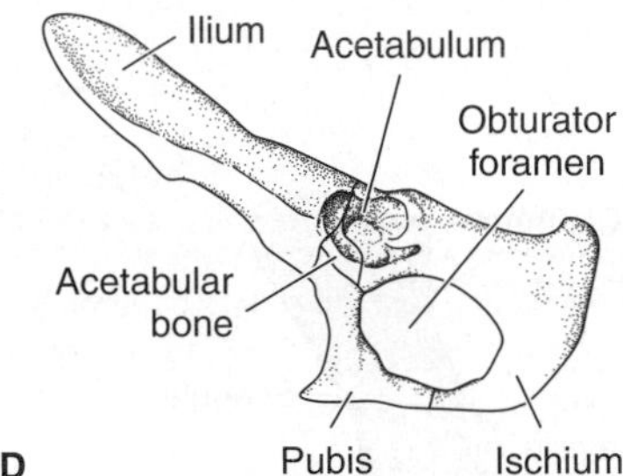

Figure 6-6

Stages in the evolution of the pelvic girdle in the line of evolution to mammals. **A,** Early tetrapod, *Dimetrodon.* **B,** Synapsid close to mammals, *Cynognathus.* **C,** Opossum, *Didelphis,* a marsupial mammal. **D,** Cat, *Felis, a placental mammal. (After Romer and Parsons, 1986.)*

united with more sacral vertebrae (Fig. 6-6**B** to **D**). This transformation, too, is correlated with the shift in limb position discussed previously.

The Pectoral Girdle and Appendage of *Necturus*

Because *Necturus* is paedomorphic, many parts of its girdles and appendages remain unossified. *Necturus* does not leave the water, so its appendicular skeleton need not be as strong as that of early tetrapods, but the position and shape of its girdles are representative of early terrestrial vertebrates.

Examine the pectoral girdle on a skeleton of *Necturus* and compare it with Figure 6-7**A.** The two halves of the pectoral girdle overlap slightly ventrally but are not united. On each side an ossified **scapula** extends dorsally cranial to the **glenoid cavity,** the articular facet for the articulation of the girdle with the humerus. A **suprascapular cartilage** caps the scapula. The ventral part of the girdle remains unossified and is called the **coracoid plate.** A **procoracoid process,** not to be confused with the anterior coracoid bone of other tetrapods, extends cranially from the coracoid plate. A **coracoid foramen,** for blood vessels and nerves, may be seen in the coracoid plate ventral to the scapula.

Study the pectoral limb skeleton, noting first the position of the different segments of the limb and the preaxial and postaxial sides (see Anatomy in Action 3-3). A single bone, the **humerus,** projects laterally nearly in the horizontal plane from the glenoid cavity to the elbow joint. Two bones of approximately equal size compose the antebrachium: a **radius** on the preaxial, or medial, side and an **ulna** on the postaxial, or lateral, side. They extend to the manus, or hand. The manus consists of a group of six cartilaginous **carpals** in the wrist, four ossified **metacarpals** in the palm of the hand, and the ossified **phalanges** of the digits, or fingers. Terms for the individual carpals are derived from their positions relative to other bones (Figs. 6-3 and 6-14), but the individual carpals usually are not distinct in dried skeletons. Notice that only four toes are present, a number characteristic of the manus of lissamphibians. Because the fossil record of lissamphibians is sparse, we cannot be certain as to which toes of amniotes they are homologous, but they may be homologous to the second through fifth toes of amniotes.

The Pelvic Girdle and Appendage of *Necturus*

Each half of the pelvic girdle has a narrow ossified **ilium** that extends dorsally from the **acetabulum,** the socket for the articulation with the femur, to attach on a single sacral rib and vertebra (Fig. 6-7**B**). Ventrally, there is a broad **puboischiadic plate,** which contains a pair of ossified **ischia** caudally and a **pubic cartilage** cranially. An **obturator foramen,** for the obturator nerve, passes through the

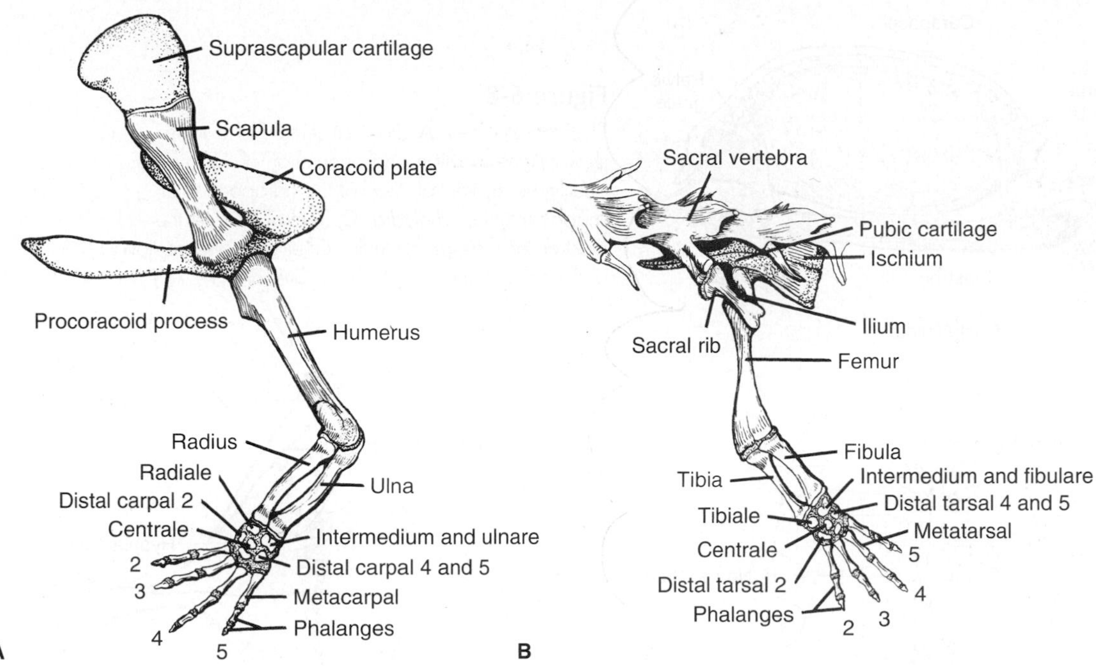

Figure 6-7
Dorsolateral views of the appendicular skeleton of *Necturus.* **A,** Pectoral girdle and appendage. **B,** Pelvic girdle and appendage.

pubic cartilage. Notice that the pelvic girdle, together with the sacral rib and vertebra, forms a skeletal ring of bone around the caudal end of the trunk. The intestine and urogenital ducts pass through this ring, which is called the **pelvic canal.**

Study the pelvic limb, noting its position and its preaxial and postaxial sides (see Anatomy in Action 3-3). The **femur** forms the proximal segment of the limb and projects laterally in, or close to, the horizontal plane. The **tibia** and **fibula** in the shank extend vertically downward to the pes, or foot. The pes consists of a group of six cartilaginous **tarsals** and ossified **metatarsals** and **phalanges** (Figs. 6-3 and 6-19). The individual tarsals cannot be distinguished in dried skeletons. Many salamanders have five toes in the hind foot, but *Necturus* has only four, with the most medial probably being homologous to the second toe of amniotes.

The Pectoral Girdle and Appendage of the Turtle

Turtles are unique in that their trunk is encased in a protective shell that is composed of a dorsal **carapace** and ventral **plastron** (Fig. 6-8**A**). When threatened, turtles usually withdraw their limbs and head into the shell. The carapace is composed of dermal plates that have fused with the underlying vertebral column (Fig. 6-8**B**). Endoskeletal ribs become incorporated into the dermal **costal plates** of the carapace. The result is that, alone among vertebrates, the pectoral girdle lies ventral and medial to the ribs. In other vertebrates the pectoral girdle lies dorsal and lateral to the ribs. Many specializations of the pectoral girdle result from its novel location (Anatomy in Action 6-1). The plastron is entirely dermal.

The endoskeletal part of the pectoral girdle has an unusual triradiate shape (Fig. 6-9**A**). The **scapula** forms a long prong that extends dorsally to articulate with the carapace lateral to the vertebral column. The **acromion** forms another prong that extends ventrally and medially to attach by a ligament to the entoplastron of the plastron. The **anterior coracoid** forms a broader and flatter plate. In dissections, an **acromiocoracoid ligament** may be seen extending between the tips of the acromion and coracoid. A **glenoid cavity** is present for articulation of the humerus.

Parts of the dermal part of the girdle are present but are incorporated in the cranial plates of the plastron. Examine a plastron in which the epidermal scutes, or **laminae,** have been removed and the dermal plates exposed (Fig. 6-8**C**). A pair of **epiplastra,** caudal to which is a median **entoplastron,** forms the front of the plastron. Three additional paired plates—the **hyoplastra,** the **hypoplastra,** and the **xiphiplastra,** lie caudal to the pair of epiplastra. All these plates represent, in part, ossifications in the

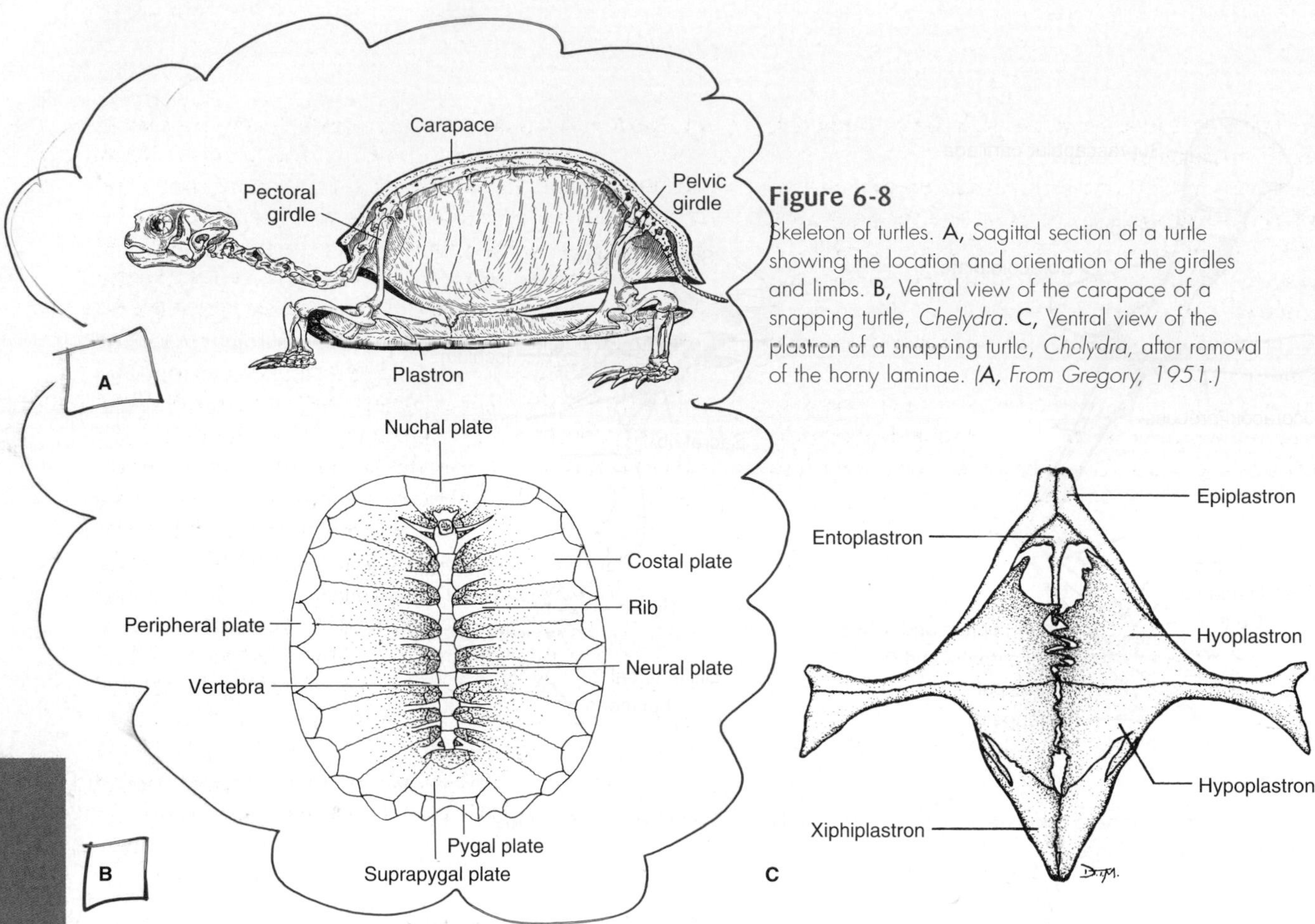

Figure 6-8
Skeleton of turtles. **A**, Sagittal section of a turtle showing the location and orientation of the girdles and limbs. **B**, Ventral view of the carapace of a snapping turtle, *Chelydra*. **C**, Ventral view of the plastron of a snapping turtle, *Chelydra*, after removal of the horny laminae. *(A, From Gregory, 1951.)*

dermis of the skin of the underside of the body. However, during embryonic development, the originally separate primordia of the clavicles and interclavicle become incorporated into the first three plates, which therefore have a compound origin. The epiplastra include the paired **clavicles;** the entoplastron includes the **interclavicle.** The remaining plastral plates may be homologous to the **gastralia** of basal amphibians and reptiles. In basal tetrapods, the gastralia were riblike rods of dermal bone found on the ventral abdominal wall. They may be remnants of the piscine dermal scales and are retained in a few extant reptiles, such as *Sphenodon,* crocodiles, and possibly turtles.

Study the pectoral appendage (Figs. 6-7**A** and 6-9**B**), identifying its preaxial and postaxial sides. The long bones of the appendage are the **humerus** in the upper arm and the **radius** and **ulna** in the forearm. The proximal end of the humerus has a round **head** that fits into the glenoid cavity and two prominent processes for the attachment of muscles. Of the forearm bones, the radius is the one on the preaxial side. This may be cranial or medial depending on limb position. The radius and ulna are about equal in size, but the ulna extends over the distal end of the humerus, whereas the radius articulates on the underside of the distal end of the humerus. As in early tetrapods,

Anatomy in Action 6-1 Pectoral Girdle Specializations of Turtles

Ann Burke (1989) demonstrated that modifications during early development of turtles induce rib precursor cells to migrate laterally just beneath the dermis of the skin on the dorsal surface of the trunk rather than taking their normal path ventrally. As a result, the ribs come to lie dorsal and lateral to the developing pectoral girdle. The attachments of the tips of the scapula and acromion to the shell allow the pectoral girdle to rotate, which moves the glenoid cavity forward and backward. This movement contributes to an increase in step length, and it may also assists in lung ventilation. The rigid shell of turtles precludes rib movements, which are important in most reptiles for changing the volume of the trunk cavity and ventilating the lungs. Turtles, therefore, change the volume of their trunk cavity by contracting and relaxing muscle sheets lining the limb pockets, assisted by rotation of the pectoral girdle.

both the radius and ulna articulate with the carpal bones, and there is no distal radioulnar joint.

The manus consists of a group of **carpals** in the wrist, a row of five **metacarpals** in the palm, and the **phalanges** in the free part of the toes. There are five toes, with the first being the most medial. The phalangeal number has been reduced in the snapping turtle from the ancestral amniote number to 2-3-3-3-3. The phalangeal number varies among chelonian species.

The carpus of the turtle is similar in many ways to that of amphibians, and its individual components should be identified (Fig. 6-9**B**). The carpals can be grouped into proximal and distal rows. The proximal row consists of three bones: an **ulnare** adjacent to the ulna; an **intermedium** lying between the distal ends of the radius and ulna; and an elongated element distal to the radius and intermedium. This elongated element has been interpreted as a fusion of two centralia with the radiale, but Burke and Alberch (1985) present embryonic evidence that it is a fusion of three **centralia,** and the radiale has been lost. The distal row consists of five distal carpals that are numbered according to the digit to which they are related: distal carpal 1, distal carpal 2, and so on. In addition there is a small **sesamoid** bone on the lateral edge of the carpus. Sesamoid bones develop in the tendons of muscles and are variable. They facilitate the movement of a tendon across a joint or alter slightly the direction of pull of a muscle. The sesamoid bone you see adjacent to the ulnare occurs in nearly all tetrapods and is called the **pisiform.**

The forelimbs of turtles are similar to those of early tetrapods in many ways. They are carried in the splayed position, and the long bones of the upper and lower limbs are relatively short and chunky. The forelimbs of turtles can rotate farther forward than in other tetrapods and are folded across the front of the shell when the animal retracts its head.

The Pelvic Girdle and Appendage of the Turtle

Study the pelvic girdle of the turtle on a mounted skeleton and from an isolated specimen (Figs. 6-8**A** and 6-10**A**). Each half of the pelvic girdle consists of a dorsal **ilium** that in-

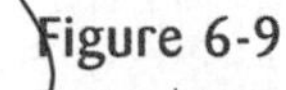

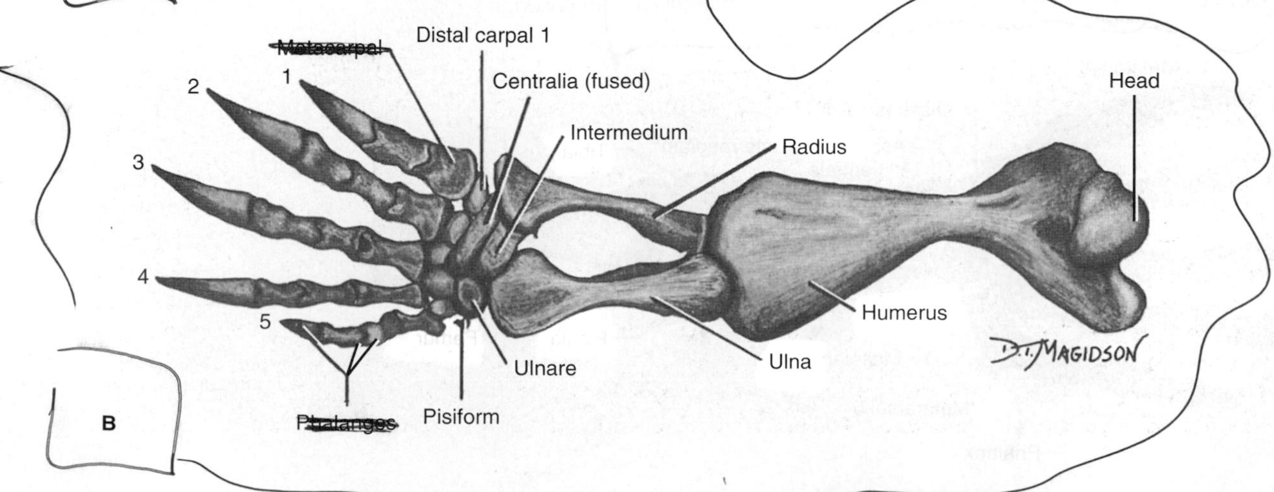

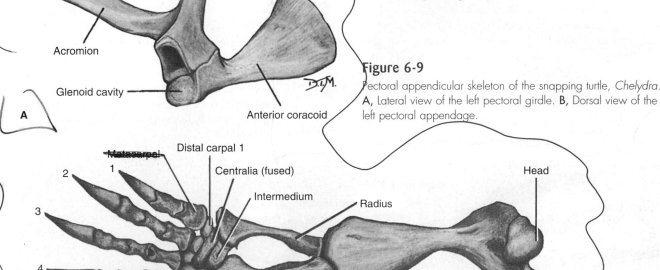

Figure 6-9

Pectoral appendicular skeleton of the snapping turtle, *Chelydra*. **A**, Lateral view of the left pectoral girdle. **B**, Dorsal view of the left pectoral appendage.

clines caudodorsally and articulates with two sacral ribs and vertebrae; a cranioventral **pubis;** and a caudoventral **ischium.** All three elements share in the formation of the **acetabulum,** which is the socket for articulating with the femur of the hind leg. The pubis and ischium of opposite sides are united by a **pelvic symphysis.** An **epipubic cartilage,** which may be partly ossified, extends cranially in the midventral line from the pubic bones. Both the pubis and ischium have a lateral process that is directed ventrally and rests on the plastron. A large **puboischiadic fenestra,** which develops in association with the origin of certain pelvic muscles, lies between the pubis and ischium on each side. The separate obturator foramen that is seen in many fishes and amphibians is not present in turtles because the obturator nerve also passes through the puboischiadic fenestra.[3]

[3]The terminology of the various pelvic openings unfortunately is confused. In most amphibians and reptiles, an *obturator foramen*, or *pubic foramen*, for an obturator nerve perforates the pubis cranial to the acetabulum (Fig. 6-6**A**). The rest of the puboischiadic plate is solid. In reptiles (lizards), an additional opening, known as the *puboischiadic fenestra*, or *thyroid fenestra*, develops between the pubis and ischium in association with the origin of certain pelvic muscles. In mammals, the fenestration of the puboischiadic plate includes both the primitive obturator foramen and the puboischiadic fenestra. Such an opening is termed an *obturator foramen*. The turtle seems to parallel this condition, but most investigators call the opening a *puboischiadic fenestra.*

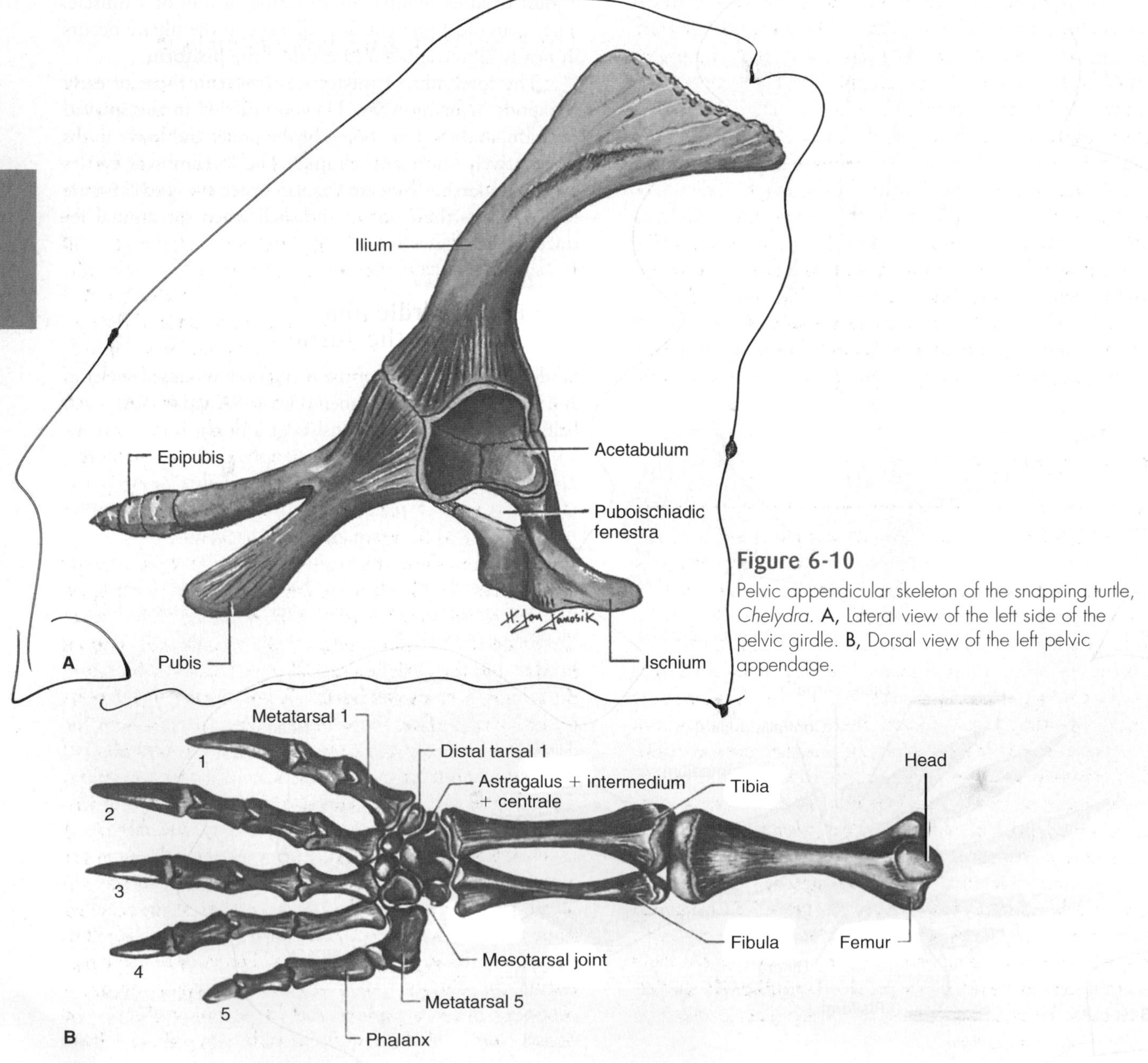

Figure 6-10
Pelvic appendicular skeleton of the snapping turtle, *Chelydra*. **A**, Lateral view of the left side of the pelvic girdle. **B**, Dorsal view of the left pelvic appendage.

Examine the pelvic limb, noting its position and its preaxial and postaxial sides (Fig. 6-10**B**). The long bone of the thigh is the **femur.** Its proximal end bears a round **head** that fits into the acetabulum and two prominent processes for the attachment of muscles. The long bones of the shank are the **tibia** and **fibula;** the former is the larger element on the preaxial side.

The pes consists of a group of **tarsals** in the ankle region, a row of five **metatarsals** in the sole of the foot, and **phalanges** in the free part of the five digits. The phalangeal formula is reduced to 2-3-3-3-3 in the snapping turtle. The metatarsal of the fifth toe is flat and broad rather than round and elongate like the others.

The individual tarsals should be studied and compared with Figures 6-10**B** and 6-3**B.** A row of four **distal tarsals** lies next to the metatarsals. The fourth distal tarsal is larger than the others and is associated with the fourth and fifth toes. All the remaining proximal and central elements of the tarsus tend to fuse into a single bone, but close inspection often reveals the lines of union between the major elements. As in the manus, the most proximal and lateral element, the fibulare, has been lost. The proximolateral element is a fusion of the intermedium and one centrale (Burke and Alberch, 1985). The proximomedial element represents the tibiale, but often it is given its mammalian name, **astragalus.** The main ankle joint of the turtle, as in other reptiles, is a **mesotarsal joint,** because it lies between the large proximal element and the distal tarsals. Rotation of the foot during locomotion occurs at this joint.

MAMMALS

The appendicular skeleton of most mammals is well suited for terrestrial locomotion. The limbs have rotated so that they lie close to the trunk, with the elbow directed caudally and the knee cranially (see Fig. 5-5 and Anatomy in Action 3-4). Limbs in this position offer good mechanical support because they are close to the body's center of gravity. The long bones of the limbs usually are longer than in species with splayed limbs, and the limbs move through longer arcs. These features increase step and stride length. The appendicular skeleton varies considerably, however, because mammals have adapted to many different modes of life and methods of locomotion. Some walk; some are adapted for running very fast; some bound; and others burrow, climb, swim, or fly.

Ancestral mammals were **plantigrade,** that is, they walked with the soles of their feet on the ground. **Digitigrade** species, such as the cat, have faster gates partly because of a longer step and stride made possible by walking on the digits with the soles of the feet off the ground. **Unguligrade** species, such as the horse and cow, carry this tendency further and walk and run on their toe tips.

We will focus on the cat's appendicular skeleton, but figures and descriptions of the rabbit's are included for comparison. Cats are quadrupedal carnivores with very flexible limbs adapted for stalking, running, and jumping as they pursue prey, and then manipulating the prey as they tear it apart. Rabbits are a gnawing herbivorous species that do not manipulate their food with their front paws, but are capable of moving fast to escape predators. Their powerful hind legs and elongated feet are adapted for bounding, that is, to give nearly simultaneous, powerful thrusts. Rabbits land on their front legs, which act as shock absorbers but are not as flexible as those of cats. Humans are bipeds and move on powerful hind legs. Human pectoral appendages and hands are used in many manipulative activities.

The Pectoral Girdle and Appendage

Study the appendages and girdles on mounted specimens of the cat or rabbit (see Figs. 5-5 and 5-6) and from disarticulated bones of the cat and other available mammals. Look for the many features of the appendicular skeleton shared by different species of mammals, and notice the modifications that reflect different modes of life. Learn to distinguish the individual girdles and the long bones of the appendages and to recognize whether they are from the left or the right side.

The pectoral girdle consists primarily of an expanded triangular **scapula** located on the side and back of the cranial end of the thorax (Fig. 6-11). Eutherian mammals have lost the anterior coracoid, and the posterior coracoid is reduced to a small, hooklike **coracoid process** that can be seen medial to the cranial edge of the **glenoid cavity,** the articular socket for articulation with the humerus. Correlated with the shift of the limbs under the body, the glenoid cavity of mammals is directed ventrally rather than laterally as in nonmammalian tetrapods.

If we envision the scapula as an inverted triangle (Fig. 6-11**A** and **C**), then the glenoid cavity is at the apex, and the curved **dorsal border** of the scapula is at the base of the triangle. The cranial edge of the scapula is its **cranial border,** and the straighter caudal edge, which is adjacent to the armpit, is its **caudal border.** A prominent ridge of bone on its lateral surface, the **scapular spine,** extends from the dorsal border nearly to the **glenoid cavity.** The ventral tip of the spine continues as a process known as the **acromion.** The clavicle articulates with this process in those mammals that have a prominent clavicle (e.g., the rabbit). A **metacromion** extends caudally from the spine dorsal to the acromion. It is exceptionally long in the rabbit. The portion of the lateral surface of the scapula caudal to the scapular spine is called the **infraspinous fossa;** the portion cranial to the spine is the **supraspinous fossa.** The scapular spine represents the primitive cranial edge of the scapula because it continues to the acromion, and the acromion lies on the cranial border of the scapula in early tetrapods and basal

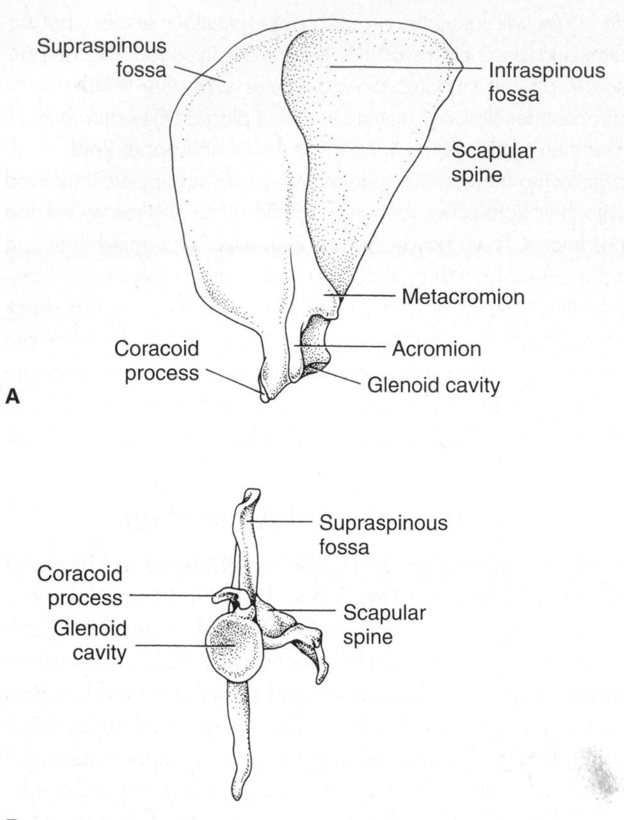

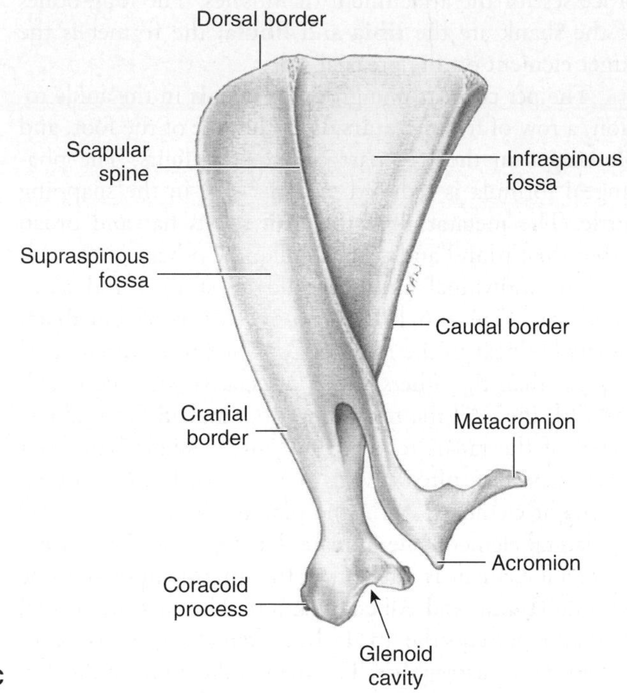

Figure 6-11
Left scapula of mammals. **A,** Lateral view of the cat scapula. **B,** Ventral view of the cat scapula. **C,** Lateral view of the rabbit scapula.

mammals (Fig. 6-5**D**). The portion of the scapula cranial to the spine appeared during the evolution of therian mammals. The medial surface of the scapula is called the **subscapular fossa.** In humans, from whom most of our anatomical terminology is derived, this surface is somewhat concave (fossalike), but in the cat and rabbit this surface is flat.

Of the elements of the dermal part of the girdle present in early tetrapods, only the **clavicle** is retained in eutherians. It is quite variable (Anatomy in Action 6-2). In many mammals, humans and rabbits included, the clavicle extends from the acromion to the manubrium of the sternum. However, in some mammals, including the cat, the clavicle

Anatomy in Action 6-2 Why Is the Clavicle Sometimes Lost?

Farrish Jenkins (1974) studied differences in locomotion between terrestrial mammals that retain the clavicle and cursorial species that have lost it, or in an individual in which it has been experimentally removed. During the swing phase of a step, the foot is lifted off the ground, the foot and leg are moved forward, and the foot is placed on the ground in front of its previous position. The glenoid cavity of the scapula also rotates forward to a limited extent. During the propulsive phase of a step, the foot retains its position on the ground and muscles extending from the trunk to the pectoral girdle and appendage pull the trunk forward relative to the girdle and appendage. If a well-developed clavicle is present, it acts as a "spoke" and fixes the distance between the acromion and manubrium of the sternum. As the trunk moves forward, the shoulder is deflected laterally, because the clavicle resists the medial component of the pull of the muscles. Relative movement between the manubrium and acromion takes the form of an arc. Terrestrial mammals that retain a clavicle tend to be those in which the legs are not rotated completely beneath the body so that the shoulder and elbow joints are not in the same parasagittal plane. The clavicle is lost in cursorial mammals, in which shoulder and elbow joints lie in the same parasagittal plane (Anatomy in Action 3-3). Without a clavicle, the shoulder is not deflected laterally during the propulsive stroke. This maximizes forward thrust as opposed to lateral thrust. Loss of the clavicle also permits the scapula to become, in effect, another limb segment because the glenoid cavity can increase its own fore-and-aft excursion. This movement increases step and stride length.

is connected to the rest of the skeleton only by ligaments and is reduced to a sliver of bone imbedded in the muscles cranial to the shoulder joint (see Figs. 5-5, 7-18, and 7-21).

The bone of the brachium is the **humerus** (Fig. 6-12). Its expanded proximal end has a smooth rounded **head** that articulates with the glenoid cavity and two processes for muscular attachment: a lateral **greater tubercle** and a medial **lesser tubercle.** An **intertubercular groove** for the long tendon of the biceps muscle lies between the two tubercles on the craniomedial surface of the humerus.

The distal end of the humerus is expanded and bears a smooth articular surface known as a **condyle.** Although there appears to be but a single condyle, it can be divided into a medial pulley-shaped portion, the **trochlea,** for the

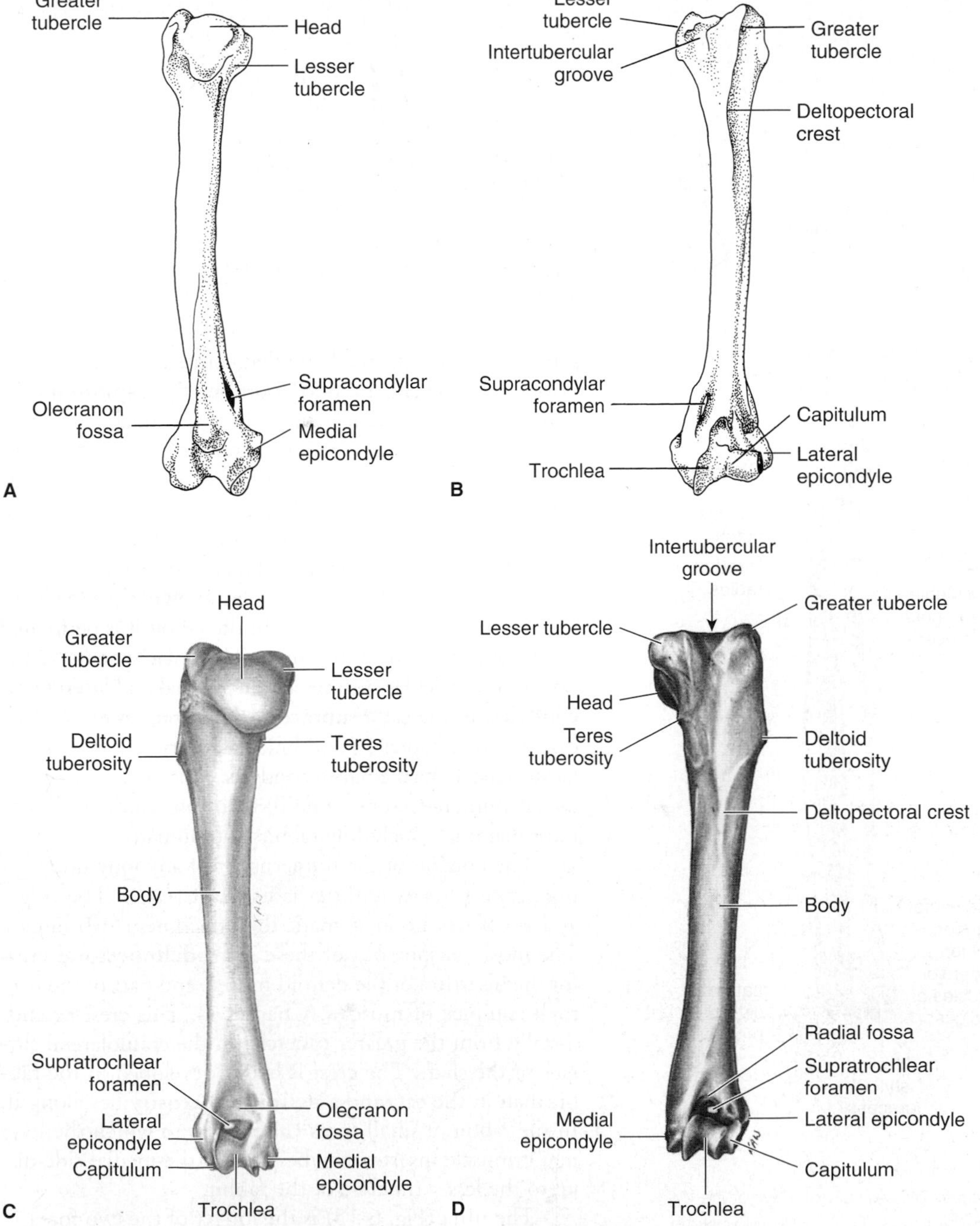

Figure 6-12

Left humerus of mammals. **A,** Caudal view of the cat humerus. **B,** Cranial view of the cat humerus. **C,** Caudal view of the rabbit humerus. **D,** Cranial view of the rabbit humerus.

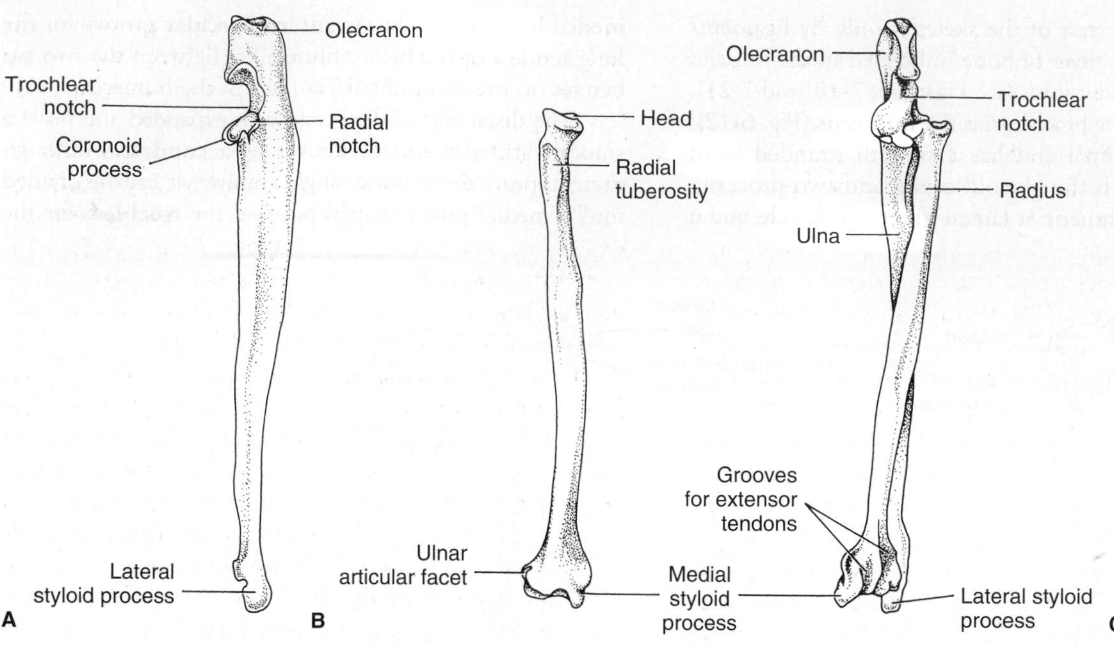

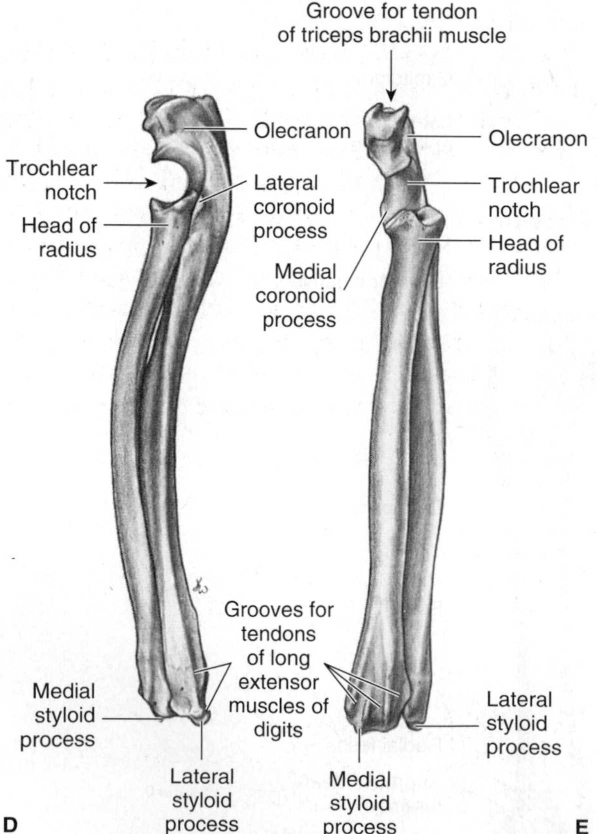

Figure 6-13
Left antebrachial bones of mammals. **A,** Lateral view of the cat ulna. **B,** Caudal view of the cat radius. **C,** Cranial view of the articulated cat radius and ulna with the hand prone. **D,** Lateral view of the rabbit radius and ulna. **E,** Cranial view of the rabbit radius and ulna.

ulna of the forearm (the bone that comes up behind the elbow), and a lateral rounded portion, the **capitulum,** on which the radius articulates. In most mammals, including the cat, the capitulum is rounded and the head of the radius rotates here as the manus changes from a prone to a supine position. You may have to articulate the ulna and radius with the humerus in order to determine the extent of the trochlea and capitulum. In the rabbit, the manus does not rotate axially. An **olecranon fossa** for the olecranon of the ulna is situated proximal to the trochlea. In the rabbit it is perforated by the **supratrochlear foramen.** The enlargements to the sides of the articular surfaces are the **medial** and **lateral epicondyles.** In the cat a **supracondylar foramen** for the median nerve (Chapter 9) and brachial artery (Chapter 11) is located above the medial epicondyles. This foramen is an ancestral amniote feature found in many amniotes but lost in most mammals, including rabbits and humans.

The portion of the humerus, or of any long bone, lying between its extremities is its **shaft** *(body).* The ridges and rugosities upon it mark the attachment of muscles. The most conspicuous of these is the **deltopectoral crest** for the insertion of the deltoid muscle and part of the pectoral complex of muscles (Chapter 7). This crest extends distally from the greater tubercle on the craniolateral surface of the shaft. The crest is better developed in the rabbit than in the cat, and a **deltoid tuberosity** lies along it. In the rabbit, a small **teres tuberosity,** to which the teres major muscle inserts, may be seen on the medial side distal to the lesser tubercle in the rabbit.

The **ulna** (Fig. 6-13) is the longer of the two forearm bones. A prominent, semilunar-shaped **trochlear notch,** for articulating with the humerus, is located near its proximal end. The end of the ulna that lies proximal to the notch is the **olecranon,** or "funny bone." In the cat a **medial coronoid process** forms the distal border of the notch, and a smaller, less conspicuous **lateral coronoid process** forms its proximal border. A **radial notch,** in which the head of the radius rotates, merges with the trochlear notch lateral to the coronoid processes. In the rabbit, the radius is fused to the ulna, hence, the radial notch is not apparent. The ulna terminates distally in a **lateral styloid process,** which lies on the lateral surface of the carpus. Note that, unlike the condition in reptiles, the ulna and radius of mammals articulate with each other distally and that the ulna plays a relatively insignificant role in the formation of the wrist joint. In some mammals, such as the horse and cow, the distal half of the ulna is lost.

The other bone of the forearm is the **radius** (Fig. 6-13). In the cat, the articular surfaces on its **head** allow the bone to rotate axially on the humerus and ulna. Recall (see Anatomy in Action 3-3) that the caudal rotation of the pectoral appendage, which brought the appendage closer to the trunk during the evolution of mammals, requires that the radius rotate on the ulna so the hand is prone, or the palm faces down. The cat can rotate its forearm to bring the palm to face up (supine) or down (prone). In the rabbit, the radius and ulna are fused with each other. No rotation is possible, so the manus is permanently in the prone position. This strengthens the forearm and helps resist stress when the rabbit lands on its pectoral appendages after a bound. In the cat, a prominent **radial tuberosity** for the insertion of the biceps muscle lies slightly distal to the head of the radius. The distal end of the radius is expanded; it bears articular facets for the ulna and carpus and a short **medial styloid process.** In the cat, there is also an articular facet for the ulna.

A generalized diagram for the components of the **carpus** (wrist) is shown in Figure 6-14. The widely used terminology for nonmammalian tetrapods is applied to the carpus of an early amniote in Panel **A.** These terms can be applied to mammals (Panel **B**), but most investigators prefer the human *Terminologia Anatomica* terms.

Study the manus of the cat or rabbit (Fig. 6-15). Its first portion, the **carpus,** or wrist, consists of two rows of small **carpal bones.** The proximal row contains three elements in many mammals, including rabbits and humans: a medial **scaphoid** (radiale), a central **lunate** (intermedium), and a lateral **triquetrum** (ulnare). In cats and other carnivores, the scaphoid and lunate are fused to form a large **scapholunate.** As in many amniotes, a **pisiform** projects caudoventrally from the triquetrum. The pisiform is one of many small **sesamoid bones** found in the appendages. They are not supporting elements; rather, they are associated with the attachments of muscle tendons, in this case with the tendon of the flexor carpi ulnaris (Chapter 7). Most sesamoid bones are not named. The four elements of the distal row of carpal bones are, from medial to lateral, the **trapezium** (distal carpal 1), **trapezoid** (distal carpal 2), **capitate** (distal carpal 3), and **hamate** (distal carpal 4).

Five **metacarpals** form the palm of the hand, and the free parts of the digits, or fingers, are composed of **phalanges.** The first digit (the thumb or **pollex**) is the most medial one. Note that the number of phalanges has been reduced from that in ancestral amniotes to 2-3-3-3-3. This correlates with the hand being directed cranially throughout the step cycle. The first digit is short in both the cat and rabbit. The thumb of humans is relatively much larger and can oppose the other digits in a precision grip. In the cat, the terminal, or **ungual,** phalanges are articulated in such a way that they, and the claws that they bear, can be either retracted and pulled back over the penultimate pha-

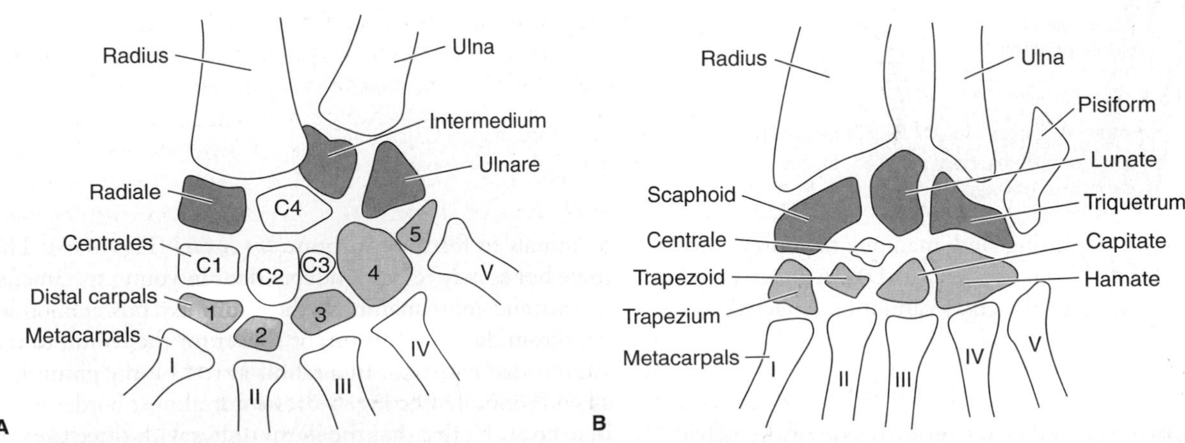

Figure 6-14
Evolution of the amniote carpus. **A,** Left carpus of an early amniote. **B,** Left carpus of a mammal.
(Modified from Romer and Parsons, 1986.)

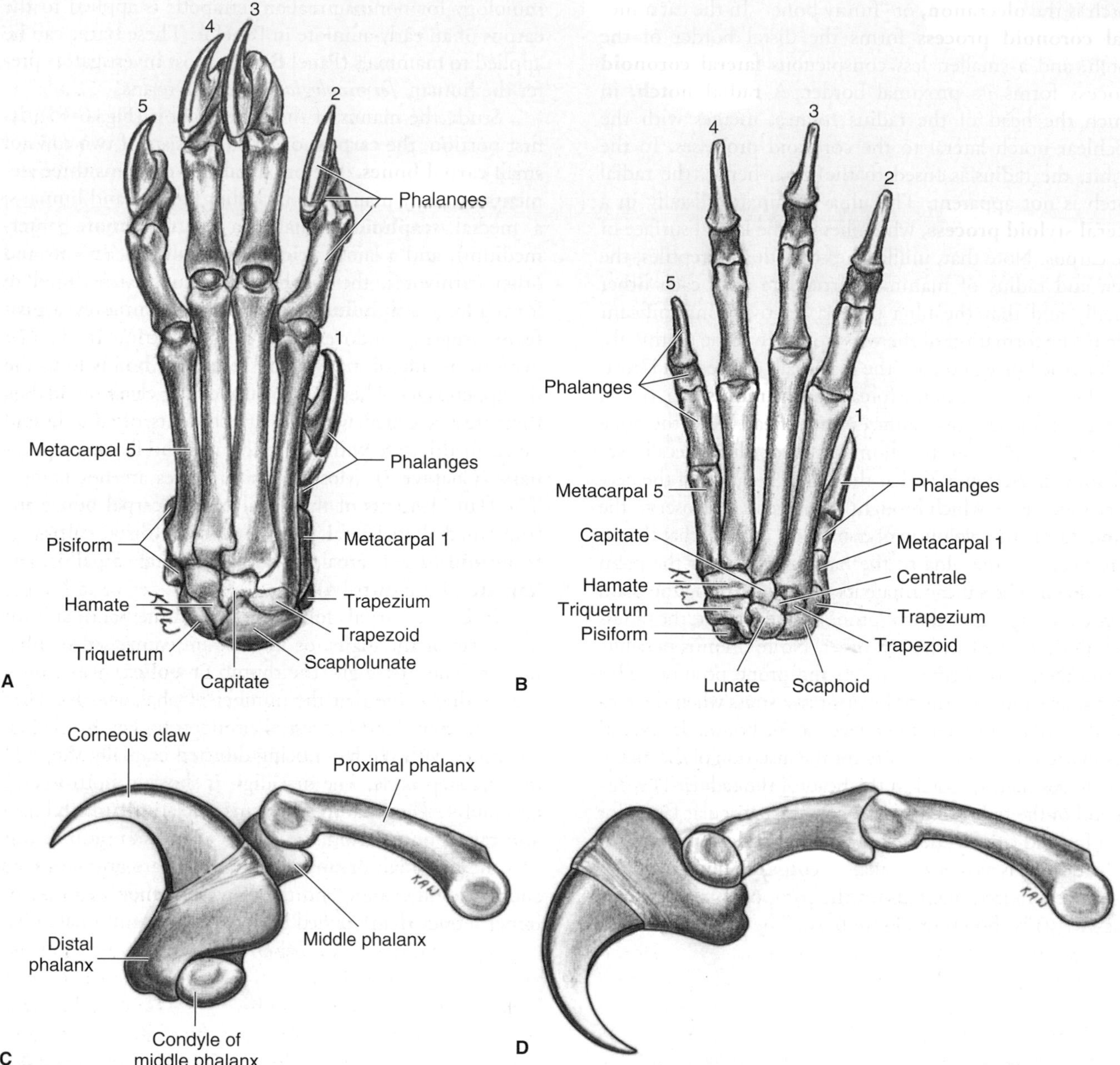

Figure 6-15

Left manus of mammals. **A**, Dorsal view of the cat manus. **B**, Dorsal view of the rabbit manus. **C**, Lateral view of the fourth digit of a cat with the claw retracted. **D**, Same digit as shown in **C** with the claw protracted.

langes or, when capturing and manipulating prey, protracted or extended (Fig. 6-15**C** and **D**). In other mammals, including rabbits, the claws cannot be retracted.

The Pelvic Girdle and Appendage

The ilium, ischium, and pubis on each side of the pelvic girdle, which are independent elements in early tetrapods and embryonic mammals, have fused together in adult mammals to form a **hip bone** *(os coxae)* (Fig. 6-16). The three bones may remain independent in young specimens. In cats and most mammals the **ilium** extends craniodorsally from the **acetabulum,** or socket for the femur, to the sacrum (see Fig. 5-5). In a rabbit at rest on the ground, it extends dorsally (see Fig. 5-6). Its dorsalmost border is the **iliac crest.** Notice that the ilium unites with three (cat) or four (rabbit) sacral vertebrae, in contrast to the one or two that connect with the ilium in amphibians and reptiles.

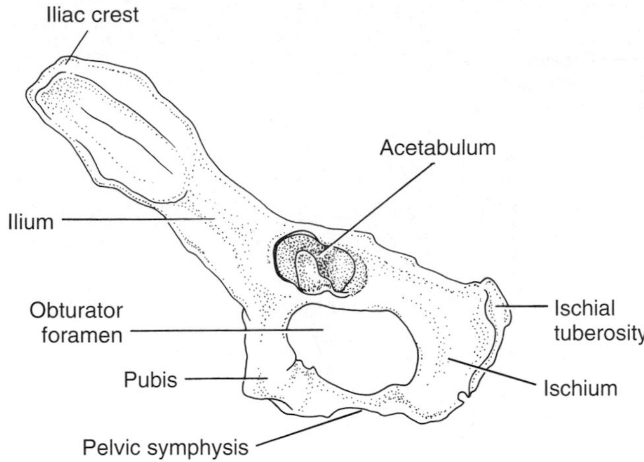

Figure 6-16
Lateral view of the left hip bone *(os coxae)* of an adult cat.

The **ischium** surrounds all but the cranial portion and some of the medial side of the large **obturator foramen** in the ventral portion of the pelvic girdle (Fig. 6-16). A sheet of connective tissue covers this foramen in life. The enlarged caudolateral portion of the ischium is called the **ischial tuberosity.** The **pubis** lies cranial to the obturator foramen and completes its medial rim. Although in most tetrapods the pubis enters the cranial portion of the acetabulum, in cats it is separated from it by a small **acetabular bone** (Fig. 6-6**D**) of unknown phylogenetic significance. The pubes and ischia of opposite sides are united by the **pelvic symphysis,** so that the pelvic girdle and sacrum form a complete ring, or **pelvic canal,** through which the digestive and urogenital tracts must pass to reach the anus and urogenital apertures.

Because the leg of mammals has rotated beneath the body, the thigh bone, or **femur,** of the cat and rabbit articulates with the acetabulum by a **head** that projects from the medial side of the proximal end of the bone (Fig. 6-17). The head is connected to the femoral shaft by a **neck.** A large, lateral **greater trochanter** and a small, medial **lesser trochanter** can also be seen on the proximal end of the femur. These processes are for muscle attachments. Rabbits have a **third trochanter** on the shaft distal to the greater trochanter, which is correlated with changes in the musculature associated with its bounding mode of locomotion. Other muscles insert along other ridges or protuberances on the shaft. A depression called the **trochanteric fossa** is situated on the caudal surface of the femur medial to the greater trochanter. The **linea aspera** for the insertion of limb muscles is the most conspicuous ridge on the femoral shaft. In the cat, it extends diagonally across the caudal side of the shaft. In the rabbit, it follows the medial side of the femoral shaft and may include a small bump. The distal end of the femur has a smooth articular facet over which the **patella,** or knee cap, glides (see Figs. 5-5, 7-31,and 7-33). The patella is a large sesamoid bone in the patellar tendon of the quadriceps femoris muscle (Chapter 7). Caudal to this are smooth **lateral** and **medial condyles** for articulation with the tibia. The rough areas above each condyle are **epicondyles,** and the depression between the two condyles is the **intercondylar fossa.**

The **tibia** is the larger and more medial of the two shank bones (Figs. 6-18). Its proximal end bears a pair of **condyles** for articulation with the femur, and a cranial, oblong bump, the **tibial tuberosity,** for the attachment of the patellar ligament. Its shaft has a relatively sharp cranial margin, the **tibial crest,** which continues distally from the tuberosity. The distal end of the tibia articulates with the tarsus, or ankle. It is elongated on the medial side and forms a process called the **medial malleolus,** which reinforces the ankle.

The more lateral **fibula** is a very slender bone (Fig. 6-18). In the cat, it is a separate and distinct element, but in the rabbit, much of its shaft fuses with the tibia. Fusion of bones reduces the chances of dislocation and resists stresses better in a bounding animal. Notice that the head of the fibula articulates with the proximal end of the tibia and does not enter the knee joint. Distally, it also articulates with the tibia and serves to strengthen the ankle laterally. Its distal end has a small process known as the **lateral malleolus.** In the cat, this process is pulleylike, and the tendon of the peroneus brevis muscle (Chapter 7) passes caudal to it. Notice that the tibia and fibula are approximately the same length as the femur in the cat, but they are distinctly longer in the rabbit. A lengthening of distal limb segments (shin and foot) provides extra leverage in a bounding animal.

The structure of joints between the long bones of the limbs is quite complex, but the generalized structure of such a joint is given in Anatomy in Action 6-3. The way long bones of the limb grow, and how this information can be used to help determine the age of a skeleton, is described in Anatomy in Action 6-4.

A generalized diagram of the amniote tarsus is shown in Figure 6-19. The elements of the early amniote tarsus are shown in panel **A,** and the bones are given their comparative anatomy names. The elements of the mammal tarsus are shown in panel **B,** and human terms are applied with one exception. Most students of comparative mammalian anatomy and paleontology use the term **astragalus** instead of *talus* (the *Terminologia Anatomica* term). This element represents the fused tibiale and intermedium and possibly includes a centrale.

Examine the foot or **pes** of the cat and rabbit (Fig. 6-20). In the rabbit, the foot is exceptionally long, which correlates with its bounding pattern of locomotion (compare Figs. 5-5

Text continues on page 113

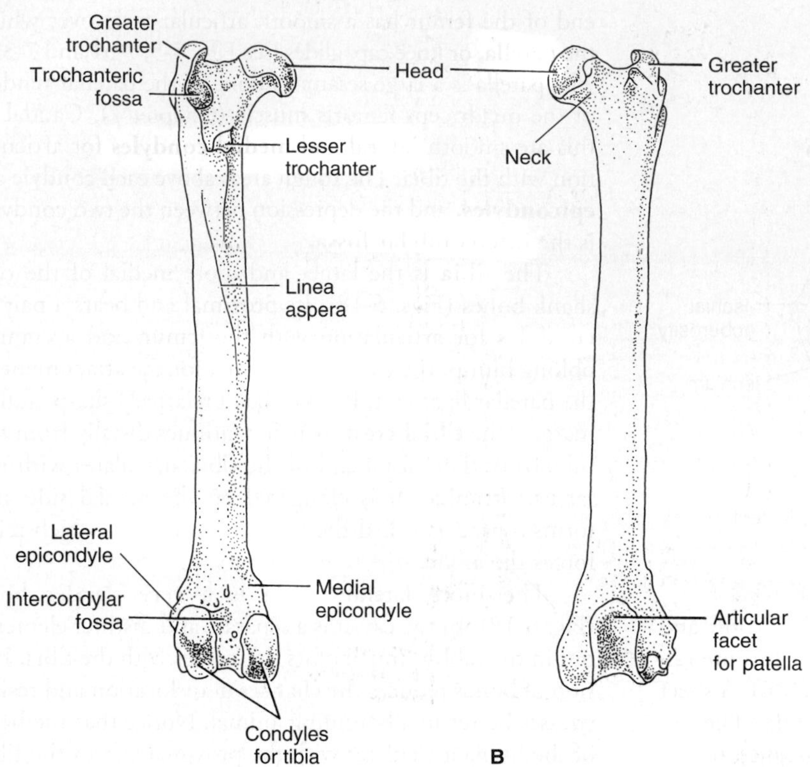

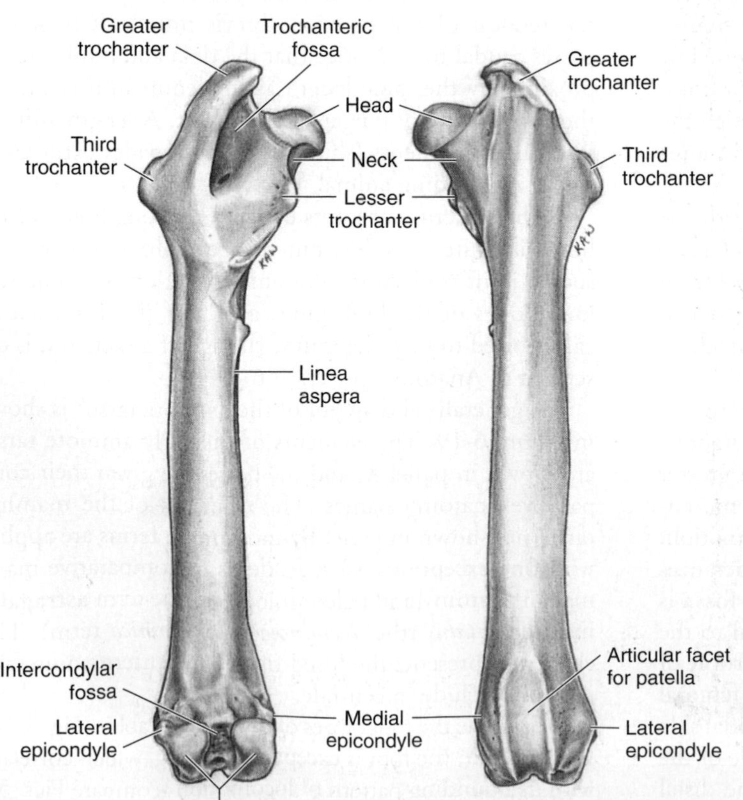

Figure 6-17

Left femur of mammals. **A**, Caudal view of the cat femur. **B**, Cranial view of the cat femur. **C**, Caudal view of the rabbit femur. **D**, Cranial view of the rabbit femur.

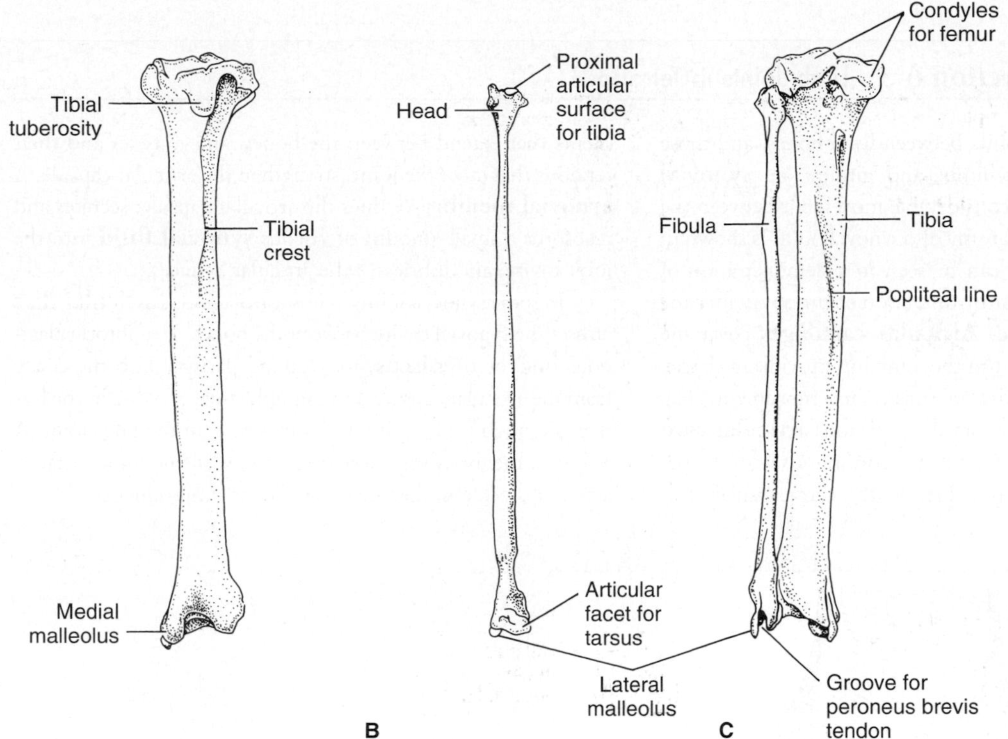

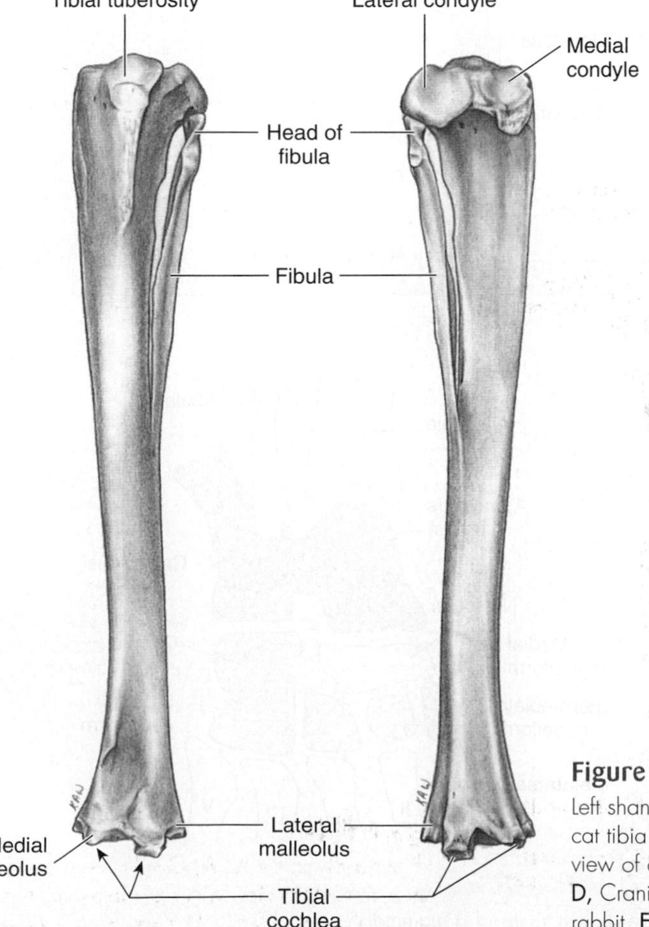

Figure 6-18

Left shank bones of mammals. **A,** Cranial view of the cat tibia. **B,** Medial view of the cat fibula. **C,** Caudal view of an articulated tibia and fibula of the cat. **D,** Cranial view of the fused tibia and fibula of the rabbit. **E,** Caudal view of the fused tibia and fibula of the rabbit.

Anatomy in Action 6-3 Limb Joints in Tetrapods

Most of the joints between limb bones, and those between the limbs and girdles, are **synovial joints** that permit considerable movement between the bones. The basic anatomy of a synovial joint is shown in the figure. Joint structure can be seen in a demonstration of the knee or another joint or by dissection of this joint after the muscles have been studied. **Articular cartilages** cover the ends of the bones that meet in the joint and provide an elastic, wear-resistant, and low-friction surface for movement. The joint is surrounded and supported by a fibrous **articular capsule** of dense connective tissue that is continuous with the periosteum covering the bones. Frequently, extracapsular ligaments that extend between the bones, and muscles and their tendons that cross the joint, strengthen the articular capsule. A **synovial membrane** lines the articular capsule; secretes and reabsorbs a small amount of viscous **synovial fluid** into the joint cavity; and lubricates the articular surfaces.

In some synovial joints, a fibrocartilaginous **articular disc** crosses the synovial cavity between the bones, or a fibrocartilaginous ring, or **meniscus,** may extend partway into the cavity from the articular capsule. For example, there is a disc in the human sternoclavicular joint and a meniscus in the knee joint. A disc or a meniscus improves the fit between the joint surfaces, acts as a shock absorber, or restricts certain movements.

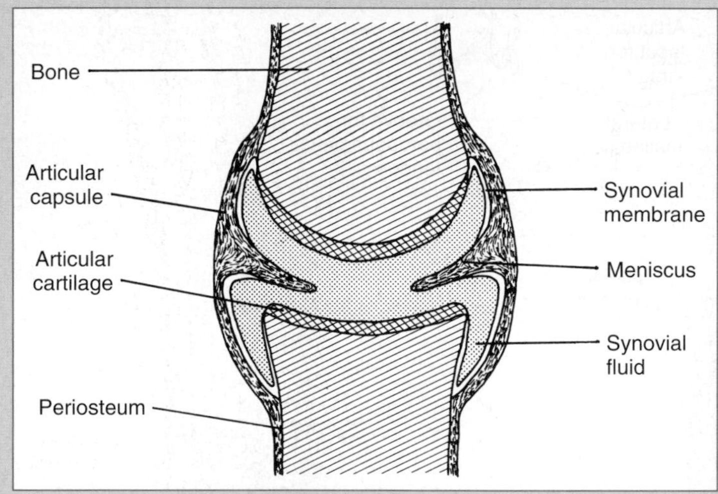

Diagram of a representative synovial joint.

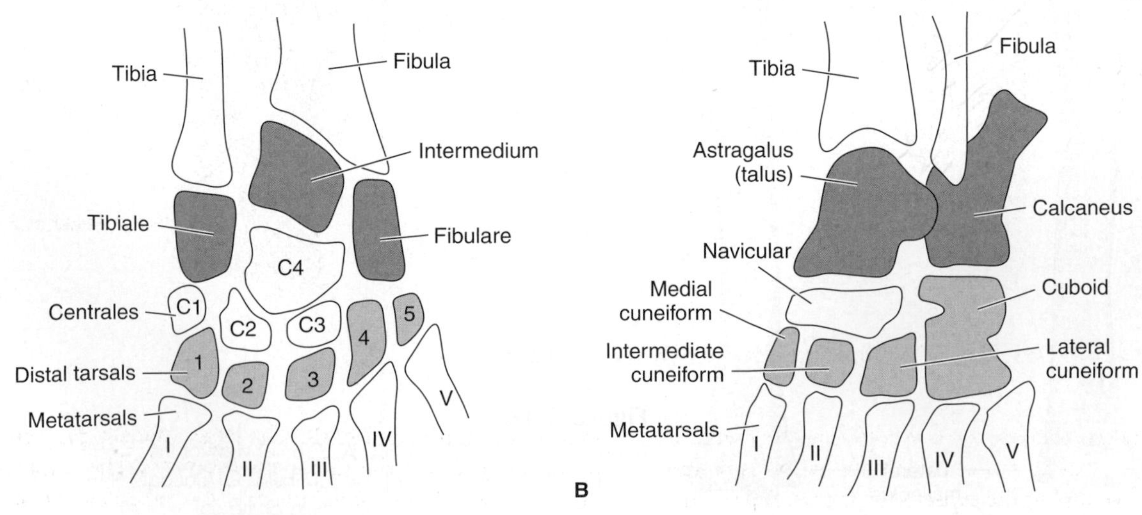

Figure 6-19

Evolution of the amniote tarsus. **A**, Left tarsus of an early amniote. **B**, Left tarsus of a mammal. *(Modified after Romer and Parsons, 1986.)*

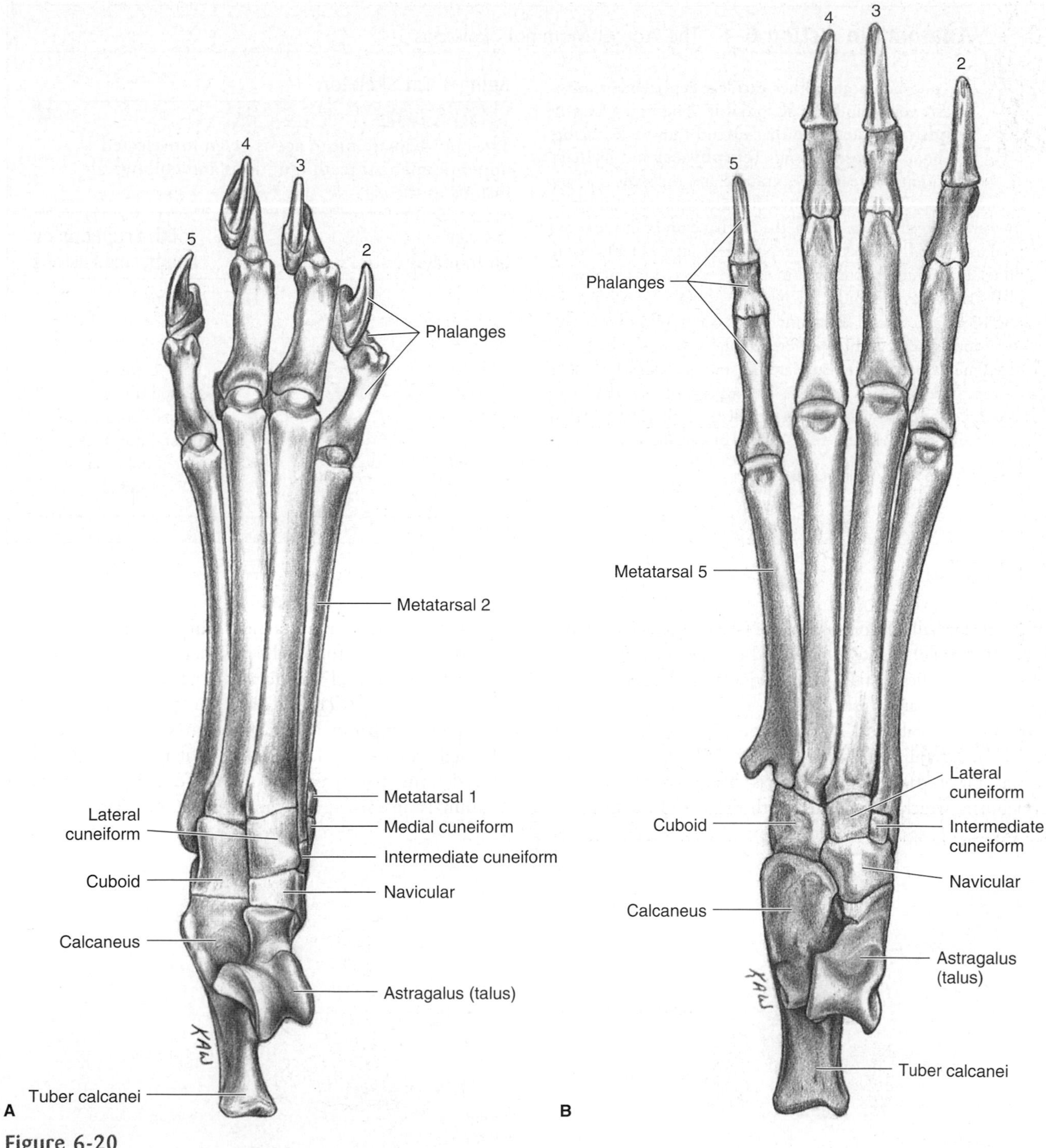

Figure 6-20

Left pes of mammals. **A,** Dorsal view of the cat foot. **B,** Dorsal view of the rabbit foot.

and 5-6). In the cat, the first toe is vestigial; in the rabbit, it is lost completely. The **tarsus** of the cat and rabbit is typical of mammals and contains the expected elements. A proximal **astragalus** *(talus)* articulates with the tibia and fibula. The ankle joint in early synapsids and mammals lies between this bone and the tibia and fibula rather than being a mesotarsal joint as in reptiles and birds. The large, caudally projecting heel bone is called the **calcaneus** (fibulare). The calcaneus extends into a large caudal process, the **tuber calcanei,** to which the calcaneal, or Achilles, tendon inserts (Chapter 7). On its plantar, or caudal, side, it widens medially and forms a grooved **sustentaculum tali** to hold in place the tendon

Anatomy in Action 6-4 The Age of Mammal Skeletons

Limb bones and other cartilage replacement bones in young individuals grow in length at plates of **epiphyseal cartilage** that extend transversely across the bone between its ends **(epiphyses)** and its shaft. The joint between the epiphysis and the shaft is a synchondrosis (p. 72). The epiphyseal plate grows by mitosis of the cartilage cells. As it grows, the cartilage on both the proximal and distal surfaces of the epiphyseal plate is replaced by bone. Bones stop increasing in length when the epiphyseal plates stop growing and have become completely ossified, so that the epiphysis and shaft unite. This occurs at different ages in various species and for different epiphyses in the body. By observing which epiphyseal plates are present and which have disappeared, one can estimate the age of the skeleton (see table). The degree to which permanent teeth have replaced milk teeth and the degree of fusion of certain skull sutures also help determine age.

Aging a Cat Skeleton

The approximate mean age is given for selected upper permanent teeth eruptions and epiphyseal fusions in the cat.

Age (in months)	Tooth Eruptions or Epiphyseal Fusions
3	Incisor 1
4	Incisor 3
5	Canine
12	Calcaneus
15	Distal ulnar
15	Distal femoral
18	Distal humeral
18	Proximal tibial
20	Distal radial

of the flexor hallucis longus muscle (see Chapter 7). A **navicular** (centrale) lies just distal to the astragalus, and a row of four bones lies distal to the navicular and calcaneus. In the cat these are, from medial to lateral, the **medial cuneiform** (distal tarsal 1), **intermediate cuneiform** (distal tarsal 2), **lateral cuneiform** (distal tarsal 3), and **cuboid** (distal tarsal 4). The rabbit has also lost the medial cuneiform, which is correlated with the loss of the first toe.

Five elongated **metatarsals,** which occupy the sole of the foot, normally follow the tarsals, but in the cat the first toe is vestigial, and its metatarsal is reduced to a small nubbin of bone articulated with the medial cuneiform. In the rabbit, even this is lost. The bones in the free part of the digits are the **phalanges.** The phalangeal formula is reduced from that of early amniotes to 2-3-3-3-3 in most mammals; in the cat and rabbit, it is 0-3-3-3-3.

The Muscular System

Chapter 7

Continuing on the general theme of organ systems whose activities support and move the body, we now will consider the muscular system. The study of this system is one of the more challenging parts of a comparative anatomy course because it makes special demands on your ability to learn and integrate a large number of different but interrelated data. Unlike most other structures of the vertebrate body, muscles are primarily defined by their attachments to skeletal elements, as well as by their muscle fiber arrangement, histological structure, embryonic development, and their action. It is important, therefore, to review these various aspects of the muscular system before starting the dissections.

There are three basic types of muscle tissue (Table 7-1), which are characterized by their histological structure, innervation, and association with organs and parts of the body. **Skeletal muscles** have a striated appearance when viewed under the microscope. They usually, but not always, are associated with skeletal structures. They are innervated by somatic nerves and usually are under voluntary control. **Smooth muscles** are associated with internal organs, blood vessels, and glands. They are innervated by visceral nerves from the autonomic nervous system and are not under voluntary control. **Cardiac muscles** are unique to the heart, and they also act as an internal pacemaker and impulse-conducting system of the heart. They are regulated by visceral nerves of the autonomic nervous system. In this chapter, we will study only the skeletal muscles. The smooth and cardiac muscles will be studied as part of the organs in which they occur (see Chapters 10 and 11).

The Structure and Function of Skeletal Muscles

The **body,** or **belly,** of a muscle is enveloped by a connective tissue sheet, the **epimysium,** and consists of numerous **muscle fiber bundles,** or **fasciculi** (singular: **fasciculus**). These muscle fiber bundles are the smallest units that are visible by unaided vision. They are surrounded by **perimysia,** which are connective tissue envelopes that are connected with the **epimysium** (Fig. 7-1). Each muscle fiber bundle consists of many **muscle fibers,** which are very long, multinucleated muscle cells. They are individually wrapped by capillaries and a network of connective tissue fibers forming an **endomysium.** The muscle fibers contain contractile **myofibrils,** which consist of series of **sarcomeres** separated from each other by transverse **Z disks,** or **Z lines.** Thin **actin myofilaments** are anchored to the Z disks and interact with the thicker **myosin myofilaments,** which are not anchored to the Z disks. When a muscle contracts, the myosin myofilaments temporarily attach to active sites on the actin myofilaments. The myofilaments ratchet over each other, thereby creating tension and shortening. The various degrees of overlap between the actin and myosin myofilaments create a striated appearance of the sarcomeres and, thus, of the myofibrils and the entire muscle cell.

Within individual muscles, the muscle fiber bundles can be arranged in various ways (Fig. 7-2). They can be oriented in a generally parallel fashion, as in **straplike, spindle-shaped,** or **sheetlike muscles,** or they can be arranged in a diverging manner, as in **fan-shaped,** or **triangular, muscles.** Parallel-fibered muscles can be subdivided by **tendinous intersections.** Muscle fiber bundles also can be circumferentially arranged in **sphincter muscles.** Within **pennate muscles,** muscle fiber bundles can assume oblique orientations. Such muscles contain external sheets of connective tissue, as well as **intramuscular septa** or **tendons,** to which the muscle fibers can attach. **Segmented muscles** are parallel fibered but are partitioned by transverse sheets of connective tissue, called **myosepta.** These myosepta mark the border lines between the embryonic myotomes (see Footnote 3 in Chapter 1) from which the muscles developed.

The ends of muscles are attached to skeletal elements or connective tissue sheets that can move relative to one another. If muscles attach only to skeletal elements, they are separated by at least one joint. Muscles that pass over two or more joints are called **biarticular muscles** or **polyarticular muscles,** respectively.

Muscles attach to skeletal elements always via **tendons,** which are made of densely and regularly arranged connective

Table 7-1 Classification of Muscles

				Embryonic Origin	Innervation	Control	Contraction Type	Histological Type
Somatic Musculature		Head M.	Extrinsic eye muscles	Somitomeres 1, 2, 3, and 5	Cranial nerves III, IV, and VI	Voluntary; but subconscious integration of synergistic muscle contractions	Usually fast and strong; fatigue	Skeletal muscle; striated appearance; long (<30 cm) multinucleated muscle fibers
		Branchiomeric Musculature	Mandibular muscles	Somitomere 4	Cranial nerve V			
			Hyoid muscles	Somitomere 6	Cranial nerve VII			
			Branchial muscles of first branchial arch	Somitomere 7	Cranial nerve IX			
			Branchial muscles of second to fifth branchial arches	Postotic somites 1–4	Cranial nerve X; cranial nerve XI			
	Axial Musculature	Cranial Axial Musculature	Epibranchial muscles	Dorsal parts of postotic and cranial trunk somites	Dorsal rami of occipital and anterior spinal nerves			
			Hypobranchial muscles	Ventral parts of postotic and cranial trunk somites	Ventral rami of occipital and anterior spinal nerves; hypobranchial nerve; hypoglossal nerve (cranial nerve XII)			
		Trunk and Tail Musculature	Epaxial muscles	Dorsal part of trunk and tail myotomes	Dorsal rami of spinal nerves			
			Hypaxial muscles	Ventral parts of trunk and tail myotomes	Ventral rami of spinal nerves			
	Appendicular Musculature		Appendicular muscles	Ventral parts of trunk and tail myotomes	Ventral rami of spinal nerves			
Visceral Musculature			Heart	Visceral layer of hypomere	Cardiac pacemaker	Involuntary; regulated by autonomic nervous system	Moderate speed; strong; no fatigue	Cardiac muscle; striated appearance; branched muscle cells
			Walls of internal organs		Muscle cells		Slow, weak; no fatigue	Smooth muscle; spindle-shaped cells, <500 μm
			Walls of blood vessels	Mesenchymal cells of endothelium	Autonomic nervous system		Moderate speed; weak; no fatigue	Smooth muscle; spindle-shaped cells, 15–200 μm
			Iris muscles Ciliary muscles	Ectoderm of eye cup			Moderate speed; weak	
			Dermal muscles	Dermatome of somites			Slow; weak	

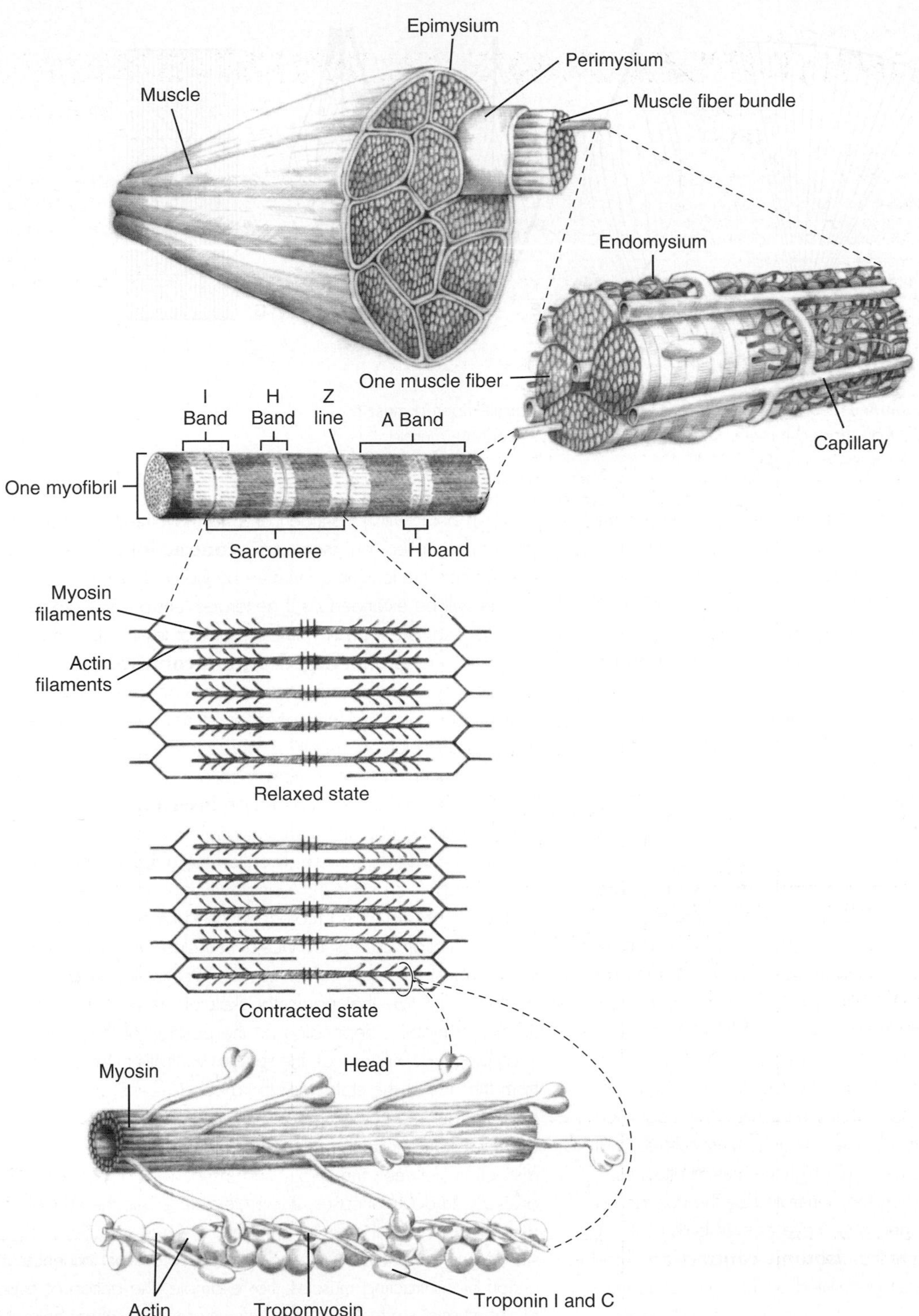

Figure 7-1
Structure of skeletal muscle and its components shown at successive levels of magnification down to the molecular level. *(Modified after Liem, Bemis, Walker, and Grande, 2001.)*

tissue consisting of mostly **collagen** fibers, but occasionally also of **elastin** fibers. The connective tissue of these tendons is continuous with the connective tissue that surrounds and pervades the muscles, such as the perimysia and endomysia. The ends of muscle fibers are anchored to the dons by **myotendinous junc** chored to skeletal elements by b fibers with those of the periosteum o

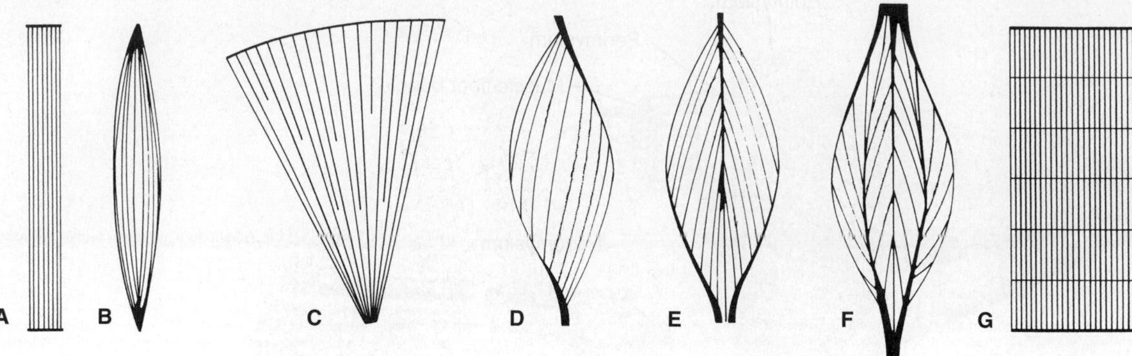

Figure 7-2
Arrangement of muscle fibers in muscles of different shape. **A,** Straplike, parallel-fibered muscle. **B,** Spindle-shaped muscle. **C,** Fan-shaped muscle. **D,** Pennate muscle. **E,** Bipennate muscle. **F,** Multipennate muscle. **G,** Segmented muscle.

etrating the matrix of the skeletal elements. Tendons can assume a great variety of shapes, depending on the shape of the muscles to which they attach. If they are sheetlike, they are called **aponeuroses.** If a tendon is so short that it is barely visible with the naked eye, we speak of a **fleshy attachment** of the muscle, in contrast to a **tendinous,** or **aponeurotic, attachment.**

There is a variety of additional types of connective tissue structures of the muscle-bone system. A **fascia** is a sheet of connective tissue that invests muscles (e.g., as epimysium), muscle groups, and other organs. It also can serve as an attachment site for muscle fibers.[1] A **ligament** usually is a band of connective tissue that connects skeletal elements to one another. **Loose connective tissue** fills spaces between muscles, nerves, and blood vessels. It often is associated with **adipose tissue,** or **fat,** and usually is picked away during the dissection process.

As we proceed now to consider the actions and functions of muscles, it is most important to keep in mind that when a muscle contracts, the muscle as a whole generates tension that is equal at both of its ends. How exactly this force affects the skeletal elements to which the muscle attaches, however, is determined by a multitude of factors in addition to the structure and physiology of the contracting muscle itself. One of these factors is the magnitude of the **load,** or external force counteracting the muscle force, that is placed on the skeletal elements to which the muscle attaches. If this load is smaller than the tension generated by the contracting muscle, the muscle will shorten and move one or both of its attachments; this process is called an **isotonic contraction.** If the load is as large as the tension generated by the contracting muscle, the muscle will still generate tension (i.e., it will become "tense"), but it will not shorten and no movement will occur; this process is called an **isometric contraction.** If the load is greater than the tension generated by the contracting muscle, the muscle will be extended as it generates tension; this process occurs, for example, in many limb muscles of tetrapods during locomotion and is called a **negative work contraction.**

Although a contracting muscle exerts an equal force on both skeletal elements to which it attaches, during isotonic muscle contractions one element usually is more mobile than the other. The muscle attachment to the more mobile element, which often is also the more distal one, is called the **insertion** of a muscle. The muscle attachment to the more stationary element, which often is also the more proximal one, is called the **origin,** or **head,** of a muscle (Figs. 7-3 and 7-4). A single muscle may have more than one place of origin and more than one place of insertion (e.g., the mammalian biceps or triceps muscles). Multiple fleshy attachments of a muscle sometimes are also called **slips.** It is important to realize that particular skeletal elements may be stationary or mobile, depending on the position of the animal. For example, in a walking cat, the serratus ventralis muscle originates from the ribs of the stationary thorax and inserts on the moving scapula, but in a heavily breathing standing cat, the scapula can be more stable than the ribs of the thorax (Fig. 7-22). Thus, the distinction between the origin and insertion of a muscle often is arbitrary. In doubtful cases, it is preferable to use the neutral term **attachment** for all ends of a muscle.

A number of other factors determine, or at least influence, the action of contracting muscles. For example, the action of a particular muscle is affected by the actions of muscles that contract at the same time. Movements of skeletal structures are rarely brought about by contraction of a single muscle. In general, several muscles contract together, or in overlapping sequences, to move a skeletal element or stabilize a joint. Muscles that work together for a particular movement, such as flexion of the elbow joint, are called **synergists.** In a group of synergistic muscles, some muscles may stabilize a joint to allow movement in only one particular direction, whereas other muscles may generate the movement.

'he difference between a fascia and an aponeurosis can be ıned as follows. An *aponeurosis* consists of tendon fibers that 'irect continuations of muscle fibers via a myotendinous ɔn; hence, the tendon and muscle fibers are oriented in the 'rection. A *fascia* provides a surface for attachment to the 'bers that are connected to muscle fibers via ıous junctions; hence, the muscle fibers and the fibers usually differ in their orientations.

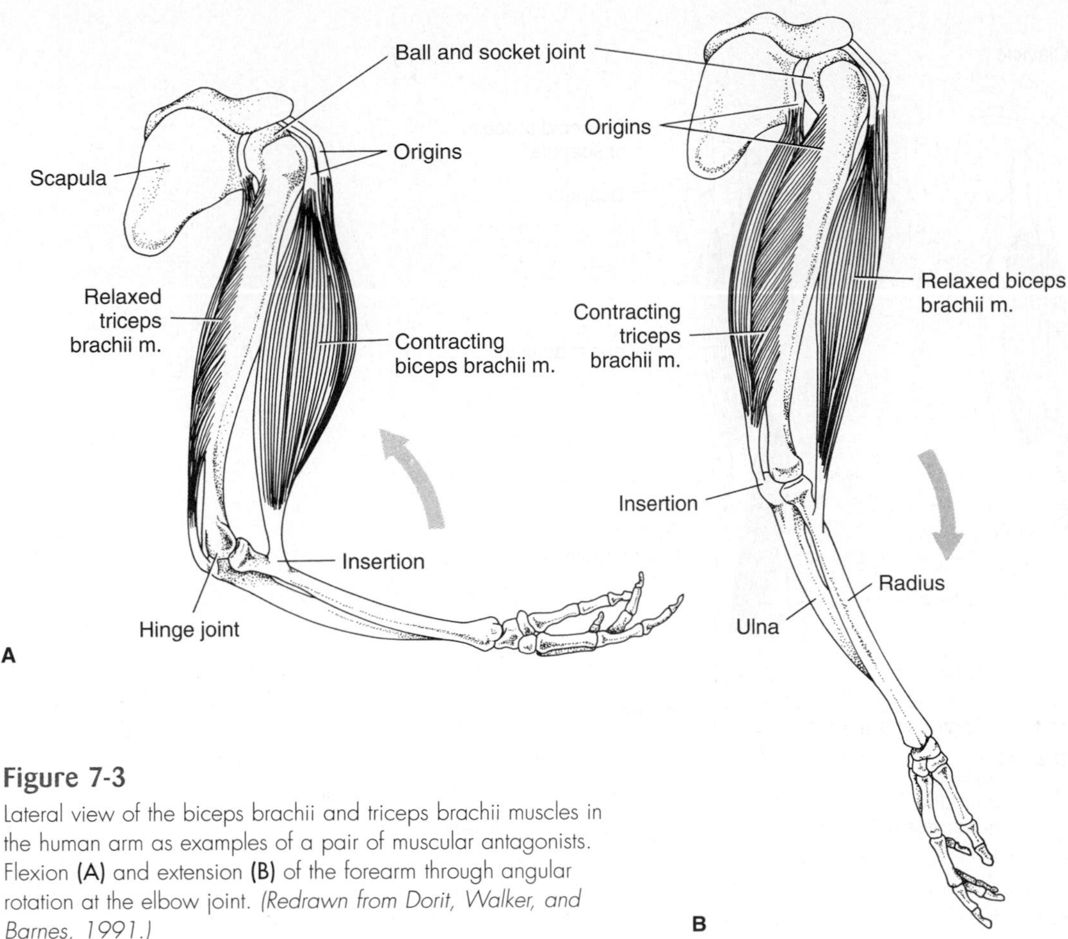

Figure 7-3
Lateral view of the biceps brachii and triceps brachii muscles in the human arm as examples of a pair of muscular antagonists. Flexion **(A)** and extension **(B)** of the forearm through angular rotation at the elbow joint. *(Redrawn from Dorit, Walker, and Barnes, 1991.)*

Some muscles may only initiate a movement, whereas other muscles may become active only as the skeletal element begins to move, and yet other muscles may be active only toward the end of a movement. Because muscles can only shorten to bring about a movement and cannot actively lengthen again, they must be extended by the contraction of other muscles, called **antagonists,** which cause the opposite movement (Figs. 7-3 and 7-4). Muscles that act antagonistically often have very different attachments and functions, as shown, for example, by the brachioradial and pronator quadratus muscles (Fig. 7-4). As with synergists, antagonists occur in groups of muscles rather than as single muscles.

All muscles generate tension as they contract, but the direction of this tensile force is determined by the position and orientation of the muscles relative to the joints they span. Whether a muscle passes over the front or the back of a joint, and whether it crosses a joint perpendicularly or diagonally, fundamentally affects its actions on the skeletal elements to which it attaches (compare Figs. 7-3 and 7-4). The action of a contracting muscle also is determined by the number of joints it spans, because a contracting muscle affects all the joints it crosses. The shape and structure of the joints themselves and their surrounding ligaments play a crucial role in controlling the direction of movements of skeletal elements. A **ball-and-socket,** or **spheroidal, joint** (e.g., the mammalian hip joint) allows a great versatility of movements. A **hinge joint,** or **ginglymus** (e.g., the humeroulnar joint in the mammalian elbow), allows movement only in one plane as an angular rotation. A **pivot,** or **trochoid, joint** (e.g., the head of the radius in the radial notch of the ulna) allows movement only as an axial rotation around the pivot's axis. There are many other types of joints that allow different degrees of mobility (see Anatomy in Action 6-3, p. 112).

Although most natural movements of the skeletal system are complex, it is possible to identify several basic types of movements that can combine to generate complex motions. When using a term describing a basic type of movement, it is necessary to be specific with respect to the identity of the moving structure and to the axis and direction of movement. A **flexion** is a movement that reduces the angle formed between skeletal elements, such as between the humerus and forearm (Fig. 7-3**A**); it is an **angular rotation** about a joint, in this case the elbow joint. Flexions can occur in different directions; for example, the head or the vertebral column can be flexed, or bent, forward (i.e., ventrally) or sideways (i.e., laterally). An **extension** is the opposite of a flexion; it increases the angle formed between skeletal elements (Fig. 7-3**B**).

An **adduction** is a movement that brings a limb closer to a point of reference (e.g., the midventral line of the body). For example, the humerus is adducted when the pectoral muscle

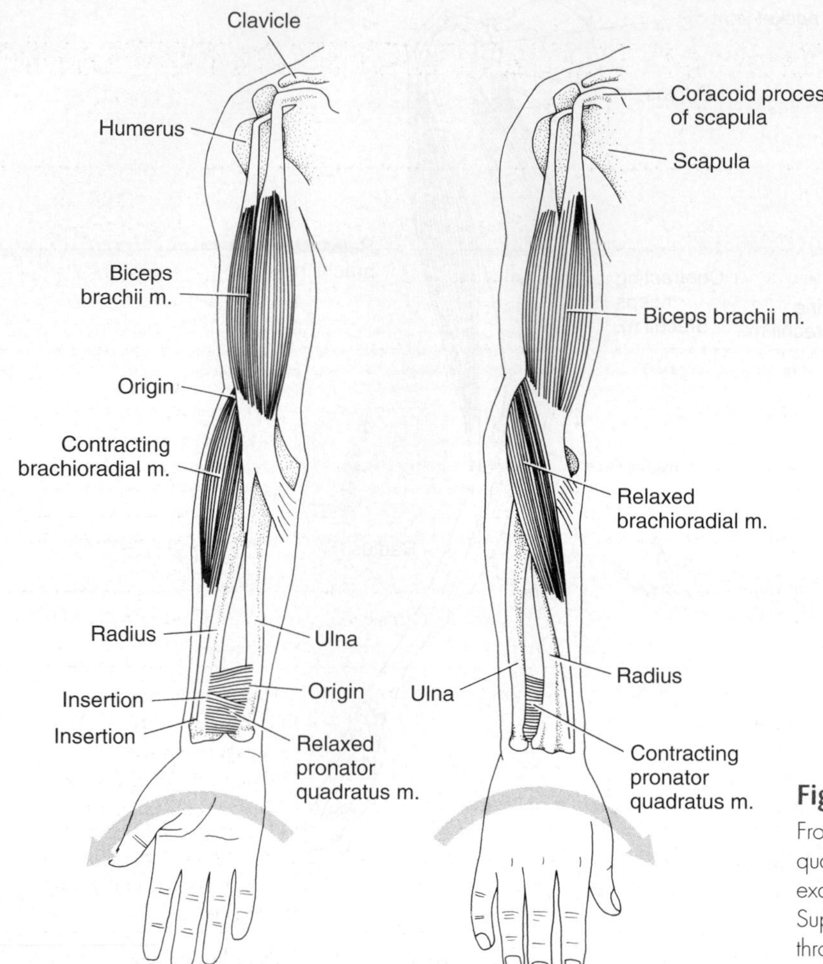

Figure 7-4
Frontal view of the brachioradial and pronator quadratus muscles in the human arm as examples of a pair of muscular antagonists. Supination **(A)** and pronation **(B)** of the hand through axial rotation of the radius at the elbow joint. *(Redrawn from Young and Hobbs, 1975.)*

pulls it toward the thorax through an angular rotation about the shoulder joint (Fig. 7-22). An **abduction** is the opposite movement that rotates a limb away from a point of reference. For example, the pectoral fin of a shark is extended when the dorsal fin muscle contracts and raises the fin away from the midventral line of the body (Fig. 7-8**B** and **C**).

A **protraction** is a movement that brings a body part forward. For example, the limbs of tetrapods are protracted when they are pulled cranially during a step by extending the limbs and rotating the humerus or femur cranially about the shoulder or hip joints, respectively (Fig. 7-10). A **retraction** is the opposite movement and pulls a limb caudally. For example, the limbs of tetrapods are retracted when they are pulled caudally during a step by first flexing and then extending the limbs and rotating the humerus or femur caudally about the shoulder or hip joints, respectively (Fig. 7-10).

An **axial rotation,** in contrast to the angular rotation discussed previously, is a circular movement about a longitudinal axis. For example, the atlas of mammals, together with the skull, rotates axially about the dens of the axis (see Fig. 5-7). Supinations and pronations are special cases of axial rotations. A **supination** is the movement that places a body part facing up in a supine position. For example, the contraction of the brachioradial muscle supinates the hand of a human, or the paw of a cat, by rotating the distal end of the radius laterally about the long axis of the forearm so that the palm of the human hand, or cat forepaw, faces upward or medially, respectively (Fig. 7-4**A**; see p. 107 in Chapter 6). A **pronation** is the opposite movement that places a body part facing down in a prone position. For example, the contraction of the pronator quadratus muscle pronates the human hand, or cat forepaw, by rotating the distal end of the radius medially about the long axis of the forearm so that the palm of the human hand, or cat paw, faces downward (Fig. 7-4**B**; see also Chapter 6 and Fig. 6-13**C**).

The Classification of Muscles into Groups

The muscular system differs significantly among vertebrates, particularly between fishes, amphibians, and mammals (Table 7-2). These differences reflect the different body constructions of the vertebrates that need to deal with very different physical conditions, depending on whether they live in water or on land. In

Text continues on page 124

Table 7-2 Comparison and Homologies of the Skeletal Muscles of Jawed Vertebrates

The table lists the main muscles described and discussed in Chapter 7 and indicates their possible homologies. For additional information, see Table 7-1.

			Innervation	Muscles of *Squalus*	Muscles of *Necturus*	Muscles of Mammals (*Felis* and *Oryctolagus*)
Axial Muscles	Extrinsic Eye Muscles		Oculomotor (III) nerve	Dorsal rectus extrinsic eye muscle	Dorsal rectus extrinsic eye muscle	Dorsal rectus extrinsic eye muscle
				Ventral rectus extrinsic eye muscle Medial rectus extrinsic eye muscle Ventral oblique extrinsic eye muscle	Ventral rectus extrinsic eye muscle Medial rectus extrinsic eye muscle Ventral oblique extrinsic eye muscle	Ventral rectus extrinsic eye muscle Medial rectus extrinsic eye muscle Ventral oblique extrinsic eye muscle
			Trochlear (IV) nerve	Dorsal oblique extrinsic eye muscle	Dorsal oblique extrinsic eye muscle	Dorsal oblique extrinsic eye muscle
			Abducens (VI) nerve	Lateral rectus extrinsic eye muscle	Lateral rectus extrinsic eye muscle	Lateral rectus extrinsic eye muscle
	Branchiomeric Muscles	Mandibular Muscles	Trigeminal (V) nerve	Adductor mandibulae muscle Levator palatoquadrati muscle Spiracular muscle Preorbital muscle	Levatores mandibulae muscles	Masseter muscle Temporal muscle Pterygoid muscles Tensor veli palatini muscle Tensor tympani muscle
				Intermandibular muscle	Intermandibular muscle	Mylohyoid muscle Rostral portion of digastric muscle
		Hyoid Muscles	Facial (VII) nerve	Levator hyomandibulae muscle Dorsal hyoid constrictor muscle	Depressor mandibulae muscle Branchiohyoid muscle	Stapedius muscle Anterior portions of platysma and facial muscles
				Interhyoid muscle Ventral hyoid constrictor muscle	Interhyoid muscle Sphincter colli muscle	Posterior portions of platysma and facial muscles Caudal portion of digastric muscle Stylohyoid muscle
		Branchial Muscles	**Anamniotes:** Glossopharyngeal (IX) nerve and vagus (X) nerve **Amniotes:** Accessory (XI) nerve and spinal nerves	Cucullaris muscle	Cucullaris muscle	Trapezius muscle complex Sternocleidomastoid muscle complex
					Levatores arcuum muscles	
				Dorsal interarcual muscles Lateral interarcual muscles Branchial adductor muscles		
				Superficial branchial constrictor muscles Interbranchial muscles	Dilatator laryngis muscle Subarcual muscles Transversi ventrales muscles Depressores arcuum muscles	Intrinsic muscles of the larynx Some muscles of the pharynx

continued

		Innervation	Muscles of *Squalus*	Muscles of *Necturus*	Muscles of Mammals (*Felis* and *Oryctolagus*)
Axial Muscles	Epibranchial Muscles	Dorsal rami of occipital and anterior spinal nerves	Epaxial and dorsal portions of hypaxial myomeres dorsal to the gill region	Anterior portion of dorsalis trunci muscle	**Anterior portion of epaxial muscles:** Multifidi muscles Splenius muscle Longissimus capitis muscle Semispinalis cervicis et capitis muscle
	Hypobranchial Muscles: Prehyoid Muscles	**Anamniotes:** Hypobranchial nerve, ventral rami of occipital and anterior spinal nerves **Amniotes:** Hypoglossal (XII) nerve and cervical nerve plexus	Coracomandibular muscle	Genioglossus muscle	Lingualis proprius muscle Genioglossus muscle
				Geniohyoid muscle	Hypoglossus muscle Styloglossus muscle Geniohyoid muscle
	Hypobranchial Muscles: Posthyoid Muscles		**Rectus cervicis muscle:** Coracoarcual muscle Coracohyoid muscle	Rectus cervicis muscle	Sternohyoid muscle Sternothyroid muscle Thyrohyoid muscle
				Omoarcual muscle Pectoriscapular muscle	
			Coracobranchial muscles 1–5		
	Epaxial Trunk Muscles	Dorsal rami of spinal nerves	Epaxial portions of myomeres	Interspinal muscles	Interspinal muscles
				Posterior portion of dorsalis trunci muscle	**Posterior portion of epaxial muscles:** Intertransversarii muscles Multifidi muscles Spinalis muscle Semispinalis muscle Longissimus dorsi muscle Iliocostal muscle
	Hypaxial Trunk Muscles	Ventral rami of spinal nerves	Hypaxial portions of myomeres	Subvertebral muscle	Longus colli muscle Psoas minor muscle Quadratus lumborum muscle
				Larva: Levator scapulae muscle **Metamorphosed adult:** Opercular muscle	Omotransversarius muscle
				Thoraciscapular muscle	Serratus ventralis muscle (part)
				External oblique muscle	Serratus ventralis muscle (part) Rhomboideus cervicis and thoracis muscle Rhomboideus capitis muscle Serratus dorsalis muscle Scalenus muscle Rectus thoracis muscle External oblique muscle External intercostal muscles
				Internal oblique muscle	Internal oblique muscle Internal intercostal muscles
				Transversus abdominis muscle	Transversus abdominis muscle Transversus thoracis muscle
				Rectus abdominis muscle	Rectus abdominis muscle Diaphragm

continued

Table 7-2 Comparison and Homologies of the Skeletal Muscles of Jawed Vertebrates *continued*

			Innervation	Muscles of *Squalus*	Muscles of *Necturus*	Muscles of Mammals (*Felis* and *Oryctolagus*)
Appendicular Muscles	Pectoral Appendicular Muscles	Dorsal Muscle Group	Brachial plexus of ventral rami of spinal nerves	Superficial and deep portions of dorsal fin muscle	Latissimus dorsi muscle	Cutaneus trunci muscle (part) Latissimus dorsi muscle Teres major muscle
					Subcoracoscapular muscle	Subscapular muscle
					Scapular deltoid muscle	Scapulodeltoid muscle Acromiodeltoid muscle
					Procoracohumeral muscle	Cleidobrachial muscle Teres minor muscle
					Triceps muscle	Triceps brachii muscle Tensor fasciae antebrachii muscle Anconeus muscle
					Extensor muscle of forearm, wrist, and digits	Extensor muscle of forearm, wrist, and digits
		Ventral Muscle Group		Cranial fin muscle Ventral fin muscle	Pectoral muscle	Cutaneus trunci muscle (part) Pectoral muscle complex
					Supracoracoid muscle	Supraspinatus muscle Infraspinatus muscle
					Coracoradial muscle	Biceps brachii muscle (part)
					Humeroantebrachial muscle	Biceps brachii muscle (part) Brachial muscle
					Coracobrachial muscle	Coracobrachial muscle
					Flexor muscles of the forearm, wrist, and digits	Flexor muscles of the forearm, wrist, and digits
	Pelvic Appendicular Muscles	Dorsal Muscle Group	Pelvic plexus of ventral rami of spinal nerves	Dorsal fin muscle	Iliotibial muscle	Sartorius muscle
					Internal puboischiofemoral muscle	Iliacus muscle Psoas major muscle Pectineus muscle Vasti muscles
					Ilioextensorius muscle	Rectus femoris muscle
					Iliofibular muscle	Gluteus superficialis muscle
					Iliofemoral muscle	Tensor fasciae latae muscle Gluteus medius muscle Gluteus profundus muscle
					Extensor muscles of the shank, ankle, and digits	Muscles on the extensor side of the shank, ankle, and digits
		Ventral Muscle Group		Proximal and distal portions of ventral fin muscle	External puboischiofemoral muscle	Obturator externus muscle Quadratus femoris muscle
					Adductor femoris muscle	Adductor femoris brevis et magnus muscle
					Pubotibial muscle	Adductor femoris longus muscle
					Ischiofemoral muscle	Obturator internus muscle Gemmelli muscles
					Caudofemoral muscle	Piriformis muscle Caudofemoral muscle (absent in some mammals)
					Puboischiotibial muscle	Gracilis muscle
					Ischioflexorius muscle	Semimembranosus muscle Semitendinosus muscle Biceps femoris muscle Abductor cruris caudalis muscle
					Flexor muscles of the shank, ankle, and digits	Muscles on the flexor side of the shank, ankle, and digits

the course of evolution, as the configurations of the skeletal elements of a fishlike body construction became modified into those of tetrapod body constructions, the structure and configuration of muscles changed accordingly.

We know from studies that assessed the extent and nature of the individual variability of muscles within and between species that muscles can change their configurations relatively easily. We infer that those variations also occurred in the past and eventually led to the differences we observe today among different vertebrates. For example, muscles can change their sizes and shapes, divide into two or more muscles, merge with other muscles, disappear altogether, or alter their internal architecture by adding, increasing, or reducing intramuscular tendons or septa. Muscles also can shift the positions of their origins or insertions and, thereby, modify their actions.

Such changes usually can be traced relatively easily in muscles from representatives of the same vertebrate class, for example, when comparing a cat and a rabbit. They are much more difficult to trace across very different vertebrate life forms, such as fishlike versus tetrapod vertebrates, especially because the number of individually identifiable muscles varies greatly between vertebrate classes (Table 7-2). In such cases, it is more meaningful to compare entire muscle groups between classes, and to do so we need to classify the muscular system into comparable parts in a manner that applies to all vertebrates.

One approach to such a classification of muscles is based on their embryonic origin. Individual muscles always develop from the same, clearly identifiable tissues in an embryo. Because these embryonic tissues are arranged in a very similar manner in all vertebrates at the early stages of embryonic development, we assume that muscles that developed from the same embryonic structure can be clustered into muscle groups that are directly comparable in all vertebrates. Muscle groups that are defined by their embryonic origin usually are innervated by nerves that arise from the same or comparable divisions of the nervous system (Tables 7-1 and 7-2). Thus, the nerve supply of a muscle often provides a useful clue to its embryonic origin and group affiliation.

In order to understand the rationale behind the classification system used almost universally for muscles in comparative anatomy, it is necessary to review the embryonic development of the musculature in a vertebrate, which is intimately tied to the differentiation of the mesoderm (Figs. 7-5 and 7-6).

In the early stages of embryonic development, the paraxial mesoderm of the trunk, which lies beside the notochord and below the neural plate, grows laterally and ventrally between the ectoderm and endoderm (Fig. 7-5A). In the trunk and starting cranially, the dorsal portion of the mesoderm gradually becomes divided into distinct segments, or **somites** (Fig. 7-5B), which define the segmental organization of the vertebrate body. A primitive coelom appears within the mesoderm. The mesoderm differentiates also from dorsal to ventral into the dorsal large **epimere,** or **somite,** comprising the segmental somites; the small intermediate **mesoderm,** or **mesomere,** which gives rise to the urogenital organs; and the long, unsegmented lateroventral **hypomere,** or **lateral plate** (Fig. 7-5C). The primitive coelom is subdivided accordingly into the dorsal **myocoel,** the intermediate **nephrocoel,** and the large coelom proper. The epimere separates from the ventral portions of the mesoderm, and its somites differentiate into the lateral **dermatome,** the intermediate **myotome,**[2] and the medial **sclerotome.** The mesomere gives rise to the segmented nephric and genital ridges. The hypomere forms the lateral **somatic layer** and the medial **visceral,** or **splanchnic, layer;** these two layers enclose the definitive **coelom** (Fig. 7-5D). The dermatomes expand under the ectoderm, lose their segmental organization, and differentiate into the dermis of the integument, including the dermal muscles (e.g., the arrector pili muscles of mammalian hair) and much of the dermal skeleton. The sclerotomes surround the notochord and neural tube. The sclerotomal cells split transversely, and each half condensate and migrate so as to be located between the original somites and eventually develop into the vertebrae. The myotomes grow ventrally between the dermatomes and the somatic layer of the hypomere (Fig. 7-5E). In gnathostome vertebrates, the myotomes subsequently are subdivided by the horizontal skeletogenous septum into the epaxial and hypaxial musculature (Fig. 7-5F); the hypaxial musculature also gives rise to the appendicular musculature. The visceral layer of the hypomere differentiates into the smooth visceral musculature of the gut and inner organs, the cardiac musculature of the heart, and the visceral serosa that covers the inner organs and forms the mesenteries suspending them in the coelomic cavities. The somatic layer of the hypomere develops into the parietal serosa that lines the coelomic cavities and provides the connective tissue associated with the appendicular musculature.

In the head, or cephalic region, of the embryo, the development of the skeletal musculature differs somewhat from that observed in the trunk region. Until a few years ago, the traditional understanding was based on the discovery of Edwin S. Goodrich (1918, 1930) that the paraxial mesoderm in the head of sharks differentiates first into eight distinct somites, although the somites that are adjacent to the expanding otic capsule of the chondrocranium are later resorbed. The somatic muscles that differentiate from the myotomes of the remaining cranial somites include the **extrinsic eye muscles** of the eyeball. The somatic **epibranchial** and **hypobranchial muscles,** which lie dorsal and ventral, respectively, to the gill, or branchial, region in the adult, differentiate from myotomes of the trunk, which grow cranially over and below the embryonic branchial region. The branchiomeric muscles first appear deeply between the visceral arches with which they are associated and next to the oropharyngeal cavity. Because of technical limitations, the embryonic origin could not be determined, and it was conjectured that the branchiomeric muscles develop from the splanchnic layer of the lateral plate mesoderm in the cephalic region. Accordingly, the branchiomeric muscles were classified as visceral muscles, comparable to those in the gut wall of the trunk, even though, unlike them, they consisted of skeletal muscles.

---

[2]See Footnote 3, Chapter 1.

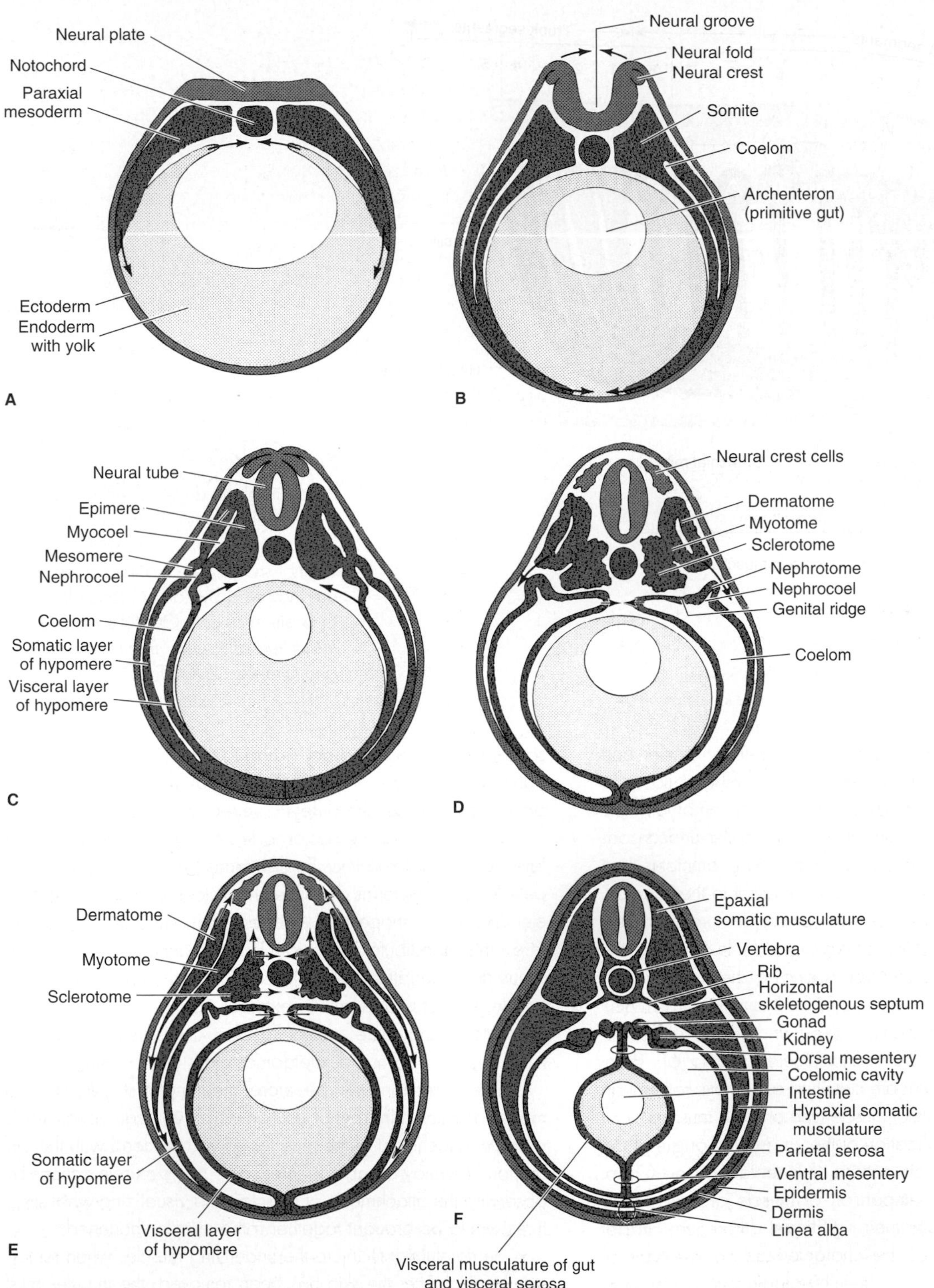

Figure 7-5

Diagrammatic cross sections through embryos with a moderate amount of yolk, such as those of amphibians, illustrating the embryonic differentiation of the mesoderm in the trunk region. Arrows indicate the direction of growth of embryonic tissues. **A,** Early stage of mesoderm formation. **B,** Early stage of coelom formation. **C,** Differentiation of the mesoderm into the epimere, mesomere, and hypomere. **D,** Differentiation of the epimere into the dermatome, myotome, and sclerotome. **E,** Differentiation of the dermatome into the dermis of the integument, of the myotome into the somatic musculature, and of the sclerotome into the axial skeleton. **F,** Final differentiation of the mesodermal organs and division of the somatic musculature into epaxial and hypaxial muscles. *(A–D, Redrawn with permission from M. Hildebrand, Analysis of Vertebrate Structure. Copyright ©1974 and 1982 by John Wiley & Sons, Inc.; **E** and **F**, modified from Hyman, 1942.)*

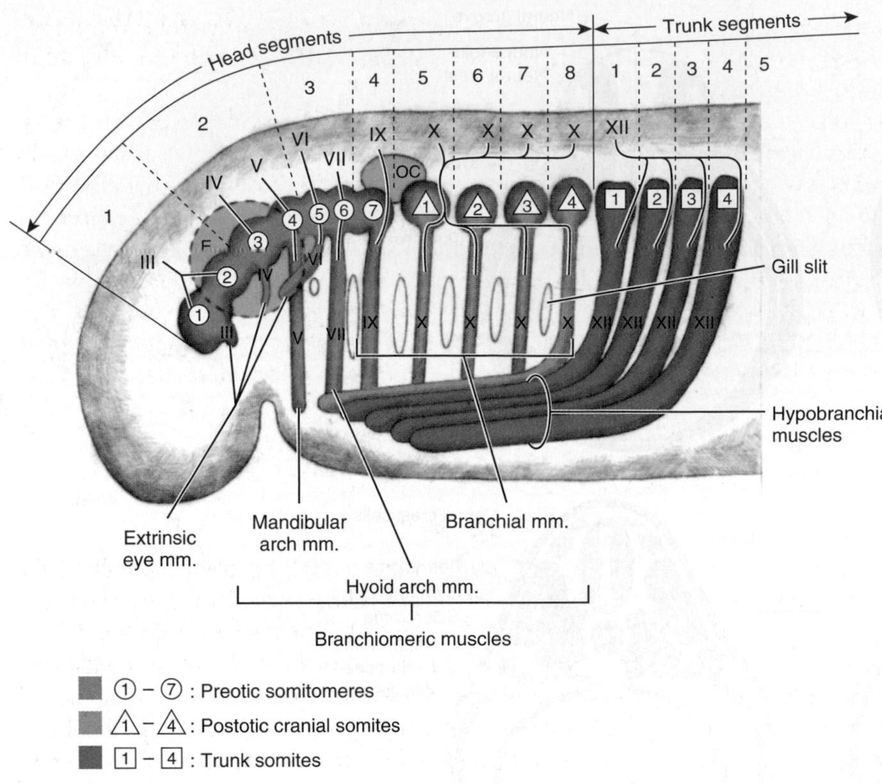

Figure 7-6
Diagram of the development of cranial muscles and nerves in an amniote vertebrate. The brain and spinal cord are not shown. The cranial nerves are indicated by Roman numerals. E, position of eye; OC, otic capsule. *(Modified after Liem, Bemis, Walker, and Grande, 2001.)*

In the 1980s, thanks to the work by Stephen Meier and Patrick Tam (1982), Drew Noden (1983, 1984), Nicole Le Douarin (1984), and Glenn Northcutt (1990), the embryonic differentiation of the head of amniotes became better understood. The head musculature is less clearly segmented in amniotes than in sharks, because the paraxial mesoderm rostral to the otic capsule differentiates into a series of seven poorly demarcated somitomeres instead of three distinct somites (Fig. 7-6). In amniotes, the first six somitomeres probably correspond to the three preotic somites of shark embryos. Furthermore, the branchiomeric musculature arises through migration of cells from the cranial somitomeres and somites into the branchiomeres (Fig. 7-6). Branchiomeric muscles, therefore, are classified as somatic rather than visceral, which fits the skeletal muscle type of these muscles.

This new conceptualization of the embryonic origin of the head muscles also fits their pattern of innervation (see Chapter 9). It facilitates the comparison of muscle groups across vertebrates classes despite their significant differences. In addition, muscle groups can be characterized by the type of their innervation and control, their histological structure, and their physiological attributes (Table 7-1).

The Study of Muscles

The anatomy of most muscle groups is described with reasonable completeness, although details of the intrinsic muscles of the hand and foot and the muscles of the tail, perineum, face, and ears have been omitted. As far as possible, the muscle descriptions are organized according to the classification outlined earlier and in Tables 7-1 and 7-2 so that muscles can be compared easily across the different vertebrate classes.

Muscle names are easier to learn if the Latin terms are related to, or translated into, English terms (see Word Roots in Appendix 1). Many terms describe characteristic features of a muscle, such as its shape (e.g., trapezius; rhomboideus); location (e.g., intermandibular, brachial); location and shape (e.g., latissimus dorsi, serratus ventralis); location and muscle fiber direction (e.g., rectus abdominis, external oblique); primary action (adductor mandibulae, tensor fasciae antebrachii); or attachment (e.g., coracohyoid, cleidomastoid).

The dissection will be more meaningful if you have a mounted skeleton in front of you on which you can visualize the points of attachment of the muscles as you proceed with the dissection. You may be able to infer some basic muscle actions by observing the attachments of muscles and visualizing what structures would be brought together if the muscle shortened.

Be careful not to injure the underlying muscles when removing the skin. After the skin has been removed, the muscles must be carefully separated from one another. This involves cleaning off the overlying connective tissue with forceps until you can see the orientation of the muscle fibers. Ordinarily, the muscle fibers of one particular muscle are held together by a sheath of connective tissue, the epimysium, and they all run in the same general direction between common attachments. The muscle fibers of adjacent muscles will be interconnected by a different sheath of connective tissue and will have different muscle fiber orientations and attachments. Such differences will provide you with clues for the identification of muscles. Separate muscles from each other by tearing the connective tissue between them with forceps, watching the muscle fiber direction as you do so. Do not use scissors or scalpels to separate muscles. If the muscles separate as units, you are dissecting correctly. If small bundles of muscle fibers are being exposed so that the result looks like a spoonful of canned tuna fish, you probably are damaging a single muscle.

It is best to expose and separate several muscles of a given region before attempting to identify them. Sometimes it will be necessary to bisect a superficial muscle to expose deeper muscles. In such cases, be sure to expose the entire muscle first by identifying and freeing its borders, which run parallel to the muscle fiber direction. Then only, cut across the middle of the muscle (not across one of its tendons or aponeuroses) at a right angle to the direction of its muscle fibers. Depending on the arrangement of the muscle fibers, you may have to cut in a curve to maintain the desired right angle to the muscle fibers. By doing this, you can reflect the bisected ends of the muscle to expose deep muscles. The bisected ends can easily be placed together again when you wish to review superficial muscles. If you need to cut through several layers of muscles, bisect them at different levels so that the various bisected ends cannot be confused.

FISHES

The muscles of *Squalus* are a good example of the condition of the musculature in ancestral piscine vertebrates. The groups of muscles can easily be recognized and studied in an adult because they have not lost their identity through migration, as they have to some extent in tetrapod vertebrates.

In order to minimize the damage to muscles and organs, follow the sequence of dissection steps indicated in Table 7-3.

When removing the skin from a shark specimen, keep in mind that the skin of sharks, unlike that of mammals, is firmly connected to the connective tissue that envelopes most muscles. The best technique may be to reflect the skin as far as possible and carefully scrape away any muscle fiber bundles from the internal surface of the skin using a scalpel. Alternatively, you can first remove the skin while leaving the connective tissue in place, and then free the muscles from superficial connective tissue.

Axial Muscles I

Muscles of the Trunk and Tail

The muscles of the trunk and tail represent the bulk of the musculature in fishes and belong to the axial group of somatic muscles (Tables 7-1 and 7-2). First remove a wide strip of skin from one side of the tail and then remove another wide strip from the same side of the trunk between the pectoral fin and the first dorsal fin (Table 7-3; Fig. 7-7). These strips should extend from the middorsal to the midventral lines of the body. It generally is easier to remove the skin from the dorsal half of the body and more challenging to do so from the ventralmost longitudinal bundle of the trunk hypaxial musculature (Fig. 7-7). If you are working carefully, you will be able to discern that the collagen fibers of the connective tissue underlying the skin are generally oriented obliquely and crosswise and, hence, are helically arranged around the body.

Notice that the trunk musculature consists of muscle segments, or **myomeres,** which develop from the embryonic myotomes. The myomeres are bent in a complex

Table 7-3 General Plan and Sequence of Dissections of *Squalus acanthias*

Left Side of Specimen	Right Side of Specimen
1. Remove transverse strips of skin on tail and trunk	**5.** Ampullae of Lorenzini; lateral line canals and pit organs (remove skin as needed)
2. Remove skin on head (ventral surface and left side to middorsal line)	**6.** Inner ear
3. Muscles of the head: Separate intermandibular and interhyoid muscles; expose constrictor and branchial muscles; separate cucullaris and epaxial muscles; open anterior cardinal sinus; separate branchial region from coracoarcual muscle; expose coracobranchial muscles	**9.** Cranial nerves VIII–X, occipital and hypobranchial nerves
4. Extrinsic eye muscles; eye bulb *in situ*	**11.** Blood vessels in the oropharyngeal cavity, including collector loops
7. Nose	
8. Cranial nerves 0–VII	
10. Open the oropharyngeal cavity	
12. Open the pleuroperitoneal cavity	
13. Open the pericardial cavity	

zigzag fashion and are separated from each other by connective tissue sheets, the **myosepta.** The myomeres can be divided, at the apices of the zigzags, into **longitudinal bundles.** A longitudinal connective tissue sheet, the **horizontal skeletogenous septum** (see Figs. 5-1 and 7-7), lies deep to the lateral line and separates the dorsal portions of the myomeres, which constitute the **epaxial musculature,** from the ventral portions, which constitute the **hypaxial musculature.** This division of the trunk musculature is found in all gnathostomes but is absent in the jawless fishes (see Fig. 2-1). Notice that the muscle fibers within the two somewhat darker longitudinal bundles directly dorsal and ventral to the horizontal skeletogenous septum are oriented longitudinally, whereas the muscle fibers in the other longitudinal bundles are oriented obliquely.

Each myomere is a complex entity, as can be seen in Figure 7-7**A.** The internal part of each V-like fold of a myomere is cone shaped and extends further cranially or caudally than the fold does at the body surface. Cutting a V-shaped notch out of the top half of several conical paper cups and then fitting them together can provide a model for the myomere arrangement. The cone-within-a-cone nature of the folds can be seen if you make a cross section of the tail. In cross section, the overlapping V-shaped folds of several myomeres appear as a series of nearly concentric rings (see Fig. 5-1). In the abdominal region, the midventral band of whitish connective tissue that ties together the myomeres of opposite sides is called the **linea alba.**

Notice that the longitudinally oriented muscle fibers near the horizontal skeletogenous septum are darker in color (see Fig. 5-1). They are called **red,** or **slow oxidative, fibers** because they contain considerable amounts of **myoglobin,** which facilitates the transfer of oxygen from the blood, and their metabolism is aerobic. They are used at normal cruising speeds, and their rate of contraction is only as fast, and the duration of action only as long, as their oxygen supply allows. They do not fatigue. When sudden bursts of speed are needed, the lighter-colored **white,** or **fast glycolytic, fibers** in the rest of the longitudinal bundles become active. They can contract very rapidly, but only for short periods of time because their metabolism is anaerobic. Eventually they fatigue and incur an oxygen debt. As a result, lactic acid accumulates, which is metabolized after the muscle fibers have ceased to contract. Myomere structure is closely related to swimming (Anatomy in Action 7-1).

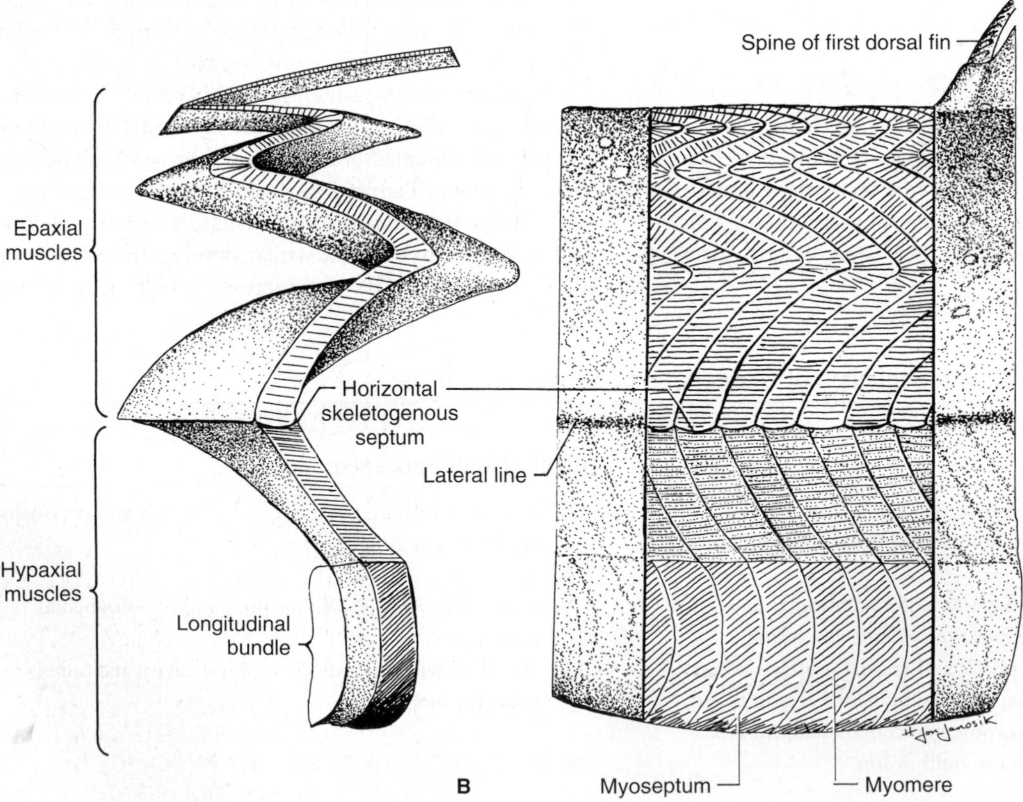

Figure 7-7

Lateral view of the trunk muscles between the pectoral and first dorsal fin of a spiny dogfish. Cranial is to the left of the figure. **A,** Diagram of an individual myomere consisting of several longitudinal bundles (only one is labeled as such). **B,** Several myomeres *in situ.*

Anatomy in Action 7-1 Swimming and Myomere Structure

In order to swim, a fish must generate a propulsive force and control its level within the water column. Most fish species propel themselves by lateral undulations of the trunk and tail. In sharks, the undulations involve primarily the caudal half of the body, and their amplitude increases as they approach the tip of the tail (Figs. **A** and **B**). As the trunk and tail push laterally and caudally against the water (F), the water pushes back against the body with a reaction force (R). This reaction force can be resolved into lateral (R_y) and forward (R_x) force components. The lateral force components at the various points of the tail largely cancel each other out, but the forward force components add up and propel the fish forward.

The trunk and tail undulations are generated by waves of contracting myomeres, which move caudally and alternate from side to side. Hence, the contraction waves on opposite sides of the body are out of phase with each other. Because the vertebral column cannot be compressed and, therefore, maintains its length, a contraction of myomeres on one side of the body pulls the myosepta together and bends the vertebral column and body, while on the opposite side of the body the myomeres relax and are extended. As a consequence of the zigzag shape of the myomeres, any given intervertebral joint is affected by several myomeres at the same time. As the most caudal myomere may just start to contract, the middle myomeres may reach their maximal tension, and the cranial myomeres may already start to relax.

As we have seen, the inextensible collagen fibers of the dermis envelop the body in a cross-wise helical pattern. This arrangement allows the connective tissue to expand on one side of the body and to be compressed on the other side, simply by changing the angles between the oblique collagen fibers that cross one another. Steven Wainwright, F. Vosburgh, and John Hebrank (1978) inferred that the dermis of the skin and the underlying connective tissue act as an "exotendon" in a manner similar to that of a tendon in the leg muscles of land vertebrates. On the convex side of a body curve, the skin and its dermis are stretched and store elastic energy. This energy is released through recoil when this body curve becomes concave so that the contracting myomeres need to shorten less.

The pectoral fins of sharks and most fishes are not used for generating thrust during swimming, in contrast to the limbs of swimming terrestrial mammals. The complex fin musculature enables the fins to assume a variety of positions and shapes and to control the direction of swimming. However, in fishes that have a boxlike, stiff body (e.g., box fishes), the pectoral fins have become the primary thrust generators. In electric fishes (e.g., knife fishes) that need to maintain an undisturbed electric field and, therefore, must not undulate their trunk and tail, the unpaired fins have been modified to form a single fin along the entire trunk and tail. These fishes can move forward and backward by undulating only this fin and keeping their trunk and tail still.

In addition to generating a propulsive force, a fish must be able to float at an appropriate depth in the water column without having to expend extra energy to prevent sinking. As described by Archimedes (ca. 287–212 B.C.), an object in a liquid experiences an upward buoyant force that is equal to the weight of the liquid that it displaces. If a fish had the same density (i.e., mass/volume) as the surrounding water, its weight would be equal to the buoyant force; therefore, it would neither sink nor rise within the water column. However, most tissues of fishes are denser than water. In order to be buoyant, a fish needs to compensate for its denser tissues with tissues or substances that are less dense than water. Actinopterygian fishes achieve buoyancy through their air-filled swim bladder. Cartilaginous fishes do not possess a swim bladder. They approximate buoyancy through their cartilaginous skeleton, which usually is less mineralized and, therefore, less dense than bone; by storing lipids, which are less dense than water, in their liver; and by maintaining a high concentration of trimethylamine oxide (TMAO), which is a very large molecule with a relatively low molecular weight, in their body fluids.

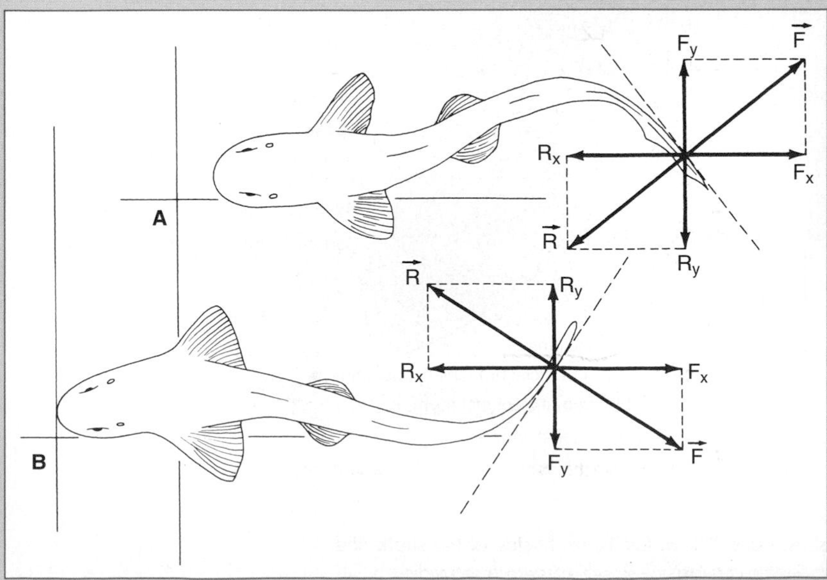

A, B Dorsal views of two stages of a swimming sequence of a spiny dogfish at an interval of 0.6 second. Graphic analysis of the forces exerted by, and acting on, one point of the tail captures a static condition when all forces are balanced and no movement takes place. In a real dynamic condition, the reaction forces by the water on the tail add up and push the fish forward. *F,* Force of the tail against the water; *Fx,* caudally directed component of the force F; F_y, laterally directed component of the force F; *R,* reaction force of the water against the tail; R_x, cranially directed component of the reaction force R; R_y, laterally directed component of the reaction force R. *(Modified from Gray, 1968; and Gans, 1974.)*

Epibranchial Muscles

As we discussed earlier, the **epibranchial musculature** represents a cranial extension of the epaxial musculature dorsal to the gill region (Fig. 7-8**A** and **B**). Remove the skin on the left half of the dorsal side of the head almost down to the gill slits (Table 7-3). Notice that the epibranchial muscles are segmented. Their deeper portions, which may be seen only later, attach to the caudal surface of the chondrocranium and vertebral column (Fig. 7-8**A**).

Hypobranchial Muscles

- Prehyoid muscles
  - Coracomandibular muscle
- Posthyoid muscles
  - Rectus cervicis muscle
    - Coracoarcual muscle
    - Coracohyoid muscle
- Coracobranchial muscles
- Also to be dissected: intermandibular and interhyoid muscles (branchiomeric muscles)

To see the hypobranchial muscles, which represent a cranial extension of the hypaxial musculature, remove the skin on the entire underside of the head from the pectoral fins rostrally to the tip of the lower jaw and laterally up to the gill slits (Table 7-3). A sheet of slightly oblique to transverse muscle fibers covers the triangular area between the mandibular cartilages (Fig. 7-8**B** and **C**). This muscle sheet is part of the branchiomeric musculature, but it must be studied now because you will need to bisect it to expose the underlying hypobranchial muscles.

The sheet of transverse musculature consists of two separate muscle layers, namely, the superficial intermandibular muscle and the deeper interhyoid muscle. The paired **intermandibular muscles** originate from the medial edges of the mandibular cartilages and meet along a collagenous midventral raphe. This raphe does not quite extend to the rostral end of the intermandibular muscles. Identify and free the most caudal muscle fiber bundles that originate from the surface of the adductor mandibulae muscle (see later) and represent the caudal border of the

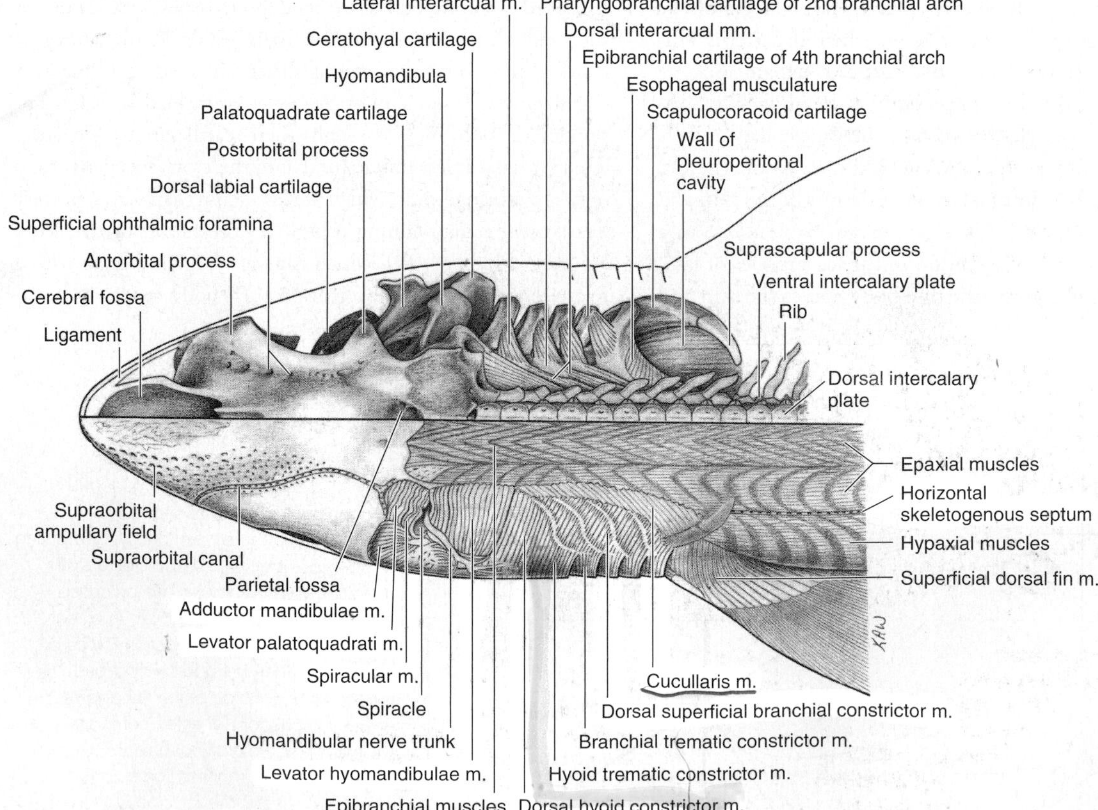

Figure 7-8

Muscles of the head and pectoral fins of *Squalus*. **A**, Dorsal view. On the right side of the shark, the gill rays, the pectoral fin, and all the muscles, except the interarcual muscles, were removed.

continued

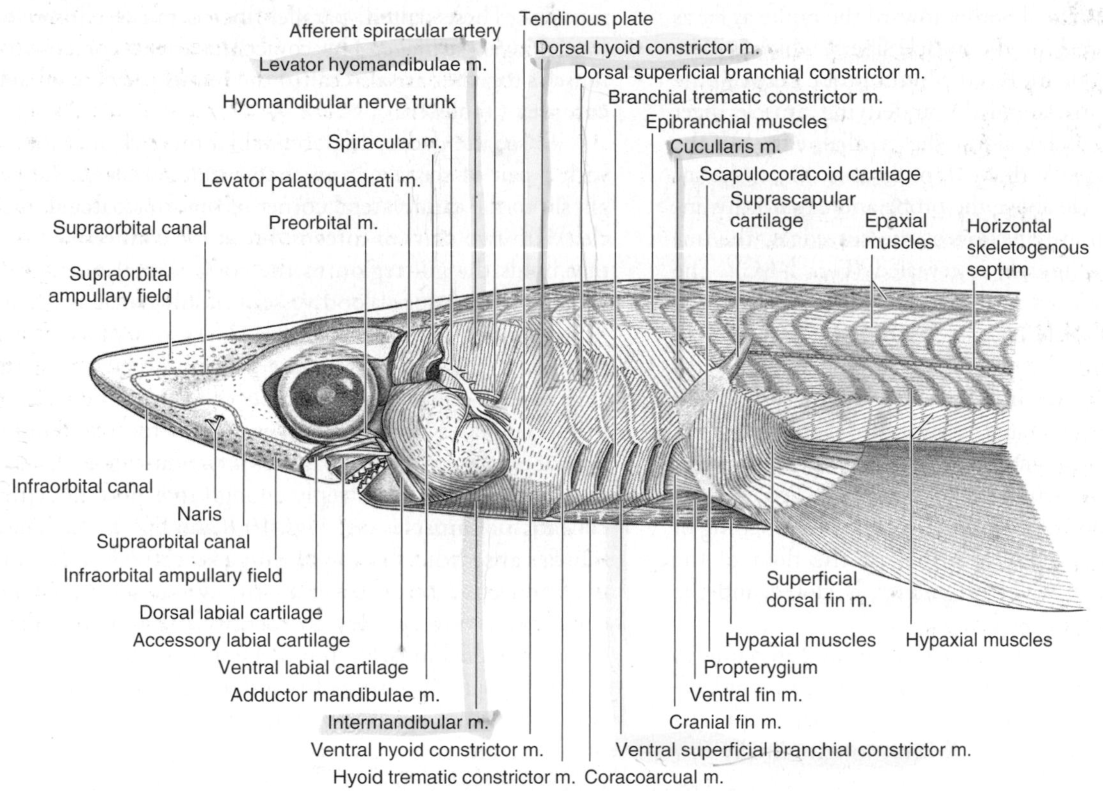

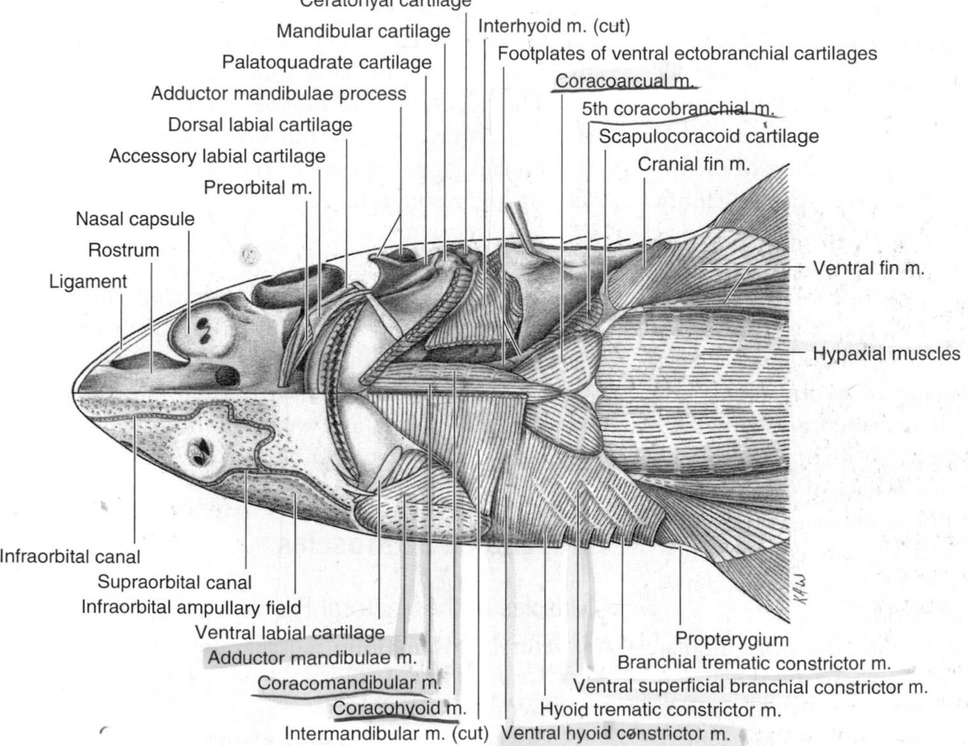

Figure 7-8 *continued*

B, Lateral view. C, Ventral view. The intermandibular and interhyoid muscles were bisected along the midline and reflected on the left side of the shark. The ventral hyoid constrictor and superficial branchial constrictor muscles were separated from the coracoarcual muscle and reflected.

muscle. Follow the caudal border toward the raphe as far as you can without breaking the muscle fibers, which will be for about 5 mm (1/5 inch). Insert a blunt probe between the intermandibular muscle and the underlying muscle sheet (the interhyoid muscle) along the medial edge of the mandibular cartilage. With a pair of scissors, bisect the intermandibular muscle above the probe and extend this incision rostrally. Upon reflection of the cut edges, the underlying interhyoid muscle is revealed (Fig. 7-8**C**). The **interhyoid muscle** does not extend as far rostrally as the intermandibular muscle because it originates laterally on the ceratohyal cartilages and inserts on the midventral raphe. Notice that the intermandibular and interhyoid muscles cannot be separated easily from each other near the raphe. The intermandibular and interhyoid muscles pull the mandibular cartilages and ceratohyal cartilages, respectively, closer toward the midline during the opening of the jaws and hyoid arch and help raise the floor of the oropharyngeal cavity when the mouth is closed and the branchial arches compressed.

Now make a longitudinal incision about 2 to 3 mm to the left of the raphe and about 2 mm deep. You will have cut deeply enough when you can see, upon separating the cut edges, an underlying muscle with longitudinally oriented muscle fibers, the coracomandibular muscle (see later). Lift and reflect the edge of the bisected muscle sheet to look at its internal surface. You will notice that the caudal border of this muscle sheet cannot be easily separated from ventral hyoid constrictor muscle (Fig. 7-8**B** and **C**) because there is extensive overcrossing of muscle fiber bundles between these muscles. Remove the connective tissue from the internal surface until the muscle fibers of the interhyoid muscle and its wide raphe are exposed. Identify the rostral border of the interhyoid muscle, as well as its mostly aponeurotic origin on the ceratohyal cartilage.

We can now proceed to study the hypobranchial muscles. The narrow midventral muscle, which is exposed after the intermandibular and interhyoid muscles have been reflected, is called the **coracomandibular muscle.** It is a prehyoid muscle (Table 7-2). It is embryonically paired, but it merges with its counterpart to form an unpaired muscle. It arises caudally from the surface of the coracoarcual muscles (Fig. 7-8**C**) and inserts on the symphysis of the mandibular cartilages. Bisect the coracomandibular muscle and reflect it. The dark or pink tissue mass dorsal to its rostral end is the **thyroid gland.**

The paired muscles deep to the coracomandibular muscle belong to the posthyoid portion of the hypobranchial musculature (Table 7-2). The caudal, paired **coracoarcual muscles** are subdivided by tendinous intersections and form a triangular midventral muscle mass. They originate from the scapulocoracoid cartilage and serve as the place of origin for the rostral **coracohyoid muscles.** These paired, parallel-fibered muscles insert on the basihyal cartilage. The coracoarcual and coracohyoid muscles together are also called the **rectus cervicis muscle complex** (Table 7-2).

To see the deeper hypobranchial muscles, make a cut with a pair of scissors from the ventral corner of the last gill slit to the caudolateral corner of the coracoarcual muscle. With two pairs of forceps, break the connective tissue that binds the gill region to the coracoarcual muscle. Be careful not to injure blood vessels unduly (such as the inferior jugular vein; see Chapter 11) as you proceed rostrally and caudally to separate the gill arches from the coracoarcual and fin muscles. You will also encounter cartilaginous plates (Fig. 7-8**C**); these are the footplates of the ventral ectobranchial cartilages, which you will see later in their entirety. Proceed deeply enough to expose the **coracobranchial muscles** (see Figs. 10-8 and Fig. 11-9). These muscles arise from the scapulocoracoid cartilage and from the connective tissue covering the dorsal surface of the coracoarcual muscle. Try to identify the five individual coracobranchial muscles, but make sure not to damage the reasonably large uninjected blood vessels that emerge between them (see Fig. 11-9). The most rostral, first coracobranchial muscle lies dorsal to the coracohyoid muscle, which you saw earlier, and inserts on the basihyal cartilage next to it. The second to fourth coracobranchial muscles insert on the three hypobranchial cartilages. The most caudal, fifth coracobranchial muscle is the largest of the five and inserts on the large caudal basibranchial cartilage.

The hypobranchial muscles are involved in opening the mouth (i.e., the jaw and hyoid arch) and expanding the oropharyngeal cavity (i.e., the branchial arches) during feeding and respiration.

Extrinsic Eye Muscles

The extrinsic muscles of the eye develop embryonically from the head somitomeres 1, 2, 3, and 5 (Table 7-1). They originate from the wall of the orbit and insert on the surface of the eyeball. They are responsible for the movements of the eyeball, and we will consider them when we study the eye (see Chapter 8).

Appendicular Muscles

Muscles of the Pectoral Fin

- Ventral division of fin musculature
  - Cranial fin muscle
  - Ventral fin muscle
- Dorsal division of fin musculature
  - Dorsal fin muscle
    - Superficial portion
    - Deep portion

The appendicular group of somatic muscles is relatively simple in most fishes compared to the limb musculature of tetrapods. In fishes, whose propulsive force for swimming is generated by lateral undulations of the trunk and tail (Anatomy in Action 7-1), the fins are used for stability, maneuverability, and breaking during swimming, and not for generating thrust. Remove the skin on the dorsal, ventral, and cranial surfaces of the base of the pectoral fin, and mobilize the fin by removing connective tissue. The **cranial fin muscle** originates from the lateral surface of the scapulocoracoid cartilage and inserts on the propterygium (Fig. 7-8**B** and **C**). It protracts the pectoral fin. The **ventral fin muscle** consists of two clearly identifiable portions and originates from the caudal process and caudal surface of the scapulocoracoid cartilage (see Fig. 4-5). It adducts the pectoral fin and makes it ventrally concave. The **dorsal fin muscle** consists of two portions. Its **superficial portion** originates from the connective tissue covering the dorsal longitudinal muscle bundle of the hypaxial musculature caudal to the scapulocoracoid cartilage. It pulls the pectoral fin dorsally and abducts it. The **deep portion** of the dorsal fin muscle can be seen only if you pull the mobilized fin cranially and look at the internal surface of the base of the fin. It originates along the caudal edge of the suprascapular process and inserts medially on the dorsal side of the pectoral fin. Together with the superficial portion, it flattens the pectoral fin.

Muscles of the Pelvic Fin

- Dorsal fin muscle
- Ventral fin muscle, with proximal and distal portions

Remove the skin from the base of the pelvic fin. In a female, the **dorsal fin muscle** arises from the surface of caudal trunk myomeres, from the iliac process of the pelvic fin girdle, and from the metapterygium of the pelvic fin. It inserts by slips onto the radial cartilages of the pelvic fin and abducts and flattens the pelvic fin. The **ventral fin muscle** is divided into proximal and distal portions. The **proximal portion** arises from the puboischiadic bar and inserts on the metapterygium. The **distal portion** arises from the metapterygium and inserts on the radial cartilages of the pelvic fin. The ventral fin muscle pulls the pelvic fin ventrally.

In a male, both the dorsal and ventral fin muscles extend onto the clasper as distinct small muscles. To see the ventral fin muscle, you need to mobilize and reflect, but not remove, the siphon, which is a long muscular sac that lies directly under the skin (see Fig. 12-8).

Axial Muscles II

Branchiomeric Muscles

- Mandibular muscles
  - Adductor mandibulae muscle
  - Levator palatoquadrati muscle
  - Spiracular muscle
  - Preorbital muscle
  - Intermandibular muscle (already dissected with the hypobranchial muscles)
- Hyoid muscles
  - Interhyoid muscle (already dissected with the hypobranchial muscles)
  - Hyoid constrictor muscle, with ventral and dorsal portions
  - Hyoid trematic constrictor muscle
  - Levator hyomandibulae muscle
- Branchial muscles
  - Superficial branchial constrictor muscles (ventral and dorsal portions)
  - Branchial trematic constrictor muscles
  - Cucullaris muscle
  - Interarcual muscles (lateral and dorsal portions)
  - Interbranchial muscles
  - Branchial adductor muscles

The branchiomeric muscles develop embryonically from somitomeres 4, 6, and 7 and from the first four postotic axial somites (Table 7-1). It is the most complex musculature in a fish.

Very carefully remove the skin from the side of the gill region, jaws, and spiracle, and from the caudal and ventral sides of the orbit. Be especially careful when removing the skin from the areas between the gill slits; in this region, the superficial musculature is very thin and is easily overlooked. In the spiracular region, remove the skin only from the surface of the head and leave the inside of the spiracle intact. Carefully remove connective tissue that may obscure the orientation of the muscle fiber bundles.

Beginning with the muscles of the mandibular arch (Table 7-2), you will discover the large **adductor mandibulae muscle** at the angle of the jaws (Fig. 7-8). It arises from the caudal part of the palatoquadrate cartilage and, especially, from the large adductor mandibulae process (see Figs. 4-4 and 4-5). It inserts on the mandibular cartilage and closes the jaws. Dorsal to the adductor mandibulae muscle and rostral to the spiracle is a muscle that arises from the side of the otic capsule and inserts on the palatoquadrate cartilage. Careful dissection will reveal that it consists of a rostral **levator palatoquadrati muscle** and a caudal **spiracular muscle.** The levator palatoquadrati muscle raises the palatoquadrate cartilage mainly when the jaws are being closed. The spiracular muscle is located on the cranial wall of the spiracle and closes the spiracle. The **preorbital muscle** will be seen later when the eye and cranial nerves are dissected (Figs. 7-8**B** and **C**, 8-7, 8-8, and 9-7). It originates from the ventral surface of the base of the rostrum of the chondrocranium and inserts on the caudal end of the palatoquadrate cartilage. It is responsible for the retraction of the chondrocranium when the jaws are being closed. The **intermandibular muscle,** the last of the mandibular arch muscles, was already observed when you dissected the hypobranchial muscles. Find it again.

To observe the muscles of the hyoid arch, first find the **interhyoid muscle** again. The **ventral hyoid constrictor muscle** is located caudal to the interhyoid muscle. The two muscles can be differentiated from each other only on the basis of their origins. The ventral hyoid constrictor muscle originates medially from a fascia and a tendinous intersection and inserts on a tendinous plate caudal to the adductor mandibulae muscle. The **dorsal hyoid constrictor muscle** lies above and behind the adductor mandibulae muscle. It inserts on the tendinous plate and originates dorsally from the surface of the cucullaris muscle (see later and Fig. 7-8**B**). The muscle fiber bundles that are located caudal to the tendinous plate and extend to the edge of the first gill slit form the **hyoid trematic constrictor muscle** (Fig. 7-8**B**). Observe the orientation of the muscle fiber bundles that attach to tendinous intersections, and confirm that they would close the flap of the gill slit if they contracted. The hyoid constrictor muscles compress the first gill chamber when the gill slit is closed and the mouth is opened (see Chapter 10 and Anatomy in Action 10-1 figures). The **levator hyomandibulae muscle** lies between the dorsal hyoid constrictor muscle and the spiracle. It arises from the surface of the epibranchial musculature and otic capsule and inserts mainly on the hyomandibular cartilage, but some muscle fiber bundles extend onto the palatoquadrate cartilage. The levator hyomandibulae muscle raises the hyomandibular cartilage when the mouth is being closed.

The remainder of the branchiomeric muscles are associated with the branchial arches. The four **dorsal** and **ventral superficial branchial constrictor muscles** follow the dorsal and ventral hyoid constrictor muscles and are separated from one another by tendinous intersections. The dorsal superficial branchial constrictor muscles originate from the surface of the cucullaris muscle (see later), and their ventral counterparts originate from the surface of the coracoarcual muscle. The muscle fiber bundles along the caudal edge of each gill slit form a **branchial trematic constrictor muscle.** The triangular muscle dorsal to the dorsal hyoid and superficial branchial constrictor muscles is the **cucullaris muscle,** which is partly homologous to the trapezius muscle of mammals (Table 7-2). To see it in its entirety, you need to separate it completely from the dorsal hyoid and dorsal superficial branchial constrictor muscles by breaking the connective tissue that anchors the dorsal hyoid and superficial branchial constrictor muscles to the surface of the cucullaris muscle (Fig. 7-8**B**). As you reflect the gill region, notice the flat cartilaginous plates on the internal surface of the dorsal superficial branchial constrictor muscles; these are the footplates of the dorsal ectobranchial cartilages, which you will see in their entirety later when you dissect the interbranchial septa. Observe that the cucullaris muscle arises from the surface of the epibranchial musculature and inserts on the scapulocoracoid cartilage and the epibranchial cartilage of the fifth branchial arch. The cucullaris muscle compresses the anterior cardinal sinus (see below) when the mouth is being opened and the branchial arches are unfolded (see Chapter 10 and Anatomy in Action 10-1).

To see the deep muscles of the branchial arches, separate the cucullaris and levator hyomandibulae muscles from the epibranchial musculature by breaking the connective tissue that binds them together. Push the branchiomeric muscles and the branchial region ventrally, and you will expose a large longitudinal cavity, the anterior cardinal sinus, which is a part of the venous system (see Chapter 11 and Fig. 11-5). Clean out any coagulated blood and remove the membrane that covers the branchial arches and their muscles. Push the epibranchial musculature as much upward and away from the branchial region as necessary to expose completely the dorsal surfaces of the branchial arches (Fig. 7-8**C**). Several nerves cross this surface; mobilize and push them out of the way to clear the view, but do not cut them. Medially, you will see the three ribbon-like **dorsal interarcual muscles** extending between the dorsal surfaces of the pharyngobranchial cartilages. Each originates from the caudal end of one pharyngobranchial cartilage and inserts on the cranial end of the next pharyngobranchial cartilage in front. The dorsal interarcual muscles adjust the distances between the pharyngobranchial cartilages when the mouth is opened and the branchial arches are expanded. The **lateral interarcual muscles** are shorter and broader, and they are located more laterally. They extend between the epibranchial and pharyngobranchial cartilages of each of the first four branchial arches. They raise the epibranchial cartilages when the branchial arches are expanded. The fifth branchial arch does not possess interarcual muscles because its pharyngobranchial cartilage is fused to the epibranchial cartilage and to the pharyngobranchial cartilage of the fourth branchial arch (see Figs. 4-4 and 4-5).

To see the intrinsic gill muscles, open all the branchial pouches by cutting through the middle of each superficial branchial constrictor muscle parallel to the tendinous intersections. Open them as wide as you can without cutting through the edges of the internal gill slits, which are communicating with the oropharyngeal cavity. In this way, you can mobilize and inspect the individual interbranchial septa (Fig. 7-9). Interbranchial septa are the plates of mostly soft tissue that are attached to the individual branchial arches and are supported by the gill rays radiat-

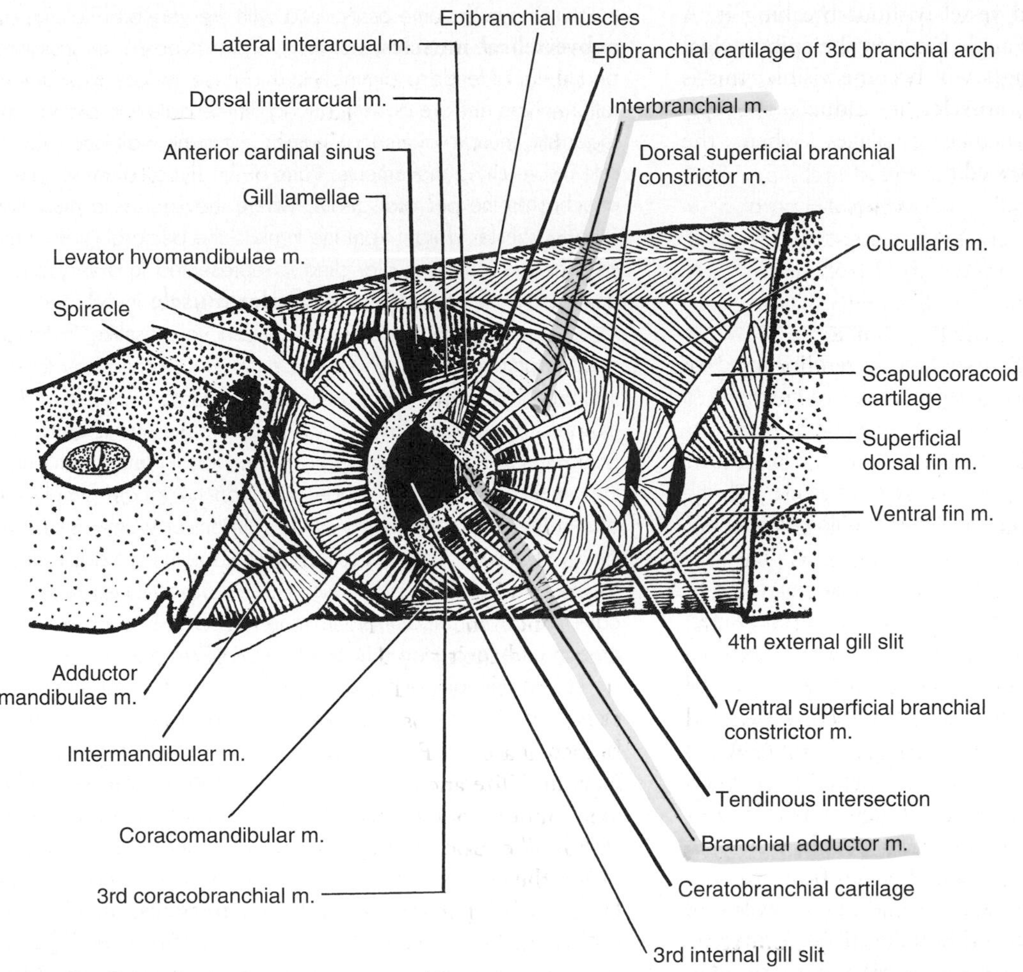

Figure 7-9
Diagrammatic lateral view of the deeper branchiomeric muscles of *Squalus*. The third definitive gill pouch has been cut and spread open. The epithelial gill lamellae were stripped from the caudal wall of the third gill pouch to reveal the interbranchial muscle and its tendinous intersections attaching to gill rays.

ing from the epibranchial and ceratobranchial cartilages of the branchial arches. Recall that the fifth branchial arch lacks gill rays, an interbranchial septum, and the associated muscles. The surfaces of the interbranchial septa are covered with a thin epithelium and a mass of capillaries forming the respiratory gill lamellae (see Fig. 10-10). Scrape away the gill lamellae and epithelium from both surfaces of the third interbranchial septum between the second and third gill slit. On the rostral (i.e., posttrematic) surface, you will discover circularly arranged muscle fiber bundles, which form the **interbranchial muscle.** This muscle represents a deep part of the branchial constrictor musculature and is subdivided by radial tendinous intersections. On the caudal (i.e., pretrematic) surface of the interbranchial septum, you will see that the cartilaginous gill rays and their connective tissue sheaths are attached to these connective tissue intersections. The interbranchial muscle straightens the interbranchial septum when the branchial pouches are compressed (see Anatomy in Action 10-1). Palpate the interbranchial septum along its outer periphery dorsal and ventral of the gill slit, and you will feel very thin cartilaginous rods, the **dorsal** and **ventral ectobranchial cartilages.** You can remove some tissue to see at least parts of them. You have already seen their footplates, which anchor them in place. The ectobranchial cartilages provide a place of attachment for the interbranchial muscles and impart stiffness to the surface of the branchial region. Return to the caudal surface of the third branchial arch, palpate and find the joint between the epibranchial and ceratobranchial cartilages, and peel away the whitish epithelium. If the specimen is injected, a red blood vessel will lie along the peripheral surface of the branchial arch. Gently mobilize and push away this blood vessel without breaking it. A short muscle spanning the angle between the epibranchial and ceratobranchial cartilages will become visible; this is the **branchial adductor muscle.** It adducts the epibranchial and ceratobranchial cartilages when the branchial arches are compressed.

AMPHIBIANS

The many morphological changes that occur in the muscular system during the evolution from a fishlike vertebrate to a tetrapod are correlated with changes in the mode of body support and locomotion, with the acquisition of a neck that allows the head to be moved independently from the trunk, with the adoption of air breathing, and with the acquisition of a mobile tongue to capture and swallow food.

Lateral undulations of the trunk and tail remain part of the locomotory motions of most amphibians, except frogs and toads, and most reptiles, except turtles and tortoises. They swing forward the limbs of alternate sides while the feet that remain planted on the ground prevent the trunk from sliding back and, thereby, ensure a forward displacement of the body (Fig. 7-10**A**). In accordance with their new role for locomotion on land, the limbs of amphibians have become slender, with a more differentiated musculature and complex hands or feet at their distal ends (Table 7-2).

Although the trunk muscles are no longer solely responsible for locomotion in amphibians, as they are in fishlike vertebrates, they remain significant and have taken on new tasks. The epaxial trunk muscles brace and tie together the elements of the vertebral column so that it can serve as a stable yet flexible axis for the attachment of the limbs and the suspension of the internal organs. The epaxial muscles also help resist torsional forces on the trunk during locomotion on land (Fig. 7-10**B**). Superficial portions of the epaxial muscles tend to fuse into long muscle bundles that cross several body segments, but the deeper portions remain segmented. Deeper parts of the hypaxial trunk musculature have become associated with the vertebral column as **subvertebral muscles,** but most of the hypaxial musculature on the flanks tends to diminish in thickness, to lose its segmentation where ribs are absent, and to differentiate into broad, antagonistic muscle layers that support the trunk wall and play a role in breathing movements. Parts of the hypaxial musculature attach onto the pectoral girdle, where they act as a muscular sling to transfer weight from the trunk to the pectoral girdle and appendage. This muscular sling is represented in a very basic configuration by the **thoraciscapular muscle** in *Necturus.*

The hypobranchial muscles remain very fishlike in larval aquatic amphibians, such as *Necturus,* but in most fully terrestrial amphibians and reptiles, they evolved into complex tongue muscles. The tongue has become more mobile and is generally responsible for capturing prey and for food transport through the mouth and pharynx, since prey can no longer simply be drawn into the mouth cavity with a current of water as in fishlike vertebrates. The hypobranchial musculature of adult metamorphosed terrestrial amphibians also has acquired a significant role for the breathing movements, but it is less crucial for opening the jaws than it is in fishlike vertebrates.

In amphibians, the cranial visceral arches and their muscles have become significantly modified. The upper jaw, or palatoquadrate bone, has become fused to the cranium. The movements of the skull and lower jaw alone are responsible for opening and closing the mouth. Whereas the jaws are closed by mandibular muscles in amphibians as in fishes, the mandible of amphibians (and of reptiles and birds) is opened primarily by the paired **depressor mandibulae muscles,** which have evolved from the musculature of the hyoid arch. Most of the hyoid musculature has lost its connection with the elements of the hyoid arch, which is moved primarily by hypobranchial muscles in amphibians. The rest of the hyoid muscles of amphibians have become associated with the auditory ossicles or have become rather superficial, forming a **sphincter colli muscle** over the neck. The more caudal visceral arches and their muscles are reduced as a consequence of the shift from branchial to pulmonary respiration in metamorphosed terrestrial amphibians. Some caudal branchiomeric muscles become associated

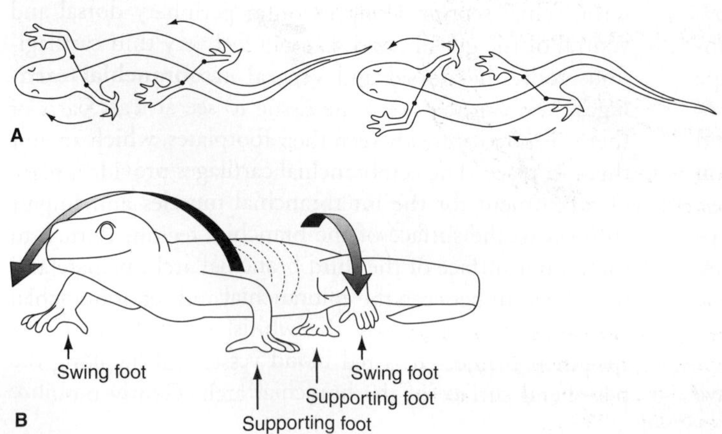

Figure 7-10
Terrestrial locomotion of a urodele amphibian (e.g., a newt or salamander) as a model of the locomotion of early tetrapods. **A,** Diagrammatic dorsal view of two stages in a walking sequence of a salamander. The orientation of the pectoral and pelvic girdle relative to the vertebral column does not change; therefore, the lateral undulations of the trunk move the swing feet forward while the supporting feet prevent the body from sliding backward. **B,** Oblique frontolateral view of a walking salamander. As the diagonally opposite front and hind legs support the body, the trunk is subjected to torsional stresses. *(From Liem, Bemis, Walker, and Grande, 2001; after Carrier, 1993.)*

with the larynx, which is a new organ that has evolved in conjunction with air breathing. The cucullaris muscle still is fishlike in larval aquatic amphibians, such as *Necturus,* but it enlarges and differentiates as the head and pectoral girdle movements become more complex in metamorphosed terrestrial amphibians (and in reptiles).

Necturus is an adequate model for an early tetrapod condition. It is, however, a permanently aquatic paedomorphic amphibian and, hence, possesses many fishlike attributes due to adaptation to its aquatic environment, such as the external gills and a more extensive branchial apparatus than would be found in a terrestrial metamorphosed adult amphibian.

Axial Muscles I

Muscles of the Trunk

- Epaxial muscles
  - Dorsalis trunci muscle
  - Interspinal muscles
- Hypaxial muscles
  - External oblique muscle
  - Internal oblique muscle
  - Rectus abdominis muscle
  - Transversus abdominis muscle
  - Subvertebral muscle (to be seen later with the iliopsoas muscle complex)

Confine your study of the muscles of *Necturus* to one side of the body, preferably the side that has not been cut open to inject the circulatory system (Table 7-4). From the middle of the trunk, remove a strip of skin that is about 5 cm long and extends from the middorsal to the midventral line. Clean off any connective tissue that may obscure the view of muscles, and note that the trunk musculature consists of **myomeres** that are separated by **myosepta** and divided into **epaxial** and **hypaxial portions.** The myomeres are not folded in a complex manner as they are in fishes.

Although the muscle fiber bundles of the epaxial musculature remain segmented by myosepta, the muscle fiber bundles on the surface tend to fuse so that a longitudinal muscle, the **dorsalis trunci muscle,** is formed (Figs. 7-11**C**

Table 7-4 General Plan and Sequence of Dissections of *Necturus maculosus*

Intact Side of Specimen	Side with Incision Through Trunk
1. Remove the skin and dissect the muscles	**4.** Open the peritoneal cavity, bisect the pectoral muscle, and open the pericardial cavity
2. Cut through angle of mouth and side of the pharynx; leave external gills on floor of the mouth	**5.** Dissect branchial blood vessels
3. Study the nose and cut open the nasal cavity	

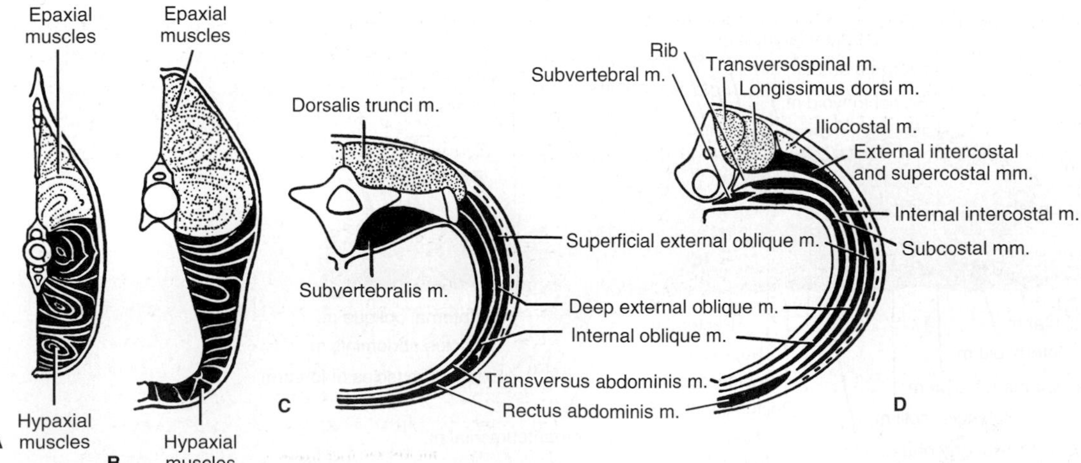

Figure 7-11
Diagrammatic transverse sections through the trunk of various vertebrates showing the evolutionary differentiation of the trunk musculature. The epaxial muscles are stippled; the hypaxial muscles are shown in black. **A,** Tail of a shark. **B,** Trunk of a shark. **C,** Trunk of a urodele amphibian. **D,** Trunk of a lizardlike reptile. *(From Romer and Parsons, 1986; mainly after Nishi, 1938.)*

and 7-12). You can see some of the deeper epaxial muscles by removing some muscle fiber bundles next to the mid-dorsal line with a pair of forceps. You will expose the tops of the vertebral arches and a series of short **interspinal muscles.** Each interspinal muscle arises from the edge of a caudal zygapophysis of one vertebra and inserts on the neural arch of the vertebra caudal to it.

Examine the hypaxial musculature and note that most of its superficial muscle fibers are obliquely oriented and run from craniodorsal to caudoventral. These constitute the **external oblique muscle.** The muscle fiber bundles next to the midventral line are oriented longitudinally and form the **rectus abdominis muscle.** This muscle is less differentiated in *Necturus* than it is in an metamorphosed terrestrial amphibian. Carefully remove the muscle fiber bundles of the external oblique muscle from two myomeres to expose the underlying **internal oblique muscle.** Its muscle fiber bundles are oriented obliquely but crosswise to those of the overlying muscle, that is, from caudodorsal to cranioventral. Midventrally, the muscle fiber bundles assume a longitudinal orientation and become part of the rectus abdominis muscle. Remove now the muscle fiber bundles of the internal oblique muscle in one of the two myomeres to uncover a third layer of the hypaxial musculature, the **transversus abdominis muscle.** Its muscle fiber bundles are oriented more transversely than those of the overlying muscle layers. If you remove some of its muscle fiber bundles, you will reach the translucent parietal serosa that lines the body cavity. The **subvertebral muscle,** another hypaxial muscle, lies on the ventral side of the vertebral bodies and transverse processes (Fig. 7-11**C**). You will see it when you observe the abdominal viscera (see Chapter 10). The subvertebral muscle interconnects the vertebrae and inserts on the individual centra and transverse processes. Several other hypaxial muscles attach onto the pectoral and pelvic girdles and will be seen later.

Hypobranchial Muscles

- Prehyoid muscles
  - Geniohyoid muscle
  - Genioglossus muscle
- Posthyoid muscles
  - Rectus cervicis muscle
  - Omoarcual muscles (to be seen with the pectoral muscles)
  - Pectoriscapular muscle (to be seen with the pectoral muscles)
- Other muscles (branchiomeric muscles)
  - Intermandibular muscle
  - Interhyoid muscle
  - Sphincter colli muscle

Remove the skin from the entire underside of the head and neck as far caudally as the pectoral region. To see the hypobranchial muscles, you will first have to dissect and observe three superficial branchiomeric muscles (Figs. 7-12 and 7-13). These muscles form sheets with a roughly transverse muscle fiber orientation and insert on a midventral raphe on the underside of the head and neck. The most rostral of these muscle sheets is the **intermandibular muscle.** It originates along the medial edges of the lower jaw.

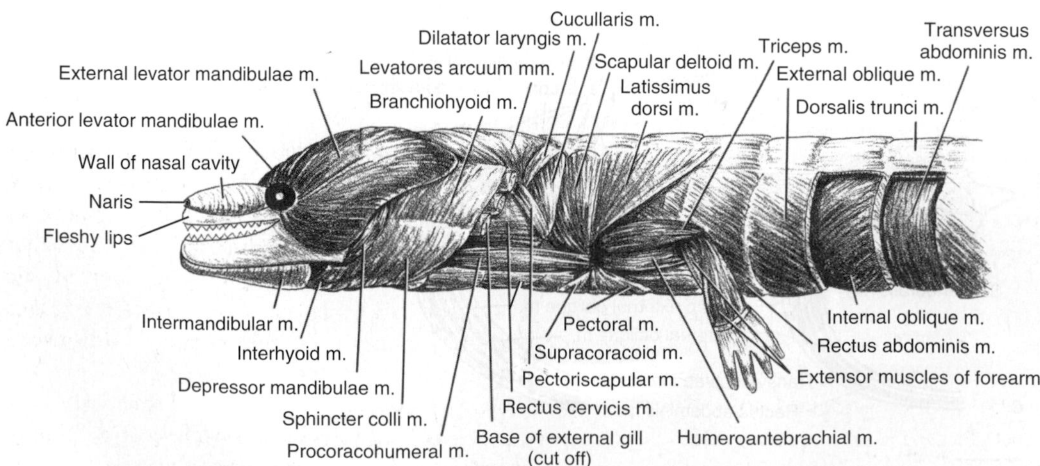

Figure 7-12
Lateral view of the head, trunk, and pectoral appendicular musculature of *Necturus*. The external oblique muscle is removed from one myomere to reveal the internal oblique muscle. In the next myomere, the internal oblique muscle is also removed to reveal the transversus abdominis muscle.

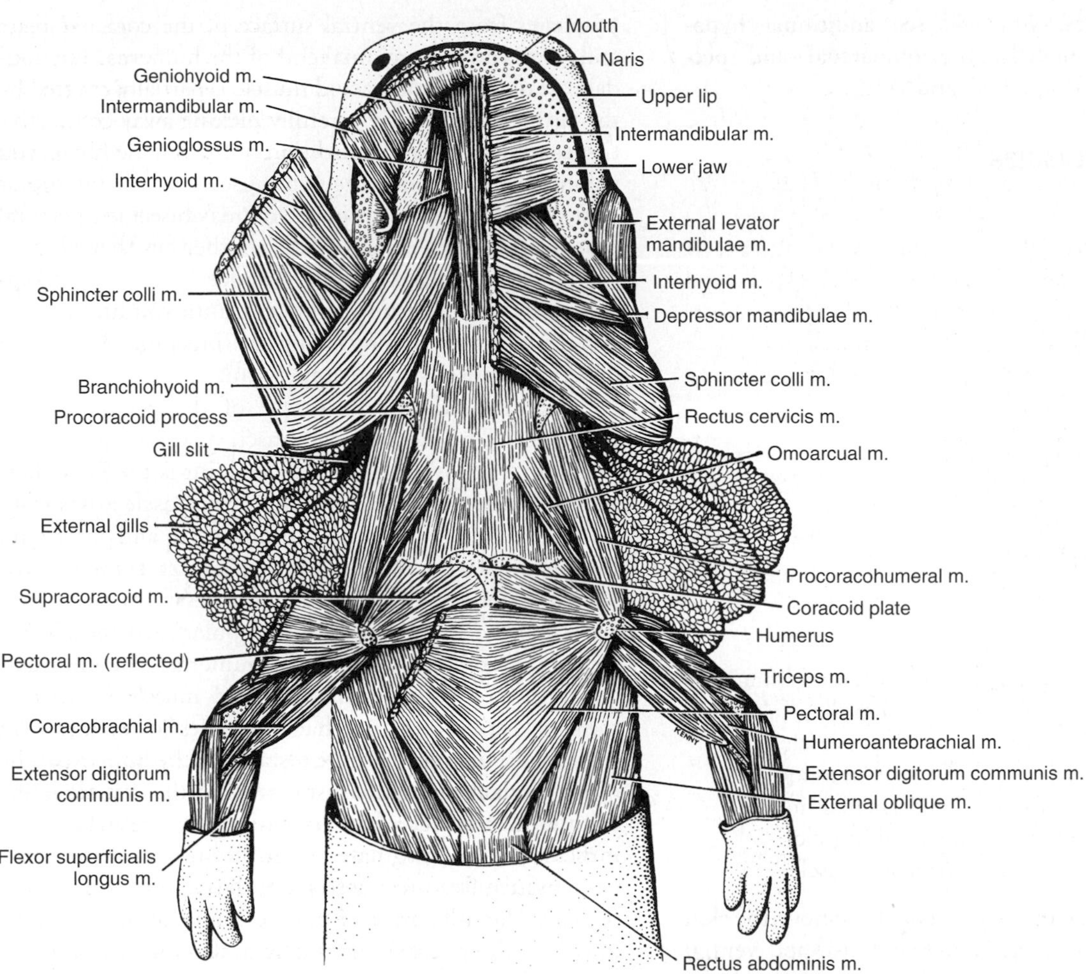

Figure 7-13
Ventral view of the head, throat, neck, and pectoral appendicular musculature of *Necturus*. Several muscles on the right side were bisected and reflected to reveal the underlying muscles.

Next, more caudal, is the **interhyoid muscle,** whose rostral end is partly covered by the intermandibular muscle. The interhyoid muscle arises from the hyoid arch. The most caudal muscle is the **sphincter colli muscle,** which sometimes is considered a part of the interhyoid muscle. It arises from a distinct connective tissue sheet that covers the branchiohyoid muscle (see later and Figs. 7-12 and 7-13), a large muscle caudal to the angle of the mouth. Bisect these three muscles slightly to one side of the midventral line to preserve the midventral raphe. A pair of midventral longitudinal muscles will be seen extending caudally from the symphysis of the lower jaw. These are the **geniohyoid muscles,** which represent the bulk of the prehyoid portion of the hypobranchial musculature (Table 7-2). Their caudal ends insert on the caudal portion of the hyoid apparatus (i.e., basibranchial cartilage 2). Deep muscle fibers that are derived from the geniohyoid muscle form the **genioglossus muscle,** a muscle that has evolved in connection with the appearance of a mobile tongue in amphibians. It can be seen by mobilizing and pushing away the geniohyoid muscle or by bisecting the geniohyoid muscle and reflecting its ends. The genioglossus muscle appears as a thin layer of essentially longitudinal muscle fibers that arise on the chin, extend caudally in the floor of the mouth, and insert on a transverse fold of the mucous membrane at the base of the tongue. Its lateral portions are developed best.

Most of the posthyoid hypobranchial musculature is represented by the **rectus cervicis muscle,** a wide muscle on the underside and lower sides of the neck. It extends caudad from the hyoid apparatus toward the pectoral girdle, where it is continuous caudally with the rectus abdominis muscle (see earlier). Note that it has tendinous intersections as a trace of segmentation. When you dissect the pectoral muscles, you will see additional hypobranchial muscles, namely, the omoarcual and pectoriscapular muscles (Figs. 7-12 and 7-13).

Appendicular Muscles

Pectoral Muscles

- Ventral appendicular muscles
  - Pectoral muscle
  - Supracoracoid muscle
  - Coracoradial muscle
  - Humeroantebrachial muscle
  - Coracobrachial muscle
  - Flexor muscles of the forearm, wrist, and fingers
- Dorsal appendicular muscles
  - Latissimus dorsi muscle
  - Scapular deltoid muscle
  - Procoracohumeral muscle
  - Subcoracoscapular muscle
  - Triceps muscle
  - Extensor muscles of the forearm, wrist, and fingers
- Other muscles acting on the pectoral girdle (Table 7-2)
  - Omoarcual muscle (a hypobranchial muscle)
  - Cucullaris muscle (a branchiomeric muscle)
  - Pectoriscapular muscle (a hypobranchial muscle)
  - Thoraciscapular muscle (a hypaxial muscle)
  - Levator scapulae muscle (a hypaxial muscle)

On the intact side of your specimen, remove the skin from the pectoral appendage and the lateral and ventral surfaces of the trunk in the pectoral region, and cut off the external gills. Be careful, because some muscles may adhere to the skin. The **procoracohumeral muscle** is the band of muscle lateral to the rectus cervicis muscle (Fig. 7-13). It arises from the external surface of the procoracoid process of the pectoral girdle and inserts on the proximal end of the humerus. Lift the medial border of the procoracohumeral muscle, and you will discover some muscle slips, the **omoarcual muscle,** which belong to the hypobranchial muscle group. One bandlike slip originates from the lateral surface of the rectus cervicis muscle and inserts on the caudal end of the procoracoid process. The other, short slip originates from the humerus next to the origin of the procoracohumeral muscle and inserts on the caudal end of the procoracoid process. A large, fan-shaped muscle, the **pectoral muscle,** will be seen on the ventral surface (Fig. 7-13). It arises from the midventral linea alba and hypaxial musculature and inserts on the proximal end of the humerus next to the procoracohumeral muscle. At the cranial end of the linea alba, you will discover the transparent or translucent, crescent-shaped cartilaginous edges of the medially overlapping coracoid plates. You will then notice the fan-shaped **supracoracoid muscle,** which originates from the ventral surface of the coracoid plate and inserts on the proximal end of the humerus. The caudal part of the supracoracoid muscle is partially covered by the pectoral muscle. By carefully picking away connective tissue and watching the direction of the muscle fibers, you will be able to separate the two muscles. To see the supracoracoid muscle in its entirety, you may bisect the pectoral muscle and reflect its ends. Still another fan-shaped muscle, the **coracoradial muscle,** lies deep to the supracoracoid muscle, but do not look for it until you find its tendon of insertion passing along the brachium when you dissect the muscles of the upper arm.

Examine the lateral surface of the pectoral girdle (Fig. 7-12). The most caudal muscle of a series of muscles that converge toward the shoulder joint is the large, fan-shaped **latissimus dorsi muscle.** This muscle arises from the surface of the dorsalis trunci muscle and inserts on the proximal end of the humerus. Next cranial is the **scapular deltoid muscle,** which arises from the lateral surface of the scapula and suprascapular cartilage and inserts on the proximal end of the humerus. The most cranial portion of the scapular deltoid muscle sometimes forms a distinct portion that originates lower down on the scapula and inserts more distally on the humerus. The relatively narrow muscle just cranial to the scapula is the **cucullaris muscle,** which is part of the branchiomeric musculature but originates from the surface of the epaxial musculature and inserts on the dorsal end of the scapula. The still narrower muscle cranial to the cucullaris muscle is the **pectoriscapular muscle,** which is a hypobranchial muscle that originates from the lateral border of the rectus cervicis muscle on the lateral side of the neck and inserts onto the distal end of the scapula.

Reflect the origin of the latissimus dorsi muscle and carefully reflect the caudal edge of the scapula cranially. You will see a band of muscle extending from the hypaxial musculature cranially to insert on the medial side of the dorsal border of the scapula. This **thoraciscapular muscle** belongs to the hypaxial muscle group (Fig. 7-14). A narrower band of muscle, the **levator scapulae muscle,** may also be seen extending cranially from its insertion on the dorsal border of the scapula in front of the insertion of the thoraciscapular muscle. You will obtain a better view of it by detaching the origin of the cucullaris muscle from the epaxial muscle surface and reflecting it. The levator scapulae muscle originates cranially from the occipital region of the skull. In metamorphosed terrestrial amphibians, this muscle becomes the **opercular muscle,** which is associated with the ear (see Anatomy in Action 8-3). Bisect the pectoral muscle, if this has not already been done, and bisect also the latissimus dorsi muscle. Pull the arm forward so you can see the medial side of the pectoral girdle. The sternal cartilages referred to in Chapter 5 can now be seen.

Clean away blood vessels and nerves dorsal to the joint with the humerus, and you will find a small muscle, the **subcoracoscapular muscle,** which arises from the caudal edge of the scapula and coracoid plate and inserts on the humerus. It passes between the two heads of the triceps muscle (see later and Fig. 7-14).

The upper arm is covered for the most part by three muscles. On the ventrolateral surface of the upper arm, you will find the **humeroantebrachial muscle** arising from the humerus and inserting on the proximal end of the radius (Figs. 7-12 and 7-13). On the ventromedial surface, the larger **coracobrachial muscle** is occasionally divided into a **brevis portion** and a **longus portion.** It arises from the coracoid plate and inserts on the shaft and distal end of the humerus. The tendon of the **coracoradial muscle,** which was mentioned earlier, can be found between the coracobrachial and humeroantebrachial muscles. After you find the tendon, trace it medially to the belly of its muscle, which lies on the ventral surface at the coracoid plate deep to the supracoracoid muscle. The **triceps muscle** covers the dorsal surface of the upper arm. Its name is derived from the comparable but three-headed muscle in the human being, but in *Necturus* only two heads are readily distinguished. They are separated proximally by the inserting subcoracoscapular muscle (see earlier and Fig. 7-14). The **coracoid head** is located on the medial side of the upper arm, and the **humeral head** is located on the dorsolateral side. The triceps muscle arises from the pectoral girdle and humerus. Both heads converge and unite, forming a common tendon that passes over the elbow and inserts on the proximal end of the ulna. The triceps muscle extends the elbow joint.

Clean the fascia from the surface of the forearm and note that the muscles are aggregated into two groups: the extensor and the flexor muscles. The wide **extensor digitorum communis muscle** covers the dorsal surface of the forearm and breaks up distally into four translucent tendons that insert on the proximal ends of each terminal phalanx. It originates from the distal end of the humerus and extends the wrist and digits. Medial to the extensor digitorum communis muscle lies the **extensor antebrachii et carpi radialis muscle.** It originates on the distal end of the humerus and inserts along the radius and on the radial carpal bones. On the opposite side of the forearm lies the **extensor antebrachii et carpi ulnaris muscle,** which also originates on the distal end of the humerus but inserts along the ulna and on the ulnar carpal bones. Various shorter extensor muscles lie deep to these long extensor muscles and along the basal phalanges of the digits. On the ventral side of the forearm, the large **flexor superficialis longus muscle** originates from the distal end of the humerus. It turns into a translucent tendon plate at the level of the wrist and breaks up into four tendons that insert at the bases of the terminal phalanges of the digits. It flexes the digits and the wrist. Medial to the flexor superficialis longus muscle lies the **flexor antebrachii et**

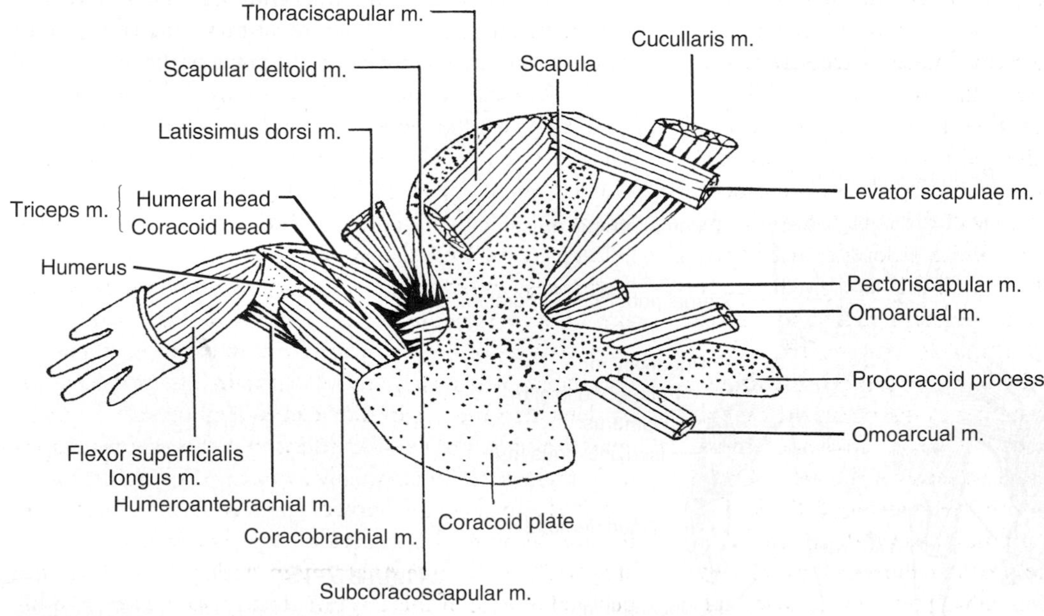

Figure 7-14

Medial view of the muscles of the isolated left pectoral girdle and limb of *Necturus*. Cranial is to the right of the figure.

carpi radialis muscle, which arises from the distal end of the humerus and inserts along the radius and on the radial carpal bones. The **flexor antebrachii et carpi ulnaris muscle** is located on the opposite side of the forearm. It also arises from the distal end of the humerus but inserts along the ulna and on the ulnar carpal bones. Various shorter flexor muscles lie deep to the long flexor muscles and along the basal phalanges of the digits.

Pelvic Muscles

- Ventral appendicular muscles
  - Pubotibial muscle
  - External puboischiofemoral muscle
  - Puboischiotibial muscle
  - Ischioflexorius muscle
  - Caudofemoral muscle
  - Ischiofemoral muscle
  - Shank flexor muscles
- Dorsal appendicular muscles
  - Internal puboischiofemoral muscle
  - Iliotibial muscle
  - Ilioextensorius muscle
  - Iliofibular muscle
  - Iliofemoral muscle
  - Shank extensor muscles
- Hypaxial muscles
  - Ischiocaudal muscle
  - Caudopuboischiotibial muscle

Remove the skin from one of the pelvic appendages and the adjacent trunk and tail on the intact side of the specimen (Table 7-4). Remove also the skin from the cloacal region and, on one side, half of the large **cloacal gland,** which lies directly under the skin. Clean the area and notice the ilium on the lateral surface dorsal to the base of the limb (Fig. 7-16). The hypaxial musculature attaches to it. Other parts of the pelvic girdle are covered by muscles.

Study first the ventral surface of the pelvic girdle and thigh. One of the most distinct muscles and, hence, one that will serve as a good landmark, is the **pubotibial muscle.** It is a narrow muscle band on the ventral surface of the thigh, which arises from the lateral edge of the pubic cartilage and inserts on the proximal end of the tibia (Fig. 7-15). The ventral surface of the pelvic girdle and the thigh medial to the pubotibial muscle are covered by a large, triangular muscle complex. The cranial two thirds represent the **external puboischiofemoral muscle,** whereas the caudal third represents the **puboischiotibial muscle.** You can detect the line of separation between these two muscles near the distal end of the thigh deep to the pubotibial muscle, then follow this line toward the midventral line of the body. Note that the cranial end of the puboischiotibial muscle overlaps the caudal end of the external puboischiofemoral muscle. The external puboischiofemoral muscle, in turn, overlaps the attachment of the rectus abdominis muscle on the cranial side of the pubic cartilage. The puboischiotibial and external puboischiofemoral muscles arise from the ventral surface of the pelvic girdle. The external puboischiofemoral muscle inserts on the femur, whereas the puboischiotibial muscle inserts on the proximal end of the tibia. The **ischioflexorius muscle** lies along the medioventral surface of the thigh caudal to the puboischiotibial muscle and, at first, appears to be a part of this muscle. The two muscles can be separated most easily at the proximal end of the shank and from the dorsal side. Try to follow the muscles medially. The ischioflexorius muscle inserts on the fascia of the distal end of

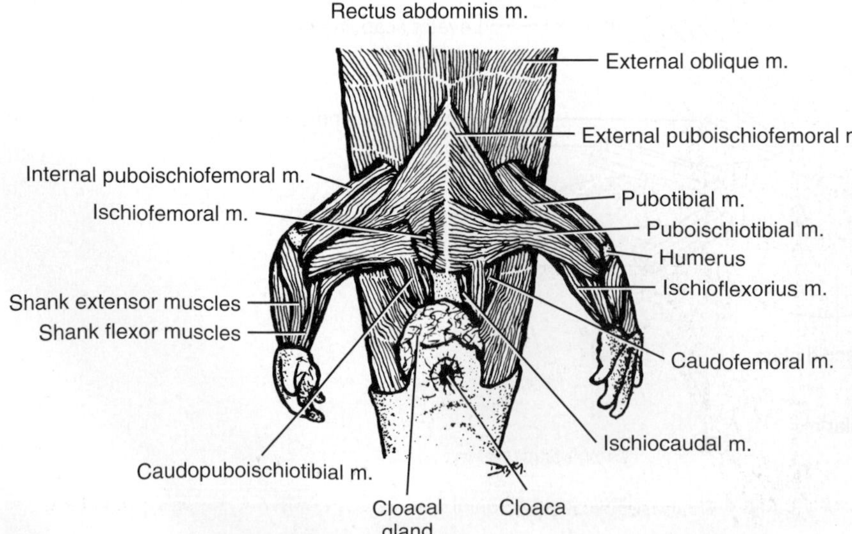

Figure 7-15
Ventral view of the pelvic muscles of *Necturus*. The puboischiotibial muscle is bisected near its origin to reveal the underlying ischiofemoral muscle.

the shank. It arises from the caudal end of the ischium. You may notice that it is subdivided into distal and proximal bellies that are separated by a short tendon.

Lateral to the cloaca you will find two longitudinal muscles. The narrower muscle beside the cloaca is the **ischiocaudal muscle.** It inserts on the caudal vertebrae and arises from the caudal end of the ischium. The more lateral muscle is the **caudopuboischiotibial muscle,** which has a similar insertion but arises from the surface of the puboischiotibial muscle. Both of these muscles are hypaxial muscles that attach to the pelvic region. They flex the tail. Dissect dorsolateral to the caudopuboischiotibial muscle, and you will find a third longitudinal muscle. This is the **caudofemoral muscle,** which is an appendicular muscle arising from the caudal vertebrae and inserting on the proximal end of the femur.

Cut across the puboischiotibial muscle and reflect its ends. The small triangular muscle lying caudal to the external puboischiofemoral muscle is the **ischiofemoral muscle.** It arises from the ischium and inserts on the proximal end of the femur. The slender muscle, arising from the underside of the femur near the insertion of the external puboischiofemoral muscle and continuing across the underside of the knee joint to the fibula, is the **popliteus muscle.** It is actually a shank muscle rather than a thigh muscle. The **adductor femoris muscle,** another deep muscle that is found cranial to the origin of the external puboischiofemoral muscle in some urodeles, is absent in *Necturus.*

The large muscle lying dorsal to the pubotibial muscle along the cranial edge of the thigh is the **internal puboischiofemoral muscle.** It originates from the internal surface of the pubic cartilage and ischium and inserts along most of the femur. To expose it, cut through the pubic attachment of the rectus abdominis muscle and dissect deeply.

The dorsal surface of the thigh is covered by two muscles that arise from the base of the ilium, extend over the knee as a tendon, and insert on the tibia (Fig. 7-16). The cranial muscle is the **iliotibial muscle;** the caudal muscle is the **ilioextensorius muscle.** These two muscles often appear as one muscle because the demarcation line between them is indistinct. Caudal to the ilioextensorius muscle, along the caudal surface of the thigh, lies the very slender **iliofibular muscle.** It, too, arises from the base of the ilium, but it diverges from the other muscles to insert on the fibula. Dissect away blood vessels and nerves caudal to the origin of the iliofibular muscle, and you will find a small, deep, triangular muscle, the **iliofemoral muscle,** which arises from the base of the ilium and inserts on the caudal edge of the femur.

The muscles of the shank fall into extensor and flexor muscles, which originate from the distal end of the femur. The **extensor muscles** lie on the dorsal surface of the shank; the **flexor muscles** lie on the ventral surface.

Axial Muscles II

Branchiomeric Muscles

- Mandibular muscles
  - Intermandibular muscle (already seen with the hypobranchial muscles)
  - Anterior levator mandibulae muscle
  - External levator mandibulae muscle
- Hyoid muscles
  - Interhyoid muscle (already seen with the hypobranchial muscles)
  - Sphincter colli muscle (already seen with the hypobranchial muscles)

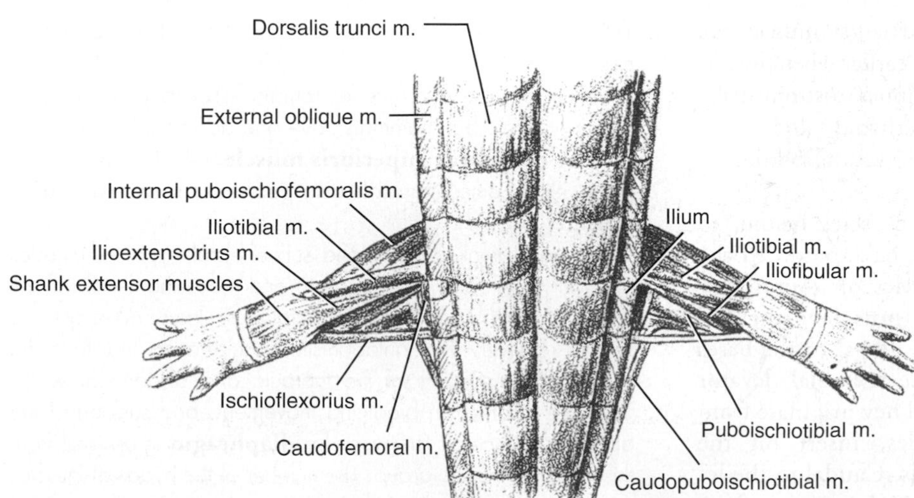

Figure 7-16
Dorsal view of the pelvic muscles of *Necturus*.

  - Depressor mandibulae muscle
  - Branchiohyoid muscle
- Branchial muscles
  - Levatores arcuum muscles
  - Dilatator laryngis muscle (already seen with the pectoral muscles)
  - Cucullaris muscle (already seen with the pectoral muscles)
  - Subarcual muscles
  - Transversi ventrales muscles
  - Depressores arcuum muscles

Remove the skin from the top and one side of the head. Be particularly careful above and below the stumps of the external gills you cut previously. Find again the intermandibular, interhyoid, and sphincter colli muscles you dissected with the hypobranchial muscles.

The mandibular arch musculature is represented by the intermandibular muscle and by the levator mandibulae muscle complex, which closes the jaws (Fig. 7-12). The **anterior levator mandibulae muscle** lies on top of the head on both sides of the middorsal line. The paired muscles with their oblique muscle fibers create a chevron pattern with the apex pointing caudally. They originate from the frontal and parietal bones of the skull and insert on the lower jaw. The massive **external levator mandibulae muscle** lies caudal to the eye. It originates from the squamosal and quadrate bones of the skull and inserts on the dentary of the lower jaw.

The hyoid musculature is represented by the interhyoid and branchiohyoid muscles and by the large muscle complex between the levator mandibulae complex and the gills. This muscle complex is divided into two muscles that arise in common from the first branchial arch in front of the gills. The more rostral muscle, the **depressor mandibulae muscle,** inserts on the caudal end, or retroarticular process, of the lower jaw and is the main muscle involved in opening the mouth. The more caudal and larger muscle, the **branchiohyoid muscle,** which we saw earlier when dissecting the hypobranchial muscles, continues rostromedially underneath the sphincter colli, interhyoid, and intermandibular muscles to insert along the ventral portion of the ceratohyal cartilage of the hyoid arch.

The remaining branchiomeric muscles belong to the branchial arches, although some have modified the position of their attachments. A series of four to five muscle slips, the **levatores arcuum muscles** (singular: levator arcus muscle), will be found dorsal to the bases of the external gills between the external levator mandibulae and cucullaris muscles. They originate from the surface of the epaxial muscles, insert on the branchial arches, and elevate the gills. Caudal to the levatores arcuum muscles and crossing the origin of the cucullaris muscle, you will see the **dilatator laryngis muscle.** The dilatator laryngis muscle looks like a levator arcus muscle, but it curves inward toward the base of the neck on its way to the lateral laryngeal cartilages instead of attaching onto the gills. It opens the glottis (i.e., the aperture of the larynx). The last dorsal branchiomeric muscle, the **cucullaris muscle,** was observed during the dissection of the pectoral muscles.

The ventral branchiomeric muscles can be found by dissecting deep between the branchiohyoid and rectus cervicis muscles. Pull the branchiohyoid muscle cranially and the rectus cervicis muscle caudally. A series of three **subarcual muscles** will be seen: a longitudinal muscle extending from the ceratohyal cartilage to the first branchial arch; another longitudinal muscle extending from the first and second branchial arches to the third branchial arch; and an oblique muscle medial to the latter and extending between the first and second branchial arches. The oblique muscle fiber bundles caudal to the subarcual muscles and deep to the rectus cervicis muscle constitute the **transversi ventrales muscles** (see Fig. 11-12). These muscle fibers interconnect the paired third branchial arches. The remaining ventral branchiomeric muscles are a series of three **depressores arcuum muscles,** which extend from the ventral portion of the three branchial arches to the bases of the three external gills. They are small and may have been removed with the skin.

MAMMALS

The morphological changes of the musculature toward the configuration in a land tetrapod, which we observed in *Necturus* as a model of an early tetrapod, continue in the evolution toward mammals. Review the general remarks on tetrapod muscle evolution on page 136, as well as Table 7-2. We will discuss a few general trends observed in the evolution of mammals at this point.

The extrinsic eye muscles remain essentially the same in all vertebrates, but mammals have one additional muscle, the **levator palpebrae superioris muscle,** which raises the upper eyelid. Some mammals also possess a **retractor bulbi muscle.**

Epaxial trunk muscles and subvertebral hypaxial muscles are well developed. They support and move the vertebral column and head. Most of the hypaxial musculature develops into several relatively thin, antagonistic muscle layers that form the muscular components of the thoracic and abdominal walls. They are involved in breathing movements and support of the abdominal viscera. The muscular **diaphragm** is derived from this hypaxial muscle group. The number of the hypaxial muscles that transfer the weight of the body from the trunk to the pectoral girdle and appendage, such as the **serratus ventralis** and **rhomboideus muscle complexes,** is increased. These hypaxial muscles also are responsible for movements between the trunk and the pectoral girdle.

The hypobranchial musculature becomes subdivided into many units that precisely control the tongue and hyoid movements during mastication and swallowing.

The appendicular muscles of mammals are much more numerous than those of early tetrapods, because the limbs are involved in supporting the body, in locomotion using various gaits, and in other activities, such as grooming, prey capture, and intraspecific and interspecific interactions. Because the feet are positioned closer to the body's central axis and center of gravity (see Anatomy in Action 3-3), the limbs can raise and maintain the trunk off the ground more easily. The limbs also can swing forward during locomotion without the need for lateral undulations of the trunk as in amphibians and reptiles. Because ventral appendicular muscles are less important for lifting the body off the ground, some of them, such as the **supraspinatus** and **infraspinatus muscles,** have shifted their positions dorsally and are involved in protracting and retracting the pectoral limb. Some muscle fiber bundles that are derived from appendicular muscles form the **cutaneus trunci muscle,** which spreads over the trunk and moves the areas of skin that cannot be reached by limbs or tails for scratching and shooing away insects.

The mandibular muscles have increased in number and complexity to ensure precise occlusion of the teeth for mastication. A small part of the mandibular musculature follows the original jaw joint into the middle ear cavity to become the **tensor tympani muscle.** A new jaw-opening muscle, the **digastric muscle,** has evolved from a combination of parts of the mandibular and hyoid musculature and replaces the depressor mandibulae muscle found in amphibians, reptiles, and birds, which have retained the original vertebrate jaw articulation. A small part of the hyoid musculature, the **stapedius muscle,** attaches to the stapes in the middle ear, and another part, the **stylohyoid muscle,** retains a connection with the hyoid. However, most hyoid muscles are superficial and form the **platysma** and the **facial muscles,** which move the skin of the neck, face, and lips. The primitive cucullaris muscle of vertebrates has enlarged and become divided into the **trapezius** and **sternocleidomastoid muscle complexes** in connection with the more complex shoulder and head movements that are independent of those of the trunk.

Many of the muscle groups, which are relatively easily observed in fishlike vertebrates and early tetrapods, have lost their clear definition in adult mammals, because some muscles have migrated from their original positions. This condition makes it more difficult to study the muscles strictly by the muscle groups established on the basis of embryonic development, but it still can be done if a few concessions are made to the mammalian topography.

Cutaneous Muscles

- Cutaneus trunci muscle
- Superficial sphincter colli muscle
- Platysma
- Facial muscles

As the first step of your dissection, it is necessary to remove the skin from your specimen. With a sharp scalpel, make incisions through the skin completely around the neck behind the skull; around the paws at the level of the wrists and ankles, respectively; and around the base of the tail, anus, and external genitals. If your specimen is a male, first locate the testicles before making the incision; do not remove the skin over the testicles and (in both sexes) around the anus and urogenital orifices. For the time being, the skin should also be left on the head, tail, and paws. Lay your specimen on its belly, find the middorsal line by palpating the spinous processes of the vertebral column, and make a longitudinal, middorsal incision through the skin that extends from the circular incision around the neck to that around the base of the tail.

Beginning on the back of your specimen, gradually separate the skin from the underlying muscles by tearing through the fibrous connective tissue of the **superficial fascia** with a pair of blunt forceps. As you separate the skin from the trunk, notice the fine, parallel brownish lines that adhere to its undersurface. They represent the muscle fiber bundles of the **cutaneus trunci muscle,** the largest of the cutaneous muscles that move the skin. This muscle is not present in humans. The cutaneus trunci muscle arises from the surface of certain appendicular muscles of the shoulder, such as the latissimus dorsi and pectoralis muscles and from the **linea alba,** a midventral band of connective tissue. It fans out over most of the trunk and inserts on the dermis of the skin. It should be removed with the skin, except for the portion attached to the shoulder muscles caudal to the armpit. Several smaller cutaneous muscles, derived from the caudal musculature, become associated with the caudal part of the cutaneus trunci muscle. They may not be noticed.

You may be able to discern two additional cutaneous muscles on the neck, namely, the superficial **sphincter colli muscle** on the ventral side and the **platysma** on the dorsal side. Both muscles are derived from the hyoid musculature. As it spreads over the face, it breaks up into many small cutaneous muscles associated with the lips, nose, eyes, and ears. They are collectively known as the **facial muscles.** You may see them later when the skin is removed from the head. These muscles also are present in humans and are responsible for facial expression.

As you continue to skin your specimen, you will come upon narrow tough cords passing to the skin. These are **cutaneous blood vessels** and **nerves** and must be cut.

Note that they tend to be segmentally arranged along the trunk. If your specimen is a pregnant or lactating female, the **mammary glands** will appear as a pair of large, longitudinal, glandular masses along the ventral side of the abdomen and thorax. They should be removed with the skin.

As you reach the limbs, try to tear the connection between the skin and the underlying muscles while leaving the skin around the limbs intact as "sleeves." Do so by prying and tearing with blunt forceps and probes under the skin, continuing from the skinned part of the trunk as well as starting from the circular cut around the wrists and ankles until you can pull out the limb from the sleevelike skin. In this way you will be able to preserve the skin as some sort of jumpsuit that can easily be pulled on and off the specimen. If your specimen is an older male cat, this may not be possible because the connective tissue may be very tough; in this case, make a longitudinal incision through the skin on the dorsal or lateral surface of the limbs to facilitate skinning.

After the specimen is skinned, clean away fat and connective tissue on the side that is to be studied (Table 7-5), but do not clean an area thoroughly until it is being dissected. If your specimen is a male, be particularly careful in removing the wad of fat in the groin because it contains on each side the proximal part of the **cremasteric pouch,** a part of the scrotum containing blood vessels and the sperm duct extending between the abdomen and scrotal skin. First find the cremasteric pouch. It is large in rabbits but rather small in cats. Clean away connective tissue deep in the groin, or inguinal region, so that you can find the actual boundary between the thigh and abdomen. In rabbits, a shiny, white **inguinal ligament** will be seen in this region, extending from the pubis to the ilium.

Caudal Trunk Muscles

The axial muscles cannot be studied all at the same time because many of them are located deep to the shoulder muscles. Those located on the trunk between the pectoral and pelvic appendages will be examined now, and the more cranial muscles will be considered after the appendages have been studied (p. 173).

Hypaxial Muscles

- Muscles of the abdominal wall
  - External oblique muscle
  - Internal oblique muscle
  - Transversus abdominis muscle
  - Rectus abdominis muscle
- Subvertebral muscles (to be studied with iliopsoas muscle complex)
  - Quadratus lumborum muscle
  - Psoas minor muscle
- Appendicular muscles also seen:
  - Latissimus dorsi muscle
  - Pectoral muscle complex

Continue to clean off the surface of the trunk between the pectoral and pelvic appendages (Table 7-5). The sheet of tough, white fascia covering the lumbar region on the back is the **thoracolumbar fascia.** The wide sheet of muscle that runs cranioventrally from the cranial part of this fascia and disappears in the armpit is the latissimus dorsi muscle, an appendicular muscle (Figs. 7-18, 7-19, 7-20 and 7-21). The large triangular muscle that covers the underside of the chest is another appendicular muscle, the pectoralis muscle. The borders of the latissimus dorsi and pectoralis muscles appear to run together caudal to the armpit. Separate the two muscles in this region by remov-

Table 7-5 General Plan of Dissections of Cats and Rabbits

Intact Side of Specimen	Specimen Side with Injection Site and Ligature in the Neck
1. Muscles of the body, neck, and limbs, except jaw, prehyoid, and branchiomeric muscles	**2.** Prehyoid and branchiomeric muscles of the floor of the mouth
4. Salivary glands and fascial nerves	**3.** Jaw or mandibular muscles
5. Tear glands and extrinsic eye muscles	**7.** Section through the floor of the mouth along the mandibular ramus, after bisecting and reflecting the stylohyoid and mylohyoid muscles and detaching the digastric muscle from its insertion on the mandible
6. External acoustic meatus	**8.** Bisect palatoglossal arches
12. Blood vessels of the neck and head, fore and hind limbs	**9.** Cut open the pharynx
13. Brachial and pelvic nerve plexus	**10.** Reveal the pterygoid muscles under the lining of the soft palate and pharynx
	11. Strip muscles off the larynx

ing the cutaneus trunci muscle and carefully trace their borders cranially. Remove any fat and loose connective tissue from beneath them. The hypaxial trunk muscles can now be studied.

As in other tetrapods (Fig. 7-11), the abdominal wall is composed of three layers of muscle, in addition to a paired longitudinal muscle along the midventral line. All these muscles serve to compress the abdomen. The **external oblique muscle** forms the outermost layer (Figs. 7-17 and 7-18). This muscle arises by slips from the surface of a number of caudal ribs and from the thoracolumbar fascia. Part of its origin lies beneath the caudal border of the latissimus dorsi muscle. Its muscle fibers extend obliquely caudoventrally to insert by an aponeurosis along the length of the linea alba. In rabbits, some of the fibers insert on the inguinal ligament. Remove fat and connective tissue from the surface of the external oblique muscle and identify its borders. The cranial border is very short, thin, and hidden under the latissimus dorsi and pectoral muscles; the caudal border is almost longitudinally oriented. Bisect the external oblique muscle perpendicularly to its muscle fiber direction. Start the cut through one of the borders by separating, as far as you can, first the sheetlike external oblique muscle from the underlying internal oblique muscle (see later and Fig. 7-17), then cut with a pair of scissors through the part of the muscle that can be lifted. Continue the separation of the two muscle layers as far as you can, cut, and so on, until you reach the other border. Make sure to cut through the middle of the muscle fibers and not through the aponeurosis; as a result the completed incision will be curved. Reflect the bisected external oblique muscle to expose the entire underlying internal oblique muscle.

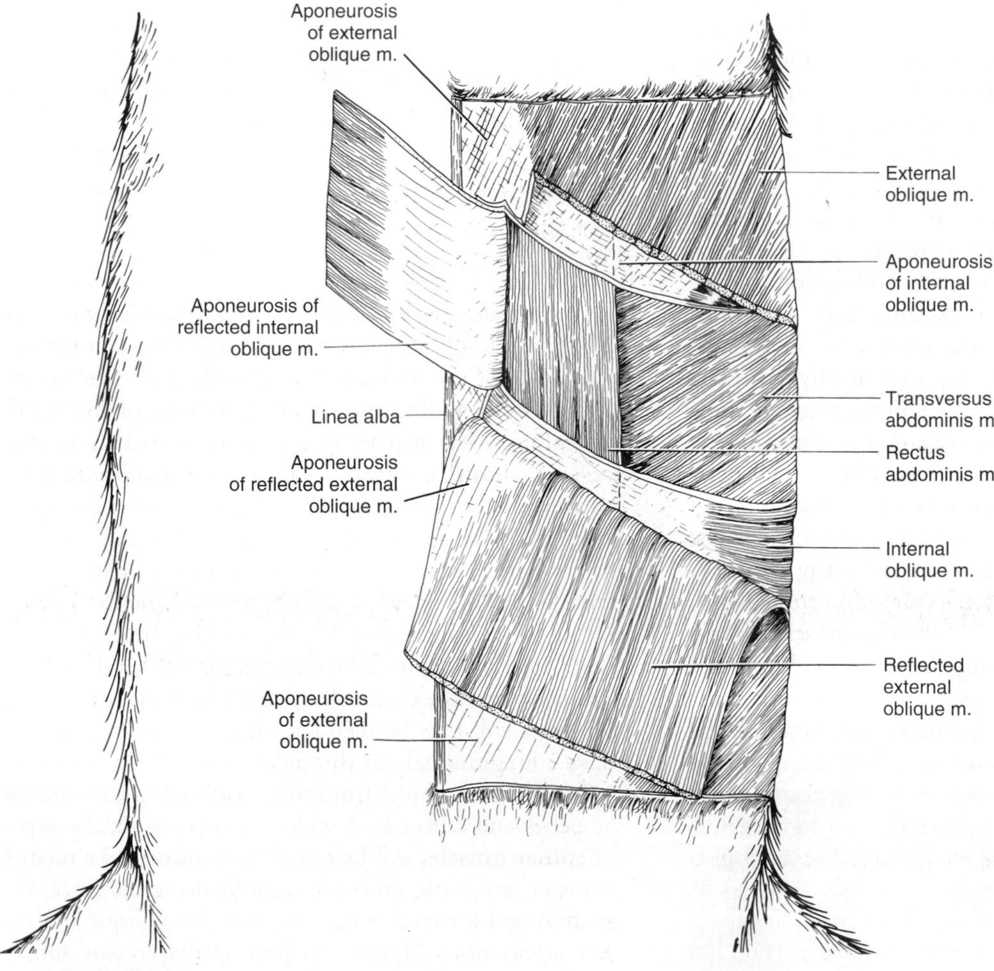

Figure 7-17
Ventral view of the abdominal muscle layers on the left side of a cat. Parts of the external oblique and internal oblique muscles are reflected to show the deeper layers. Note that this figure is not representative of the dissection that you will perform when following the instructions in the text.

The **internal oblique muscle** lies underneath the external oblique muscle. It is most apparent high up on the side of the abdomen near its main origin from the thoracolumbar fascia. The muscle fibers of the internal oblique muscle extend ventrally and slightly cranially at right angles to the muscle fibers of the external oblique muscle and soon lead into a long aponeurosis that inserts along the linea alba. Only the dorsal half of the muscle is fleshy. The ventral half consists of an aponeurosis, but this may not be readily apparent because this aponeurosis is transparent so that the muscle fibers of the underlying transversus abdominis muscle (see later) are visible. The cranial border of the internal oblique muscle is situated slightly caudal to the rib cage, and the caudal border reaches far caudad deeply into the groin region. Most of the caudal part of the internal oblique muscle passes over an underlying longitudinal muscle that inserts on the femur (i.e., the iliopsoas muscle; see p. 168), except for a small caudal muscle slip, which passes under it.

To see the **transversus abdominis muscle,** make a longitudinal cut through the fleshy portion of the internal oblique muscle, in the same manner as described for the external oblique muscle, and reflect the bisected muscle. The transversus abdominis muscle arises primarily from the medial surface of the more caudal ribs and from the transverse processes of the lumbar vertebrae. The latter portion of the origin lies in a furrow deep to the epaxial musculature. Note that this furrow has a location comparable to that of the horizontal skeletogenous septum of fishes. The muscle fibers of the transversus abdominis muscle extend ventrally, and slightly caudally, to insert along the linea alba by a short aponeurosis. Separate some of the muscle fibers of the transversus abdominis muscle from one another, and you will expose the **parietal peritoneum** lining the abdominal cavity. There is an **internal fascia** between the transversus abdominis muscle and the parietal peritoneum, but it cannot be seen easily because it usually merges with the transversus abdominis muscle.

The reflection of the internal oblique muscle also exposes a longitudinal band of muscle that lies on both sides of the midventral line. This is the **rectus abdominis muscle.** Transverse tendinous intersections sometimes can be seen in it, but these do not correspond to the myosepta that delimit primitive muscle segments. The rectus abdominis muscle arises from the pubis and inserts on the sternum and cranial costal cartilages. In cats, its cranial portion lies directly under the external oblique muscle; its middle portion lies between two layers of the aponeurosis of the internal oblique muscle; and its caudal portion lies between two layers of the aponeurosis of the transversus abdominis muscle. In rabbits, most of the rectus abdominis muscle lies between the internal oblique and transversus abdominis muscles.

In addition to the muscular layers of the abdominal wall, the caudal hypaxial musculature includes a subvertebral muscle group (Fig. 7-11), which lies along the ventral surfaces of the lumbar and caudal thoracic vertebrae. This muscle group, which includes the **quadratus lumborum** and **psoas minor muscles,** is associated with certain pelvic muscles and is described in connection with the iliopsoas muscle complex (p. 168).

As mentioned earlier, the entire body is enveloped by the **superficial fascia,** which lies directly under the skin. It is a thin layer of loose connective tissue that cannot be isolated easily because it blends with the dermal connective tissue or fuses with the **deep fascia.** This layer of connective tissue is thicker and tougher, and it is reinforced in particular places, such as the thoracolumbar fascia. There is often a layer of fat, the **panniculus adiposus,** embedded between the superficial and deep fascias. Over the trunk, the superficial and deep fascias together form the **external fascia of the trunk** *(fascia trunci).* The **internal fascia of the trunk** *(fascia transversalis),* in contrast, lies between the serosa of the body cavities (see Chapter 10) and the innermost muscle layer of the abdominal and thoracic body walls.

Epaxial Muscles

- Multifidi muscles
- Erector spinae muscle

Lift up the thoracolumbar fascia with a pair of forceps and make a longitudinal incision through it about 0.5 cm to one side of the middorsal line (Table 7-5). Identify its location by palpating the spinous processes of the vertebral column. When making this incision, make sure that you cut through only the most superficial layer, or **lamella,** of the thoracolumbar fascia; if you see muscle fibers, you have cut too deeply. If this is the case, look at the cut edges and separate the two fascial lamellae. Slip a probe under the most superficial fascial lamella to avoid cutting too deeply and extend the incision from the latissimus dorsi muscle to the sacral region. Reflect the superficial lamella of the thoracolumbar fascia. A deeper layer of this fascia will now be seen encasing the epaxial muscles. Make a longitudinal cut through it about 1 cm (in a cat) or 0.5 cm (in a rabbit) from the middorsal line cranial to the pelvis and reflect it. A wide muscular band, the **erector spinae muscle,** will be revealed. It consists of a medial narrower part with almost longitudinal muscle fiber orientation and a lateral wider part with an oblique muscle fiber orientation. This lateral part of the erector spinae muscle is subdivided by the deep layer of the thoracolumbar fascia, which dips into it. The muscle fibers of the more medial subdivision run obliquely from caudolateral to craniomedial, whereas the muscle fibers of the more lateral subdivision run from caudomedial to craniolateral. These subdivisions, however, do not correspond to the subdivisions that will be seen later on the cranial part of the trunk (p. 174).

Mobilize the medial border of the erector spinae muscle and push it laterally. The **multifidi muscles** will be revealed as a narrow band of muscle next to the spinous processes of the vertebrae and lie under a translucent layer of connective tissue, which you can remove. The multifidi muscles represent a series of short muscles that interlace the vertebrae and whose diagonally oriented muscle fiber bundles run from caudolateral to craniomedial. It is difficult to distinguish them individually.

Pectoral Appendicular Muscles

Most of the muscles of the pectoral region are, of course, appendicular muscles, but a number of axial muscles and several branchiomeric muscles have become associated with the pectoral girdle and appendage.

Pectoral Muscle Complex

- Pectoralis superficialis muscle
  - Pectoralis descendens muscle
  - Pectoralis transversus muscle
- Pectoralis profundus muscle
  - Xiphihumeral muscle (in cats)
- Other muscles also seen:
  - Cleidobrachial muscle (an appendicular muscle)

The large triangular muscle complex covering the chest is the **pectoral muscle complex.** It arises from the sternum and inserts primarily along the humerus. Its major actions are to adduct the humerus toward the chest and to retract it, but in standing animals, it also transfers body weight from the trunk to the pectoral girdle and appendage (Fig. 7-22). Clean the surface of this muscle complex well enough so that you can see the orientation of its fibers.

The pectoral muscle complex of quadrupedal mammals is divided into superficial and deep portions, which are homologous to the pectoralis major and pectoralis minor muscles of humans, respectively, but the deep portion is the larger muscle in quadrupeds. Each portion may be further subdivided.

The **pectoralis superficialis muscle** arises from approximately the cranial one third of the sternum. Its muscle fibers extend more or less laterally to insert on the humerus. In cats, the insertion extends to the distal end of the humerus and onto the antebrachium. The superficial pectoral muscle can be divided into two parts (Fig. 7-18).

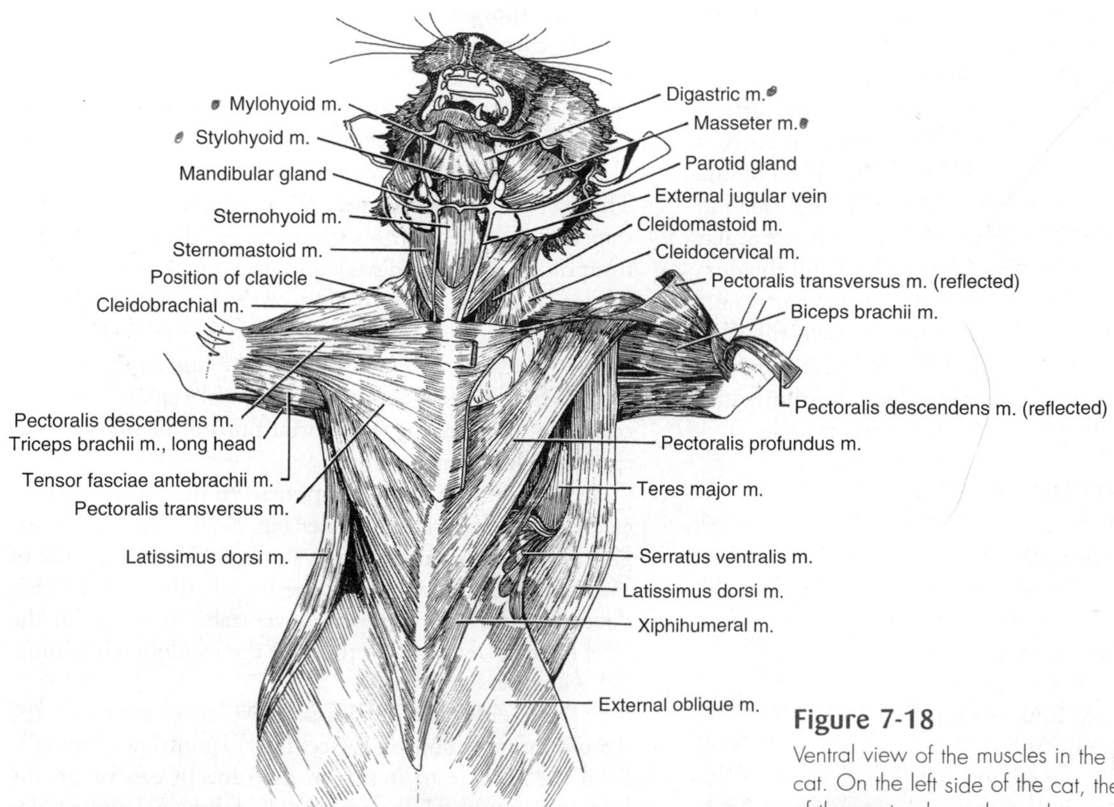

Figure 7-18

Ventral view of the muscles in the pectoral region of a cat. On the left side of the cat, the superficial portions of the pectoral muscle are bisected and reflected to reveal the deeper muscles.

A narrow band of very superficial muscle fibers, the **pectoralis descendens muscle,** extends in cats from the cranial end of the sternum to the **antebranchial fascia** *(fascia antebrachii),* or, in rabbits, to the middle of the humerus. The rest of the superficial pectoral muscle is called the **pectoralis transversus muscle.**

The **cleidobrachial muscle** covers the front of the shoulder and part of the pectoralis transversus muscle (Figs. 7-18 and 7-41). Separate the borders of the cleidobrachial and pectoralis transversus muscles by breaking and picking away the connective tissue that binds them together. In cats, the clavicle is imbedded on the underside of the cleidobrachial muscle at the base of the neck; you can feel it if you palpate the muscle between your thumb and forefinger. Cut through the connective tissue that ties the medial end of the clavicle to the manubrium of the sternum so that you can push the cleidobrachial muscle cranially and expose the entire pectoral muscle. Be careful not to cut through the jugular vein and cranial border of the pectoral muscle.

The **pectoralis profundus muscle** originates from the caudal two thirds, or from the full length, of the sternum. Its muscle fibers extend craniolaterally and disappear beneath those of the superficial pectoral muscle. Toward the armpit, the lateral border of the pectoralis profundus muscle is strongly tied to the ventral border of the latissimus dorsi muscle and cannot be separated easily. You may need to cut the two muscles apart with a pair of scissors. In cats the insertion is confined to the humerus, but in rabbits some muscle fibers also attach onto the clavicle, and a group of deep muscle fibers pass dorsal to the clavicle and sweep over the front of the shoulder to insert on the scapular spine. This portion will be seen later in more detail (p. 153). This scapular muscle portion of rabbits probably helps to pull the dorsal border of the scapula cranially, an action that occurs in bounding. In cats, the most caudal muscle fibers of the pectoralis profundus muscle form a distinct narrow band, the **xiphihumeralis muscle,** which continues cranially underneath the pectoralis profundus muscle.

Trapezius and Sternocleidomastoid Muscle Complexes

- Trapezius muscle complex
  - Cleidocervical muscle (or clavotrapezius muscle) (in cats)
  - Cervical trapezius muscle
  - Thoracic trapezius muscle
- Sternocleidomastoid muscle complex
  - Sternomastoid muscle
  - Sterno-occipital muscle (in cats)
  - Cleidomastoid muscle
  - Cleido-occipital muscle (in rabbits)
- Other muscles also seen:
  - Cleidobrachial muscle (a deltoid muscle)
  - Brachiocephalic muscle (cleidobrachial muscle and cleidocervical muscle) (in cats)
  - Posterior depressor conchae muscle (a cutaneous muscle) (in rabbits)

The muscles belonging to the trapezius and sternocleidomastoid muscle complexes are branchiomeric muscles that have become associated with the pectoral girdle, because they evolved from the cucullaris muscle of fishes and amphibians. The trapezius muscle in humans is a large undivided muscle, but it is subdivided in the mammals being considered here. Most muscle portions act on the scapula, abducting it toward the middorsal line, as well as protracting and retracting it. In cats, the most cranial part of the trapezius muscle (i.e., the cleidocervical muscle, see later) inserts on the clavicle and, together with the cleidobrachial muscle, protracts the arm. The sternocleidomastoid muscle complex acts primarily upon the head, turning it to the side and flexing it downward. Clean off connective tissue from the ventral, lateral, and dorsal surfaces of the neck and from the dorsal part of the shoulder region. You may have to remove more skin from the back of the head. Do not injure the large external jugular vein located superficially on the ventrolateral surface of the neck.

The **thoracic trapezius muscle** is a thin sheet of muscle covering the cranial part of the latissimus dorsi muscle, from which it should be separated (Figs. 7-19 and 7-20). From their origin on the middorsal line of the thorax, the muscle fibers of the thoracic trapezius muscle converge to insert on the dorsal part of the scapular spine. The **cervical trapezius muscle** lies cranial to the thoracic trapezius muscle. It arises from the middorsal line of the front of the thorax and neck and from an aponeurosis that interconnects the left and right cervical trapezius muscles. In cats its origin extends to the base of the neck, but in rabbits it extends as far forward as the nuchal crest of the skull. The muscle fibers of the cervical trapezius muscle converge to insert on the ventral portion of the scapular spine and its metacromion.

Cats possess a third component to the trapezius muscle, the **cleidocervical muscle** (Fig. 7-19). From their origin on the nuchal crest of the skull and middorsal line of the cranial part of the neck, the muscle fibers of the cleidocervical muscle extend caudoventrally to merge, at the level of the clavicle, with those of the cleidobrachial muscle, which was seen earlier.

The **cleidobrachial muscle,** which is a part of the deltoid muscle complex (see later), continues caudally from the clavicle to insert on the ulna in cats or on the humerus in rabbits. In cats, the cleidobrachial and cleidocervical muscles tend to merge because the clavicle is re-

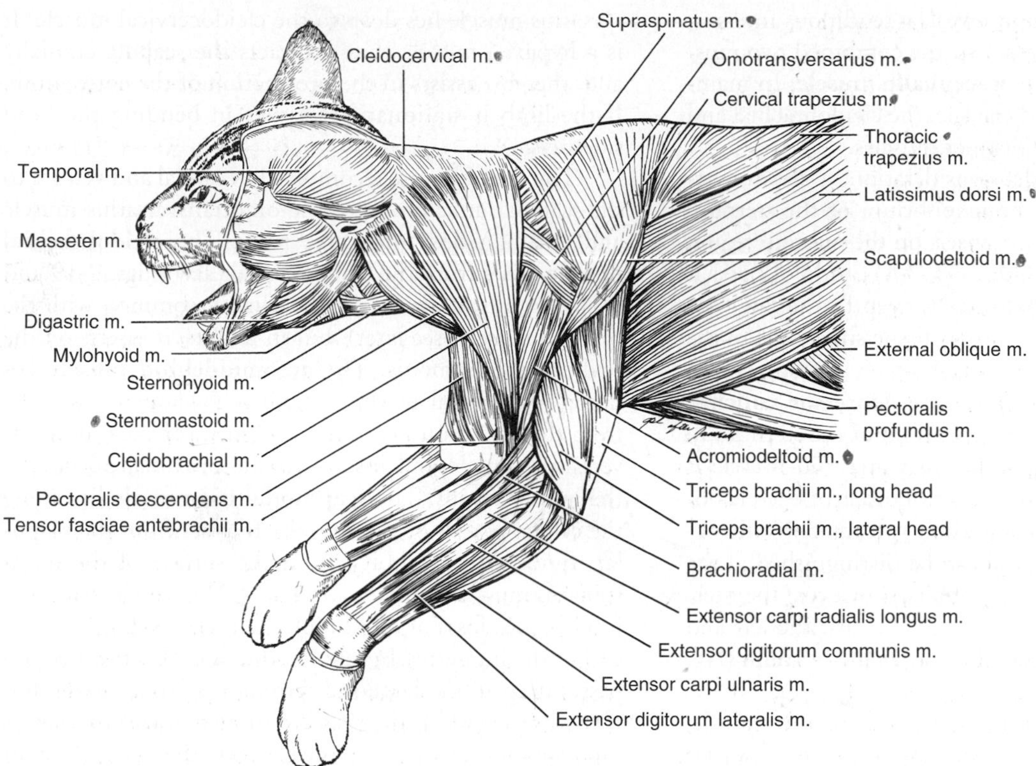

Figure 7-19
Lateral view of the head, neck, shoulder, and pectoral muscles of a cat.

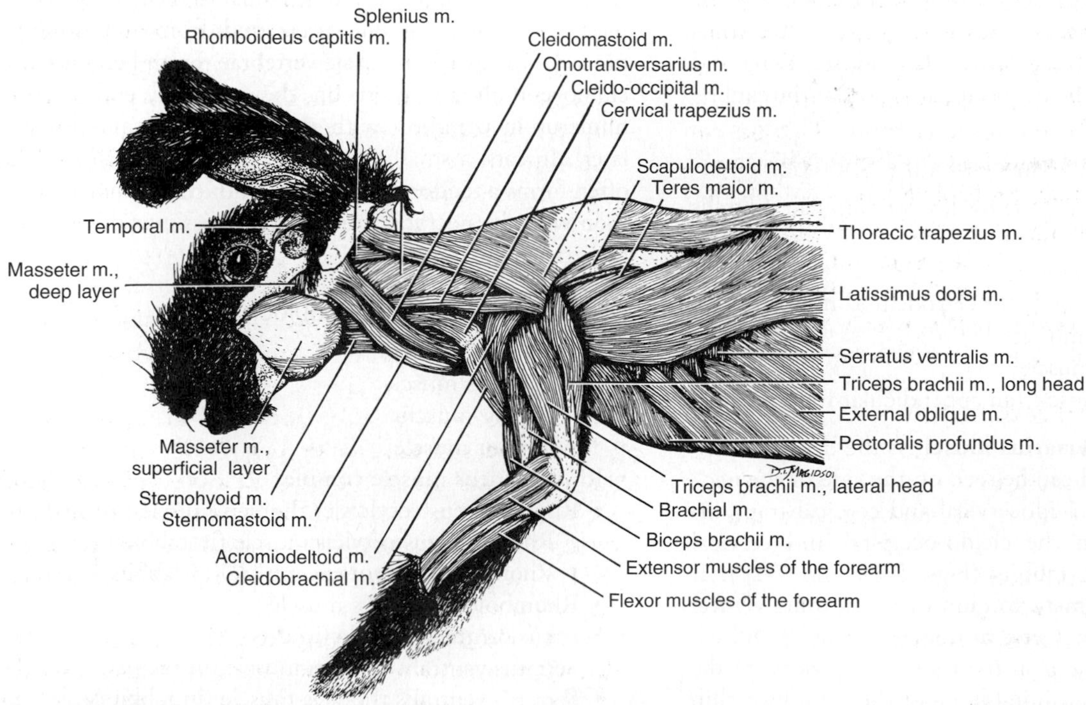

Figure 7-20
Lateral view of the head, neck, shoulder, and pectoral muscles of a rabbit.

duced, but careful dissection reveals a tendinous intersection between them. Nevertheless, the combined two muscles are also called the **brachiocephalic muscle.** In mammals with a well-developed clavicle, the cleidobrachial and cleidocervical muscles are distinct muscles.

In the mammals under consideration, a **sternomastoid muscle** arises from the manubrium of the sternum and extends craniodorsally to insert on the mastoid region of the skull (Figs. 7-19, 7-20, 7-38, 7-41, and 7-42). As it extends cranially, the muscle passes deep to the large external jugular vein. In cats, muscle tissue superficial to the vein is part of the platysma; in rabbits, it is a specialized part of the platysma, the **posterior depressor conchae muscle** extending from the manubrium of the sternum to the base of the ear. In cats, the sternomastoid muscle is wide and parallels the cranioventral border of the cleidocervical muscle; part of its insertion spreads onto the occipital region of the skull and can be distinguished as the **sterno-occipital muscle.** In rabbits, however, the sternomastoid muscle is a narrow band. In cats, the left and right sternomastoid muscles merge near the sternum (Fig. 7-18) and may need to be cut apart.

The **cleidomastoid muscle** attaches to the clavicle, where it joins other muscles attaching on the clavicle and extends to the mastoid region of the skull (Figs. 7-18 and 7-20). In cats, much of it lies deep to the wide sternomastoid and cleidocervical muscles, but in rabbits, it lies dorsal to the sternomastoid muscle. In rabbits, a **cleido-occipital muscle** arises from the clavicle dorsolateral to the origin of the cleidomastoid muscle (Fig. 7-20). It extends cranially deep to the cleidomastoid and sternomastoid muscles to insert on the basioccipital region of the skull.

Remaining Superficial Muscles of the Shoulder

- Omotransversarius muscle (or levator scapulae ventralis muscle) (a hypaxial muscle)
- Deltoid muscle complex
  - Cleidobrachial muscle (or clavodeltoid muscle)
  - Acromiodeltoid muscle
  - Scapulodeltoid muscle
- Latissimus dorsi muscle (an appendicular muscle)

The **omotransversarius muscle** is the band of muscle whose caudal end can be seen on the side of the neck of cats between the cleidocervical and cervical trapezius muscles, or between the cleido-occipital and cervical trapezius muscles in rabbits (Figs. 7-19 and 7-20). It originates from the metacromion of the scapula ventral to the insertion of the cervical trapezius muscle and extends cranially to insert on the transverse process of the atlas and to the basioccipital bone of the skull by a thin tendon that merges with the longissimus capitis muscle (Figs. 7-39 and 7-40). In cats, much of the omotransversarius muscle lies deep to the cleidocervical muscle. It is a hypaxial muscle that protracts the scapula cranially and, thereby, assists in the protraction of the entire limb. If the limb is stationary, it assists in bending the head sideways.

The deltoid muscle complex lies caudal and ventral to the scapular attachment of the omotransversarius muscle and is subdivided into three parts. The **cleidobrachial muscle** arises from the clavicle, as we saw (Figs. 7-19 and 7-20). In cats, it inserts on the ulna in common with the brachial muscle (see later), but in rabbits, it inserts on the body of the humerus. The **acromiodeltoid muscle** lies dorsal and caudal to the cleidobrachial muscle. It arises from the acromion near the attachment of the omotransversarius muscle and inserts on the proximal portion of the humeral body. The **scapulodeltoid muscle** lies along the caudal border of the scapula. It arises from the scapular spine and, in rabbits, from the surface of the infraspinatus muscle (see later and Fig. 7-21), which covers the infraspinous fossa of the scapula. It inserts on the proximal end of the humerus. In rabbits, the muscle is thinner and passes deep to the large metacromion of the scapula, but if you break and lift the metacromion, you will see that its topographical relationship is essentially the same as that of cats. The cleidobrachial muscle protracts the arm, but the caudal parts of the deltoid muscle complex retract and abduct the humerus.

The **latissimus dorsi muscle** has already been observed on the side of the trunk caudal to the arm. It arises from the thoracolumbar fascia and from the spinous processes of the last thoracic vertebrae. From here, it runs cranioventrally to insert on the proximal end of the humerus in common with the teres major muscle (see later). In cats, a small part of the latissimus dorsi muscle often forms a tendon that inserts on the humerus in common with the pectoral muscle. The latissimus dorsi muscle retracts the humerus.

Deeper Muscles of the Shoulder

- Supraspinatus muscle
- Infraspinatus muscle
- Teres major muscle
- Teres minor muscle
- Rhomboideus muscle complex
  - Rhomboideus cervicis et thoracis muscle (in cats)
    - Rhomboideus cervicis muscle (in rabbits)
    - Rhomboideus thoracis muscle (in rabbits)
  - Rhomboideus capitis muscle
- Serratus ventralis muscle (in cats)
  - Serratus ventralis cervicis muscle (in rabbits)
  - Serratus ventralis thoracis muscle (in rabbits)
- Subscapular muscle
- Also seen: Pectoralis profundus muscle (in rabbits)

Cut across the middle of the latissimus dorsi muscle at right angles to its muscle fibers. Also cut across the middle of the thoracic and cervical trapezius muscles. Reflect the ends of these muscles, and clean out any fat and loose connective tissue from underneath them in order to expose the deeper muscles of the shoulder. In rabbits, the **pectoralis profundus muscle** can be seen sweeping over the front of the shoulder and scapula and inserting on the scapular spine. It should be bisected near its insertion and reflected.

The **supraspinatus muscle** occupies the supraspinous fossa of the scapula (Fig. 7-21). It inserts on the greater tuberculum of the humerus and protracts the humerus. The **infraspinatus muscle** occupies the infraspinous fossa. In rabbits, most of this muscle is covered by the scapulodeltoid muscle. The infraspinatus muscle inserts on the greater tuberculum of the humerus and rotates the humerus outward.

The **teres major muscle** arises from the caudal border of the scapula caudoventral to the infraspinatus muscle. It runs cranioventrally to insert on the proximal end of the humerus in common with the latissimus dorsi muscle. It rotates the humerus inward and retracts it.

Lift up the cranioventral border of the scapulodeltoid and infraspinatus muscles and dissect deeply between the infraspinatus muscle and the long head of the triceps brachii muscle, which is the large muscle on the caudal surface of the upper arm (see later and Fig. 7-19). Eventually you will come upon a very small triangular muscle arising by a tendon from the middle of the caudal border of the scapula and inserting on the greater tuberculum of the humerus. This is the **teres minor muscle,** which assists the infraspinatus muscle in rotating the humerus outward.

Examine the shoulder region from a dorsal view. The large muscle that arises from the spinous processes of the posterior cervical and anterior thoracic vertebrae and inserts along the dorsal border of the scapula is the **rhomboideus cervicis et thoracis muscle.** In cats, it forms a continuous sheet of muscle bundles, but in rabbits, it is clearly divided into a cranial **rhomboideus cervicis muscle** and a caudal **rhomboideus thoracis muscle** (Fig. 7-24). The most cranial muscle bundle of the rhomboideus muscle complex of

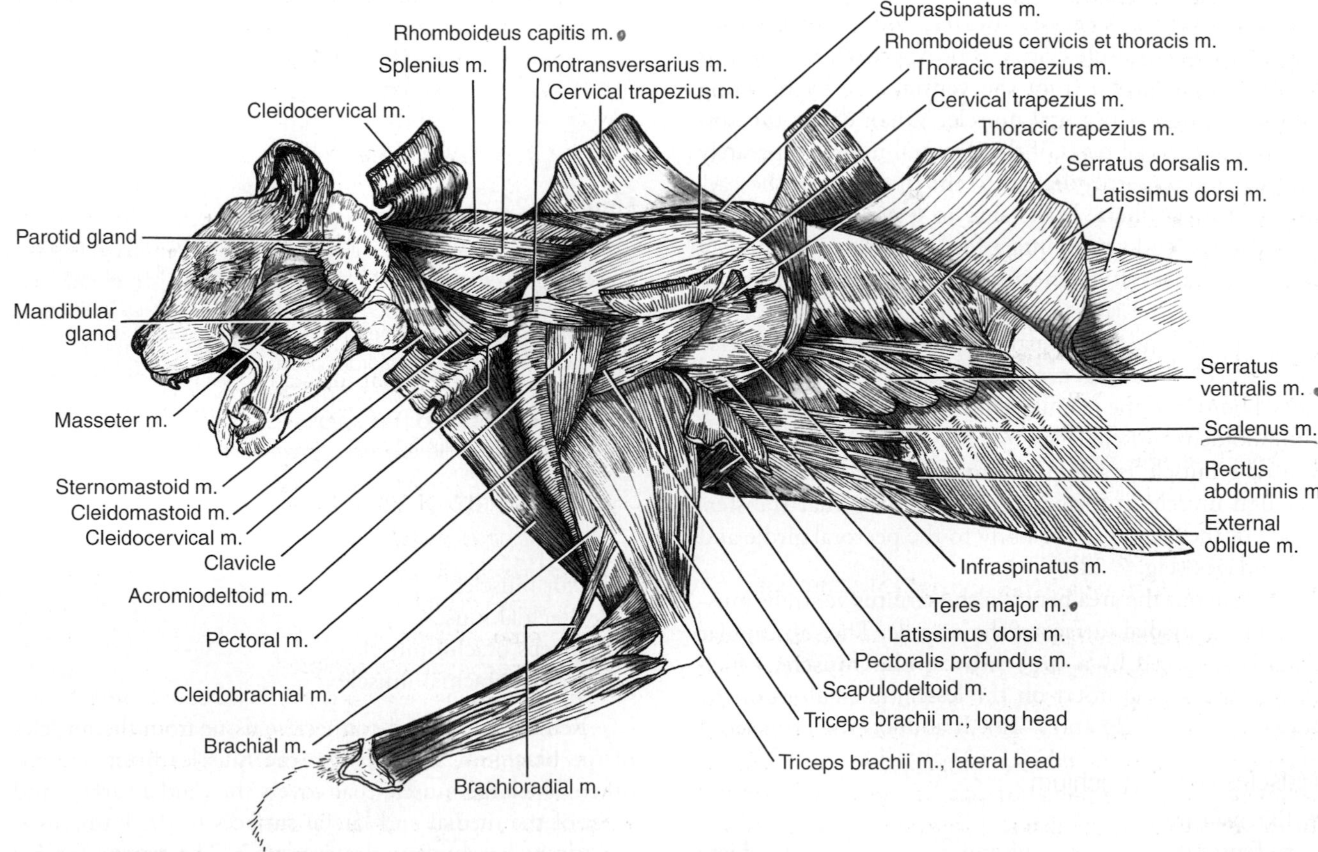

Figure 7-21
Lateral view of the neck, shoulder, and pectoral muscles of a cat. The trapezius muscle complex and the latissimus dorsi muscles are bisected and reflected to reveal the deeper muscles.

both cats and rabbits extends farther cranially to insert on the back of the skull and is called the **rhomboideus capitis muscle.** The rhomboideus muscle complex is part of the hypaxial musculature that pulls the scapula toward the vertebral column, helps to hold it in place, and assists in its protraction and retraction.

Muscles that lie deep to the rhomboideus muscle complex and extend from the dorsal border of the scapula to the ribs and the transverse processes of the cervical vertebrae constitute the serratus ventralis muscle complex. To see it, first carefully free and mobilize the ventral border of the rhomboideus capitis muscle and separate it from the cranial border of the underlying serratus ventralis muscle. Bisect each of the three parts of the rhomboideus muscle complex at a different level so that you can distinguish the different parts even after they have been bisected. When you do this, make sure not to cut through the serratus ventralis muscle, otherwise the entire pectoral limb will fall off your specimen. Pull the top of the scapula laterally, and clean away fat and loose connective tissue from the exposed area. The large fan-shaped muscle that you see is the **serratus ventralis muscle** proper (Figs. 7-21 and 7-22). In rabbits, it is divided into a **serratus ventralis cervicis muscle** and a **serratus ventralis thoracis muscle** (Fig. 7-39). Approach the serratus ventralis muscle also from the ventral side by bisecting and reflecting the pectoral muscle. When doing this, bisect the individual parts of the pectoral muscle separately in order to cut perpendicular to the direction of the particular muscle fibers. Make sure to leave the underlying nerves and blood vessels intact for later study (Chapters 9 and 11). The serratus ventralis muscle arises by a number of slips from the ribs just dorsal to the junction of the bony ribs and the costal cartilages and from the transverse processes of the caudal cervical vertebrae. It inserts on the dorsal border of the scapula ventral to the insertion of the rhomboideus muscle (Figs. 7-23 and 7-24). The serratus ventralis muscle is a hypaxial muscle. Together with the pectoral muscle, it forms a muscular sling that transfers much of the weight of the body to the pectoral girdle and appendages (Fig. 7-22).

Clean out the area between the serratus ventralis muscle and the medial surface of the scapula. The subscapular fossa is occupied by a large **subscapular muscle,** which passes ventrally to insert on the lesser tuberculum of the humerus (Figs. 7-23 and 7-24). It adducts the humerus.

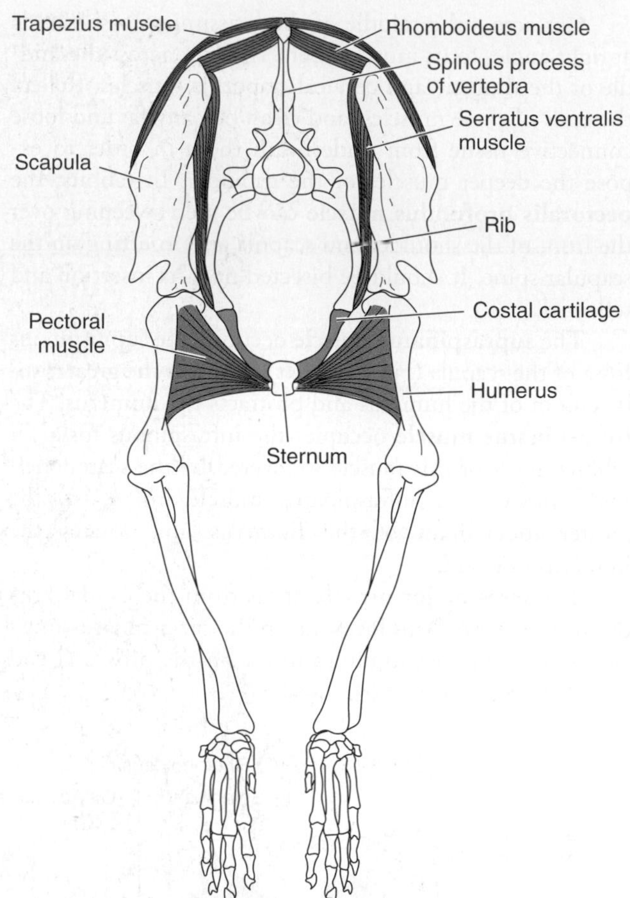

Figure 7-22

Diagrammatic cranial view of the thorax and pectoral appendages of a cat. The muscles shown form a muscular sling that suspends the thorax between the pectoral appendages. The trapezius and rhomboideus muscles pull the proximal end of the scapula mediodorsally, and the pectoral muscles pull the humerus and, thereby, the distal end of the scapula medially. In this manner, the scapula is stabilized. The serratus ventralis muscles suspend the thorax between the scapulae. *(From Liem, Bemis, Walker, and Grande, 2001; after Walker and Homberger, 1992.)*

Muscles of the Brachium

- Extensor muscles
  - Tensor fasciae antebrachii muscle (or dorsoepitrochlear muscle)
  - Triceps brachii muscle
  - Anconeus muscle (in cats)
- Flexor muscles
  - Brachial muscle
  - Biceps brachii muscle
  - Coracobrachial muscle

Remove superficial connective tissue from the muscles of the brachium and separate the muscles from one another. The large muscle that covers the caudal surface and most of the medial and lateral surfaces of the humerus is the triceps brachii muscle (see later). The **tensor fasciae antebrachii muscle** is closely associated with the medial surface of the triceps brachii muscle and should be studied first. It arises primarily from the lateral surface of the latis-

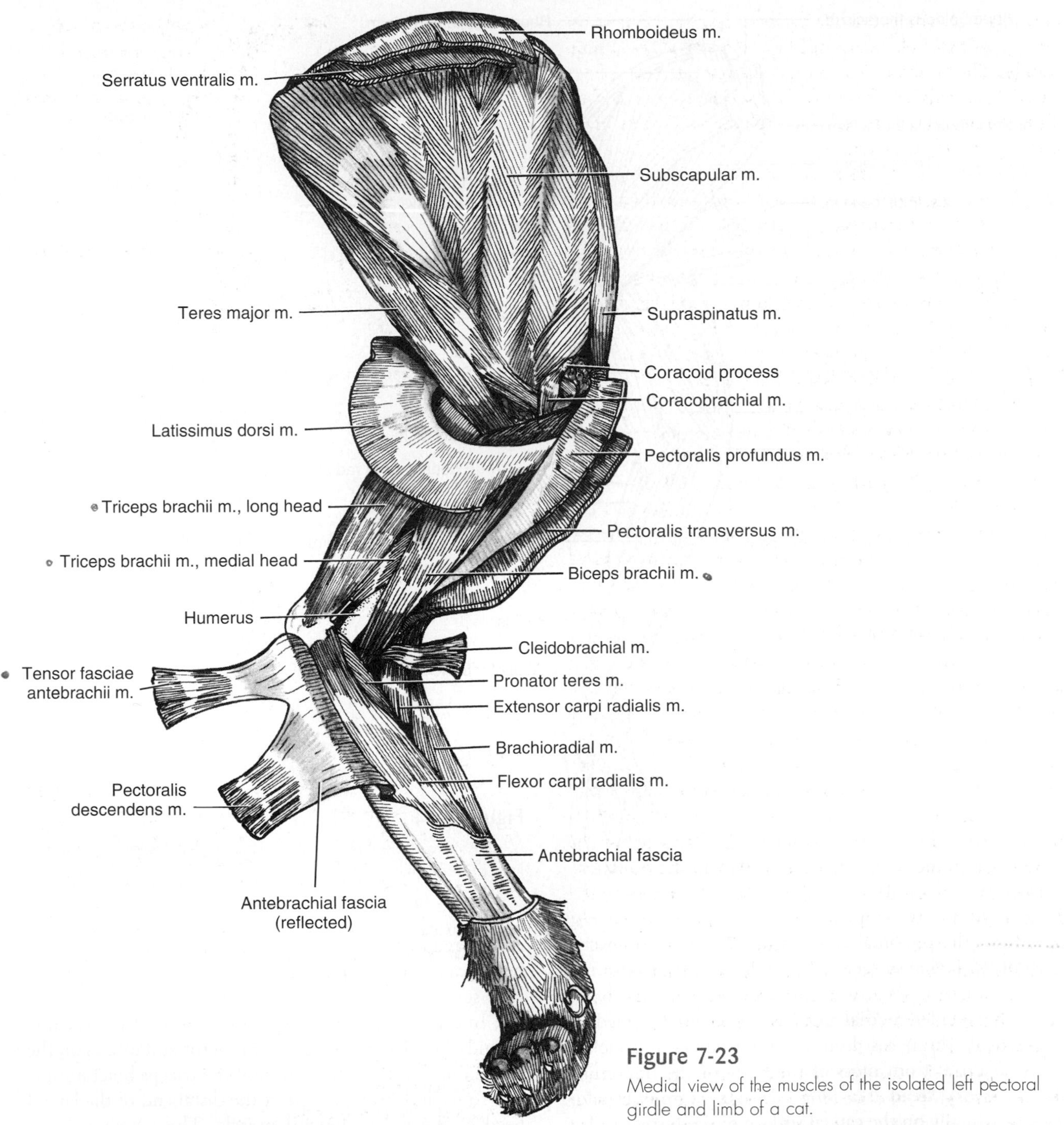

Figure 7-23
Medial view of the muscles of the isolated left pectoral girdle and limb of a cat.

simus dorsi muscle and extends distally along the medial surface of the arm to insert on the tendon of the triceps brachii muscle and on the antebrachial fascia (Figs. 7-23 and 7-24).

Bisect the tensor fasciae antebrachii muscle and reflect its halves. The **triceps brachii muscle** is now exposed (Figs. 7-19, 7-23, and 7-24). Contrary to its name, which is derived from its condition in humans, it has five heads in cats and four in rabbits. All heads insert in common on the olecranon of the ulna. The **long head,** which is located on the caudal surface of the humerus, is the largest. Note that it arises from the distal third of the caudal border of the scapula and, therefore, acts on the shoulder joint as well as on the elbow joint. The large **lateral head**

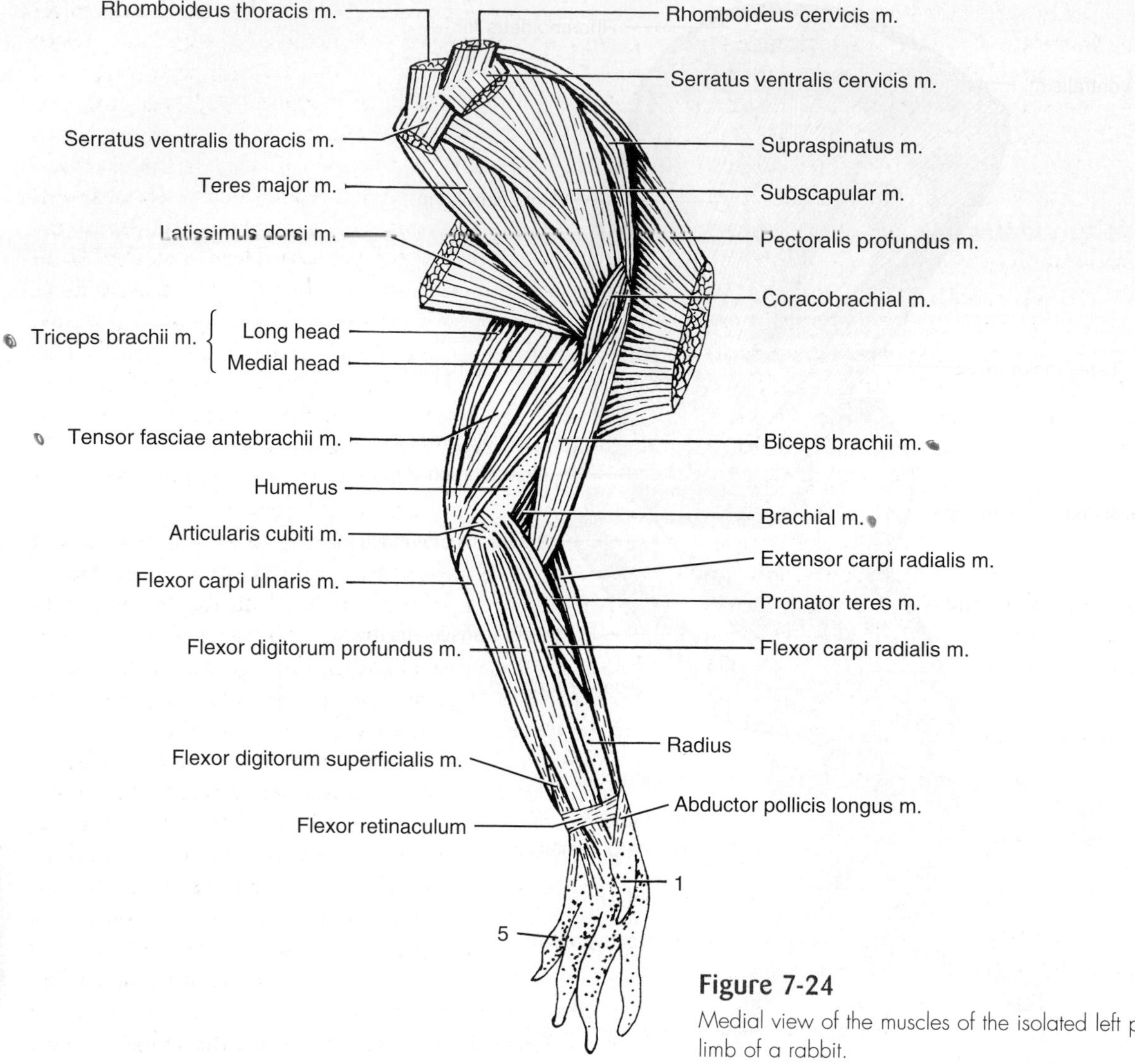

Figure 7-24

Medial view of the muscles of the isolated left pectoral girdle and limb of a rabbit.

arises from the proximal end of the humerus and covers much of the lateral surface of this bone. In cats, it is quite distinct, but in rabbits, it is partly united with the long head. The smaller **medial head** is found on the medial surface of the humerus deep to several nerves and blood vessels. It arises from most of the body of the humerus. The **accessory head** is a long and narrow muscle and arises proximally on the caudal surface of the humerus. In cats, it is largely fused with the medial head. In rabbits, only its proximal portion is muscular and lies between the long, medial, and lateral heads. Distally, the long tendon of the accessory head merges with the medial head. In cats, a very small **short head** arises from the medial surface of the distal end of the humerus just cranial to the supracondylar foramen (see Fig. 6-12**A** and **B**) and passes medially across the proximal ends of the flexor muscles of the antebrachium (see later). The short head usually is considered to be an unnamed division of the medial head, but because it is a distinct muscular head in all cats we have seen, we prefer to give it its own name.

In cats, a small **anconeus muscle** arises from the distal end of the lateral side of the humerus and inserts on the olecranon next to the insertion of the triceps brachii muscle. To see it, bisect and reflect the distal end of the lateral head of the triceps brachii muscle. The triceps brachii, tensor fasciae antebrachii, and anconeus muscles extend the forearm.

The craniolateral surface of the humerus is covered by the **brachial muscle,** which arises from the body of the humerus. In cats, it inserts on the proximal end of the ulna (Fig. 7-21), but in rabbits, it inserts on the ulna and radius in common with the biceps brachii muscle (Fig. 7-20).

The **biceps brachii muscle** is a prominent muscle on the craniomedial side of the humerus (Figs. 7-18, 7-23, and 7-24). To see it clearly, reflect the pectoral muscle near its insertion on the humerus. The biceps brachii muscle, which derives its name from the condition in humans (Figs. 7-3 and 7-4), originates by a single tendon from the dorsal border of the glenoid cavity on the scapula. This tendon lies in the intertubercular groove between the greater and lesser tubercles of the humerus. In cats, it inserts by a tendon on the radial tuberosity, which is located on the caudal surface of the radius (see Chapter 6 and Fig. 6-13**B**); therefore, the biceps brachii muscle also supinates the forearm. In rabbits, the biceps brachii muscle inserts on both the radius and ulna, which are immovably fused (see Chapter 6). The biceps brachii and brachialis muscles are the major flexors of the elbow joint.

Clean off the connective tissue from the medial side of the shoulder joint, and you will find a short band of muscle arising from the coracoid process and inserting on the proximal end of the humerus. This is the **coracobrachial muscle** (Figs. 7-23 and 7-24). In some individuals, a long head extends to the distal end of the humerus shaft. The coracobrachial muscle assists in the adduction of the humerus.

Muscles of the Antebrachium

- Extensor muscles
  - Brachioradial muscle (in cats)
  - Extensor carpi radialis muscle
    - Extensor carpi radialis longus muscle
    - Extensor carpi radialis brevis muscle
  - Extensor carpi ulnaris muscle
  - Extensor digitorum communis muscle
  - Extensor digitorum lateralis muscle
  - Extensor pollicis muscle (or extensor digiti I muscle)
  - Extensor indicis muscle (or extensor digiti II muscle)
  - Abductor pollicis longus muscle
  - Supinator muscle (in cats)
- Flexor muscles
  - Articularis cubiti muscle (in rabbits)
  - Pronator teres muscle
  - Flexor carpi radialis muscle
  - Flexor carpi ulnaris muscle (humeral and ulnar heads)
  - Flexor digitorum superficialis muscle
  - Flexor digitorum profundus muscle
  - Pronator quadratus muscle (in cats)
  - Lumbrical muscles
  - Abductor muscle of the fifth digit (in cats)

Before you can study the muscles of the forearm and hand, you will have to remove the very extensive antebrachial fascia. When removing this fascia, be careful not to damage the brachioradial muscle (see later and Figs. 7-21 and 7-25). The deeper layer of the antebrachial fascia is continuous with the tendons of the tensor fasciae antebrachii, triceps brachii, and pectoralis descendens muscles. It forms a dense fibrous sheet that dips down between many of the muscles that attach onto the ulna and radius. At the level of the wrist, part of the antebrachial fascia forms ligaments that encircle the wrist and hold the longitudinal tendons in place. A band of dense collagenous fibers, the **extensor retinaculum,** bridges the tendon grooves on the dorsal side of the radius, and a comparable **flexor retinaculum** lies on the palmar side of the wrist (Fig. 7-24). The antebrachial fascia extends into the hand and, on the palmar side, contributes to the formation of fibrous sheaths, the **vaginal ligaments,** through which the long flexor tendons run toward their attachments on the fingers. After you have removed the antebrachial fascia, separate the major muscles from one another before attempting to identify them.

The muscles of the forearm and hand can be sorted into an extensor muscle group located on the craniolateral surface of the forearm and the back of the hand and a flexor muscle group located on the caudomedial surface of the forearm and palm of the hand. In the elbow region, the insertions of the biceps brachii, brachial, and cleidobrachial muscles pass between these groups at the cranial border of the arm, and the ulna and the insertion of the triceps brachii muscle separate them caudally.

Most extensor muscles arise from or near the lateral epicondyle of the humerus. You can recognize the following superficial muscles at the level of the elbow, starting from the medial, or thumb, side of the forearm: brachioradial muscle, which is absent in rabbits; extensor carpi radialis muscle complex; extensor digitorum communis muscle; extensor digitorum lateralis muscle; and extensor carpi ulnaris muscle (Fig. 7-25). The **brachioradial muscle** arises more proximally from the humerus than the other extensor muscles and inserts on the styloid process of the radius. The extensor carpi radialis muscle complex can be divided into the more superficial and cranial **extensor carpi radialis longus muscle** and a deeper and more caudal **extensor carpi radialis brevis muscle.** Both muscles originate near the lateral epicondyle of the humerus, and the tendons of both muscles pass deep to the tendon of the abductor pollicis muscle (see later and Fig. 7-25) to insert on the proximal ends of the second and third metacarpals, respectively. These three extensor muscles act to extend the paw at the wrist. The brachioradial muscle also assists the supinator muscle (see later) in supinating the paw. The **extensor carpi ulnaris muscle** is positioned on the ulnar side of the forearm, and its broad tendon inserts on the base of the fifth metacarpal. In cats, this muscle is a flexor *(sic!)* and abductor of the wrist; it was named for its condition in humans.

The long extensor muscles of the digits are more complex. The **extensor digitorum communis muscle** originates from the lateral epicondyle of the humerus. Its dis-

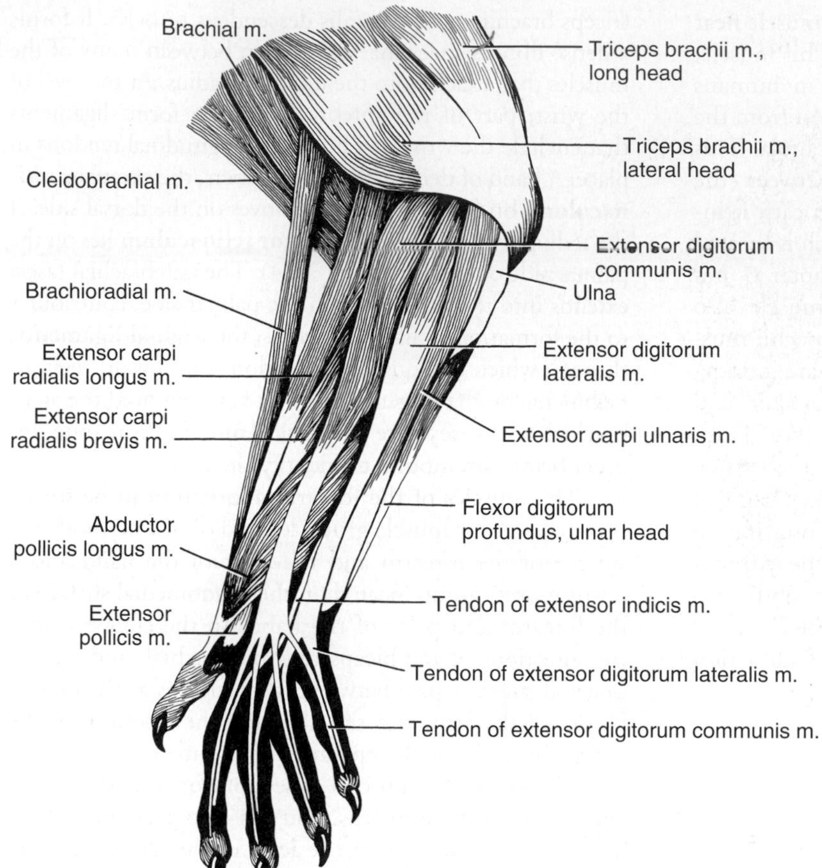

Figure 7-25

Lateral view of the muscles of the left forearm of a cat. The antebrachial fascia and extensor retinaculum are removed to reveal the underlying muscles.

tal tendon passes through the lateral tendon groove on the distal end of the radius and breaks up into four tendons, which run along the dorsal side of digits 2 to 5. The tendons are held in place on each phalanx by connective tissue bands and insert on the terminal phalanges. The **extensor digitorum lateralis muscle** originates from the humerus near the origin of the extensor digitorum communis muscle, and its distal tendon passes through the tendon groove on the distal end of the ulna. At the level of the wrist, the tendon breaks up into three or four tendons, which run along the dorsolateral surface of digits 2 to 5, or digits 3 to 5. These tendons merge with their counterparts of the extensor digitorum communis muscle near the distal end of the proximal phalanges. Occasionally, the tendon to the fifth digit arises from a separate muscular head. The extensores digitorum communis et lateralis muscles extend the wrist and the second to fifth digits.

Four muscles lie deep to the preceding ones on the extensor surface of the forearm (Fig. 7-26). Expose them by bisecting and reflecting the muscular portions of the two long extensor muscles of the digits and the extensor carpi ulnaris muscle. The narrow **extensor pollicis muscle** and the even narrower **extensor indicis muscle** lie deep to the extensor carpi ulnaris muscle and arise from the proximal three fourths of the lateral surface of the ulna. The extensor pollicis muscle often is absent. If present, it turns into a tendon proximal to the extensor retinaculum of the wrist and adheres to the cranial surface of the extensor indicis muscle, which turns into a tendon only after passing under the extensor retinaculum. The tendons of both muscles merge with the tendons of the long extensor muscles of the digits at mid-length of the digits. Both muscles contribute to the extension of the first two digits.

The powerful **abductor pollicis longus muscle** arises from much of the lateral surface of the ulna and adjacent parts of the radius. Its muscle fibers converge, continue underneath the tendon of the extensor digitorum communis muscle, and form a tendon that crosses over the extensor carpi radialis tendons to insert upon the radial side of the first metacarpal. This muscle abducts the thumb.

In cats, the **supinator muscle** crosses obliquely over the proximal half of the radius underneath the belly of the extensor digitorum communis muscle and proximal to the belly of the abductor pollicis longus muscle (Fig. 7-26). It arises from the lateral epicondyle of the humerus and el-

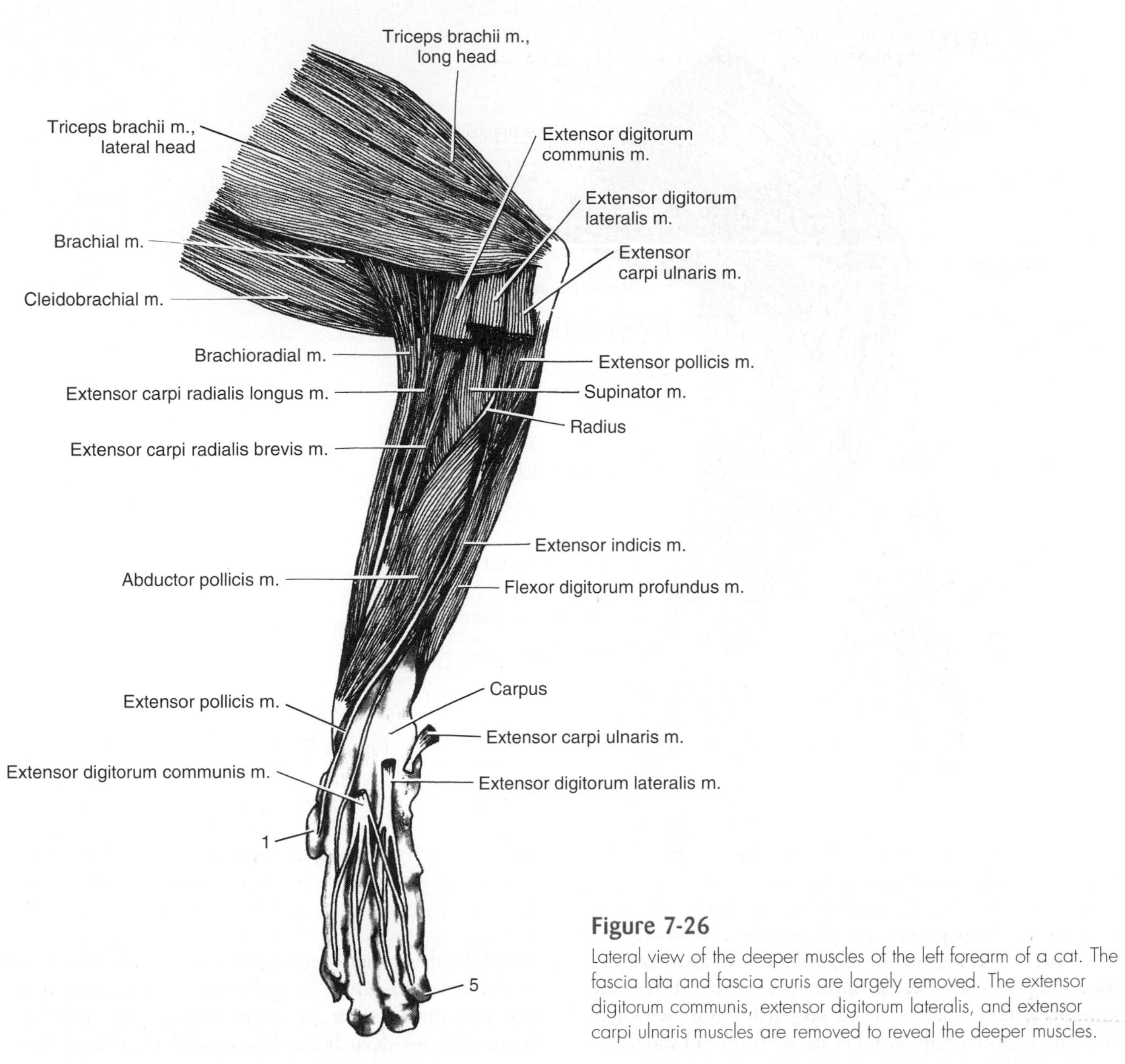

Figure 7-26
Lateral view of the deeper muscles of the left forearm of a cat. The fascia lata and fascia cruris are largely removed. The extensor digitorum communis, extensor digitorum lateralis, and extensor carpi ulnaris muscles are removed to reveal the deeper muscles.

bow ligaments and inserts upon the radius. Its diagonal muscle fibers enable it to act as a powerful supinator of the forepaw.

Most flexor muscles arise from or near the medial epicondyle of the humerus. In rabbits, a small **articularis cubiti muscle** strengthens the medial side of the humeroulnar joint. The remaining superficial muscles are, from cranial to caudal, the pronator teres, flexor carpi radialis, flexor digitorum superficialis, and flexor carpi ulnaris muscles (Figs. 7-24 and 7-27). The **pronator teres muscle** extends diagonally from the medial epicondyle of the humerus to the medial border of the radius. It pronates the forearm and forepaw.

The **flexor carpi radialis muscle** extends distally along the radial border of the forearm and forms a long tendon that passes deep to other tendons in the hand to insert on the proximal ends the second and third metacarpals. To see the insertions, it is necessary to cut through the overlying tough connective tissue with a pair of small scissors. The **flexor carpi ulnaris muscle** is positioned on the ulnar side of the forearm. It inserts on the pisiform bone of the carpus and arises by two heads. The more superficial **humeral head** arises from the medial epicondyle. The **ulnar head,** which is largely fused to the medial and deep surfaces of the humeral head, arises from the caudomedial side of the ulnar olecranon. Both carpal flexor muscles flex the paw at the wrist.

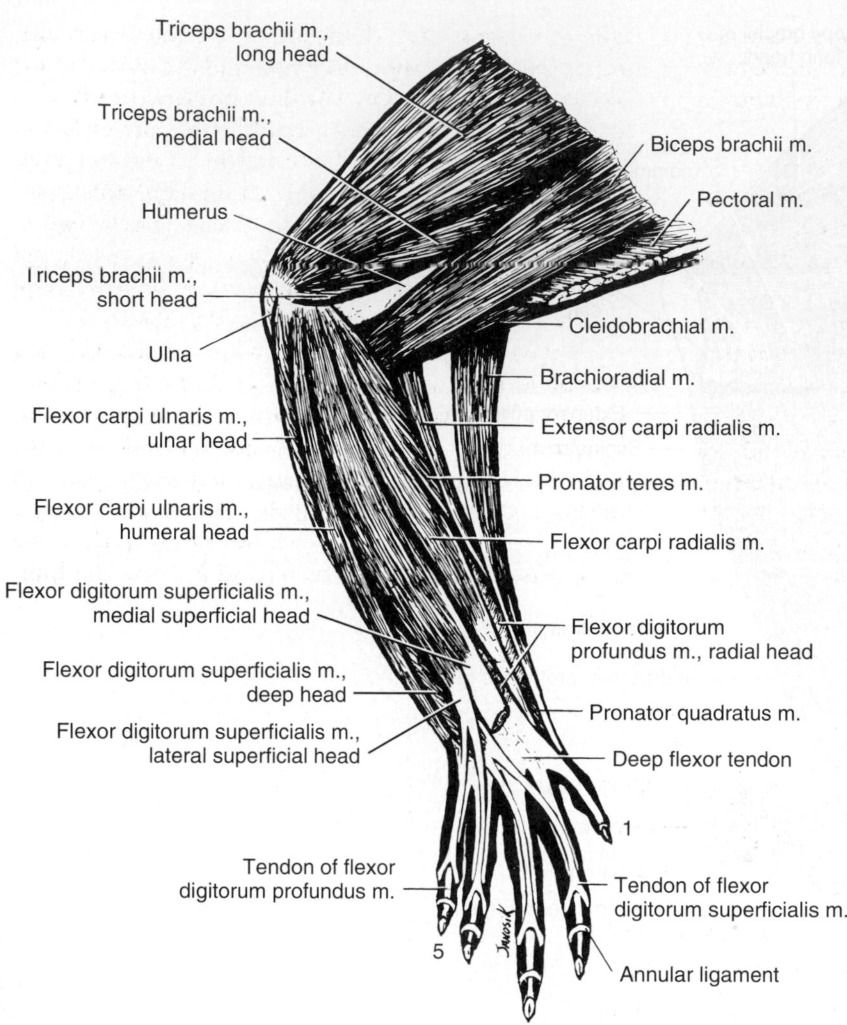

Figure 7-27
Medial view of the muscles of the left forearm of a cat. The antebrachial fascia, flexor retinaculum, palmar aponeurosis, and part of the flexor digitorum superficialis muscle are removed to reveal the deeper muscles.

The **flexor digitorum superficialis muscle** lies between the two carpal flexor muscles. It is the widest of the superficial flexor muscles and arises partly from the medial epicondyle of the humerus and partly from the surface of the flexor digitorum profundus muscle (see later). At the level of the wrist, it becomes tendinous and runs underneath the flexor retinaculum. Some superficial tendons branch off and attach to the footpads (these superficial tendons should be cut near their attachments when the skin is removed), but the several main tendons insert on the middle phalanx of each digit. Clean the flexor digitorum superficialis muscle and remove the connective tissue surrounding the individual tendons so that you can see their insertions on the phalanges. In cats, the flexor digitorum superficialis muscle is very large. It must be mobilized and pushed to the side to expose its various heads, as well as the underlying single wide and thick tendon of the flexor digitorum profundus muscle (see later). You can do so by cutting through the flexor retinaculum along the medial border of the flexor digitorum superficialis muscle proximally to the first digit. Three heads of the flexor digitorum superficialis muscle can be distinguished: two superficial heads and one deep head. The **medial superficial head** forms a strong tendon, which splits into four relatively thick superficial tendons to digits 1 to 4. The **lateral superficial head** is smaller and splits into two thin, deeper tendons to digits 3 and 4. The **deep head** arises from the surface of the underlying flexor digitorum profundus muscle and sends three thin tendons to digits 2 to 4 (Fig. 7-28). Thus, the flexor digitorum superficialis muscle sends one tendon each to digits 1 and 5; two tendons each to digits 2 and 3; and three tendons to digit 4. Near its insertion, each tendon splits into two branches and attaches to the sides of the middle phalanx of each digit. The tendons of the flexor digitorum profundus muscle (see later) pass through, and are held in place by, the arches formed by the splitting of the superficial tendons. The flexor digitorum superficialis muscle flexes the wrist and the digits up to their middle phalanges. In rabbits, the flexor digitorum superficialis muscle is much smaller. It consists of only one muscle belly that originates from the medial epicondyle of the humerus and sends three tendons to digits 2 to 4.

Push the flexor digitorum superficialis muscle to the side and identify the thick wide tendon of the **flexor digitorum profundus muscle.** Put a blunt probe under this tendon and isolate it from the tendons of the flexor digitorum superficialis muscle. The five (in cats) or four (in rabbits) muscle heads of the flexor digitorum profundus muscle all converge toward the common deep flexor tendon (Fig. 7-28). If necessary, bisect the muscle heads of overlying muscles. In cats, identify first the very large ulnar and radial heads of the flexor digitorum profundus muscle. The **ulnar head** arises from most of the length of the outer edge of the ulna, and part of it is visible from the extensor side of the forearm (Fig. 7-25). The **radial head** is the most medial and deepest of the five heads; it arises from the middle third of the radius, the interosseus ligament stretching between the radius and the ulna, and the adjacent parts of the ulna. The three **humeral heads** lying between the radial and ulnar heads arise more or less in common from the medial epicondyle of the humerus. Two of these heads lie deep to the flexor digitorum superficialis muscle and the flexor carpi radialis muscle, respectively. The central humeral head is more superficial, and part of it can be seen occasionally between the flexor carpi radialis and flexor digitorum superficialis muscles.

The common deep flexor tendon extends into the palm of the forepaw and breaks up into five strong tendons, which run along the ventral surfaces of the digits, pass underneath annular ligaments, perforate the tendons of the flexor digitorum superficialis muscle, and finally insert on the terminal phalanges of the digits. The flexor digitorum profundus muscle flexes the wrist and all segments of the digits. Some small intrinsic muscles of the paw, the **lum-**

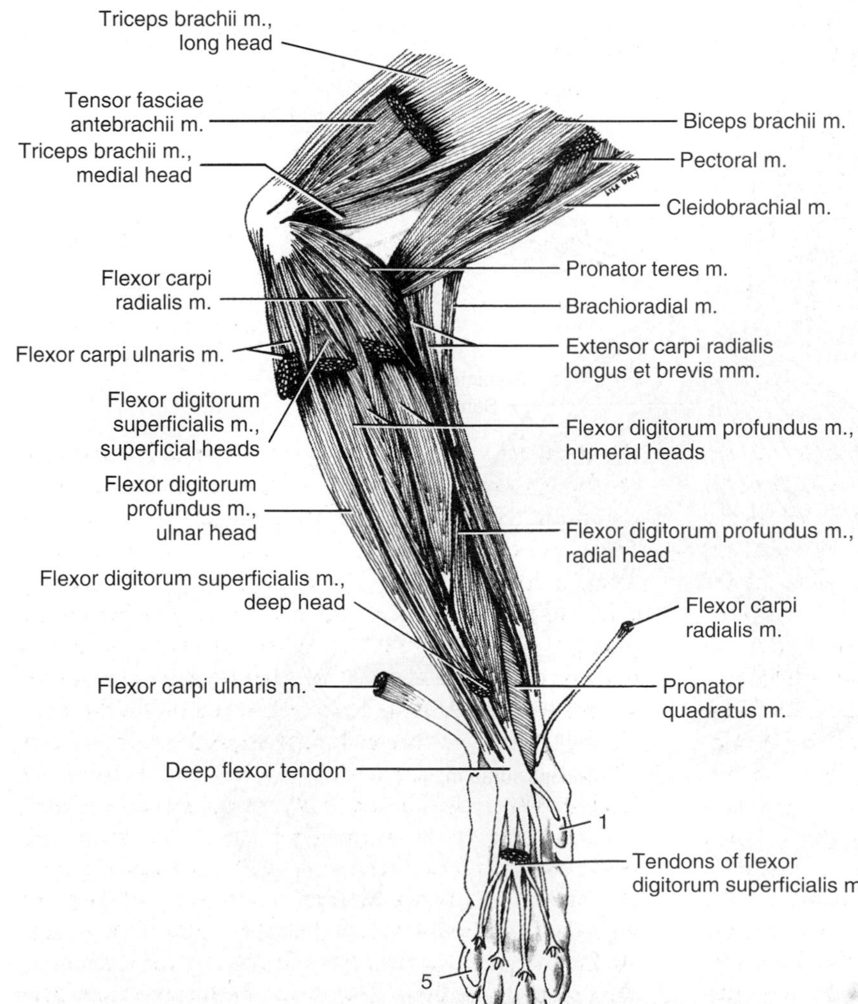

Figure 7-28
Medial view of the deeper muscles of the left forearm of a cat. The flexor digitorum superficialis, flexor carpi ulnaris, and flexor carpi radialis muscles are removed to reveal deeper muscles.

brical muscles, arise superficially from the common deep flexor tendon and insert on the joints of the proximal phalanges of digits 2 to 5. They flex the proximal phalanges. A small but fairly powerful and bulbous muscle, the **abductor muscle of the fifth digit,** arises distal to the flexor carpi ulnaris muscle on the pisiform and fifth metacarpal bones. It inserts by a long tendon on the proximal phalanx of digit 5. Other intrinsic muscles of the forepaw are more difficult to identify without magnification.

Spread apart the radial head, the common tendon of the flexor digitorum profundus muscle, and the tendon of the flexor carpi radialis muscle. In cats, a bluish, shiny fascia will be revealed. Make a longitudinal incision through this fascia and expose the **pronator quadratus muscle** (Fig. 7-28). Its muscle fibers extend diagonally from their origin on the distal third of the ulna to their insertion on the outer edge of the radius. It pronates the forepaw (see Fig. 6-13**C**). Rabbits lack this muscle along with the brachioradial and supinator muscles on the extensor surface of the forearm, because the configuration of their radius and ulna does not permit axial rotation of the forearm (see Fig. 6-13**D** and **E**).

Pelvic Appendicular Muscles

Although the configuration of thigh muscles of all mammals is basically the same, there are distinct differences in the topographical relationships between certain muscles of cats and rabbits. Clear away any fat and loose connective tissue from the surfaces of the pelvic region and thigh as a preliminary to the dissection. When doing so, take care not to damage the aponeuroses of the gluteal muscles and of the tensor fasciae latae and biceps femoris muscles (Fig. 7-29). Also remove the large pads of fat lateral to the base of the tail and in the **popliteal fossa,** the depression behind the knee. Start to separate the more obvious muscles.

Lateral Thigh and Adjacent Muscles

- Sartorius muscle
- Tensor fasciae latae muscle

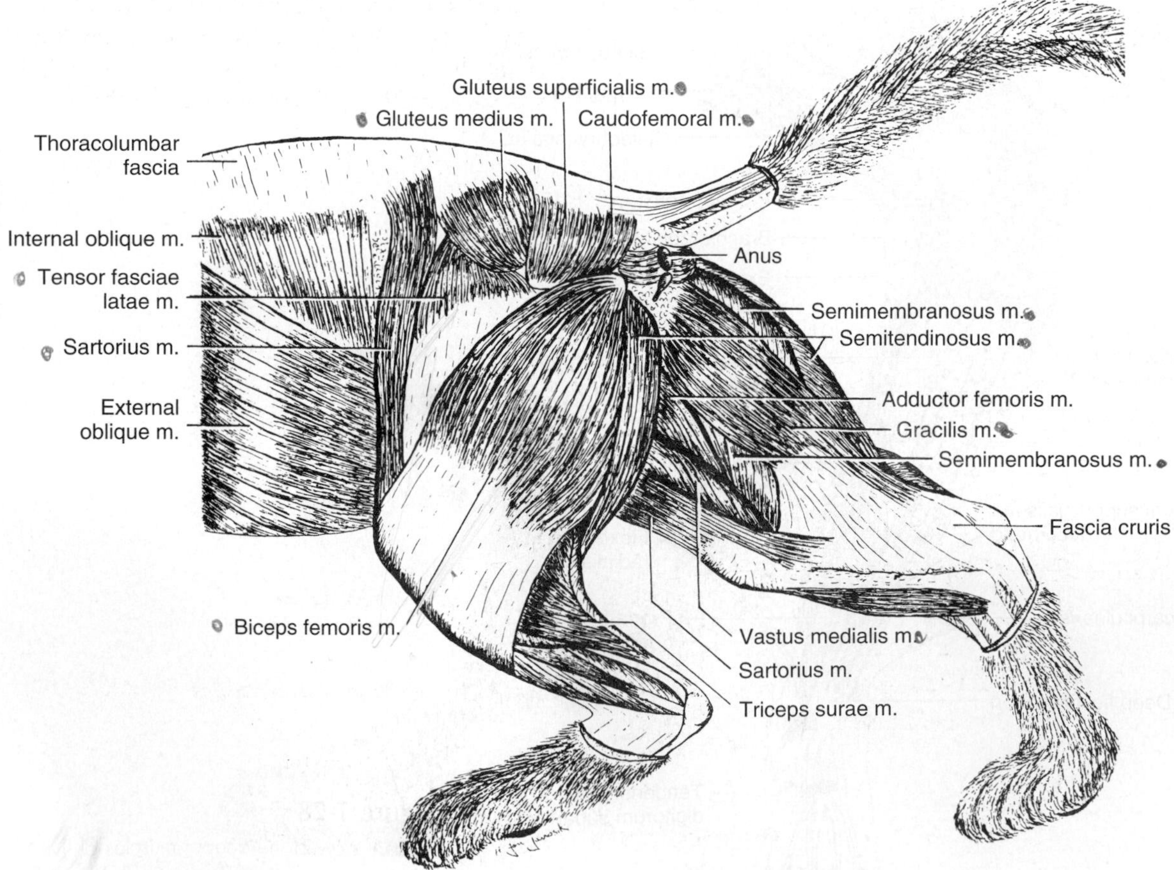

Figure 7-29

Superficial pelvic and thigh muscles of a cat. The left leg shows a lateral view of the thigh muscles. The right leg shows a medial view of the thigh muscles.

- Biceps femoris muscle
- Semitendinosus muscle
- Caudofemoral muscle (or coccygeofemoral muscle, or gluteobiceps muscle)
- Abductor cruris caudalis muscle

The most cranial muscle on the thigh of cats is the **sartorius muscle** (Fig. 7-29). It is a muscle band that extends from its origin on the crest and ventral border of the ilium to its insertion on the patella and medial surface of the thigh, but part of it can be seen laterally. This lateral part of the sartorius muscle arises from the thoracolumbar fascia. In rabbits, the muscle lies entirely on the medial surface (Fig. 7-34) and originates from the inguinal ligament. The sartorius muscle inserts on the medial condyle of the tibia and merges distally with the gracilis muscle (see later and Fig. 7-29). The sartorius muscle adducts and protracts the thigh and contributes to the extension of the shank at the knee.

In cats, the tough, whitish **fascia lata** lies on the lateral surface of the hip and thigh caudal to the sartorius muscle, but in rabbits, it lies on the craniolateral surface of the thigh. The **tensor fasciae latae muscle** arises from the ventral border of the ilium and cranially from the surface of underlying muscles, and it inserts on the fascia lata. In cats, the tensor fasciae latae muscle is subdivided into two portions by a dorsal extension of the fascia lata. The cranial portion is longer and thicker and covers most of the cranial surface of the thigh. The caudal portion is shorter and sheetlike and tends to merge caudally with the gluteus superficialis muscle (see later). In rabbits, the tensor fasciae latae muscle spreads over the medial side of the thigh and partly unites with the vastus medialis muscle of the quadriceps muscle complex (see later and Fig. 7-34).

The lateral surface of the thigh caudal to the fascia lata is covered by the very broad **biceps femoris muscle** (Figs. 7-29 and 7-30). It has a narrow origin from the ischial tuberosity of the ischium and fans out to insert by a broad aponeurosis on the patella and much of the tibia. The biceps femoris muscle forms the lateral wall of the **popliteal fossa,** the depression on the caudal side of the knee. It flexes the shank and abducts the thigh.

The more slender **semitendinosus muscle** lies caudal to the origin of the biceps femoris muscle and contributes to the medial wall of the popliteal fossa. It also arises from the ischial tuberosity, but in rabbits, a part of it arises from the fascia overlying the biceps femoris muscle. It inserts on the medial side of the proximal end of the tibia. In rabbits, part of its insertion joins the calcaneus tendon of several shank flexor muscles (see later). The semitendinosus muscle flexes the shank and retracts the thigh.

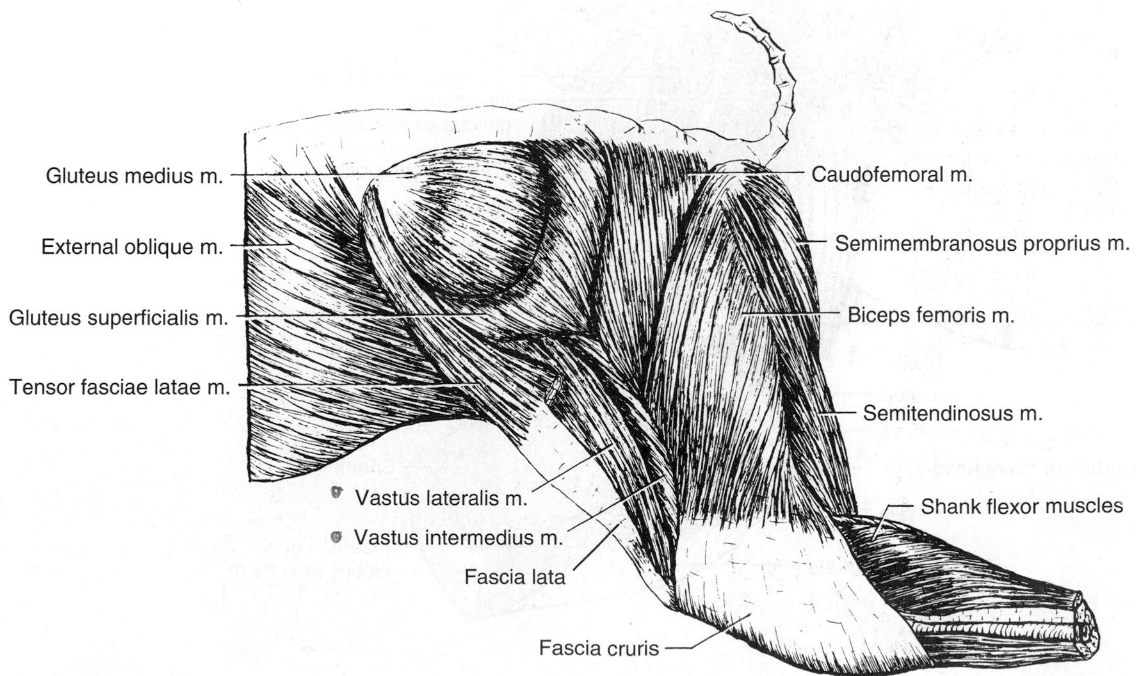

Figure 7-30

Lateral view of the superficial pelvic and thigh muscles of a rabbit. Part of the fascia lata has been removed to reveal the underlying muscles.

The **caudofemoral muscle** forms a band of muscle craniodorsal to the origin of the biceps femoris muscle. It arises from the more caudal sacral and anterior caudal vertebrae, passes underneath the cranial border of the biceps femoris muscle, and forms a narrow tendon that inserts on the patella in common with part of the biceps femoris muscle. Lift up the biceps femoris muscle to follow the caudofemoral muscle. The caudofemoral muscle abducts and retracts the thigh and extends the shank. It is closely associated with the gluteal muscles (see later).

Free and mobilize the caudal border of the biceps femoris muscle. As you remove the superficial connective tissue, a very thin muscle band, the **abductor cruris caudalis muscle,** becomes visible projecting from under the caudal border of the biceps femoris muscle in the popliteal fossa. Bisect and reflect the biceps femoris muscle without damaging the abductor cruris caudalis muscle (Fig. 7-31). This muscle arises from a sacral or caudal vertebra and inserts on the tibia and adjacent crural fascia *(fascia cruris)* with the biceps femoris muscle. Because it contains a disproportionally high number of muscle spindles, it probably serves as a tension sensor rather than as a thigh abductor as its name implies.

Gluteal Muscle Complex and Deeper Pelvic Muscles

- Gluteus muscle complex
  - Gluteus superficialis muscle
  - Gluteus medius muscle
  - Gluteus profundus muscle
  - Gluteus accessorius muscle (in rabbits)
- Piriformis muscle
- Articularis coxae muscle (in cats)
- Gemellus cranialis muscle
- Gemellus caudalis muscle
- Obturator externus muscle
- Obturator internus muscle
- Quadratus femoris muscle
- Also seen: Coccygeus muscle (an axial muscle)

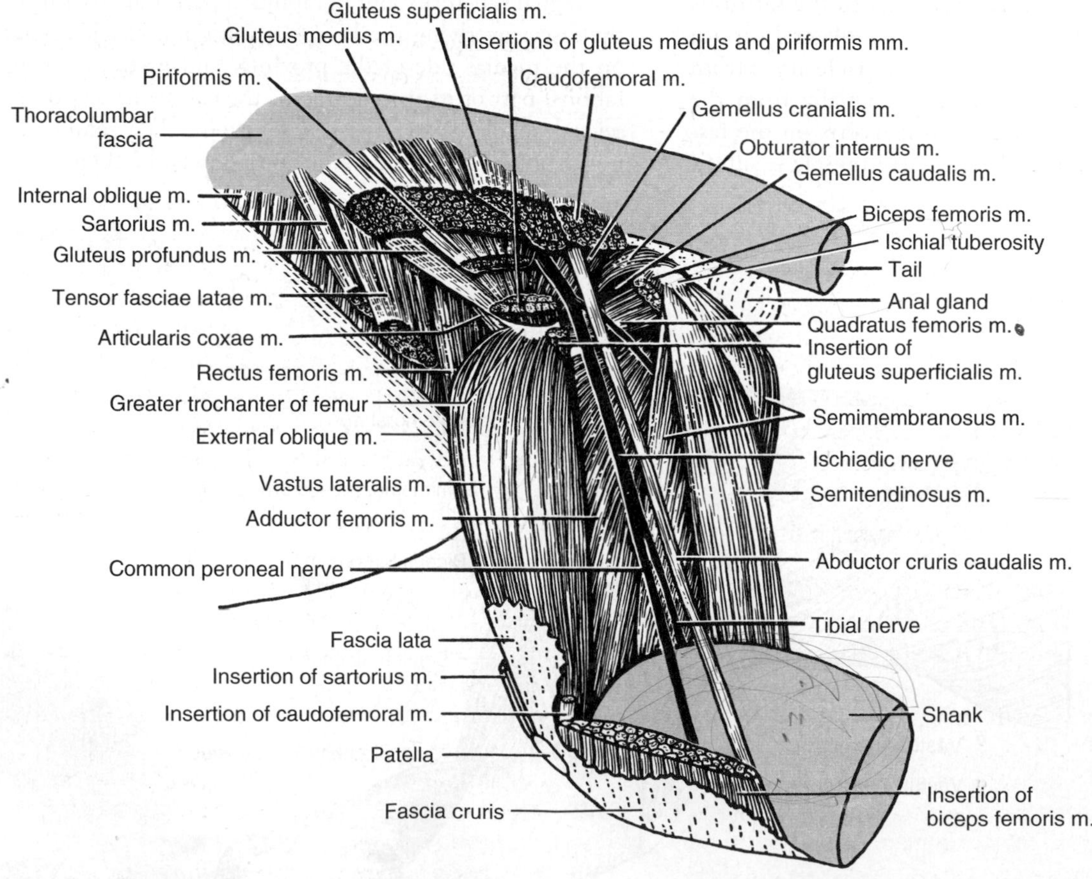

Figure 7-31

Lateral view of deep muscles of the pelvis and thigh of a cat. The sartorius, tensor fasciae latae, biceps femoris, gluteus medius, gluteus superficialis, caudofemoral, and piriformis muscles are largely cut away to expose the deeper muscle, but the origins and insertions of these muscles are shown as points of reference.

The gluteus muscle complex lies underneath the tensor fasciae latae muscle (Figs. 7-29 and 7-30). To see it, you must first transect and reflect the tensor fasciae latae muscle. In cats, start to cut through the cranial border of the muscular part of its cranial portion, which tends to merge proximally with the underlying gluteus medius muscle (see later; you do not need to separate these two muscles now, because later they can be reflected together). It may not be easy to identify the caudal border of the tensor fasciae latae muscle because it tends to merge with the gluteus superficialis muscle. If this is the case, you will have to separate the two muscles later on the basis of their different insertions, and you will have to bisect the sartorius muscle. In rabbits, start to bisect the tensor fasciae latae muscle at its medial border after you have separated it from the vastus medialis muscle on the medial side of the thigh (Fig. 7-34).

The **gluteus superficialis muscle** is the most superficial muscle of the gluteus muscle complex. It arises from the sacral portion of the thoracolumbar fascia and from the spinous processes of sacral and anterior caudal vertebrae. In rabbits, a cranial and more superficial portion also arises from the ventral border of the ilium, deep to the ventral border of the gluteus medius muscle. In cats, the muscle fiber bundles converge to insert slightly distal to the greater trochanter of the femur so that you can now clearly distinguish it from the tensor fasciae latae muscle because of its different insertion. In rabbits, the gluteus superficialis muscle inserts on the third trochanter of the femur (see Fig. 6-17**C** and **D**).

Bisect and reflect the gluteus superficialis muscle. The **gluteus medius muscle,** which lies partly under the gluteus superficialis muscle, arises from the sacral portion of the thoracolumbar fascia and from the crest and lateral surface of the ilium and from adjacent vertebrae. It inserts upon the greater trochanter of the femur. Both the gluteus superficialis and gluteus medius muscles abduct the thigh.

Bisect the gluteus medius muscle with a scalpel by slowly cutting deeper and deeper. In cats, the gluteus medius muscle tends to merge with the underlying **piriformis muscle;** as soon as you have reached a muscle layer that is heavily interspersed with tendons, you have reached the piriformis muscle. Separate now the gluteus medius muscle from the piriformis muscle by reflecting the gluteus medius; if this is not possible, proceed to cut through the piriformis muscle. A prominent ischiadic nerve emerges from underneath the caudal border of the piriformis muscle. For now, do not cut deeper than the level of the ischiadic nerve. The piriformis muscle arises from the last sacral and first caudal vertebrae and inserts on the greater trochanter of the femur. It abducts the femur. Reflect the two halves of the piriformis muscle.

The **gluteus profundus muscle** is the deepest muscle of the gluteus muscle complex. In cats, it consists of a cranial part, which is cylindrical with a shiny tendinous surface, and a caudal part, which is fan shaped and extends caudally under the piriformis muscle. The gluteus profundus muscle arises from the lateral surface of the ilium and inserts on the greater trochanter of the femur, skirting the distal side of the inserting gluteus medius and piriformis muscles. It abducts the thigh. In rabbits, the gluteus profundus muscle lies cranial to the piriformis muscle and is associated with the **gluteus accessorius muscle.**

If you are dissecting a cat, look deep between the gluteus profundus muscle and the origins of the vastus lateralis and rectus femoris muscles (Fig. 7-31). The thin muscle observed is the **articularis coxae muscle.** It arises from part of the lateral surface of the ilium and inserts on the proximal end of the femur. Because of its small size and richness in muscle spindles, it probably serves as a tension sensor rather than as a thigh protractor. Rabbits do not possess such a muscle.

Bisect the ischiadic nerve and the abductor cruris caudalis muscle (if it is not already torn) to have a clear view of three small muscles on the lateral surface of the ischium. The **gemellus cranialis muscle** is the most cranial one of these three muscles and arises from the palpable dorsal rim of the ilium and ischium. The **obturator internus muscle** arises from the medial surface of the ischium and the connective tissue sheet covering the obturator foramen (see Chapter 6) and emerges over the dorsal rim of the pelvis to lie between the two gemelli muscles. The **gemellus caudalis muscle** arises from the lateral surface of the ischium just cranial to the ischial tuberosity. Occasionally, the origins of the gemellus caudalis and obturator internus muscles grade into each other if the obturator internus muscle extends its origin all the way to the ischial tuberosity. The three muscles abduct the thigh. In cats, the three muscles insert on the greater trochanter of the femur along the proximal border of the insertion of the gluteus medius and piriformis muscles. In rabbits, the three muscles converge into a thick tendon and insert in the trochanteric fossa of the femur.

The **coccygeus muscle** is a deep axial muscle (Fig. 7-32). At first it may look like a dorsal extension of the gemellus cranialis muscle, but it originates from the dorsal rim of the ilium and ischium, extends dorsally to the origin of the gemellus cranialis muscle, and inserts on the tail vertebrae. It contributes to the formation of the wall of the pelvic canal.

The **quadratus femoris** muscle is the rather thick band of muscle distal to the gemellus caudalis muscle. It arises from the ischial tuberosity deep to the origin of the biceps femoris muscle and inserts on the femur at the bases of the greater and lesser trochanters. It is primarily a thigh retractor.

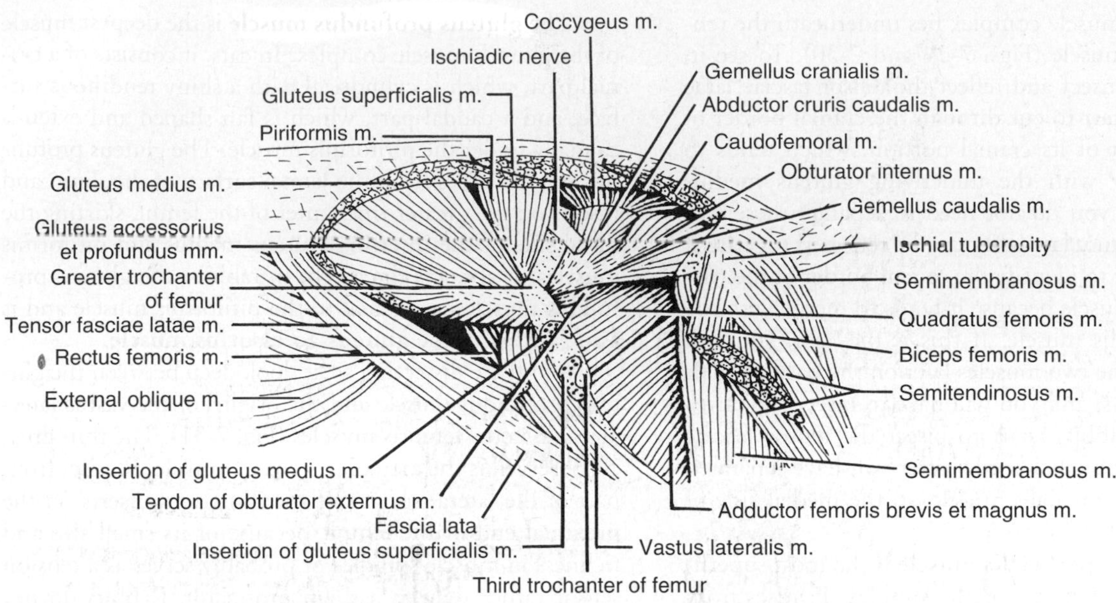

Figure 7-32

Lateral view of the deep pelvic muscles of a rabbit. The caudofemoral, abductor cruris caudalis, gluteus superficialis, gluteus medius, and biceps femoris muscles, as well as the fascia lata, are largely cut away to expose the deeper muscle, but the origins and insertions of these muscles are shown as points of reference.

Separate the gemellus caudalis and quadratus femoris muscles. Bisect the quadratus femoris muscle to reveal the **obturator externus muscle.** In cats it is muscular, but in rabbits it is tendinous. It arises from the lateral surface of the pelvis around the obturator foramen and inserts deep in the trochanteric fossa of the femur. It adducts and retracts the thigh.

Quadriceps Femoris Muscle Complex

- Quadriceps femoris muscle complex
  - Rectus femoris muscle
  - Vastus lateralis muscle
  - Vastus medialis muscle
  - Vastus intermedius muscle

The front portion of the thigh of mammals is covered by a group of four muscles that insert in common on the **patella,** which is a sesamoid bone (see Chapter 6, p. 109). The patella is anchored to the tibial tuberosity via the **patellar tendon.** It contributes to the formation of the cranial surface of the knee joint and permits the common tendon of the quadriceps muscle to slide easily across the joint. The muscle complex in its entirety is called the **quadriceps femoris muscle complex,** and it is the primary shank extensor.

Reflect and separate the tensor fasciae latae muscle from adjacent muscles and clean the exposed area. The large muscle on the craniolateral surface of the thigh, which was largely covered by the tensor fasciae latae muscle, is the **vastus lateralis muscle** (Figs. 7-30 and 7-31). It arises from the greater trochanter and adjacent parts of the femur. The **rectus femoris muscle** lies on the cranial thigh surface medial to the vastus lateralis muscle (Figs. 7-33 and 7-34). Because it arises from the ilium just cranial to the acetabulum, it also acts across the hip joint and assists in protracting the thigh. The **vastus medialis muscle** lies on the medial surface of the thigh caudal to the rectus femoris muscle and will be seen later after the overlying gracilis muscle has been bisected and reflected. It arises from the body of the femur (Figs. 7-29 and 7-34). In cats, the **vastus intermedius muscle** can be found by dissecting deeply between the vastus lateralis and rectus femoris muscles. It arises from the femur lateral to the origin of the vastus medialis muscle and often merges with the vastus medialis muscle. In rabbits, the vastus intermedius muscle is larger so that it can also be seen partly on the lateral side of the thigh, if the vastus lateralis and caudofemoralis muscles are pulled apart slightly (Fig. 7-30).

Caudomedial Thigh Muscles

- Gracilis muscle
- Semimembranosus muscle
  - Semimembranosus proprius muscle (in rabbits)
  - Semimembranosus accessorius muscle (in rabbits)
- Pectineus muscle

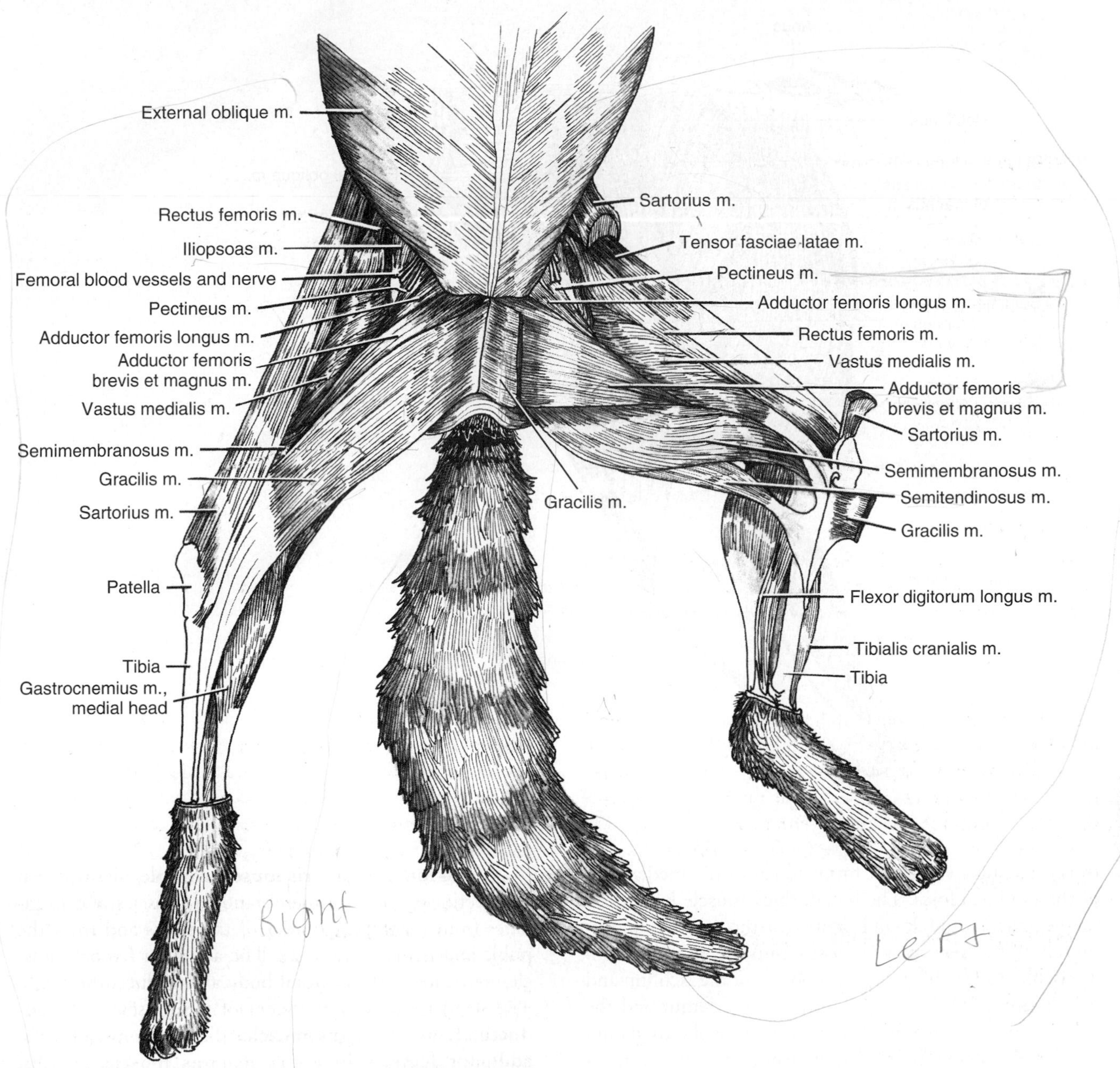

Figure 7-33
Medial view of the thigh muscles of a cat. The right leg shows the superficial muscles. On the left leg, the sartorius and gracilis muscles are largely cut away to reveal the deeper muscles, but the origins and insertions of these muscles are shown as points of reference.

- Adductor femoris muscle complex
  - Adductor longus muscle
  - Adductor brevis et magnus muscle

The medial surface of the thigh caudal to the sartorius and quadriceps femoris muscles is largely covered by the **gracilis muscle,** a wide, thin muscle that arises from the pubic and ischial symphyses (Figs. 7-29, 7-33, and 7-34). It inserts by an aponeurosis onto the tibia and crural fascia. In rabbits, it merges with the distal part of the sartorius muscle. The gracilis muscle adducts and retracts the thigh and flexes the shank.

Bisect the gracilis muscle and reflect its ends. Find again the semitendinosus muscle, which you saw earlier

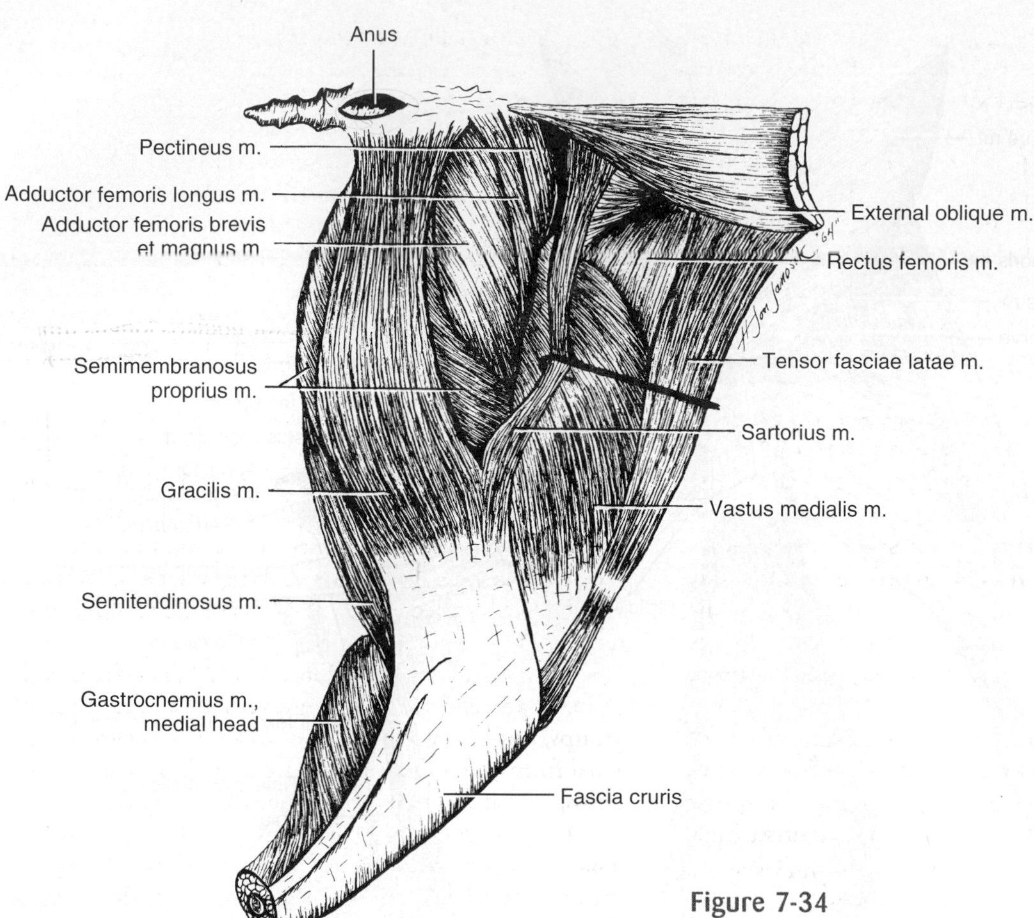

Figure 7-34
Medial view of the left thigh muscles of a rabbit.

from the lateral surface. It forms much of the medial wall of the popliteal fossa. The broad, thick muscle lying deep to the gracilis muscle and cranial to the semitendinosus muscle is the **semimembranosus muscle.** It arises from the ischial tuberosity and caudal border of the ischium and inserts upon the medial epicondyle of the femur and the adjacent part of the tibia. It is a retractor of the thigh and flexor of the shank. In rabbits, the semimembranosus muscle can be divided into two parts: the **semimembranosus proprius muscle,** which you have exposed, and the **semimembranosus accessorius muscle,** which is imbedded within the semimembranosus proprius muscle. In order to see the latter, carefully cut through the belly of the semimembranosus proprius muscle until you come to a darker cylindrical portion. This is the semimembranosus accessorius muscle. Its origin is limited to the ischial tuberosity and its insertion to the medial condyle of the tibia.

The small, triangular **pectineus muscle** arises from the cranial border of the pubis just caudal to the point where the femoral blood vessels emerge from the body wall. It inserts on the femoral body beside the origin of the vastus medialis muscle. It is a thigh adductor.

The **adductor femoris muscle complex** lies between the pectineus and semimembranosus muscles and originates from the cranial border of the pubis and from the pubic and ischial symphyses. The adductor femoris muscle inserts along the femoral body and adducts the thigh. The slender cranial component of the muscle is the **adductor femoris longus muscle;** the rest constitutes the **adductor femoris brevis et magnus muscle.** In cats, these subdivisions are less clear than in rabbits.

Iliopsoas Muscle Complex and Adjacent Muscles

- Iliopsoas muscle complex
  - Psoas major muscle
  - Iliacus muscle
- Subvertebral muscles (hypaxial trunk muscles)
  - Psoas minor muscle
  - Quadratus lumborum muscle

Observe the thick bundle of muscle that emerges from the body wall medial to the origin of the rectus femoris muscle. This is the iliopsoas muscle complex. You may study it now or postpone it until you have dissected the abdominal viscera (see Chapter 10). If you choose to dissect it now, trace the muscle bundle cranially by cutting through the muscle layers of the abdominal wall. The iliopsoas muscle complex lies in a retroperitoneal position; you need to remove the lining of the body cavity to see them. The main part of the muscle bundle represents the **psoas major muscle** (see Fig. 11-22). This muscle portion arises primarily from the centra of the last two or three thoracic vertebrae and from the centra of all lumbar vertebrae. It inserts on the lesser trochanter of the femur and protracts and rotates the thigh.

Lateral and slightly dorsal to the extreme caudal portion of the psoas major muscle, you will see a group of muscle fiber bundles arising from the ventral border of the ilium. In rabbits, some muscle fiber bundles also arise from the adjacent vertebrae. These muscle fiber bundles represent the **iliacus muscle.** They insert and act in common with the psoas major muscle. The psoas major and iliacus muscles are more intimately united in cats than in rabbits; together they are called the **iliopsoas muscle complex.**

The thin muscle medial to the psoas major muscle is the **psoas minor muscle** (see Fig. 11-22). In cats, it arises from the centra of the last one or two thoracic and first three or four lumbar vertebrae, but in rabbits, it arises from the centra of the caudal lumbar vertebrae. It inserts on the pelvic girdle near the origin of the pectineus muscle. The psoas minor muscle is one of the subvertebral hypaxial muscles referred to on page 146, rather than an appendicular muscle. It flexes the back.

Lift up the lateral border of the psoas major muscle near its middle. The thin muscle that lies on the ventral surface of the transverse processes of vertebrae is the **quadratus lumborum muscle,** another subvertebral hypaxial muscle. It arises primarily from the centra and transverse processes of the lumbar and last several thoracic vertebrae. In rabbits, it extends farther cranially and some of its muscle fibers also arise from the ribs. It inserts on the ilium cranial to the origin of the iliacus muscle. It flexes the back and, if active unilaterally, bends the vertebral column laterally.

Muscles of the Shank

- Muscles on the flexor side of the shank
  - Shank flexor muscle complex (or calf muscle)
    - Triceps surae muscle
      - Gastrocnemius muscle (medial and lateral heads)
      - Soleus muscle
    - Flexor digitorum superficialis muscle
    - Flexor digitorum brevis muscle
  - Popliteus muscle
  - Flexor digitorum profundus muscle complex
    - Flexor digitorum longus muscle
    - Flexor hallucis longus muscle
  - Tibialis caudalis muscle
- Muscles on the extensor side of the shank
  - Tibialis cranialis muscle
  - Extensor digitorum longus muscle
  - Extensor digitorum lateralis muscle
  - Extensor hallucis longus muscle
  - Extensor digitorum brevis muscle
  - Peroneus muscle complex
    - Peroneus longus muscle (or fibularis longus muscle)
    - Peroneus brevis muscle (or fibularis brevis muscle)

The shank is covered by the tough **crural fascia,** which is partly united with the tendons of some thigh muscles, including the biceps femoris and gracilis muscles. Remove the crural fascia; bisect and reflect the gracilis muscle, if you have not already done so; and reflect the sartorius, semitendinosus, and biceps femoris muscles, which were bisected earlier. Try to leave the tendons of the reflected muscles intact.

As was the case in the forearm and hand, the muscles of the shank and foot fall into extensor and flexor muscle groups, but the two muscle groups are not as clearly separated from each other as in the forearm. In general, the extensor muscles lie on the craniolateral surface of the shank; the flexor muscles lie on the caudomedial surface. They are separated by an exposed strip of the tibia on the medial side (Fig. 7-36). On the lateral side, their separation is indicated by the position of the fibula, although the fibula is not exposed at the surface (Fig. 7-35).

The large calf muscle on the caudal surface of the shank is a functional unit, the **shank flexor muscle complex,** which is composed of the gastrocnemius, flexor digitorum superficialis, and soleus muscles (Figs. 7-35 and 7-36). The **lateral head** of the **gastrocnemius muscle** arises from the lateral epicondyle of the femur, lateral surface of the patella, and adjacent parts of the tibia. In cats, it also arises by a small slip from the crural fascia. Its **medial head** arises from the medial epicondyle of the femur. Although the **soleus muscle** is visible on the lateral surface of the shin as it protrudes cranially from under the lateral head of the gastrocnemius muscle (Fig. 7-35), most of it lies deep to the lateral head of the gastrocnemius muscle. In cats, the soleus muscle arises from the proximal one third of the fibula, but in rabbits, it arises by a thin tendon from the proximal end of the fibula. Because the tendons of the soleus muscle and the two heads of the gastrocnemius muscle and soleus muscle converge, these muscles also are given the collective name of **triceps surae muscle.** Dissecting from the caudal surface of the shank, separate the two heads of the gastrocnemius muscle. The muscle lying between them is the **flexor digitorum superficialis muscle.** It originates deep to the lateral head of the gastrocnemius

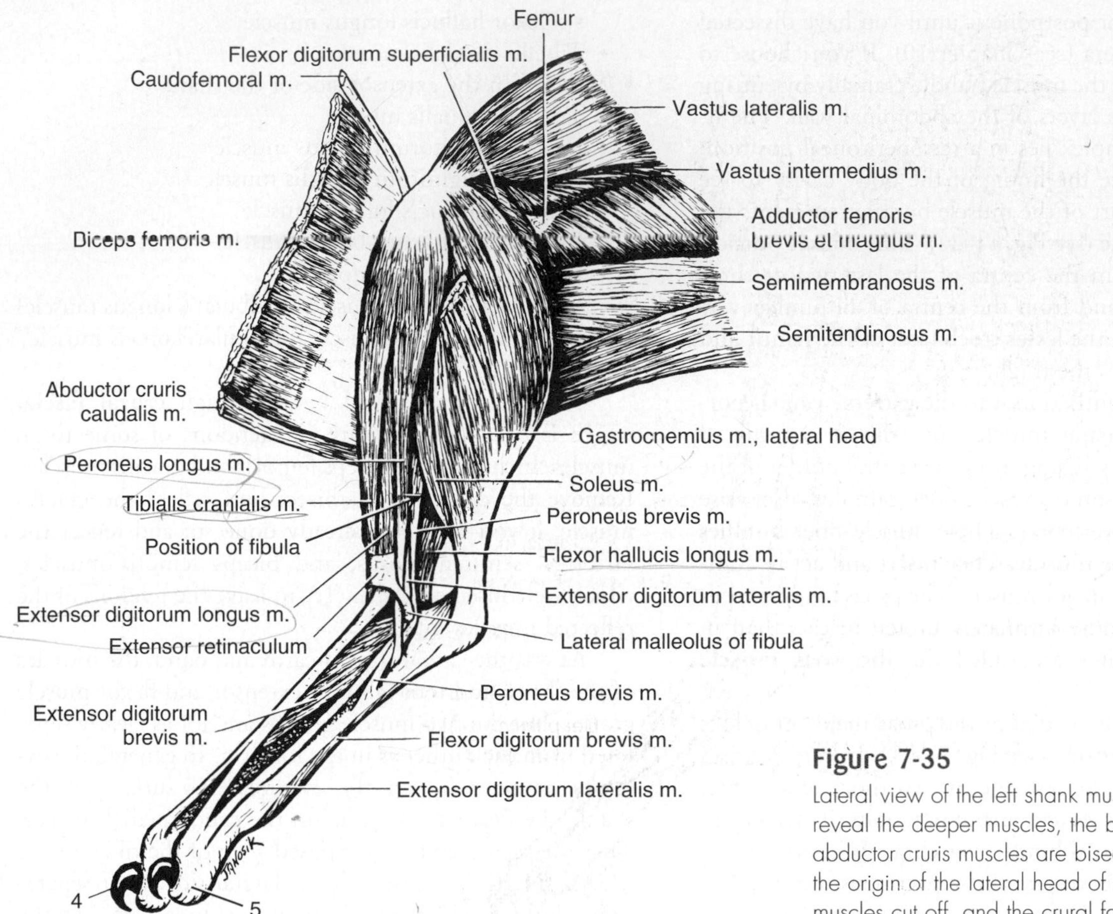

Figure 7-35
Lateral view of the left shank muscles of a cat. To reveal the deeper muscles, the biceps femoris and abductor cruris muscles are bisected and reflected, the origin of the lateral head of the gastrocnemius muscles cut off, and the crural fascia removed.

muscle from the lateral epicondyle of the femur and adjacent part of the patella. Distally, its tendon joins the tendon of the triceps surae muscle to form a large common tendon, the **calcaneal tendon,** or Achilles tendon, which inserts on the tuber calcanei. In rabbits, a branch of the tendon of the semitendinosus muscle also joins the calcaneal tendon. Remove the connective tissue around the calcaneal tendon so that you can separate the individual tendons of the gastrocnemius, soleus, and flexor digitorum superficialis muscles. You will see that the tendon of the flexor digitorum superficialis muscle forms a cap over the tuber calcanei, the **calcaneal cap,** and extends distally to the tips of the toes (see later). From an evolutionary point of view, these muscles belong to the flexor muscles of the ventral limb musculature, and their action on the foot is one of plantar flexion. This action is, however, less confusingly called an extension of the ankle, or an "extension of the foot" in human anatomy. These shank flexor muscles are particularly important in thrusting the foot upon the ground and pushing the body upward during the propulsive phase of a step, hence their large size.

The remaining four muscles on the flexor side of the shank can be seen from the medial side, where they lie between the tibia and the muscle group just described (Fig. 7-36). To see them from the lateral side (Fig. 7-35), first identify and separate, then bisect and reflect, the muscle bellies of the gastrocnemius, soleus, and flexor digitorum superficialis muscles (Fig. 7-37). A triangular **popliteus muscle** arises by a narrow tendon from the lateral epicondyle of the femur, extends toward the medial side of the shank, passes caudal to the knee joint, and fans out to insert upon the proximal one third of the tibia. It assists in flexing the knee and slightly rotates the tibia medially, pointing the foot toward the midventral line.

Return for a moment to the calcaneal cap of the flexor digitorum superficialis muscle and follow the superficial flexor tendon distally. A thin, broad muscle sheet, the **flexor digitorum brevis muscle,** attaches to and surrounds much of the superficial flexor tendon. It flexes the digits. Distally, the superficial flexor tendon splits into four tendons. These tendons give off branches to the foot pads, but their main portions continue to the second to

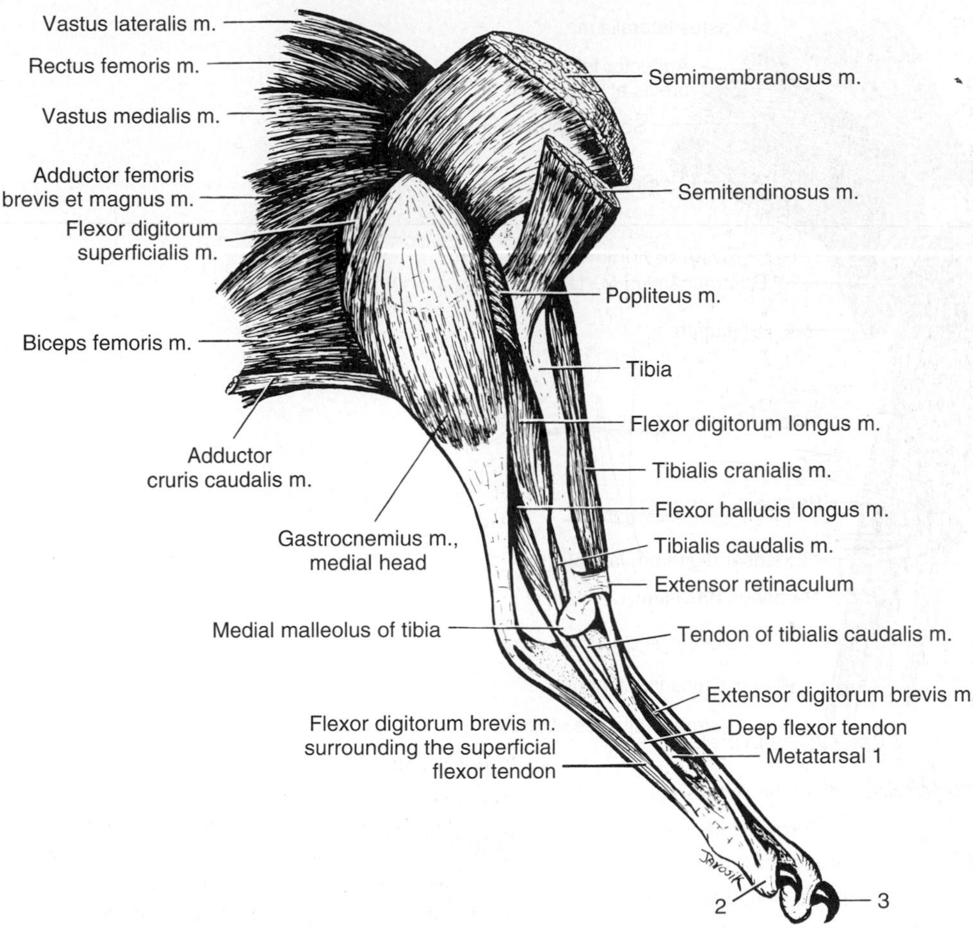

Figure 7-36

Medial view of the left shank muscles of a cat. To reveal the deeper muscles, the tensor fasciae latae, gracilis, semitendinosus, and semimembranosus muscles are bisected and reflected, and the crural fascia is removed.

fifth toes. Each tendon splits into two branches near its insertion and attaches to the sides of the middle phalanx of each digit.

The **flexor digitorum profundus muscle complex** is composed of the flexor digitorum longus and flexor hallucis longus muscles. The **flexor hallucis longus muscle** is the largest and most lateral of the three muscles. It arises from the caudal surface of the fibula and caudolateral surface of the tibia next to the origin of the soleus muscle. Near the tarsus, it turns into a strong tendon. The **flexor digitorum longus muscle** is smaller and lies medial to the flexor hallucis longus muscle and next to the popliteus muscle. It originates from the head of the fibula and the popliteal line of the tibia. It turns into a tendon at mid-length of the tibia and passes caudally over a groove between the cochlea and the medial malleolus of the tibia. This smaller tendon joins the tendon of the flexor hallucis longus muscle in the tarsal region to form the **deep flexor tendon,** which passes over the sustentaculum tali (see Chapter 6) to the sole of the foot deep to the superficial flexor tendon and the flexor digitorum brevis muscle. The flexor digitorum longus and flexor hallucis longus muscles extend the ankle joint and flex the toes.

The **tibialis caudalis muscle** is the smallest muscle of the flexor digitorum profundus muscle complex. In cats, it originates deep between the flexor hallucis longus and flexor digitorum longus muscles from the head of the fibula and the caudal surface of the proximal end of the tibia. It turns into a tendon at mid-length of the tibia and runs caudally over a groove on the medial malleolus of the tibia, then crosses medially underneath the tendon of the flexor digitorum longus muscle before inserting proximally on the intermediate cuneiform bone of the tarsus (see Fig. 6-20). It extends the ankle joint.

To expose the deep flexor tendon down to the tips of the toes, free the medial border of the flexor digitorum

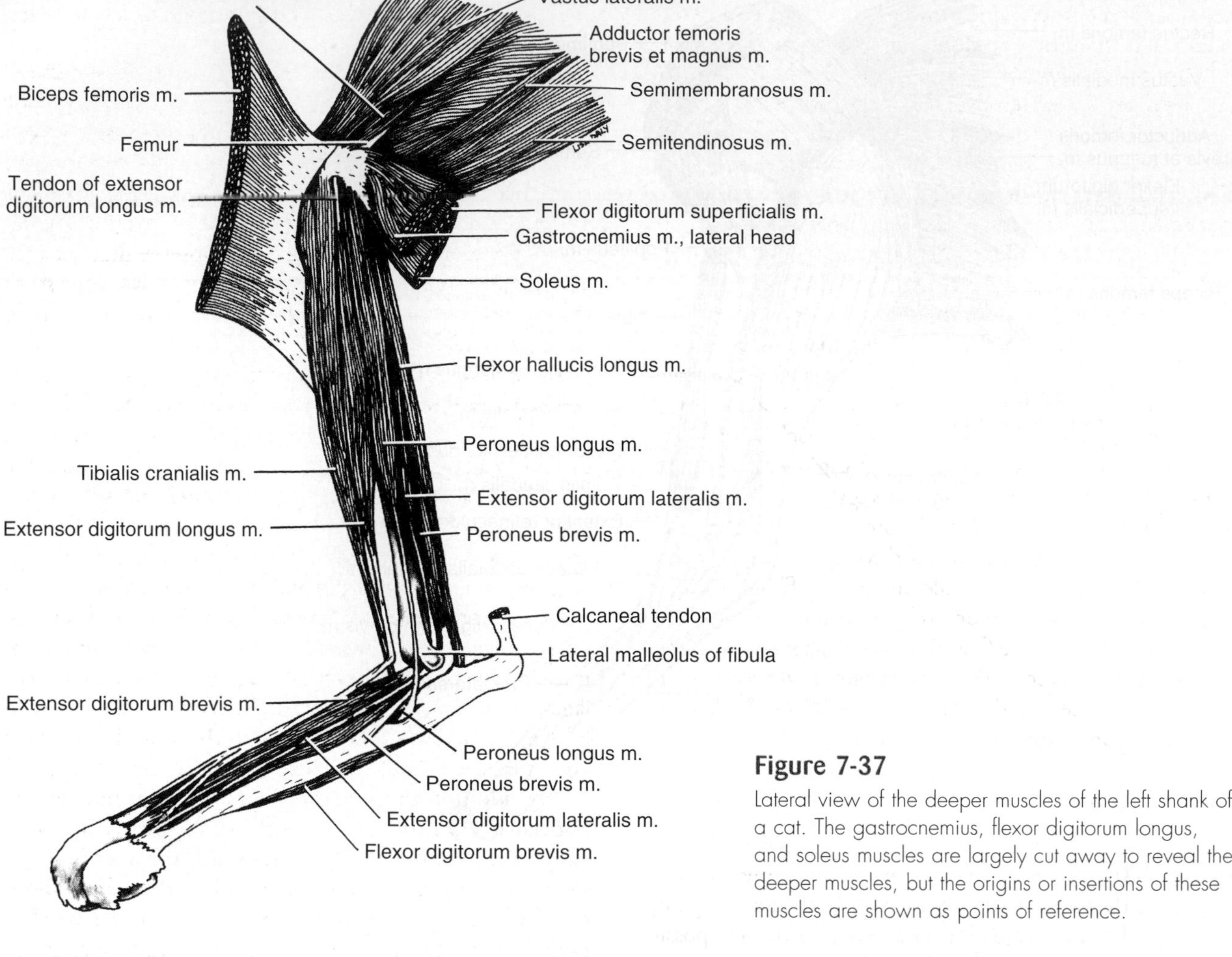

Figure 7-37
Lateral view of the deeper muscles of the left shank of a cat. The gastrocnemius, flexor digitorum longus, and soleus muscles are largely cut away to reveal the deeper muscles, but the origins or insertions of these muscles are shown as points of reference.

brevis muscle from underlying connective tissue and push the muscle with the superficial flexor tendon aside. The deep flexor tendon breaks up into four tendons that perforate the tendons of the superficial flexor tendon at the level of the middle phalanges and extend along the flexor sides of digits 2 to 5 to the terminal phalanges.

The musculature on the extensor side of the shank is much less massive than the musculature on the flexor side, because it is involved mainly in the recovery movements of the limbs as opposed to the pushing forward of the body during locomotion. The most cranial extensor muscle is the **tibialis cranialis muscle** (Figs. 7-36 and 7-37). It arises from the proximal one third of the fibula and adjacent parts of the tibia, passes underneath the **extensor retinaculum** of the ankle joint, and inserts on the second metatarsal bone. It flexes the ankle joint, a motion also called "dorsiflexion of the foot" in human anatomy.

The **peroneus muscle complex** lies caudal to the extensor digitorum longus muscle (see later and Figs. 7-35 and 7-37). Its two components are most distinct at their insertions, although they arise from different parts of the fibula. The **peroneus longus muscle** arises from the proximal one third of the fibula. Its tendon passes through a groove on the cranial surface of the lateral malleolus of the fibula and continues through a diagonal groove deep in the sole of the foot across the bony surface of the tarsus to insert onto the proximal ends of the second to fourth metatarsal bones. It abducts and everts the foot and flexes the ankle joint. The **peroneus brevis muscle** originates from the distal half of the fibula. Its tendon passes through a groove on the caudal border of the lateral malleolus of the fibula, together with the tendon of the extensor digitorum lateralis muscle (see later). The peroneus brevis muscle inserts by a stout tendon on the lateral side of the fifth metatarsal. It abducts the foot and extends the ankle joint. The remaining muscles lie on the lateral side of the shank (Fig. 7-37). The **extensor digitorum longus muscle** is located caudal and deep to the tibialis cranialis muscle. It arises from the lateral epicondyle of the femur and passes over the knee joint. Toward the distal end of the shank, it turns into a tendon that passes underneath the extensor retinaculum of the ankle joint and breaks up into four tendons. These tendons run along the dorsal sides of the digits and insert on the terminal phalanges of digits 2 to 5. The extensor digitorum longus muscle extends the digits and flexes the ankle joint. Intrinsic muscles of the foot include the **extensor digitorum brevis muscle.**

The **extensor hallucis longus muscle** of cats lies deep to the extensor digitorum longus and tibialis cranialis muscles. It originates from the middle portion of the cranial surface of the fibula, and its tendon usually merges with that of the tibialis cranialis muscle. Occasionally, it originates only from the proximal end of the fibula and completely merges with the tibialis cranialis muscle. It assists in flexing the ankle joint. The **extensor digitorum lateralis muscle** lies between the peroneus longus and flexor hallucis longus muscles. It originates from the proximal part of the fibula, and its tendon passes through a groove on the caudal side of the lateral malleolus of the fibula together with the tendon of the peroneus brevis muscle. It inserts on the lateral surface of the first phalanx of the fifth digit after blending with the tendon of the extensor digitorum longus muscle. The extensor digitorum lateralis muscle extends the fifth toe and extends the ankle joint.

Cranial Trunk Muscles

The caudal trunk muscles were described before the appendages were dissected, and some other trunk muscles were seen during the dissection of the shoulder region. Now that the appendages have been examined, it is possible to resume the study of the cranial trunk muscles.

Hypaxial Muscles

- Muscles acting on pectoral girdle (seen previously)
  - Omotransversarius muscle
  - Rhomboideus cervicis et thoracis muscle
  - Rhomboideus capitis muscle
  - Serratus ventralis muscle
- Muscles forming part of the thoracic wall
  - Rectus abdominis muscle
  - External intercostal muscles
  - Internal intercostal muscles
  - Transversus thoracis muscle
  - Rectus thoracis muscle
  - Scalenus muscle
  - Serratus dorsalis muscle (cranial and caudal parts)
- Subvertebral muscle
  - Longus colli muscle

All the trunk muscles that become associated with the pectoral girdle belong to the hypaxial muscle group. They are the omotransversarius, rhomboideus cervicis et thoracis, rhomboideus capitis, and serratus ventralis muscles. Find them again.

Lay your specimen on its back, reflect the pectoral muscle, and examine the muscles on the ventrolateral side of the thoracic wall. The **rectus abdominis muscle** will be seen running cranially to its insertion on the sternum and cranial costal cartilages (Figs. 7-17, 7-21, and 7-38). The thoracic wall is composed of three muscle layers comparable to those of the abdominal wall. Observe that the outermost layer, the **external intercostal muscles,** consists of muscle fibers that extend from one rib caudoventrally to the next caudal rib. This layer does not extend all the way to the midventral line. Cut through and reflect one external intercostal muscle, and you will expose one of the underlying **internal intercostal muscles.** Its muscle fibers extend from one rib cranioventrally to the next caudal rib. The third layer, the **transversus thoracis muscle,** is incomplete and is found only near the midventral line. To see it, lift the rectus abdominis muscle, and cut through and reflect the ventral portion of an internal intercostal muscle. The transversus thoracis muscle arises from the dorsal surface of the sternum and inserts by a number of muscle slips onto the internal surfaces of the costal cartilages; its muscle fibers run parallel to the ribs. You will have a better view of the transversus thoracis muscle when you open the thorax (see Chapter 10 and Fig. 10-17).

In addition to these muscle layers, other muscles are associated with the thoracic wall (Fig. 7-38). The diagonally oriented muscle, which arises near the middle of the sternum and crosses the cranial end of the rectus abdominis muscle to insert on the first rib, is the **rectus thoracis muscle.** Dorsal to it is a fan-shaped muscle complex that extends between the cervical vertebrae and the ribs. This is the **scalenus muscle.** It arises from the transverse processes of most of the cervical vertebrae in cats, or the last four cervical vertebrae in rabbits, and has multiple insertions on various ribs. In cats, one portion of its insertion extends as far caudad as the ninth or tenth rib. In rabbits, much of the insertion passes between the cranial and caudal halves of the serratus ventralis muscle.

Turn your specimen on its side and reflect the latissimus dorsi muscle. Observe that the dorsal border of the external oblique muscle continues cranially deep to the latissimus dorsi muscle and that its fascia is fused to the underside of the latissimus dorsi muscle. Carefully break this connection to mobilize the cranial end of the dorsal border of the external oblique muscle. With a pair of scissors, cut the external oblique muscle through the middle of its mobilized dorsal border perpendicularly to its muscle fibers, all the way to the first section through the muscle you made earlier, and reflect the parts. Pull the top of the scapula away from the trunk and carefully clean away any fat without

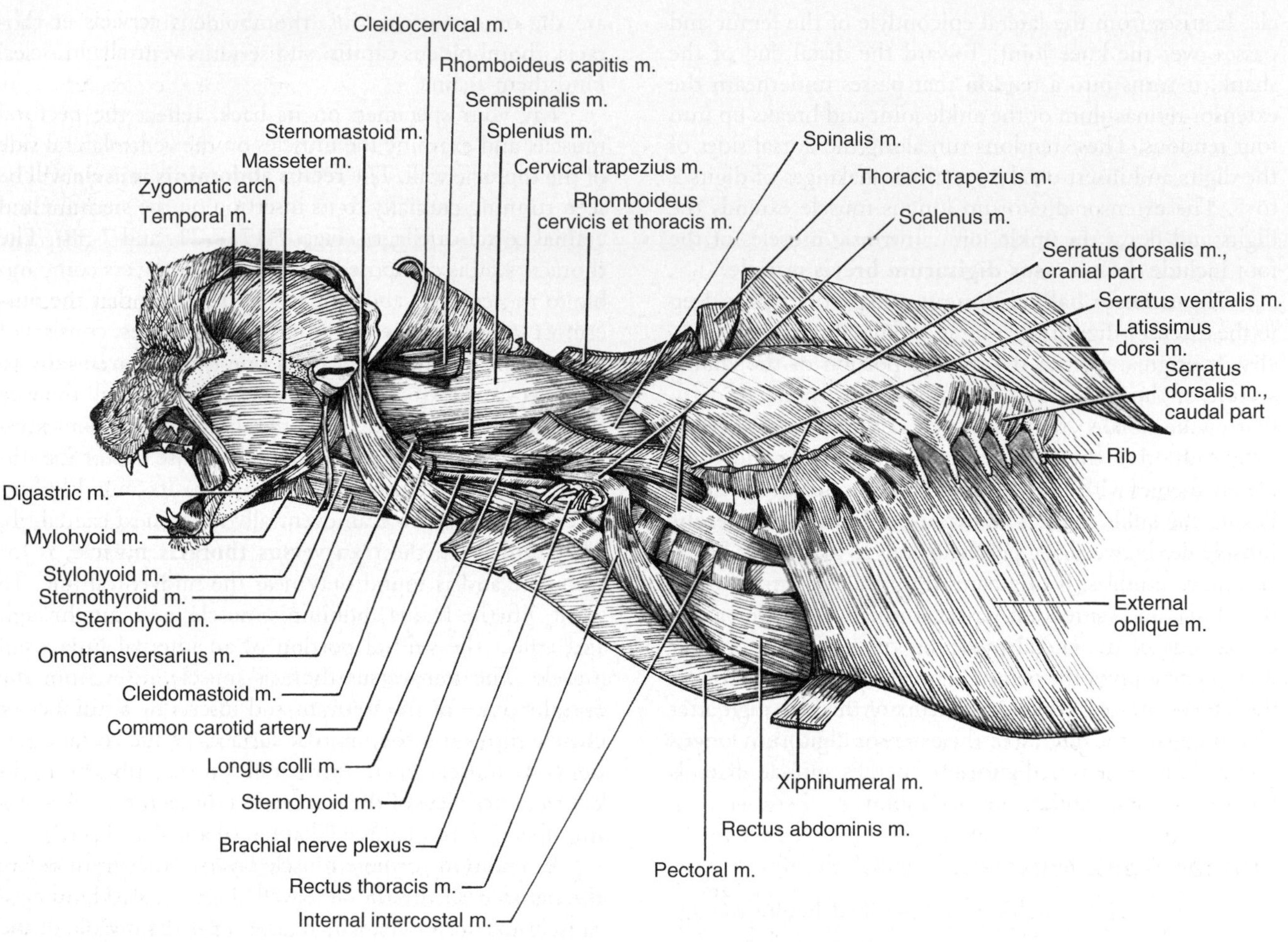

Figure 7-38

Lateral view of the muscles of the neck and thorax of a cat. The shoulder and limb are removed, and the superficial muscles are bisected and reflected.

damaging the underlying fascia and muscle layer. Medial to the scapula and serratus ventralis muscle, a number of short muscular slips arise from the thoracolumbar fascia and insert on the proximal portion of the ribs. These constitute the **serratus dorsalis muscle** (Fig. 7-38). The **cranial part** with obliquely oriented muscle fiber bundles can be distinguished from the **caudal part,** whose muscle fiber bundles are oriented more transversely. The last slip of the caudal part inserts on the last rib. Its caudal border is adjacent to the cranial border of the internal oblique muscle. Between the reflected latissimus dorsi muscle and the caudal portion of the serratus dorsalis muscle, you will discover a layer of connective tissue. Bisect it from the caudal end of the muscular part of the latissimus dorsi muscle to the first slip of the caudal part of the serratus dorsalis muscle. Now extend your section through the muscular part of the cranial portion of the serratus dorsalis muscle. All these thoracic muscles, together with the muscular diaphragm and the abdominal muscles, are involved with respiratory movements.

The subvertebral portion of the hypaxial musculature in the neck and thoracic regions is represented primarily by the **longus colli muscle** (Fig. 7-38). It is a muscle band in the neck ventromedial to the origin of the scalenus muscle and is comprised of multiple muscle slips. You will be able to see only the portion along the cervical vertebral column, unless you have already opened the thoracic cavity (see Chapter 10). It arises from the ventral surfaces of the first six thoracic vertebrae and extends cranially, incorporating additional muscle slips that originate from the transverse processes and centra of cervical vertebrae. It inserts on the centra and transverse processes of each cervical vertebra. It assists in ventral and lateral flexions of the neck.

Epaxial Muscles

- Splenius muscle
- Multifidi muscle
- Spinalis muscle

- Erector spinae muscle
- Longissimus dorsi muscle
- Longissimus capitis muscle
- Iliocostal muscle
- Semispinalis cervicis et capitis muscle
  - Semispinalis cervicis muscle (in cats)
  - Semispinalis capitis muscle (in cats)

Some of the epaxial muscles were observed earlier. Find again the erector spinae and multifidi muscles, the latissimus dorsi muscle, and the serratus dorsalis muscle. Bisect the entire serratus dorsalis muscle while making sure to cut through its muscular portion, not its aponeurosis. Reflect it and notice a shiny transparent sheet of connective tissue covering the underlying muscles in the cervical region. This is the **spinocostotransversal fascia;** bisect and reflect it. The epaxial musculature should now be well exposed. Trace the multifidi and erector spinae muscles caudally and cranially.

The **multifidi muscles** can be seen most clearly in the lumbar region, although they interconnect vertebrae along the entire length of the vertebral column. The individual muscles extend from the mammillary processes, transverse processes, and zygapophyses of vertebrae to the spinous processes of more cranial vertebrae. Most multifidi muscles connect two or three adjacent vertebrae between their origin and insertion. In rabbits, some also attach to the fascia covering the muscle. More cranial multifidi muscles lie deep to the spinalis muscle (see later and Fig. 7-38).

As the **erector spinae muscle** continues cranially from its origin on the iliac crest and dorsal surfaces of the more caudal trunk vertebrae, it splits into a **longissimus dorsi muscle** lying lateral to the spinalis muscle (see later) and a more lateral **iliocostal muscle.** The muscle fiber bundles of these muscles extend diagonally from caudomedial to craniolateral. The muscle fiber bundles of the iliocostal muscle insert on the ribs, and those of the longissimus dorsi muscle insert chiefly on the transverse processes of the thoracic and cervical vertebrae.

The **spinalis muscle** lies lateral to the spinous processes of the thoracic and caudal cervical vertebrae. In cats, most of it arises from the inner and external surfaces of the deep layer of the thoracolumbar fascia covering the erector spinae muscle slightly cranial to where it splits into two portions. For this reason, the spinalis muscle sometimes is considered a division of the erector spinae muscle. Deeper parts of the spinalis muscle arise from the dorsal surface of vertebrae. The spinalis muscle inserts on the spinous processes of vertebrae. The muscle fiber bundles of the spinalis muscle, like those of the multifidi muscles, extend diagonally from caudolateral to craniomedial. In rabbits, the spinalis muscle is similar, except that it is not as intimately associated with the erector spinae muscle.

As mentioned earlier, the thoracolumbar fascia is part of the deep fascia and external fascia of the trunk. Like many fasciae, it provides a surface for the attachment of muscles. In the course of the previous dissections, you probably noticed that the thoracolumbar fascia is a very complex structure consisting of two main layers and several subdivisions called lamellae. The thoracolumbar fascia is attached to the spinous processes of the thoracic and lumbar vertebrae and to the crest of the ilium. It also is anchored to the transverse processes of the lumbar vertebrae and to the ribs.

Before we can study the rest of the epaxial muscles, we need to reflect some muscles of the sternocleidomastoid muscle complex. Find again the cleidocervical muscle and its insertions on the skull and neck, then find the cleidomastoid muscle and its attachment on the clavicle, deep to the insertion of the cleidocervical muscle. Separate the two muscles. Bisect the cleidocervical muscle between the clavicle and its attachment on the neck and skull. Because the cleidobrachial muscle was bisected earlier, only the middle portion of the brachiocephalicus muscle and the origin of the cleidomastoid muscle will remain attached to the clavicle. Reflect the dorsal half of the cleidocervical muscle after separating its cranial border from the caudal border of the sternomastoid muscle. Clean the surface of the sternomastoid muscle and mobilize the salivary glands from its surface, without damaging or removing them, in order to see its attachment to the mastoid process of the skull. Bisect the sternomastoid muscle and reflect its halves, then reflect the exposed cleidomastoid muscle with the clavicle. Now the entire splenius muscle is revealed.

The most superficial of the cranial epaxial muscles is the **splenius muscle** (Figs. 7-39 and 7-40). It is a thin but broad triangular muscular sheet that covers the back of the neck. The splenius muscle arises from the middorsal line of the neck and passes craniolaterally to insert on the nuchal crest and occipital region of the skull and on the transverse process of the atlas. The splenius muscle assists in flexing the head laterally. If it contracts bilaterally, it elevates the head. Identify the ventral border of the splenius muscle and separate it from the adhering **longissimus capitis muscle,** which is a cranial extension of the longissimus dorsi muscle, which you saw earlier.

Bisect the splenius muscle perpendicularly to the orientation of its muscle fiber bundles. There is practically no dorsal border because of the triangular shape of the muscle. Reflect the two flaps of the splenius muscle and expose the underlying **semispinalis cervicis et capitis muscle.** In cats, the **semispinalis cervicis muscle,** which has a parallel arrangement of muscle fiber bundles, can be distinguished from the **semispinalis capitis muscle,** which is more fan shaped. The semispinalis cervicis et capitis muscle arises from the vertebrae between the cranial ends of

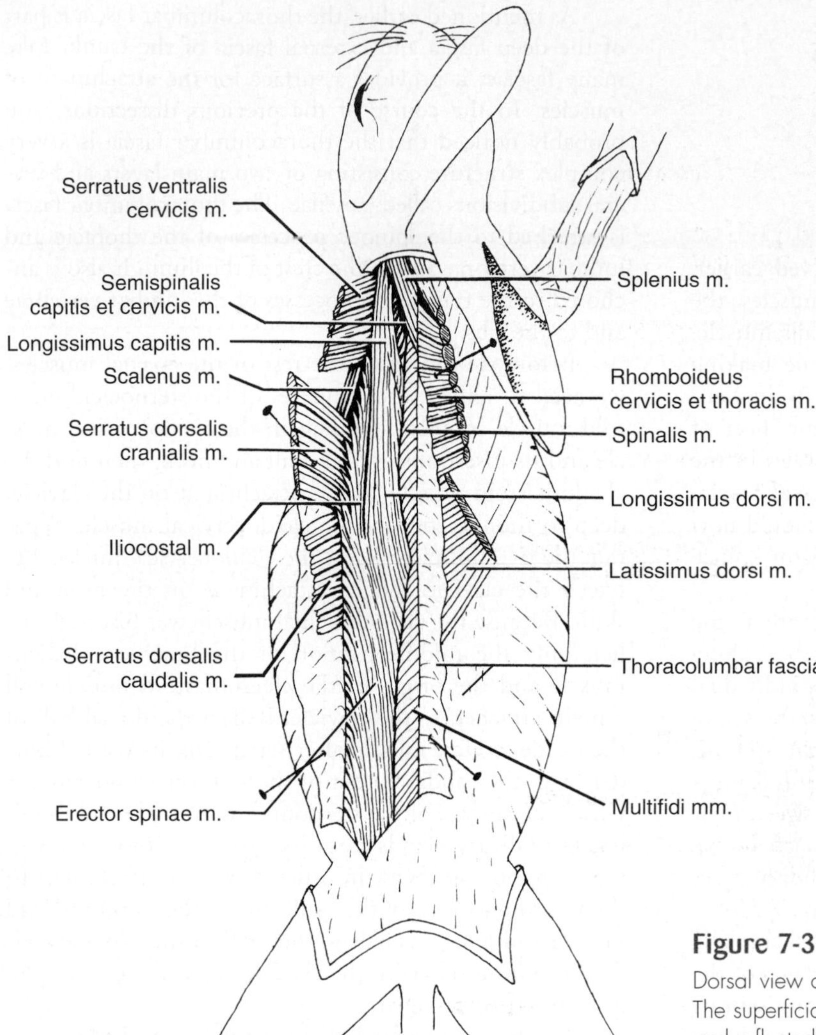

Figure 7-39
Dorsal view of the epaxial muscles on the left side of a rabbit. The superficial muscles and the thoracolumbar fascia are bisected and reflected.

the spinalis and longissimus capitis muscles and inserts on the back of the skull.

Deeper epaxial muscles are not described here. Collectively, the epaxial muscles extend and laterally flex the back, neck, and head.

Although the epaxial muscles of mammals have become quite complex, the original configuration seen in reptiles can still be recognized. As shown in Figure 7-11, the epaxial muscles of reptiles fall into three muscle groups: the transversospinalis, longissimus, and iliocostalis muscle complexes. The reptilian transversospinalis muscle complex is represented in mammals by several deep muscles, such as the interspinal, intertransversarii, and occipital muscles, and the multifidus spinae, spinalis, and semispinalis muscles. The reptilian longissimus muscle complex is represented in mammals by the longissimus and splenius muscles, and the reptilian iliocostal muscle complex is represented by the iliocostal muscle in mammals.

Hypobranchial Muscles

The hypobranchial muscles are a hypaxial muscle complex that is located on the ventral side of the neck and throat. The hypobranchial muscles move the larynx, hyoid apparatus, and tongue. They play a crucial role in opening the mouth, manipulating food in the oral cavity, and swallowing.

Completely remove the skin of this region as far cranially as the chin (Table 7-5). Be careful not to damage the thin and fragile stylohyoid muscle (see later and Fig. 7-18), which lies directly under the platysma. Clean away any loose connective tissue and fat. You may reflect and push aside, but not destroy, salivary glands and ducts, lymph nodes, and prominent blood vessels and nerves.

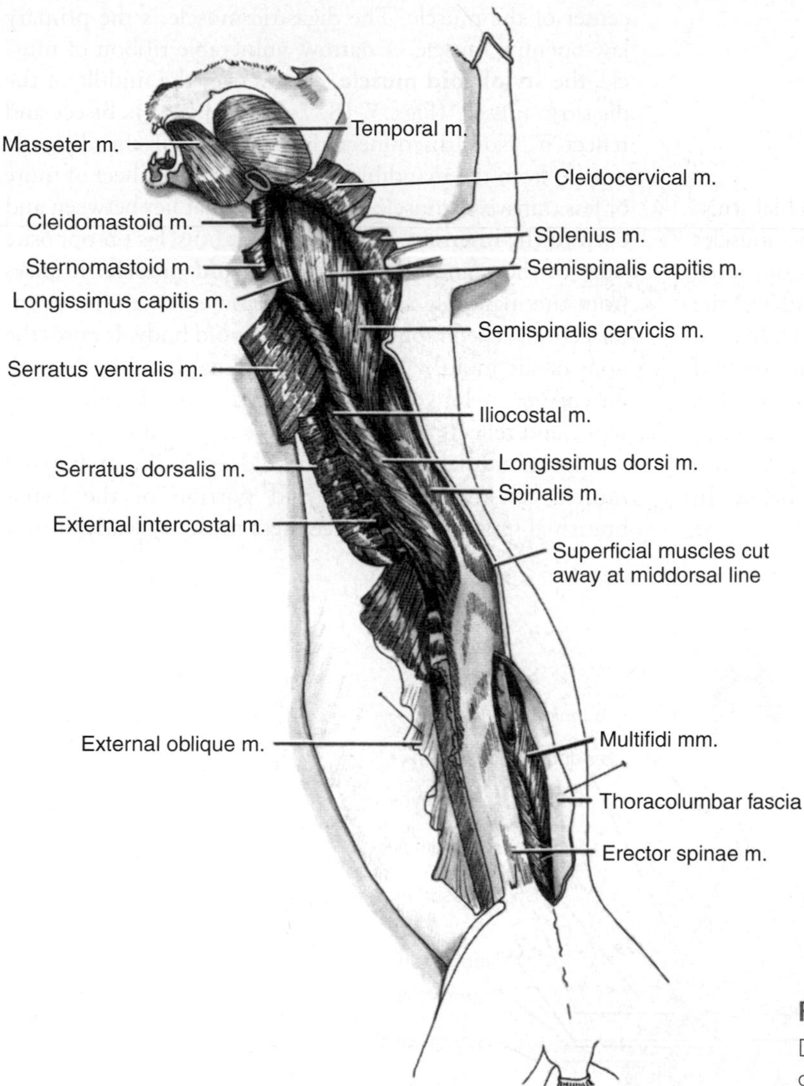

Figure 7-40

Dorsal view of the epaxial muscles on the left side of a cat. The superficial muscles and the thoracolumbar fascia are bisected and reflected.

Posthyoid Hypobranchial Muscles

- Sternohyoid muscle
- Sternothyroid muscle
- Thyrohyoid muscle

Find again the sternomastoid muscles and reflect them. The **sternohyoid muscle** is a thin, midventral longitudinal band of muscle that covers the windpipe, or trachea, and extends from its origin on the cranial end of the sternum to its insertion on the hyoid body (Figs. 7-41 and 7-42). The sternohyoid muscle actually is paired, but in cats it usually is fused into an unpaired midventral muscle.

Carefully separate the sternohyoid muscle from another band of muscle that lies dorsolateral to it. This muscle band, the **sternothyroid muscle,** has a similar origin, but runs cranially to insert on the thyroid cartilage of the larynx. The larynx, or "Adam's apple," is located caudal to the hyoid body. Therefore, the sternothyroid muscle is not as long as the sternohyoid muscle. A short band of muscle, the **thyrohyoid muscle,** lies on the lateral surface of the larynx. It arises next to the insertion of the sternothyroid muscle and runs rostrally to insert on the hyoid body. The sternothyroid and thyrohyoid muscles appear as a single muscle band unless their attachments on the larynx are carefully exposed.

Prehyoid Hypobranchial Muscles

- Geniohyoid muscle
- Hyoglossus muscle
- Genioglossus muscle
- Styloglossus muscle

- Lingualis proprius muscle
- Branchiomeric muscles also seen:
  - Digastric muscle
  - Stylohyoid muscle
  - Mylohyoid muscle

In order to expose the prehyoid hypobranchial muscles (Tables 7-1 and 7-2), three branchiomeric muscles must be studied and reflected first. Notice the stout muscle band that is attached to the ventromedial border of the mandible. This is the **digastric muscle** (Fig. 7-41). In cats, it has a fleshy origin from the paracondyloid and mastoid processes of the skull. In rabbits, its origin is by a tendon from just the paracondyloid process. The digastric muscle derives its name from its condition in humans, in which it is divided by a central tendon into two distinct bellies. In cats, there is an occasional tendinous intersection in the center of the muscle. The digastric muscle is the primary jaw-opening muscle. A narrow, vulnerable ribbon of muscle, the **stylohyoid muscle,** crosses over the middle of the digastric muscle (Figs. 7-18, 7-41, and 7-42). Bisect and reflect it. Now disconnect the insertion of the digastric muscle from the mandible and reflect it. The sheet of more or less transverse muscle fiber bundles that lies between and deep to the insertions of the digastric muscles on opposite sides of the mandible is the **mylohyoid muscle.** It arises from the mandible and inserts on a medial tendinous intersection, called raphe, and on the hyoid body. It raises the floor of the mouth. Make a longitudinal incision through one of the mylohyoid muscles slightly to one side of the raphe and reflect it (Table 7-5). It is rather thin.

Longitudinal muscles that lie deep to the mylohyoid muscle constitute the prehyoid portion of the hypobranchial musculature. To see these muscles clearly, it may

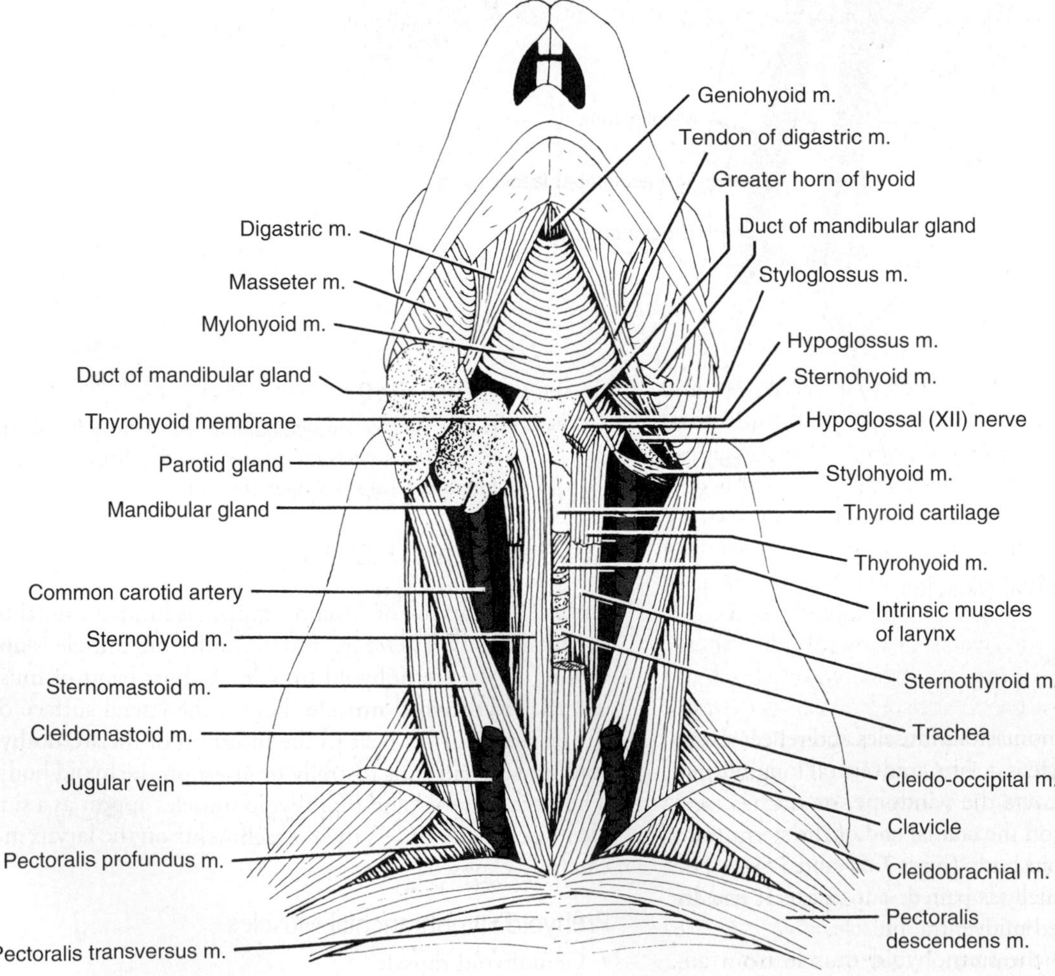

Figure 7-41

Ventral view of the muscles in the neck and floor of the mouth of a rabbit. The salivary glands and sternohyoid muscle are removed from the left side of the rabbit to reveal the deeper structures.

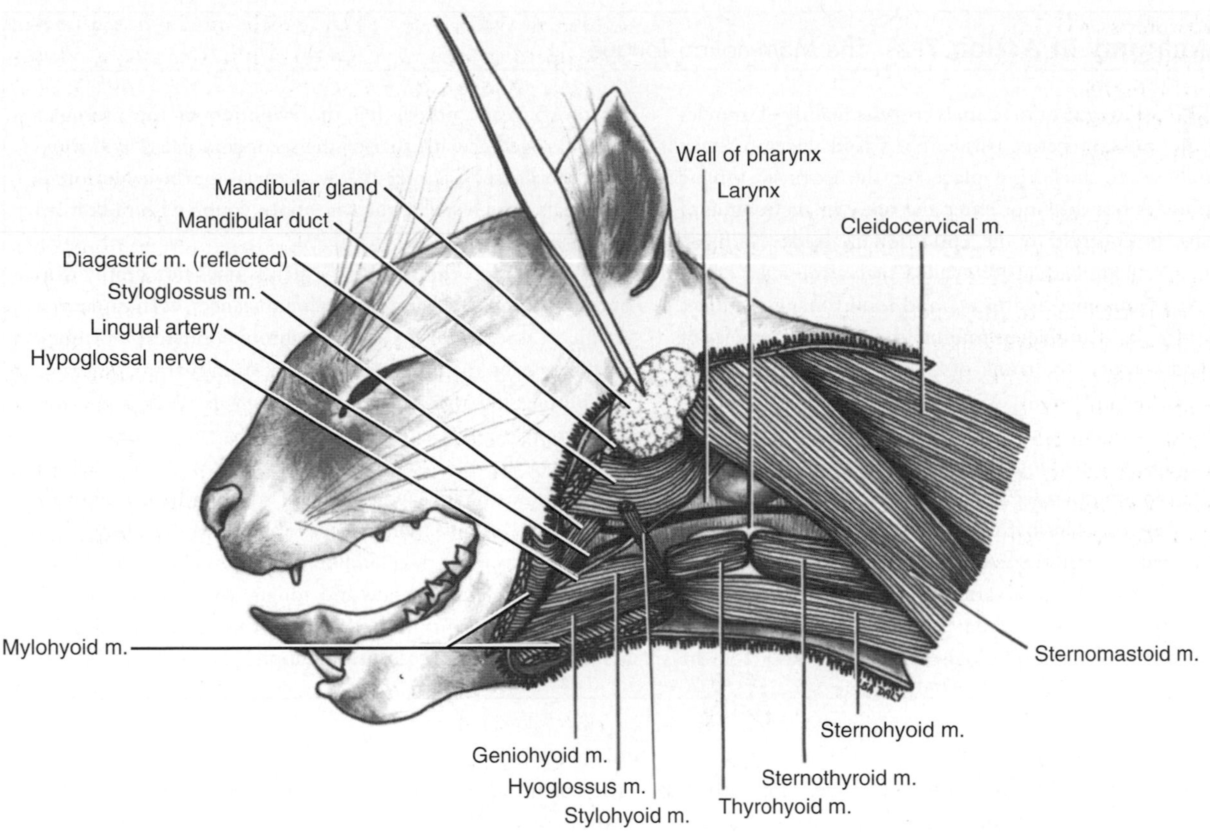

Figure 7-42
Lateral view of the hypobranchial and adjacent muscles of a cat.

be necessary to transect the mandibular symphysis and spread the two mandibular rami apart. The midventral band of muscle is the **geniohyoid muscle** (Figs. 7-41 and 7-42). It consists of a pair of muscles that have merged. It arises from the intermandibular symphysis and inserts on the hyoid body. Bisect the geniohyoid muscle and reflect its ends. The band of muscle that arises from the chin deep to the origin of the geniohyoid muscle is the **genioglossus muscle.** It runs caudodorsally into the tongue. The muscle that arises from the hyoid body lateral and deep to the insertion of the geniohyoid muscle is the **hyoglossus muscle.** It runs cranially into the tongue. Pull the tip of the tongue, and you will notice that the muscle also moves. The muscle band that arises from the mastoid process of the skull and runs rostrally into the tongue is the **styloglossus muscle.** It is located lateral to the rostral portion of the hyoglossus muscle. The hyoglossus and styloglossus muscles are distinct at their origins, but they merge as they enter the tongue. The glossus muscles (i.e., the genioglossus, hyoglossus, and styloglossus muscles), together with intrinsic muscle fiber bundles within the tongue, which are collectively called the **lingualis proprius muscle,** form the substance of the tongue and are responsible for its movements (Anatomy in Action 7-2).

Branchiomeric Muscles

Mandibular Arch Muscles

- Mylohyoid muscle
- Digastric muscle (rostral portion)
- Masseter muscle (deep and superficial parts)
- Temporal muscle
- Pterygoid muscle
- Tensor tympani muscle
- Tensor veli palatini muscle
- Hyoid arch muscles also seen: cutaneous muscles
  - Platysma
  - Facial muscles
  - Scutuloauricular muscle (in rabbits)
  - Frontoscutular muscle (in rabbits)

The **mylohyoid muscle** and the rostroventral half of the **digastric muscle,** which were seen during the dissection of the hypobranchial muscles, are branchiomeric

Anatomy in Action 7-2 The Mammalian Tongue

The tongue of mammals consists mainly of muscles and connective tissue. The hyoid skeleton serves only as an anchoring place for the extrinsic tongue muscles but does not enter the body of the tongue of mammals, in contrast to the condition in birds. Kathleen Smith and William Kier (1989) called such a muscular boneless organ a "muscular hydrostat" and found many examples of it besides the mammalian tongue, such as the tentacles of squids and octopus, the trunk of elephants, and the tongues of turtles and many lizards. When the musculature of muscular hydrostats contracts, the tissues inside the organ are compressed and their internal pressure rises. As a consequence, the tissues become rigid under this compression and act as a hydrostatic skeleton. The hydrostatic skeletons, in contrast to skeletons made of hard tissues, such as bone and cartilage, can be bent and molded into a variety of shapes so that they can adapt to complex space requirements. Dietrich Starck (1978), the eminent German morphologist, who died in 2001 at the age of 93 years, noted that the evolution of the muscular tongue, together with that of the secondary palate and muscular cheeks (see Chapter 10), was crucial for the evolution of mammals, which suckle their newborn young with milk from their mammary glands.

Milk within the mammary glands does not simply drip out of the nipple under normal circumstances, even under the action of the "milk-releasing" neurohypophyseal hormone oxytocin. Baby mammals extract milk from mammary glands by compressing the mammary glands with their paws and snouts and by sucking on the nipple. As the studies of Sue Herring (1985) and Rebecca German and W. W. Crompton (2000) show, the lips and tongue of suckling baby mammals curl or wrap around the mother's nipple, and the baby's nasal passage between the nasal and oral cavities is closed off. At the same time, rhythmical jaw and tongue movements create a negative pressure within the sealed-off oral cavity so that milk is sucked and pressed out of the nipple.

muscles of the mandibular arch. To see the other mandibular arch muscles, remove the skin from the top of the head and the cheek region (Table 7-5). Mobilize the auricle by removing fat and loose connective tissue around its base. The platysma and the facial muscles, both of which belong to the hyoid muscle group (see later) need to be removed, but be careful not to injure salivary glands, lymphatic nodes, nerves, and blood vessels in this region. Exercise special care in removing the skin, fat, and loose connective tissue from the cheek region, because the duct of a salivary gland crosses the cheek just underneath the facial muscles (see later and Fig. 10-14). Find the zygomatic arch. The powerful **masseter muscle** arises from the zygomatic arch and extends ventrally. Its **deep part** inserts on the masseteric fossa on the lateral surface of the mandibular ramus. Its **superficial part,** however, extends farther and curves around the ventral edge of the mandibular ramus to attach on the medial surface of the mandibular ramus (Fig. 7-19 and 7-20). In rabbits, the deep and superficial parts are quite distinct because the deep part arises from the caudal end of the zygomatic arch and extends rostroventrally at nearly right angles to the superficial part.

Another mandibular muscle, the **temporal muscle,** is located dorsal to the zygomatic arch. In cats, it is a sizable muscle and covers the large temporal fossa of the skull. It arises primarily from the surface of the cranium, but some fibers originate from the dorsal edge of the zygomatic It passes medial to the zygomatic arch and inserts on the coronoid process of the mandible. In rabbits, the temporal fossa and the temporal muscle are relatively small. To see the temporal muscle, bisect the powerful facial **scutuloauricular** and **frontoscutular muscles,** which extend from the top of the skull to the base of the auricle and erect the auricles. The origin of the temporal muscle extends from a point dorsal to the base of the auricle to the back of the orbit (Fig. 7-20). It inserts by a tendon that follows the caudal wall of the orbit to the coronoid process of the mandible.

The **pterygoid muscle** extends from the base of the skull to the medial side of the lower jaw. You will be able to see it in its entirety later, when you observe the oral and pharyngeal cavities (see Chapter 10). At that time, you will be able to see also the **tensor veli palatini muscle,** which extends from the base of the skull into the soft palate. It is impractical at this time to find the **tensor tympani muscle,** because it attaches to the malleus within the middle ear (see Chapter 8, Anatomy in Action 8-3).

Hyoid Arch Muscles

- Platysma
- Facial muscles
- Digastric muscle (caudal portion)
- Stylohyoid muscle
- Stapedius muscle

The major hyoid arch muscles are the **platysma,** the **facial muscles,** the **caudal half of the digastric muscle,**

Anatomy in Action 7-3 Jaw Mechanics in Carnivores and Herbivores

The skulls, dentitions, and jaw muscles of cats and rabbits are quite different as a result of their different diets and behaviors. In a cat, as an example of a carnivore, the temporal muscle is much larger than the masseter muscle; the canine teeth are used for stabbing and killing prey; and the carnassial teeth are used for cutting flesh (Fig. **A**). The hingelike jaw articulation restricts fore-and-aft movements of the mandible but allows slight lateral movements so that the carnassial teeth of one side of the upper and lower jaws can be aligned to cut meat. The mandible is characterized by a low ramus and position of the jaw joint and by a tall, large coronoid process (Fig. **B**). This configuration results in a relatively small moment arm for the force of the masseter muscle and a relatively large moment arm for the temporal muscle. Both muscles generate a clockwise moment around the jaw joint to counteract any force acting on the canine or carnassial teeth. Because the force of a prey item acting on the canine teeth has the mechanical advantage of a longer moment arm, the jaw muscles may need to generate great forces to compensate for their smaller moment arms, to enable the canine teeth to puncture the prey. However, the same difference between the moment arms of the jaw muscles and the force acting on the tip of the mandible ensures that the jaw muscles need to shorten relatively little while causing a relatively large change in the gape. Furthermore, the force acting on the carnassials, when they cut through meat, has a smaller moment arm than that acting on the canine teeth and is, therefore, more easily exceeded by the force generated by the jaw muscles.

To understand the size difference between the temporal and masseter muscles in a cat, we need to analyze the forces that the muscles exert on the mandible (Fig. **C**). (For the sake of simplicity, the actions of the pterygoid muscles are not considered here.) The force of the temporal muscle comprises a dorsal and a caudal force component; the force of the masseter muscle comprises a dorsal and a rostral force component. The dorsal force components of the two muscles add up to create the force closing the jaw. The rostral and caudal force components may cancel each other. However, the temporal muscles, being the larger one, has the potential to generate a greater force and, hence, a greater caudal force component than the rostral force component of the masseter muscle. This apparent mismatch of the horizontal force components of the two muscles is resolved when we consider the behavior of cats (pussy cats as well as tigers). A captured prey usually struggles to pull away from the jaws of a predator and thereby generates rostrally directed forces on the lower jaw. If that happens, the temporal muscle can contract more forcefully and generate a greater caudally directed force component to prevent a luxation of the jaw joint.

In a rabbit, as an example of a gnawing herbivore, the temporal muscle is much smaller than the masseter muscle, the incisor teeth are used for gnawing hard materials, and the cheek teeth are used for grinding fibrous plant materials (Fig. **D**). The jaw joint allows fore-and-aft movements of the mandible, as well as lateral movements, for grinding food between the upper and lower cheek teeth. The mandible is characterized by a relatively tall ramus, a very high jaw joint, and a very low coronoid process (Fig. **E**). This configuration results in a small moment arm for the force of the temporal muscle and a large moment arm for the force of the masseter muscle. Both muscles generate a clockwise moment around the jaw joint to counteract any force acting on the incisor or cheek teeth. Because the force that is acting on the teeth has the mechanical advantage of a longer moment arm, the jaw muscles may need to generate greater forces to compensate for their smaller moment arms. However, the same difference between the moment arms of the jaw muscles and the force exerted by food or a prey ensures that the jaw muscles need to shorten relatively little to cause a relatively greater gape. Furthermore, the force acting on the cheek teeth, when they grind fibrous vegetal matter, has a smaller moment arm than that acting on the incisors and, therefore, is more easily exceeded by the force generated by the jaw muscles.

To understand the size difference between the temporal and masseter muscle in a rabbit, we need to analyze the forces that the muscles exert on the mandible (Fig. **F**). We also need to recall that the masseter muscle of rabbits comprises two distinct parts with opposite muscle fiber directions. (For the sake of simplicity, the actions of the pterygoid muscles are not considered here.) The force of the temporal muscle comprises a dorsal and a caudal force component. The force of the superficial part of the masseter muscle (Ms) comprises a dorsal and a rostral force component; the deep part of the masseter muscle (Mp) comprises a dorsal and a caudal force component. The dorsal force components of the two muscle layers add up to generate crushing forces on the food. The rostral and caudal force components may cancel each other. However, as Geerling Langenbach and T.M.G.J. Eijden (2001) discovered, the two muscle layers also can contract by alternating between the two sides of the jaw so that one side of the jaw is pulled rostrally while the opposite side is pulled back. These movements are responsible for the cyclic chewing motions that are observed in herbivores and are needed to grind down fibrous food between the cheek teeth.

continued

Anatomy in Action 7-3 Jaw Mechanics in Carnivores and Herbivores *continued*

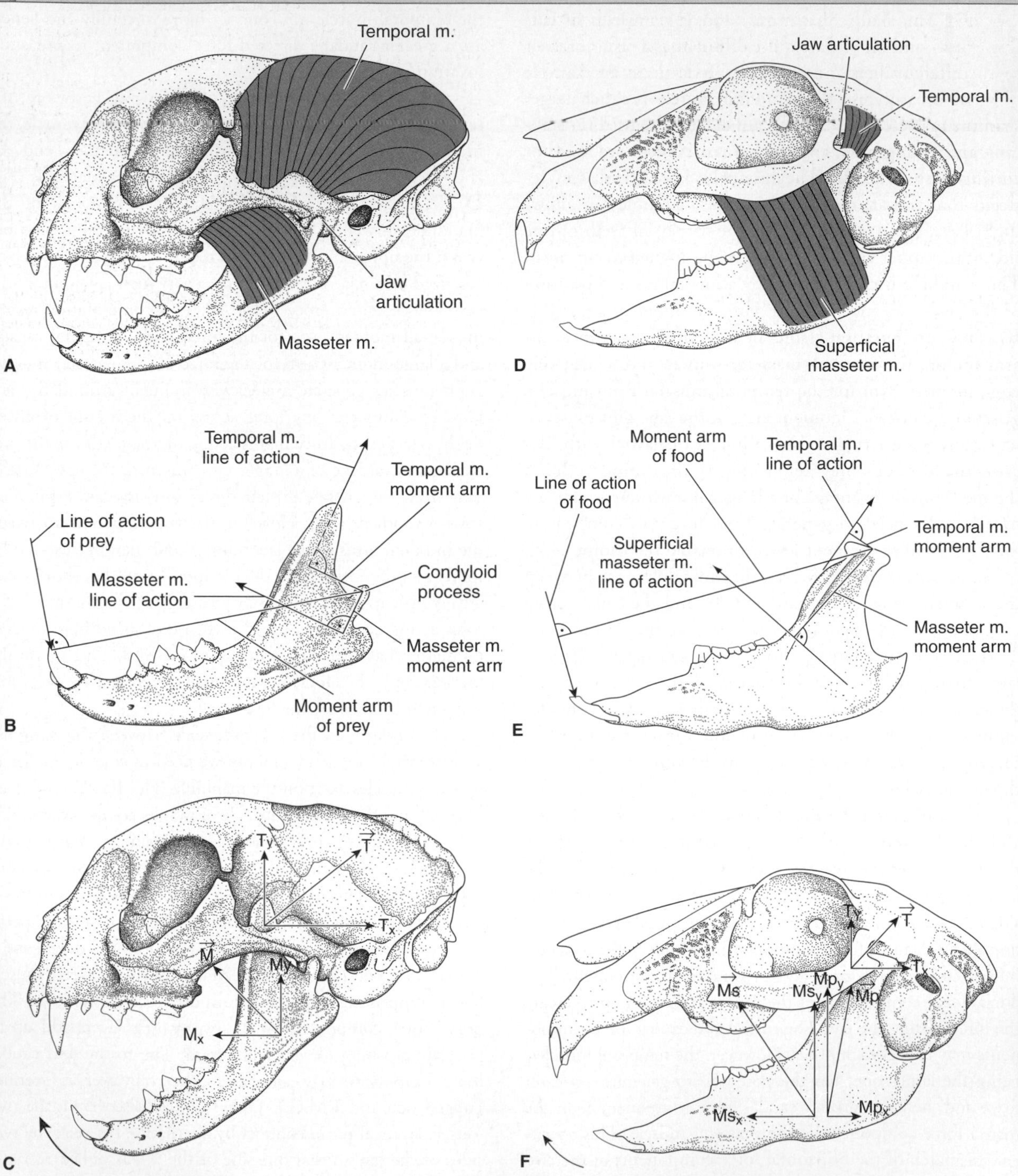

A, Diagrammatic lateral view of the superficial jaw muscles of a cat, B, Moment diagram of the mandible of a cat. C, Force diagram of the mandible of a cat. D, Diagrammatic lateral view of the superficial jaw muscles of a rabbit. E, Moment diagram of the mandible of a rabbit. F, Force diagram of the mandible of a rabbit. *(**A** and **D** from Liem, Bemis, Walker, and Grande, 2001.)*

and the **stylohyoid muscle.** Find the stylohyoid muscle again; it is a small ribbon of muscle. In cats, it is located lateral to the caudal portion of the digastric muscle; in rabbits, it is located caudal to the caudal portion of the digastric muscle (Fig. 7-42). In cats, it arises from the stylohyal ossicle of the chain of hyoid ossicles and inserts on the hyoid body. In rabbits, it arises from the base of the skull and inserts onto the greater horn of the hyoid bone. It moves the hyoid body. The **stapedius muscle,** which has followed the stapes into the middle ear, can be seen later (see Chapter 8).

Caudal Branchiomeric Muscles

- Trapezius muscle complex
- Sternocleidomastoid muscle complex
- Intrinsic laryngeal muscles
- Thyroarytenoid muscle
- Cricoarytenoid muscle
- Cricothyroid muscle

Much of the caudal branchiomeric musculature is lost during the morphological transformation from a fishlike vertebrate to a mammal in the course of evolution, but some parts become associated with the pectoral girdle as the **trapezius** and **sternocleidomastoid muscle complexes.** These were already described in connection with the musculature of the shoulder region. Some remaining parts of the caudal branchiomeric musculature form the **intrinsic muscles of the larynx,** such as the **thyroarytenoid, cricoarytenoid,** and **cricothyroid muscles**, and some other parts contribute to the formation of the wall of the pharynx. Some intrinsic muscles of the larynx can be seen on the ventral surface of the larynx deep to the rostral end of the sternohyoid muscle (Fig. 7-41).

The Sense Organs

The sense organs and nervous system integrate the activities of all parts of the body. The sense organs, the central nervous system, and the basic pattern of distribution of the peripheral nerves are the topics of this and the following chapter. It is appropriate to consider them at this time before examining the visceral organs because the most conspicuous effector organs are the muscles described in the previous chapter. If separate heads cannot be provided for studying the brain and cranial nerves of *Squalus* and separate sheep brains for studying the mammalian nervous system, instructors may wish to postpone this unit of work until the end of the course (see also Tables 7-3 and 7-5).

Types of Receptors

To survive in a changing world, organisms must monitor changes in their external and internal environments and react appropriately. The individual cells of many simple invertebrates, such as hydras and jellyfishes, can both receive and respond to stimuli. More specialized animals have **receptors** that receive stimuli; nerve cells, or **neurons,** that transmit and integrate information; and specialized muscle cells, gland cells, ciliated cells, and other **effectors** that mediate the responses to the stimuli.

Chapter 8

Receptors may be the ends of neurons themselves or specialized cells that in turn activate sensory neurons. In vertebrates, free sensory neuron endings, which sometimes are encapsulated in specialized layers of connective tissue, detect various cutaneous sensations such as pain, touch, temperature, and pressure. Odoriferous particles also are detected by specialized neuronal endings in the nose. Most other sensations are detected by specialized receptor cells that are particularly sensitive to a single type of environmental stimulus or modality, such as light or mechanical deformation. Receptor cells are transducers, which are devices that transform minute amounts of the energy to which they are attuned into another type of energy, the nerve impulse. A **sense organ** is an aggregation of receptor cells or neuron endings together with tissues that support and protect the cells and frequently direct or amplify the environmental stimulus; for example, the lens of the eye helps to focus light rays onto the receptor cells in the retina.

Receptor cells and receptive neurons often are classified according to the sensory modality to which they are attuned. Mechanoreceptors are activated by a physical deformation of a part of the cell, photoreceptors by a change in light intensity or wave length, thermoreceptors by temperature changes, and so forth. As with the skeleton and muscles, it is convenient to sort this array of receptors, and of the sensory neurons that arise from them, into somatic and visceral groups. **Somatic sensory receptors** lie in the "outer tube" of the body, which consists of the skin and the somatic muscles. Receptors associated with somatic muscles and tendons are known as **proprioceptors,** and they monitor tension generated by the muscles themselves. Proprioceptors also occur in branchiomeric muscles. **Visceral sensory receptors** are associated with the "inner tube" of the body, that is, the viscera.

Before studying the nervous system, we will study the macroscopically visible sense organs, because they are more superficial (Tables 7-3 and 7-5). The central nervous system and nerves cannot be studied without damaging the sense organs.

THE NOSE

Animals detect many chemical changes both within their bodies and in the external environment. The ability to detect chemical changes in the environment is known as **chemoreception;** it occurs in all animals. **Gustatory,** or **taste, receptors** are chemoreceptors that respond to chemical changes in the immediate environment of the animal. Usually the chemicals are in direct contact with a body surface, such as the tongue. **Olfactory,** or **smell, receptors** are chemoreceptors that respond to stimuli that emanate from a greater distance. The distinction between taste and smell may be clear to us, but becomes blurred in aquatic animals. The structures for the detection of gustatory and olfactory stimuli, however, are quite distinct. Gustatory receptors are microscopic groups of specialized epithelial cells known as **taste buds.** Because taste buds develop from endodermal cells, they are considered special visceral sensory receptors. They are innervated by the facial (VII), glossopharyngeal (IX), and vagus (X) cranial nerves. Most are confined to the lining of the mouth and pharynx, but in some fishes, including catfishes, many taste buds migrate onto the body surface. The tentacles, or barbels, around the mouth are particularly rich in taste buds. Taste buds are too small to study in dissections, and we will not consider them further.

Olfactory receptors are specialized neurons, often called **neurosensory cells** because they directly receive environmental stimuli and transmit nerve impulses. They develop, along with other parts of the nervous system, from ectoderm and are considered special somatic sensory receptors. They are innervated by the olfactory cranial nerve (I). The thickened ectoderm that forms the olfactory receptors and the nasal sac is one of several **neurogenic placodes**[1] in the head of embryonic vertebrates. The olfactory cells are too small for us to see, but we will consider the nasal sacs and cavities to which they are confined.

Gustation and olfaction are the most important senses in many vertebrates, even in some animals that have well-developed eyes and ears. It is through these senses that many animals find food and mates and often gain important clues used in navigation. We may find this difficult to appreciate because our olfactory sense is somewhat reduced. Experiments have shown that smell is a major sense that guides a shark to the general location of prey, but other senses, particularly electroreception, are important in directing the strike. Recordings from individual olfactory cells show that the odor of cutaneous mucus and blood of marine animals most frequently elicits responses. Smell also appears to play a role in reproductive behavior.

In most fishes, the nose consists of a pair of **olfactory sacs,** each of which connects to the surface by a pair of external nostrils, called **incurrent and excurrent nares,** through which water carrying odoriferous particles circulates. In lungfishes and tetrapods, each olfactory sac, now called a **nasal cavity,** has but one external nostril, called a **naris,** but each opens into the mouth cavity through an internal nostril, or **choana.** In addition to its original olfactory function, the nose of tetrapods serves as an air passage.

Although most olfactory receptors of tetrapods are located in the dorsal parts of the nasal cavities, in many species some olfactory receptors remain ventral where they form the **vomeronasal organ** (Jacobson's organ). The vomeronasal organ is a blind-ending tube that opens on the roof of the mouth cavity (Chapter 10). The receptive cells of this organ resemble other olfactory cells. However, their neurons travel in a special bundle of the olfactory nerve and terminate in an accessory olfactory bulb of the brain. The vomeronasal organ is a part of a communication system based on the secretion of species-specific chemical messengers known as **pheromones** into the external environment. Pheromones may warn conspecifics of danger, establish territories, indicate social status, and signal sexual readiness.

In order for a substance to stimulate the olfactory epithelium, it must be in solution. Tetrapods have solved this problem by evolving glands whose secretions keep the epithelium moist. The secretions of these glands, and the mucosa of the nasal passages as a whole, also condition the air that passes to the lungs by moistening, cleaning, and, in birds and mammals, warming it. In some reptiles and in birds and mammals, the mucosal surface is increased through the evolution of **conchae,** or turbinate bones (Fig. 4-21). In mammals, the secondary palate (see Chapters 4 and 10) prolongs the respiratory passages.

[1]*Placodes* are ectodermal thickenings that give rise to the lens of the eye and many integumentary structures, including scales, teeth, glands, and hair follicles. Neurogenic placodes also form receptor cells and neurons.

Fishes

Study the nose of *Squalus.* Note that each **naris** is divided into two openings (Figs. 7-8**C** and 8-1**A**) by a superficial flap of skin and a deeper ridge. The lateral opening, which also faces rostrally to a slight extent, is the **incurrent naris,** through which a current of water enters the olfactory sac; the medial opening is the **excurrent naris.** A thin, flaplike valve along its caudal margin prevents water from entering the excurrent naris. Remove the skin and other tissue from around the **olfactory sac** on the side of the head to be used for dissecting the sense organs (Table 7-3). The nasal capsule and antorbital process of the chondrocranium, which nearly surround the olfactory sacs, must be cut away. You now can see that the olfactory organ is a round sac that has no connection with the oropharyngeal cavity. The **olfactory tract** of the brain extends to the caudomedial surface of the olfactory sac and there expands into an **olfactory bulb.** Cut open the olfactory sac and notice that its internal surface area is increased by many septalike folds, called **olfactory lamellae,** which in turn bear minute secondary folds (Fig. 8-1**B**). The olfactory cells are concentrated at the bases of the troughs between the lamellae.

Amphibians

Note the pair of widely separated **nares** on your specimen of *Necturus.* Remove the skin between the naris and the eye on the side of the head used for dissecting the muscles (Table 7-4). You will see the long, tubular nasal cavity after picking away the surrounding connective tissue (Fig. 7-12). Notice that the walls of the nasal cavity are not ossified. In order to see the internal nostril, or choana, you need to open the mouth cavity by cutting through the angle of the mouth. To do so without damaging the structures that you dissected in Chapter 7 and that you will want to see in Chapter 10, identify the levator mandibulae externus, levatores arcuum, dilatator laryngis, cucullaris, scapular deltoid and latissimus dorsi muscles (Fig. 7-12). Detach the origins of the levatores arcuum and cucullaris muscles from the epibranchial musculature and bisect the latissimus dorsi muscle. Gently reflect the suprascapular cartilage (Chapter 6) away from the trunk and bisect the levator scapulae and thoraciscapularis muscles, which insert on the medial side

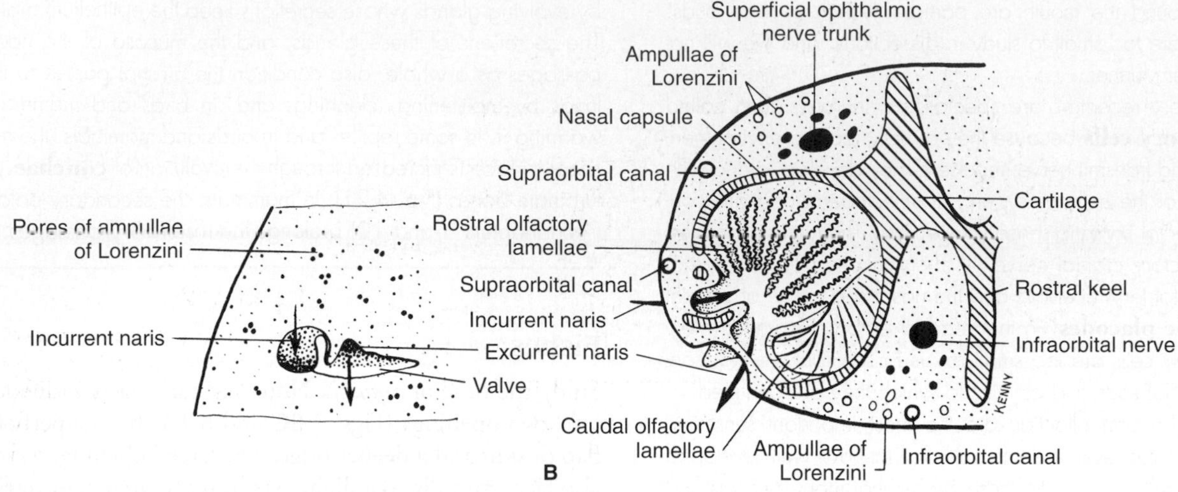

Figure 8-1

A, Ventral view of the nares on the right side of *Squalus*. Rostral is toward the top of the figure. B, Diagram of the left olfactory sac as seen in a caudal view of a cross section taken just caudal to the nares. Arrows indicate the direction of the water current. The ampullae of Lorenzini and supraorbital and infraorbital canals are parts of the lateral line system.

of the suprascapular cartilage, so that the scapula is mobilized. Insert now a blunt probe between the jaws and pry them slightly open to facilitate the insertion of one blade of a pair of scissors into the angle of the mouth. Proceed to cut through the angle of the mouth along the caudoventral edge of the levator mandibulae externus muscle. Slightly reflect the mobilized muscles and scapula and extend your cut caudally through the wall of the neck and dorsal to the external gills. Notice the large superficial blood vessels at the base of the external gills. Cut as far caudal as the level of the carpus of the adducted forelimb, then proceed transversely through the body wall to intersect the incision through the body wall, which usually was made by the biological company during preservation of the specimen. The deeper part of your incision should pass dorsal to the lung and into the esophagus. Swing open the floor of the mouth and pharynx. Spread apart the cut edges and notice that only the wide oropharyngeal cavity has been cut open. Find the slitlike **choana** lateral to the most caudal and shortest row of teeth, called the pterygoid teeth (Figs. 4-10 and 10-13). Cut open the nasal cavity and find the **olfactory lamellae** within. The vomeronasal organ is absent in the permanently aquatic *Necturus,* but is present in metamorphosed terrestrial amphibians and in terrestrial reptiles, except turtles.

Mammals

Study the nose of a mammal in sagittal sections of the head cut in such a way that one half shows the nasal septum and the other half shows the inside of the nasal cavity. The nose should be studied from demonstration preparations, unless this unit of work has been postponed to the end of the course or heads from specimens of a previous class have been saved for the purpose. This dissection should be correlated with the description of the sagittal section of the skull (Fig. 4-21). The following account is based on the cat but is applicable to many other mammals.

The **nares,** which lie close together in mammals, lead into paired **nasal cavities.** The nasal cavities occupy the area of the skull rostral to the cribriform plate of the ethmoid bone and dorsal to the hard palate. On the larger section, you can see that they are completely separated from each other by a **nasal septum** (Fig. 8-2). The vomer bone forms the ventral portion of the nasal septum; the **perpendicular plate of the ethmoid** forms the caudodorsal portion; and cartilage forms the rest of the septum.

On the section in which the nasal septum has been removed, you can see that each nasal cavity is filled to a large extent by three folded **conchae,** or turbinate bones (see Figs. 4-21, 8-2, and 10-15). These are covered with the nasal mucosa. A simple fold that extends from the dorsal edge of the naris caudoventrally to about the middle of the hard palate forms the **ventral concha** (maxilloturbinate). The nasolacrimal duct that drains tears from the eye enters lateral to the ventral concha. The duct entrance is best seen on a skull in which this concha has been removed. The **dorsal concha** (nasoturbinate) is represented by a single longitudinal fold in the dorsal part of the nasal cavity. The area between and caudal to these two conchae is filled by the complexly folded **middle concha** (ethmoturbinate). Most of the olfactory epithelium is limited to

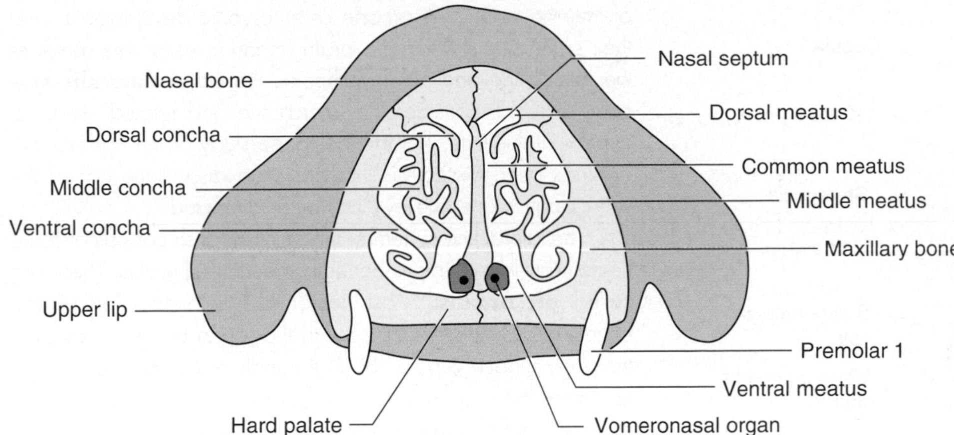

Figure 8-2
Semidiagrammatic cross section through the nasal cavity of a cat at the level of the first premolar tooth.

the more caudal parts of the conchae. With low magnification, you may see **olfactory nerves** passing through the cribriform foramina of the cribriform plate to the olfactory bulb of the brain.

Air passages lead from the naris between the conchae. The most prominent of these is the **ventral meatus,** which lies between the ventral and middle conchae on the one hand and the hard palate on the other hand. It opens by the **choana** into the nasopharynx. The choanae are located at the caudal border of the hard palate (see Chapter 10). The **dorsal meatus** lies dorsal to the dorsal concha. The **middle meatus** is represented by the spaces formed by the complex folds of the middle concha. A **common meatus** lies between the nasal septum and the conchae. Certain air passages communicate with air sinuses in adjacent bones, including the frontal and sphenoid bones (see Fig. 10-15). The air sinuses lighten the skull and in some species act as resonating chambers that modify or amplify sounds produced primarily in the larynx.

Paired **vomeronasal organs** are present in the cat. The entrance to one can be seen on the roof of the mouth cavity just caudal to the first incisor tooth (see Figs. 10-15 and 10-16). From here an **incisive duct** leads through the palatine fissure (Fig. 4-20) to the organ. You can find the vomeronasal organ by carefully dissecting away the rostroventral portion of the nasal septum. It appears as a blind-ending tube with a cartilaginous wall lying on the hard palate and against the nasal septum. It extends approximately 1 cm caudal to the incisor teeth.

THE LATERAL LINE SYSTEM AND ELECTRORECEPTORS

Fishes, larval amphibians, and a few aquatic adult amphibians have a superficial sensory system known as the **lateral line system** for detecting currents and other water movements. Water is a good conductor of electricity, and in many species parts of the lateral line system are specialized **electroreceptors** that detect weak electric currents. The lateral line system also is closely related to the ear. The receptive cells in both the lateral line and ear are **hair cells** (Figs. 8-3A and 8-4A). In the lateral line system, clusters of hair and supporting cells form small organs called **neuromasts,** which usually are located within canals in the skin that open to the surface by pores (Fig. 8-3B). The surface of each hair cell bears a single modified cilium, called a **kinocilium,** and a cluster of microvilli known as **stereocilia** (Fig. 8-3A). The stereocilia are graded in length, with the longest located next to the kinocilium. A gelatinous matrix, called a **cupula,** caps the kinocilium and stereocilia. Low-frequency vibrations and water movements displace the cupula, and this displacement bends the hairs.

The hair cells in the lateral line system are tonic receptors, that is, they have a certain base level of activity at all times. This activity activates sensory, or afferent, neurons that synapse with the cells. If the movement of the cupula is toward the kinocilium, then the rate of activity of the cells increases. Movement of the cupula in the opposite direction decreases the rate of activity of the hair cells. Motor, or efferent, neurons modulate the level of sensitivity of the hair cells.

The hair cells in both the lateral line and ear develop embryonically from a series of neurogenic placodes. Both the lateral line and ear are supplied by neurons that terminate in adjacent and closely related centers in the brainstem. The neurons from the ear form the statoacoustic (VIII) cranial nerve. The **lateralis neurons** from the lateral line system travel in six **lateral line,** or **lateralis, nerves** that develop from six distinct neurogenic placodes. These neurons enter the brain very close to the bases of the facial (VII), glossopharyngeal (IX), and vagus (X) nerves (see Chapter 9). Traditionally they were considered to be parts of these nerves, but we treat them separately because of their distinctive origin and function. Lateralis nerves are absent in amniotes so they were missed when the conventional numbering system and terminology of cranial nerves were developed in amniotes. Because of the similarities in the structure

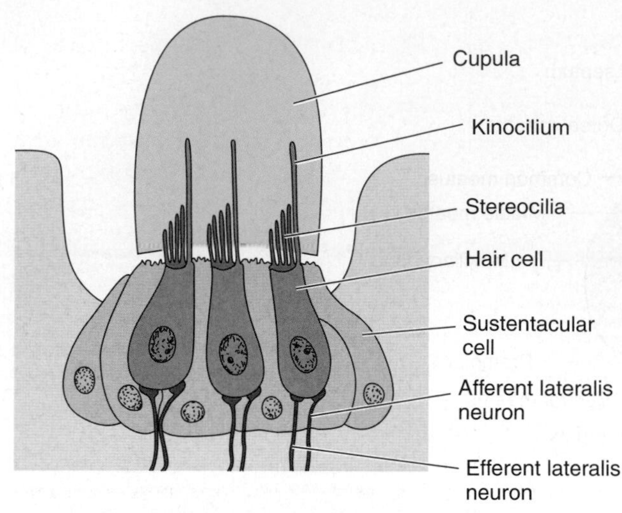

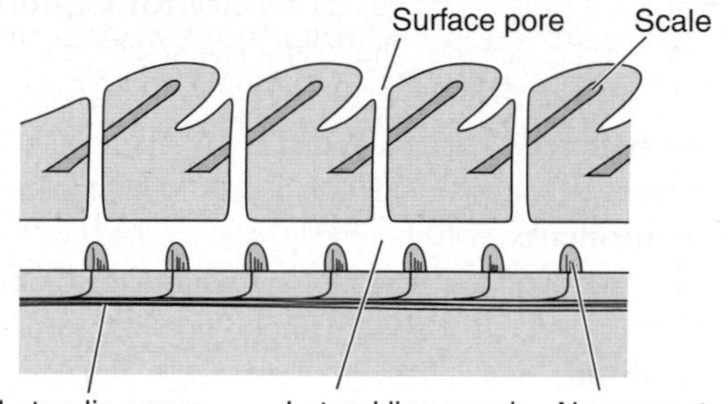

Figure 8-3

Lateral line receptors. **A,** Receptor hair cells are grouped with sustentacular cells into neuromasts. **B,** Most neuromasts lie within canals in the skin that open to the surface through pores. The figure is of a teleost fish in which the pores perforate scales. *(After Liem, Bemis, Walker, and Grande, 2001.)*

of their hair cells, their mode of embryonic development, and their connections within the brain, many investigators speak of the lateral line and ear together as the **octavolateralis system.** We will consider the lateral line and related electroreceptors at this time, but will postpone study of the ear until the eye has been dissected. It is easier to dissect the ear of the spiny dogfish after the eye has been dissected.

The lateral line system of the spiny dogfish consists of some neuromasts that lie in pits beneath specialized scales. These are called **pit organs.** Most neuromasts lie within lateral line canals in or just beneath the skin that open by many pores to the body surface (Fig. 8-3**B**). The canals are filled with sea water. Experiments on a variety of fishes have shown that the lateral line system enables them to localize objects at a distance, even in turbid water, either by the disturbance produced by a moving object or by the reflected waves set up by the fish's own movements. Because the neuromasts in the canals have different polarities (for example, the kinocilium may be located toward the cranial or the caudal ends of the body), the fish can distinguish the direction of water movements and the source of the disturbance. The lateral line system has been called one of "distant touch."

All cartilaginous fishes, other nonteleostean fishes, and a few amphibians (caecilians and some salamanders) have groups of modified lateral line receptors known as **ampullary organs** or, in cartilaginous fishes, **ampullae of Lorenzini.** The receptor cells are neuromasts in which the hair cells have only one hairlike process and no cupula is present (Fig. 8-4**B**). Efferent neuron fibers are absent. The receptors lie in small swellings (ampullae) at the base of jelly-filled tubules that open on the body surface (Fig. 8-5**C**). The tubules lie just beneath the skin and tangential to it. They vary in length and collectively extend in many directions, that is, some groups extend rostrally,

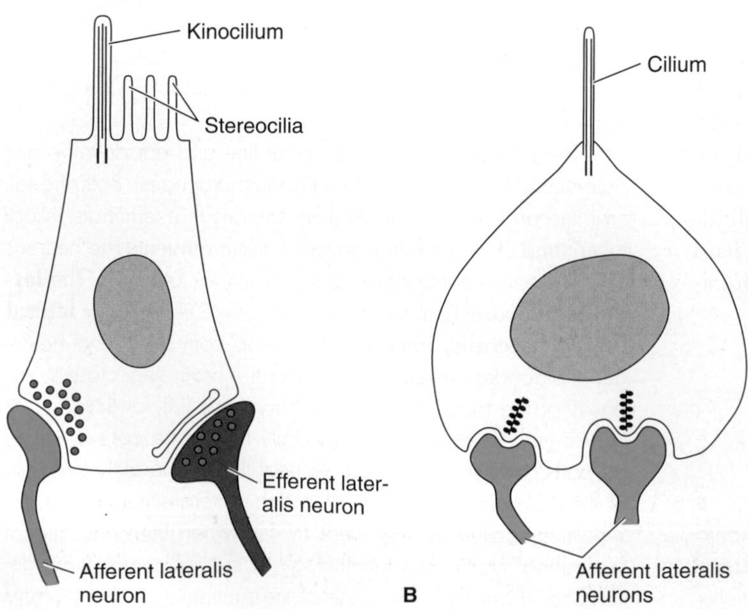

Figure 8-4

Comparison of the structure of a hair cell from a canal neuromast organ **(A)** and a sensory cell of an ampullary organ **(B)**. *(From Liem, Bemis, Walker, and Grande, 2001.)*

some caudally, some medially, and so forth. The function of the ampullary organs is electroreception. The skin and wall of the ampullary organs have such high electrical resistance that a small voltage charge in the external environment does not affect deeper tissues directly through them. The jelly in the ampullary organs, however, acts as an electrical capacitor. It can hold a voltage charge for a moment and transmit it without a voltage drop to the receptive cells at the base of the organ. The length and direction of a tube determine the stimulus intensity at the receptive cells. Tubes directly in line with the voltage gradient are most affected, enabling the animal to detect the source of the electric field. The ampullae are sensitive enough to detect the direct current and low-frequency bioelectric fields generated by the activity of respiratory and other muscles of nearby prey. The electroreceptive system can lead a shark to its prey in the absence of other sensory clues, but field experiments have shown that sharks are attracted into an area by olfactory clues. The electroreceptive system is the most important sensory system in directing their final attack. Attacks can be provoked by a voltage gradient as low as 5 nV/cm.

Interactions between the earth's magnetic field and a swimming shark, or one passively drifting in a current, induce electrical fields. Experiments have shown that cartilaginous fishes can detect such fields. It has been suggested that they use this information in orientation and navigation.

The lateral line system occurs in all fishes and larval amphibians; the electroreceptive system occurs in chondrichthyans and certain amphibians. Although lost in early actinopterygians, it has reevolved in a few teleosts. Both sensory systems were lost during the evolution of reptiles, birds, and mammals and were never reacquired, not even in such aquatic tetrapods as the Cetacea. Parts of the systems have been noted during the study of the external features of *Petromyzon*, *Squalus*, and *Necturus* (Chapter 3). We will consider them in detail only in *Squalus*.

The Ampullae of Lorenzini

Examine your specimen or a large head of *Squalus*. The pores on the snout through which a jellylike substance exudes when the area is squeezed are the openings of the **ampullae of Lorenzini.** Note their distribution. They are very abundant on the rostral part of the snout (Fig. 8-5**A** and **B**). Especially conspicuous patches of ampullary organs, which are called **ampullary fields,** are visible as a pair of patches on the dorsal side of the snout and on most of the ventral side of the snout. Smaller clusters of ampullary organs can be seen over the outer surface of the adductor mandibulae muscle. Other smaller groups of pores are located caudal to the angle of the mouth. Carefully

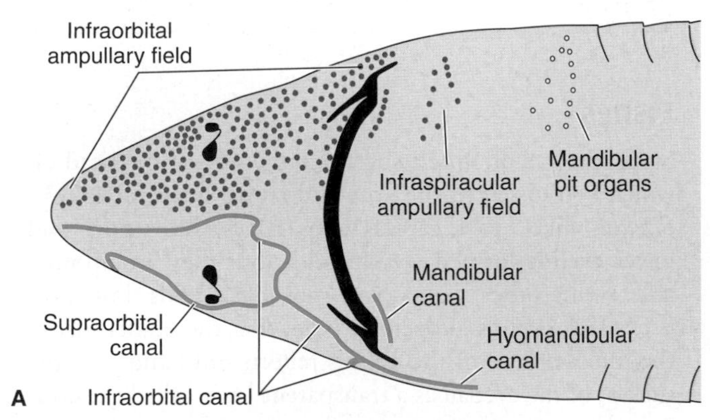

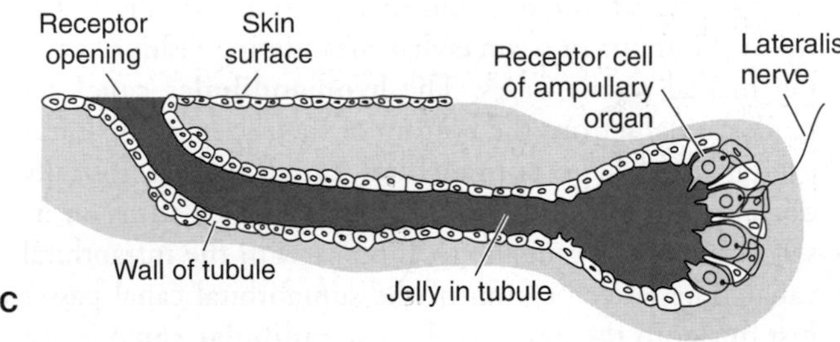

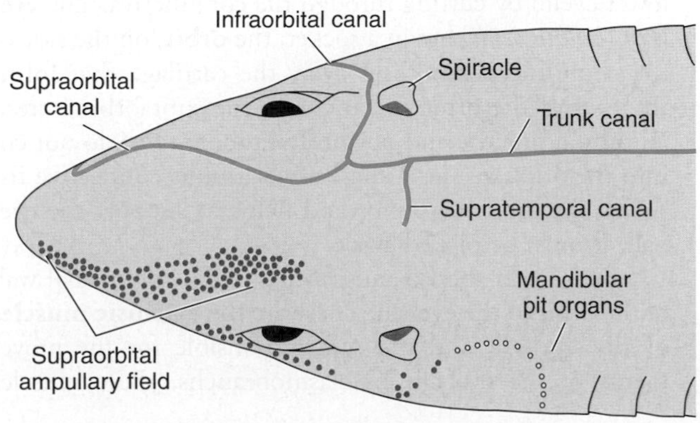

Figure 8-5

A and **B**, Ventral and dorsal views of the lateral line canals, ampullary organs, and pit organs on the head of *Squalus*. **C**, Longitudinal section through an ampullary organ. *(From Liem, Bemis, Walker, and Grande, 2001.)*

reflect the skin from the right half of the snout and head as far back as the first gill slit. Make sure to reflect only the skin and to leave any connective tissue on the specimen. Note that each pore leads into a long, jelly-filled tubule that extends tangentially under the skin and ends in a round bulb, the ampulla proper. Each ampulla is supplied with a branch of one of the lateral line nerves.

The Lateral Line Canals and Pit Organs

Observe the intact skin over the lateral line on the sides of the trunk (see Fig. 3-2), and you will see several less pigmented spots at intervals of about 1 to 2 inches. If you study the center of one of these spots with a stereomicroscope, you will discover a pore that leads into the **trunk lateral line canal** under the skin. Reflect a section of skin above the trunk lateral line canal in the same manner as you have done to study the ampullary organs on the head. Notice part of the canal on the underside of the reflected skin and part on the specimen.

Look now at the underside of the skin flaps you have reflected from the snout and head; also observe the exposed surface of the underlying connective tissue on the specimen. A **supratemporal canal** crosses the top of the head caudal to the endolymphatic pores (Figs. 7-8 and 8-5**B**). The **infraorbital canal** passes ventral to the eye, zigzags beneath the eye, and then extends rostrally to the tip of the snout in a somewhat meandering fashion, passing medial to the nares. The **hyomandibular canal** extends caudally from the bottom of the zigzag of the infraorbital canal. The **supraorbital canal** extends rostrally dorsal to the eye and onto the snout. It then turns on itself and extends caudally to connect with the infraorbital canal. The latter portion of the supraorbital canal passes just dorsal to the nares. A short **mandibular canal** is not connected to the others. It appears as a row of pores overlying the mandible caudal to the labial pocket.

Pit organs of the lateral line system are difficult to discern grossly. There is a pair rostral to each endolymphatic pore, a mandibular row on the lower jaw, and scattered pit organs on the flank dorsal to the lateral line canal. Dermal denticles adjacent to them usually are specialized and overlap the organs.

THE EYEBALL AND ASSOCIATED STRUCTURES

The eyes are somatic sensory organs whose photoreceptive cells develop embryonically as outgrowths from the diencephalon of the brain. All vertebrates have a pair of **image-forming lateral eye,** unless they have been secondarily lost. In addition, many fishes, some amphibians, and some reptiles have a median **pineal eye,** a **parietal eye,** or both (Fig. 8-6). Recall the pineal eye complex of the lamprey (Chapter 2). Experiments on larval lampreys and other vertebrates show that the pineal and parietal eye generate a low level of nerve impulses in dim light, but that strong light inhibits this activity. In some species, the median eye complex is a neuroendocrine transducer that translates light signals into chemical messengers. When light levels are low, the median eye complex produces enzymes that lead to the synthesis of the hormone **melatonin.** The median eye complex is transformed into the endocrine **pineal gland** in birds and mammals. We will return to this gland when we study the brain.

We will study the lateral image-forming eyes at this time. Although the structure of the eyeball is similar in all vertebrates, there are differences in the role of the cornea and lens in light refraction, the ways and speed with which the eye adapts to changes in light intensity, and methods of accommodation, or focusing. Many of these differences are associated with the different refractive properties of water and air and problems associated with vision in these media. The needs for protecting and cleansing the surface of the eyeball also are different. The surface of the fish eyeball is bathed in water, which keeps the eyeball moist and clean. Tetrapods have evolved movable eyelids of various types, as well as tear glands and ducts that protect, cleanse, and moisten the surface of the eyeball.

Fishes

Study the eye of *Squalus* on the same side of the head on which you dissected the nose and lateral line (Table 7-3). *Squalus,* like other cartilaginous fishes, has upper and lower **eyelids** formed of skin folds, but they are immovable. Some other sharks have movable eyelids, but most fishes lack eyelids altogether. Note that the epidermis on the inner surface of the eyelids reflects onto and over the surface of the eyeball as a transparent layer called the **conjunctiva.**

Remove the upper eyelid and free the eye from the lower eyelid by cutting through the conjunctiva. The **eyeball** *(bulbus oculi)* lies in a socket, the **orbit,** on the side of the chondrocranium. Cut away the cartilage that forms the roof of the orbit and includes the supraorbital crest, antorbital process, and postorbital process, but do not cut into the otic capsule. A mass of gelatinous connective tissue in the orbit surrounds and helps to support the eyeball. It must be picked away.

Six ribbon-shaped muscles pass from the medial wall of the orbit to the eyeball. These are the **extrinsic muscles of the eyeball,** and they are responsible for the movements of the eyeball. In elasmobranchs, these muscles

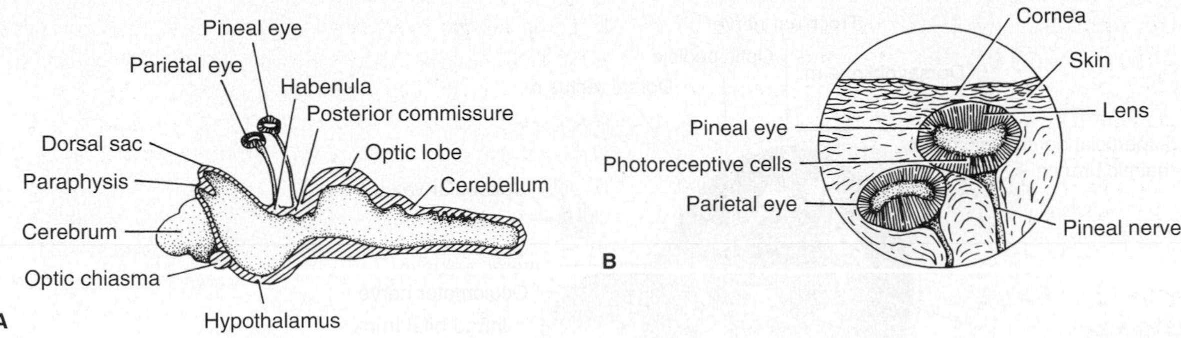

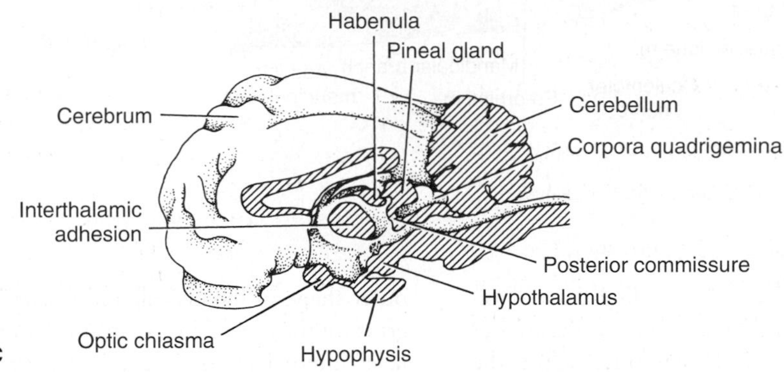

Figure 8-6

Diagrams of the pineal-parietal complex shown in diagrammatic longitudinal section. **A**, Relationship of the median eyes to the brain in a lamprey. **B**, Enlargement of the median eyes of a lamprey. **C**, Pineal gland of a sheep and its relationship to the brain.

contain both white muscle fibers, which presumably are used in rapid eyeball movements, and red muscle fibers, which are used in slower, more sustained actions. As explained in Chapter 7, the extrinsic eye muscles belong to the axial subdivision of the somatic muscles. Although they develop from the first three myotomes or comparable somitomeres (see Tables 7-1 and 7-2), they form two anatomical groups in the adult: an oblique group and a rectus group. Push the eyeball caudally and note the two muscles that arise from the rostromedial corner of the orbit and insert on the eyeball (Fig. 8-7). The muscle that inserts on the dorsal surface is the **dorsal** (superior) **oblique;** the muscle that inserts on the ventral surface is the **ventral** (inferior) **oblique.** The four muscles that arise from the caudomedial corner are the recti muscles. Three can be seen in a dorsal view. The muscle that inserts on the dorsal surface of the eyeball adjacent to the insertion of the dorsal oblique is the **dorsal** (superior) **rectus;** the muscle that lies medial to the eye is the **medial rectus;** and the muscle that lies caudal to the eye is the **lateral rectus.** Lift up the eye and look at its ventral surface. The **ventral** (inferior) **rectus** passes to insert on the ventral surface of the eyeball beside the insertion of the ventral oblique muscle.

Most of the other strands of tissue that pass to the muscles and through the orbit are nerves and nerve trunks and will be considered later; at this time, find the **optic nerve.** It is the large nerve that passes from the eyeball caudal to the ventral oblique muscle (Fig. 8-7). Do not confuse it with the **optic pedicle,** which is a stalk of cartilage that passes to the back of the eyeball between the four recti muscles. The pedicle is shaped like a golf tee and anchors the eyeball to the skull.

The outermost layer of the wall of the eyeball is the tough **fibrous tunic,** which is modified into the transparent **cornea** on the front of the eyeball. The cornea is covered by, and fused to, the equally transparent **conjunctiva.** The conjunctiva is a modified portion of the skin. Observe how the skin surrounding the eye becomes thin and soft on the inside of the eyelids and folds over the front of the eyeball as the transparent conjunctiva.

Cut off the dorsal third of the eyeball using a pair of sharp, fine-point scissors. Note the large spherical **lens,** but

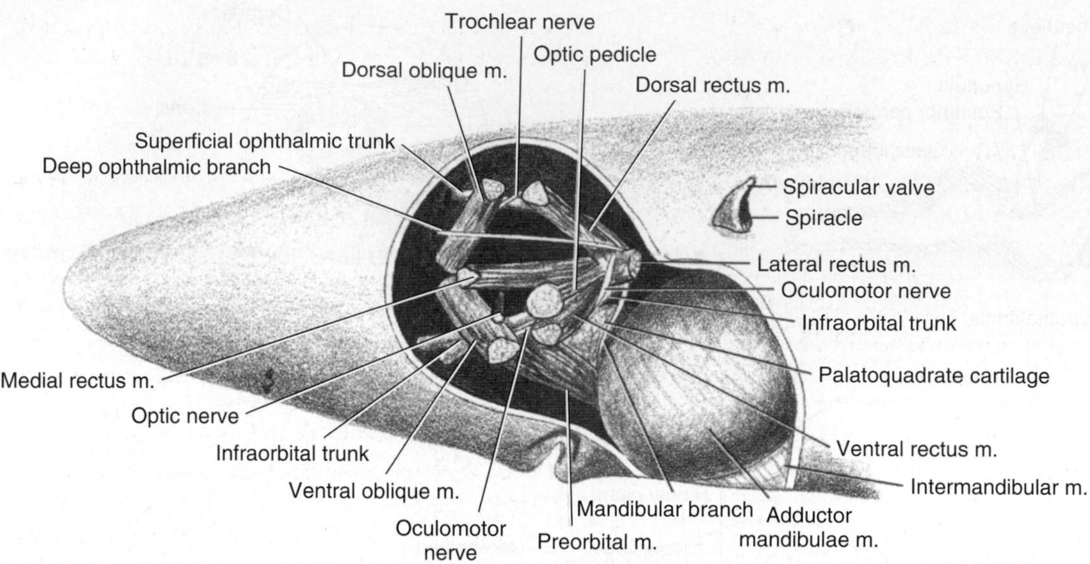

Figure 8-7
Lateral view of the orbit of *Squalus* after the eyeball is removed.

try not to break it away from the surrounding tissues. You can see the three layers, or tunics, that make up the wall of the eyeball at the back of the eye. As mentioned before, the outermost layer is the fibrous tunic. Its front portion is the cornea; the rest of it constitutes the **sclera** (Fig. 8-8). Much of a shark's sclera is reinforced by cartilage to maintain the outer shape of the eyeball and isolate it from the forces exerted on the eyeball by extrinsic eye muscles. The pigmented layer internal to the sclera is the **vascular tunic,** most of which forms the **choroid.** The **iris** is a modified portion of the vascular tunic, and it projects between the cornea and the lens. The **pupil** is the hole in the center of the iris. The

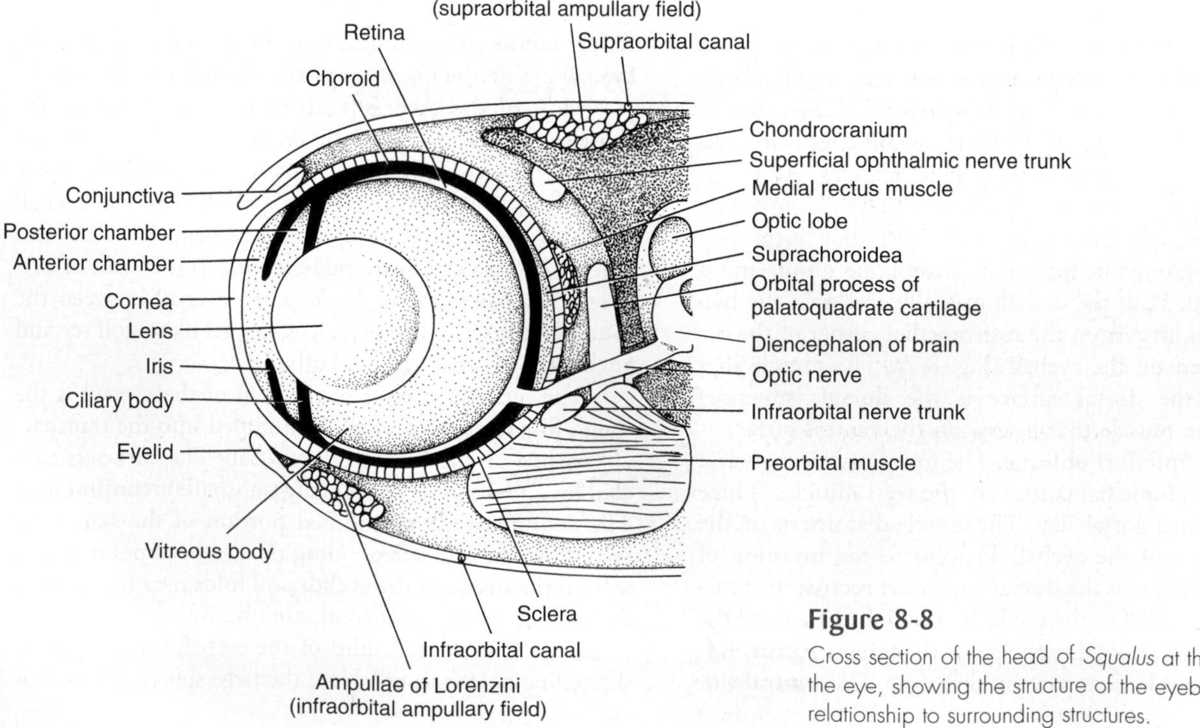

Figure 8-8
Cross section of the head of *Squalus* at the level of the eye, showing the structure of the eyeball and its relationship to surrounding structures.

whitish layer internal to the choroid is the **retina,** which is an incomplete layer that disappears near the lens. The retina is not firmly attached to the vascular tunic; it is held in place by the internal pressure exerted by the **vitreous body** in the intact eyeball. The vitreous body is the gelatinous material that fills the large cavity of the eyeball, which is called the **chamber of the vitreous body.** Once the eyeball is opened, the retina tends to fall off the vascular tunic and crumple up in the bottom of the chamber of the vitreous body. The point at which the optic nerve connects with the retina is the **optic disc,** sometimes called the **blind spot** because there are no receptive cells at this location.

The lens is held in place by the **ciliary body,** a ring of black tissue that consists of minute, radially arranged folds and which is a modified part of the vascular tunic near the base of the iris. A delicate fibrous material, the **zonule,** which cannot be discerned with the naked eye, extends between the ciliary body and the lens. A suspensory ligament, which cannot be seen with the naked eye, also assists in anchoring the lens. The cavity in front of the lens is filled with the watery **aqueous humor.** This cavity is subdivided by the iris into the very small **posterior chamber** between the lens and the iris and the **anterior chamber** between the iris and the cornea.

Cut across the back of the eyeball where the optic pedicle attaches. The relatively thick layer of material between the sclera and choroid is known as the **suprachoroidea.** It is a vascular connective tissue that is found only in chondrichthyans that have an optic pedicle.

Cut into the lens and observe that it is composed of multiple layers of modified epithelial cells arranged concentrically like the skins of an onion.

The sclera is the supporting layer of the eyeball, and the presence of cartilage within it is not surprising, because the sclera develops in part from the optic capsule of the embryonic chondrocranium (Chapter 4). The **sclerotic bones** that are present in some large-eyed vertebrates, such as birds, ossify in this cartilage. The choroid performs several functions. It is vascular and helps to nourish the light-sensitive, avascular retina. Its pigment, some of which is derived from the outer pigmented layer of the embryonic retina, prevents internal reflections and scattering of light. In addition, the choroid of many vertebrates, including the spiny dogfish, reflects some light back onto the retina. Such a reflecting device, known as a **tapetum lucidum,** is found in vertebrates that live under conditions of varying light intensity—certain fishes and nocturnal tetrapods. The elasmobranch tapetum is remarkable among vertebrates because it depends on silvery **guanine plates** within certain cells. Light passes through the retina to the choroid. The guanine plates are set at such an angle that light is reflected directly back to the receptive cells of the retina and does not scatter and blur the image. Adjacent **choroid chromatophores** can extend or retract, thereby carrying their pigment across the guanine plates, or retracting it, and thus adapting the eye to light or dark conditions (Fig. 8-9). These chromatophores are called

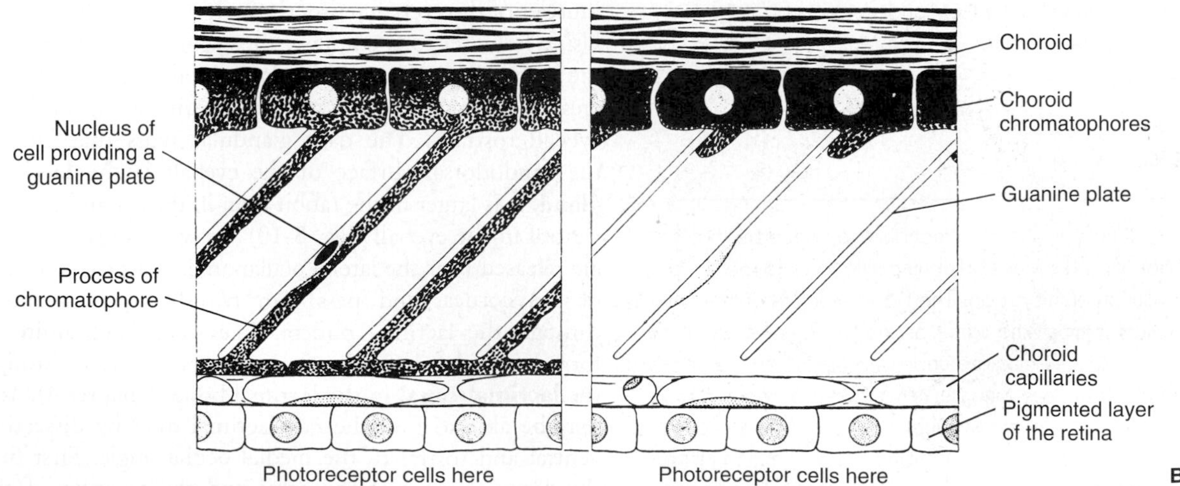

Figure 8-9
Diagrams of sections through the choroid of the smooth dogfish *Mustelus mustelus* showing the tapetum lucidum adapted for light **(A)** and dark **(B)** conditions. In the light-adapted eye, processes of the chromatophores cover the guanine plates. In the dark-adapted eye, they are withdrawn and allow the guanine to reflect light back to the receptor cells of the retina. *(From Walls, 1942, after Franz, 1934.)*

independent effectors because they detect changes in illumination and respond directly. The circular and radial muscle fibers within the iris, which narrow and widen the pupil, respectively, further adjust the eye to light and dark. Although in elasmobranchs these muscles are under some degree of neural control, they also respond directly to the level of ambient light.

Because the refractive index of the cornea and humors of the eye is nearly the same as that of water, these structures do not take part in the refraction of light rays in aquatic vertebrates such as *Squalus*. Nearly all the refraction occurs at the thick, nearly spherical lens in elasmobranchs and most other fishes, although some variation in lens shape does occur (Sivak, 1991). In elasmobranchs, the lens is held in such a position that moderately distant objects are in focus on the retina. In bright light, images are automatically brought into sharp focus by contraction of the pupil, regardless of the distance of the object. This is the principle used in an old-fashioned pinhole camera. To accommodate for very close objects, the lens can be pulled forward by the contraction of a small **protractor lentis muscle** that is located ventrally in the ciliary body. The pressure of the aqueous humor pushes the lens back to its resting position when this muscle relaxes.

The retina contains the photoreceptive cells. Both rods and cones are present in elasmobranchs. The ratio of rods to cones varies with the species, the habitat it occupies, and its mode of life. Bottom-dwelling species have few cones; pelagic species have more. The ratio of rods to cones is 50:1 in *Squalus*. The rods are particularly sensitive in low-intensity light; cones require higher light intensity. It is unclear to what extent elasmobranchs have color vision. As divers know, water absorbs red, orange, and other colors that have longer wavelengths, so that the underwater world is mostly blue. In summary, the well-developed eyes of elasmobranchs vary with the environment in which they live, and vision is an important sense at intermediate distances.

Mammals

We turn directly to the mammalian eye, because that of the aquatic *Necturus* contributes little to an understanding of the evolution of this organ. The basic structure of the eyeball in mammals is much the same as that in *Squalus*, but the details differ. The iris muscles are under the control of the autonomic nervous system (see Chapter 9) and respond quickly to the rapid changes in light intensity that occur on land. Because the mammalian cornea has a much higher refractive index than air, it is important in bending light rays. The mammalian lens is less important in refraction and is more oval in shape than the lens of *Squalus*. Accommodation occurs through changes in the curvature of the lens.

Tear Glands and Extrinsic Muscles of the Eyeball

Examine the eye of your specimen on the side that you used to study the muscles (Table 7-5). Movable **upper** and **lower eyelids** *(palpebrae)* are present. The slitlike opening between them is called the **palpebral fissure.** The corners of the eye where the eyelids unite are the **ocular angles.** Cut through the lateral ocular angle and pull the eyelids apart. You now can clearly see the **nictitating membrane** *(semilunar fold).* It is attached at the medial ocular angle, but its lateral edge can spread over the surface of the eye if the eyeball is retracted slightly. The nictitating membrane in humans is reduced to a small semilunar fold that can be seen covering the medial corner of the eyeball. With a hand lens, examine the edge of each eyelid of your specimen 3 or 4 mm from the medial ocular angle. If you are fortunate, you will see on each a minute opening, the **lacrimal punctum,** that leads into a **lacrimal canaliculus.** If you cannot find them in your specimen, look for one on a human eye by pulling down the lower eyelid and examining its edge near the most medial eyelash.

Cut off the upper and lower eyelids, leaving a bit of skin around the medial ocular angle. As you remove the eyelids, note that the **conjunctiva** on the underside of the eyelids reflects over the **cornea.** If the cornea is not too opaque, then the **pupil** and **iris** can be seen. A facial muscle, the **orbicularis oculi,** encircles the eyelids and will be cut off with them. It closes the eyelids.

Free the eyeball and associated glands from the bony rims of the orbit by picking away connective tissue. Do not dissect deeply in the region of the medial ocular angle, and try not to destroy a loop of connective tissue attached to the rostrodorsal wall of the orbit. One of the extrinsic muscles of the eyeball passes through this loop (Fig. 8-11). Using bone scissors, cut away the zygomatic arch beneath the eyeball; also cut away the postorbital processes and the crest of bone above the orbit (supraorbital arch). Push the eyeball rostrally. The dark glandular mass that lies on the caudodorsal surface of the eyeball is the **lacrimal gland.** It is larger in the rabbit than in the cat and extends ventral to the eyeball (Fig. 8-10). Its secretions, the tears, are released near the lateral ocular angle, bathe the surface of the cornea, and pass into two lacrimal canaliculi through the lacrimal puncta. These canaliculi unite to form a **nasolacrimal duct** that enters the nose through the **lacrimal canal** in the lacrimal bone (Chapter 4). You may be able to find the nasolacrimal duct by dissecting ventral and rostral to the medial ocular angle. First find the rostral border of the orbit and the position of the lacrimal canal. Do not injure an extrinsic eye muscle that is attached to the orbital wall near the lacrimal bone.

A second tear gland, the **gland of the nictitating membrane** (Harderian gland), lies just rostral to the nic-

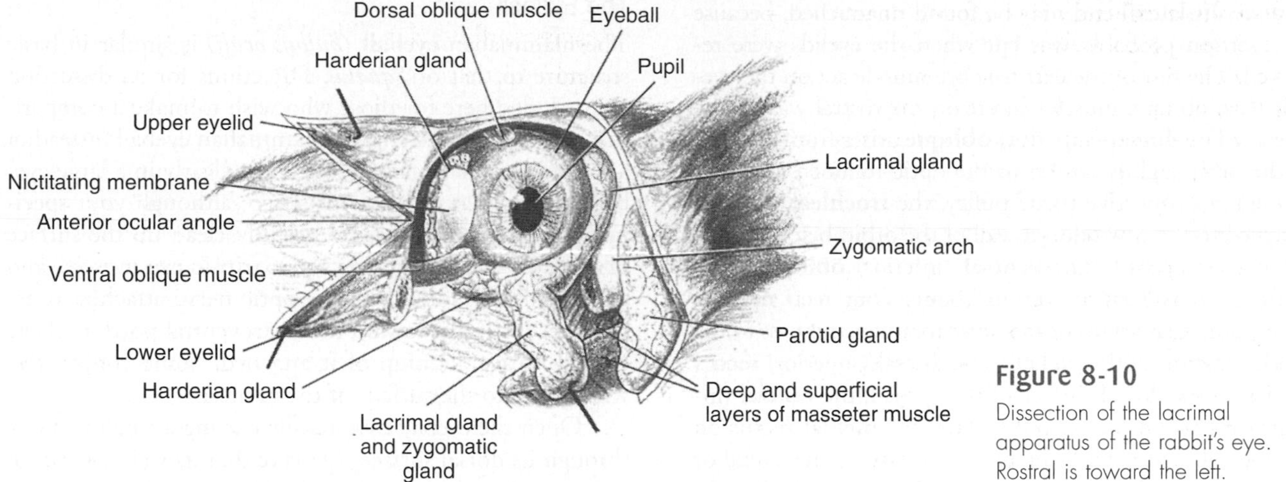

Figure 8-10
Dissection of the lacrimal apparatus of the rabbit's eye. Rostral is toward the left.

titating membrane. It is small in carnivores but large in the rabbit. In the rabbit, part of this gland extends ventrally, goes deep to the ventral oblique muscle, and then appears on the underside of the eyeball.

The cat and rabbit have a **zygomatic gland** located in the floor of the orbit ventral to the eyeball. In the rabbit, it is difficult to separate this gland from the ventral portion of the lacrimal gland. The zygomatic gland is not part of the lacrimal apparatus; rather, it secretes saliva that is discharged into the mouth at a point near the last upper molar.

Remove the glands and connective tissue from around the eyeball of your cat or rabbit specimen and expose the extrinsic muscles of the eyeball (Fig. 8-11). The pattern of the muscles is much the same as in *Squalus,* except for two additional muscles. One of these, the **levator palpebrae superioris,** arises from the medial wall of the orbit dorsal to the optic foramen and inserts on the upper eyelid, which

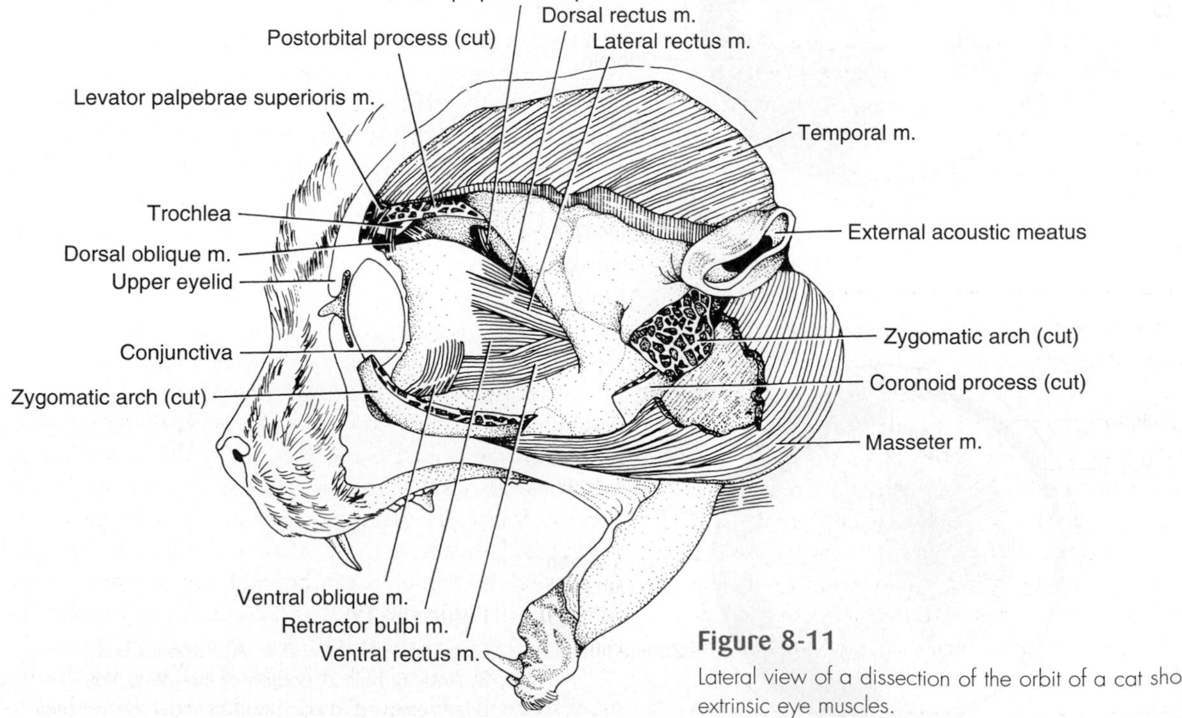

Figure 8-11
Lateral view of a dissection of the orbit of a cat showing the extrinsic eye muscles.

it raises. Its lateral end may be found unattached, because its insertion probably was cut when the eyelids were removed. The rest of the extrinsic eye muscle act on the eyeball. Two oblique muscles insert on the rostral wall of the eyeball. The **dorsal** (superior) **oblique** arises from the wall of the orbit slightly rostral to the optic foramen and goes through a connective tissue pulley, the **trochlea,** which is attached to the rostrodorsal wall of the orbit, before inserting on the eyeball. The **ventral** (inferior) **oblique** arises from the maxillary or lacrimal bone. Four recti muscles arise from the margins of the optic foramen and pass to the caudal portion of the eyeball. The **dorsal** (superior) **rectus** inserts on the dorsal surface of the eyeball, the **ventral** (inferior) **rectus** on the ventral surface, the **medial rectus** on the medial surface, and the **lateral rectus** on the lateral or caudal surface. Another muscle that is not seen in *Squalus* is the **retractor bulbi,** which can be divided into four parts that pass to the eyeball deep to the recti muscles. The embryonic derivations of the extrinsic muscles of the eyeball are listed in Tables 7-1 and 7-2.

The Eyeball

The mammalian **eyeball** *(bulbus oculi)* is similar in basic structure to that of *Squalus.* Directions for its dissection are included here for those who wish to make a comparison or who prefer to study a mammalian eyeball instead of a shark's eyeball. It can be seen most clearly in a large eyeball such as that of a cow or sheep, although your specimen's eyeball can be used. Carefully clean up the surface of the eyeball by removing the extrinsic eye muscles and associated fat. Notice that the optic nerve attaches somewhat eccentrically, toward the rostroventral portion of the eyeball. Leave a stump of it attached. Some conjunctiva will adhere to the surface of the cornea.

Open the eyeball by carefully cutting a small window through its dorsal surface. Observe that its wall consists of the same three layers as in a shark's eye: an outer **fibrous tunic;** a middle dark layer, the **vascular tunic;** and the inner, whitish **retina** (Fig. 8-12). The fibrous tunic is a dense, supporting connective tissue. Approximately the medial two thirds of the fibrous tunic form the opaque **sclera;** the

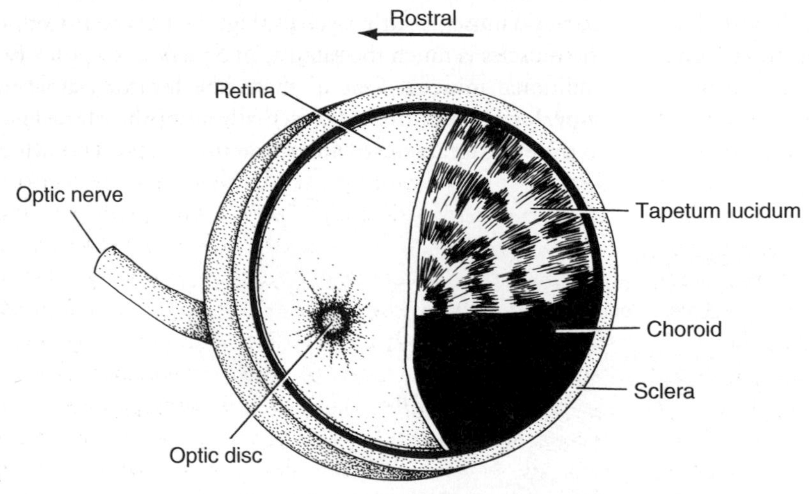

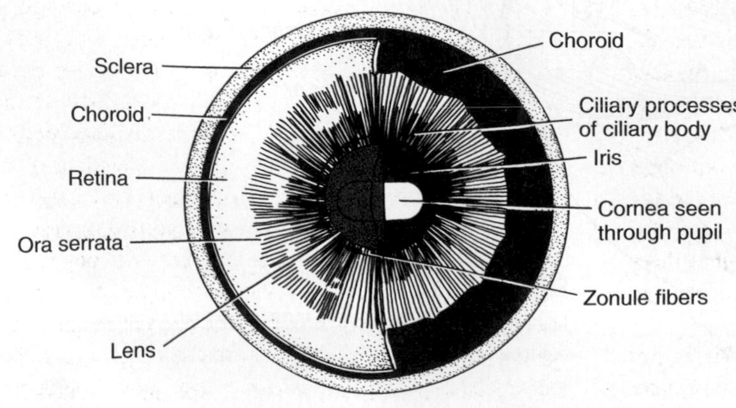

Figure 8-12
Dissection of a cow eye. **A,** Posterior half. **B,** Anterior half. A portion of the retina has been removed in each half to show the vascular tunic.

lateral one third forms the transparent **cornea,** through which light enters the eye. The vascular tunic is rich in blood vessels; most of this layer forms the **choroid,** which lies behind the retina and helps to nourish it. The portion of the retina that you see is the nervous layer, which contains the receptive rods and cones on its choroid surface. In the embryo, the retina has a distinct pigmented layer, but in the adult this layer becomes associated with the choroid. The pigment cells reduce light reflections within the eye and play a role in the biochemistry of vision.

With a pair of fine scissors, extend a cut from the window that you made on its dorsal side around the equator of the eyeball, thus separating the eyeball into an anterior half containing the lens and cornea and a posterior half containing most of the retina. Cut through all the layers. The jellylike mass filling the eyeball between the lens and retina is the **vitreous body,** a material that holds the retina in place, supports the lens, and refracts light entering the eye. The space in which it lies is the **chamber of the vitreous body.** Keep the vitreous body with the anterior half of the eyeball. Submerge both halves in a dish of water.

Examine the posterior half (Fig. 8-12**A**). The retina probably has become partly detached from the choroid, but it can be floated back into its normal position. Note the round spot, called the **optic disc,** where the optic nerve attaches to the retina. This region is devoid of rods and cones and hence is often called the blind spot. Remove the retina and observe that an extensive section of the choroid dorsal to the optic disc is quite iridescent. This is the **tapetum lucidum,** an area that reflects some of the light passing through the retina back onto the rods and cones and hence facilitates the animal's ability to see in dim light. This tapetum depends on a compact layer of endothelial cells in the choroid, not on guanine plates as in a spiny dogfish. Many mammals, including humans, lack a tapetum lucidum.

Carefully remove the vitreous body from the anterior half of the eyeball. Notice that the white nervous layer of the retina does not extend far into this half of the eyeball. The line of demarcation between it and the dark choroid (plus embryonic pigment layer of retina) is the **ora serrata** (Fig. 8-12**B**). The portion of the vascular tunic that you see extending from the ora serrata toward the **lens** is the **ciliary body.** The portion of the ciliary body next to the lens has a pleated appearance; the individual folds are the **ciliary processes.** While observing the area with a dissecting microscope, carefully stretch the region between the lens and the ciliary processes. You will see many delicate **zonule fibers** passing from the ciliary body to the equator of the lens. The zonule fibers and ciliary body with its ciliary muscle fibers play an important role in accommodation in mammals (Anatomy in Action 8-1).

Carefully remove the lens and notice that it is not spherical, as in the spiny dogfish, but somewhat flattened. The vascular tunic continues in front of the lens to form the **iris.** The **pupil** is the opening through the iris. Its diameter, and the amount of light it permits to pass, is regulated by circular and radial muscle fibers within the iris. The space between the lens and iris is the **posterior chamber;** the space between the iris and cornea is the **anterior chamber.** Both chambers are filled with a watery **aqueous humor** secreted by the ciliary processes. This liquid maintains the intraocular pressure. Excess liquid is drained off by a microscopic **scleral venous sinus** (canal of Schlemm), which encircles the eye between the base of the cornea and the iris. If you make a vertical cut through the anterior half of the eyeball and examine it under a dissecting microscope, you may be able to see this canal and the ciliary muscles within the ciliary body.

THE EAR

The **inner ear** of vertebrates, as we have discussed (p. 187), is closely related to the lateral line system. Nineteenth-century investigators first proposed that the inner receptive part of the ear evolved by the invagination of a portion of the lateral line system, although there is no direct evidence to confirm or refute this hypothesis. In whatever way it evolved, the inner ear is isolated from surface disturbances and is specialized to detect internal disturbances caused by changes in the orientation and movements of the body. The inner ear consists of a series of thin-walled ducts and sacs filled with a fluid known as the **endolymph.** These ducts and sacs are collectively called the **membranous labyrinth** (Figs. 8-13 and 8-14). They may lie within a single chamber in the otic capsule, but usually they are imbedded within a series of parallel canals and chambers within the otic capsule known as the **cartilaginous,** or **osseous, labyrinth.** The membranous labyrinth and osseous labyrinth are separated from each other by spaces filled with fluid and criss-crossed by minute strands of connective tissue. This fluid is the **perilymph.**

Only an inner ear is present in fishes. As in terrestrial vertebrates, it is an organ of equilibrium. In many species, parts of the inner ear are sensitive to sound waves. A sound generated under water has two components. First, there is physical disturbance of water particles analogous to the ripples generated when a stone is dropped into a pond. Water particles are displaced with a certain velocity, and this **displacement wave** travels outward from the sound source as vibrations of low frequency and high amplitude. These "near-field" waves do not travel far or fast. Second, **pressure,** or **sound, waves** are generated that have a higher frequency and lower amplitude. They propagate very rapidly and travel considerable distances as the "far-field" component of sound. The displacement wave can be detected by the lateral line system and by parts of the inner ear in some species. The sound waves moving through water

Anatomy in Action 8-1 Accommodation in the Eye of Mammals

Refraction and accommodation differ in mammals and fishes. In mammals, intraocular pressure causes the wall of the eyeball to bulge. As a consequence, the zonule fibers exert a radial pull on the elastic lens, so that the lens is under tension and somewhat flattened (see figure). Under these conditions, the lens has minimal refractive powers, and the eye is focused on distant objects. On land, the greatest refraction of the light occurs at the cornea because of the significant difference between the refractive indices of the cornea and air. The lens refracts light less strongly because the differences between the refractive indices of the aqueous humor, lens, and vitreous body are relatively small. The main role of the lens is to focus the eye on close objects, for which the light needs to be refracted more strongly. To do so, the lens needs to increase its curvature. This occurs when ciliary muscle fibers in the ciliary body contract and pull the sclera closer towards the lens. This relieves tension on the zonule fibers and allows the lens to bulge. When the ciliary muscles relax, the sclera returns to its resting position and the zonule fibers pull again on the lens so that it returns to its flatter configuration.

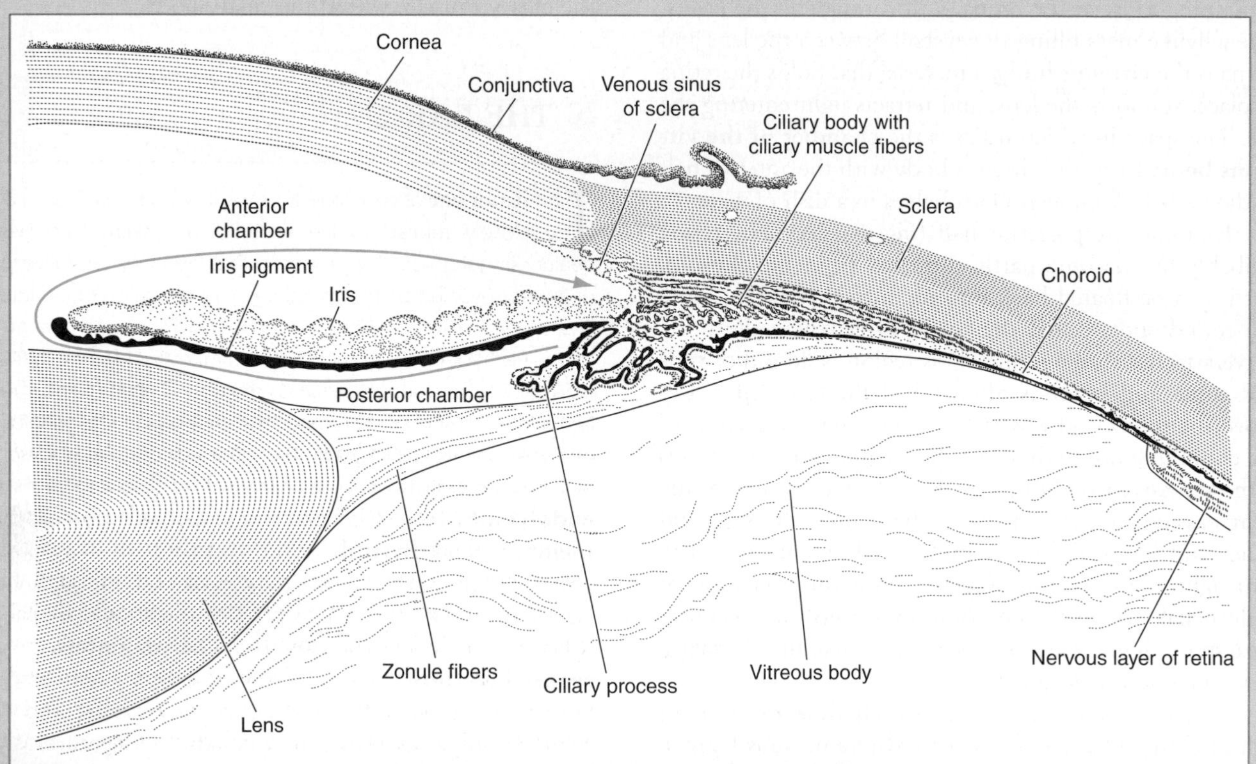

Detailed structure of the anterior portion of the mammalian eyeball. The blue line shows the path of the aqueous humor from the ciliary body, where it is secreted, to the venous sinus of the sclera, where it enters the circulatory system. *(From Liem, Bemis, Walker, and Grande, 2001, after Fawcet, 1994.)*

easily travel through the head and inner ear of a fish because the body tissues consist mostly of water. In a sense, a fish is transparent to sound waves. Fishes cannot detect these pressure waves unless they have a mechanism to transduce them into displacement waves that can affect certain hair cells of the inner ear. Connections between the swim bladder and the inner ear do this in some species of teleosts.

Fishes

You can study the inner ear of *Squalus* on demonstration preparations, or, if time permits, you may dissect it on the side of the head used for the study of other sense organs (Table 7-3). The basic structure of the membranous labyrinth can be seen better in cartilaginous fishes than in any other vertebrates, because the labyrinth is relatively large and can be freed more easily from cartilage than from

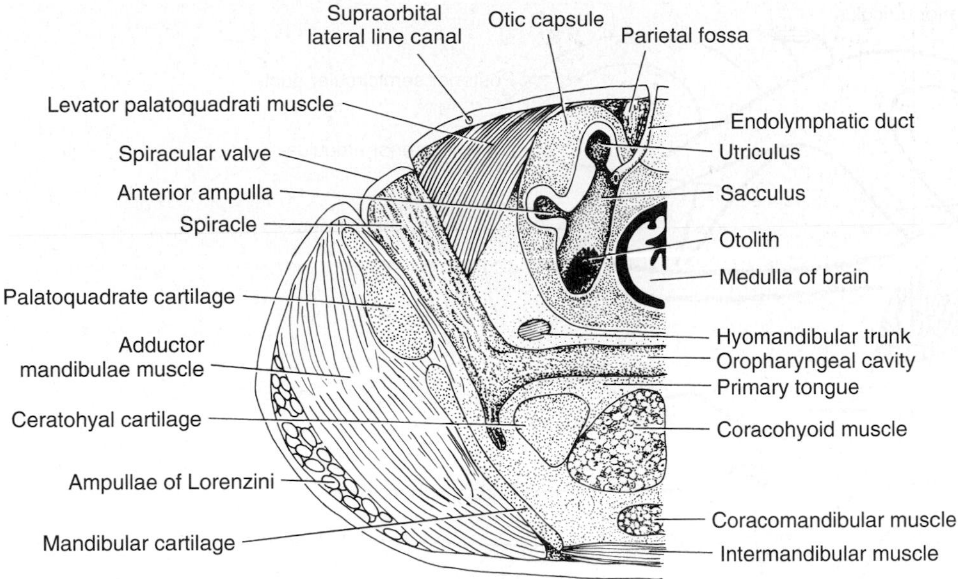

Figure 8-13

Cross section through the head of *Squalus* at the level of the spiracle showing the relationship of the otic capsule and inner ear to surrounding structures.

bone. First expose the otic capsule by removing the skin and muscles from its dorsal, lateral, and caudal surfaces. You can also cut away the spiracle and adjacent parts of the mandibular and hyoid arch (Fig. 8-13). Note the **endolymphatic duct** that leads from one of the two endolymphatic pores in the skin to the parietal fossa on the dorsal side of the chondrocranium. It ultimately connects with the sacculus of the membranous labyrinth (Fig. 8-14).

To expose the **membranous labyrinth** of the inner ear, you must carefully break away the surrounding cartilage of the otic capsule piece by piece with forceps, beginning on the dorsal and lateral surfaces and gradually working ventrally and medially. You usually can see the various ducts and chambers through the translucent cartilage of the skull shortly before you reach them. Use special care as you dissect the cartilage from around the parts of the membranous labyrinth, and try not to break them. You will first see the three semicircular canals of the **cartilaginous labyrinth,** namely, the **anterior vertical**, **posterior vertical,** and **lateral horizontal semicircular canals.** The **semicircular ducts** of the membranous labyrinth lie within the semicircular canals (Fig. 8-14). The ventral portion of each semicircular duct bears a round swelling, the **ampulla,** which contains a sensory patch of hair cells and a cupula known as a **crista.** Branches of the **statoacoustic (VIII) nerve** can be seen leaving each ampulla, but the cristae may not be visible. The ends of each semicircular duct connect to one of the two membranous utriculi. The ampullae of the anterior vertical and lateral horizontal semicircular ducts are connected to the ventrolateral end of the **anterior utriculus,** and the opposite ends of these semicircular ducts are connected to the dorsomedial end of the anterior utriculus. The posterior vertical semicircular duct connects to the **posterior utriculus.** The utriculi connect to the **sacculus** through inconspicuous openings. The sacculus lies in a large cavity medial to the lateral horizontal canal of the cartilaginous labyrinth. If the liquid has drained from the sacculus, its lateral wall often is collapsed over the bottom of the sacculus and covers the otolith (see below). Ventrally and medially, the wall of the sacculus is attached to the surrounding cartilage of the otic capsule. Expose the connection between the endolymphatic duct and the sacculus by removing cartilage. Dissect away as much of the cartilage as possible from the ventral side of the sacculus without destroying it and observe its short caudoventral extension or bulge. This extension is the **lagena.** The sacculus and lagena contain large sensory patches of hair cells called **maculae.** Each macula is overlaid by a mass of minute mineral concretions imbedded in a gelatinous matrix to form an **otolith.** In many sharks, the otolith is composed of secreted crystals of calcium carbonate, but in *Squalus,* sand grains that enter through the endolymphatic ducts form the otolith. In preserved specimens, the matrix that holds the sand grains together tends to disintegrate. Hearing in sharks is discussed in Anatomy in Action 8-2.

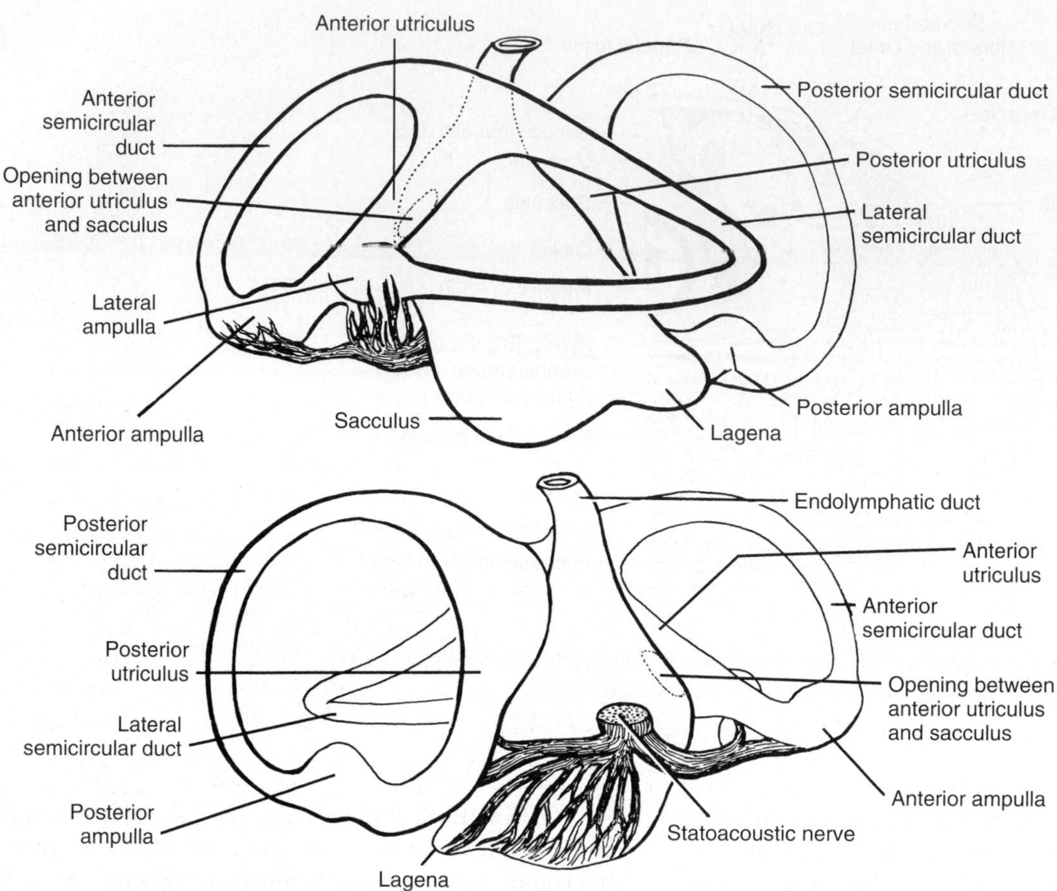

Figure 8-14

Left inner ear (membranous labyrinth) of the shark, *Heptanchus maculatus*. **Top,** Lateral view. **Bottom,** Medial view. *(Redrawn from Daniel, 1934.)*

Amphibians and Reptiles

Some terrestrial vertebrates lack an eardrum and respond primarily to low-frequency vibrations that are transmitted to the inner ear through certain skull bones. In those with ears sensitive to air-borne sounds, higher-frequency sound pressure waves impinge upon an eardrum or **tympanic membrane.** The tympanic membrane constitutes the **external ear.** Sound waves are transmitted from this membrane across an air-filled middle ear cavity, or **tympanic cavity,** by one or more auditory ossicles (Fig. 8-15). The tympanic cavity and auditory ossicles are the **middle ear.** The footplate of the **columella,** which is the only auditory ossicle of amphibians, reptiles, and birds, fits into the **fenestra vestibuli,** or oval window, on the side of the otic capsule. (In mammals, as we shall see later, the columella is called the stapes and it is the innermost of three ossicles.) The columella often also has a connection to the quadrate bone, so its topographical relationships are essentially the same as those of the hyomandibula, to which it is homologous. The difference in size between a large tympanic membrane and a small footplate of

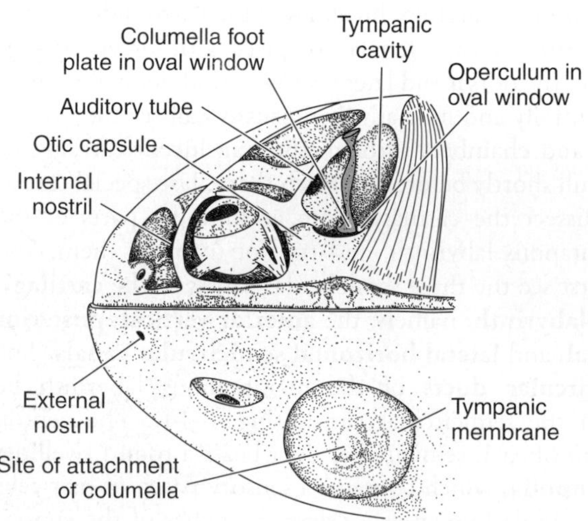

Figure 8-15

Tympanic membrane and middle ear of a bullfrog in a dorsal surface view **(bottom)** and dissected **(top).** *(From Liem, Bemis, Walker, and Grande, 2001, after Walker, 1981.)*

Anatomy in Action 8-2 Hearing in Sharks

No studies have been done on the acoustic properties of the ear of *Squalus,* but some have been done in larger reef and lemon sharks, whose ears are structurally similar to those of the spiny dogfish (for review, see Corwin, 1981). Behavioral studies and recordings of action potentials in the statoacoustic nerve demonstrate that the larger sharks are aroused by low-frequency vibrations of the type that would be generated by the movements of a wounded and struggling fish. Their highest sensitivity is for frequencies between 40 and 160 Hz. Moreover, the sharks can localize the source of a sound coming from as far away as 250 meters.

The primary area of sensitivity in the ear is the inconspicuous and previously little-studied **crista neglecta,** which is located in the **posterior duct canal** (figure). This duct canal is the equivalent of the dorsal part of the posterior utriculus of *Squalus,* but instead of connecting ventrally with the posterior semicircular duct, the posterior duct canal terminates ventrally directly in the sacculus and on a **lateral membrane** adjacent to the large perilymphatic space. Low-frequency vibrations reaching the fish from above and the side cause water particles to be displaced as shown by arrows in the figure. Because of their low frequency, few of those vibrations impinging on the cartilage of the otic capsule enter the ear, but those hitting the skin over the **parietal fossa** travel easily through the loose, watery connective tissue filling the parietal fossa and impinge upon the posterior duct canal, which reaches the parietal fossa through the **perilymphatic foramen.** This foramen is analogous to the fenestra ovalis of tetrapods. Particle displacement waves travel directly through the posterior duct canal, move the cupula overlying the very dense population of hair cells in the crista neglecta, and finally are released through the lateral membrane into the perilymphatic space. Notice from the figure that vibrations coming from above and the side affect primarily the contralateral ear. When the shark turns toward the sound source, both ears will be affected equally, and the shark is able to hone in on the target.

Electrical recordings from parts of the statoacoustic nerve show that the sacculus also responds to low-frequency vibrations but to a lesser extent than does the crista neglecta. Low-frequency vibrations, with their large amplitude, cause the entire fish to vibrate to some extent. Because the heavy otolith particles are not anchored and have more inertia than the endolymph, their movement lags, and this stimulates the underlying hair cells of the macula sacculus. Because the distribution of stereocilia on the hair cells determines the direction of the displacements to which they will respond, groups of cells within the macula with different patterns of distribution of these processes will detect displacements coming from different directions. Sharks presumably use this system to detect sounds coming from behind or beneath them.

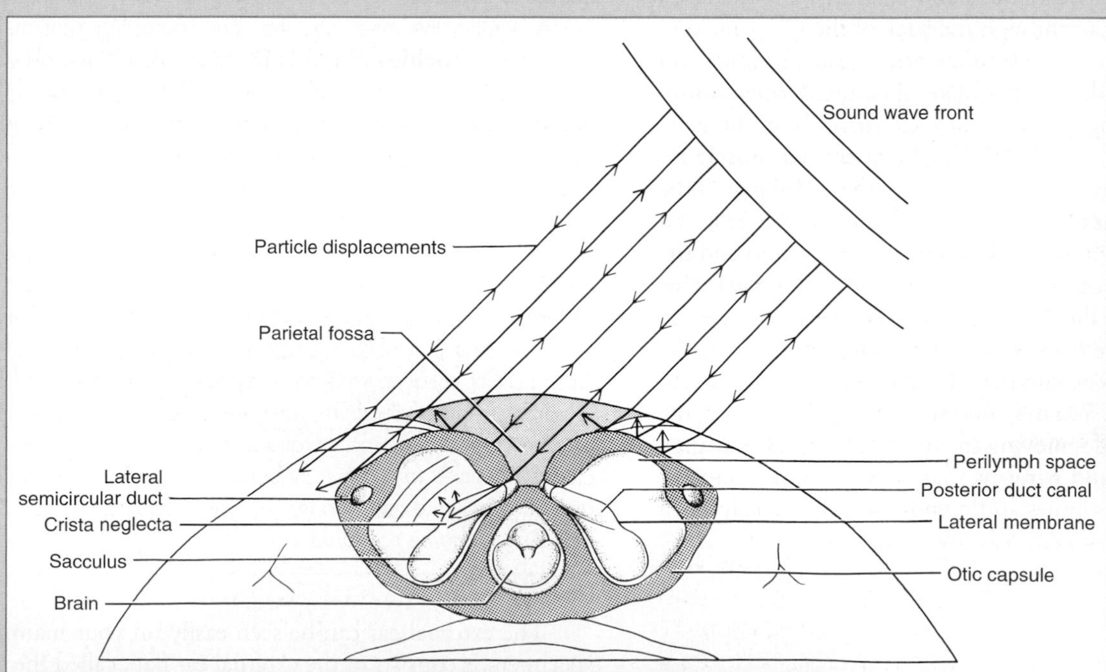

Diagram of a cross section through a shark's head indicating the path by which particle displacements induced by a low-frequency wavefront reach the inner ear. *(From Corwin, 1981.)*

the columella amplifies the pressure sufficiently to overcome the inertia of the liquids in the inner ear. The sound pressure wave is transduced into a displacement wave that travels through parts of the perilymph to the receptive part of the membranous labyrinth. Vibrations finally are released through the **fenestra cochleae,** or round window, back into the middle ear cavity or into the cranial cavity. The location of the tympanic membrane, the relationship between the columella and certain nerves, the number of auditory ossicles, the location of the receptive cells within the membranous labyrinth, and other aspects of the tetrapod ear vary so much among groups that investigators now believe that tympanic ears sensitive to air-borne pressure waves evolved independently of one another several times in the course of tetrapod evolution (Lombard and Bolt, 1979).

The urodele ear lacks the tympanic membrane and tympanic cavity, and its components are hard to dissect. Therefore, you should study the structure of the ear of a frog on demonstration dissections of a bullfrog or by making your own dissection of a frog. The large external tympanic membrane of the frog is easily seen (Fig. 8-15). It is located high on the skull. Remove it on one side and you will expose the **tympanic cavity**. Open the mouth and note that the tympanic cavity communicates with the caudal part of the oropharyngeal cavity via the **auditory** (Eustachian) **tube.** A long rodlike **columella** crosses the middle ear.

Expose the **otic capsule,** which contains the inner ear, by removing the overlying skin and muscle. Also reflect the superficial, laterally extending musculature lying caudal to the middle ear. Cut through the back of the tympanic cavity and trace the columella to the otic capsule. Its inner end is associated with the knoblike specialized **operculum,** which is homologous to the urodele structure of the same name (Fig. 4-10). A tiny muscle, the **opercular muscle** extends caudally from the operculum to the dorsal part of the scapula just beneath the superficial muscle you reflected (Anatomy in Action 8-3). Remove the operculum and the footplate of the columella and you will see the **oval window** into which they fit. Vibrations enter the inner ear at this point and are released through a fenestra, which, in lissamphibians, opens into the cranial cavity dorsal to the statoacoustic nerve. You may find this fenestra by opening the cranial cavity and removing the brain. This fenestra is analogous to the **round window** that opens into the tympanic cavity of certain reptiles and mammals and is usually given this name, even though it is not homologous.

Mammals

Although many reptiles have a tympanic membrane and a middle ear located caudal to the quadrate bone (Fig. 8-16A and B), early synapsids appear to have had a different type of sound-transmitting system (Allin, 1975). Their skulls did not have enough space for a tympanic membrane large enough to amplify sound waves, and the columella was a heavy bone that retained a broad connection with the quadrate bone. Early synapsids probably detected low-frequency ground-borne vibrations via their lower jaw, and these were most likely transmitted to the inner ear via the articular and quadrate bones that form the jaw joint, as well as the columella. More advanced synapsids had an unusual flange, the reflected lamina, on the angular bone of the lower jaw (Fig. 8-16C), which extended ventrally and caudally around an air-filled space that probably lodged a relatively large "mandibular" tympanic membrane. Such a mechanism may have detected higher-frequency air-borne vibrations and thus supplemented sound transmission through bones. The quadrate, articular, and angular bones were small, and, although involved with jaw mechanics, they were not bound tightly to adjacent bones, so they could vibrate to a limited extent. With the further reduction in the size of the bones forming the jaw joint and other adjacent bones, and with the shift in jaw joint that occurred during the transition to mammals (Chapter 4), the ear could become specialized for detecting air-borne sound waves. The columella became much smaller and because of its stirrup shape is called the **stapes.** The quadrate became the **incus;** the articular became the **malleus** (Fig. 8-16E). The retroarticular process of the articular bone, which was no longer needed for the attachment of jaw-opening muscles, was in a strategic position to become a lever arm or the handle of the malleus. The angular became the **tympanic bone** supporting the tympanic membrane.

An **auricle** evolved around the entrance to the **external acoustic meatus,** the canal leading to the tympanic membrane. Within the inner ear, the sound-detecting portion became a long **cochlea** (Fig. 8-16D). Its spiraled shape allows it to fit within the otic region. The cochlea consists of the **cochlear duct,** which is the part of the membranous labyrinth containing the receptive cells; and two perilymphatic ducts, the **scala vestibuli** and **scala tympani,** by which pressure waves enter and are released from the cochlear duct.

According to this hypothesis, ears sensitive to air-borne vibrations evolved independently in certain lissamphibians and reptiles and in later synapsids. In the line of evolution toward mammals, the columella, quadrate, and articular were at all times part of a sound-transmitting system. They were used first to transmit ground vibrations, then were used for a combination of ground- and air-borne vibrations, and finally served as specialized ossicles to transmit air-borne vibrations. The quadrate and articular were not fitted into an ear already specialized for detecting air-borne sound waves.

The external ear can be seen easily on your mammal specimen. It consists of the external ear flap, called the **auricle,** and of the **external acoustic meatus,** a canal with a cartilaginous wall leading inward to the skull. The eardrum, or **tympanic membrane,** lies at the base of the

Anatomy in Action 8-3 Hearing in Frogs and Salamanders

Lissamphibians have evolved unique auditory systems (Capranica, 1976). Their inner ear has two areas sensitive to sound waves. The **basilar papilla** is sensitive to the high-frequency sounds associated with the mating and release calls of male frogs. The mating call identifies species, and the release call identifies sex because it signals a male to release a male frog it may have inadvertently grasped. The **amphibian papilla** is sensitive to low-frequency vibrations of the type that are generated and probably transmitted through the body tissues when a male grasps a female. It probably is sensitive to other low-frequency environmental sounds as well. Sound waves reach the inner ear of a frog by way of the tympanic membrane, columella, and operculum. The columella and operculum are notched in such a way that they lock together when tension is exerted on the slender **opercular muscle,** which extends from the operculum to the scapula. The two ossicles then form a functional unit of considerable size and inertia that responds only to low-frequency sounds (Hetherington et al., 1986). When the opercular muscle relaxes, the two ossicles uncouple. The smaller size and lower mass of the columella by itself allow it to respond to high-frequency sounds such as mating calls. By this means a frog can tune its ear to either low-frequency or high-frequency sounds.

Salamanders lack a tympanic membrane and tympanic cavity, and they have no mating calls. Salamanders respond only to low-frequency environmental noises and ground vibrations. Their larvae have a columella that attaches to superficial skull bones, but this system is replaced in the adults by the opercular system (Figs. **A** and **B**).

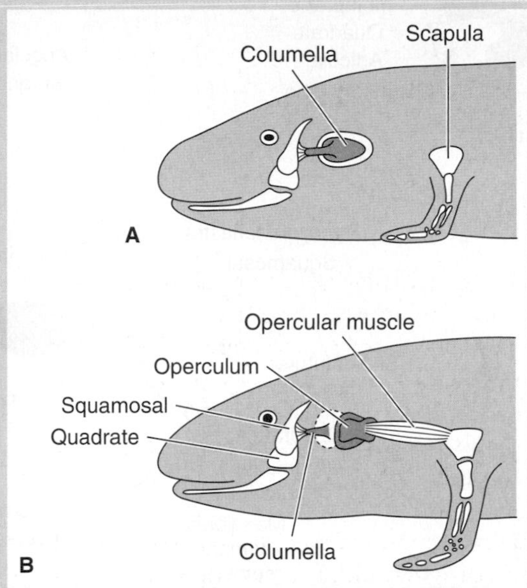

Ear of salamanders. **A**, The larva has a columella connected to superficial skull bones. **B**, The adult retains the columella, but most of its hearing occurs by way of the opercular apparatus.

meatus. You can find it on the side of the head on which you removed the auricle by cutting away as much of the external acoustic meatus as possible and shining a light into the hole. Notice that the tympanic membrane is set at an angle; its rostral portion extends more medially than its caudal portion. The opaque line seen through the membrane is the handle of the malleus.

The rest of the ear is difficult to dissect and should be observed on demonstration preparations or models, or studied from a diagram (Fig. 8-16**D** and **E**). If desired, dissections can be prepared from the sagittal sections of the head that were used to study of the nose. The following account is based on the cat but is applicable to many other mammals. Remove the muscles and other tissue from around the tympanic bulla except at its rostromedial corner. The middle ear cavity, or **tympanic cavity,** lies within the bulla and opens into the nasopharynx by the **auditory tube** (Eustachian tube). The opening of the auditory tube appears as a slit in the lateral wall of the nasopharynx (Fig. 10-15). Careful dissection between this slit and the rostromedial corner of the bulla will reveal the auditory tube; part of its wall is cartilaginous and part is bony.

The rest of the dissection will be much easier if you first decalcify the specimen by placing it in a weak solution (0.06%) of nitric acid for a few days. Rinse the decalcified specimen in running water for several hours before handling it. Break away the caudomedial portion of the bulla, the mastoid and paracondyloid processes, and adjacent parts of the nuchal crest, exposing the caudomedial chamber of the tympanic cavity. Note that it is largely separated from a smaller rostrolateral chamber by an approximately vertical plate of bone. A hole through the dorsolateral portion of this plate connects the two chambers of the middle ear cavity. You can see the **fenestra cochleae,** or round window, through this hole (see Fig. 4-22). Carefully break away this plate of bone and open the rostrolateral chamber. You can see the handle of the **malleus** on the inside of the tympanic membrane, and you will notice that the fenestra cochleae is situated on a round promontory of bone. A fingerlike process of cartilage extends from the caudolateral wall of the tympanic cavity between the handle of the malleus and this bony promontory. The tiny nerve that runs along it and leaves its tip is the **chorda tympani,** a branch of the facial nerve that supplies the taste buds of the tongue and certain

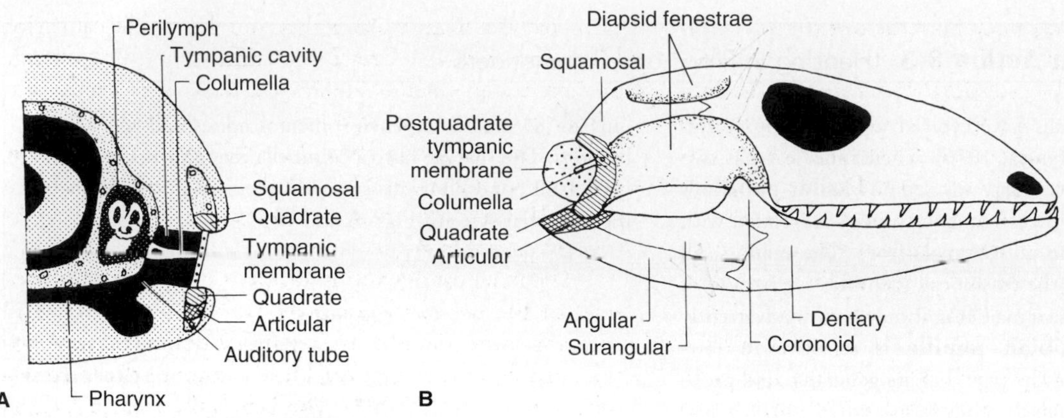

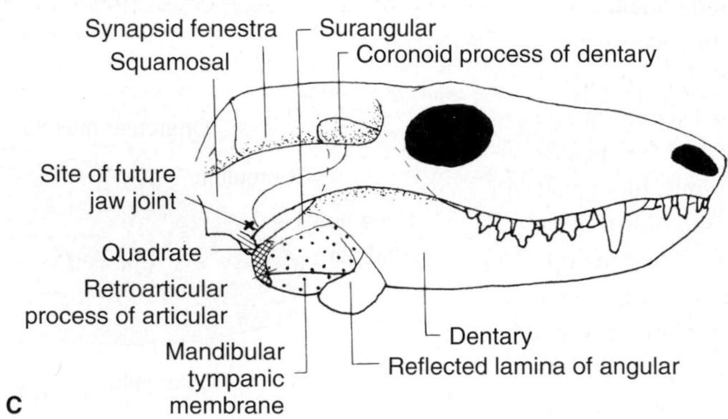

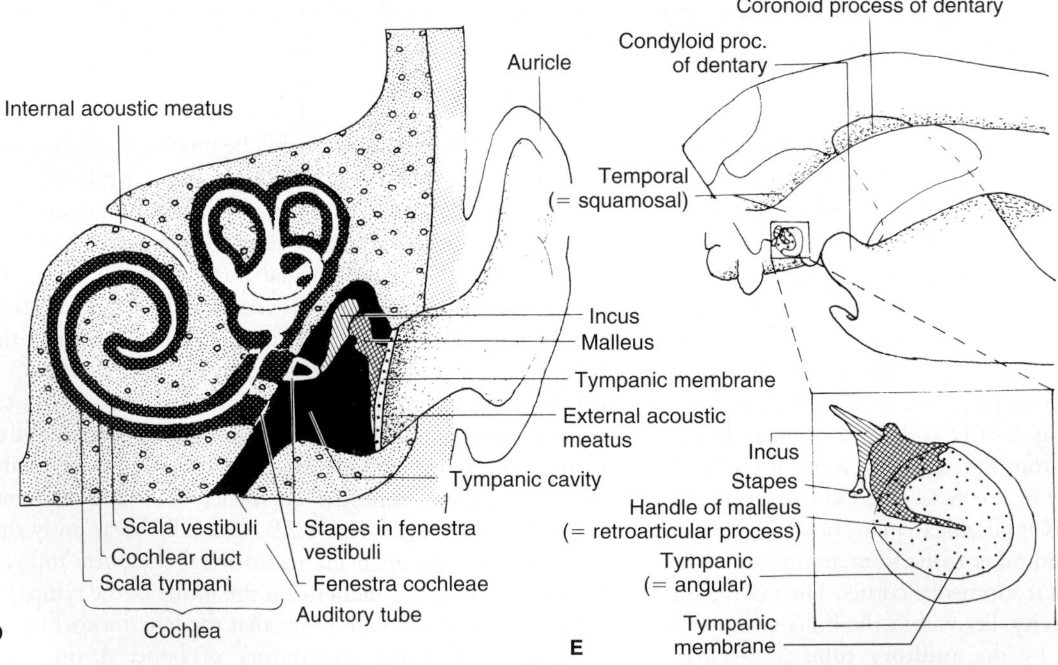

Figure 8-16

Diagrams of some types of ears found in terrestrial vertebrates. **A**, Transverse section through the otic region of a lizard. **B**, Lateral view of the skull and ear of a generalized lizard. **C**, Lateral view of the skull and ear of a synapsid close to the mammal transition. **D**, Transverse section through the otic region of a human skull. **E**, Lateral view of the posterior portion of an opossum skull, with the otic region shown enlarged below. *(B and C, redrawn from Allin, 1975; E, redrawn from Crompton and Parker, 1978.)*

salivary glands. Break away bone from the rostromedial corner of the tympanic bulla and find the entrance of the auditory tube. The other auditory ossicles, **incus** and **stapes,** and the **fenestra vestibuli,** or oval window, in which the stapes fits, lie dorsal to the fenestra cochleae. To see them, you must remove a piece of bone caudal and dorsal to the external acoustic meatus without damaging the plate of bone supporting the tympanic membrane. You also will notice two small muscles attaching to certain of the auditory ossicles. The **stapedius muscle** arises from the medial wall of the tympanic cavity caudad to the fenestra cochleae and inserts on the stapes. The **tensor tympani muscle** arises from the medial wall rostral to the fenestra vestibuli and inserts on the malleus. These muscles are parts of the hyoid and mandibular musculature, respectively, which have followed parts of the hyoid and mandibular arches into the middle ear cavity. Their contractions increase the stiffness of the tympanic membrane and chain of auditory ossicles to reduce the amplitude of loud sounds and noises, whose vibrations would damage the delicate receptive hair cells in the cochlea. The evolution of these complex middle ear structures allowed mammals to tune their middle ear to perceive high-frequency, low-energy sound waves.

The inner ear lies within the petrous portion (otic capsule) of the temporal bone. By removing the brain and chipping away pieces of the temporal bone, you may notice portions of the inner ear, but it cannot be dissected satisfactorily by this method. The **internal acoustic meatus** for the vestibulocochlear and facial nerves lies in the cranial cavity on the caudomedial surface of the petrous portion.

The Nervous System

Chapter 9

Receptors alert an animal to changes in its internal and external environments. Often, an animal combines and sorts out sensory signals from different sources and draws upon memories of past experiences before it makes a motor response appropriate to the changed conditions. The motor response frequently involves the activation (or inhibition) of many muscles, glands, and other effectors. We refer to the totality of these processes as **integration** or **coordination.** Locomotion and other rapid integrative responses are mediated by nerve impulses traveling along nerve cells, or **neurons,** to a combination of specific effectors. Most slower responses that regulate continuing processes that affect many parts of the body, such as metabolism, growth, and reproduction, are mediated by chemical messengers called **hormones,** which are secreted by **endocrine glands** and distributed through the blood to their targets. A hormone reaches all parts of the body, but only target cells with matching biochemical receptors on their surfaces will respond to it. Although there is a distinction between nervous and endocrine integration, the dichotomy is not absolute. Neurons secrete chemical messengers called **neurotransmitters** that transfer signals from one neuron to another neuron or to an effector. A few neurons secrete hormones into the blood, a phenomenon called **neurosecretion,** and both nerve impulses and hormones together affect some physiological processes, including color changes in many fishes.

Because the endocrine glands are widely scattered, we will not discuss them in one place but as they are observed during the dissection of other organ systems. We consider the nervous system at this time because it is the last of the group of organ systems that deals directly with the general function of body support and movement.

Neurons and Glial Cells

The nervous system is composed of several cell types. **Neurons** are the functional units of the nervous system. They process and transmit nerve impulses, but many other cells, which are far more abundant than the neurons, surround, protect, and maintain the neurons. These cells are known as **glial cells** in the brain and spinal cord and as **neurilemma cells** in the nerves. Many glial and neurilemma cells wrap around the axons of neurons many times, surrounding them with a fatty **myelin sheath,** which increases the speed of nerve impulse transmission. Beyond this, glial cells help to maintain the composition of the extracellular fluid, they have a phagocytic role, and they have been shown to have a role in inducing and stabilizing the junctions, or **synapses,** between neurons (Ullian et al., 2001).

A neuron can be divided into four regions. (1) An enlarged cell body, or **trophic region,** contains the nucleus, synthesizes materials needed by the cell, and is primarily responsible for maintaining the cell. (2) A **receptive region** receives stimuli from receptors or other neurons and initiates the nerve impulse. This region consists of relatively short, branching processes known as **dendrites** that connect with the cell body, or it may consist of just the cell body itself. (3) The **conductive region** is composed of a long slender process, known as the **axon,** which rapidly transmits the nerve impulse to the end of the cell. (4) A **transmitting region** at the end of the axon consists of branching processes that transmit the nerve impulses across synapses to other neurons or to effector cells. Usually the cell body is located near the receptive end of the neuron, in which case the dendrites are short processes that lead toward the cell body and the axon is a long process that leads away from it, but the cell body can be located at any point along a neuron. In sensory neurons, for example, the cell body is set off to the side of a long axon.

Aggregates of nerve cell bodies within the vertebrate brain are known as **nuclei;** aggregates of cell bodies outside the brain and spinal cord are called **ganglia. Tracts** are groups of axons (often simply called fibers) that run together within the brain and spinal cord. **Nerves** are groups of axons (fibers) ensheathed by connective tissue that lie peripheral to the brain and spinal cord.

Divisions and Components of the Nervous System

Macroscopically, the nervous system can be divided into central and peripheral portions. The **central nervous system** consists of the brain and spinal cord (Fig. 9-1). Both the brain and spinal

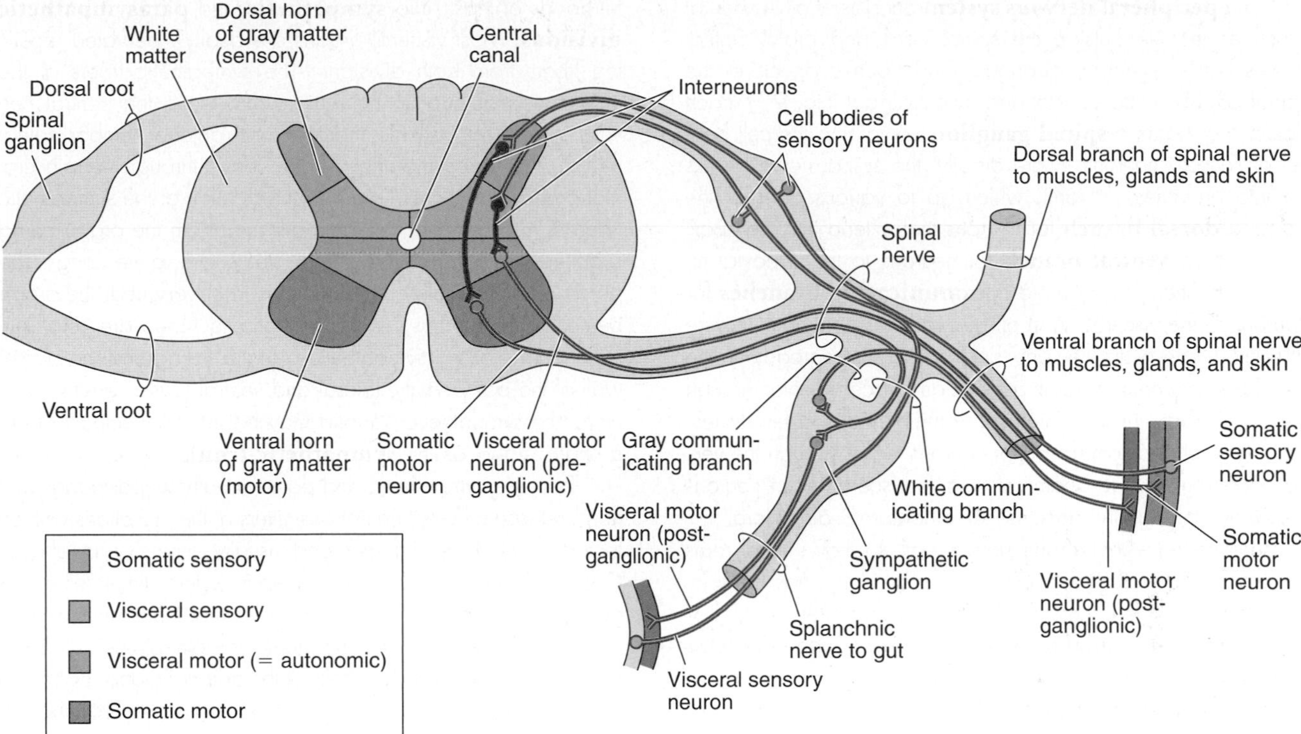

Figure 9-1

Diagram of a cross section through the thoracic part of the spinal cord and a pair of spinal nerves of a mammal showing the organization and functional components of the nervous system. The sensory and motor regions of the gray matter in the spinal cord and the types of neurons are color coded. Two interneurons are shown in the gray matter between the sensory and motor neurons. *(From Liem, Bemis, Walker, and Grande, 2001.)*

cord are hollow; they contain a **central canal,** which expands to form large chambers called **ventricles** in many regions of the brain. Recall that a single, dorsal, tubular nerve cord is one of the diagnostic characteristics of a chordate.

Neurons are distributed within the central nervous system as **gray matter** and **white matter.** The gray matter consists primarily of the cell bodies of neurons and unmyelinated fibers; the white matter consists primarily of myelinated fiber tracts. Most of the gray matter is centrally located. In the spinal cord it forms a continuous column whose cross section in amniotes usually resembles the letter H or a butterfly. **Sensory,** or **afferent, neurons** enter the **dorsal horn** of the gray matter at the level at which they attach to the spinal cord, or they may ascend or descend in the white matter of the spinal cord before entering the dorsal horn or nuclei in the brain. Their cell bodies usually are located in a peripheral ganglion. On entering the gray matter, some sensory neurons may extend directly to motor neurons, but most terminate on **interneurons** whose cell bodies contribute to the gray matter of the dorsal horn. The interneurons transmit the impulse to the motor neurons, or they may enter the white matter and ascend to the brain. During their ascent, they normally cross, or **decussate,** to the opposite side of the central nervous system. The cell bodies of **motor,** or **efferent, neurons** lie in the **ventral horn,** and their axons extend to the effectors. They may receive impulses directly from sensory neurons, but usually they receive impulses from interneurons coming from the dorsal horn or the brain. Interneurons descending from the brain decussate on their way caudad.

Sensory and motor neurons, and the horns of gray matter where they end or begin, are divided into somatic and visceral columns (Fig. 9-1). Somatic sensory neurons are stimulated primarily by cutaneous somatic receptors and enter the somatic sensory column of the dorsal horn. Somatic motor neurons begin in the somatic motor column and transmit impulses to somatic muscles. Visceral sensory and motor neurons and columns lie between the somatic columns and serve the visceral receptors and effectors. Thus, we can speak of **somatic sensory, visceral sensory, visceral motor,** and **somatic motor columns.** Further subdivisions sometimes are made.

The gray columns continue rostrally into the brain, where they break up into distinct islands of gray matter, or nuclei. The topographical relationships among nuclei are similar to the relationships among columns. The most dorsal nuclei are somatic sensory nuclei; the most ventral nuclei are somatic motor nuclei. Nuclei are closely associated with the terminations and origins of neurons that travel in the cranial nerves. In mammals, some gray matter migrates to parts of the brain surface and forms a conspicuous **cortex.**

The **peripheral nervous system** consists of all the neural structures that lie outside the spinal cord and brain. Spinal nerves are segmentally arranged, and each connects to the spinal cord by a dorsal root and a ventral root (Fig. 9-1). Each dorsal root bears a **spinal ganglion** containing the cell bodies of sensory neurons. More distally, the spinal nerve breaks up into branches, or rami, which go to various parts of the body: a **dorsal branch** for neurons that extend into the epaxial region, a **ventral branch** for neurons to the hypaxial region, and often one or more **communicating branches** for neurons to the viscera. Most of the branches contain a mixture of sensory and motor neurons, but the neurons segregate in the roots. Sensory neurons (both somatic and visceral) enter through the dorsal roots in all craniates. In amniotes and a few fishes, both visceral and somatic motor neurons leave through the ventral roots, but in most anamniotes some visceral motor neurons also leave through the dorsal roots. In lampreys all visceral motor neurons leave though the dorsal root, so their ventral roots contain only somatic motor fibers. This condition has been interpreted to be the ancestral condition for craniates.

Most of the **cranial nerves** are believed to be serially homologous to either a dorsal or a ventral root of a spinal nerve of the type seen in lampreys. Dorsal root nerves, therefore, would be expected to contain all the sensory neurons, both somatic from cutaneous receptors and visceral from receptors in the mouth and pharynx, as well as visceral motor neurons. Ventral root nerves would contain only somatic motor neurons. Cranial nerves considered to be serially homologous to dorsal and ventral roots contain the expected types of neurons, but the head also contains structures not found in the trunk. The branchiomeric muscles, which are somatic (Table 7-1), are supplied by somatic motor neurons that exit the brainstem via the dorsal roots. Because branchiomeric muscles are confined to the head, this group of neurons is not found in spinal nerves. The head also contains many unique receptors, including the nose, eye, ear, and (in aquatic anamniotes, including amphibian larvae) the lateral line and electroreceptive systems. Special somatic sensory neurons from these organs travel in unique sensory cranial nerves that have no serial homologues in the trunk.

The visceral motor neurons that travel in certain spinal and cranial nerves supply the smooth muscles in the walls of visceral organs and glands and constitute the **autonomic nervous system.** Visceral sensory neurons returning from the viscera usually travel in the same nerves as the visceral motor fibers but are not considered to be a part of the autonomic nervous system, which is defined as a motor system. A morphologically distinctive feature of the autonomic nervous system is that its neurons do not extend from the central nervous system directly to effectors (as do somatic motor neurons); rather, there is always a relay in a ganglion (Fig. 9-1). Thus, we can distinguish between **preganglionic** and **postganglionic neurons** in the autonomic nervous system. Another unique feature of the autonomic nervous system is its subdivision, in tetrapods at least, into **sympathetic** and **parasympathetic divisions.** Most visceral organs are dually innervated receiving fibers from both divisions. The sympathetic fibers of the mammalian autonomic nervous system leave the central nervous system through the thoracic and anterior lumbar spinal nerves; the parasympathetic fibers leave through certain cranial and sacral nerves. The peripheral relay of the sympathetic fibers is in a ganglion at some distance from the organ being supplied, so its postganglionic fibers are quite long. The parasympathetic relay, in contrast, is in or very near the organ being supplied; thus, its postganglionic fibers are relatively short. Many of the sympathetic ganglia lie against the dorsal wall of the body cavity, lateral and ventral to the vertebral column. The sympathetic ganglia may be interconnected to form a chain known as the **sympathetic trunk.**

Because sympathetic and parasympathetic postganglionic fibers release different neurotransmitters at their junctions with effectors, they have different and usually antagonistic effects. One division of the autonomic nervous system stimulates an organ; the other inhibits it. The effect of sympathetic stimulation is similar to the effect of the hormone norepinephrine, which is produced by the medullary cells of the adrenal gland. Both help a vertebrate adjust to stress by mobilizing resources that increase energy output. Cardiac muscles contract with greater force and speed; blood pressure rises; blood flow to cardiac and skeletal muscles increases; and blood sugar levels rise. Gut peristalsis and digestive functions are inhibited. In contrast, parasympathetic stimulation activates processes that produce and store energy. More blood is directed to the gut, digestive activities increase, and glucose is stored.

Development and Divisions of the Brain

The brain is the most complex part of the nervous system. We can begin to understand its many regions by considering its development. As shown in Figure 9-2, the brain arises as an enlargement of the cranial end of the neural tube. As it develops, it differentiates into three regions: a rostral forebrain, or **prosencephalon;** a middle midbrain, or **mesencephalon;** and a caudal hindbrain, or **rhombencephalon.** The mesencephalon does not become subdivided further, but the other two brain regions do. In most vertebrates, paired vesicles that will become the **telencephalon** evaginate and grow rostrolaterally from the rostral end of the prosencephalon. The part of the prosencephalon remaining in the midline becomes the **diencephalon.** The rhombencephalon divides into a rostral **metencephalon** and a caudal **myelencephalon.** Each of these five regions further differentiates. The telencephalon gives rise to the cerebral hemispheres and olfactory bulbs; the diencephalon gives rise to the thalamus, hypothalamus, and epithalamus. The dorsal part of the mesencephalon forms a roof, or tectum, that differentiates in mammals into the paired superior colliculi (optic lobes) and paired inferior colliculi. The metencephalon gives rise to the

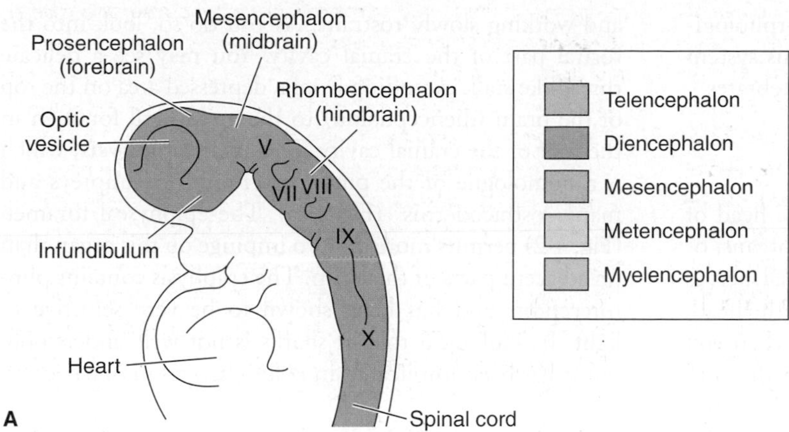

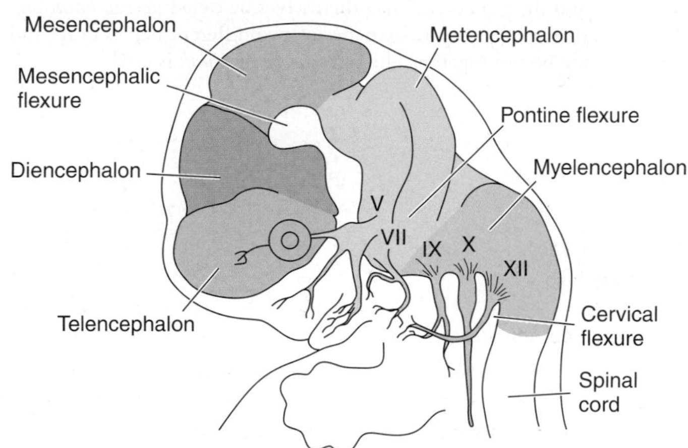

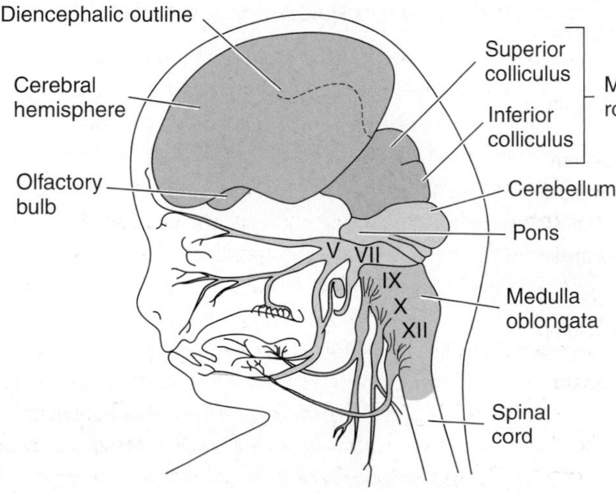

Figure 9-2

Diagrams of three stages in the development of the human brain showing the differentiation of the principal brain regions. **A,** Human embryo at 33 mm in length. **B,** Human brain from an embryo at 7 weeks of age. **C,** Human brain and cranial nerves from a human embryo at 3 months of age. Brain regions are color coded. Certain cranial nerves are identified by Roman numerals. *(From Liem, Bemis, Walker, and Grande, 2001, after Corliss, 1976.)*

cerebellum and, in mammals, to the pons; the myelencephalon gives rise to the medulla oblongata. The **brainstem** is the brain exclusive of the cerebellum, diencephalon, and telencephalon.

FISHES

The nervous system of *Squalus* should be studied carefully because it shows exceptionally well the basic structure of the vertebrate nervous system. The cartilage that encases it is easy to remove, and the brain is close to a morphologically ancestral and generalized stage. The nervous system of *Squalus* is a good prototype for that of all vertebrates.

The Dorsal Surface of the Brain

You may dissect the nervous system on the large head of *Squalus* that you used for the study of the sense organs, or you may dissect it in your specimen. The cranial nerves should be studied primarily on the intact side of the head, because some of them probably were destroyed when you dissected the sense organs (Table 7-3). Remove the skin and underlying tissue from the dorsal surface of the chondrocranium and from around the eye. Be careful not to cut the large dorsal nerve, called the **superficial ophthalmic trunk,**[1] that lies on the dorsal side of the orbit and lateral to the rostrum (Fig. 9-7). Cut away the cartilaginous roof of the cranial cavity by starting between the otic capsules and working slowly rostrally. As you do so, look into the rostral part of the cranial cavity. You may see a delicate threadlike stalk extending from a depressed area on the top of the brain (diencephalon) to the epiphyseal foramen in the roof of the cranial cavity. This is the **epiphysis,** which is a homologue of the pineal eye found in lampreys and many "ostracoderms" (Fig. 9-3). The epiphyseal foramen (Fig. 4-2) permits more light to impinge on this organ than on adjacent parts of the brain. The epiphysis contains photoreceptors and has been shown to be very sensitive to light. Its biological role in sharks is not well understood, but it has been implicated in color changes in one species.

[1]The term *nerve* can be applied to any collection of axons peripheral to the central nervous system. Nerves often branch, and these branches may themselves be called *nerves, branches,* or *rami.* When the branches of two different nerves unite and are bound together, the term *nerve* or *trunk* is used.

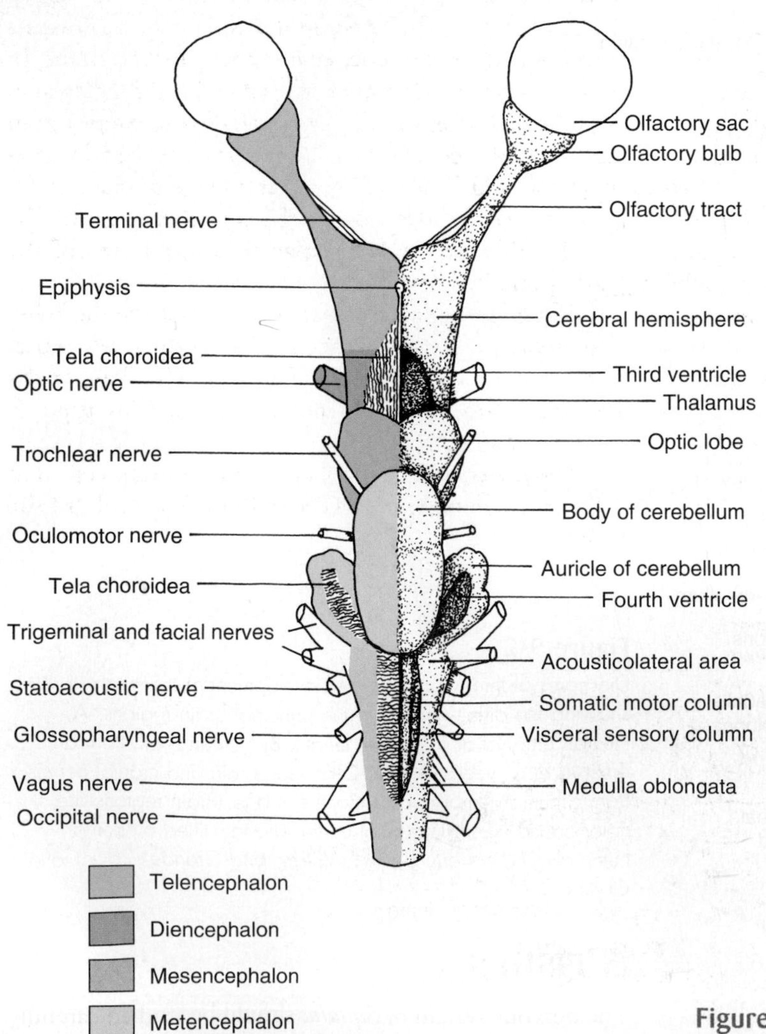

Figure 9-3

Dorsal view of the brain of *Squalus.* The tela choroidea has been removed on the right side.

Next, cut away the supraorbital crest and as much of the lateral walls of the cranial cavity as possible without injuring nerves. Much of the inner ear on the intact side may have to be cut away. Be particularly careful not to break the small **trochlear nerve** that leaves the brain dorsally and passes to the dorsal oblique muscle of the eyeball (Fig. 9-7). The brain now should be well exposed. Its surface is covered with a delicate, vascular connective tissue, the **primitive meninx.** Strands of connective tissue pass from the primitive meninx to the connective tissue lining of the cranial cavity, which is called the **endochondrium.** In life, a protective, mucoid perimeningeal tissue lies between the brain and endochondrium. It sometimes coagulates with preservation and must be washed out carefully.

The paired **olfactory bulbs** form the most rostral part of the brain (Figs. 9-3 and 9-7). They are the lateral enlargements in contact with the olfactory sacs, and they receive the primary olfactory neurons originating in the olfactory epithelium. Secondary olfactory neurons originate in the olfactory bulbs and form the **olfactory tracts** that extend caudally to the cerebral hemispheres. Most of these neurons terminate in a ventrolateral portion of the cerebral hemisphere that is essentially homologous to the mammalian piriform lobe. Olfactory bulbs, olfactory tracts, and cerebral hemispheres together constitute the telencephalon.

The **diencephalon** is the depressed area, often with a darkly colored roof, situated caudal to the cerebral hemispheres. (The rich vascular supply causes the roof to be dark.) The roof of the diencephalon is the **epithalamus,** its lateral walls form the **thalamus,** and its floor, which will be seen later, is the **hypothalamus.** The epiphysis may be seen attaching to the caudal part of the epithalamus. Most of the roof of the diencephalon is very thin, vascular, and membranelike, forming a **tela choroidea.** The tela choroidea consists of only the ependymal epithelium that lines the central canal of the central nervous system and the delicate vascular primitive meninx that covers it. Carefully remove the tela choroidea on one side of the epithalamus and, while doing so, note that part of it extends onto the caudal surface of the cerebral hemispheres (Fig. 9-4). This part of the tela choroidea actually does not belong to the diencephalon but forms a thin-walled sac, the **paraphysis,** which is considered part of the telencephalon. A prominent fold extends from the tela choroidea down into the large cavity **(third ventricle)** within the diencephalon. This fold, the **velum transversum,** represents the rostral end of the diencephalon. The velum transversum often can be discerned as an opaque transverse line visible through the translucent tela choroidea of the third ventricle. A vascular tuft from the tela choroidea extends rostrally from the velum transversum into the lateral ventricle within each cerebral hemisphere. Each vascular tuft constitutes a **choroid plexus** of the lateral ventricle. Choroid plexuses secrete and absorb a lymphlike **cerebrospinal fluid** that slowly circulates through the ventricles of the brain. The cerebrospinal fluid helps maintain a proper nutritive and ionic environment for the brain. In sharks, only a small amount of cerebrospinal fluid escapes from the ventricles to circulate in meningeal spaces, but in mammals, the central nervous system is bathed in cerebrospinal fluid. The small transverse bulge of nervous tissue at the very caudal part of the epithalamus under the tela choroidea is the **habenular region,** an area of the brain specialized for olfactory integration.

A large pair of **optic lobes** lies just caudal to the habenular region. The optic lobes develop in the roof, or **tectum,** of the mesencephalon (Fig. 9-3). The floor of the mesencephalon, which cannot be seen at this time, is known as the **tegmentum.**

The metencephalon lies caudal to the mesencephalon and consists dorsally of the **cerebellum.** The **body of the**

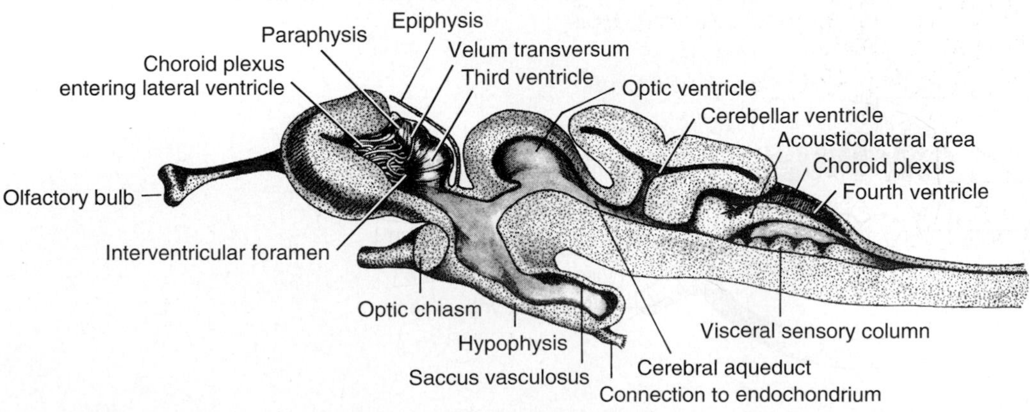

Figure 9-4
Sagittal section of the brain of *Squalus.*

cerebellum is the large, median, oval mass whose rostral end overhangs the optic lobes. Note that a longitudinal groove and a transverse groove partially subdivide it into four parts. The pair of earlike flaps that lie on either side of the caudal part of the body of the cerebellum are the **auricles of the cerebellum.** The ventral part of the metencephalon contributes to the medulla oblongata in anamniotes.

The myelencephalon lies caudal to the metencephalon and forms the greater part of the **medulla oblongata,** which is the elongated region of the brain that is continuous caudally with the spinal cord. Most of the roof of the medulla oblongata consists of a thin, vascular tela choroidea, which covers a large brain cavity, the **fourth ventricle.** Carefully remove the tela choroidea and its rostral extension onto the dorsal surface of the auricles of the cerebellum. Vascular folds, the **choroid plexus of the fourth ventricle,** can be seen protruding into the ventricle. Lift up the caudal end of the body of the cerebellum and notice that the auricles are continuous with each other across the dorsal midline of the brain.

The columns of gray matter that continue from the spinal cord into the brain can be seen in the ventral and lateral walls of the fourth ventricle after removing the tela choroidea. The pair of midventral longitudinal folds that lie on the floor of the ventricle are the **somatic motor columns** (Figs. 9-3 and 9-5). They contain the cell bodies of somatic motor neurons. There is a deep longitudinal groove lateral to each somatic motor column. The **visceral motor column,** a narrow column containing the cell bodies of visceral motor neurons, is situated laterally deep in this groove. A longitudinal row of small bumps lies dorsal to the visceral motor column. This is the **visceral sensory column,** which receives and relays impulses from the visceral sensory neurons. The longitudinal groove between the visceral motor and visceral sensory columns is the **sulcus limitans,** a landmark separating the ventral motor portion from the dorsal sensory portion of the spinal cord and brainstem. The dorsolateral rim of the fourth ventricle constitutes the **somatic sensory column,** which receives and relays impulses from the somatic sensory neurons. You can visualize the relationships among these columns particularly well in a cross section, which you can make after the brain and nerves have been studied (Fig. 9-5).

The rostral part of the somatic sensory column is enlarged. This portion, known as the **acousticolateral area** (Fig. 9-3), receives neurons from the ear and lateral line organs, but neurons from the ear, lateral line organs, and ampullae of Lorenzini each terminate in their own nuclei within this area. These nuclei cannot be seen grossly. The acousticolateral area is continuous rostrally with the auricles of the cerebellum. Although it cannot be seen in a gross dissection, these sensory and motor columns continue rostrally through the metencephalon and into the mesencephalon. The columns break up into nuclei.

Cranial and Occipital Nerves

We must consider the cranial and occipital nerves before examining the ventral and internal parts of the brain. You will have to remove more of the lateral wall of the chondrocranium as you study the nerves. Cranial nerves, as well as most anatomical structures, were first described in humans. Twelve nerves were recognized by early human anatomists. In addition to being named, each was given a Roman number based on its sequential position. These numbers are universally accepted and apply well to all amniotes, although the sequence is not always identical to the condition in humans. Anamniotes lack the last two cranial nerves found in amniotes, although they have comparable neurons. Anamniotes have some cranial nerves that are not present in humans and other amniotes therefore are not included in the numbering system.

Table 9-1 summarizes the cranial nerves and their components, or the types of neurons that they contain. We recognize six functional components that are potentially

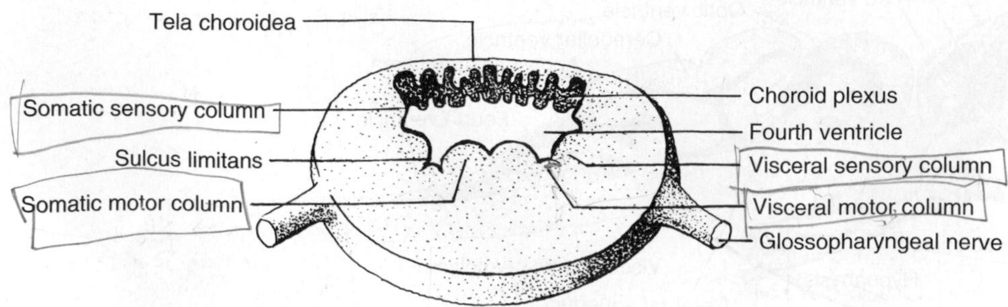

Figure 9-5

Cross section through the medulla oblongata of *Squalus* made at the level of the glossopharyngeal nerve.

present in a cranial nerve: **general somatic sensory, special somatic sensory** (olfaction, vision, lateral line/electroreception), **general visceral sensory, special visceral sensory** (taste), **visceral motor** (autonomic nervous system), and **somatic motor.** Based on their components and our knowledge of the evolution of the spinal nerves, the cranial nerves can be sorted into three groups. **Dorsal root nerves** appear to be serially homologous to dorsal roots of ancestral spinal nerves; **ventral root nerves** appear to be serially homologous to the ventral roots of ancestral spinal nerves; and a group of **unique sensory nerves** supply the special sense organs on the head: nose, eye, ear, lateral line, and electroreceptors. This organization of the nerves makes them easier to learn (Fig. 9-6). In the following dissection instructions, we will work through the nerves from rostral to caudal.

Terminal Nerve

Squalus and nearly all vertebrates, except birds, have a minute **terminal nerve** that was discovered after the numbering system was established and so was not included. Sometimes it is given the designation **Zero (0).** The terminal nerve (Fig. 9-3) in *Squalus* lies along the medial surface of the olfactory tract and extends between the olfactory sac and cerebral hemisphere. It is seen best at the medial angle formed by the junction of the olfactory tract with the cerebral hemisphere because it separates slightly from the olfactory tract in this region. It is closely associated with the olfactory system but appears to be restricted to detecting pheromones. It is classified as a special somatic sensory nerve, as are other nerves from the special sense organ of the head (Fig. 9-6 and Table 9-1).

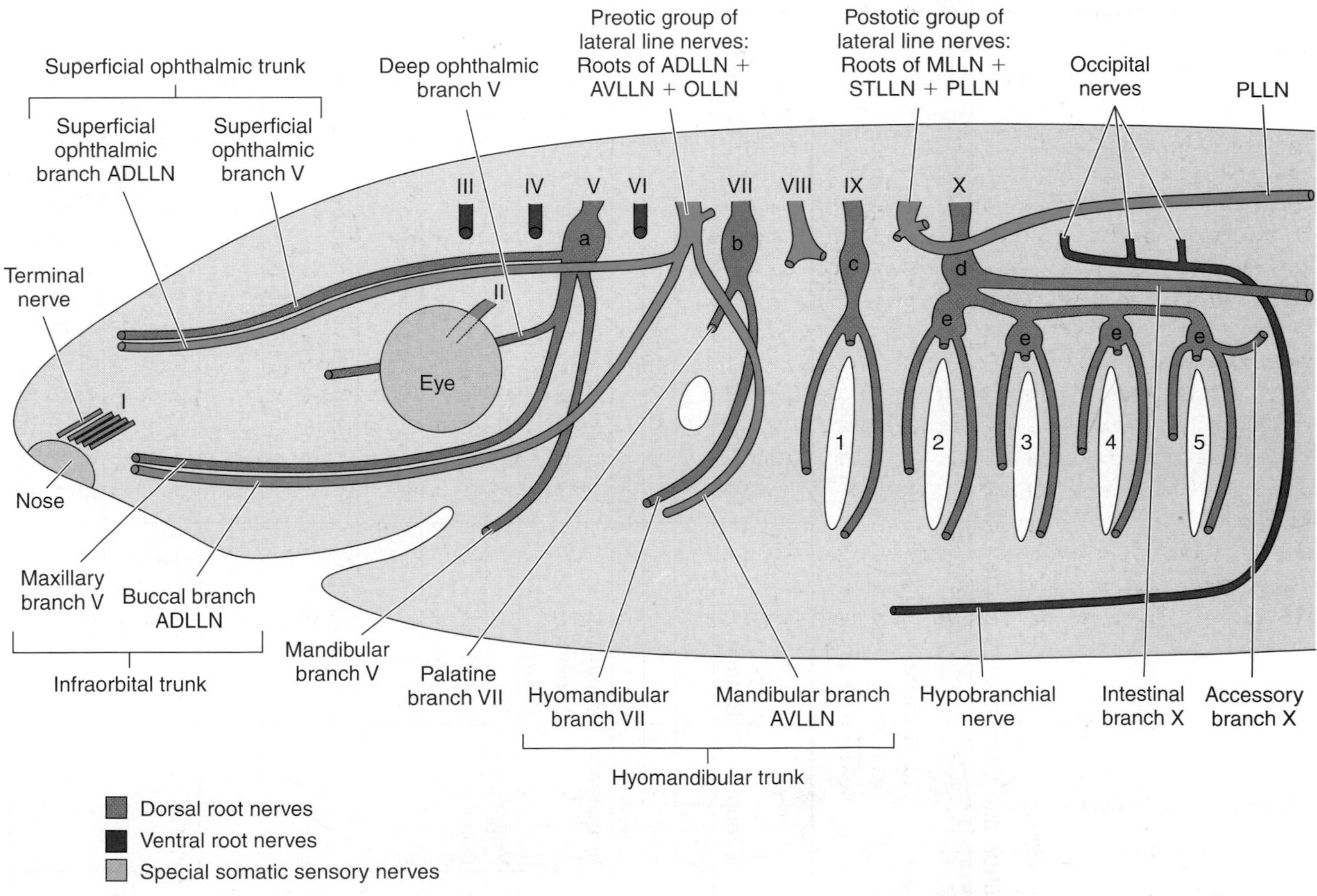

Figure 9-6

Diagram in lateral view of the cranial nerves and their major branches of a shark. Cranial nerves are identified by their Roman numerals. Notice that some branches of dorsal root nerves and lateral line nerves run together and form nerve trunks. Lateral line nerves are identified by their initials: *ADLLN*, anterodorsal lateral line nerve; *AVLLN*, anteroventral lateral line nerve; *OLLN*, otic lateral line nerve; *MLLN*, middle lateral line nerve; *STLLN*, supratemporal lateral line nerve; *PLLN*, posterior lateral line nerve. Important sensory ganglia are identified by the following letters: *a*, semilunar ganglion; *b*, geniculate ganglion; *c*, petrosal ganglion; *d*, jugal ganglion; *e*, nodose ganglia. Gill slits are indicated in Arabic numerals. *(Modified after Liem, Bemis, Walker, and Grande, 2001.)*

Table 9-1 Distribution and Components of the Cranial Nerves of Sharks

The cranial and occipital nerves of selachians, together with the group of nerves to which they belong, their major branches or rami, and their functional components are summarized in this table. An X indicates the presence of a component, and an (X) that the component is present in some but not all species. A few motor fibers, not shown in the table, modulate the sensitivity of the neurons in many special sense organs of the head.

Nerve	Group*	Branch	Distribution	Somatic Sensory		Visceral Sensory		Visceral Motor	Somatic Motor
				General†	Special	General‡	Special¶		
Terminal	Special		Olfactory sac		X				
I. Olfactory	Special		Olfactory epithelium		X				
II. Optic	Special		Retina		X				
III. Oculomotor	Ventral root	4 muscular branches Ciliary	Ventral oblique; ventral, dorsal, and medial recti muscles Muscles in eyeball					X	X
IV. Trochlear	Ventral root		Dorsal oblique muscle						X
V. Trigeminal	Dorsal root	Superficial ophthalmic	Skin on top and side of head	X					
		Maxillary	Skin on lower jaw, underside of rostrum	X					
		Deep ophthalmic (profundus)	Skin on top and side of rostrum, sensory fibers from eye, sometimes motor fibers to eye	X				(X)	
		Mandibular	Mandibular muscles, skin over lower jaw, inside of mouth	X					X
Preotic group of lateral line nerves: ADLLN, AVLLN, OLLN	Special	Superficial ophthalmic	Accompanies same branch of V; supplies lateral line neuromasts and ampullae of Lorenzini		X				
		Buccal	Accompanies maxillary branch of V; supplies lateral line neuromasts and ampullae of Lorenzini	X					

VI. Abducens	Ventral root		Lateral rectus muscle						X
VII. Facial	Dorsal root	Hyomandibular	Hyoid muscles, mouth lining			X	X		X
		Palatine	Mouth lining			X	X		
Postotic group of lateral line nerves: MLLN, STLLN, PLLN	Special		Supplies postotic lateral line neuromast. PLLN continues along trunk supplying trunk canal. Does not supply any ampullae of Lorenzini.		X				
VIII. Statoacoustic	Special		Hair cells of inner ear		X				
IX. Glossopharyngeal	Dorsal root	Pretrematic	Rostral wall of first gill pouch			X	X		
		Posttrematic	Muscles of third visceral arch, caudal wall of first gill pouch			X	X		X
		Pharyngeal	Adjacent pharyngeal lining			X	X		
X. Vagus	Dorsal root	Visceral	Gives rise to 4 branchial branches, each with pretrematic, posttrematic, and pharyngeal branches that supply remaining gill pouches and muscles of remaining visceral arches.			X	X		X
		Accessory	Cucullaris muscle						X
		Intestinal	Postpharyngeal gut wall, pericardial cavity, stomach, intestine			X		X	
Occipital nerves	Ventral roots		Epibranchial and hypobranchial musculature. The first several spinal nerves contribute to them.	X					X

*We interpret cranial nerves as serially homologous to the dorsal or ventral roots of cranial nerves or as special somatic sensory nerves that are unique to the head.
†Most of the general somatic sensory fibers in the trigeminal nerve come from cutaneous receptors; a few are propioceptive fibers.
‡Neurons from gustatory receptors primitively may have formed unique cranial nerves.
¶These are preganglionic fibers of the parasympathetic part of the autonomic nervous system.

Olfactory Nerve

The **olfactory nerve (I)** carries olfactory stimuli (special somatic sensory) from the olfactory sac to the olfactory bulb. Free the olfactory bulb completely by removing some of the cartilage and tough perichondrial connective tissue surrounding it. Continue to do so until you reach the olfactory sac and can see that the olfactory bulb and olfactory sac are adjacent to each other. The extremely short olfactory nerve is not recognizable as a separate structure at the gross level, but it consists of many minute bundles of olfactory neurons that pass between the olfactory sac and the olfactory bulb. You may see the region where the olfactory sac and olfactory bulb are connected by making a longitudinal cut through the olfactory sac and olfactory bulb. The neurons of the olfactory epithelium both receive stimuli (from odorant molecules) and transmit nerve impulses.

Optic Nerve

The **optic nerve (II)** carries sensory impulses (special somatic sensory) from the retina. Find it in the orbit and trace it medially (Fig. 9-7). It is a thick nerve. Push the brain away from the cranial wall and notice that the optic nerve

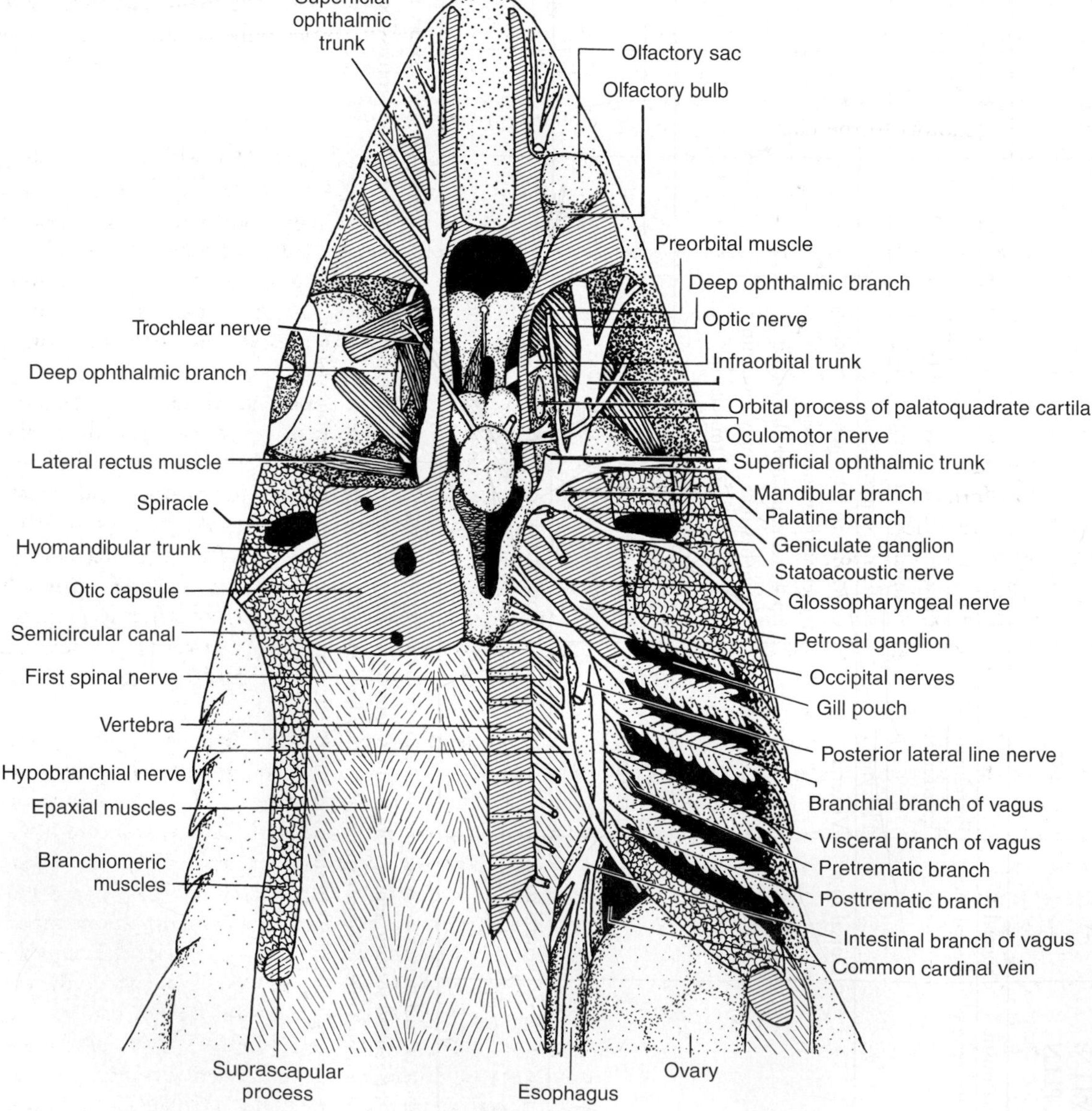

Figure 9-7

Semidiagrammatic dorsal dissection of the brain and cranial nerves of *Squalus*. Deeper structures are shown on the right side.

attaches to the ventral surface of the diencephalon. Because the retina of the eye develops embryonically from an outgrowth of the brain, many investigators consider the retina and optic nerve to be parts of the brain. The optic nerve from the retina, therefore, is technically a brain tract rather than a peripheral nerve. Fibers in the optic nerve are the axons of ganglion cells whose cell bodies lie in the retina. The ganglion cells receive impulses from the photoreceptive rods and cones via short bipolar cells. In addition, other cells within the retina make many horizontal interconnections, so the retina does considerable processing of visual information before it sends signals to other parts of the brain.

Oculomotor Nerve

The **oculomotor nerve (III)** is a ventral root nerve. As such it carries somatic motor impulses to four of the six extrinsic muscles of the eyeball (Table 9-1). It also carries autonomic fibers (visceral motor) to the ciliary ganglion and then to the ciliary body of the eye. To see it, mobilize the eye on the intact side of the head in the manner described for the dissection of the eye (Chapter 8), remove the gelatinous connective tissue filling the orbit, and look on the ventral surface of the eyeball. The branch of the oculomotor nerve going to the ventral oblique muscle will be apparent (see Fig. 8-7). Follow it caudally and medially. It passes ventral to the ventral rectus muscle, and at the caudal margin of this muscle it crosses a small and often whitish blood vessel. This artery follows the margin of the ventral rectus and enters the eyeball. The autonomic fibers of the oculomotor nerve form a small **ciliary branch** that travels along this artery, but these fibers seldom can be seen grossly. The branch of the oculomotor nerve that goes to the ventral rectus muscle lies between the ventral rectus muscle and the small artery.

Turn your specimen over and find the oculomotor nerve from the dorsal side. Cut the lateral rectus muscle near its insertion on the eyeball and cut the superficial ophthalmic trunk. Reflect them to see the oculomotor nerve clearly. It extends dorsal to the origin of the dorsal rectus muscle and enters the cranial cavity. Just before it enters, it gives off one branch to the dorsal rectus muscle and another to the medial rectus muscle. (Do not confuse the oculomotor nerve with another nerve of about the same size, the deep ophthalmic nerve, which crosses the base of the oculomotor nerve and extends along the medial surface of the eyeball.) Carefully push the brain away from the cranial wall and note the attachment of the oculomotor nerve on the ventral surface of the mesencephalon.

Trochlear Nerve

You observed the **trochlear nerve (IV)** crossing an optic lobe in the dorsal dissection of the brain (p. 211). Lift the rostral end of the body of the cerebellum and note where it attaches to the brain. The trochlear nerve passes through the cranial wall; goes ventral to, or perforates, the large superficial ophthalmic trunk; and extends to the dorsal oblique muscle (Fig. 9-7). Like the oculomotor nerve, it is a ventral root nerve and carries somatic motor fibers, but it does not carry autonomic fibers.

Abducens Nerve

Skip the fifth nerve for a moment and consider the **abducens nerve (VI),** which is a ventral root nerve that carries somatic motor fibers to the lateral rectus muscle. You can see it on the ventral surface of this muscle. You will see its attachment on the ventral surface of the medulla oblongata later when you remove the brain (Fig. 9-8). Like the trochlear nerve, it carries only somatic motor fibers and no autonomic fibers.

Trigeminal Nerve

Return to the **trigeminal nerve (V),** which is the first of the dorsal root nerves (Table 9-1 and Fig. 9-6). The name suggests that it has three main branches, which it does in amniotes. These branches sometimes are numbered V_1, V_2, and V_3. *Squalus* and other anamniotes have four branches. The trigeminal nerve carries somatic motor fibers to the muscles of the mandibular arch and returns general somatic sensory fibers from cutaneous and other receptors on most of the head and within the oropharyngeal cavity. The trigeminal nerve emerges from the medulla oblongata just caudal to the auricles of the cerebellum. Its roots lie very close to the preotic group of lateral line nerves (Fig. 9-6), as well as the facial nerve (VII) and the statoacoustic nerve (VIII). The trigeminal and facial nerves exit the chondrocranium of *Squalus* through the trigeminofacial foramen (see Fig. 4-2**C**). It is not practical to try to separate these cranial nerves grossly near the brain, but they can be distinguished by their peripheral branches.

The **superficial ophthalmic branch** of the trigeminal nerve (V_1 in part) and the superficial ophthalmic branch of the anterodorsal lateral line nerve (see later) form the large **superficial ophthalmic trunk** (Fig. 9-6), which already was noted passing through the dorsal part of the orbit (see Fig. 8-7, and Fig. 9-7). Observe how several branches of this trunk emerge from the row of superficial ophthalmic foramina (Fig. 4-2) to the skin and the lateral line organs in the supraorbital canal above the eye (see Fig. 8-5). After emerging from the rostralmost and largest superficial ophthalmic foramen in the chondrocranium, the superficial ophthalmic trunk sends off a lateral branch that passes first through the cartilage of the antorbital process and supplies the skin and ampullae of Lorenzini caudal to the naris on the underside of the snout. Follow the main part of the superficial ophthalmic trunk

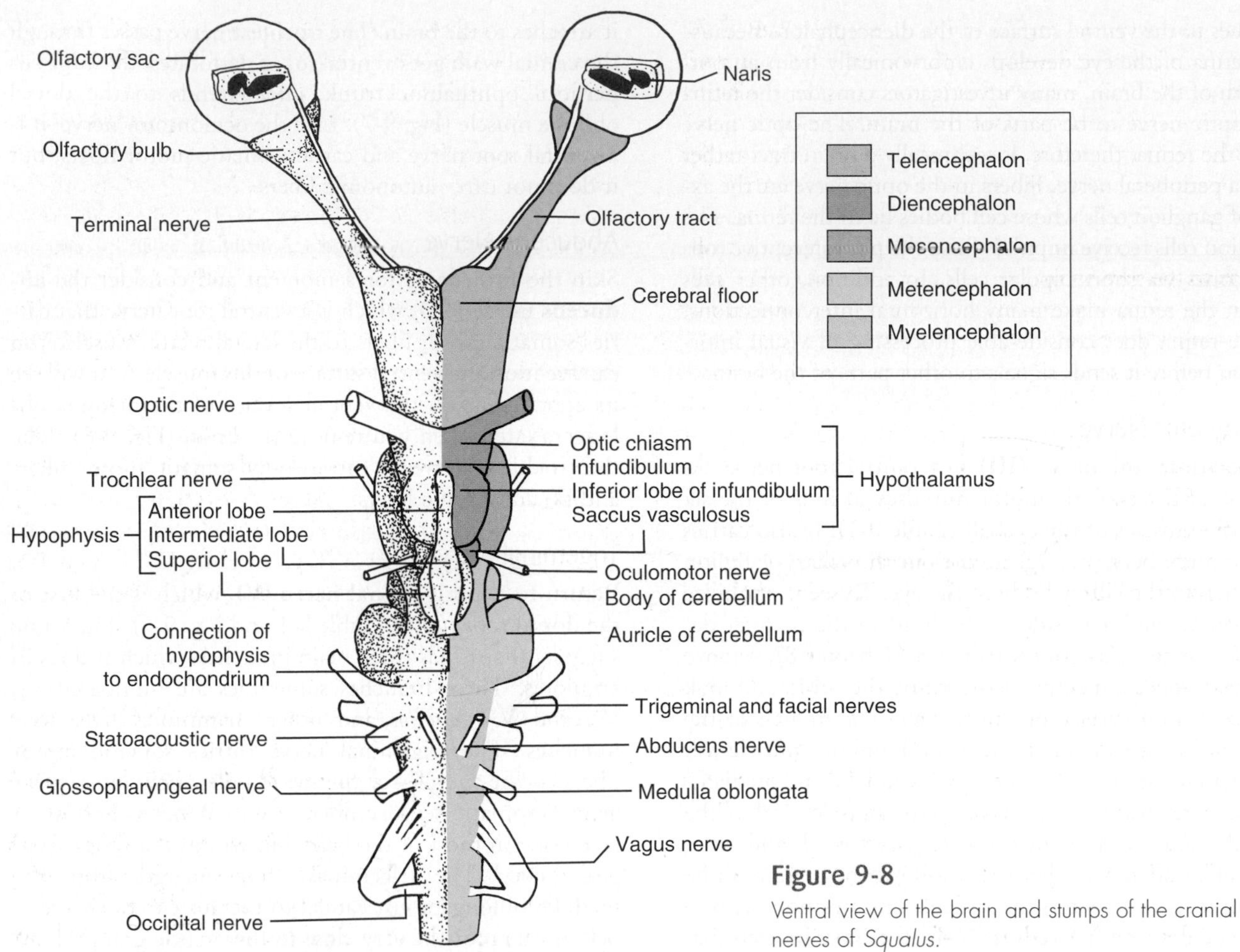

Figure 9-8

Ventral view of the brain and stumps of the cranial nerves of *Squalus*.

rostrally and observe how it sends tiny fibers to the ampullae of Lorenzini on the dorsal side of the snout. The nerve fibers supplying the ampullae of Lorenzini and lateral line organs are derived from the superficial ophthalmic branch of the anterodorsal lateral line nerve.

A **deep ophthalmic branch** (V_1 in part) of the trigeminal nerve, which also is called the **profundus nerve,** enters the orbit dorsal to the oculomotor nerve; adheres to the connective tissue on the medial surface, or back, of the eyeball; and leaves the front of the orbit through a small foramen to run with the superficial ophthalmic trunk (Fig. 9-7). Trace the deep ophthalmic branch on the side of the head in which the eyeball and extrinsic muscles of the eyeball were dissected (Table 7-3). The deep ophthalmic branch of the trigeminal nerve returns general somatic sensory impulses from cutaneous sense organs (but not lateral line organs) in the skin on the top and side of the head. In addition, the deep ophthalmic branch has several minute and inconspicuous branches, which return general somatic sensory fibers from parts of the eyeball. In some sharks they may also carry a few autonomic fibers (visceral motor) to the eye (Norris and Hughes, 1920).

You can find the **mandibular branch** (V_3) of the trigeminal nerve by dissecting away the connective tissue on the caudal wall of the orbit (Fig. 9-7). It is a fairly thick nerve that lies caudal to the lateral rectus muscle. The mandibular branch carries somatic motor fibers to the branchiomeric muscles of the mandibular arch and returns some general somatic sensory fibers from cutaneous sense organs in the skin and the oropharyngeal cavity in the regions that it supplies.

The **maxillary branch** (V_2) of the trigeminal nerve and the buccal branch of the anterodorsal lateral line nerve (see later) form the large **infraorbital trunk** (Fig. 9-6). This trunk extends rostrally across the **preorbital muscle** on the floor of the orbit (see Fig. 7-8, Fig. 8-7, and Fig. 9-7). Clean the surface of the preorbital muscle, mobilize the infraorbital trunk, and reflect the skin and underlying connective tissue to reveal the entire extent of the muscle and infraorbital trunk on the lateral and ventral sides of the snout. The preorbital muscle follows the palatoquadrate cartilage and dorsal labial cartilage medially and attaches to the ventral surface of the chondrocranium. It inserts caudally by a long tendon to the jaw joint, deep to the mandibular adductor muscle (see Fig. 7-8), and pulls the chondrocranium down during the closing of the mouth (Chapter 7). The infraorbital trunk bifurcates into a smaller branch to the skin on the lateral part of the underside of the snout and a larger medial branch, which dips down rostral to the preorbital muscle to innervate the neuromasts of the infraorbital lateral line canal and the ampullae of Lorenzini of the central part of the underside of the snout. You will have to make a longitudinal, midventral cut through the skin and the underlying connective tissue to see the entire extent of this large nerve branch. The maxillary branch of the infraorbital trunk returns general somatic sensory fibers from cutaneous sense organs; the buccal branch of the anterodorsal lateral line nerve supplies the ampullae of Lorenzini and lateral line organs.

Cut away enough cartilage and connective tissue from the caudomedial corner of the orbit to see where the four branches of the trigeminal nerve come together, and again note the attachment of the trigeminal nerve to the medulla oblongata. The main part of the trigeminal nerve bears a slight enlargement, the **semilunar ganglion** (Fig. 9-6), which contains the cell bodies of the sensory neurons; however, it is unlikely that you can distinguish this ganglion.

Lateral Line Nerves (unnumbered)

Sharks, like most other aquatic anamniotes, including many amphibian larvae, possess lateral line and electroreceptive organs that are innervated by special somatic sensory nerves. These sensory organs and their nerves, as well as their central connections within the brain, are lost in terrestrial vertebrates. Until recently, these special sensory nerves, known as the **lateral line nerves,** were considered components of the facial, glossopharyngeal, and vagus nerves, because their roots emerge from the brainstem adjacent to these cranial nerves. Because of their embryonic origin from distinct neurogenic placodes, their distinct function, and their termination in discrete nuclei in the brain, we recognize six lateral line nerves as separate cranial nerves, comparable to the other special somatic sensory cranial nerves. The six lateral line nerves can be classified into preotic and postotic groups (Fig. 9-6). The preotic group consists of the **anterodorsal lateral line nerve** (ADLLN), the **anteroventral lateral line nerve** (AVLLN), and the **otic lateral line nerve** (OLLN). These three nerves carry both mechanosensory information from neuromasts in the lateral line canals and electroreceptive information from the ampullae of Lorenzini. The postotic group consists of the **middle lateral line nerve** (MLLN), the **supratemporal lateral line nerve** (STLLN), and the **posterior lateral line nerve** (PLLN). These three nerves carry only mechanosensory information.

We have already identified the **superficial ophthalmic branch** of the anterodorsal lateral line nerve that joins the superficial ophthalmic branch of the trigeminal nerve to form the large superficial ophthalmic trunk, as well as the **buccal branch** of the anterodorsal lateral line nerve that joins the maxillary branch of the trigeminal nerve to form the large infraorbital trunk. Later, you will identify the **mandibular branch** of the anteroventral lateral line nerve that joins the hyomandibular branch of the facial nerve to form the **hyomandibular trunk.** The medial and supratemporal lateral line nerves join dorsal branches of the glossopharyngeal and vagus nerves respectively (see later). You will see the large posterior lateral line nerve as you dissect the vagus nerve. It runs along the lateral line canal of the trunk, innervating the many neuromasts of the trunk canal (see Fig. 8-5).

Facial Nerve

The **facial nerve (VII),** which is a dorsal root nerve, is the nerve of the hyoid arch and spiracle. One branch of the facial nerve, the **hyomandibular branch,** together with the mandibular branch of the anteroventral lateral line nerve form the **hyomandibular trunk** (Figs. 7-8**A** and **B,** and 9-6). This trunk emerges caudal to the spiracle and crosses the levator hyomandibulae muscle dorsal to the adductor mandibulae muscle (Fig. 9-7; see also Fig. 7-8**B**). Follow it peripherally, noting that it is distributed to the hyoid muscles (somatic motor fibers) and the lining of the oropharyngeal cavity (general visceral sensory fibers and special visceral sensory fibers that carry information from taste buds). The nerve fibers that innervate the lateral line organs of the mandibular canal and ampullary organs in the infraspiracular ampullary field (see Fig. 8-5) are derived from the buccal branch of the anteroventral lateral line nerve. Follow the hyomandibular trunk proximally to the brainstem, where its roots enter very close to the trigeminal and the statoacoustic nerves (VIII). You may have to cut away some of the spiracle, surrounding muscles, and otic capsule as you go, because the hyomandibular trunk passes ventral to the inner ear. About 1 cm from the brain, the facial nerve bears a slight enlargement, the **geniculate ganglion,** which contains the cell bodies of sensory neurons (Figs. 9-6 and 9-7). Another, and smaller, **palatine branch** of the facial nerve leaves from the rostroventral surface of this ganglion. It returns general visceral sensory neurons from the lining of the oropharyngeal cavity and special visceral sensory fibers from taste buds in the area it supplies.

Statoacoustic Nerve

You may have noted a part of the **statoacoustic nerve (VIII)**[2] coming from the ampullae of the anterior vertical and lateral semicircular ducts of the inner ear during the dissection of the hyomandibular trunk. On the other side of the specimen (Table 7-3), continue to remove the cartilage of the otic capsule and note another, and longer, part of this nerve coming from the ampulla of the posterior vertical semicircular duct, as well as from the sacculus, and parts of the utriculus. The statoacoustic nerve thus consists of special somatic sensory fibers.

Glossopharyngeal Nerve

The **glossopharyngeal nerve (IX),** which is a dorsal root nerve, is the nerve of the third visceral arch and the first of the five definitive gill pouches. It can be seen crossing the floor of the otic capsule caudal to the sacculus (Fig. 9-7). It passes ventral to the caudal branch of the statoacoustic nerve and at first may be confused with this nerve. Bisect this part of the statoacoustic nerve and find the attachment of the glossopharyngeal nerve on the side of the medulla oblongata. Trace the glossopharyngeal nerve laterally; this will be easier if you open the first gill pouch by cutting through the skin and muscle dorsal and ventral to the first external gill slit. Cut all the way to, but not through, the internal gill slit, which is the opening between the gill pouch and pharynx (Chapter 7). As the glossopharyngeal nerve leaves the otic capsule, it bears an oval swelling, the **petrosal ganglion,** which contains the cell bodies of sensory neurons. Three branches leave from the petrosal ganglion. Follow them distally and expose them. A large **posttrematic branch** passes down the caudal face of the first gill pouch to carry somatic motor fibers to the branchiomeric muscles of the third visceral arch and return general and special (gustatory) visceral sensory fibers from this region. A smaller **pretrematic branch,** which consists entirely of general and special visceral sensory fibers, passes down the cranial face of the first gill pouch. A still smaller **pharyngeal branch** follows the pretrematic branch a short distance. It then curves medially around a conspicuous white tendon that attaches to the dorsal tip of the hyomandibular cartilage and extends caudomedially toward the vertebral column dorsal to the first internal gill slit. The fibers of the pharyngeal branch are distributed to the wall of the oropharyngeal cavity. You may be able to see this pharyngeal branch by mobilizing the bifurcation of the pretrematic and posttrematic branches, pulling it laterally, and looking at its medial surface. The pharyngeal branch also consists entirely of visceral sensory fibers. A small dorsal branch of the middle lateral line nerve, which carries special somatic sensory fibers from the lateral line organs in the canals of this region, leaves near the point where the glossopharyngeal nerve divides, but it is impractical to find it.

Notice the overall pattern of the branches of the glossopharyngeal nerve (Fig. 9-6). Its branches are not accompanied by any lateral line nerves, so it is a good prototype for the dorsal root nerves supplying the visceral arches and pouches. Its primary branch is the posttrematic branch, which contains the somatic motor fibers to the branchiomeric muscles and visceral sensory fibers returning from the region. Pretrematic and pharyngeal branches carry only visceral sensory fibers from the area.

You will notice later and in Figure 9-6 that this pattern is repeated with some modifications for other dorsal root nerves. The pattern of branches for the four parts of the vagus nerve supplying the remaining visceral arches and pouches is almost the same as that of the glossopharyngeal nerve. The hyomandibular branch of the facial nerve and the mandibular branch of the trigeminal nerve are equivalents to posttrematic branches. The facial nerve lacks a pretrematic branch, although it has a small pharyngeal branch (called the **palatine branch**). The trigeminal nerve lacks a pharyngeal branch but has an extensive pretrematic portion represented by the superficial ophthalmic and maxillary branches.

Vagus Nerve

The **vagus nerve (X)** is a dorsal root nerve and the nerve of the remaining visceral arches. Find its connection on the dorsolateral surface of the caudal end of the medulla oblongata and follow it caudally out of the otic capsule (Fig. 9-7). To see the rest of the nerve, you must cut open the remaining gill pouches in the manner described for the first one. If you cut as far as you should, then you will cut into a large space, the anterior cardinal sinus, which lies dorsal to the internal gill slits. If you did not expose this sinus during the dissection of branchial muscles (Chapter 7), open it now by separating the cucullaris muscle from the epaxial musculature (see Fig. 7-8). A large **visceral branch** of the vagus nerve lies beneath the connective tissue on the dorsomedial wall of the anterior cardinal sinus. It gives off four **branchial branches** that cross the floor of the sinus and then are distributed to the last four visceral arches and pouches. Each branchial branch follows the pattern of the glossopharyngeal nerve, having a sensory **nodose ganglion** from which **posttrematic, pretrematic,** and **pharyngeal branches** arise. The pharyngeal branches of the branchial branches of the vagus nerve split off the posttrematic branch (unlike the pharyngeal branch of the glossopharyngeal

[2]This nerve is homologous to the vestibulocochlear nerve of birds and mammals, which receives stimuli from the vestibular apparatus associated with equilibrium and the cochlea that detects sound. We use the term *statoacoustic nerve* for all other vertebrates because they lack a cochlea in their inner ear.

nerve, which splits off the pretrematic branch). Visceral sensory fibers in the posttrematic branch of the last branchial branch are distributed to the caudal surface of the last gill pouch, but there are no branchial muscles here. Somatic motor fibers in the last branchial branch form a small **accessory branch** that goes to the cucullaris muscle. This nerve branch is difficult to find. The vagus nerve also gives rise to an **intestinal branch,** which runs with the visceral branch, and then continues along the wall of the esophagus to the viscera. The intestinal branch also sends a branch to the pericardial cavity. The intestinal branch contains visceral motor fibers and general visceral sensory fibers.

Look for the **posterior lateral line nerve** just before the visceral branch of the vagus enters the rostral end of the anterior cardinal sinus. It lies medial to the visceral branch and runs with it a short distance before separating. It then runs caudally between the epaxial and hypaxial musculature, receiving special somatic sensory fibers from neuromasts in the trunk canal.

The cell bodies of the sensory neurons of the vagus lie in several ganglia (Fig. 9-6). The **jugal ganglion** is the most proximal, and four smaller **nodose ganglia** lie near the origin of each branchial nerve.

Occipital and Hypobranchial Nerves

Free the intestinal branch of the vagus nerve from the wall of the anterior cardinal sinus. A **hypobranchial nerve** emerges from the epaxial musculature, ventral to the posterior lateral line nerve, crosses the visceral branch of the vagus nerve at about the level of its last branchial branch, and curves ventrally in the wall of the common cardinal vein (Fig. 9-7). It carries somatic motor fibers to the hypobranchial musculature and returns a few proprioceptive and cutaneous somatic sensory fibers (Table 7-2).

Trace the hypobranchial nerve craniomedially toward the vertebral column and chondrocranium. It becomes progressively narrower and more difficult to trace through the musculature, but if you are successful you will see that it is formed by the confluence of several spinal nerves and two or three **occipital nerves.** The **first spinal nerve** emerges between the chondrocranium and the first vertebra. The last occipital nerve can be seen between this point and the large root of the vagus nerve. More rostral occipital nerves lie deep to the root of the vagus nerve. The occipital nerves resemble accessory rootlets of the vagus nerve because they appear to join it. Actually, they travel with the vagus nerve only a short distance as they leave the chondrocranium; then they separate and, together with several spinal nerves, form the hypobranchial nerve.

The hypobranchial nerve, and its occipital and spinal components, is homologous to the twelfth cranial nerve of amniotes, which is known as the hypoglossal nerve (Table 7-2). By convention, the hypobranchial nerve is not considered a cranial nerve in anamniotes because its origin relative to the caudal end of the skull varies in different groups. Frequently, as in *Squalus,* the origin is partly from the back of the brain (the occipital nerves) and partly from the front of the spinal cord (the spinal nerves). The occipital nerves are serially homologous to the ventral roots of spinal nerves. The dorsal roots of those spinal nerves, whose ventral roots form the occipital nerves, constitute the vagus nerve.

The Ventral Surface of the Brain

It is now possible to return to the brain, remove it from the braincase, and study its ventral surface. Before lifting the brain out of the braincase, cut across the caudal end of the medulla oblongata, the olfactory tracts, and the roots of the cranial nerves. Try to leave long stumps of the nerves attached to the brain. After you cut across the trigeminal, facial, and statoacoustic nerves, push the brain to one side and notice the small abducens nerve leaving from the ventral surface. It too must be cut. Lift up the caudal end of the brain and carefully pull it rostrally. You will soon see a part of the brain extending into a recess, called the **sella turcica,** in the floor of the cranial cavity. It probably will be necessary to cut away some of the floor of the chondrocranium caudal to this recess to remove the brain intact.

After the brain has been removed, examine its ventral surface (Fig. 9-8). Identify the roots of the cranial nerves and the major regions of the brain. Several new structures can be seen in the region of the hypothalamus, which is the floor of the diencephalon. The optic nerves attach at the rostral end of the hypothalamus, and their fibers decussate at this point, forming an X-shaped structure known as the **optic chiasm.** In most nonmammalian vertebrates, the decussation is complete so that the left eye projects to the right side of the brain and *vice versa.* In some amphibians and reptiles, a few fibers do not decussate and remain ipsilateral. In many mammals, especially those species in which the visual fields of the two eyes overlap to a large extent and stereoscopic vision is present, many fibers remain ipsilateral and the rest decussate. The visual field of each eye in such mammals projects to each side of the brain. After passing through the optic chiasm, the optic fibers form a band, the **optic tract,** which leads caudodorsally to the optic lobes. You will have to remove the primitive meninx to distinguish the tract. Most of the rest of the hypothalamus consists of a large, caudally projecting **infundibulum.** Just caudal to the optic chiasm, the infundibulum bears a pair of prominent lateral lobes known as the **inferior lobes of the infundibulum.** The **hypophysis,** or pituitary gland, lies partly between and partly caudal to the inferior lobes. The infundibulum finally forms a thin-walled, dark **saccus vasculosus,** which lies caudal to the inferior lobes and dorsal to the hypophysis. The saccus vasculosus of sharks is a specialized structure that is composed of neural epithelium and contains an extensive vascular plexus. Its function is unknown, but one hypothesis is that the saccus vasculosus detects water pressure and, hence, depth. Another hypothesis is that it helps control the composition of cerebrospinal fluid.

If the brain was carefully removed, then you can recognize the various lobes of the hypophysis (Fig. 9-8). Two conspicuous lobes lie in the midventral line: an **anterior lobe** between the inferior lobes of the infundibulum and an **intermediate lobe** caudal to this. A pair of **superior lobes** lies between the intermediate lobe and the saccus vasculosus.

The hypophysis is an endocrine gland of dual embryonic origin. Most of its anterior lobe, which attaches to the infundibulum, develops as an outgrowth of the diencephalon and, therefore, is homologous to the mammalian neurohypophysis. The rest of the hypophysis develops from an embryonic mouth or hypophyseal pouch, which is an invagination from the embryonic stomodaeum, and hence is homologous to the mammalian adenohypophysis. Functions of the hypophysis of sharks are not fully understood, but available evidence suggests that they are very similar to those of mammals, which we discuss later. The neurohypophysis stores and releases hormones involved in osmoregulation. As in mammals, these hormones are synthesized by neuron cell bodies in the hypothalamus and reach the neurohypophysis by traveling along the axons of the cells from which they are released. The adenohypophysis produces a hormone that helps regulate growth, and others that interact with the thyroid, adrenal, and gonadal hormones in the regulation of body metabolism, dispersion of pigment in chromatophores, mineral metabolism, and reproduction.

The Ventricles of the Brain

The central cavity that characterizes the neural tube of chordates expands to form large chambers, or ventricles, in the brain (Fig. 9-4). You have already observed some of these. To see the other cavities, make a sagittal section of the brain and cut into one of the cerebral hemispheres. A **lateral ventricle** lies in each of the paired cerebral hemispheres. Each connects with the **third ventricle** of the diencephalon by a narrow passage known as the **interventricular foramen** (foramen of Monro). The third ventricle connects with the **fourth ventricle** of the hindbrain by a narrow ventral passage known as the **cerebral aqueduct,** or aqueduct of Sylvius. **Optic ventricles** in the optic lobes and a **cerebellar ventricle** in the body of the cerebellum lie dorsal to the cerebral aqueduct. In life, all these cavities are filled with **cerebrospinal fluid.**

The Spinal Cord and Spinal Nerves

The spinal cord lies in the vertebral canal of the vertebral column (Fig. 5-2). To see it and the spinal nerves, remove several centimeters of the muscles overlying the vertebral column caudal to the chondrocranium and carefully shave away the dorsal part of the neural arch. You will soon see the **spinal cord** and the **dorsal roots** of the spinal nerves through the translucent cartilage. A **spinal ganglion** is present on each dorsal root but is rather small. Each dorsal root lies slightly caudal to its corresponding ventral root. To find the ventral root you must cut away the lateral wall of the neural arch cranial to the dorsal root. The **ventral root** arises from the spinal cord by a fan-shaped group of rootlets and passes out of the vertebral canal through a foramen in the neural plate. In *Squalus* the two roots unite in the musculature lateral to the vertebral column, but this union, and the subsequent splitting of the spinal nerve into dorsal, ventral, and communicating branches, is difficult to find.

The components in the roots of the *Squalus* spinal nerves appear to be similar to those in the nerves of amniotes, that is, sensory neurons enter the spinal cord via the dorsal roots, and motor neurons leave the spinal cord via the ventral roots. Preganglionic autonomic fibers present in the dorsal roots of many anamniotes have not been positively identified in sharks, but some may be present. Certainly most of the preganglionic visceral motor fibers of the autonomic system in the trunk extend through the ventral roots to synapse with postganglionic fibers in a series of ganglia lying dorsal to the posterior cardinal sinus and kidneys. The autonomic system of *Squalus* thus has cranial and spinal contributions, but the visceral organs apparently do not have the dual innervation found in mammals. Also, autonomic fibers do not go to the skin in *Squalus.* Functions of the nervous system of sharks are summarized in Anatomy in Action 9-1.

AMPHIBIANS

In its organization, the nervous system of amphibians has not evolved much beyond the condition seen in sharks. This is particularly the case for salamanders such as *Necturus,* although its cerebellum is not as well developed as in active fishes such as sharks. *Necturus* can be studied as an example of the amphibian condition, but it need not be studied in detail because the differences from *Squalus* are few.

Anatomy in Action 9-1 Spinal Cord and Brain Functions in Sharks

Much can be learned about basic central nervous system organization and function from *Squalus,* in which structures and activities are simpler than in mammals. Although locomotion is affected by impulses from the brain, much of it is initiated in the **spinal cord.** Experiments in which the connection between the spinal cord and brain has been severed have shown that the spinal cord of sharks is more than a center for reflex action. Such a "spinal" shark still displays normal, rhythmic swimming movements. These experiments suggest the presence of a **central pattern generator** within the spinal cord that integrates swimming because electrical recordings from individual motor neurons show a spontaneous rhythmic activity. This activity is maintained even when curare poisoning blocks muscle activity. Other studies have demonstrated that sensory feedback from the skin and active muscles plays a role and at least modifies any inherent rhythmic activity (for review, see Williamson and Roberts, 1986).

Central to integration within the brain is a network of short interconnected neurons known as the **reticular formation.** The reticular formation lies deep in the medulla oblongata and extends rostrally through the floor of the metencephalon and mesencephalon. The reticular formation receives afferent input from the sensory nuclei of the cranial nerves that attach to this region, that is, from all except for the olfactory and optic nerves. It also receives input from many other parts of the central nervous system, such as the spinal cord, deep nuclei in the cerebrum (the striatum), optic lobes, and cerebellum. Efferent fibers lead to motor nuclei of the cranial nerves and to the motor columns of the spinal cord. Neuron pools in the reticular formation integrate many motor activities, including eyeball movements, feeding movements, respiratory movements, heart rate, and swimming. Efferent impulses traveling down the fibers of the **reticulospinal tract** are the only impulses to reach the spinal cord from the brain in fishes. There are no direct fibers from the cerebrum. Integration that occurs partly in the reticular formation makes it possible for a fish to engage in complex patterns of behavior, such as finding food, swimming toward it, and capturing it.

Sensory neurons from the inner ear, lateral line, and electroreceptors terminate in adjacent but distinct nuclei in the acousticolateral area of the medulla oblongata. Fibers from these nuclei extend into the cerebellum, particularly the auricles of the cerebellum. The body of the cerebellum also receives visual, tactile, and proprioceptive inputs. Efferent fibers go to the reticular formation, from which they can be sent to many motor centers. As in other vertebrates, the cerebellum of elasmobranchs monitors the body's position in space, the degree of muscle contraction, and the general environment, or "landscape," in which the fish is living. It also initiates appropriate motor adjustments.

Most of the optic lobes form an **optic tectum** or roof that contains six layers of neuron cell bodies and fibers, which makes it anatomically the most complex region of the elasmobranch brain. Although it receives its major sensory input from the eyes, the tectum also receives projections that bring in most other sensory modalities. Fibers from the inner ear, lateral line, and electroreceptors enter a more caudal and deeply located nucleus within the optic lobes called the **torus semicircularis.** Although some efferent fibers pass rostrally to dorsal thalamic nuclei (see below), most efferent fibers lead to the reticular formation. The tectum and torus semicircularis can affect motor activity at many levels by means of these pathways. The coordination of eye movements with swimming movements is one example. Some investigators believe that the optic tectum is the primary integration center of the elasmobranch brain, analogous to the mammalian cerebral hemispheres.

The diencephalon contains many important centers. The **habenula** in its roof is an olfactory center. The **thalamus,** which forms its lateral walls, consists primarily of a group of **dorsal thalamic nuclei** that relay ascending sensory impulses from the reticular formation, optic tectum, and torus semicircularis to parts of the cerebrum. Efferent fibers lead from a ventral part of the thalamus back to the optic tectum and reticular formation. The **hypothalamus** in the floor of the diencephalon receives projections from olfactory, gustatory, and general visceral senses. Most of its efferent fibers lead by way of the reticular formation to visceral motor neurons, but some of its neurons produce neurosecretions that affect the activity of the hypophysis. The hypothalamus of elasmobranchs appears to affect the level of body activity, such as rest versus wakefulness, and helps regulate water balance, blood sugar levels, and other aspects of homeostasis. In conjunction with parts of the cerebrum, it controls aspects of behavior related to feeding, aggression, and sexual activity.

The olfactory bulbs are the first olfactory integration centers. Additional olfactory integration occurs in other parts of the cerebral hemispheres to which the olfactory bulbs project. Much of the cerebrum, however, is not olfactory. Instead, this region receives projections via the dorsal thalamic nuclei from visual and other sensory centers. Its efferent fibers lead from a group of deep cerebral nuclei, called the **striatum,** to the hypothalamus and ventral part of the thalamus. Impulses can go from these structures to the optic tectum and reticular formation. The cerebrum has neuronal connections that imply an important role in integration, but little is known about its functions in elasmobranchs.

The Dorsal Surface of the Brain

Expose the dorsal surface of the cranium by removing the skin and muscles overlying it, then carefully chip away the bone that forms its roof. In amphibians the primitive meninx of fishes is represented by two layers of connective tissue applied to the surface of the brain: a tough outer **dura mater** and an inner, vascular **pia-arachnoid mater.** If you did not remove the dura mater with the skull bones, then you must remove it now in order to see the brain. The rostral region of the brain is formed by the paired **cerebral hemispheres** and **olfactory bulbs** (Fig. 9-9). The olfactory bulbs lie rostral to the hemispheres but are not clearly separated from them. At most, only a slight lateral indentation may be seen between bulbs and hemispheres. Because the olfactory bulbs are adjacent to the rest of the telencephalon rather than to the olfactory sacs, there is no long, narrow olfactory tract as there is in *Squalus.* The bands of nervous tissue that extend between the olfactory sacs and the olfactory bulbs are the **olfactory nerves.** The small dark body that lies between the caudal ends of the cerebral hemispheres is the **paraphysis,** which is an evagination from the telencephalon.

The **epithalamus** of the diencephalon appears as a small, triangular, light-colored area caudal to the paraphysis. As in other vertebrates, the **epiphysis** and **habenula** are present in this region, but they are difficult to distinguish. From the **tela choroidea** in this region, a **choroid plexus** has invaginated into the third ventricle and has extended caudad into the cavity of the midbrain. You can see it by cutting open the epithalamus.

The **optic lobes** lie caudal to the epithalamus. They are relatively small in *Necturus* and do not bulge dorsally to the extent seen in *Squalus.* The line of separation between optic lobes and epithalamus is not sharp.

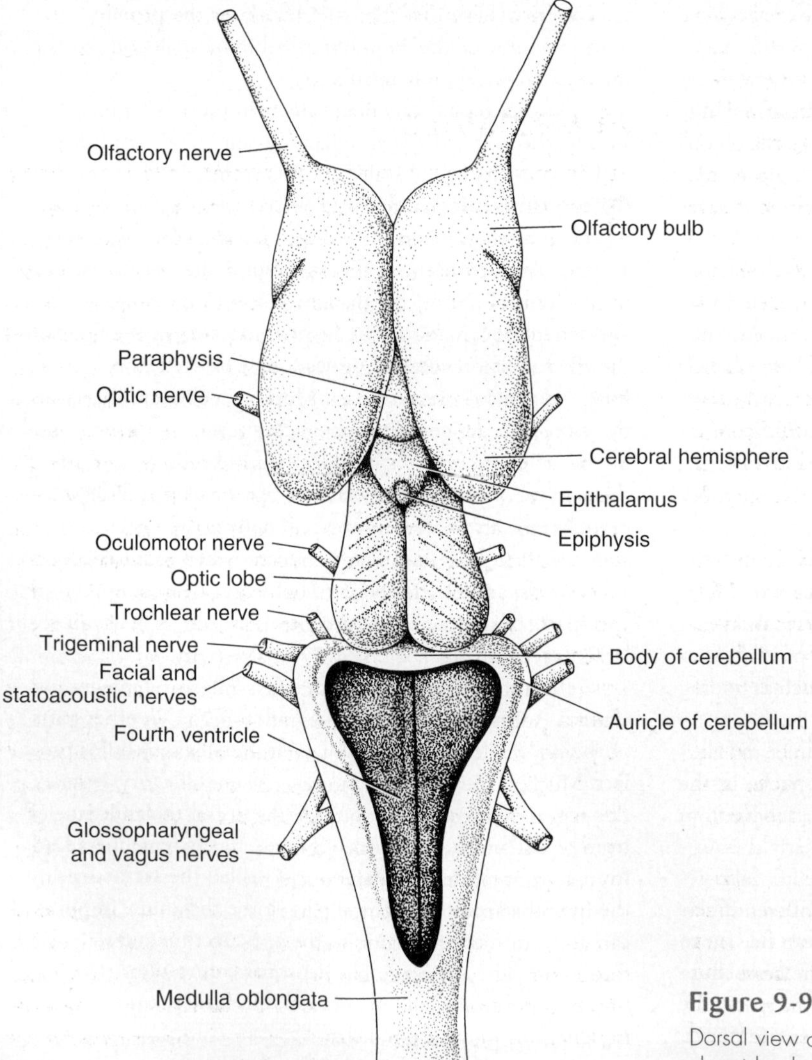

Figure 9-9

Dorsal view of the brain and cranial nerves of *Necturus.* The tela choroidea has been removed.

The **medulla oblongata** lies caudal to the optic lobes; you can recognize it by the very dark **tela choroidea** that forms most of its roof. Remove the tela choroidea with its **choroid plexus** and note the heart-shaped **fourth ventricle.**

As mentioned at the outset, the **cerebellum** is very small in amphibians. Its body consists of only a narrow transverse band of nervous tissue that lies between the optic lobes and the medulla oblongata. This fold may be difficult to distinguish from the optic lobes. The tissue that surrounds the anterolateral corners of the fourth ventricle represents the **auricles of the cerebellum.**

The Cranial and Occipital Nerves

The **olfactory nerve (I)** was observed coming into the brain from the olfactory sac. The **optic nerve (II)** from the eye crosses the cranial cavity and attaches to the brain at the rostral end of the diencephalic floor (hypothalamus).

The nerves to the extrinsic eye muscles [the **oculomotor nerve (III),** the **trochlear nerve (IV),** and the **abducens nerve (VI)**] are present but can be seen only under magnification. They have the same topographical relationships as in *Squalus* and supply the same muscles. The abducens nerve, however, also goes to the retractor bulbi muscle. This muscle is present only in tetrapods, although its homologue is present in extant coelacanths, a surviving group of early sarcopterygian fishes.

The large nerve arising from the lateral surface of the rostral end of the medulla is the **trigeminal nerve (V);** the nerve trunk caudal to it is the common attachment of the **facial (VII)** and **statoacoustic nerves (VIII).** *Necturus* has a well-developed lateral line system in the head, and the same three preotic lateral line nerves seen in *Squalus* innervate the neuromasts of the more anterior canals. These lateral line nerves are difficult to distinguish grossly from the other cranial nerves.

The roots caudal to the facial and statoacoustic nerves represent the origin of the **glossopharyngeal nerve (IX),** the **vagus nerve (X),** and the postotic **lateral line nerves.** These nerves run together until they leave the cranium, then they separate. The distribution and composition of these nerves are substantially the same as in *Squalus,* but there are some modifications in the gill region. The lateral line nerve fibers and receptors present in *Necturus* and aquatic larvae of other salamanders are lost in completely metamorphosed salamanders.

Like other salamanders, *Necturus* has a **hypobranchial nerve** to the hypobranchial musculature. This nerve is formed primarily of fibers running with the first spinal nerve, but it also receives contributions from the second spinal nerve and, in some salamanders, a small twig from the glossopharyngeal-vagus trunk. Certain of these nerve fiber contributions would obviously be homologous to the occipital nerves of *Squalus,* but just which ones is uncertain. It is of interest in this connection that the first, and sometimes the second, spinal nerve consists of only the ventral root.

The Ventral Surface of the Brain

Cut across the caudal end of the medulla oblongata and across the cranial nerves and remove the brain. Try to lift the hypophysis from the **sella turcica** without breaking it off. Note the major regions of the brain and the stumps of the cranial nerves, and study the floor of the diencephalon, or **hypothalamus,** in more detail. The optic nerves decussate to form an **optic chiasm** at the rostral end of the hypothalamus, but the chiasm is small and inconspicuous in *Necturus.* As in *Squalus,* the greater part of the hypothalamus consists of a large, posteriorly projecting **infundibulum** to which the **hypophysis** is attached. A small **saccus vasculosus** is present at the base of the roof of the infundibulum but is difficult to see grossly.

MAMMALS

The many changes that occurred in the central nervous system during the evolution of amniotes are related in large measure to the increased activity and flexibility of response in these animals. A major change, especially in birds and mammals, is the great increase in the amount of sensory information projected forward to the cerebrum and the speed with which it is transmitted in these endothermic vertebrates. The cerebrum becomes the dominant integration center in the brain. To appreciate the areas of the cerebrum affected, it is necessary to consider the distribution of the gray matter in the cerebrum (Fig. 9-10A). The gray matter of the cerebrum of all vertebrates can be divided into a dorsal **pallium,** or roof, and a ventral **subpallium** that consists of a group of nuclei called the **striatum** and another group of nuclei called the **septum.** The pallium is divided into a **lateral pallium** (the olfactory lobes or primary olfactory region), a **medial pallium** (another olfactory center), and a **dorsal pallium** located between them. The dorsal pallium is very small in anamniotes, but it expands in amniotes because it is the region that receives the new projections of largely cutaneous somatosensory, visual, and acoustic information from dorsal thalamic nuclei. The dorsal pallium receives no direct olfactory projection from the olfactory lobes.

The evolution of the dorsal pallium in the various amniote groups is not entirely clear, but we follow a hypothesis developed by Butler and Hodos (1996). They recognize two groups of dorsal thalamic nuclei: (1) a **collothalamus,** which receives auditory, visual, and some somatosensory impulses from the optic tectum, which is called **colliculi** in mammals; and (2) a

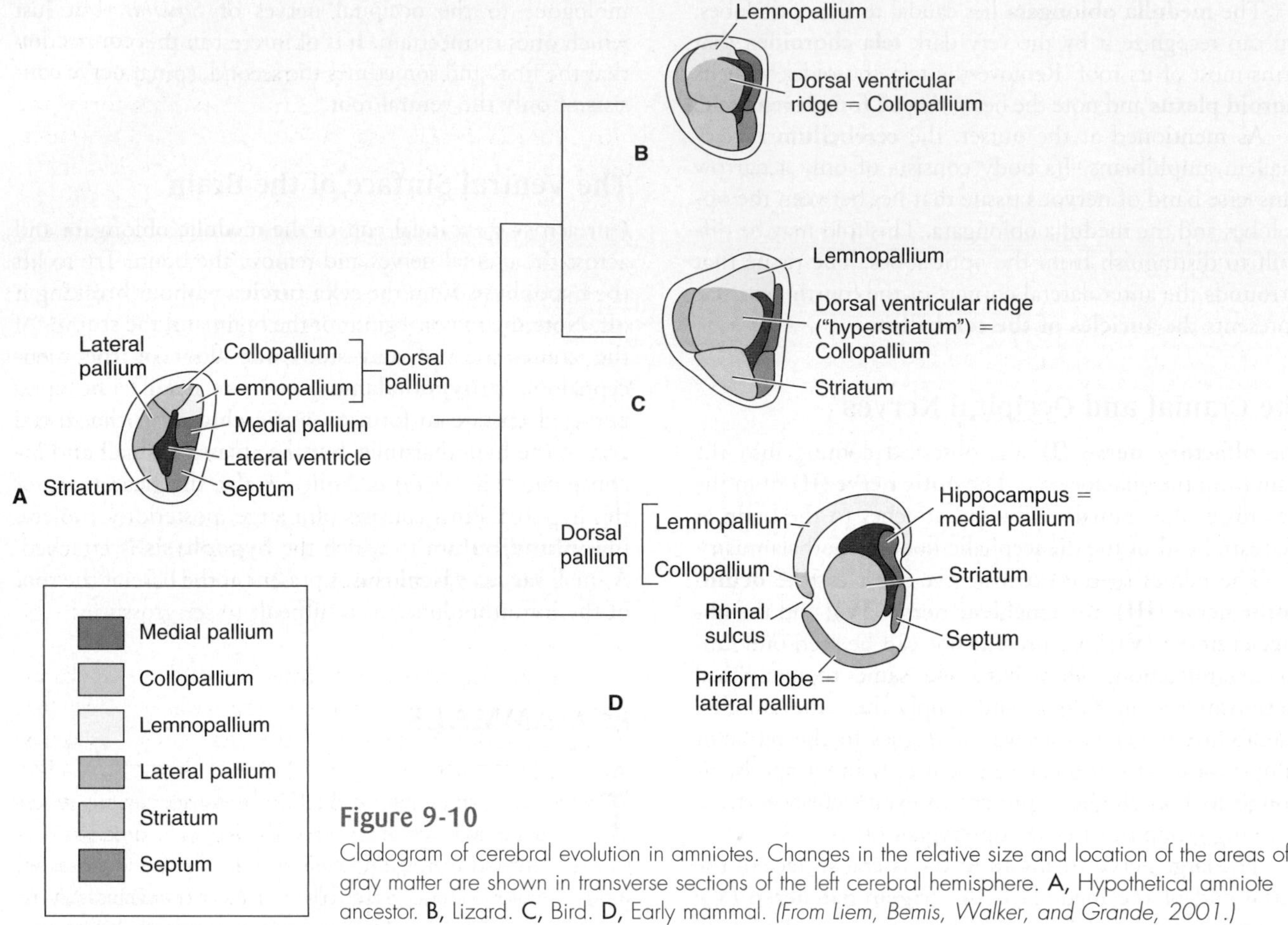

Figure 9-10

Cladogram of cerebral evolution in amniotes. Changes in the relative size and location of the areas of gray matter are shown in transverse sections of the left cerebral hemisphere. **A**, Hypothetical amniote ancestor. **B**, Lizard. **C**, Bird. **D**, Early mammal. *(From Liem, Bemis, Walker, and Grande, 2001.)*

lemnothalamus, which primarily receives somatosensory and visual information on tracts (called **lemnisci**) that bypass the optic tectum and ascend directly from centers where these impulses enter the central nervous system. These parts of the thalamus project to regions of the dorsal pallium called the **collopallium** and **lemnopallium,** respectively.

In the sauropsid line of amniote evolution to reptiles and birds, the primary expansion of the dorsal pallium occurs in the collopallium (Fig. 9-10**B** and **C**), which remains in its ancestral location deeply seated next to the lateral ventricle. In the synapsid line of evolution to mammals, the primary expansion occurs in the lemnopallium (Fig. 9-10**D**), which migrates to the cerebral surface where it forms a gray cortex called the **isocortex** or **neopallium.** This expansion is so great in mammals that the cerebral hemispheres grow caudally over the roof of the brain and contact the expanding cerebellum. As a consequence, the diencephalon and mesencephalon of a mammalian brain are hidden beneath the expanded cerebral hemispheres. A series of folds called **gyri** (Fig. 9-12) further increase the surface area of the isocortex in many eutherian mammals. As the isocortex expands, it pushes the lateral and medial pallia apart. The lateral pallium, called the **piriform lobes** in mammals, comes to lie on the ventrolateral surface of the cerebrum. It remains the primary olfactory area. The medial pallium is pushed medially and inward to bulge into the lateral ventricle forming the **hippocampus,** another important olfactory and visceral integration center.

Much sensory (afferent) information is projected forward to the cerebrum more rapidly by bypassing the reticular formation and optic tectum and going directly to the dorsal thalamic nuclei on a **spinothalamic tract** from the spinal cord and a **bulbothalamic tract** from the brainstem. During mammalian evolution, the dorsal thalamic nuclei increased in size and number. The optic tectum and reticular formation lost some of the functions observed in sharks but retained many others.

The striatum in the cerebrum, subthalamus, and reticular formation continues to participate in transmitting efferent signals back to motor centers, but speed of transmission is increased by the evolution of direct pathways from the cerebral cortex to motor centers, such as the **corticobulbar tract** to the brainstem and the **corticospinal tract** to the spinal cord.

With the loss of the lateral line and electroreceptive systems, which probably occurred with the loss of aquatic larvae in the earliest amniotes, the **cerebellum** of tetrapods no longer receives input from these systems, but other sensory input is extensive. The auricles of the cerebellum, called **flocculonodular lobes** in mammals, and its body, now represented by most of the **vermis,** remain. In mammals, a pair of large **cerebellar hemispheres** evolved and have extensive connections with other brain regions, including the isocortex. The cerebellum continues to be very important in motor coordination and in motor learning.

The basic pattern of the cranial nerves of mammals remains much the same as you saw in sharks (Table 9-1), but a few changes were superimposed upon this plan during the evolution of mammals. (1) The lateral line nerves are lost. (2) Mammals retain the autonomic visceral motor fibers present in the oculomotor nerve of many anamniotes and, in addition, acquire autonomic visceral motor fibers in the facial and glossopharyngeal nerves. Those in the facial nerve go to most of the salivary glands; those in the glossopharyngeal nerve go to the parotid salivary gland. (3) With the elaboration of the cucullaris muscle to form the trapezius and sternocleidomastoid complex of muscles, the somatic motor fibers that supply these muscles separate from the vagus nerve to form a new cranial nerve, the **accessory nerve (XI).** This change is seen in both sauropsid and synapsid lines so must have occurred at the level of the first amniotes. (4) The major change in ventral root nerves occurs when the caudal limit of the cranium becomes fixed, and the occipital nerves, plus one or two rostral spinal nerves, form a definite cranial nerve, the **hypoglossal nerve (XII).** Among extant vertebrates, this change is first seen at the level of amniotes, but it probably occurred earlier, because there is a foramen for such a nerve in the skulls of basal tetrapods. In lissamphibians, the hypobranchial musculature is supplied by fibers that travel in the first spinal nerve.

Meninges

You should study the mammalian brain and the stumps of the cranial nerves on isolated sheep brains. As explained earlier, the peripheral distribution and composition of the cranial nerves are, with a few exceptions, essentially the same as in fishes. The foramina through which they leave the cranium are listed in Table 4-3, and you should review them as you study the nerves. During later dissections, you will see peripheral parts of some of these nerves. If it is not possible to study the brain from isolated specimens, then you can remove it from your own specimen, but if this is to be done, then postpone this study until the end of the course. To remove the brain, first make a sagittal section of the head and then carefully loosen the halves of the brain and pull them out. Leave on the tough membrane, called the dura mater, covering the brain as you take it out. The cranial nerves will have to be cut, but leave stumps as long as possible attached to the brain.

The primitive meninx of sharks has differentiated in mammals into three layers: dura mater, arachnoid, and pia mater (Fig. 9-11). Because the brain fills the cranial cavity in mammals, the tough connective tissue that forms the outer **dura mater** has been pushed against the periosteum, lining the cranial cavity and fuses with it. If the dura mater is still on your specimen, then carefully remove it. As you do so, note that it sends an extension down between the cerebrum and cerebellum and another between the two cerebral hemispheres. The former extension is called the **tentorium;** the latter is the **falx cerebri** (see Fig. 10-15). These extensions of the dura mater help to stabilize the brain within the cranial cavity and prevent it from being distorted during sudden rotational movements of the head. The tentorium is ossified in the cat (see Fig. 4-21). Cut across the falx cerebri and notice that it contains the **superior sagittal venous sinus.** Irregular pits evident on the cerebral surface of the falx cerebri lead into this venous sinus. In life, small **arachnoid villi** project from the next deeper meninx (the arachnoid) through these pits into the venous sinus.

The arachnoid and pia mater are difficult to distinguish grossly. The **pia mater** is the vascular layer that closely invests the surface of the brain. The avascular **arachnoid** lies between the pia and dura mater. These layers are most easily distinguished on the cerebrum in the region overlying the grooves, called **sulci,** on the brain surface because the arachnoid does not dip into the sulci whereas the pia mater does. In life, a narrow **subarachnoid space,** which is crisscrossed by delicate strands of connective tissue, which are called **arachnoid trabeculations,** lies between the pia mater and arachnoid (Fig. 9-11**A**). The name *arachnoid* is derived from these spider web–like strands. The subarachnoid space expands into a large **cerebellomedullary cistern** where the arachnoid leaves the caudal border of the cerebellum and extends over the medulla oblongata (Fig. 9-11**B**). **Cerebrospinal fluid** fills the subarachnoid space as well as the ventricular system of mammals.

Most cerebrospinal fluid is secreted into the ventricles of the brain by the choroid plexuses (Fig. 9-11**B**). After slowly moving through ventricles, the fluid leaves the cavities of the brain through minute pores in the roof of the fourth ventricle and slowly circulates in the subarachnoid space. It reenters the blood vessels through the arachnoid villi, which protrude into the superior sagittal and certain other cranial venous sinuses. The cerebrospinal fluid forms a liquid cushion around the central nervous system that protects and supports the exceedingly soft and delicate nervous tissue. Cerebrospinal fluid gives the brain a great deal of buoyancy. It also helps to provide the brain with carefully selected nutrients and other substances. The immediate environment of brain cells must be more

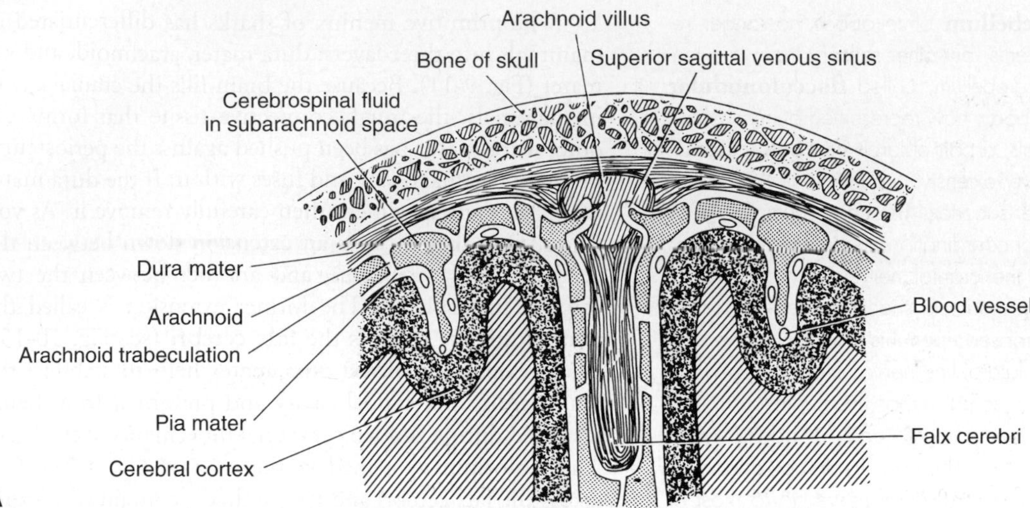

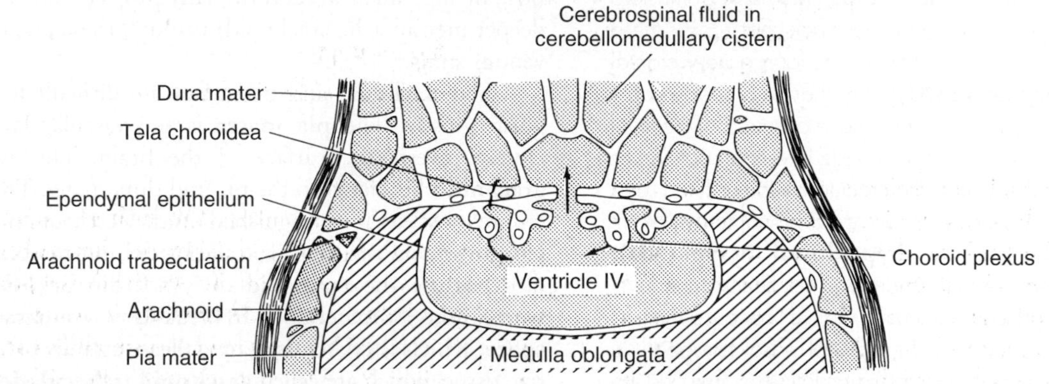

Figure 9-11

Diagram of the meninges and cerebrospinal fluid circulation in a mammal. **A**, Transverse section through the dorsal surface of the cerebral hemisphere showing an arachnoid villus, which is a major point where cerebrospinal fluid enters the venous system. **B**, Transverse section through the medulla oblongata showing secretion of the cerebrospinal fluid into the fourth ventricle and its escape into an enlargement of the subarachnoid space, the cerebellomedullary cistern.

carefully regulated than that of other cells because the activity of brain cells is adversely affected by the accumulation of waste products and by fluctuating levels of sugars, other nutrients, hormones, and certain ions. Tight junctions between brain capillary cells and between the cells of the choroid plexuses form a **blood-brain barrier** that controls the passage of many substances. Gases and some other materials pass easily, whereas other substances are held back. Glial cells and the cells of the choroid plexuses actively transport some substances into or out of the cerebrospinal fluid.

External Features of the Brain and the Stumps of the Cranial Nerves

The sheep brain is widely studied as a model for the mammalian brain, and good online atlases and dissection guides are available (Johnson et al. and Wheeler et al. 1996).

The Telencephalon

The paired cerebral hemispheres and the cerebellum are so large that little else is apparent at first in surface views of the brain (Figs. 9-12 and 9-13). Notice that a deep longitudinal furrow known as the **longitudinal cerebral fissure**

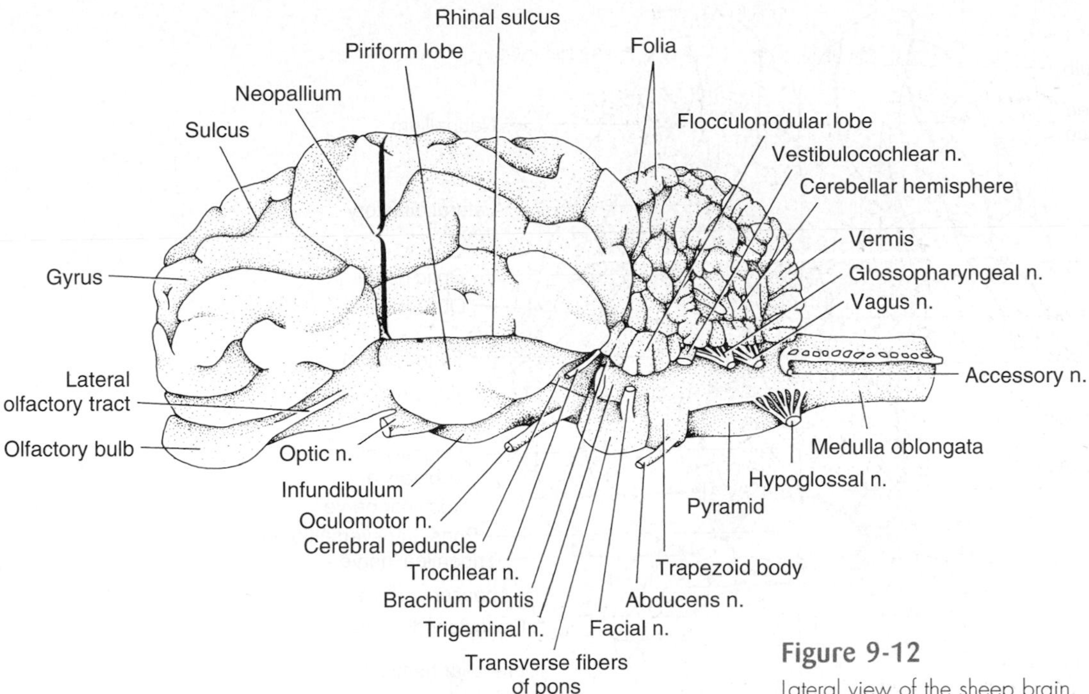

Figure 9-12
Lateral view of the sheep brain.

separates the cerebral hemispheres dorsally from each other. Spread the dorsal parts of the cerebral hemispheres apart and note the thick transverse band of nerve fibers that connects them. This is the **corpus callosum,** an isocortex **commissure**[3] characteristic of eutherian mammals. The surface of each hemisphere is thrown into many folds, called **gyri,** which are separated from each other by grooves called **sulci.** A pair of **olfactory bulbs** project from the rostroventral portion of the cerebrum. They can be seen best in a lateral or ventral view. The olfactory bulbs lie over the cribriform plate of the ethmoid bone and receive the bundles of olfactory neurons from the nose through the **cribriform foramina** (Chapter 4). These neurons, which were pulled off the olfactory bulb during brain removal, constitute the **olfactory nerve (I).** A whitish band, the **lateral olfactory tract,** extends at an angle caudally and laterally from each olfactory bulb. The ventral portion of the cerebrum, to which each olfactory tract leads, is known as the **piriform lobe.** It is separated laterally from the rest of the cerebrum by the **rhinal sulcus.** As explained earlier, the piriform lobe represents the lateral pallium, or primary olfactory cortex. Because the medial pallium is pushed medially and internally (Fig. 9-10**D**), all of the cortex that you see lying dorsal to the rhinal sulcus is the isocortex, the major nonolfactory integration area of the mammalian brain.

[3]A *commissure* is a fiber tract that interconnects comparable structures on opposite sides of the brain.

The Diencephalon

The telencephalon has enlarged to such an extent that it has grown caudally over and covers the diencephalon and much of the mesencephalon. To see the dorsal portion of the diencephalon, the **epithalamus,** it is necessary to spread the cerebral hemispheres apart carefully, cutting the corpus callosum. Do not destroy the delicate tela choroidea and its choroid plexus that forms part of the roof of the diencephalon (Fig. 9-16). Gently push the folds of the tela choroidea rostrally and laterally to expose the knoblike **pineal gland,** which lies caudal to the tela choroidea. Pick away the tela choroidea and its choroid plexus. The longitudinal slit that is exposed is the **third ventricle** (Fig. 9-14). The narrow transverse band of tissue between the pineal gland and the ventricle is the **habenular commissure,** and the tissue forming the posterolateral rim of the ventricle is the **habenula,** an olfactory area.

Turn the brain over and examine the ventral surface of the diencephalon, which is known as the **hypothalamus** (Fig. 9-13). The hypothalamus is a very important visceral integration area. The **optic nerves (II),** which enter the skull through the **optic canal** (Chapter 4), undergo a partial decussation at the rostral border of the hypothalamus, forming the prominent **optic chiasm.** The rest of the hypothalamus is the oval area lying caudal to the optic chiasm. A narrow stalk, the **infundibulum,** extends from the hypothalamus to the **hypophysis,** or pituitary gland. If the gland is still attached, then remove it in

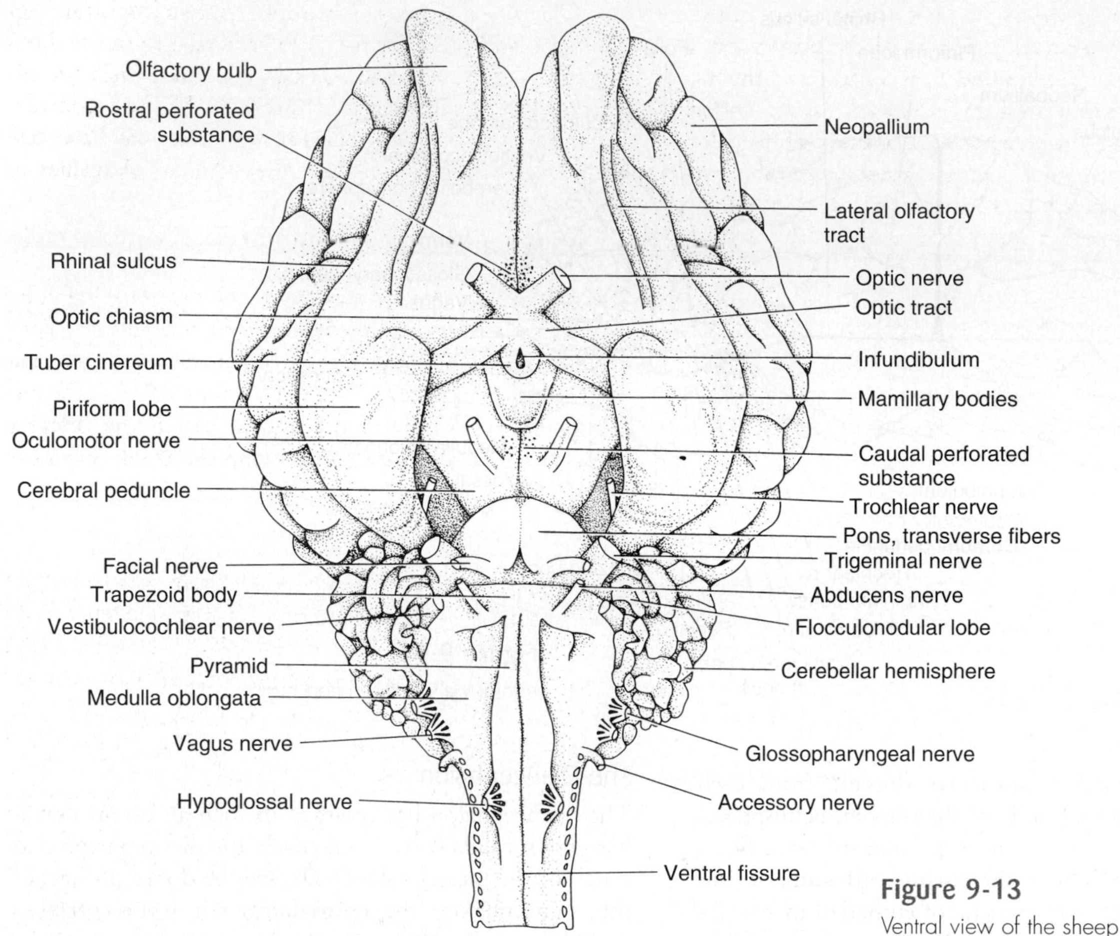

Figure 9-13

Ventral view of the sheep brain.

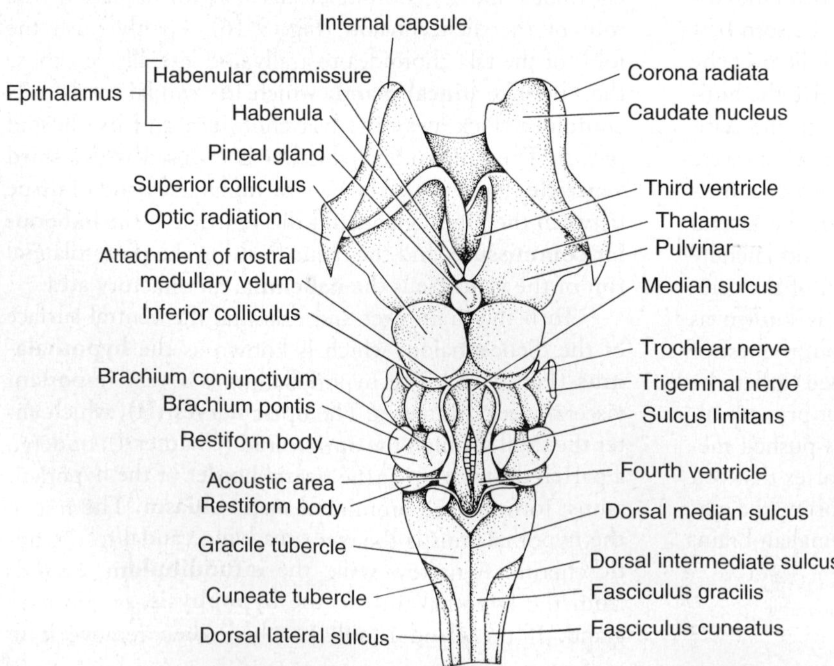

Figure 9-14

Dorsal view of the sheep brainstem. The telae choroideae have been removed. *(After Ranson and Clark, 1959.)*

order to get a clearer view of the region. The cavity in the infundibulum represents an extension of the third ventricle. The portion of the hypothalamus adjacent to the attachment of the infundibulum is known as the **tuber cinereum.** A pair of rounded **mamillary bodies** forms the caudal end of the hypothalamus. Macroscopically, the mamillary bodies are not as obviously paired in the sheep brain as they are in the human brain, but nuclei within them are paired.

In order to glimpse the **thalamus,** or lateral wall of the diencephalon, you must carefully pull one of the cerebral hemispheres rostrally and look beneath it. You will see this area more clearly after you have removed the cerebrum (p. 237), so postpone further study of the thalamus until then.

The pineal gland and hypophysis, two important endocrine glands that develop at least in part from the diencephalon, are functionally closely integrated with the diencephalon (Anatomy in Action 9-2). We will consider other functions of the diencephalon after we have dissected the cerebrum and have seen the lateral surface of the thalamus.

The Mesencephalon

You can see the roof, or **tectum,** of the mesencephalon by spreading apart the cerebrum and cerebellum. Four prominent, round swellings, the **corpora quadrigemina,** characterize this region. The larger, rostral pair is the **superior colliculi;** the smaller, caudal pair is the **inferior colliculi.** The inferior colliculi evolved from the more

Anatomy in Action 9-2 Endocrine Glands Associated with the Diencephalon in Mammals

Two endocrine glands whose functions are regulated partly by the nervous system develop from the roof and the floor of the diencephalon. The **pineal gland** of birds and mammals evolved from the pineal/parietal median eye complex found in many fishes and early tetrapods (see Fig. 8-6). The posterior lobe of the hypophysis, called the **neurohypophysis,** develops as an outgrowth from the floor of the diencephalon, the hypothalamus. The anterior and intermediate lobes, called the **adenohypophysis,** develop as an evagination from the roof of the embryonic mouth, or stomodaeum (Fig. 2-2).

The pineal gland of mammals continues to synthesize the hormone **melatonin,** first produced by the median eye complex. Because melatonin synthesis is inhibited by light, its levels in the blood are highest at night or during times of the year when days are short. Light affects the pineal gland directly in some species of birds, as it does in lizards and anamniotes. In mammals the cells that produce melatonin are under neural control through a complex pathway that leads from a **suprachiasmatic nucleus** that lies above the optic chiasm in the floor of the hypothalamus. This is an example of a **neuroendocrine transducer** because neural signals control the synthesis of a hormone. Because the blood levels of melatonin have a daily or circadian rhythm, and in some species a seasonal rhythm, this hormone appears to be involved in regulating certain circadian physiological cycles, such as sleep and some hormone levels. It also may be involved with seasonal reproductive cycles. Melatonin is known to have an inhibiting effect on gonadal development in some rodents. The suprachiasmatic nucleus appears to be the biological clock that regulates the production of melatonin, but the clock must be adjusted periodically to changes in day length. Recent discoveries have shown that the suprachiasmatic nucleus receives light signals from a specialized group of ganglion cells in the retina (Hatter et al., 2002).

The neurohypophysis synthesizes two hormones. **Antidiuretic hormone** promotes the reabsorption of water from kidney tubules and hence prevents diuresis, or the production of copious dilute urine. **Oxytocin** stimulates the contraction of smooth muscles in certain organs, most typically the uterus during childbirth and the mammary glands when an infant suckles (Anatomy in Action 7-2). Both of these hormones are synthesized by neurons with cell bodies located in the hypothalamus and axons that travel through the infundibulum to the neurohypophysis, where the hormones are stored. Appropriate stimuli cause the hormones to be released into blood capillaries. This is an example of **neurosecretion.**

The adenohypophysis synthesizes seven major hormones that affect the level of metabolism **(thyroid stimulating hormone),** reproductive maturation and cycles **(follicle stimulating hormone** and **luteinizing hormone),** growth **(growth hormone** and **prolactin),** secretion of a hormone by the adrenal cortex **(adrenocorticotropic hormone),** and color change in certain species **(melanocyte stimulating hormone).** The synthesis and release of these hormones are regulated primarily by **releasing hormones** or, in at least one case, by a **release inhibiting hormone,** that are produced in the hypothalamus and travel to the adenohypophysis *via* a group of small **hypophyseal portal veins.** Here, again, the nervous system has an important influence on the endocrine system.

deeply seated tori semicircularis of fishes (Fig. 9-14; Anatomy in Action 9-1). Note that the **trochlear nerves (IV)** arise slightly caudal to the inferior colliculi. Most of the integrative functions of the optic tectum of earlier vertebrates are transferred to the isocortex in mammals, but some optic and auditory fibers are still projected to the superior and inferior colliculi, which remain significant visual and auditory reflex centers, respectively.

A pair of **cerebral peduncles** lie along the ventrolateral surface of the mesencephalon (Figs. 9-12 and 9-13). Each emerges from beneath the **optic tract,** which extends dorsally and caudally from the optic chiasm. The peduncles are large bundles of fibers that extend caudally from the cerebral hemispheres and carry efferent corticobulbar and corticospinal fibers directly from the motor part of the cortex to the brainstem and spinal cord.

An **oculomotor nerve (III)** arises from the surface of each cerebral peduncle. The third, fourth, and sixth cranial nerves (described later) converge and leave the skull together through the **orbital fissure** (Chapter 4) to enter the base of the orbit, where they are distributed to the extrinsic eye muscles. The depression between the two cerebral peduncles is the **interpeduncular fossa.** If you strip the meninges from this region, you may be able to see small holes through which blood vessels enter the brain. This region constitutes the **caudal perforated substance.** A comparable **rostral perforated substance** lies rostral to the optic chiasm.

The Metencephalon

The dorsal portion of the metencephalon forms the **cerebellum.** Numerous platelike folds called **folia,** which are separated from each other by **sulci** (Fig. 9-12), increase its surface area. The median part of the cerebellum, which has the appearance of a segmented worm bent nearly in a circle, is called the **vermis;** the lateral parts are the **cerebellar hemispheres.** The lobe of each hemisphere that lies ventral to the main part of the hemisphere, and lateral to the region where the cerebellum attaches to the rest of the brain, is known as the **flocculonodular lobe.** These lobes are homologous to the cerebellar auricles of *Squalus;* most of the vermis is homologous to the body of the cerebellum in *Squalus;* and most of the cerebellar hemispheres are new additions, known as the **neocerebellum,** with cerebral connections.

The cerebellum is connected to other parts of the brain by three prominent fiber tracts, or peduncles (Figs. 9-14 and 9-15). Most of the **brachium pontis,** or middle peduncle, lies medial to the rostral half of the flocculonodular lobe. You will have to dissect off this lobe on one side in order to see the brachium pontis clearly. Note that the brachium pontis connects ventrally

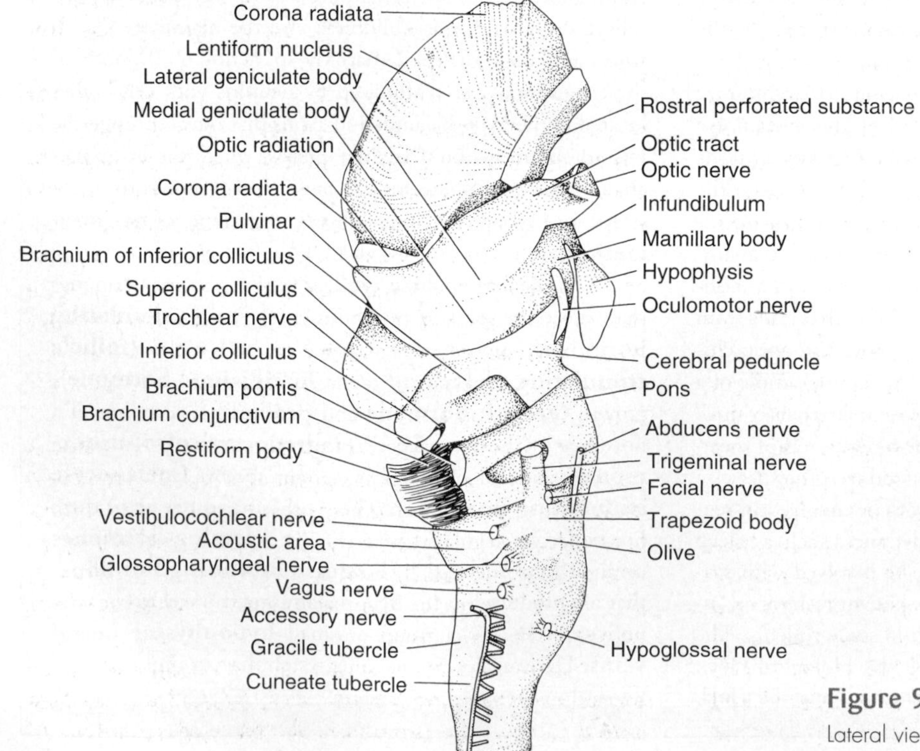

Figure 9-15

Lateral view of the sheep brainstem. *(After Ranson and Clark, 1959.)*

with a transverse band of fibers known as the **transverse fibers of the pons** (Fig. 9-12). Trace the brachium pontis dorsally into the white matter of the cerebellum, the **arbor vitae.** The arbor vitae will be seen more clearly in the sagittal section (p. 237). The tissue caudal and slightly medial to the brachium pontis constitutes the **restiform body,** or caudal peduncle. It continues along the dorsolateral margin of the medulla oblongata. The **brachium conjunctivum,** or rostral peduncle, lies medial to the brachium pontis; you can see it by looking in the area between the cerebellum and caudal colliculi. An important auditory pathway, the **lateral lemniscus,** forms a bundle of fibers that emerges between the brachium conjunctivum and brachium pontis.

In mammals, the ventral portion of the metencephalon has differentiated sufficiently from the medulla oblongata to be considered a distinct region—the **pons** (Fig. 9-13). Macroscopically, the pons includes the region of transverse fibers bordered rostrally by the interpeduncular fossa and caudally by the trapezoid body, which is a part of the myelencephalon. The large **trigeminal nerve (V)** arises from the lateral portion of the pons and extends rostrally across the base of the brachium pontis. The trigeminal nerve of mammals consists of three primary branches: (1) the ophthalmic branch (V_1), which returns sensory fibers from much of the head above the level of the orbit; (2) the maxillary branch (V_2), which returns sensory fibers from the skin over the upper jaw; and (3) the mandibular branch (V_3), which returns sensory fibers from the lower jaw area and carries motor fibers to the mandibular muscles. These branches leave the cranial cavity through the three foramina that lie just caudal to the optic canal: the **orbital fissure** (ophthalmic branch with cranial nerves III, IV, and VI), the **foramen rotundum** (maxillary branch), and the **foramen ovale** (mandibular branch). The cerebellum is a major center for equilibrium and motor coordination (Anatomy in Action 9-3).

The Myelencephalon

The rest of the brain belongs to the myelencephalon and forms the **medulla oblongata.** In order to see the parts of the medulla oblongata clearly, you must strip off the meninges on at least one side, but before doing this you should identify the remaining cranial nerves (Figs. 9-12 and 9-13). The **abducens nerve (VI)** and the **facial nerve (VII)** lie at the border between the pons and the medulla oblongata. The abducens nerve lies more medial and extends rostrally across the pons. The facial nerve extends laterally and dorsally across the base of the trigeminal nerve. The stump of the **vestibulocochlear nerve (VIII)** lies dorsal to the cut end of the facial nerve. Facial and vestibulocochlear nerves leave the cranial cavity together through the **internal acoustic meatus** (Fig. 4-21). The vestibulocochlear nerve begins at the inner ear, which is lodged within the petrous part of the temporal bone, but the facial nerve continues through the temporal bone to emerge on the skull surface at the **stylomastoid foramen** (Fig. 4-22).

The **glossopharyngeal nerve (IX)** and the **vagus nerve (X)** are represented by a number of fine rootlets caudal to, but in line with, the vestibulocochlear nerve. In most specimens, these rootlets have been cut so short that they cannot be traced into the peripheral parts of the nerves. It is impossible to say more than that the rostral rootlets belong to the glossopharyngeal nerve and the caudal rootlets to the vagus nerve. If possible, examine a demonstration dissection in which the dura mater has been removed in such a way that you can trace the rootlets into the glossopharyngeal and vagus nerves. The **accessory nerve (XI)** is the large longitudinal nerve caudal to the vagus nerve. It arises by a number of fine rootlets from the caudal end of the medulla oblongata and the rostral end of the spinal cord. The rostral end of the accessory nerve lies close to the vagus nerve; indeed, the accessory nerve evolved from a part of the vagus nerve of anamniotes. Glossopharyngeal, vagus, and accessory nerves leave the cranial cavity together through the **jugular foramen** (Figs. 4-20**A** and 4-21) in company with the internal jugular vein. The **hypoglossal nerve (XII)** is represented by the rootlets on the caudoventral portion of the medulla oblongata. Notice how close these are to the level of the foramen magnum; such proximity is to be expected, because the nerve evolved from the spinooccipital nerves of fishes. The hypoglossal nerve leaves through the **hypoglossal canal** (Fig. 4-21).

Now strip off the meninges on one side of the medulla oblongata, and carefully remove the **tela choroidea** with its **choroid plexus** that forms much of the roof of the medulla oblongata. Pull the cerebellum rostrally and note the large **fourth ventricle** extending forward into the metencephalon. The caudal part of the roof of the ventricle was formed by the tela choroidea, but a thin layer of fibers termed the **rostral medullary velum,** which you can see by pushing the cerebellum caudally, forms the rostral part of the roof. The trochlear nerve decussates in this velum.

Note the enlargement on the dorsal rim of the medulla oblongata just caudal to the point at which the restiform body turns into the cerebellum and dorsal to the vestibulocochlear nerve. Observe that it extends medially to an oval enlargement in the ventrolateral part of the floor of the fourth ventricle. This enlargement constitutes the **acoustic area** (Fig. 9-14).

Examine the dorsal surface of the caudal end of the medulla oblongata, which extends as far caudad as the **first spinal nerve.** The prominent **dorsal median sulcus** of the spinal cord (Fig. 9-14) continues into the medulla

Anatomy in Action 9-3 The Functions of the Cerebellum in Mammals

Although the cerebellum of amniotes has lost its sensory input from lateral line organs, it remains an important center for maintaining equilibrium and coordinating muscle activity. It has become particularly important in motor coordination because the range and complexity of motor activity increase in birds and mammals. It also becomes closely interrelated with the isocortex in mammals.

Most of the gray matter of the cerebellum has moved outward to form a complex multilayered cortex where afferent impulses terminate and efferent ones are initiated (figure). Impulses enter and leave the cerebellum through its peduncles: the brachium conjunctivum, brachium pontis, and restiform body. Efferent signals leave on the axons of large **Purkinje cells** and relay in several **deep cerebellar nuclei.** The Purkinje cells have an inhibitory effect upon their targets, but the degree of inhibition is altered by connections within the cerebellar cortex. Some impulses strengthen the degree of inhibition; others reduce it, a process called disinhibition. The fiber tracts within the cerebellum, which you will see later in the sagittal section of the brain, form a picturesque branching pattern called the **arbor vitae,** or tree of life.

The flocculonodular lobes of mammals, which represent the cerebellar auricles of fishes, remain important vestibular centers and receive input from the vestibular nucleus to which the inner ear projects. The vestibular nucleus is a part of acoustic area in the medulla oblongata (figure). The vermis of mammals, which is homologous to most of the body of the cerebellum of fishes, continues to receive an extensive input from most sensory modalities, but particularly from proprioceptors—for example, fibers from the cuneate tubercle of the medulla oblongata.

The newly evolved cerebellar hemispheres of mammals have extensive interconnections with other parts of the brain, but particularly with the newly evolved isocortex. "Carbon copies," so to speak, of many motor impulses initiated in the motor part of the isocortex go to pontine nuclei in the pons of the metencephalon, relay, decussate, and ascend to the cerebellar cortex. Some efferent fibers from the cerebellum return impulses to the isocortex *via* a relay in the dorsal thalamus. Other efferent fibers from the cerebellum go to phylogenetically older motor centers in the reticular formation of the brainstem, both directly and *via* a relay in the red nucleus in the floor of the mesencephalon. Centers where relays occur may also receive input from the colliculi and other centers in the brain so that further modulation of the signals may occur.

The cerebellum, in summary, receives a great deal of information about the environmental landscape in which the animal is operating, the current orientation of the body in space, and the degree of contraction of the muscles of the body. It monitors motor directives that are initiated in other centers and may influence the activity of nearly all motor pathways by adjusting the extent, duration, and timing of muscle contraction. Beyond this, there is evidence that the cerebellum plays a role in motor learning, that is, the ability to repeat precise, carefully timed, and complex motor sequences such as those involved in shooting a basketball or playing a musical instrument.

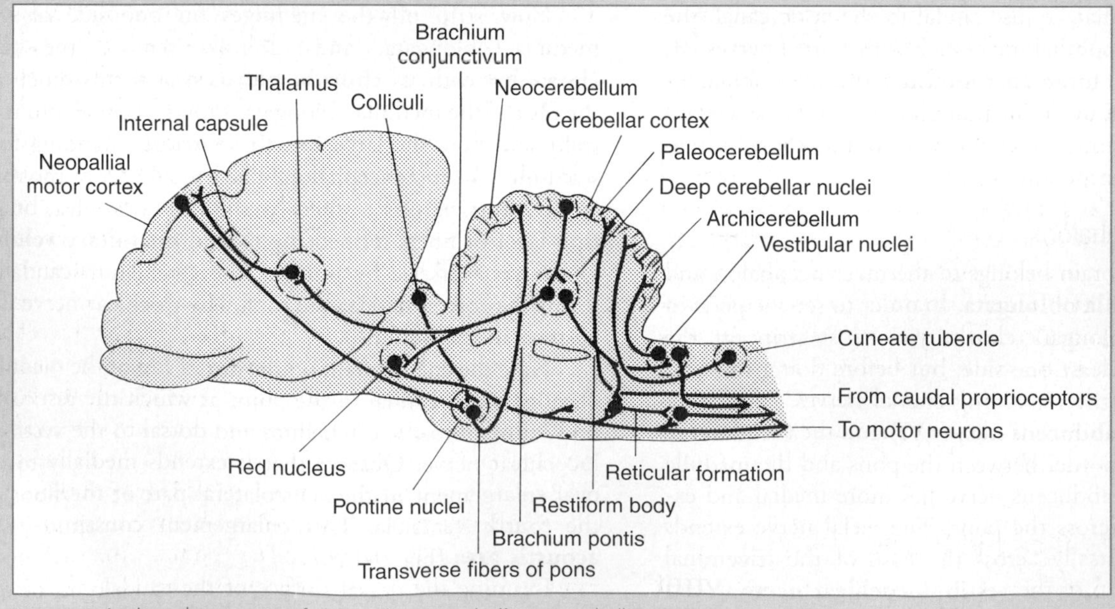

Diagram of a lateral projection of major afferent and efferent cerebellar connections in the sheep brain.

oblongata nearly to the fourth ventricle. A less distinct **dorsal intermediate sulcus** lies about 0.25 cm lateral to the dorsal median sulcus in a sheep brain, and a **dorsal lateral sulcus** lies slightly lateral to the dorsal intermediate sulcus. These grooves outline two longitudinal fiber tracts, a dorsal **fasciculus gracilis** and a more lateral **fasciculus cuneatus.** The rostral end of the former tract expands slightly to form a structure called the **gracile tubercle.** The fasciculus cuneatus has a comparable enlargement known as the **cuneate tubercle.** These tubercles are the outward manifestations of nuclei. Sensory fibers entering the spinal cord and carrying information about proprioception, pressure, and tactile discrimination, which is the ability to distinguish between one touch and two, terminate and are relayed to the dorsal thalamus in these nuclei rather than in the sensory columns of the spinal cord where they entered the spinal cord. As already stated, the restiform body forms the dorsolateral rim of the medulla oblongata rostral to these tubercles. It then turns dorsally to enter the cerebellum.

Turn the brain over and examine the ventral surface of the medulla oblongata. The narrow transverse band of fibers immediately caudal to the pons is the **trapezoid body,** an acoustic decussation (Fig. 9-13). Notice that its fibers can be followed dorsolaterally into the acoustic area. The midventral groove extending the length of the medulla oblongata is the **ventral median fissure.** The longitudinal bands of tissue on either side of it that are approximately 0.5 cm wide in the sheep brain are known as the **pyramids.** Note that some of the pyramidal fibers lie superficial to the trapezoid body.

The medulla oblongata is a transitional region between the rostral parts of the brain and the spinal cord. As in sharks (Anatomy in Action 9-1), in mammals it contains part of the reticular formation and the sensory and motor nuclei of cranial nerves that attach in this region. Numerous reflex activities occur in this region, and many important visceral activities are controlled here, including respiratory movements, salivation, swallowing, rate of heartbeat, and blood pressure. Most of the nuclei cannot be seen grossly, but some form bulges on the surface. The acoustic area contains the vestibular and cochlear nuclei, where most of the sensory neurons in the vestibulocochlear nerve terminate and relay. Some of the relays extend vestibular impulses to the cerebellum, and some extend auditory impulses through the trapezoid body. Much of the white matter of the medulla represents fiber tracts passing through the region.

Sagittal Section of the Brain

Cut the brain in half as close as possible to the sagittal plane. Be especially careful to cut through the median plane of the pineal gland. If you deviate from the plane, take the larger half and dissect away enough tissue to be able to see clearly the median cavities of the brain (Fig. 9-16). Many of the features described for the entire brain can be seen in this view. Note in particular the way in which the cerebral hemispheres extend caudally over the diencephalon and mesencephalon. The **corpus callosum** of the cerebrum shows particularly well. Its expanded rostral end is known as its **genu;** its expanded caudal end is its **splenium.** The thinner region between is called the **trunk of the corpus callosum.** A pair of thin vertical septa of gray matter lies ventral to the rostral part of the corpus callosum. Although each officially is called the **septum pellucidum** in mammals, each is homologous to the **septum** of nonmammalian amniotes (Fig. 9-10) and often goes by this name.

The lateral ventricle within a cerebral hemisphere lies lateral to the septum. You can observe this ventricle by breaking the septum. A band of fibers called the **fornix** lies within the diencephalon caudal to the septum pellucidum. The **body of the fornix** begins near the splenium, and then the band curves rostrally and ventrally as the **column of the fornix.** It passes out of the plane of the section caudal to a small round bundle of fibers that is a cross section of the **anterior commissure,** an olfactory decussation. The thin ridge of tissue that extends ventrally from the anterior commissure to the optic chiasm is the **lamina terminalis**—a landmark representing the rostral end of the embryonic neural tube. The cerebral hemispheres are evaginations that extend laterally and rostrally from the rostral end of the neural tube. The third ventricle and diencephalon lie caudal to the column of the fornix, anterior commissure, and lamina terminalis. Note that the **third ventricle** is very narrow but has a considerable dorsoventral and rostrocaudal extent. It is lined by a shiny epithelial membrane called the **ependymal epithelium,** as are all the cavities within the neural tube. The thalamus lies lateral to the third ventricle, but a portion of it, the **interthalamic adhesion,** extends across the third ventricle and appears as a dull circular area not covered by the shiny ependymal epithelium. The **interventricular foramen** (foramen of Monro), through which each lateral ventricle communicates with the third ventricle, lies in the depression rostral to the interthalamic adhesion. The hypothalamus lies ventral to the third ventricle, and the epithalamus lies dorsal to it. Note again the **pineal gland,** the **habenular commissure** (which shows in cross section), and the **habenula.** These features show very well in the section. The epithalamus also includes a **posterior commissure.**

Carefully dissect away tissue between the anterior commissure and the mamillary bodies. You will soon find a band of fibers, the **postcommissural fornix,** which is continuous with the column of the fornix and leads to the mamillary bodies (Fig. 9-17). This is one of the main connections between the cerebrum and hypothalamus. Dissecting just rostral to the anterior commissure, you will find a thin layer of fibers leading from the fornix into the area of gray matter just under the septum pellucidum. This is the **septal region,** and these fibers rostral to the anterior commissure constitute the **precommissural fornix.**

A narrow **cerebral aqueduct,** or aqueduct of Sylvius, leads through the mesencephalon to the fourth ventricle of the hindbrain (Fig. 9-16). The paired superior and inferior colliculi show nicely on the dorsal surface of the mesencephalon.

The cerebellum lies dorsal to the fourth ventricle. Note that most of its **gray matter** is in the form of a gray

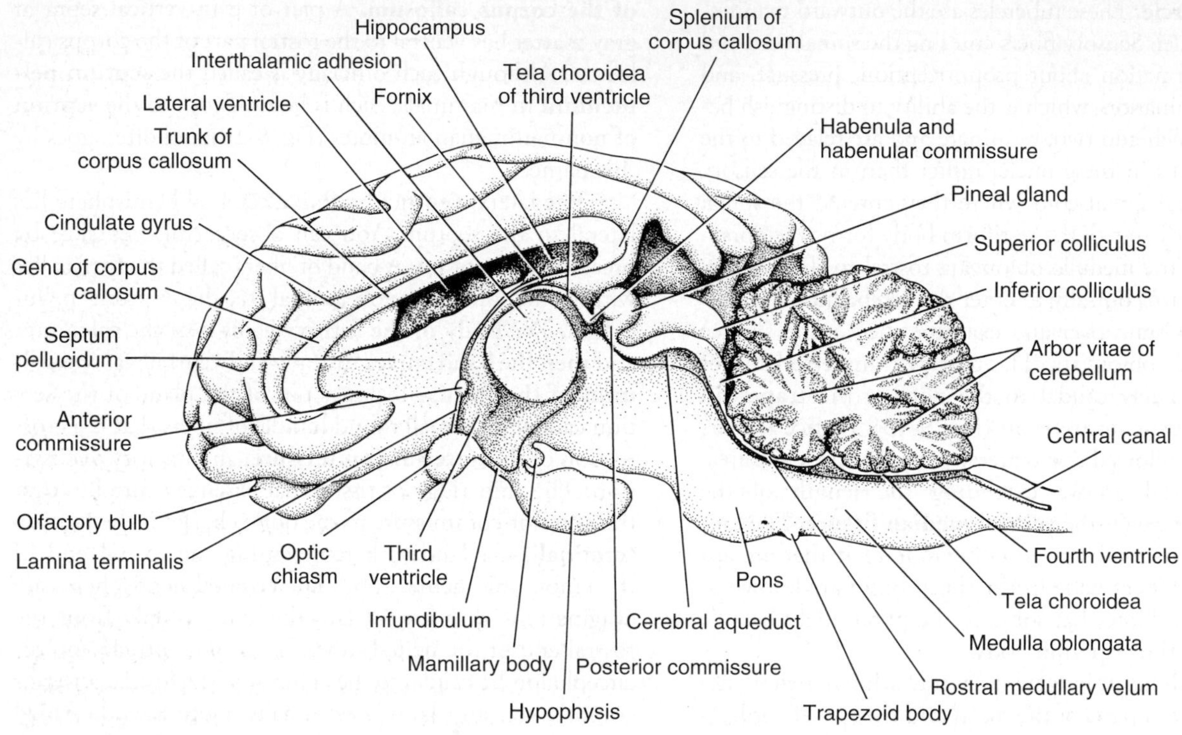

Figure 9-16
Sagittal section of the sheep brain.

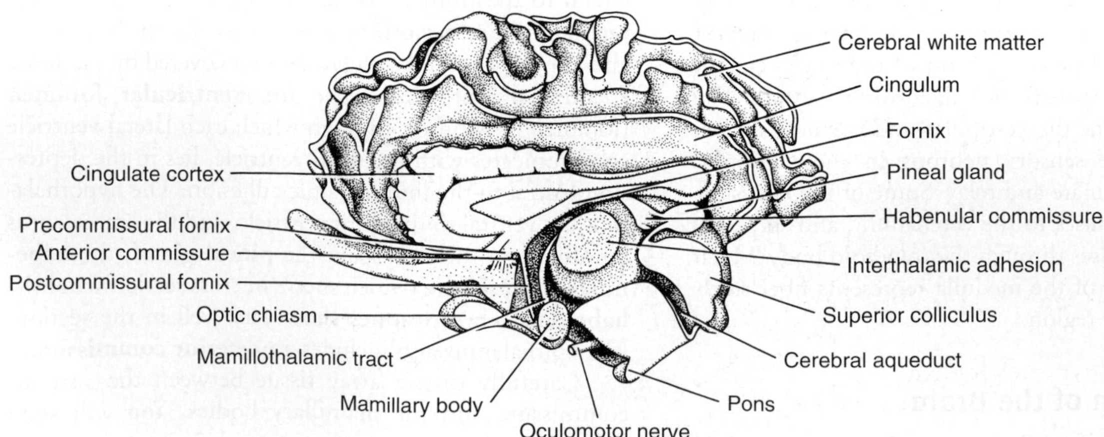

Figure 9-17
Dissection of a sagittal section of the cerebrum and diencephalon of a sheep brain.

cortex over the surface of the **folia,** whereas the **white matter** is centrally located. The white matter, which represents fiber tracts extending between the cerebellum and other parts of the brain, has the appearance of a tree and is called the **arbor vitae,** or tree of life.

Dissection of the Cerebrum

The Cerebral Cortex

Most of the gray matter of the mammalian cerebrum forms a **gray cortex** on the surface. As we have pointed out, all of the cortex lying dorsal to the rhinal sulcus belongs to the isocortex, and it is the primary nonolfactory integrating area of the mammalian brain. You can gain an appreciation of the complexity of the cerebrum by dissecting one half of your sheep brain specimen, preferably the one in which the fornix has been dissected, in one of two ways. (1) Using a blunt instrument, such as an orangewood manicure stick or the handle of a pair of forceps, scrape deeply into each sulcus and shuck out the cortex from the deeper white matter. The cortex is soft and scrapes out easily, whereas the underlying white fibers are quite firm. If time permits, do this over all of the isocortex except for the insular cortex, which is the area lying dorsal to the rhinal sulcus about the level of the optic chiasm. (2) If less time is available, then you can see the cortex and white fibers by slicing off the top 1 or 2 cm of the cerebrum in the plane of the corpus callosum.

The isocortex consists microscopically of six layers of interconnected neuron cell bodies, whereas the phylogenetically older olfactory cortex, located in the piriform lobe, consists of only three layers. The white matter that is exposed is a tangle of three types of fibers that cannot be distinguished macroscopically. (1) Some **association fibers,** whose cell bodies lie in the cortex, interconnect adjacent gyri; others connect more distant parts of the same cerebral hemisphere. (2) **Commissural fibers,** whose cell bodies also lie in the cortex, interconnect the two cerebral hemispheres and form pathways for the exchange of information. Those extending between the isocortex on each side form the corpus callosum. (3) The remaining fibers are **projection fibers** that interconnect the cerebrum with other parts of the brain or spinal cord. Projection fibers are named according to the location of their cell bodies and the destination of their axons. **Thalamocortical fibers** ascend from the thalamus; **corticobulbar** and **corticospinal fibers** extend from the cortex to the brainstem and to the spinal cord, respectively.

The Internal Capsule and Striatum

A group of nuclei interspersed with fibers is situated deeply in the base of the cerebrum and constitutes the **striatum** (Fig. 9-10). Carefully scrape away the insular cortex lying dorsal to the central portion of the rhinal sulcus (Fig. 9-18) and you will come upon a thin layer of white fibers, the **external capsule.** Expose about 2 cm of this capsule, then carefully peel it off to expose a large mass of gray matter beneath. This is the **lentiform nucleus.** Neuroanatomists divide this nucleus into a lateral **putamen** and a medial **globus pallidus,** but they cannot be distinguished grossly in the sheep brain.

Cut off the top of the cerebral hemisphere in the plane of the corpus callosum at the level of the dorsal surface of the corpus callosum. In doing this, you will probably cut through part of the roof of the **lateral ventricle** (Fig. 9-19). Continue to cut away the roof and lateral wall of this ventricle, but leave a strip of corpus callosum near the sagittal plane. Part of the ventricle extends laterally and ventrally into the piriform lobe just caudal to the

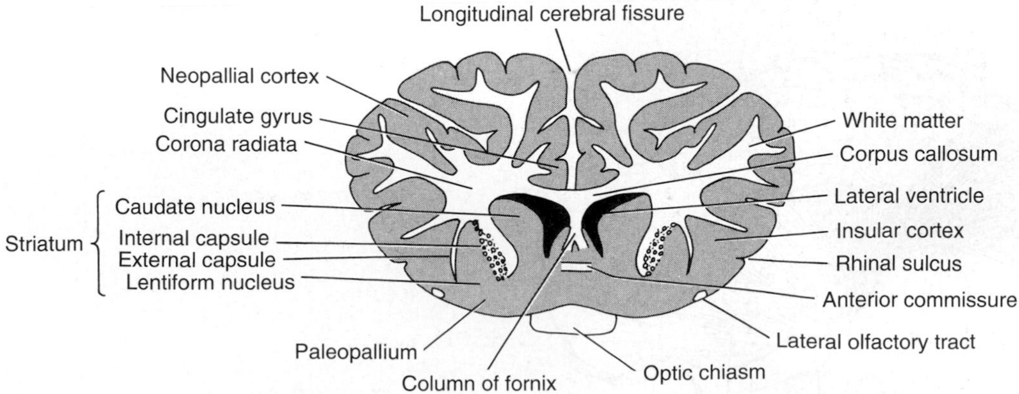

Figure 9-18

Transverse section through the cerebrum at the level of the optic chiasm of the sheep brain. White and gray matter are differentiated in this drawing.

region you have been dissecting. Cut away brain tissue to expose it. You can see the dark **choroid plexus** of the lateral ventricle extending inward from the thin medioventral wall of the cerebrum. Remove it. Another major nucleus of the striatum lies in the floor of the rostral part of the lateral ventricle. It is called the **caudate nucleus** because of its thick rostral end and tapering tail that extends caudally and ventrally (Figs. 9-18 and 9-19).

The fibers that form the white matter between the lentiform and caudate nuclei constitute the **internal capsule** (Fig. 9-18). You can expose the internal capsule by carefully scraping away the lentiform nucleus until you reach it (Fig. 9-20). As these fibers pass beyond the striatum, they fan out to form a **corona radiata** that extends to various parts of the isocortex. This radiation probably has been cut away. The internal capsule is a critical part of the brain because an important motor pathway, known as the **pyramidal system** (see below), and all sensory projection fibers that ascend from dorsal thalamic nuclei to the isocortex pass through it. A blood clot or other injury in this region can cause extensive damage. You now have seen the major parts of the brain by which signals from cutaneous receptors reach the brain (Anatomy in Action 9-4).

The nuclei of the striatum are a major motor center in the brains of nonmammalian tetrapods, and they continue to be important for much motor activity in mammals (Anatomy in Action 9-5). Mammals also have evolved a new, more direct, and faster motor pathway to motor centers. This pathway, the **corticobulbar** and **corticospinal tracts,** or **pyramidal system,** is used primarily for voluntary actions. The pathway begins with pyramid-shaped cell

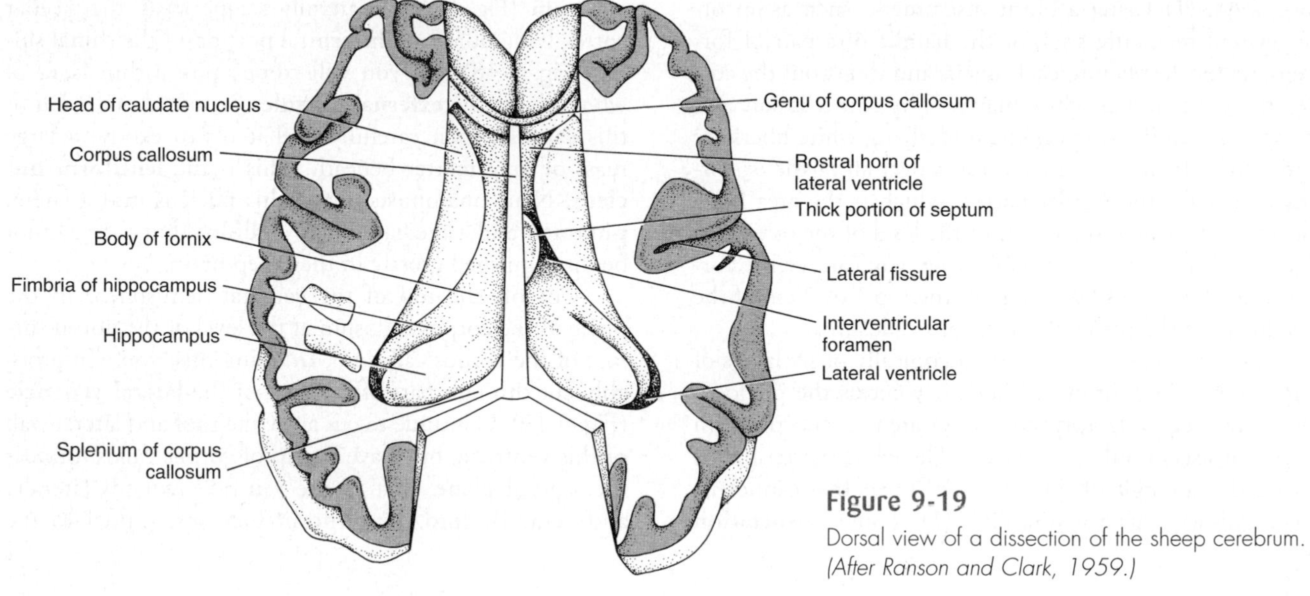

Figure 9-19
Dorsal view of a dissection of the sheep cerebrum. *(After Ranson and Clark, 1959.)*

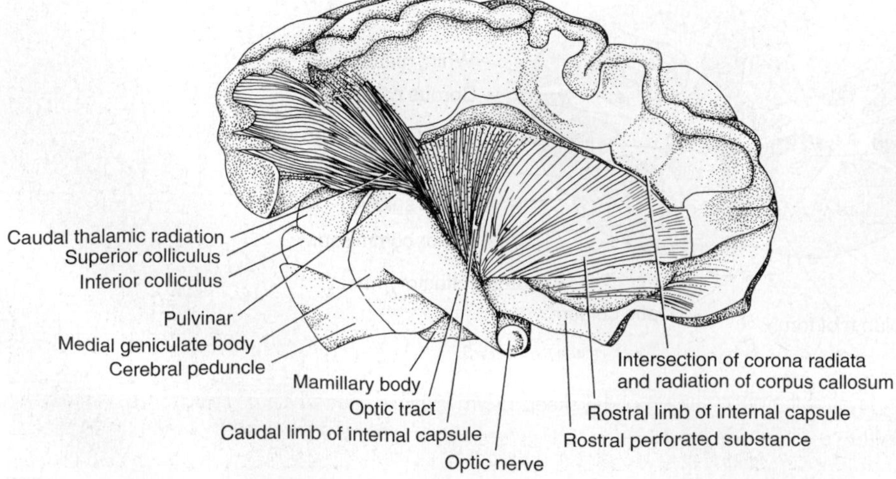

Figure 9-20
Lateral view of a dissection of the sheep cerebrum showing the internal capsule. The lentiform nucleus has been removed. *(After Ranson and Clark, 1959.)*

Anatomy in Action 9-4 The Somatosensory Pathway of Mammals

Although mammals retain some of the phylogenetic older sensory pathways through the reticular formation, they also evolved more direct sensory pathways to higher brain centers. Cutaneous impulses such as tactile, thermal, and pain (collectively called **somatosensory impulses**) from all parts of the body ascend to the somatosensory area of the isocortex in pathways that are very similar (figure). On entering the dorsal horn of the spinal cord or dorsal nuclei in the brainstem, the primary sensory neuron in the spinal or cranial nerve synapses with an interneuron that at some point decussates and continues on to nuclei in the lemnothalamus (a part of the dorsal thalamus). Some processing of sensory information occurs here. Extraneous information ("noise") may be filtered out and the meaningful signal enhanced. Neurons beginning in the lemnothalamus then continue through the internal capsule and corona radiata to the somatosensory isocortex. The body is essentially projected upside down on the somatosensory cortex. The amount of cortex devoted to any body region is not proportional to the surface area in this body region but to the number and density of receptors in the region. Thus, the human face occupies as much cortical area as the entire trunk.

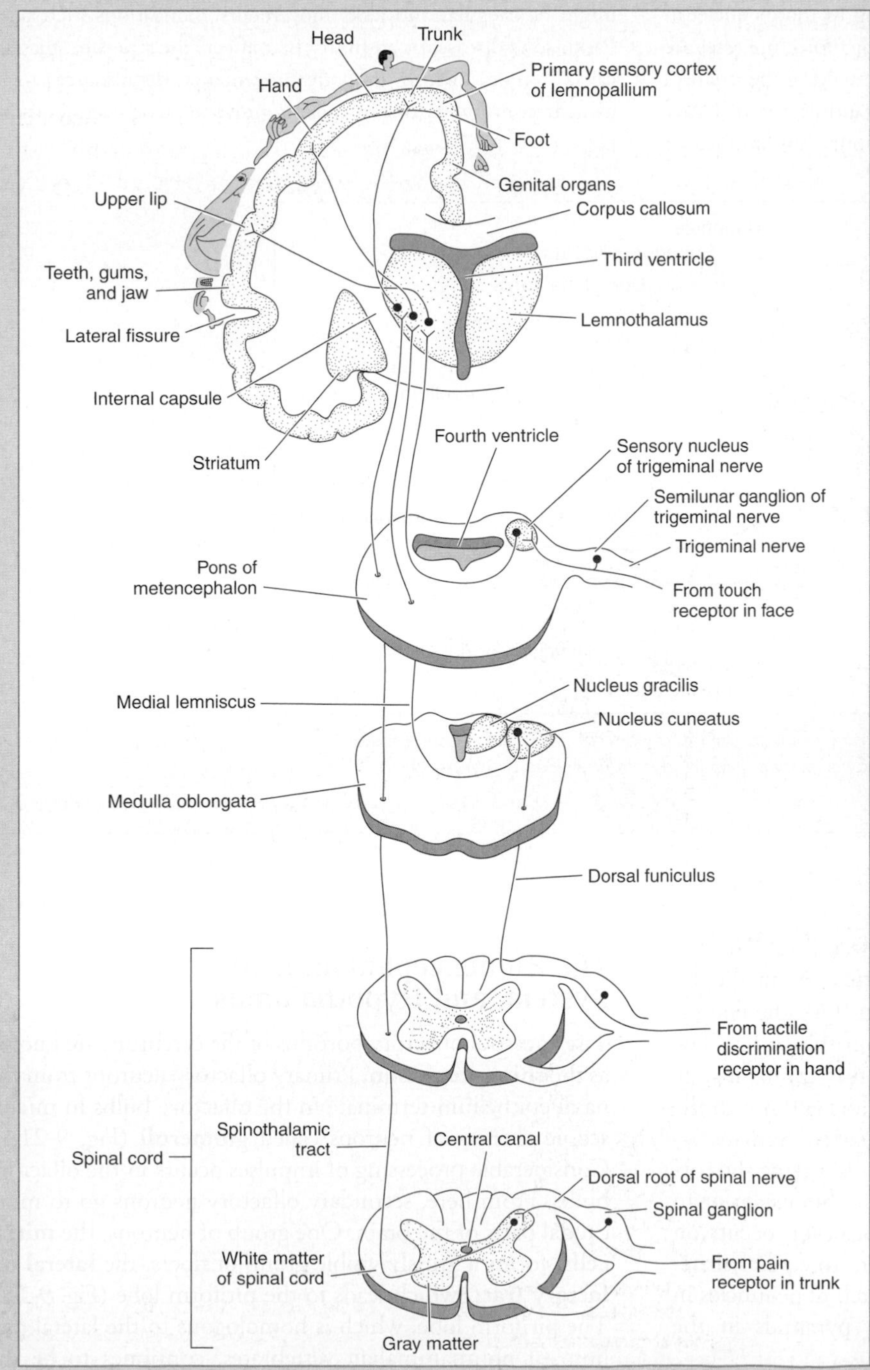

Diagram of major cutaneous sensory pathways in a mammal as seen in transverse sections of parts of the spinal cord and brain. The extent of the cortical area receiving impulses in a human is indicated by the relative size of the cartoon beside the cortex. *(From Liem, Walker, Bemis, and Grande, 2001.)*

Anatomy in Action 9-5 Motor Functions of the Mammalian Striatum

The most conspicuous motor pathways in mammals are those that go directly from the motor cortex to motor nuclei in the brainstem (corticobulbar tract) or to motor nuclei in the spinal cord (corticospinal tract). Parts of the latter are shown in the figure. In addition, many motor impulses, especially involuntary impulses, are initiated in the motor part of the isocortex and travel to the striatum. Before continuing to motor nuclei in the brainstem and reticular formation, the motor impulses are subject to many interactions among the nuclei of the striatum and many other parts of the brain, including the olfactory cortex and limbic system, dorsal thalamus, subthalamus or ventral thalamus, nuclei in the substantia nigra, and tegmental nuclei in the floor of the mesencephalon (figure). Many different types of neurotransmitters are released in the synapses in these nuclei, especially in the substantial nigra, that allow for many excitatory and inhibitory reactions as the striatum processes motor signals. In general, the striatum makes corrections that tend to smooth out what otherwise might be irregular and jerky movements. Conditions such as Parkinson's disease, or an injury that affects the striatum, may cause tremors in limb or body movements, disturbances in gait, or continuous uncontrollable movement of some body part.

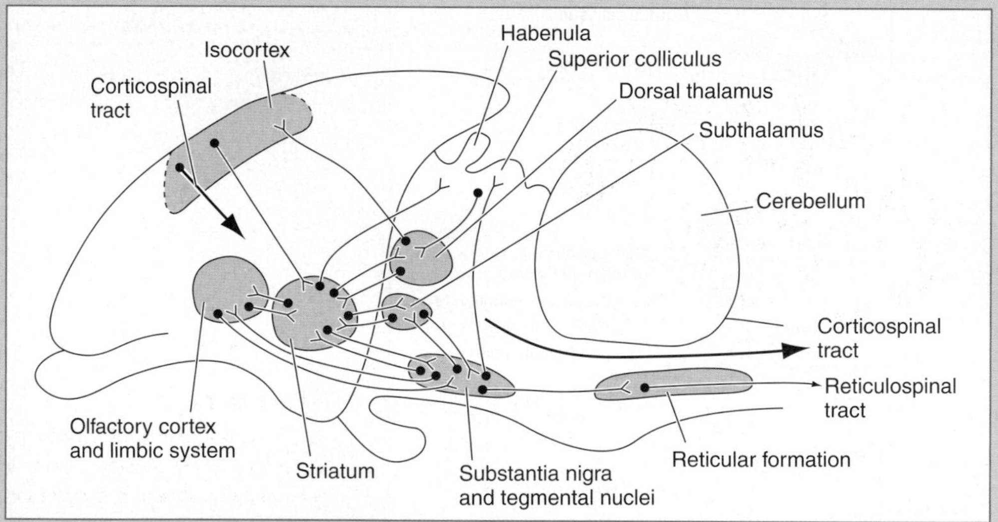

Diagram in lateral view of the major connections of the striatum in a mammal. Parts of the corticospinal tract, or pyramidal system, are shown by heavier arrows. *(Modified from Liem, Bemis, Walker, and Grande, 2001.)*

bodies in the motor part of the isocortex. As in the somatosensory cortex (Anatomy in Action 9-4), the body is projected approximately upside down on the cortex. The amount of cortex devoted to each region is approximately proportional to the number of motor units and not to the size of the area. Thus, the thumb and fingers of humans occupy a disproportionately large area. Fibers leaving the cortex go directly to the motor nuclei in the brainstem or to the motor columns of the cord. A decussation occurs on the way. These fibers form or contribute to several structures you have seen: internal capsule, cerebral peduncles in the floor of the mesencephalon, and pyramids in the medulla oblongata.

The Rhinencephalon, Limbic System, and Hypothalamus

The ancestral olfactory portions of the cerebrum are known as the **rhinencephalon.** Primary olfactory neurons from the nasal epithelium terminate in the olfactory bulbs in microscopic clusters of neurons called **glomeruli** (Fig. 9-21**A**). Considerable processing of impulses occurs in the olfactory bulbs. From here, secondary olfactory neurons go to more caudal parts of the brain. One group of neurons, the **mitral cells,** forms a grossly visible group of fibers, the **lateral olfactory tract,** which leads to the piriform lobe (Fig. 9-13). The piriform lobe, which is homologous to the lateral pallium of nonmammalian vertebrates, continues to be the

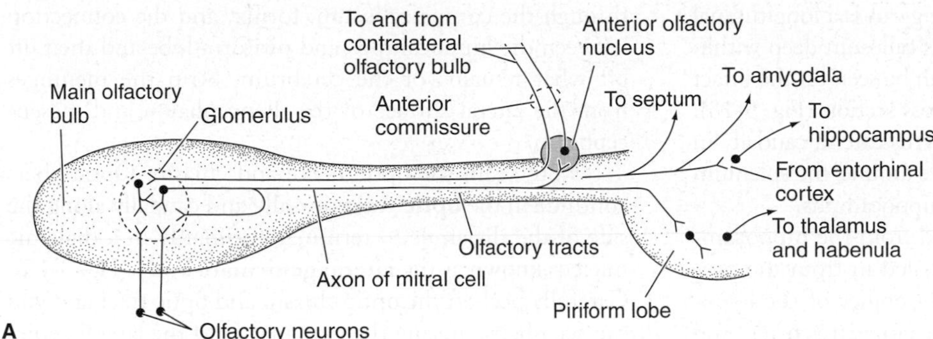

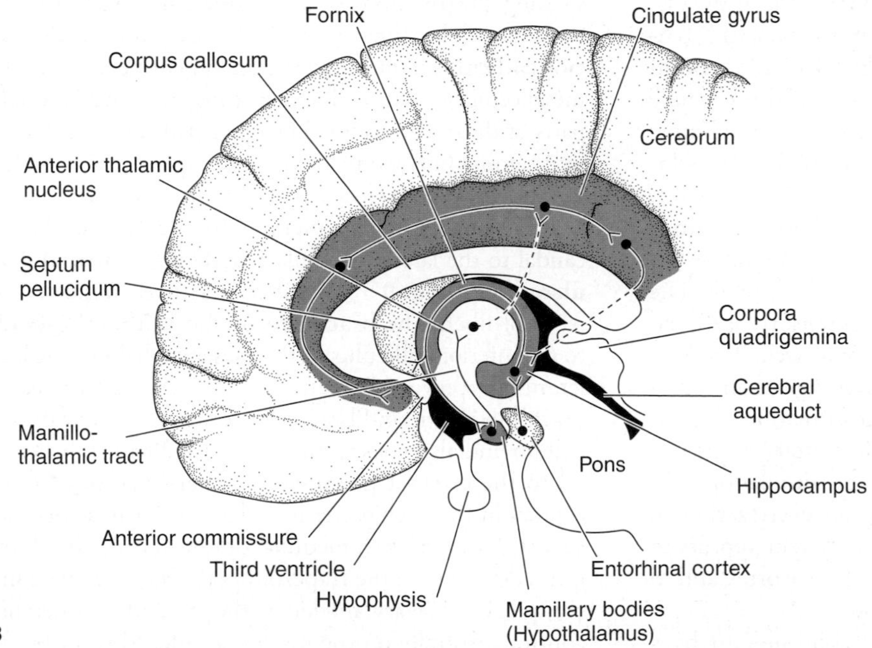

Figure 9-21
A, Diagram in lateral view of the major projections of the olfactory bulb to the cerebrum in a sheep. B, Diagram in lateral projection of the major parts of the limbic system of the sheep brain.

primary olfactory integration region. The piriform lobe projects to the dorsal thalamus, habenula, and limbic system (see below). Another group of neurons, which form the **medial olfactory tract,** extends through the anterior commissure to the contralateral olfactory bulb (Fig. 9-21**A**). This tract, most of which is deeply situated, can be followed most easily by dissecting from the anterior commissure toward the olfactory bulb. Other tracts that are not visible grossly lead from the olfactory bulbs to other parts of the brain, including the hippocampus and amygdala, a part of the limbic system described below.

The **hippocampus** represents the ancestral medial pallium, which has been rolled inward and medially (Fig. 9-10) and now bulges into the caudal part of the floor of the lateral ventricle where you can see it (Fig. 9-19). Nick the hippocampus with a scalpel in order to verify that it contains gray matter. Neurons that have their cell bodies in the hippocampus emerge on its surface to form the thin covering of white matter—the **alveus**—that you just nicked, and flow together to form the free rostral border of the hippocampus, which is called the **fimbria of the hippocampus.** These same neurons continue into the body of the fornix, which you saw in the sagittal section that you dissected (Fig. 9-16), turn ventrally and caudally as the columns of the fornix and postcommissural fornix, and terminate in the mamillary bodies of the hypothalamus (Fig. 9-21**B**).

Neurons whose cell bodies lie in the mamillary bodies extend dorsally as the **mamillothalamic tract** to terminate in the anterodorsal part of the thalamus. Expose this tract in the sagittal section that you have partly dissected by carefully scraping away thalamic tissue caudal to the columns of the fornix (Fig. 9-17). Another group of neurons goes from the anterodorsal part of the thalamus through the internal capsule and corona radiata to the **cingulate gyrus** of the isocortex (Fig. 9-21**B**). The cingulate gyrus is a longitudinal fold that lies just dorsal to the corpus callosum deep within the longitudinal cerebral fissure. It can be seen on the intact sagittal section (Fig. 9-16) and in cross section (Fig. 9-18). Fibers that initiate in the cingulate gyrus extend caudally in the white matter beneath it and curve around the splenium of the corpus callosum to enter the hippocampus.

The feedback loop you followed from the hippocampus—where olfactory information is fed in from the piriform lobe—to the fornix, mamillary bodies of the hypothalamus, thalamus, internal capsule, cingulate cortex, and back to the hippocampus constitutes most of the limbic system. It is also called the **Papez circuit** ("Papez" rhymes with "tapes"). This circuit (Fig. 9-21) interconnects the piriform lobe and hippocampus (medial pallium), and it represents a primary way in which olfactory information reaches the thalamus, hypothalamus, and isocortex, where other connections can be made. Other parts of the limbic system include the **amygdala,** a deep cerebral nucleus located lateral to the optic chiasm, which you will not see, and the septum. Olfactory input into the limbic system is of prime importance in nonmammalian vertebrates, but additional sensory information reaches it in mammals. The hippocampus receives gustatory, visceral sensory, auditory, visual, and somatic sensory information as well. The limbic system acts, in part at least, by inhibiting certain activities of the hypothalamus and reticular formation. Electrical stimulation of parts of the limbic system suppresses behavior that is occurring at the time of stimulation, and destruction of limbic centers leads to an overreaction to various stimuli. Beyond this, the limbic system appears to be involved in arousal or activation of the isocortex and in the formation of short-term memories.

The hypothalamus of mammals continues to be a very important visceral center. Its major afferent impulses come from the limbic system, other olfactory centers, and gustatory and other visceral centers. Its neural output is to the motor nuclei in the brainstem and to the reticular formation, but it also has an endocrine output that affects the activity of the hypophysis, as we have seen (Anatomy in Action 9-2). The hypothalamus helps regulate the level of body activity (sleep and wakefulness), water balance, blood sugar levels, body temperature, and other aspects of homeostasis. It interacts with the limbic system to affect drinking, feeding, fighting, reproduction, and other emotional and motivational behaviors that relate to self and species preservation.

The Surfaces of the Diencephalon and Mesencephalon

Remove the cerebrum from one half of the brain to expose the surfaces of the diencephalon and mesencephalon. If you dissected the cerebrum, then all you need do is to cut through the corpus callosum, fornix, and the connection between the hippocampus and piriform lobe and then lift off what remains of the cerebrum. Strip the meninges from the lateral surface of the diencephalon and mesencephalon.

Note again the optic nerve and chiasm. Optic fibers continue in the **optic tract** dorsally and caudally along the side of the thalamus to terminate primarily in a thalamic nucleus known as the **lateral geniculate body** (Fig. 9-15). Carefully peel off the optic chiasm and optic tract and you can see fibers entering the gray matter of the lateral geniculate body. Continue to peel off the optic tract and notice that some of the more superficial fibers terminate in a swelling, the **pulvinar,** located on the dorsolateral part of the thalamus. Other fibers travel as the **brachium of the superior colliculus** to terminate in the superior colliculus of the mesencephalic tectum. The structures you have exposed are parts of the optic system. The optic pathways are discussed in Anatomy in Action 9-6 as an example of the complexity of special somatic sensory pathways in mammals.

The smaller enlargement of the side of the thalamus caudal to the lateral geniculate body is the **medial geniculate body** (Fig. 9-15). Notice that it is connected by a band of fibers, the **brachium of the inferior colliculus,** to the inferior colliculus of the mesencephalon. The lateral lemniscus, previously seen (p. 233), emerges between the brachium pontis and brachium conjunctivum of the cerebellum and also goes to the inferior colliculus. It lies dorsal to the cerebral peduncles. Primary neurons from receptors in the cochlea terminate in cochlear nuclei in the acoustic area of the medulla oblongata. Most of these fibers decussate in the trapezoid body and terminate in inconspicuous olivary nuclei in the medulla. Other fibers continue rostrally on the ipsilateral side. Most fibers from the olivary nuclei extend through the lateral lemniscus to terminate in the inferior colliculus. Most fibers originating in the inferior colliculus continue *via* the brachium of the inferior colliculus to the medial geniculate body. From the medial geniculate body, acoustic fibers extend through the internal capsule to the **primary auditory cortex,** which is located in the temporal region. Although processing of auditory signals occurs at all levels, the synthesis and interpretation of auditory signals occur at the cortical level.

Some fibers beginning in the inferior colliculus project to the reticular formation and to motor nuclei. These mediate various acoustic reflexes, including those that enable a mammal to position its head so that it can localize the source of sound.

You now have seen how specific dorsal thalamic nuclei relay olfactory, optic, and acoustic impulses. In a similar way, other, more deeply situated dorsal thalamic nuclei relay other sensory modalities to the cerebral cortex (Anatomy in Action 9-4). However, the thalamus is more than a simple relay station. It acts as a subcortical integrating

Anatomy in Action 9-6 The Optic Pathways of Mammals

Because of the partial decussation in the optic chiasm of optic fibers from the ganglion cells of the two eyes, some fibers from each eye go to each side of the brain (figure). Each side of the brain receives an image from the binocular field of vision but from slightly different angles. Interpretation of these differences within the cerebral cortex results in stereoscopic vision and depth perception.

In most vertebrates, the optic fibers extend through the optic tract to the optic lobes, which are the homologues of the mammalian superior colliculi, where an interneuron then carries the information to the thalamus. However, in mammals most of the fibers in the **optic tract** bypass the superior colliculi and go directly to thalamic nuclei in the **lateral geniculate body.** Most fibers project from here through a part of the internal capsule known as the **optic radiation** to the **primary visual cortex** in the occipital pole of the isocortex. The processing of visual signals occurs at all levels from (1) the retina to (2) the primary visual cortex and on to (3) the extrastriate cortex to which the visual cortex projects. The extrastriate cortex is essential for many interpretations of images, such as locating their position relative to other objects.

Some optic fibers follow the ancestral route to the superior colliculi. From here some relay fibers go to thalamic nuclei from which they too are projected by other neurons to the extrastriate cortex. Fibers beginning in the superior colliculi also project to motor nuclei in the brainstem and mediate autonomic reflexes that control pupillary size, accommodation, and coordination of the movements of the eyeballs.

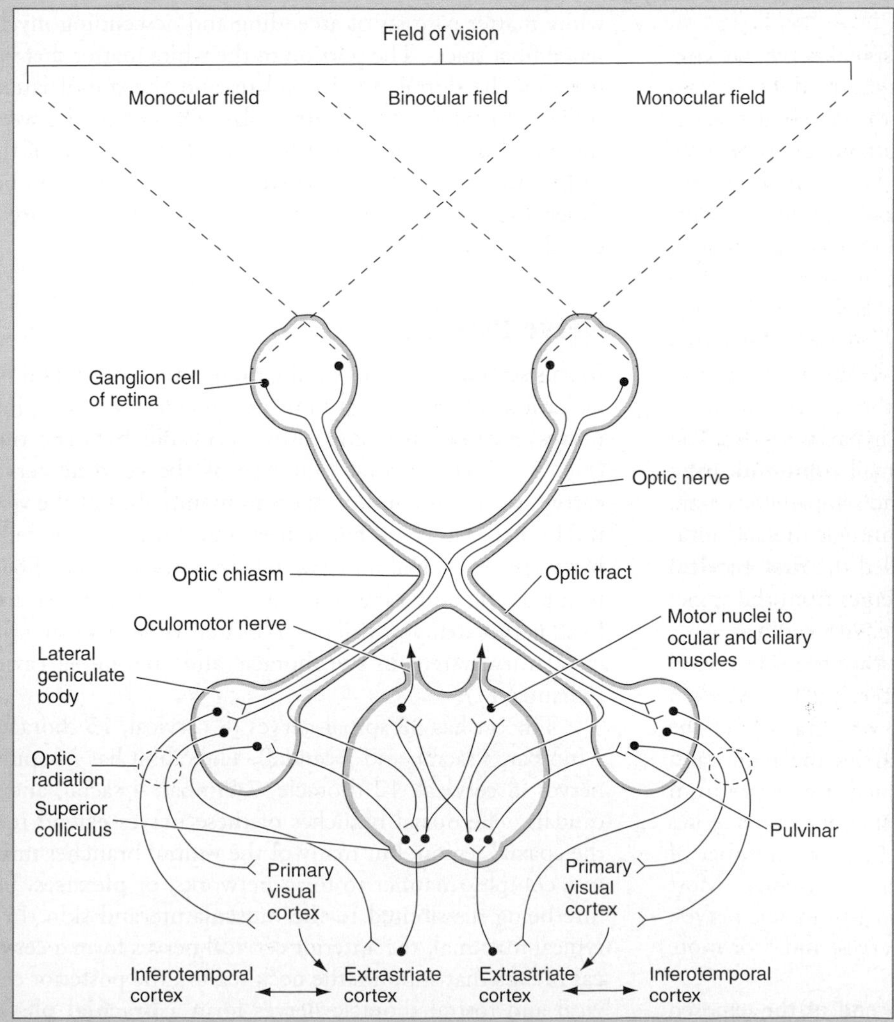

Diagram of the major parts of the optic system. *(From Liem, Bemis, Walker, and Grande, 2001.)*

center, and its numerous interconnections with higher cortical centers implicate it in many cortical functions.

The Spinal Cord and Spinal Nerves

The **spinal cord** *(medulla spinalis)* is a nearly cylindrical cord that lies within the vertebral canal of the vertebral column. It is not uniform in diameter. It bears **cervical** and **lumbosacral enlargements** from which the nerves to the appendages arise, and the caudal end tapers as a fine **terminal filament** to end in the base of the tail. You need to study only a short portion of the spinal cord. To approach it, remove the epaxial muscles from the caudal part of the thoracic region of your specimen so as to expose several centimeters of the vertebral column. With bone scissors, carefully cut across the pedicles of the vertebrae and remove the tops of the neural arches. The spinal cord will be seen lying in the vertebral canal. Continue chipping away bone and removing fat from around the spinal cord until you have satisfactorily exposed it, along with several roots of the spinal nerves.

The spinal cord and roots of the spinal nerves are covered by the tough **dura mater.** Note that the dura mater is not fused with the periosteum lining the vertebral canal as it is in the cranial cavity. Leave the dura mater on for now and examine the roots of the spinal nerves. At each segmental interval, there is a pair of dorsal and ventral roots (Fig. 9-1). Trace a dorsal and a ventral root laterally on one side. The roots pass into the intervertebral foramen before uniting to form a **spinal nerve.** Just before uniting, the dorsal root bears a small round enlargement—the **spinal ganglion.** If you trace the spinal nerve laterally, you may see it divide into a **dorsal branch** to the epaxial regions of the body and a **ventral branch** to the hypaxial regions. You probably will not be able to see the small **communicating branches** to the sympathetic ganglia and sympathetic trunk.

The first spinal nerve emerges through the alar foramen of the atlas (Fig. 5-7) and is called the **first cervical nerve.** The second cervical nerve emerges from the spinal cord just caudal to the atlas and the eighth cervical nerve leaves the spinal cord just caudal to the seventh cervical vertebra. Therefore, there are eight cervical nerves, even though there are only seven cervical vertebrae. After the cervical region, each spinal nerve carries the name and number of the vertebra caudal to which it emerges from the spinal cord. For example, the first thoracic nerve lies just caudal to the first thoracic vertebra. The number of spinal nerves varies with the number of vertebrae. Most mammals have 8 cervical nerves, 12 to 14 thoracic nerves, 6 to 7 lumbar nerves, 3 to 4 sacral nerves, and 5 or more caudal nerves.

Slit open the dura mater at one end of the exposed area. This opens the **subdural space.** Observe that the roots of the spinal nerves do not have a simple attachment on the spinal cord but unite by a spray of fine **rootlets.** Cut out a segment of the spinal cord and strip off the remaining meninges—**arachnoid** and **pia mater.** The spinal cord has a deep ventral furrow known as the **ventral median fissure,** a less distinct dorsal furrow called the **dorsal median sulcus,** and a more prominent furrow slightly lateral to the middorsal line, the **dorsal lateral sulcus.** Note that the dorsal rootlets enter along the dorsal lateral sulcus (see below). Recall that these same grooves extend onto the medulla oblongata (p. 233).

Make a fresh cross section of the cord with a sharp instrument such as a razor blade. Examine the cut surface and compare it with Figure 9-1. Also study demonstration slides if possible. The tiny **central canal** generally can be seen grossly. If you are fortunate, you may be able to distinguish the butterfly-shaped central gray matter from the peripheral **white matter.** As explained in the introduction to this chapter, the gray matter consists of unmyelinated fibers and the cell bodies of motor and interneurons; the white matter consists of ascending and descending myelinated fiber tracts. The portion of the white matter that lies between the dorsal median sulcus and the dorsal lateral sulcus is called the **dorsal funiculus.** The portion between the dorsal lateral sulcus and the line of attachment of the ventral roots is the **lateral funiculus,** and the portion between the ventral roots and the ventral median fissure is called the **ventral funiculus.**

Nerve Plexuses

In this section, we focus on the branching pattern and innervations of the ventral branches, or rami, of the spinal nerves because they are easily accessible by dissection (Fig. 9-1). The ventral branches of the cervical nerves carry only somatic motor fibers in mammals, but the ventral branches of the thoracic nerves and some of the lumbar and sacral spinal nerves also carry visceral motor fibers (sympathetic and parasympathetic fibers), which leave to form nerves that you will observe later while exploring the circulatory system in the thoracic and abdominal cavity (Chapter 11).

The cat has 38 spinal nerves (8 cervical, 13 thoracic, 7 lumbar, 3 sacral, and 7 caudal). The rabbit has 36 spinal nerves (8 cervical, 12 thoracic, 7 lumbar, 4 sacral, and 5 caudal). The dorsal branches of these nerves extend into the epaxial region, but many of the ventral branches unite in a complex manner to form networks, or **plexuses,** before being distributed to the musculature and skin. In a typical mammal, the anterior cervical nerves form a **cervical plexus** that supplies the neck region; the posterior cervical and rostral thoracic nerves form a **brachial plexus** that supplies the pectoral appendage; and the lumbar, sacral, and anterior caudal nerves form a **lumbosacral plexus** that supplies the pelvic appendage. Considerable variation occurs in the manner in which spinal nerve branches unite, branch, reunite, and branch again within the plexuses, but despite this complexity at the gross anatomical level, the targets of neurons from individual spinal segments remain very similar in different individuals. You may dissect either the brachial or lumbosacral plexus as examples of a plexus, or, if time permits, both.

The Brachial Plexus

The brachial plexus lies medial to the shoulder and rostral to the first rib; you should approach it from the ventral surface. If it is still intact on the side on which the muscles were dissected, then study it there (Table 7-5); otherwise, carefully cut through the pectoral muscle complex on the other side. Dissection of the brachial plexus involves meticulously picking away the fat and connective tissue from around the nerves and the accompanying blood vessels. If you find it necessary to cut any of the larger vessels, then do so in such a way that you will be able to appose the cut surfaces when you study the circulatory system. Clean off the nerves from a point as near as you can reach to the vertebral column to the point at which the nerves disappear into the shoulder and forelimb muscles.

The union of the ventral branches of the sixth to eighth cervical nerves and the first thoracic nerve forms the brachial plexus of the cat. A common pattern for the brachial plexus of the cat and rabbit is shown in Figure 9-22. The ventral branches of the nerves that enter the plexus are called the **roots of the plexus.** They emerge between the longus colli and scalenus muscles. Note that each root tends to split into two **divisions.** The divisions of different nerves unite to form **trunks** from which peripheral nerves arise. Also note that the splitting of roots into divisions, and the union of

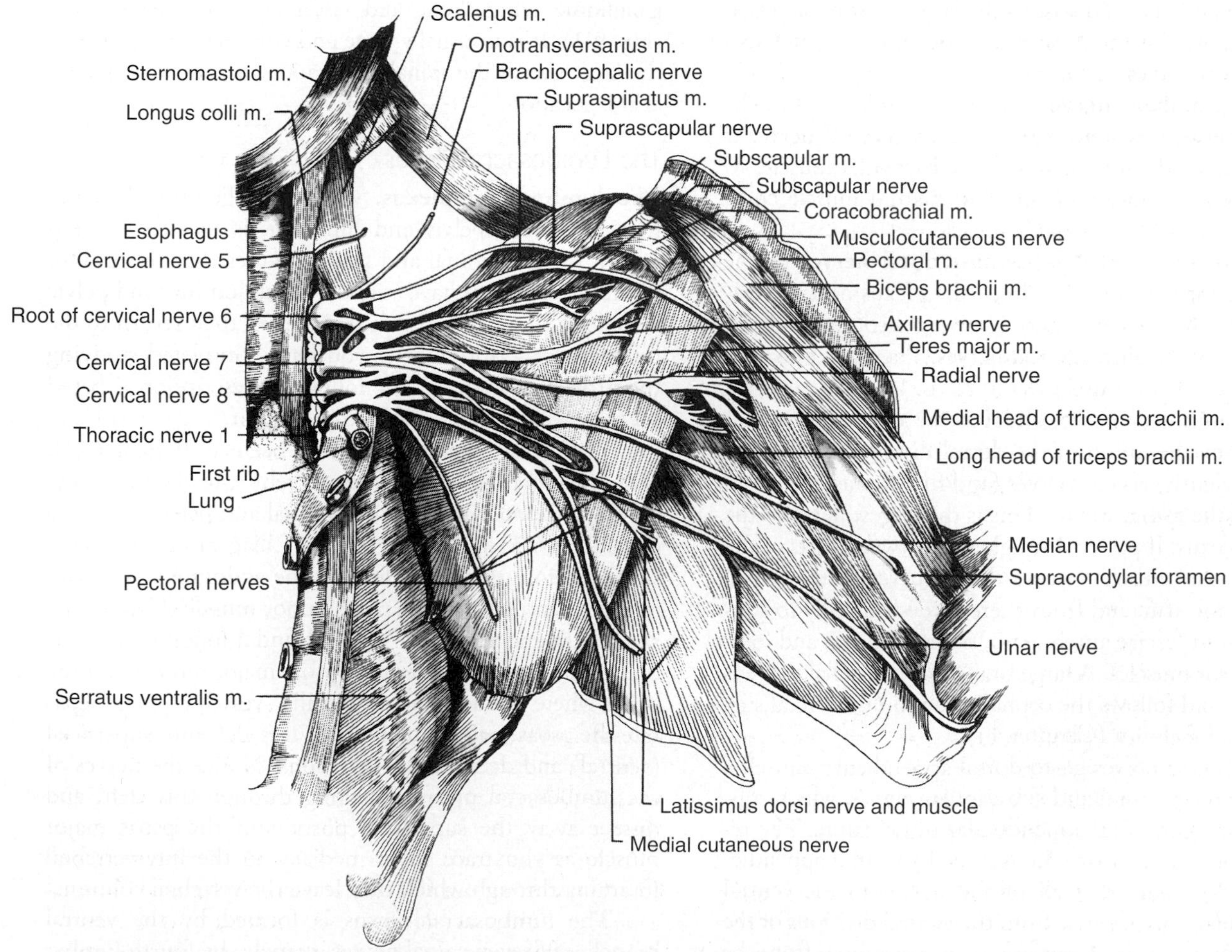

Figure 9-22
Ventral view of a dissection of the left brachial plexus of a cat.

divisions to form trunks, occurs in such a way that the nerves supplying the dorsal appendicular muscles tend to segregate early from those supplying the ventral appendicular muscles. Many of the nerves also carry cutaneous fibers that are distributed to the skin, but we will describe only the major cutaneous branches.

The most ventral nerves of the brachial plexus are several **pectoral nerves.** These small nerves arise from the ventral divisions of the plexus. They may unite with each other, and they pass to the pectoral muscle complex.

A large **suprascapular nerve** leaves the cranial end of the plexus, where it arises for the most part from the sixth cervical nerve. It passes between the subscapular and supraspinatus muscles to supply the supraspinatus and infraspinatus muscles, and some of the skin over the shoulder and brachium.

A small **brachiocephalic nerve** arises from the sixth cervical nerve and turns cranially to innervate the brachiocephalic muscle.

One or more **subscapular nerves** leave caudal to the suprascapular nerve and pass to the large subscapular muscle. They arise, for the most part, from the sixth and seventh cervical nerves.

A large **axillary nerve,** which lies caudal to the subscapular nerve, arises from the seventh cervical nerve. It passes between the teres major and subscapular muscles to supply the teres minor muscle and deltoid muscle complex and often the teres major muscle.

One or more small **latissimus dorsi nerves** to the latissimus dorsi muscle arise from the plexus near the origin of the axillary nerve and receive contributions from the seventh and eighth cervical nerves. They may also supply the teres major muscle.

The large, deep nerve caudal to the axillary nerve that is formed by the union of the dorsal divisions of the seventh and eighth cervical nerves (and first thoracic nerve in the cat) is the **radial nerve.** This is the largest nerve of the brachial plexus. It passes through and between parts of the triceps brachii muscle, crosses the humerus to the lateral surface of the arm, and from there extends distally to supply the tensor fasciae antebrachii, triceps brachii, and forearm extensor muscles. A large branch of the radial nerve is cutaneous and follows the cephalic vein on the lateral surface of the brachium (Chapter 11).

All of these nerves go to dorsal appendicular muscles, except for the pectoral and subscapular nerves, which supply parts of the ventral appendicular musculature. The remaining nerves innervate the rest of the ventral appendicular muscles. Note that all of the nerves to the ventral appendicular muscles arise from the ventral divisions of the plexus. A small **musculocutaneous nerve** springs from the ventral branch of the sixth and seventh cervical nerves and enters the biceps brachii muscle. It generally branches before reaching the biceps brachii muscle, and it supplies the biceps brachii, coracobrachialis, and brachialis muscles, and some of the skin over the forearm.

The ventral divisions of the seventh and eighth cervical nerves and the first thoracic nerve combine to form two prominent nerves that extend distally along the medial side of the brachium. The more cranial of these is the **median nerve;** the more caudal nerve is the **ulnar nerve.** They are distributed to the forearm flexor muscles and skin of the hand. In the cat, the median nerve passes through the supracondylar foramen of the humerus (Fig. 6-12**B**).

A small **medial cutaneous nerve,** which arises from the first thoracic nerve, runs parallel with and caudal to the ulnar nerve. It supplies some of the skin over the forearm.

If you have already dissected the internal organs and the circulatory system in the thoracic cavity, then find again the stellate ganglion (Chapter 11). You will discover a fine nerve strand connecting with the first thoracic nerve at the base of the stellate ganglion. This is a **white communicating branch,** or **white ramus communicans,** carrying preganglionic sympathetic and visceral sensory nerve fibers (Fig. 9-1). It is unusually large and conspicuous because of the large size of the spinal nerve and sympathetic ganglion at this location.

The Lumbosacral Plexus

The lumbosacral plexus, which supplies the skin and muscles of the pelvis and hind leg, is located so deep within the abdominal and pelvic cavities that you cannot dissect it until you have studied the abdominal and pelvic viscera. If you are to dissect this plexus, then return to the following description after you have completed studying the urogenital system. The following description is based on the cat but is applicable to the rabbit.

The pelvic symphysis will have been cut (Chapter 11). Push the two hind legs dorsally, thereby spreading open the pelvic canal, and push the abdominal and pelvic viscera to one side. Also bisect the external iliac artery and vein shortly before they pass through the abdominal wall, and reflect them. Identify the psoas minor muscle (Chapter 7), and bisect and reflect it. You will find a longitudinal cleft on the lateroventral part of the psoas major muscle near the point where the deep circumflex iliac vessels cross it. Separate the psoas major muscle along this cleft into superficial (ventral) and deep (dorsal) portions. Notice the nerves of the lumbosacral plexus emerging through this cleft, and dissect away the superficial portion of the psoas major muscle as you trace them medially to the intervertebral foramina through which they leave the vertebral column.

The lumbosacral plexus is formed by the ventral branches of seven spinal nerves, namely the fourth lumbar nerve to the third sacral nerve. As with the brachial plexus, some variation occurs in the way the seven roots of the plexus split to form divisions and in the way the divisions unite to form the trunks from which the peripheral nerves arise. Figure 9-23 shows a common pattern.

Lumbar nerve 4 splits soon after emerging from the intervertebral foramen into a **genitofemoral nerve** and a branch that passes caudad to join the divisions of the fifth lumbar nerve. The genitofemoral nerve continues caudad close to the external iliac vessels and passes through the body wall with the external pudendal vessels to supply the skin in the groin and on parts of the external genitalia. In a male, small branches also supply the cremasteric muscle.

One division of **lumbar nerve 5** unites with the caudal branch of the fourth lumbar nerve to form the **lateral cutaneous femoral nerve,** which extends laterally near the deep circumflex iliac vessels, perforates the body wall, and supplies the skin over the lateral surface of the hip and thigh. The other division of lumbar nerve 5 continues caudad to unite with the divisions of the sixth lumbar nerve. Nerve branches to the psoas muscles may arise from this division or from the lateral cutaneous femoral nerve.

Lumbar nerve 6 is a large nerve. Its largest division unites with a branch of the caudal division of the fifth lumbar nerve to form the large **femoral nerve.** After perforating the body wall with the femoral artery and vein, the femoral nerve dips deeply into the thigh musculature between the sartorius and pectineus muscles. It innervates the quadriceps femoris muscle and certain other extensor thigh muscles. It also gives rise to the **saphenous nerve,** a prominent cutaneous branch that follows the caudal border of the sartorius muscle and supplies the skin on the medial side of the thigh and shin.

In the cat, a second division of lumbar nerve 6 receives another division of the fifth lumbar nerve to form the **obturator nerve.** In the rabbit, divisions of the sixth and seventh lumbar nerve form the **obturator nerve.** This nerve extends caudolaterally, perforates the obturator internus muscle near the brim of the pelvis, and goes through the obturator foramen of the pelvic girdle to supply primarily the gracilis, obturator externus, and adductor femoris muscles. These muscles belong to the ventral appendicular group.

In the cat, a third division of lumbar nerve 6 continues caudad to join **lumbar nerve 7,** forming a large nerve. As this nerve continues caudolaterally, it soon receives a division from sacral nerve 1 and, a bit farther distally, a contribution from sacral nerves 2 and 3. The resulting nerve is called the **lumbosacral nerve trunk.** In the rabbit, the

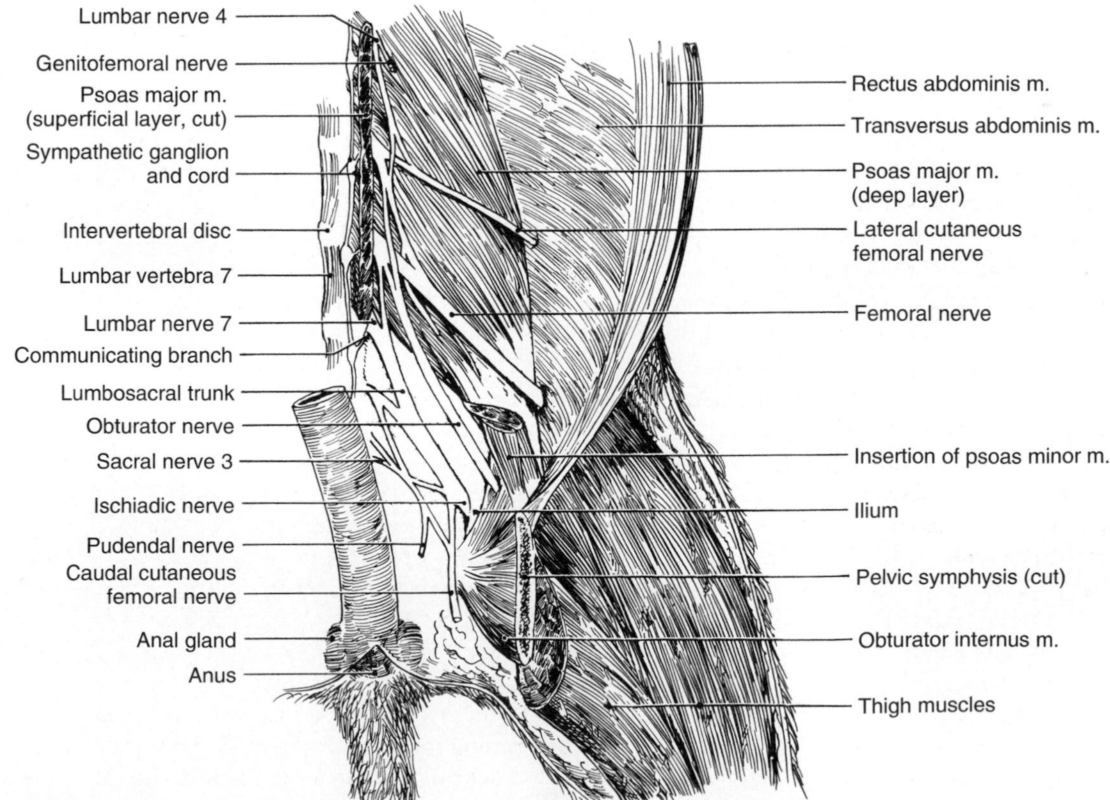

Figure 9-23

Ventral view of a dissection of the left lumbosacral plexus of a cat.

lumbosacral nerve trunk is formed mainly by a large division of sacral nerve 1, a large nerve formed by sacral nerves 2 and 3, and a small division of lumbar nerve 7. The lumbosacral nerve trunk leaves the pelvic canal by passing between the ilium and sacrum laterodorsally to the pelvis and breaks up into several branches. **Gluteal nerves** supply the gluteal and other laterodorsal hip muscles, and a large **ischiadic** (sciatic) **nerve,** which is the main continuation of the lumbosacral trunk, continues distally down the lateral surface of the thigh (see Figs. 7-31 and 7-32), innervating the biceps femoris, semimembranosus, semitendinosus, and other flexor muscles of the thigh. It bifurcates near the distal end of the thigh into **tibial** and **common peroneal** (fibular) **nerves.** These innervate the flexors and extensor muscles of the shank, respectively. Cutaneous branches of the ischiadic, tibial, and common peroneal nerves help to supply adjacent skin.

Sacral nerves 1, 2, and 3 also interconnect with one another to form a network. To see the first two, you need to bisect and reflect the obturator internus muscle. From the sacral network, several small nerves arise; the most conspicuous are the pudendal nerve and the caudal cutaneous femoral nerve, which can be seen dorsolateral to the coccygeus muscle. The **pudendal nerve** contains motor fibers to striated muscles in the anal region and sensory fibers returning from the anal region and from the penis or clitoris. The **caudal cutaneous femoral nerve** helps supply the skin in the anal area and adjacent parts of the thigh.

During this dissection, you can see a portion of the **sympathetic trunk** and **ganglia** lying on the ventral surface of the lumbar vertebrae (Fig. 9-23). Each ganglion receives a **communicating branch** from adjacent lumbar spinal nerves. Pelvic viscera receive their sympathetic innervation by minute branches from the pelvic extension of the sympathetic trunk that follow the blood vessels to the organs. A **pelvic nerve** formed by very small branches from the sacral nerves supplies parasympathetic innervation. These nerves are seldom seen in dissections.

The Coelom and the Digestive and Respiratory Systems

Chapter 10

We now turn from the organ systems that support, move, and integrate the body's activities to organ systems that sustain metabolism. The digestive system ingests water and food; breaks down food into substances that can be absorbed; absorbs water, minerals, and nutrients; and eliminates the indigestible parts of the ingested food. The respiratory system exchanges carbon dioxide against oxygen. The circulatory system transports substances, molecules, and ions to and from the cells. The excretory system excretes nitrogenous waste products of the cell metabolism and maintains the water and ion balance of the body. In teleost bony fishes, the gill epithelium excretes additional nitrogenous waste products and participates in maintaining the water and ion balance. These various systems are functionally distinct, but it is convenient to study the digestive and respiratory systems together because they are topographically closely associated. This close association has its roots in the embryonic development of the respiratory system as outgrowths from the digestive system.

Structure and Function of the Coelomic Cavity and Its Contents

The body cavity, called the **coelom** or the **coelomic cavity,** is completely filled with the inner organs, or **viscera.** Typically, organs within the coelom alter their shape and volume rhythmically over short periods of time or cyclically over longer periods of time. This is true for the heart, lungs, stomach, intestine, ovaries, and uterus. The liver, pancreas, and spleen, however, are located in the coelomic cavity because of their embryonic derivation from the primitive gut. As the viscera expand, contract, and shift their positions accordingly, their surfaces must be able to glide past one another and across the walls of the coelomic cavity. This mobility is ensured by the **serosa,** which lines the walls of the coelom and envelops all viscera. The serosa consists of a thin, smooth, flat-celled epithelium, the **mesothelium,**[1] and an underlying layer of loose connective tissue. It secretes small amounts of a serous fluid that keeps the mesothelial surfaces moist and lubricated. Each inner organ is anchored to the wall of the coelom by a **mesentery** (Fig. 10-1B and C). A mesentery consists of two layers of serosa or, in other words, a central layer of connective tissue sandwiched between two layers of mesothelium. The serosa of the mesenteries is continuous with the serosa that lines the body cavity and envelops the viscera. Mesenteries anchor the viscera but allow them some degree of movement. They also provide avenues for the nerves and the blood and lymph vessels that supply the viscera and travel within the connective tissue layer of the mesenteries.

In vertebrates, the coelom is subdivided into at least two cavities: the **pericardial cavity** and the **pleuroperitoneal cavity.** In mammals and some reptiles, the latter is subdivided further into the **peritoneal cavity,** or **abdominal cavity,** and into the paired **pleural cavities.** It is customary to name the serosa according to its location in a particular subdivision of the coelom, namely, the **pericardium, peritoneum,**[2] and **pleura.** It also is customary to distinguish the **parietal serosa,** which lines the walls of the coelom and its subdivisions, from the **visceral serosa,** which envelops the viscera.

As we have already seen (see Chapter 7), formation of the primitive embryonic coelom and mesenteries is a process that is relatively easily understood. However, in the course of both embryonic and evolutionary morphological transformations, the inner organs grow, differentiate, and shift their positions to fit within the available space, and the primitive coelom is subdivided by the formation of septa. As a result, the anatomy of the body cavity with its subdivisions and mesenteries can be very complex and confusing, unless one is familiar with its differentiation during embryonic development.

[1]The mesothelium of the serosa is derived from mesoderm.

[2]In many textbooks, the term *peritoneum* is used as a synonym for the general term *serosa.* This can be confusing unless it is specified whether the term *peritoneum* is being used in its broad or strict sense. In this book, we will use the term *peritoneum* strictly for the serous lining the peritoneal cavity. For the lining of the pleuroperitoneal cavity, we will use the general term *serosa* instead of the technically correct, but longer, term *pleuroperitoneum.*

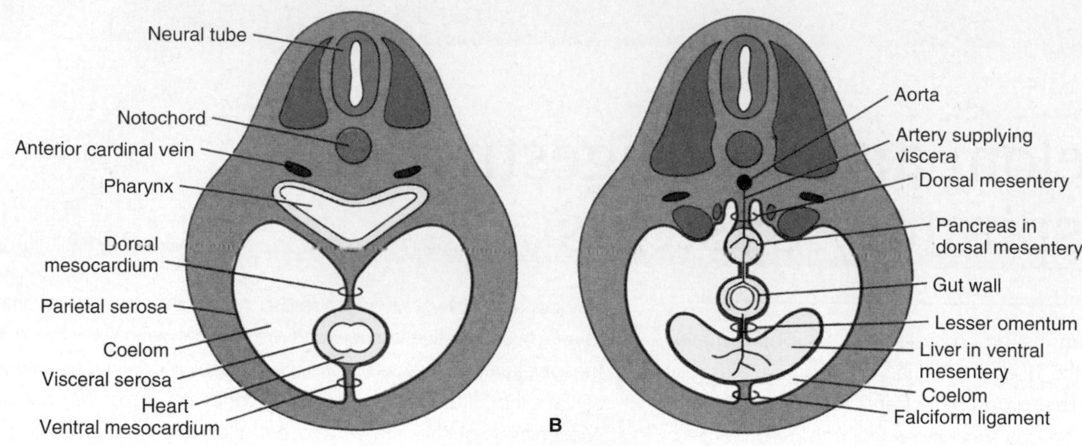

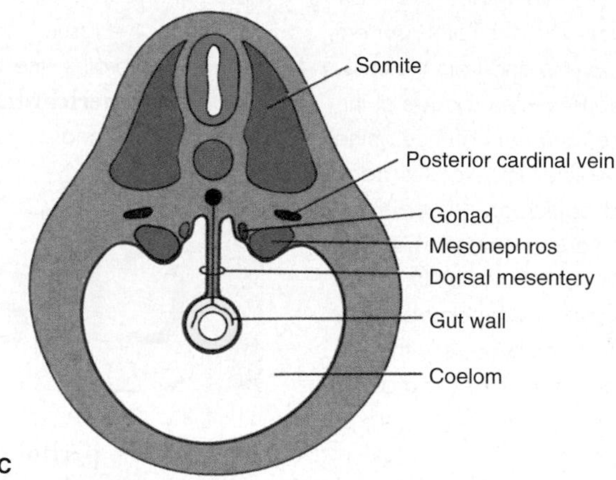

Figure 10-1

Diagrammatic cross sections through an embryo illustrating the topographical relationships of the coelom and mesenteries to the visceral organs. **A**, Cross section at the level of the heart, which develops within the ventral mesentery ventral to the pharynx and cranial to the liver. **B**, Cross section at the level of the liver and pancreas, both of which are outgrowths of the primitive gut. **C**, Cross section at the level of the intestine caudal to the liver, where the ventral mesentery has disappeared. *(Modified after Corliss, 1976.)*

The Embryonic Development of the Coelom and Its Subdivisions

As we have already seen, the coelom appears embryonically as paired cavities within the paired lateral plates (see Fig. 7-5). A mesentery is formed where the expanding paired coelomic cavities meet along the midline; the **primitive gut,** or **archenteron,** is enclosed between them. The serosa develops from the innermost mesoderm layer of the lateral plate, or mesomere, that surrounds the coelom. The parietal serosa forms from the somatic layer and the visceral serosa from the visceral layer.

At an early embryonic stage, the primitive gut is straight and undifferentiated and divides the mesentery into the **dorsal mesentery** and the **ventral mesentery** (see Fig. 7-5F). As the primitive gut differentiates, the liver grows into the ventral mesentery and the pancreas into the dorsal mesentery (Fig. 10-1**B** and Anatomy in Action 10-3, Fig. **A** and **C**). The heart lies far cranial and ventral to the pharynx (Fig. 10**A** and Anatomy in Action 10-3, Fig. **A**). It develops within the ventral mesentery (Figs. 10-3 and 10-4). Caudad to the liver, most of the ventral mesentery disappears so that the coelom becomes one confluent cavity (Fig. 10-1**C** and Anatomy in Action 10-3, Fig. **A**).

At first, the heart and liver lie adjacent to each other within the ventral mesentery (Fig. 10-2**A**), but later, the **transverse septum** forms and separates the **pericardial cavity,** which surrounds the heart, from the rest of the coelom, which subsequently is called the **pleuroperitoneal cavity** (Fig. 10-2**D**). The ventral part of the transverse septum is formed in connection with the growth and differentiation of the liver. As the liver expands, it reaches the wall of the coelom at the level of the pectoral girdle, where the visceral serosa of the liver comes into contact and fuses with the parietal serosa and, thereby, subdivides the coelom (Fig. 10-2**B**). In the course of its further differentiation,

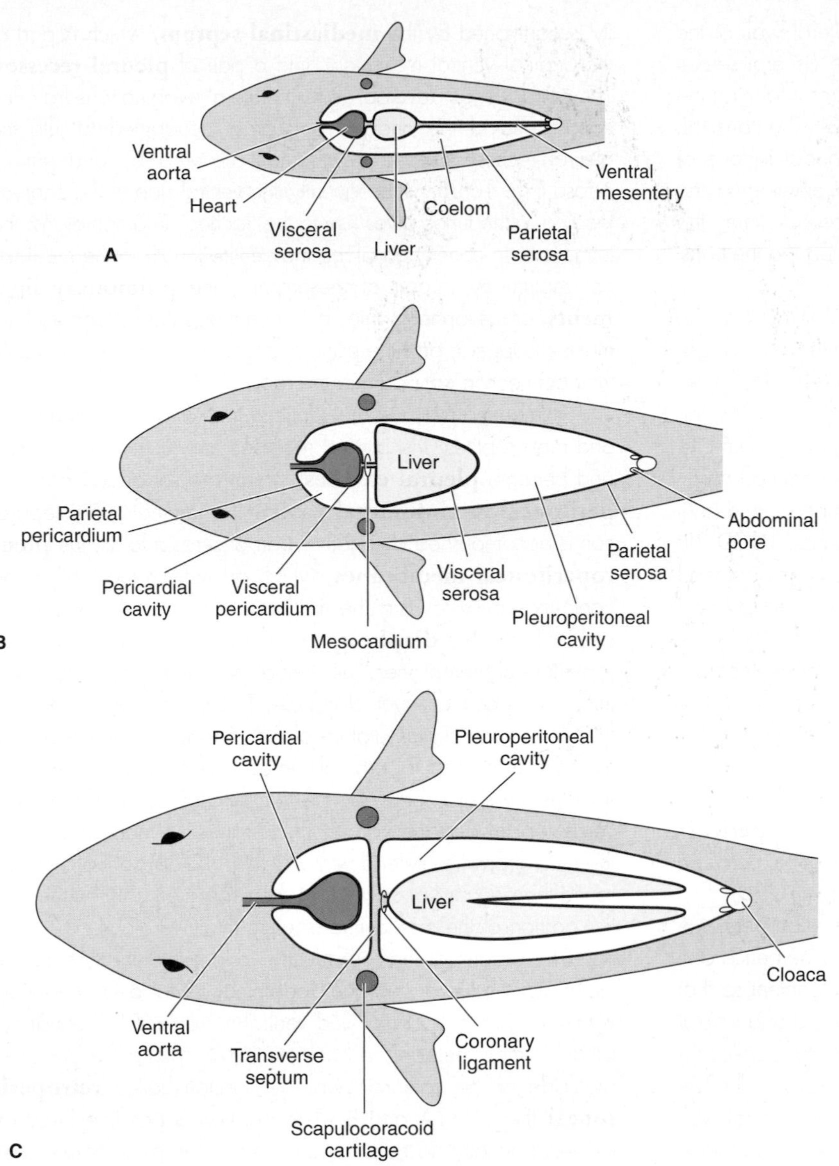

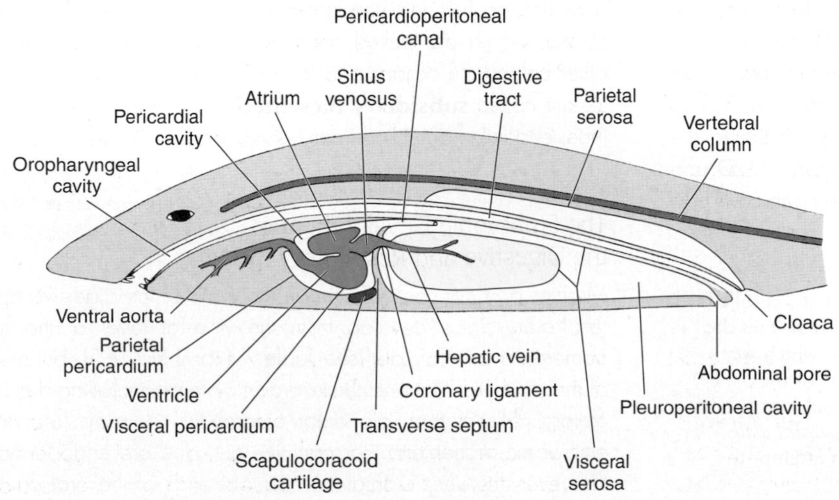

Figure 10-2

Diagrams illustrating the formation of the transverse septum and the subdivision of the coelom as a function of the expansion of the liver. A–C, Horizontal sections through the coelom of a shark ventral to its digestive tract at different developmental stages. A, Relatively early embryonic stage, in which the ventral mesentery caudal to the liver is still present. B, Later embryonic stage, in which the liver expands laterally and the visceral serosa of the liver fuses to the parietal serosa of the body wall. C, Final stage, in which the liver contracts and elongates, the serosa doubles back toward the center to form the transverse septum, and the mesocardium disappears. D, Longitudinal section through an adult shark showing the topographical and functional relationships between the pericardial and pleuroperitoneal cavities and the viscera.

the liver elongates and pulls away from the lateral wall of the pleuroperitoneal cavity. During this process, the visceral serosa of the liver doubles back toward the center, thereby forming the ventral part of the transverse septum (Fig. 10-2**C**). The transition zone between the parietal serosa lining the caudal surface of the transverse septum and the visceral serosa enveloping the liver forms the **coronary ligament.** At the same time, the **mesocardium,** the mesentery connecting the heart to the transverse septum, disappears (Fig. 10-3).

The dorsal part of the transverse septum is formed as the heart grows and differentiates. At an early embryonic stage, the heart is a longitudinal tube, but as it continues to elongate, its growth soon outstrips that of the surrounding body region. Because the heart is anchored to the parietal pericardium cranially by the ventral aorta and caudally by the common cardinal and hepatic veins, it is folded into an S-shaped curve to fit into the given space of the pericardial cavity (Fig. 10-2**D**). In the course of this folding process, the sinus venosus is displaced dorsally. At the same time, the paired, transversely oriented common cardinal veins, or ducts of Cuvier, which drain the blood from the body walls (see Fig. 11-5), are partly incorporated into the sinus venosus. As the common cardinal veins are pulled mediodorsally in the process, they entrain the serosa so that a pair of serosal folds form the dorsal part of the transverse septum. In chondrichthyan fishes, the dorsal part of the transverse septum does not close completely and the **pericardioperitoneal canal** remains as a communication between the pericardial and pleuroperitoneal cavities (Fig. 10-2**D**).

In amniote tetrapods, a distinct neck forms between the head and the pectoral girdle, and the heart within the pericardial cavity migrates caudally during embryonic development so that in an adult it is situated caudal to the pectoral girdle and ventral to the pleuroperitoneal cavity. As a consequence, the dorsal part of the transverse septum is oriented obliquely to horizontally. The pair of serosal folds, which grow medially from the lateral wall of the coelomic cavity and form the dorsal part of the transverse septum, called the **pleuropericardial membrane,** fuse with the ventral mesentery dorsal to the heart (Fig. 10-3**A** and **B**). The **pericardial wall,**[3] hence, consists of three layers: the parietal or mediastinal pleura, the parietal pericardium, and a fibrous connective tissue layer formed by and sandwiched between the two serosae. Soon the mesocardia, which originally anchored the heart within the pericardial cavity, disappear (Figs. 10-2**D** and 10-3**C**). The pleuroperitoneal cavity dorsal to the pericardial cavity is partitioned by the **mediastinal septum,** which is part of the original ventral mesentery, into a pair of **pleural recesses.** The paired lungs develop as ventrolateral evaginations from the caudal end of the pharynx and grow caudomediad into the pleural recesses, while they remain covered by and entrain serosa (Fig. 10-3**B**). Subsequently, the ventral side of the pharynx leading to the lungs gives rise to the trachea. In amphibians, the lungs remain connected along their entire length to the mediastinal septum by a pair of mesenteries, the **pulmonary ligaments,** or mesopneumonia, but in amniotes the pulmonary ligaments disappear, and the parietal and visceral serosae maintain their connection only around the roots of the lungs.

In mammals and some reptiles, in contrast to amphibians and most reptiles, the pleural recesses are closed off caudally and become **pleural cavities,** which are separated from the **peritoneal** or **abdominal cavity.** In mammals, this separation is accomplished by another pair of serosal folds, the **pleuroperitoneal membranes,** which grow from the wall of the coelomic cavity toward the midline, together with some other serosal folds. The **diaphragm** of mammals consists of the pleuroperitoneal membranes, the ventral part of the transverse septum, and a central layer of muscle. This muscle layer develops from the hypaxial musculature of the cervical region (recall that the heart is located far cranially in early embryonic stages) and invades the space between the two serosal layers of the transverse septum and the various membranes. In many mammals, the pleural cavities extend ventrally (Fig. 10-3**B**) so that they surround the pericardial cavity and are separated only by the ventral portion of the mediastinal septum (Fig. 10-3**C**).

The organs of the reproductive and excretory systems, although located in or near the abdominal cavity, do not develop within the primitive dorsal and ventral mesenteries. The kidneys usually remain outside the coelomic cavity, along the dorsomedial side of the coelomic wall, a position called **retroperitoneal** (Fig. 10-1**A** and **B**). This position is possible because kidneys normally do not change in size. The gonads and their ducts, however, expand and contract cyclically in both sexes, and they project into the coelomic cavity. The two layers of serosa, which are pulled after these organs, adhere to each other behind the organs and form mesenteries. These mesenteries are called **subsidiary mesenteries,** because they develop independently from the primary dorsal and ventral mesenteries.

The Embryonic Development of the Digestive and Respiratory Systems

Muscles and connective tissue in the walls of the digestive and respiratory tracts develop from the visceral layer of the hypomere, or lateral plate (see Table 7-1 and Fig. 7-5), but most of the epithelia that line these organ systems, including the secretory cells of their glandular outgrowths, develop from the embryonic archenteron, or primitive gut, and are endodermal. However, the very cranial and caudal ends of the oral cavity

[3]The three-layered pericardial wall of tetrapods usually is called *pericardium.* This is confusing because the serosa that lines the pericardial cavity also is called the *pericardium,* of which we distinguish the parietal pericardium, which is part of the pericardial wall, and the visceral pericardium, which surrounds the heart (Fig. 10-3**C**). Our use of the term *pericardial wall* instead of the widely used term *pericardium* is an attempt to clarify the situation for didactic purposes.

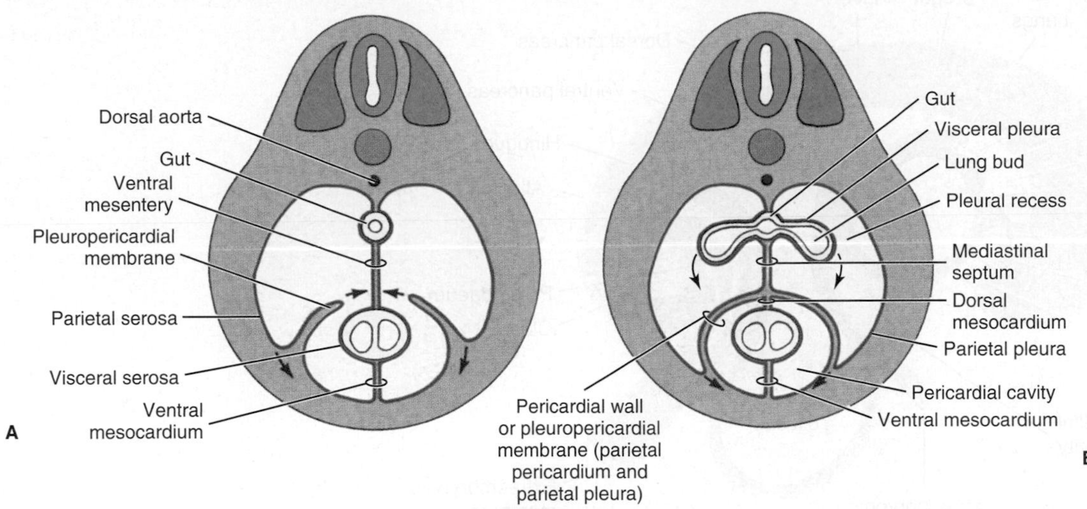

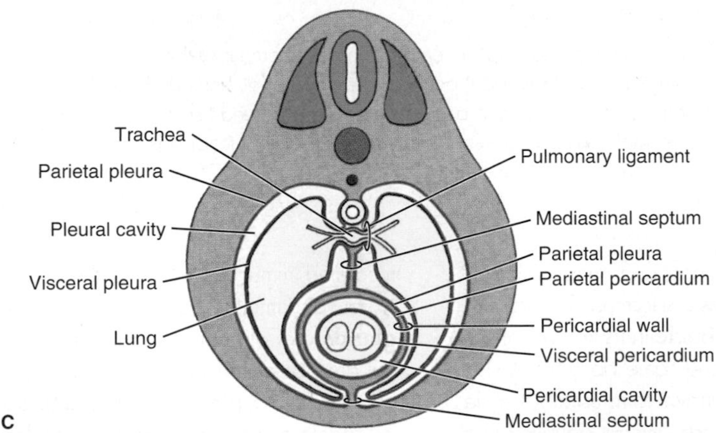

Figure 10-3

Diagrammatic cross sections through mammalian embryos at different stages illustrating the development of the pericardial and pleural cavities. **A**, Formation of the pleuropericardial membranes from the parietal serosa. **B**, Fusion of the pleuropericardial membranes with the mediastinal septum, separation of the paired pleural recesses from the pericardial cavity, ventromedial extension of the pleural recesses, and appearance of lung buds. **C**, Final stage, in which the pericardial sac is built in the mediastinal septum, the dorsal and ventral mesocardia have disappeared, and the pleural cavities with the lungs completely surrounding the pericardial cavity.

and the cloaca, respectively, are lined by ectodermal epithelium because they are formed by ectodermal invaginations, namely, the **stomodaeum** and the **proctodaeum,** respectively (Fig. 10-4). At first, these ectodermal invaginations are separated from the archenteron by plates of tissue, but these eventually break down. The borderlines between ectodermally and endodermally derived epithelia do not remain distinct in the oral cavity and cloaca of adult vertebrates.

The archenteron is divided into the **foregut,** which eventually differentiates into the pharynx, esophagus, and stomach; and the **hindgut,** which forms the intestine and much of the cloaca (Fig. 10-4). In most mammals, the cloaca is present only in early embryos. It soon becomes divided; the dorsal part contributes to the rectum and the ventral part to the urogenital passages. The egg-laying monotreme mammals retain a cloaca also as adults.

In all vertebrates, a series of paired endodermal **pharyngeal pouches** grows out of the sides of the pharynx (Fig. 10-4). There are six of these pouches in most fishes, fewer in tetrapods. The tissues between the pouches constitute the **branchiomeres,** or **pharyngeal bars.** The first branchiomere lies cranial to the first pharyngeal pouch. The skeletal visceral arches, branchial muscles, certain nerves, and the aortic arches grow into these branchiomeres. In fishes, the endodermal pharyngeal pouches meet comparable ectodermal **pharyngeal furrows** that develop on the surface of the embryo. When the

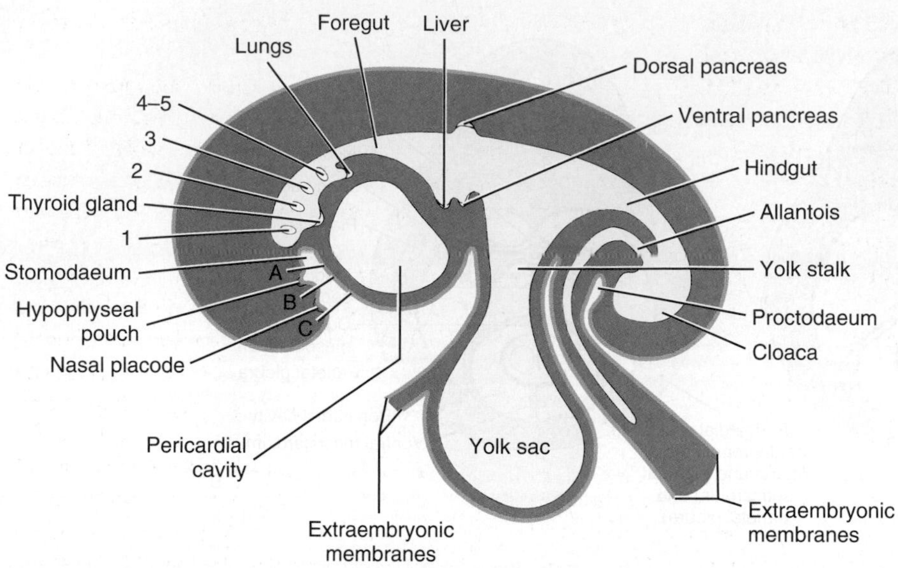

Figure 10-4

Diagrammatic sagittal section of a mammalian embryo showing the embryonic development of the digestive and respiratory systems. The openings to the pharyngeal pouches are numbered. Lines *A, B,* and *C* indicate the comparative position of the mouth opening in a jawless fish *(A)*, a jawed fish without internal nostrils *(B)*, and fishes with internal nostrils (e.g., sarcopterygian fishes) and tetrapods *(C)*. For a more complete view of the extraembryonic membranes, see Figures 11-1 and 12-25.

tissue between the pharyngeal pouches and furrows breaks through, the **gill slits** and **gill pouches** are formed. The first pharyngeal pouch and furrow form the **spiracle** in fishes with ancestral characters, such as many sharks and some bony fishes. The rest of the pharyngeal pouches and furrows form usually five definitive gill slits and pouches. In tetrapods, except aquatic larval amphibians, the pharyngeal pouches and furrows normally do not break through, although they are formed in the embryo. The first pharyngeal pouch gives rise to the **tympanic cavity** and **auditory tube** in reptiles (except snakes, which lack these structures), birds, and mammals. The second pharyngeal pouch of mammals develops into the **tonsillar fossa** containing the palatine tonsils, which consist of lymphoid tissue (see Chapter 11).

The rest of the pharyngeal pouches give rise to certain endodermal glands and organs and are later dismantled. The **thymus,** as one such endodermal organ, develops as epithelial thickenings of the dorsal part of most, or all, pharyngeal pouches in fishes and urodele amphibians, but from the ventral part of only the third and fourth pharyngeal pouches in mammals. These thymal primordia develop as individual glandlike structures or coalesce into a single glandular body. The thymus is a lymphoid organ and plays a crucial role in the development of the immune system because it hosts one particular population of lymphocytes, the **T lymphocytes**, while they mature and acquire the capacity to respond to certain invading antigens (see Chapter 11). Later, these lymphocytes travel to lymph nodes and other lymphoid organs and are involved in cell-mediated immune reactions of the body. The thymus is very large in young mammals, but it atrophies in older ones. **Parathyroid glands** are endocrine glands that are present in all tetrapods but are absent in fishes. They develop as epithelial thickenings from the ventral part of the third and fourth pharyngeal pouches in urodele amphibians, but they develop from the dorsal part of these pouches in mammals and become embedded in the thyroid gland (see below). They produce **parathyroid hormone,** which reacts to dropping calcium levels in the blood. It mobilizes Ca^{2+} from the bones, decreases Ca^{2+} excretion by the kidneys, and, together with vitamin D (cholecalciferol), increases the uptake of Ca^{2+} by the intestine. **Ultimobranchial bodies** are also endocrine glands and develop in all vertebrates from the caudal surface of the last pharyngeal pouch. In mammals, they become incorporated into the thyroid gland as the parafollicular **C cells.** These cells produce **calcitonin** (thyrocalcitonin) in reaction to elevated calcium levels in the blood. In mammals, its main role is to suppress excessive Ca^{2+} loss from the bones during pregnancy and lactation.

In addition to the lateral pharyngeal pouches, a couple of median evaginations develop from the floor of the embryonic pharynx. The endocrine **thyroid gland** develops from one such evagination at the level between the first and second pharyngeal pouches, but later it loses its connection to the pharynx. In fishes, the thyroid gland tends to remain in this cranial position, but in tetrapods, it migrates caudad to various extents. In mammals, it migrates to the level of the larynx and cranial end of the trachea. The thyroid gland may remain an unpaired single gland, or it may bifurcate and form a pair of glandular bodies. Under the influence of the thyroid-stimulating hormone (TSH) from the adenohypophysis (see Chapter 8), the thyroid gland produces and stores thyroglobulin, which is broken down and released into the blood as **thyroxin** and triiodothyronine. These hormones increase the rate of oxidative metabolism. As discussed earlier, the development of the thyroid gland of the lamprey from a part of the endostyle-like subpharyngeal gland of the ammocoete (see Chapter 2) suggests that the thyroid gland of vertebrates may have evolved from the endostyle of protochordates.

The **lungs** represent another example of organs that develop from a median evagination of the floor of the pharynx (Fig. 10-4 and Anatomy in Action 10-3). In sarcopterygian bony fishes and in tetrapods, the lungs develop from a bilobed evagination just caudal to the pharyngeal pouches. In the embryos of some amphibians, the early pulmonary primordia resemble a pair of pharyngeal pouches, which raises the possibility that they may have evolved from a ventrally displaced pair of caudal pharyngeal pouches. Lungs appeared at a very early point of vertebrate evolution as the fossils of some early fishes (Fig. 3-1) show evidence of lunglike structures. Lungs that are used for air-breathing to supplement gill respiration are also found in the lungfishes among extant sarcopterygian fishes and in some actinopterygian fishes that have retained a number of ancestral anatomical conditions. Most actinopterygian fishes, however, do not possess lungs; they possess a gas-filled **swim bladder,** which is located dorsal to the gut and regulates buoyancy (see Anatomy in Action 7-1). The swim bladder usually develops from an evagination of the roof of the caudal pharyngeal region, but in at least one actinopterygian species, it develops from an evagination of the lateral wall of the pharynx. It is unclear whether the swim bladder evolved from a primitive lung by shifting its location dorsal to the gut or whether lungs and, hence, swim bladders, evolved more than once in the course of evolution.

Additional outgrowths from the digestive tract of most vertebrates are found only after the transition from the foregut to the hindgut, where the liver and pancreas develop. The **liver** arises embryonically as a prominent ventral diverticulum (Fig. 10-4 and Anatomy in Action 10-3), which, as mentioned earlier, grows into the ventral mesentery caudal to the heart, and by its expansion forms the ventral part of the transverse septum. Functionally, the liver is a very diverse organ. It secretes bile, which is a mixture of excretory products from the breakdown of hemoglobin and fat-emulsifying bile salts, and its cells come into intimate contact with blood that the hepatic portal system brings to it from the stomach and intestinal region. Many metabolic conversions occur here. Excess sugars are stored, largely in the form of glycogen. Glycogen is broken down into glucose so that the glucose content of the blood is kept at a constant level. Amino acids are deaminated, and their amino groups converted to urea. Toxins may be removed, and many plasma proteins are synthesized.

The **pancreas** is an organ that is large enough to be visible with the unaided eye in most vertebrates, but in lampreys it is represented only by scattered cells in the walls of the intestine and liver (see Chapter 2). The pancreas arises embryonically from one or more intestinal outgrowths near the liver primordium. Frequently there is both a dorsal evagination (forming the **dorsal pancreas**) and a ventral evagination (forming the **ventral pancreas**). The latter often is paired and is associated with the base of the liver primordium and, hence, with the future bile duct (Fig. 10-4). The ventral pancreas grows around the intestine and merges with the dorsal pancreas; together the two pancreatic primordia extend into the dorsal mesentery (Fig. 10-1**B**). The stalks of the pancreatic primordia persist as ducts, although some stalks may be lost in the adult. The ventral pancreatic duct can be identified by its entrance into the intestine in common with the bile duct. The dorsal pancreatic duct can be identified by its independent entrance on the opposite side of the intestine. Most pancreatic cells are exocrine and secrete digestive enzymes, which are discharged into the intestine via pancreatic ducts and break down proteins, carbohydrates, fats, and nucleic acids. Small clusters of endocrine cells, the **islets of Langerhans,** are scattered among the exocrine cells and produce the hormones **insulin** and **glucagon,** which play a crucial role in the carbohydrate metabolism.

Farther caudally along the hindgut, the **yolk stalk** connects the archenteron with the **yolk sac** in the embryos of amniotes and certain fishes (Fig. 10-4). The yolk stalk and sac shrink as the embryo grows and uses up the yolk.

The caudalmost outgrowth of the hindgut in most tetrapods is the **urinary bladder.** It develops from the embryonic cloaca near the caudal end of the hindgut. In embryos of amniotes, this structure expands and eventually projects beyond the embryo as the **allantois,** an extraembryonic membranous sac (Anatomy in Action 10-3, Fig. **A**; see Chapter 12 and Anatomy in Action 12-4).

FISHES

The coelom and the digestive and respiratory systems of *Squalus* are good models of the condition of these structures in ancestral jawed fishes. However, bony fishes ancestral to tetrapods also had lungs that developed from embryonic outgrowths of the caudal part of the pharynx. The differentiation of the gut into a stomach and intestine may have evolved in connection with the evolution of a jaw apparatus. Filter feeders, such as tunicates and amphioxus, which feed more or less continuously on minute food particles, and jawless parasites and scavengers, such as lampreys and hagfishes, do not need a stomach (see Chapters 1 and 2). However, with the evolution of jaws, large chunks of food can be ingested at irregular time intervals, and the stomach is needed to hold food and apportion its passage through the intestine.

Pleuroperitoneal Cavity and Its Contents

The Body Wall and Pleuroperitoneal Cavity

We will study the caudal parts of the digestive system before the oropharyngeal area. Open the **pleuroperitoneal cavity** by a longitudinal incision slightly to the left side of the midventral line, which is on the side where the muscles were dissected (see Table 7-3). Extend the incision as far forward as the scapulocoracoid cartilage and as far caudad as the base of the tail. In doing the latter, cut through the pelvic girdle and continue caudally on one side of the cloacal aperture. Make a transverse incision through the body wall at mid length and extend it to the lateral line. You now have four flaps of the body wall, which you can fold back to expose the body cavity and its contents, but do not break any tissues that extend between the cranial part of the liver and the ventral body wall.

Note the layers of the body wall through which you have cut. The outermost layer is the **skin.** Underneath is a thin layer of **connective tissue** comparable to the external fascia of mammals, the hypaxial musculature, and finally the shiny coelomic epithelium, the **serosa,**[4] which lines the pleuroperitoneal cavity. The portion of the serosa lining the musculature of the body wall is the **parietal serosa;** the portion covering the inner organs, or viscera, is the **visceral serosa.** The portion of serosa that extends from the body wall to the viscera contributes to the formation of the **mesenteries** (for more details on the mesenteries, see later). As a result of their embryonic development, mesenteries consist of two layers of mesothelium with connective tissue sandwiched between them. Blood and lymphatic vessels and nerves that supply the viscera are embedded in this connective tissue (Fig. 10-1).

In most vertebrates, the coelom is a closed cavity that has no direct communication with the outside. However, in vertebrates with ancestral features, such as the spiny dogfish, the coelom communicates with the exterior by a pair of **abdominal pores** (Figs. 10-2**B**, 10-5, and 12-8). You can find these by probing the most caudal recess of the pleuroperitoneal cavity beside the **cloaca,** the chamber receiving the intestine and genital ducts, on the side of the body that is still intact. Each pore opens through the lateral wall of the cloacal aperture, but sometimes the lips of the pore have grown together. The abdominal pores allow the elimination of excess fluid from the coelom, but it is not understood under which circumstances this would be necessary.

The Visceral Organs

We will first get an overview of the contents of the pleuroperitoneal cavity and then will study the visceral organs in more detail after we have studied the mesenteries. A large **liver** with a pair of long pointed lobes occupies most of the cranioventral portion of the pleuroperitoneal cavity. If you need to cut off the ends of these lobes, make sure not to remove too much of them and, thereby, damage the blood vessels that supply the liver or the bile duct that leads from the liver to the intestine. The **bile duct** accompanies these blood vessels for most of its length but goes its own separate way near the beginning of the intestine at the pylorus (see later and Fig. 10-5). Spread the lobes of the liver apart, and you will see the **esophagus** and **stomach** in a more dorsal position. Both have about the same diameter, so there is no external line of demarcation separating them. Any discernible constriction represents a peristaltic contraction that was fixed at death and during preservation of the animal, not a line of demarcation. The caudal end of the stomach curves cranially and gives the organ a J-shape. The **pylorus** marks the end of the stomach; its wall contains a thick sphincter muscle to prevent backflow of food stuff from the intestine to the stomach. The digestive tract then turns caudally and forms the straight **valvular** or **spiral intestine,** which continues to the **cloaca.** If part of the intestine has been everted through the cloaca, you need to pull it back into the body cavity, but before doing so, observe the deep spiral fold, the **spiral valve,** in its lining (Fig. 10-7).

The large triangular organ that is attached to the caudal edge of the stomach where it curves back cranially is the **spleen,** a lymphoid organ responsible for the production and storage of blood cells (see Chapter 11). The elongate **dorsal lobe of the pancreas** is located dorsal to the right side of the spleen and runs along the medial edge of the recurved part of the stomach. The oval **ventral lobe of the pancreas** is applied to the surface of the intestine near the pylorus. The very narrow, thin part of the pancreas that connects the dorsal and ventral lobes is called the **isthmus of the pancreas.** Farther caudally, the **digitiform gland** (rectal gland) is located along the dorsal side of the caudal end of the intestine. It is an osmoregulating gland that excretes surplus ions, such as Cl^- and NO_3^- and, thereby, also Na^+ and K^+.

Other organs within the pleuroperitoneal cavity are parts of the urogenital system. We will consider them in more detail later (see Chapter 12), but you should identify them at this time. A large **gonad,** either a **testis** or an

[4] See Footnote 1.

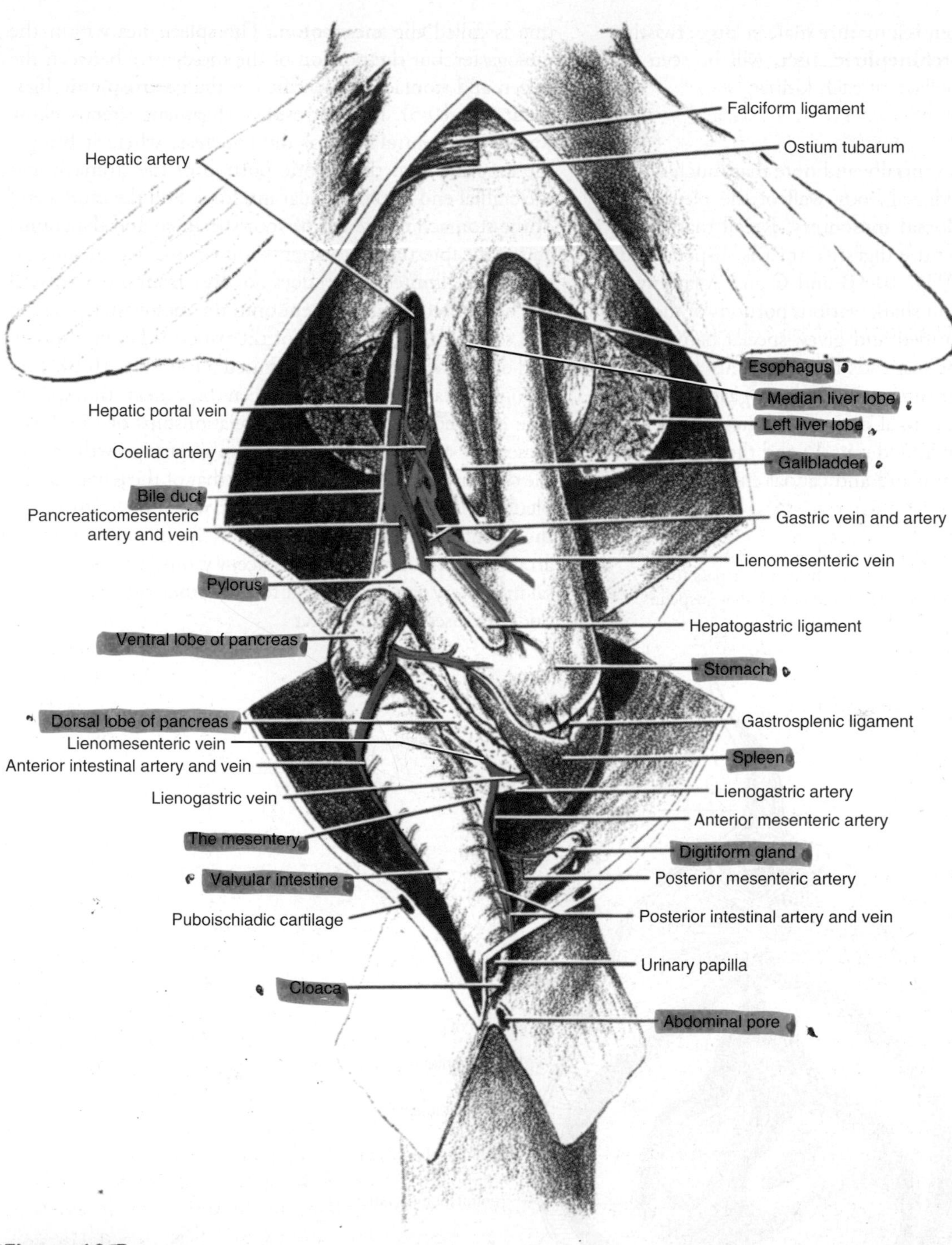

Figure 10-5
Ventral view of the abdominal viscera and major blood vessels of a female *Squalus*. The distal parts of the lateral liver lobes have been removed.

ovary, is situated on each side of the cranial end of the esophagus and stomach. If your specimen is a mature female, a pair of prominent oviducts will be seen dorsal to the ovaries and extending to the cloaca. The caudal portions of the oviducts are enlarged, greatly so in pregnant females. The paired **kidneys** are long bands of slightly darker tissue that are flattened against the dorsal side of the pleuroperitoneal cavity on either side of the dorsal mesentery (see later) and are covered by the parietal serosa. Their caudal portion is much wider than the cranial portion. If the specimen is a mature male, a large, twisting excretory duct, the **archinephric duct,** will be seen apposed to the ventral surface of each kidney.

The Mesenteries

Pull the digestive tract ventrally and note that much of it is connected to the middorsal body wall of the pleuroperitoneal cavity by the **dorsal mesentery.** Recall that, at the embryonic stage, the entire digestive tract is suspended by the dorsal mesentery (Fig. 10-1**B** and **C** and Anatomy in Action 10-3). In an adult shark, various portions of the dorsal mesentery are identified and given special names. The mesentery portion that suspends the esophagus and stomach is called the **greater omentum,** or **mesogaster;** the portion that suspends the cranial part of the intestine is called **the mesentery** (in the limited sense)[5]; and the portion that suspends the digitiform gland and caudal end of the intestine is called the **mesocolon.** The spleen lies within the mesogaster, but the portion of the mesogaster between the spleen and stomach is identified as the **gastrosplenic ligament** (Fig. 10-5). The gastrosplenic ligament extends all the way to the ventral lobe of the pancreas, where it bridges the loop between the pyloric portion of the stomach and the cranial end of the valvular intestine. Pull the caudal end of the stomach to the left of your specimen and the cranial end of the intestine to the right. You will see that the mesentery (in the limited sense) does not arise from the body wall but has shifted its attachment onto the mesogaster. You can also see that the dorsal lobe of the pancreas lies in a special fold of the mesentery (in the limited sense) (Fig. 10-6). This topographical relationship is seen best near the spleen. The complicated topographical relationships of the dorsal mesentery are a consequence of the greater growth rate of the digestive tract in comparison to that of the length of the pleuroperitoneal cavity during embryonic development. As the stomach and cranial portion of the intestine were rearranged to fit into the given space, various parts of the dorsal mesentery came to lie next to each other and became secondarily fused to each other.

[5]Unfortunately the term *mesentery* is used in two ways: in a broad sense for all double-layered serosal membranes that pass to the viscera, and in a limited sense for the mesentery that suspends most of the small intestine.

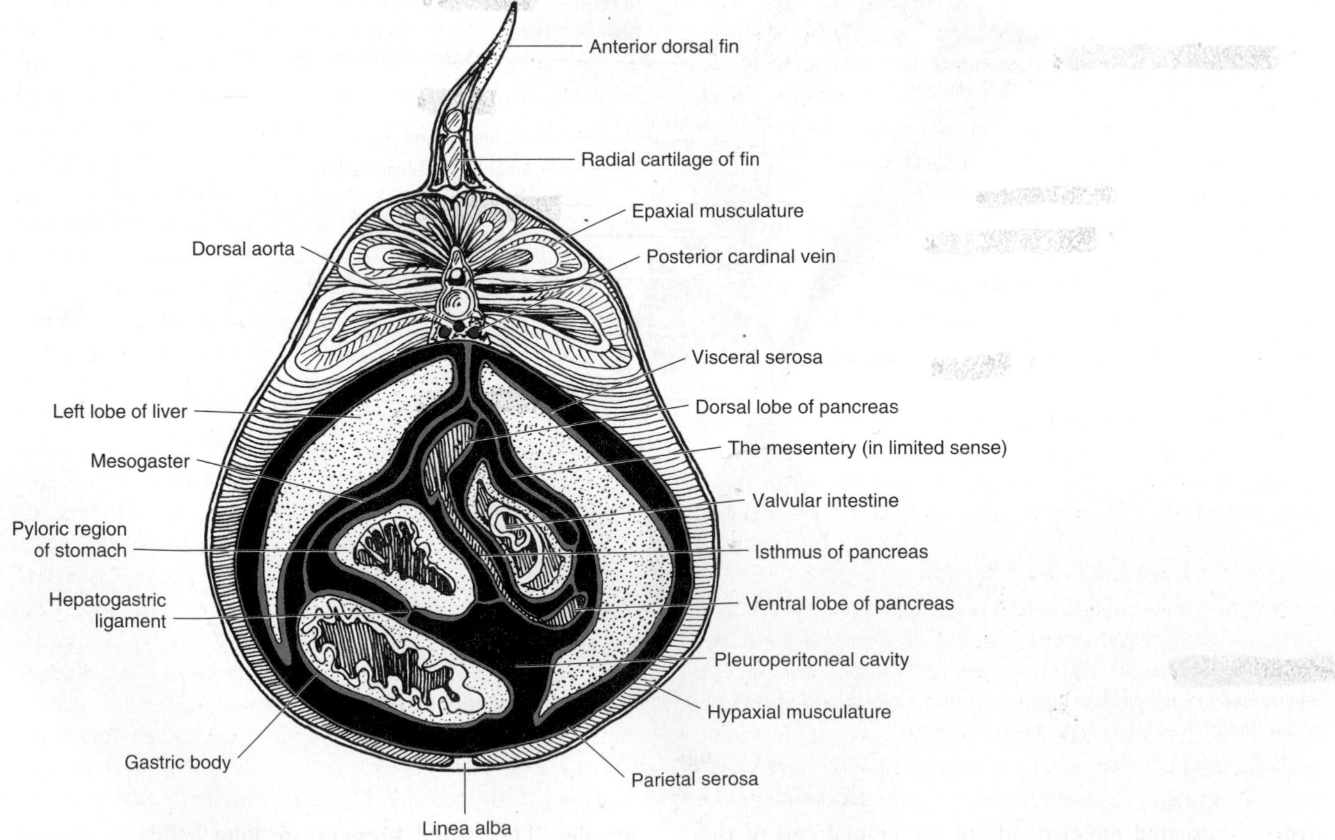

Figure 10-6

Cross section through the trunk of *Squalus* at the level of the isthmus of the pancreas, viewed from behind. The serosa is highlighted in red.

The original ventral mesentery has largely disappeared, except for the portion into which the liver has grown. The part of the ventral mesentery that extends between the ventral surface of the cranial end of the liver and the midventral body wall is the **falciform ligament.** Identify its free edge and try now to probe the **ostium tubarum,** the single entrance to the paired oviducts, because the falciform ligament may inadvertently become damaged before you are able to study the reproductive system (see Chapter 12). The portion of the ventral mesentery between the liver on the one hand and the caudal portion of the stomach and the cranial portion of the intestine on the other hand is the **lesser omentum,** or **gastrohepatoduodenal ligament,** whose structure is very complex in the spiny dogfish. The gastrohepatoduodenal ligament encloses the bile duct and the blood vessels supplying the liver. It is a unit near the liver, but it divides toward the digestive tract. The **hepatogastric ligament** extends to the curvature of the stomach, and the **hepatoduodenal ligament** carries the bile duct to the beginning of the intestine. Note that a part of the mesentery (in the limited sense) that wraps the dorsal lobe of the pancreas in a special fold protrudes ventrally between these two limbs of the gastrohepatoduodenal ligament. This particular fold of the mesentery (in the limited sense) also contains the pancreaticomesenteric artery and hepatic portal vein (see Chapter 11). To see these topographical relationships clearly, hold this particular fold of the mesentery (in the limited sense) between your thumb and forefinger and follow it caudally and dorsally over the pylorus. You will then be able to see that it envelops the dorsal lobe, isthmus and, though less obviously, the ventral lobe of the pancreas. Note also how the ventral lobe of the pancreas is anchored to the pylorus and valvular intestine by very short mesenteries that are derived from the dorsal mesentery.

Subsidiary mesenteries anchor the genital organs to the dorsal body wall. Each testis is suspended by a **mesorchium,** each ovary by a **mesovarium,** and each oviduct by a **mesotubarium.** The latter develops only in mature females.

Structure of the Digestive Organs

Cut open the esophagus and stomach by making a longitudinal incision that extends all the way to the intestine. Remove and wash out any content. You can distinguish the esophagus from the stomach by the structure of the internal lining. The lining of the esophagus is strewn with **papillae** and the wall of the stomach is folded into longitudinal **rugae.** If the stomach was greatly distended at the time of death, the rugae will have been stretched out and the stomach lining will be smooth. Three regions can be identified for the stomach: the **cardiac region** at the cranial end of the stomach, which is adjacent to the esophagus; the main portion called the **gastric body;** and the **pyloric region** of the recurving portion of the stomach. Note that the pyloric region has a thicker muscular wall, especially just before the intestine, where it forms a muscular sphincter, the **pylorus.** These topographical regions of the elasmobranch stomach do not correspond to those of the same name in the mammalian stomach, which are based on the type of gastric glands. Sharks possess only one type of gastric glands. The side of the stomach to which the mesogaster and gastrosplenic ligament attach is called the **greater curvature** of the stomach; the opposite side is called the **lesser curvature.** The greater curvature of the adult stomach corresponds to the original dorsal side of the embryonic stomach.

Look for the bile and pancreatic ducts as they enter the cranial end of the valvular intestine. The digitiform gland empties into the caudal end of the intestine. The rest of the valvular intestine is undifferentiated. It contains a complex spiral fold, the **spiral valve,** which is also found in some other fishes with ancestral characters but is absent in most fishes and tetrapods (Fig. 10-7). The spiraling line of attachment of the spiral valve to the intestinal wall is discernible through the translucent wall of the intact intestine. Make a longitudinal incision through the wall of the intestine where there are no blood vessels and spread the cut edges apart to observe the internal structure of the valvular intestine. The spiral valve increases the internal surface area of the intestine to maximize the digestive and absorptive processes while maintaining a compact shape of the intestine. It is not useful to try to find corresponding parts between the valvular intestine of certain fishes and the much longer, tubular intestine of other vertebrates.

Examine the liver in more detail. It consists of long **right** and **left lobes** and a smaller **median lobe.** The elongate, thin-walled **gallbladder** is imbedded within the median liver lobe, but a small part of it is exposed on the surface (Fig. 10-5). Make an incision in the exposed wall of the gallbladder and insert a probe to find the cranial end of the gallbladder. By scraping away superficial liver tissue from the cranial end of the gallbladder, you will find the origin of the bile duct. In the spiny dogfish, the bile leaves the liver tissue through a number of inconspicuous **hepatic ducts** and accumulates in the gallbladder. The bile is expelled by the contracting gallbladder and flows through the **bile duct** when food reaches the intestine. The bile duct travels for some distance within the wall of the valvular intestine before it empties into the lumen of the intestine. You can trace the bile duct by removing the visceral serosa and longitudinal muscle layer of the wall of the valvular intestine. The liver of sharks, like that of other vertebrates, is a central organ for metabolic processes and the storage of certain metabolites. In addition, it contains large amounts of oil, called **squalene,** which is metabolically inert and is

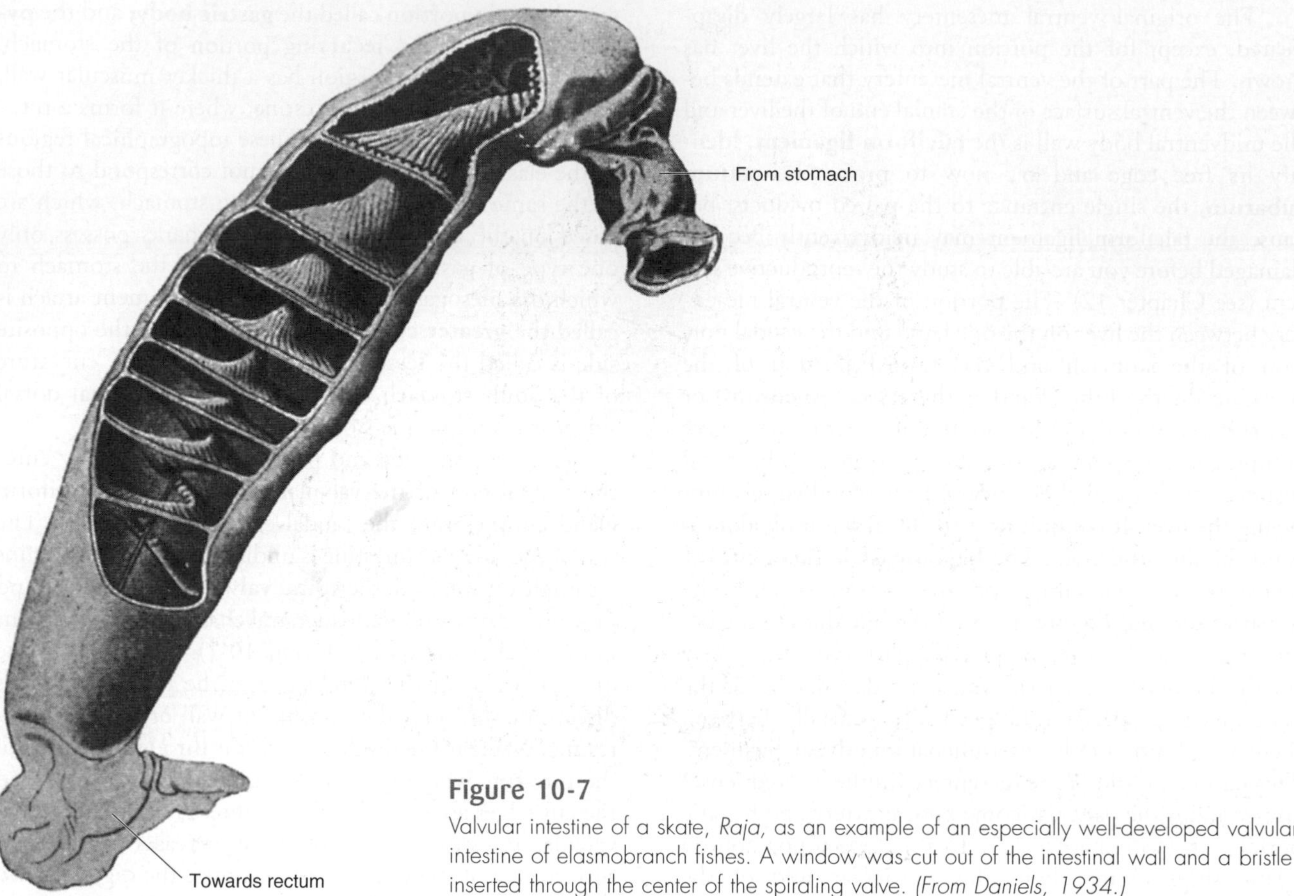

Figure 10-7
Valvular intestine of a skate, *Raja*, as an example of an especially well-developed valvular intestine of elasmobranch fishes. A window was cut out of the intestinal wall and a bristle inserted through the center of the spiraling valve. *(From Daniels, 1934.)*

partly responsible for making sharks buoyant. In this respect, the liver of sharks can be said to be analogous to the swim bladder of actinopterygian fishes (see Anatomy in Action 7-1).

Return to the pancreas and find again the dorsal and ventral lobes and the interconnecting isthmus. Both lobes drain by a common **pancreatic duct** that leaves the ventral lobe and travels obliquely caudad within the wall of the valvular intestine for a short distance before it empties into the lumen of the intestine. To see the pancreatic duct, it is necessary to bisect the mesentery that anchors the ventral lobe of the pancreas to the intestine. Slowly scrape away the soft tissue from the caudal part of the ventral lobe of the pancreas to expose the thin, whitish pancreatic duct. Follow it by removing the visceral serosa and longitudinal musculature from the wall of the intestine. Even though the pancreas of the spiny dogfish is subdivided into a dorsal and a ventral lobe, it is derived entirely from the embryonic dorsal primordium of the pancreas, because the ventral primordium of the pancreas is lost in sharks. Nevertheless, the pancreatic duct enters on the lateral surface of the intestine. This perhaps confusing topographical relationship arose because this portion of the digestive tract has undergone a rotation of nearly 180 degrees around its longitudinal axis during embryonic development so that the dorsal mesentery of the adult shark attaches along the ventral surface of the intestine (Fig. 10-6).

The Pericardial Cavity

The second division of the coelom, the **pericardial cavity,** of fishes is located in the head region cranial to the scapulocoracoid cartilage, or pectoral fin girdle, directly ventral to the basibranchial cartilages of the branchial arches and dorsal to the scapulocoracoid cartilage and hypobranchial musculature (Figs. 10-8 and 11-9). Keep in mind that fishes do not have a neck and that sharks do not have lungs. To open the pericardial cavity and see the heart, return to your original incision through the ventral body wall and extend it cranially. Make sure to cut to the left of the falciform ligament (from the point of view of the specimen, see Table 7-3) in order not to destroy it. Veer toward the midline between the paired coracoarcual muscles when you cut through the scapulocoracoid cartilage, but avoid damaging the coracomandibular muscle. Extend this longitudinal incision to the very rostral end of the

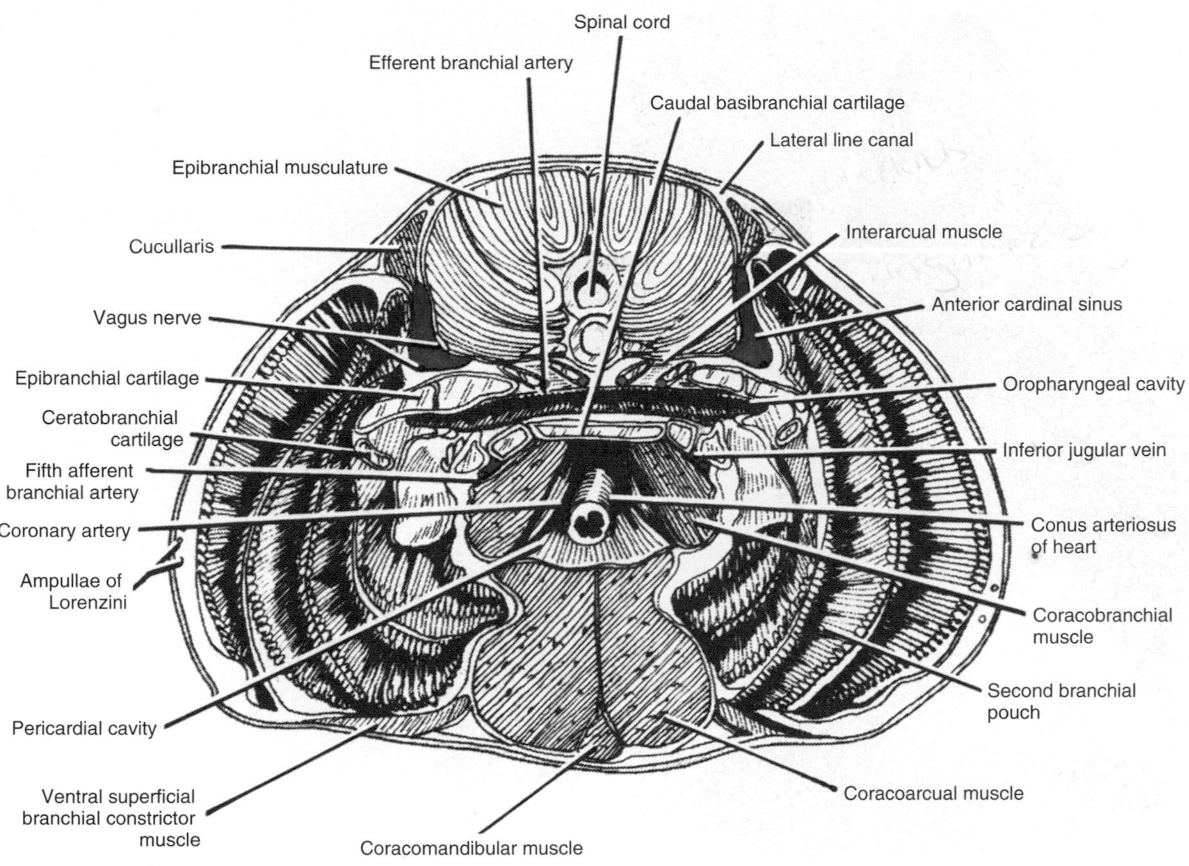

Figure 10-8

Cross section through the branchial region of *Squalus* at the level of the rostral end of the pericardial cavity in a caudal view.

pericardial cavity. Finally, make transverse cuts through the origin of the coracoarcual muscles just cranial to the scapulocoracoid cartilage so that you can reflect flaps to reveal the contents of the pericardial cavity.

The only organ within the pericardial cavity is the **heart.** The portion of the serosa that lines the wall of the pericardial cavity is the **parietal pericardium;** the portion that covers the heart is the **visceral pericardium.** Recall the embryonic development of the heart in sharks (Figs. 10-1 and 10-2) so that you can understand why the heart of an adult shark does not have any mesenteries and is anchored to the pericardial wall only through the major blood vessels entering and leaving the pericardial cavity.

The transverse, vertical septum separating the pericardial and pleuroperitoneal cavities is the **transverse septum.** You can now see that the liver is attached to the caudal face of the transverse septum by the coronary ligament. The coronary ligament does not look like a typical mesentery; it is formed just by the serosa bridging the gap between the liver and the transverse septum. It also interconnects the visceral serosa covering the liver and transverse septum on the one hand and with the cranial part of the falciform ligament on the other hand. As in many fishes with ancestral characters, the pericardial and pleuroperitoneal cavities are interconnected by the **pericardioperitoneal canal** in the spiny dogfish (Fig. 10-2**D**). You will be able to see it when you study the heart (see Chapter 11).

The Oral Cavity, Pharynx, and Respiratory Organs

The objective of the following operation is to separate the dorsal half of the shark head from its ventral half in order to gain access to the oropharyngeal cavity (Figs. 10-9 and 11-7). If we assume that the digestive tract forms the central axis of the body, this is a relatively straightforward task because most organs are located either dorsal or ventral to the digestive tract. For example, the heart and liver are situated ventral to the digestive tract, whereas the

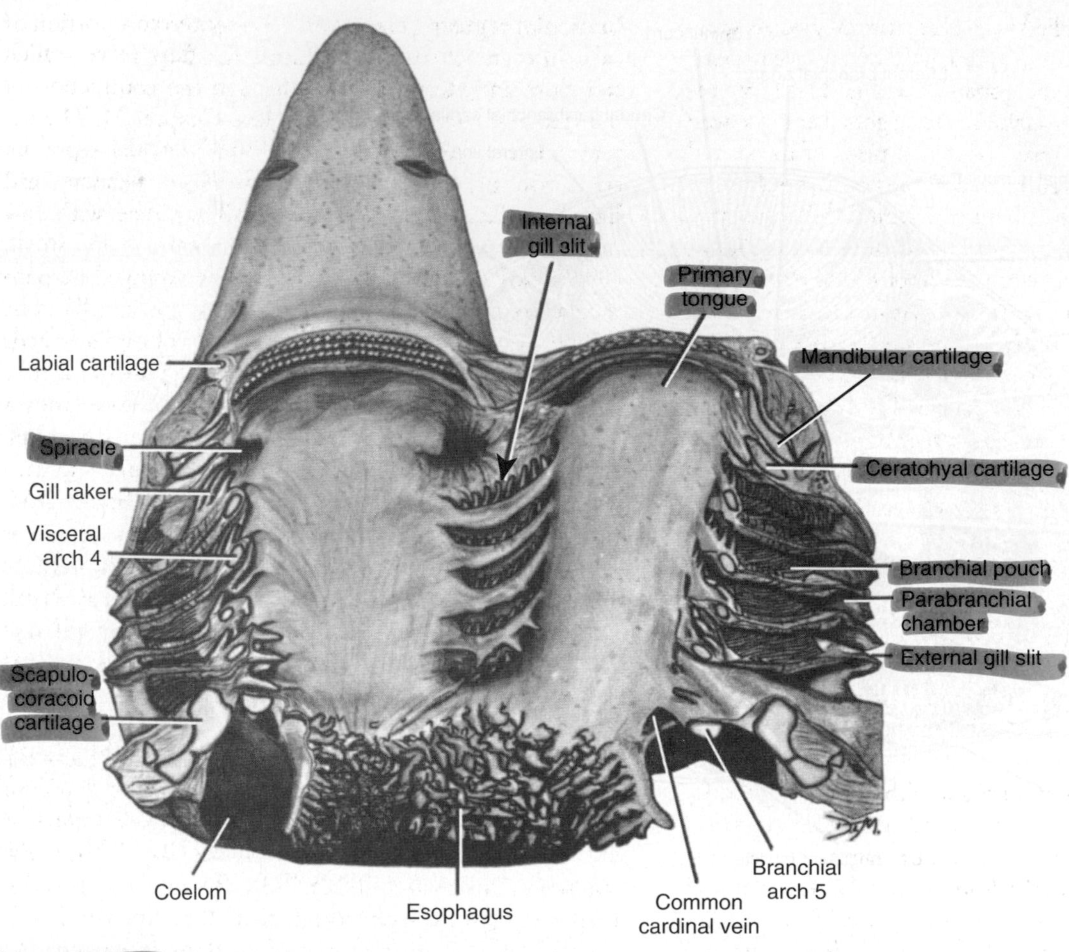

Figure 10-9
Oropharyngeal cavity of *Squalus*. The floor of the oropharyngeal cavity has been reflected to the left of the specimen to show an internal view.

brain and the vertebral column are situated dorsal to it. Some organs, however, cross the midline from dorsal to ventral, such as the visceral arches, the scapulocoracoid cartilage, and the gonads. These organs need to be bisected when the oropharyngeal cavity is opened, but by doing it systematically we can avoid undue damage to the specimen.

On the side of the specimen where you dissected the muscles (see Table 7-3), use a strong pair of scissors to cut horizontally through the mandibular and hyoid arches at the corner of the mouth, as well as through the middle of the adductor mandibulae and hyoid constrictor muscles. When extending your cut caudally through the branchial region, cut through the middle of each interbranchial septum individually and apply your cut at the level of the joint between the epibranchial and ceratobranchial cartilages. Also cut through the last (fifth) branchial arch that is apposed to the scapulocoracoid cartilage at the base of the pectoral fin. Extend your cut caudally along the ventral side of the base of the pectoral fin, through the coracoid bar, and for about 1 inch farther caudad through the muscular body wall. You now have reached the pleuroperitoneal cavity caudal to the pericardial cavity (see Fig. 11-9 for general orientation). Now proceed to cut medially, first only through the skin and hypaxial musculature toward the longitudinal midventral section that was made when the pleuroperitoneal cavity was opened.

Reflect the body wall with the pectoral fin and the dorsal half of the head and look into the pleuroperitoneal cavity. Find again the large liver and the left gonad that is anchored to the dorsal body wall by a short subsidiary mesentery. Cranially, the gonad is also anchored to the wall of the esophagus. With a pair of scissors, extend the longitudinal section through the lateral wall of the esophagus for about 1.5 inches beyond the scapulocoracoid cartilage. To do so, you need to cut also through the posterior cardinal sinus, which usually is filled with blue latex in injected specimens and lies against the dorsal wall of the pleuroperitoneal cavity lateral to the gonads (see Fig. 11-5). Subsequently, cut medially through the entire gonad and the wall of the esophagus. As a result, a cranial piece of the gonad will be attached to the ventral half of the body, and the caudal rest of the gonad will remain *in situ.* In this manner, you will keep the heart and sinus venosus intact for later dissection. You can now swing open the floor of the oropharyngeal cavity, as shown in Figure 10-9. At first, this will meet with resistance because of the stiffness of the preserved tissues, but these will become more pliable as you continue to work on this region. If the esophagus has everted into the oropharyngeal, you need to pull it back.

The demarcation between the **oral cavity** and the **pharynx** is not clearly definable in adult sharks; therefore, the cavity between the mouth and the esophagus is called the **oropharyngeal cavity.** The floor of the oropharyngeal cavity is formed by the **primary tongue,** which is supported by the basihyal, basibranchial, and hypobranchial cartilages (see Chapter 4).

You will see the internal opening of the **spiracle** at the rostral end of the roof of the oropharyngeal cavity and the five **internal gill slits** along the lateral wall. A number of papilla-like **gill rakers** project across the internal gill slits; they act as strainers and keep food and other particles from entering the gill pouches and damaging the delicate gills. Each internal gill slit leads into a large **branchial pouch.** The external portion of the branchial pouch lateral to the gill lamellae is the **parabranchial chamber** (Fig. 10-9 and Anatomy in Action 10-1); it opens on the body surface by the **external gill slit.** The tissue layer between each branchial pouch constitutes the **interbranchial septum** (Fig. 10-10). The outermost portion of each interbranchial septum forms the **flap valve,** which can close an external gill slit through the contraction of the trematic constrictor muscle (see Chapter 7). The actual gills on the surfaces of the interbranchial septa are composed of platelike, radially arranged **primary gill lamellae.** Examine the primary gill lamellae with low magnification and note that each bears many small, closely packed **secondary gill lamellae** arranged perpendicular to the surface of each primary gill lamella. The secondary gill lamellae consist essentially of capillary beds covered with a very thin, gas-permeable epithelium; they are the site of respiratory gas exchange as water moves across them (Anatomy in Action 10-1). Interbranchial septa that bear gill lamellae on both their cranial and caudal surfaces constitute a complete gill, or **holobranch** (Fig. 10-10). The first interbranchial septum, which is supported by the hyoid arch, is a **hemibranch,** because gill lamellae are present only on its caudal surface. Recall that the fifth branchial arch does not bear any gill rays and, hence, does not support gills. As a consequence, the posttrematic wall of the fifth branchial pouch does not bear gill lamellae.

Cut open the spiracle on the side on which the oropharyngeal cavity was cut open. On the cranial wall of the spiracle, you can find a minute hemibranch, known as the **pseudobranch,** on the valvelike flap, called the **spiracular valve** (see Table 7-3). Because the pseudobranch is supplied with oxygen-rich blood (see Fig. 11-7), and because the spiracle is used to draw water into the oropharyngeal cavity, this pseudobranch is considered to be a vestigial gill, but it probably has some as yet unknown function.

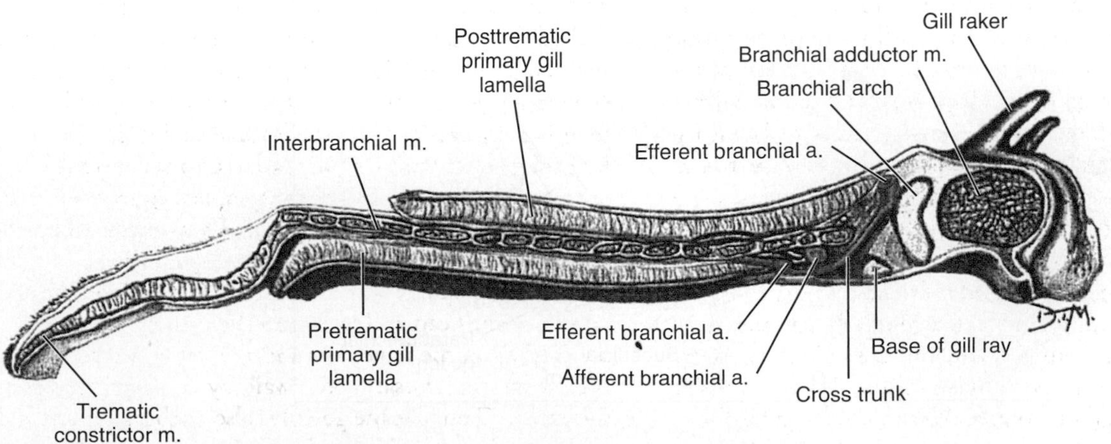

Figure 10-10

Transverse section through a left interbranchial septum and holobranch of *Squalus.* The cranial posttrematic surface is toward the top of the figure. It is shown at twice its natural size.

Anatomy in Action 10-1 Gill Ventilation, Respiration, and Feeding in Elasmobranch Fishes

In order to ensure the uptake of oxygen (O_2) and the release of carbon dioxide (CO_2) by the blood in the capillaries of the gill lamellae, a fish must maintain a flow of oxygen-rich water through its gill pouches. In *Squalus,* this flow is driven by a double-pump mechanism. When the jaw, mouth opening, and spiracle are being closed and the oropharyngeal cavity compressed (Figs. **A** and **B**), water in the oropharyngeal cavity is forced through the gill pouches, and the flap valves of the external gill slits are forced open by the water pressure to let the water flow through the parabranchial chamber. Then the gill slits are shut by the contracting trematic constrictor muscles so that suction can be generated while the mouth and spiracle are still closed (Fig. **D**). The pressure is the lowest in the parabranchial chambers, and the flap valves can be seen being sucked inward. Subsequently, several actions take place to again fill the oropharyngeal cavity with water. The spiracle opens and water is sucked into the oropharyngeal cavity. The mouth opens by slightly raising the chondrocranium and stabilizing the mandible in place; the rostral part of the oropharyngeal cavity expands by retracting the basihyal-ceratohyal cartilages like a piston; and the rest of the oropharyngeal cavity expands by unfolding the branchial arches (Figs. **C** and **D**), except the fifth branchial arch. As a result, water enters through the mouth opening and spiracle and fills the oropharyngeal cavity. Experiments using colored water have shown that water entering through the spiracle preferentially exits the first two gill slits.

The heartbeat in *Squalus* is loosely coupled with the ventilatory movements as the movements of the oropharyngeal cavity affect the pericardial cavity (see Anatomy in Action 11-4). This coordination ensures that oxygen-depleted blood is delivered to the gill capillaries when water is moved across the gill lamellae.

Elasmobranchs display a great variety in their morphology and behavior, including their mechanism of gill ventilation. For example, bottom-living skates and rays draw water into their oropharyngeal cavity only through their enlarged spiracles, which are situated dorsally. In some rapidly swim-

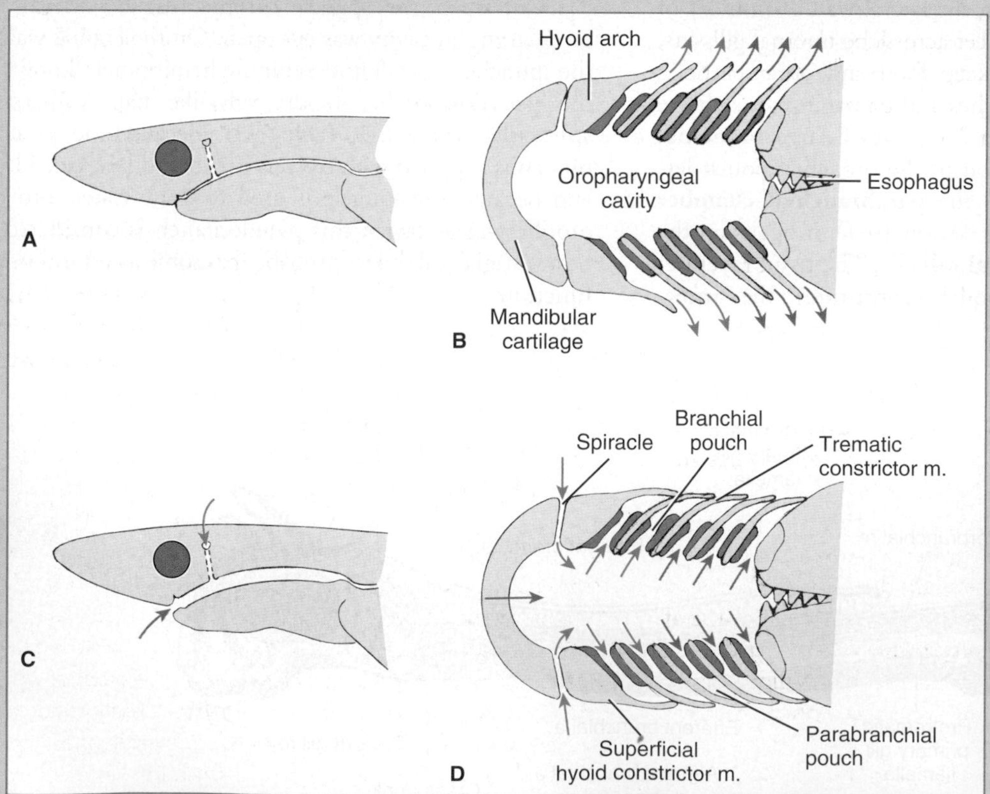

Diagrams showing the mechanics of gill ventilation in *Squalus*. **A,** Lateral view with closed mouth and spiracle. **B,** Horizontal section through the oropharyngeal cavity with closing mouth and spiracle. **C,** Lateral view with open mouth and spiracle and slightly expanded oropharyngeal cavity. **D,** Horizontal section through the oropharyngeal cavity with opening mouth and spiracle. *(B and D, after Hughes, 1963.)*

ming pelagic species that do not have spiracles, such as great white sharks, the entire water for gill ventilation enters through the mouth. Some very large sharks that swim slowly, such as whale sharks and basking sharks, swim with their mouth open and their oropharyngeal and internal gill slits expanded so that their forward motion drives water through their gill pouches. This type of ventilation is called **ram ventilation.**

The movements involved in the ingestion of prey in principle are very similar to those involved in ventilation, except that the mouth and oropharyngeal cavity, including the fifth branchial arch, need to be expanded much more to accommodate the food item (see Anatomy in Action 11-4). Cheryl Wilga and Phil Motta (Motta and Wilga, 2001; Wilga and Motta, 1998) have studied the functional morphology of feeding in *Squalus* and other elasmobranch species. They observed three different types of feeding mechanisms. During **suction feeding,** suction is created inside the oropharyngeal cavity, and the prey rushes into the oropharyngeal cavity as the mouth is opened. Any water that is ingested with the prey exits through the external gill slits. During **biting feeding,** sharks, such as the great white shark, open and close their jaws directly onto their prey and gouge chunks off their prey. During **ram feeding,** fast swimming sharks overtake their prey with widely opened mouth and engulf it. *Squalus* feeds by using a combination of ram and suction feeding.

Examine the cut surface of a holobranch and note its composition (Fig. 10-10). The **branchial arch** provides the base for the cartilaginous **gill rays** (see Figs. 4-4**B**, 4-5**B**, and 7-9), which are the skeletal support of the interbranchial septum. The **gill rakers** along the internal surface of the branchial arch have a cartilaginous core. The **branchial adductor muscle** attaches on the medial surfaces of the epibranchial and ceratobranchial cartilages. The **interbranchial muscle** with its circularly arranged muscle fiber bundles is draped over the rostral, or posttrematic, side of the gill rays, and the trematic constrictor muscle forms the core of the flap valve of the interbranchial septum. Close examination will reveal several blood vessels (see also Chapter 11). The **afferent branchial artery** is located near the center of the interbranchial septum and runs distal to the external surface of the branchial arch. It brings oxygen-depleted blood from the heart and ventral aorta to the capillaries in the gill lamellae and usually is not injected. A pair of **efferent branchial arteries,** which usually are filled with red latex in injected specimens, run along the external surface of the branchial arch just proximal to the primary gill lamellae. They drain the gill capillaries and carry oxygen-rich blood to the dorsal aorta. Recall that each interbranchial septum consists of a **pretrematic** and a **posttrematic branch** of a cranial nerve. Which cranial nerve is associated with the interbranchial septum you have been studying?

A variety of endocrine glands, such as the thyroid gland, a series of thymus bodies, and an ultimobranchial body, develop from the embryonic pharyngeal pouches but are difficult to see. You may have seen the thyroid gland during the dissection of the hypobranchial muscles (see Chapter 7).

AMPHIBIANS

The partitioning of the coelom of amphibians is comparable to that of the spiny dogfish, except that the pericardial cavity is situated slightly more caudal, so that the transverse septum assumes a more oblique orientation.

More conspicuous changes are seen in the respiratory and digestive systems, because amphibians, as their name implies, can interact simultaneously or sequentially with aquatic as well as terrestrial environments.

In regard to respiration, larval and paedomorphic amphibians, such as *Necturus,* possess gills as well as lungs (Anatomy in Action 10-2). The gills, though, are **external gills,** which are supported by skeletal elements of the hyobranchial apparatus (see Chapter 4). Small gill slits are located between the bases of the external gills. External gills are typical for urodele and frog larvae, although they also occur in some fish larvae. In frog larvae, the external gills are replaced by internal gills in the course of embryonic development; it is unclear whether these are homologous to the internal gills of adult fishes. Aquatic larvae inhale air into the oropharyngeal cavity and lungs at the water surface to satisfy between 40% and 70% of their oxygen needs (Anatomy in Action 10-2). Additional gas exchange takes place across the relatively thin, richly vascularized skin and across the epithelial lining of the oropharyngeal cavity. At metamorphosis, the larval gills are dismantled, and the lungs become the primary site of gas exchange, significantly supplemented by respiration across the thin and moist skin, particularly for the release of CO_2. Skin respiration is so significant in amphibians that some metamorphosed terrestrial salamanders were able to dismantle their lungs completely, while their hyoid and projectile tongue apparatus increased in size and complexity. In its resting position, this hyobranchial ap-

Anatomy in Action 10-2 Amphibian Ventilation and Feeding in Water and on Land

As we have seen earlier (Anatomy in Action 10-1), the ventilation of gills in fishes proceeds through a one-way flow from the mouth and, if present, the spiracle, through the oropharyngeal cavity, into the branchial pouches, and from there through the external gill slits. The external gills of aquatic amphibian larvae, however, are exposed to fresh, oxygen-rich water simply when they are moved through the water; no flow through the oropharyngeal cavity is necessary. The same is true for respiration through the skin surface. The situation for air breathing, is more complicated because the lungs are blind-ending sacs; hence, the flow of the respiratory medium cannot be unidirectional. Nevertheless, as we will see, at least the process of inspiration in gill-breathing fishes and air-breathing aquatic larval amphibians is comparable. (For gas exchange through the gills and lungs, see Anatomy in Action 11-5).

Elizabeth Brainerd and coworkers (Brainerd, 1998; Brainerd et al., 1993; Simons et al., 2000) have studied the mechanics of **air breathing** in urodele amphibians, such as *Necturus* and *Ambystoma.* The aquatic larva of *Ambystoma* is very similar to that of *Necturus,* but *Ambystoma* eventually metamorphoses into a terrestrial adult so that it lends itself to a comparative study of air breathing at different stages of its life history. Before inhalation starts, the lungs are full of air, and the glottis, gill slits, and mouth are closed (Fig. **C**). As the mouth opens and the oropharyngeal cavity expands, the oropharyngeal cavity fills mostly with inhaled fresh air, but also with air that is exhaled from the lungs as the hypaxial trunk muscles contract and compress the pleuroperitoneal cavity and the lungs (Fig. **A**). Because the air in the lung is exhaled slowly and only partly, the air in the oropharyngeal cavity consists of 80% of fresh air and only 20% exhaled air from the lung. Subsequently, the mouth and gill slits close, the oropharyngeal cavity is compressed, and the air is forced into the lungs (Fig. **D**). Because the volume of the oropharyngeal cavity is larger than that of the incompletely emptied lungs

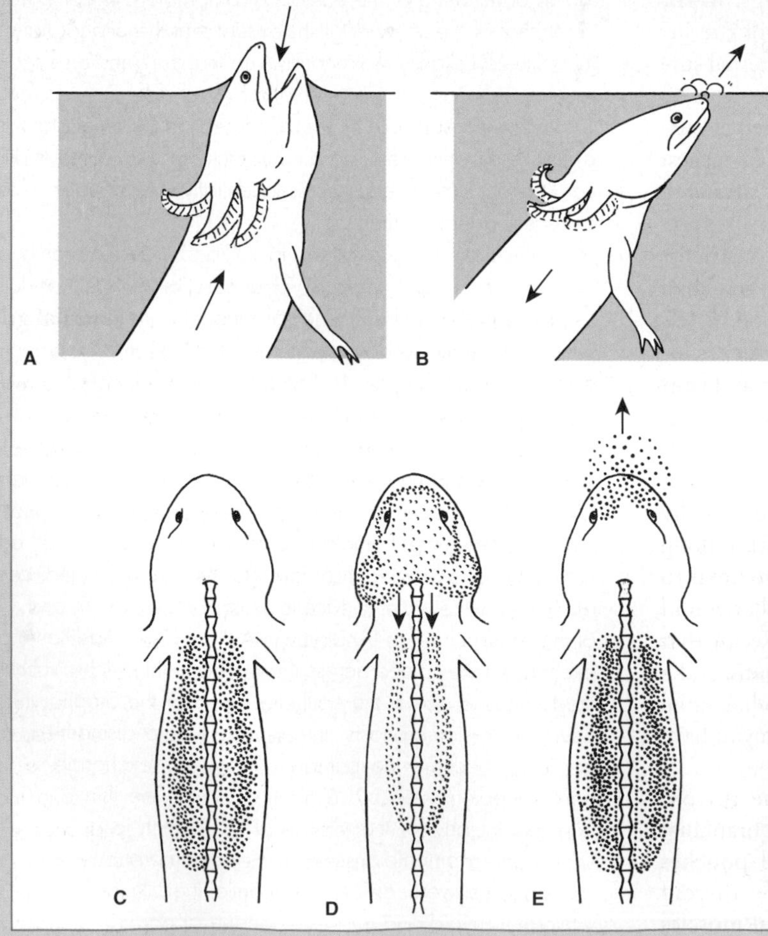

Mechanics of lung ventilation in *Ambystoma,* a larval urodele amphibian similar to *Necturus.* Arrows indicate the direction of air flow. **A** and **B**, Lateral views. **A,** Opening of the mouth and expansion of the oropharyngeal cavity. Air is inhaled into the oropharyngeal cavity and exhaled from the lungs. **B,** Closing of the mouth and compression of the oropharyngeal cavity. Air is pushed into the lungs and residual air in the oropharyngeal cavity is released through the gill slits and the mouth. **C–E,** Drawings from x-ray films of dorsal views of *Ambystoma.* **C,** The lungs are inflated, the glottis and gill slits are closed, and the oropharyngeal cavity is compressed. **D,** The mouth is closed, and the oropharyngeal cavity is filled with air. The lungs are being filled with air through compression of the oropharyngeal cavity. **E,** After the lungs have been filled with air, the oropharyngeal cavity is further compressed and residual air in the oropharyngeal cavity is released through the mouth. *(A and B, after Simons, Bennett, and Brainerd, 2000; C–E,* after Brainerd, 1998.)

(i.e., the tidal volume of the lungs), the residual air in the oropharyngeal cavity is released through the gill slits and mouth opening (Fig. **B** and **E**).

The ventilation process of metamorphosed adult *Amphioxus* on land differs from that of the larval stage in some key aspects. It consists of two components. During **buccal oscillation,** air is first inhaled through the nares, while the floor of the oropharyngeal cavity is depressed and the oropharyngeal cavity expands. Exhalation through the nares proceeds by raising the floor of the oropharyngeal cavity. **Lung ventilation** is initiated by a more pronounced expansion of the oropharyngeal cavity, which fills with fresh air that is inhaled through the nares, as well as with a small amount of exhaled air from the lungs. As the oropharyngeal cavity subsequently is compressed, the air is forced into the lungs. At rest, exhalation from the lungs appears to be a result of elastic recoil of the lungs and surrounding viscera as the glottis opens. During exertions, however, the hypaxial trunk muscles contract and compress the pleuroperitoneal cavity for a more complete emptying of the lungs.

The lung ventilation of metamorphosed adults that return to an aquatic life mode during the mating season resembles that of larvae in certain aspects. On reaching the water surface, they fill their oropharyngeal cavity with air by opening their mouth in a gulping motion, and the lungs are almost emptied by a forceful compression of the pleuropericardial cavity through the contracting hypaxial trunk muscles.

Not surprisingly, the feeding mechanisms of larval aquatic and metamorphosed terrestrial *Amphioxus* differ significantly, too, as shown by Bradley Shaffer and George Lauder (1988). Aquatic larvae ingest food items by using a mechanism comparable to that used by fishes (Anatomy in Action 10-1). They create suction in the oropharyngeal cavity by closing the gill slits and mouth and by depressing the floor of the oropharyngeal cavity. When they open the mouth, water and food rushes into the oropharyngeal cavity; the food is swallowed; and the water is expelled through the open gill slits. Metamorphosed adult salamanders on land cannot use the same mechanism as the larvae, because they would have to generate unrealistic suction forces to create an air stream powerful enough to carry food items into their mouth cavity. Terrestrial amphibians, in general, propel their sticky tongue out of their mouth cavity onto the food item and pull it into the mouth cavity by retracting their tongue. Fully metamorphosed *Ambystoma* salamanders that have returned to the water during the mating season capture prey again by suction, but because their gill slits have closed during metamorphosis, any surplus water, which may have been ingested with the food items, must leave through the mouth opening. Hence, while aquatic larval amphibians feed mostly like fishes, aquatic adults feed more like whales, which also lack gill slits.

paratus is retracted into the trunk. Cutaneous respiration is especially effective in extant amphibians because of their small size and the resulting favorable surface-to-volume ratio, but there is a potential trade-off in evaporative loss of water through the skin. Most amphibians mitigate this by seeking out wet and humid habitats, although some have managed to thrive in deserts and other xeric habitats through special adaptations.

Major morphological and physiological transformations in the digestive tract are necessary during the metamorphosis of aquatic larval amphibians into terrestrial adult ones (Anatomy in Action 10-2). Food acquisition and the morphology of the oropharyngeal cavity of aquatic larval amphibians in principle resemble those of fishes, but those of terrestrial adults are modified to deal with the different physical conditions on land, where gravity is the dominant force. During metamorphosis, the primary tongue acquires the **gland field,** a swelling that develops between the hyoid and mandibular arches, and prehyoid hypobranchial muscles insert on the tongue apparatus. Metamorphosed adult amphibians possess oral **salivary glands,** which produce a slightly sticky secretion, and a muscular **tongue,** which can be projected out of the oral cavity to capture and swallow prey.

In conjunction with the evolution of lungs for breathing air, a **laryngotracheal chamber** has evolved as a gatekeeper to and from the lungs. The intestine is tubular and long, and its internal lining is enlarged through folds and papillae to maximize the surface area for the digestion and absorption of nutrients. The intestine has differentiated into a **small** and a **large intestine.** The **urinary bladder** opens into the cloaca on its ventral side. It is very large and reabsorbs water from the dilute urine generated by the kidneys.

Necturus is an excellent model that exemplifies how the drastic move from an aquatic to a terrestrial environment could have been possible, because it combines features for an aquatic mode of life with features that allow a terrestrial mode of life after metamorphosis. The terrestrial features are present, even though *Necturus* is paedomorphic and, therefore, does not metamorphose; it remains permanently aquatic. One has to keep in mind, however, that the extant amphibians, including *Necturus,* belong to the highly derived group of Lissamphibia, which differ in multiple aspects from the ancestral amphibians (see Fig. 3-1). Because *Necturus* is permanently aquatic, it is not an ideal model of a tetrapod, whose body has been transformed to deal with the terrestrial environment.

Furthermore, *Necturus,* like all urodele amphibians, also lacks an auditory tube and a tympanic cavity (see Anatomy in Action 8-3).

The Pleuroperitoneal Cavity and Its Contents

The Body Wall and Pleuroperitoneal Cavity

If the blood vessels of your specimen were injected with colored latex, a partial incision will have been made through the body wall on one side of the midventral line (see Table 7-4). Extend this incision cranially up to the pectoral girdle and caudally through the pelvic girdle and along one side of the cloacal aperture. The basic organization of the body wall is comparable to that in fishes, such as *Squalus,* and comprises the **skin, connective tissue, hypaxial muscles,** and the **parietal serosa.** Recall the various hypaxial muscles (see Chapter 7). The part of the coelom you have opened is the **pleuroperitoneal cavity.** It is lined with **parietal serosa,** and the inner organs, or viscera, are covered with **visceral serosa.**

The Visceral Organs

The **liver** is the largest of the visceral organs and is located cranioventrally (Fig. 10-11). Notice that it is displaced toward the right side of the pleuroperitoneal cavity. It is quite flat and not as clearly subdivided into lobes as it is in *Squalus.* A portion of the original ventral mesentery, the **falciform ligament,** extends from the midventral line of the pleuroperitoneal cavity to attach along the entire length of the ventral surface of the liver. The **coronary ligament,** which anchors the cranial end of the liver to the transverse septum, is much longer in *Necturus* than in sharks and mammals. You will eventually have to bisect these mesenteries to see all the viscera, but do this in a manner that allows you to reconstruct the large uninjected veins that pass through it, such as the ventral abdominal vein, caudal vena cava, and left hepatic vein (see Chapter 11). Move the liver away from the left body wall, and you will see the elongate **stomach.** It is straight, because it is in its larval stage; in fully metamorphosed terrestrial urodele amphibians it is J-shaped. The **esophagus** is very short and interconnects the oropharyngeal cavity and stomach; you will see it more clearly later. The **spleen** is the elongate, oval organ that is anchored by a short mesentery to the left side of the stomach. It is a lymphoid organ involved in the production and storage of blood cells (see Chapter 11). The paired, very long, left **lung** lies dorsal to the spleen; the similar right lung lies in a comparable position on the other side of the body. In injected specimens, they are completely red because of the density of the injected blood vessels.

The **small intestine** begins as the **duodenum** at the pylorus at the caudal end of the stomach (Fig. 10-11). It loops back cranially before continuing in a number of convolutions of the rest of the small intestine. The whitish, irregularly shaped **pancreas** is anchored to the ascending portion of the duodenum by a mesentery that becomes very short toward the apex of the loop of the duodenum. Another portion of the pancreas adheres to the dorsal surface of the liver that is adjacent to the apex of the loop of

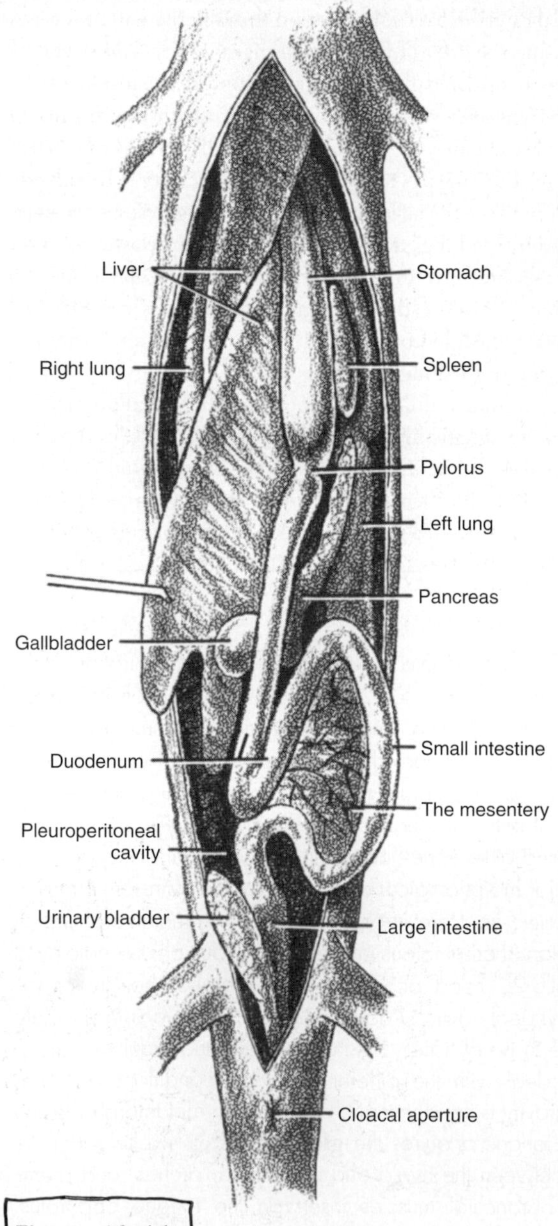

Figure 10-11
Ventral view of the digestive tract and associated viscera of *Necturus.* The caudal half of the liver is reflected so that its dorsal side is visible.

the duodenum. A third portion, the **tail of the pancreas,** is enclosed in the mesentery that connects the spleen and left lung (i.e., the mesogaster; see later) and reaches almost the caudal end of the spleen. Shortly before entering the **cloaca,** the small intestine widens and forms a short **large intestine.** The elongate **urinary bladder,** which usually is collapsed and shriveled in preserved specimens, is located ventral to the large intestine and opens into the cloaca independently from the urogenital and digestive tracts.

If your specimen is a female, the large, paired granular **ovaries** may almost fill the pleuroperitoneal cavity, and the long, convoluted **oviducts** lie dorsal to each ovary and ventral to the **kidneys** (see Fig. 12-11). If your specimen is a male, a pair of large, elongate, oval **testes** are located dorsolateral to the small intestine. The **kidneys** lie dorsal and lateral to the gonads. They project into the pleuroperitoneal cavity, and each is anchored to the body wall by a subsidiary mesentery. In males, each kidney is located in a special fold of the subsidiary mesentery anchoring the testis. Each kidney is accompanied and drained by a conspicuous, convoluted, often pigmented **archinephric duct** along its lateral border.

The Mesenteries

First find the spleen and the mesentery that anchors it to the stomach; this is the **gastrosplenic ligament** (Fig. 10-12). Now look at the dorsal side of the spleen and notice a longer mesentery that may appear to attach the spleen to the left lung. In reality, however, this mesentery continues to the middorsal wall of the pleuroperitoneal cavity, and the left lung is enclosed in a special fold of this mesentery. The entire mesentery is called the **mesogaster;** the gastrosplenic ligament is a part of it. The very short mesentery anchoring the left lung to the mesogaster is the **left pulmonary ligament** *(left mesopneumonium).* The mesogaster is a special portion of the original dorsal mesentery. The portion of the dorsal mesentery that gathers the various convolutions of the small intestine and anchors them to the middorsal body wall is **the mesentery** (in the limited sense) (Fig. 10-11). The portion that anchors the straight large intestine is the **mesocolon.**

You have already inspected and bisected the **falciform ligament.** It is part of the primitive ventral mesentery. Most of it was dismantled in the course of embryonic development, but in *Necturus,* it is much more extensive than in sharks or mammals. Its caudalmost portion is still present as the **median ligament of the urinary bladder,** which anchors the urinary bladder to the midventral body wall. The **hepatogastric ligament** interconnects the cranial end of the stomach to the right side of the liver. The short and narrow **hepatoduodenal ligament** interconnects the liver to the apex of the loop of the duodenum. It encloses a portion of the pancreas. Notice that the tail of the pancreas is enclosed in the mesogaster.

Return to the remainder of the dorsal mesentery on the right side of the pleuroperitoneal cavity. Find the right lung and observe that it lies in a fold of a mesentery arising from the dorsal side of the liver. This large mesentery is the **hepatocavopulmonary ligament,** so called because its attachment to the liver follows the course of the underlying vena cava (Fig. 10-12). Dorsally, it fuses to the mesogaster. The short mesentery that connects the right lung to the hepatocavopulmonary ligament is the **right pulmonary ligament** *(right mesopneumonium).*

The subsidiary genital mesenteries of a male comprise the **mesorchium,** which anchors the testis, and its special

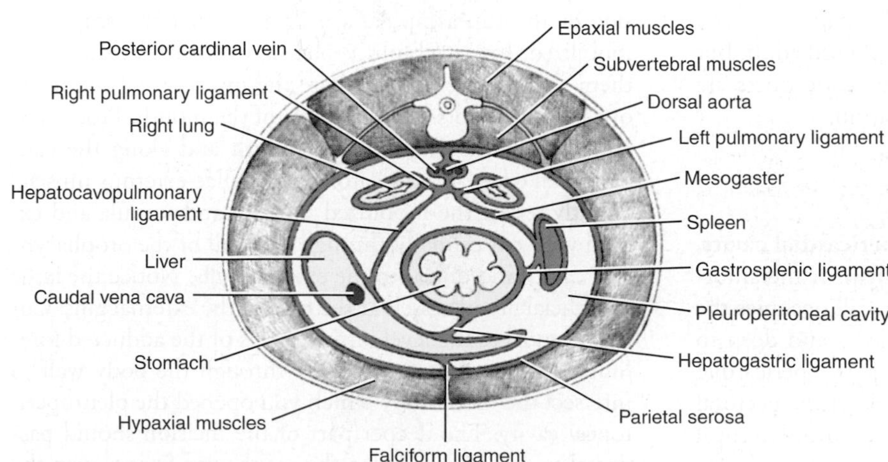

Figure 10-12

Diagrammatic transverse section through a male of *Necturus* at the level of the spleen showing the mesenteric relationships, as viewed from cranial.

fold for the kidneys. In a female, the subsidiary mesenteries comprise the **mesovarium,** which anchors the ovary; the **mesotubarium,** which anchors the oviduct; and a separate mesentery that anchors the kidney.

Structure of the Digestive Organs

Cut open the stomach and wash it out. Its lining is thrown into a number of irregular and longitudinal folds called **rugae.** The stomach comprises the **cardiac region** next to the esophagus and closest to the heart; the central **gastric body,** or **gastric corpus;** and the caudal **pyloric region,** but these regions are not clearly demarcated. The **pylorus** separates the stomach from the small intestine.

Cut open the small intestine at several points and notice that there is no spiral valve. Instead, its internal surface area is increased through many small, wavy, longitudinally arranged folds called **plicae.** The tubular intestine of *Necturus* is much longer than the valvular intestine of the shark and, therefore, is thrown into loops and convolutions that allow it to fit within the pleuroperitoneal cavity. This elongation of the tubular intestine has the same effect as the spiral valve in the valvular intestine of shark, namely, to increase the surface area for absorptive and digestive processes.

Most of the bile is drained from the liver by several **hepatic ducts;** these unite to form a **common bile duct,** which opens into the duodenum. Some of the bile backs up through a **cystic duct** into the large **gallbladder,** where it is temporarily stored. You can find the gallbladder projecting from the dorsal surface of the liver near the apex of the loop of the duodenum, but most of the hepatic ducts are imbedded in the pancreatic tissue within the hepatoduodenal ligament. Some hepatic ducts can be exposed by careful dissection.

The pancreas of urodele amphibians develops embryonically from a dorsal primordium and a pair of ventral primordia. All parts merge in a metamorphosed adult, but each part retains its own duct. The pancreatic ducts are too small to be found without magnification.

The Pericardial Cavity

The second division of the coelom, the **pericardial cavity,** is located farther rostrally in *Necturus* than in metamorphosed terrestrial amphibians, because it still occupies the larval position cranial to the pectoral girdle and deep to the hypobranchial musculature. To open the pericardial cavity and expose the heart within, first bisect the pectoral muscle on the side where you opened the pleuroperitoneal cavity (see Table 7-4). Move the coracoid plate with the supracoracoid muscle to the side so as not to damage it, before extending the incision through the wall of the pleuroperitoneal cavity cranially. In doing so, you will also bisect the vertical **transverse septum,** which completely separates the pleuroperitoneal cavity from the pericardial cavity. The **coronary ligament,** which is long in *Necturus,* extends from the caudal surface of the transverse septum to the cranial end of the liver. A large vein, the caudal vena cava (see Chapter 11) passes through this ligament.

The pericardial cavity is lined with **parietal pericardium** and contains only the **heart,** which is covered with **visceral pericardium.**

The Oropharyngeal Cavity and Respiratory Organs

Because *Necturus* is paedomorphic, its major respiratory organs are three pairs of **external gills** that project from the back of the head; the lungs are used occasionally (Anatomy in Action 10-2). If your specimen's blood vessels were injected, the gill filaments will be red because of the high density of blood vessels showing underneath the thin translucent gill epithelium. Spread the external gills apart and find the two **gill slits** between their bases. If a live specimen can be watched in an aquarium, notice the numerous **gill filaments** and observe how the gills are slowly moved back and forth through the water.

In order to open the oropharyngeal cavity (unless you have not already done so in Chapter 8), turn to the side of the head where you dissected the muscles (see Table 7-4). Identify the levator mandibulae externus, levatores arcuum, dilatator laryngis, cucullaris, scapular deltoid, and latissimus dorsi muscles (see Fig. 7-12). Detach the origins of the levatores arcuum and cucullaris muscles from the epibranchial musculature and bisect the latissimus dorsi muscle. Gently bend the suprascapular cartilage (see Chapter 6) away from the trunk and bisect the levator scapulae and thoraciscapular muscles, which insert on the medial side of the suprascapular cartilage so that the scapula is mobilized. Insert a blunt probe between the jaws and pry them slightly open to facilitate the insertion of one blade of a pair of scissors into the angle of the mouth. Proceed to cut through the angle of the mouth and along the caudoventral edge of the levator mandibulae externus muscle. Slightly reflect the mobilized muscles and scapula and extend your cut caudally through the wall of the oropharyngeal cavity and dorsal to the external gills. Notice the large superficial blood vessels at the base of the external gills. Cut as far caudal as the level of the carpus of the adducted forelimb, then proceed transversely through the body wall to intersect the incision by which you opened the pleuroperitoneal cavity. The deeper part of the incision should pass dorsal to the lung and into the esophagus. Swing open the floor of the oropharyngeal cavity (Fig. 10-13). Spread apart the cut edges and notice that only the wide oropharyngeal cavity has been cut open. Clear the lateral side of the esophagus and stomach as far back as the level of the carpus of the adducted pectoral limb by mobilizing the lungs and larger blood vessels. You may have to break some mesenteries when doing so. Now extend the longitudinal cut through the wall of the esophagus and stomach. At the level of the carpus of the adducted pectoral limb, make a transverse cut through the ventral wall of the gut, but leave the lung intact, if possible, by extending the incision underneath it. You will now be able to spread open the oropharyngeal cavity without tearing tissues (Fig. 10-13).

In vertebrates with gills and gill slits, it is not possible to draw a sharp line between the rostral **oral cavity** and the more caudal **pharynx.** This is why it is better to call the entire cavity the **oropharyngeal cavity.** Caudally, it leads to a slightly constricted, short passage, the **esophagus,** which opens into the stomach. The longitudinal folds in the lining of the esophagus tend to be smaller than those in the stomach.

The **tongue** is a fold of tissue supported by the hyoid arch. It is located in the floor of the oropharyngeal cavity (Fig. 10-13). It is very similar to the primary tongue of fishes, but a few hypobranchial muscle fibers insert on its base. The third to fifth visceral arches lie caudal to the hyoid arch and can be palpated. The two gill slits are located between the third and fourth and the fourth and fifth visceral arches. Small **gill rakers** can be seen on the oropharyngeal side of the gill slits.

Necturus can breathe air and possesses all the structures that are needed for pulmonary respiration (Anatomy in Action 10-2). If you did not dissect the nasal cavity when you studied the sense organs (see Chapter 8), expose it now. A **choana,** or **internal nostril,** can be seen in the roof of the mouth lateral to the pterygoid teeth, the most caudal and shortest of the tooth rows (see Chapter 4). The **glottis** is a very short, inconspicuous longitudinal slit in the floor of the oropharyngeal cavity at the level of the caudal gill slits. The glottis leads into a median **laryngotracheal chamber,** which leads to the lungs through a pair of openings at its caudal end and the very short **bronchi.** Make an incision into one lung and insert a blunt probe towards the glottis. A pair of small cartilages, called **lateral laryngeal cartilages,** support the wall of the laryngotracheal chamber. They are difficult to discern, although they can be palpated. They probably are derived from the sixth and seventh visceral arches. Cut open one of the lungs and observe that it is an empty sac with a smooth internal surface. In most metamorphosed terrestrial amphibians, the internal surface area is increased by pocketlike folds. If possible, examine a preparation of the lung of a frog.

The **thyroid gland,** a series of **parathyroid glands,** and the **thymus** develop embryonically from the pharyngeal epithelium, but they are difficult to find. The thyroid gland may be seen during the dissection of the arteries (see Fig. 11-12).

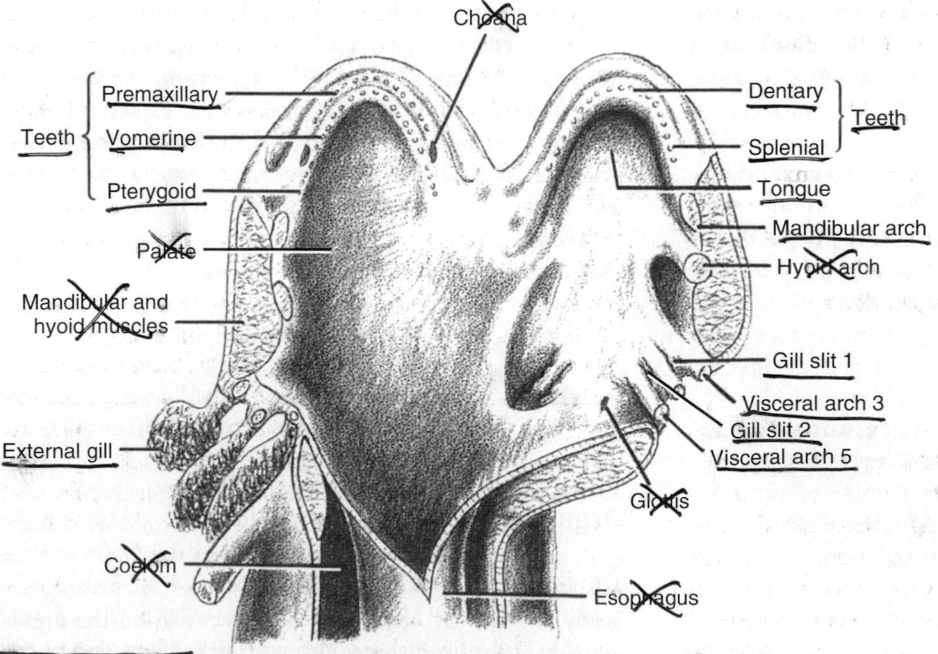

Figure 10-13

Oropharyngeal cavity of *Necturus*. The floor of the oropharyngeal cavity has been reflected to the left of the specimen.

MAMMALS

The evolution of tetrapods from ancestral amphibians to ancestral amniotes and on to mammals has been one of elaboration upon terrestrial adaptations and even multiple returns to an aquatic life mode. Here, we concentrate on terrestrial mammals. Mammals are active, warm-blooded animals. Most maintain a high and narrowly determined rate of metabolism, that is, they are **endothermic** as well as **homoiothermic.** It is important to realize that homoiothermy-endothermy (warm bloodedness) by itself is not necessarily better than cold bloodedness (poikilothermy-ectothermy). For example, homoiothermy-endothermy requires frequent provisions of energy simply to keep up the continuously high rate of metabolism. Cold-blooded animals, such as amphibians and most reptiles, can survive with fewer and relatively smaller amounts of nutrition because their rate of metabolism is tied to the environmental conditions. Several evolutionary changes in the digestive, respiratory, circulatory, and excretory systems of mammals have enabled mammals to develop and sustain their high level of metabolism.

In conjunction with the evolution of a more mobile and longer neck, the heart and pericardial cavity have become situated further away from the head and in the thorax caudal to the pectoral girdle. The paired pleural recesses of terrestrial amphibians and many reptiles became fully separated as the paired **pleural cavities** containing the lungs. The rest of the former pleuroperitoneal cavity became the **peritoneal cavity** containing most of the remaining viscera. The serosal folds that separate the pleural cavities from the peritoneal cavity were invaded by muscles and formed the muscular **diaphragm,** whose movements, together with those of the ribs, provide the motor for the ventilation of the lungs.

With the evolution of a longer neck, the laryngotracheal chamber of amphibians differentiated into a **larynx,** and the **trachea** negotiated the longer distance between the larynx and the lungs. The evolution of a **hard** and a **soft palate** resulted in a separation of the nasal passage from the oropharyngeal cavity. This construction allows simultaneous respiration and processing of food within the mouth cavity. It also braces the upper tooth rows during mastication. Food and air passages cross each other only in the caudal part of the pharynx. Normally, food and air are prevented from entering the wrong passages by reflexes that coordinate the swallowing motions and the opening and closing of the glottis.

In most reptiles and in mammals, the lungs are the only respiratory organs. In mammals, the respiratory passage branches extensively within the lungs and terminates in grapelike clusters, or **acini,** of minute, thin-walled, highly vascularized **alveoli,** where gas exchange occurs. This great increase of the respiratory surface area enables mammals to maintain their high levels of metabolism.

Many changes have occurred in the digestive system of mammals. Large **salivary glands** produce saliva to lubricate food items during mastication. Some salivary glands produce **ptyalin,** a digestive enzyme that initiates the hydrolysis of starch. The tongue, which manipulates food within the mouth cavity and initiates the swallowing process, has evolved into a muscular hydrostat through the invasion of hypobranchial muscles (see Anatomy in Action 7-2). The mammalian tongue has evolved from the primary tongue, which is typical of fishes and larval amphibians, through the addition of a pair of **lateral lingual swellings** and a gland field, or **tuberculum impar,** to the rostral end of the embryonic tongue. The hyobranchial skeleton was reduced to the hyoid body, which anchors the lingual muscles, and the hyoid ossicles, which suspend the lingual apparatus from the skull (see Chapters 4 and 7). The lingual muscles are innervated by the hypoglossal nerve, which is derived from the hypobranchial nerve of fishes (see Table 7-1), but the sensory innervation of the tongue is provided by the trigeminal (V), facial (VII), and glossopharyngeal (IX) nerves. These sensory nerves supply comparable parts of the oropharyngeal cavity in fishes and amphibians. Specifically, the trigeminal nerve provides general sensory nerve fibers to approximately the rostral two thirds of the tongue; the facial nerve provides gustatory nerve fibers to the taste receptors in the rostral two thirds of the tongue; and the glossopharyngeal nerve supplies both general sensory and gustatory fibers to the caudal third of the tongue.

The digestive tract is characterized by a great increase in internal surface area. This is accomplished through an elongation of the intestine and through numerous minute, fingerlike **villi.** The cloaca of therian mammals has become divided into dorsal and ventral portions. The ventral portion of the cloaca contributes to the formation of the urogenital passages; the dorsal portion contributes to that of the **rectum.** Superimposed upon these general features are numerous modifications of the digestive tract in adaptation to a great variety of diets and modes of life among mammalian species. The digestive tract of cats is an excellent example of a digestive tract adapted to a carnivorous diet, whereas that of rabbits exemplifies a digestive tract adapted to a diet in which fibrous vegetal matter predominates. The stomach of sheep exemplifies the multichambered stomach of herbivorous mammals specialized on a fiber-rich, protein-poor diet.

The Digestive and Respiratory Organs of the Head and Neck

The Salivary Glands

Study the salivary glands of your specimen on the opposite side of the head that you used for the dissection of the jaw, prehyoid and branchiomeric muscles (see Table 7-5). Carefully remove the skin overlying the cheek, throat, and side of the neck ventral to the auricle. You must also reflect and remove the superficial cutaneous facial muscles and platysma (see Chapter 7). You may notice that the cutaneous facial muscles over the cheek comprise two layers with muscle fibers oriented crosswise to each other.

Pick away any connective tissue ventral to the base of the ear, and you will expose the large **parotid gland.** You can recognize it by its lobulated texture. This salivary gland is approximately oval in cats (Fig. 10-14) but is dumbbell shaped in rabbits. The **parotid duct** emerges from the rostroventral edge of the parotid gland, crosses the large masseter muscle, and perforates the epithelium lining of the upper lip near the molar teeth to empty into the oral vestibule (see later). Frequently, individual accessory lobules of glandular tissue are found along the parotid duct. Two nerve branches, the **dorsal** and **ventral buccal rami** of the **facial nerve,** emerge from underneath the parotid gland, cross the masseter muscle dorsal and ventral to the parotid duct, and supply the facial muscles. Do not confuse them with the parotid duct. The parotid duct can be traced into the parotid gland, whereas the facial nerve extends deep to the parotid gland toward the stylomastoid foramen through which it leaves the skull (see Fig. 4-22).

The **mandibular gland** is located caudal to the angular process of the jaw and deep to the ventral edge of the parotid gland. It is a large, oval salivary gland with a lobulated texture. The individual lobules are larger than those of the parotid gland. Do not confuse it with the smaller, smoother-textured **mandibular lymph nodes** of variable sizes in this region. The **mandibular duct** emerges from the rostral side of the mandibular gland and runs rostrally, first passing lateral to the digastric muscle. The digastric muscle is the large muscle that arises from the base of the skull and inserts along the ventral border of the lower jaw (see Chapter 7). Detach the origin of the digastric muscle from the mandibular ramus and reflect it (if you have not already done so), so that you can see the mandibular duct continuing deep to the caudal border of the mylohyoid muscle. The mylohyoid muscle is the thin muscular sheet with transverse muscle fibers located between the paired digastric muscles. Bisect and reflect the mylohyoid muscle (if you have not already done so) and follow the mandibu-

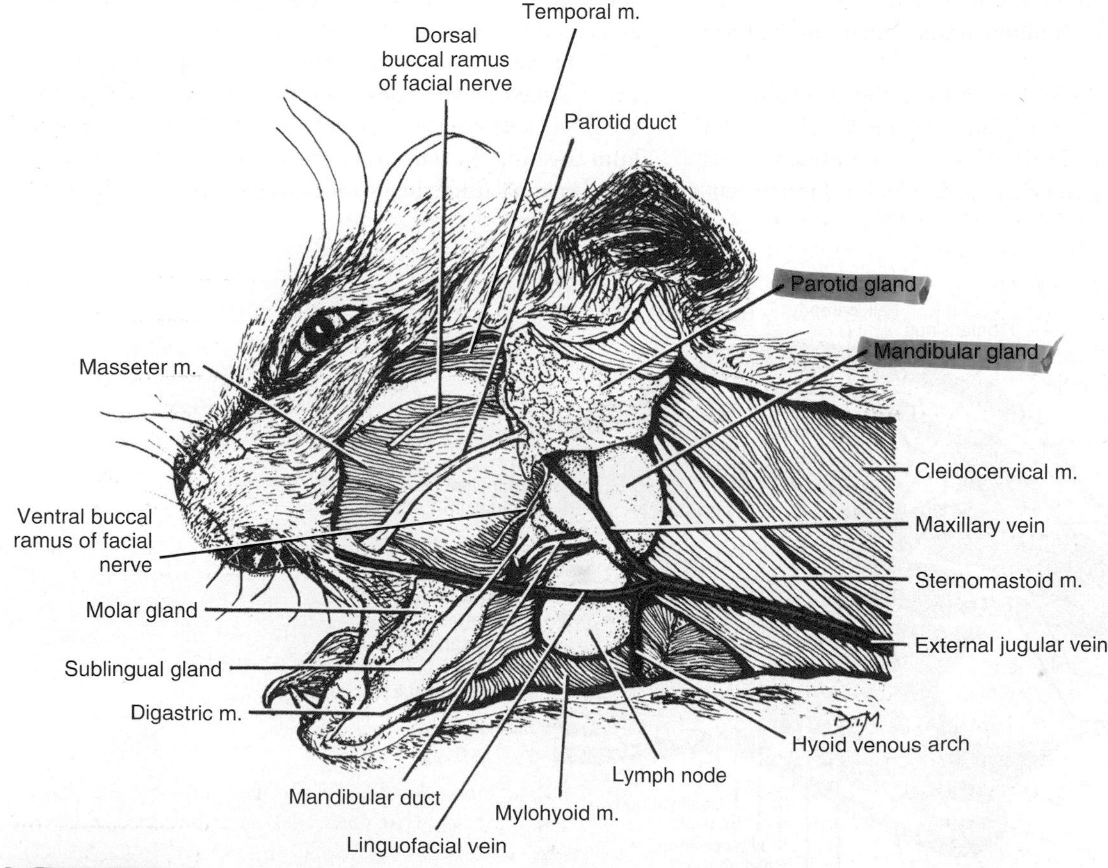

Figure 10-14

Lateral view of the salivary glands and surrounding superficial structures of the head and neck of a cat.

lar duct rostrally as far as you can. It is crossed rostrally by the **lingual nerve** returning sensory nerve fibers from the tongue. The lingual nerve contains mostly general sensory cutaneous nerve fibers of the trigeminal nerve, but it also comprises some the gustatory nerve fibers from the facial nerve. The **hypoglossal nerve,** which carries motor fibers to the tongue muscles, crosses the ventral side of the mandibular duct on its way from the jugular foramen through the hypoglossal canal in the otic region of the skull (see Chapter 4) before it dips rostrally into the musculature of the tip of the tongue. The mandibular duct continues rostrally until it reaches the mandibular ramus. Here it dips into the muscular floor of the mouth and opens just rostral to the **lingual frenulum,** the midventral septum of tongue. In cats, the paired openings of the mandibular ducts are hidden under flattened papillae, which can be lifted with a pair of forceps.

The small, elongated **sublingual gland** is situated beside the mandibular duct. In cats, this salivary gland lies along the caudal one third of the mandibular duct and usually abuts against the mandibular gland. In rabbits, it lies along the rostral one third of the mandibular duct. The sublingual gland is drained by a minute duct that parallels the mandibular duct, but it is hard to identify without magnification. The sublingual duct opens on the floor of the mouth.

The parotid, mandibular, and sublingual glands are the most common salivary glands in mammals, but in some species, additional salivary glands are present. Cats and rabbits have a **zygomatic gland,** which is located ventral to the eyeball; you have already seen it if you dissected the eye (see Fig. 8-10). Cats possess a small, elongate **molar gland** situated between the skin and epithelium of the caudal half of the lower lip. It empties on the inside of the lower lip through several small ducts that cannot be seen without magnification. Rabbits possess elongated **buccal glands** beneath the skin of the upper and lower lips.

The Oral Cavity

Because it is not possible to pry open the jaws in preserved specimens, you need to gain access to the **oral cavity** by separating the soft floor of the oral cavity from the mandible. On the side on which the stylohyoid and mylohyoid muscles were bisected earlier (see Chapter 7; Table 7-5), separate the digastric muscle from the mandible at its origin and reflect it. With a scalpel, cut through the entire length of the muscular floor of the mouth as close as possible to the mandibular ramus. The scalpel blade will emerge within the oral cavity next to the tongue. Cut through the symphysis of the mandible with a pair of bone cutters. Pry the two mandibular rami apart, and pull the tongue and the floor of the mouth ventrally through the gap so that you can look into the oral cavity. If the tissues are too stiff, you may have to cut partly through the floor of the mouth also along the opposite mandibular ramus.

The rostral part of the roof of the oropharyngeal cavity is formed by the bony **hard palate,** the caudal part by the fleshy **soft palate** (Figs. 10-15 and 10-16). The epithelium covering the hard palate is heavily cornified and bears transverse palatine ridges called **rugae.** In cats, the rugae

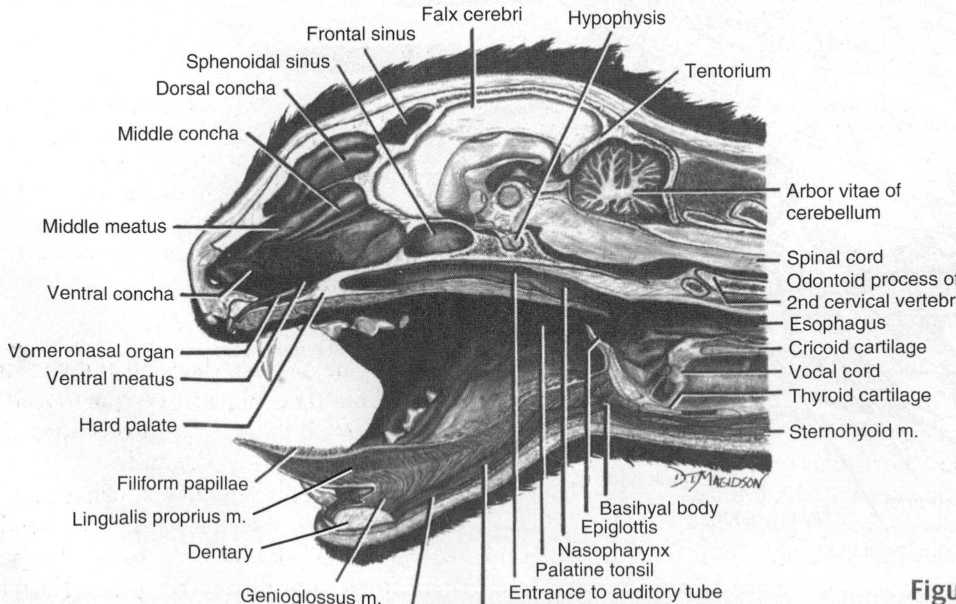

Figure 10-15
Sagittal section through the head of a cat.

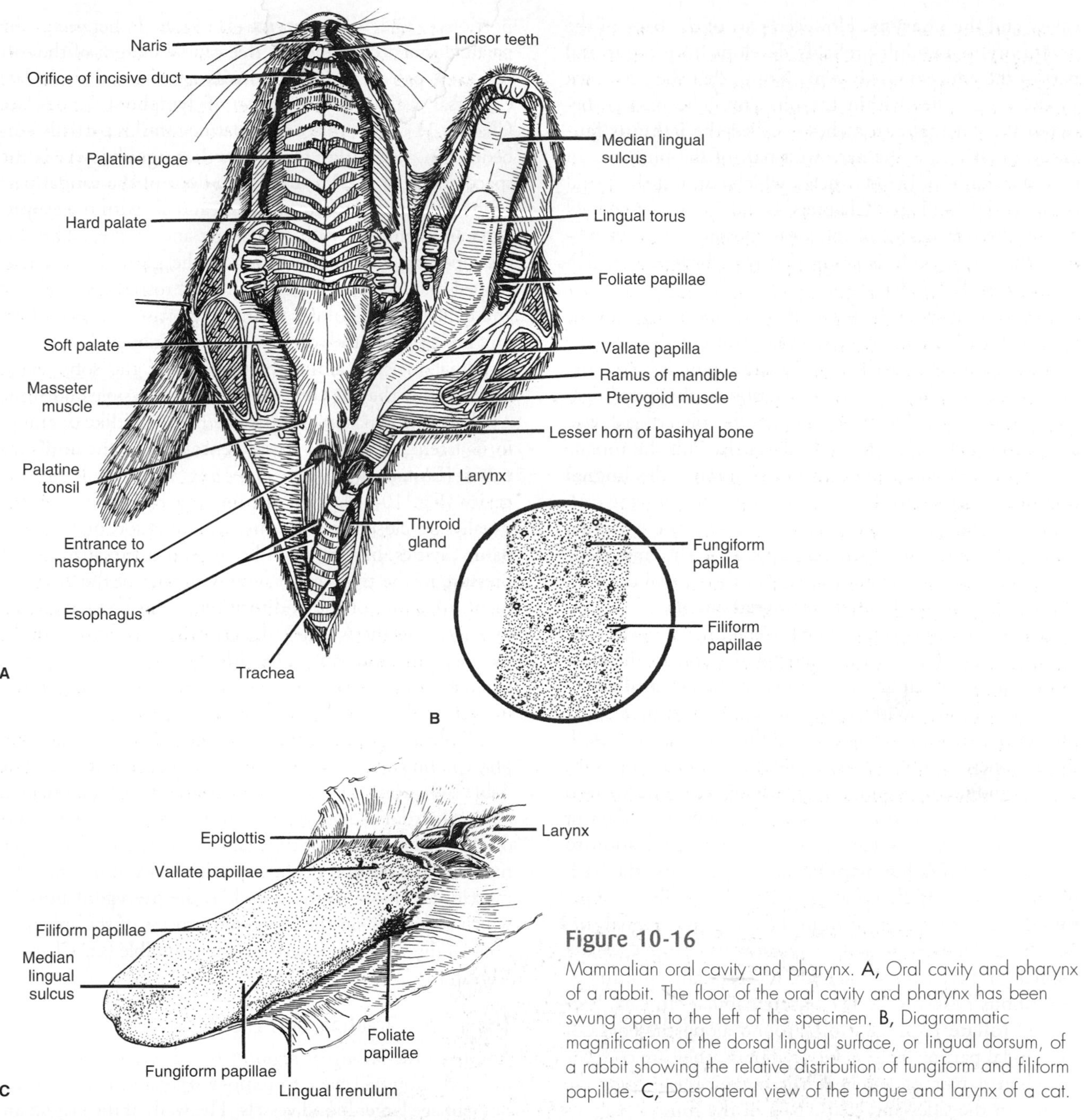

Figure 10-16

Mammalian oral cavity and pharynx. **A,** Oral cavity and pharynx of a rabbit. The floor of the oral cavity and pharynx has been swung open to the left of the specimen. **B,** Diagrammatic magnification of the dorsal lingual surface, or lingual dorsum, of a rabbit showing the relative distribution of fungiform and filiform papillae. **C,** Dorsolateral view of the tongue and larynx of a cat.

carry cornified **papillae.** The rugae and papillae point caudally and assist the tongue in moving the food toward the pharynx during mastication. You will see a pair of small openings at the very front of the hard palate just caudal to the incisor teeth. These are the openings of the **incisive ducts,** which pass through the palatine fissures (see Chapter 4) and lead to the vomeronasal organs in the floor of the nasal cavities (see Chapter 8). The lateral walls of the oral cavity are bounded by the teeth, lips, and cheeks. The portion of the oral cavity lying between the teeth and cheeks is called the **oral vestibule.** The muscular **tongue** *(lingua)* lies in the floor of the oral cavity to which it is connected by the vertical lingual frenulum seen previously.

Pull the tongue far enough ventrally to tighten and bring into prominence a pair of lateral folds that extend from the sides of the caudal portion of the tongue to the soft palate. These folds constitute the **palatoglossal arches,** and they represent the boundary between the oral cavity and the **pharynx.** However, part of the back of the oral cavity in an adult probably develops from the rostral part of the embryonic pharynx. Notice that the very back of the tongue lies within the pharynx. The passage between the palatoglossal arches is called the **isthmus faucium,** because it is the narrowest part of the pharynx. In cats, the chains of hyoid ossicles, which connect the hyoid body to the skull (see Chapters 4 and 7), are imbedded within the soft tissues of the palatoglossal arches. In rabbits, the hyoid body is suspended by a ligament, which may not easily be identified.

In a cat, palpate the chain of hyoid ossicles within the palatoglossal arch on the side where you cut through the floor of the mouth (see Table 7-5) and expose it by removing the surrounding soft tissue. In rabbits, the hyoid ligament is not palpable. Cut through the pharyngeal wall and, in cats, through the epihyal ossicle. Pull the tongue farther down and examine its dorsal surface, the **lingual dorsum** (Fig. 10-16). In rabbits, but not in cats, the caudal half of the lingual dorsum is raised and forms a **lingual torus.** The lingual torus has developed from the embryonic tuberculum impar. The rostral part of the lingual dorsum, which is divided by a **median lingual sulcus,** has developed from the embryonic paired lateral lingual swellings.

The lingual dorsum is covered with papillae, the most numerous of which are the pointed **filiform papillae.** In cats, the rostral filiform papillae are stiff and bear spiny projections that assist in grooming the fur or rasping flesh from bones, but the caudal filiform papillae are soft. Small, rounded **fungiform papillae** are interspersed among the filiform papillae, especially along the edges of the tongue. You may have to use some magnification to see them well. **Vallate papillae** are located near the back of the tongue. Each vallate papilla is a relatively large, round patch that is set off from the rest of the lingual surface by a circular moat. Cats possess four to six vallate papillae arranged along the limbs of a caudally pointing V, but rabbits possess only two vallate papillae. In cats, leaf-shaped **foliate papillae** can be found along the edges of the caudal part of the tongue. In rabbits, they are concentrated in a pair of patches. Microscopic taste buds are found on the sides and at the base of the fungiform, vallate, and foliate papillae.

The Pharynx

With a pair of scissors, cut caudally through the lateral wall of the pharynx on the side on which the palatoglossal arch was bisected (see Table 7-5). Follow the contour of the tongue to the laryngeal region, then extend the cut dorsal to the larynx and caudally into the esophagus. Do not cut into the soft palate or larynx. Swing open the floor of the oral cavity and pharynx. For descriptive purposes, the pharynx can be divided into oral, nasal, and laryngeal portions. The **oropharynx** is located between the palatoglossal arches and the free caudal margin of the soft palate. A pair of **palatine tonsils** are situated in its laterodorsal walls. They consist of lymphoid tissue (see Chapter 11). Note that each palatine tonsil is partially embedded in a **tonsillar fossa.** The **laryngopharynx** is the space dorsal to the **larynx** in the floor of the caudal part of the pharynx. It communicates caudally with the **esophagus** and ventrally with the larynx and trachea. The slitlike opening within the larynx is the **glottis.** A trough-shaped fold, the **epiglottis,** is situated rostral to the glottis and deflects food boluses from the glottis. The **nasopharynx** lies dorsal to the soft palate. Open it by making a longitudinal incision along the midline of the soft palate. Spread apart the cut edges as widely as possible and shine a light into the nasopharynx. The pair of slitlike openings in its laterodorsal walls are the entrances of the **auditory tubes** (Eustachian tubes) emerging from the tympanic cavity (Fig. 10-15). The openings to the auditory tubes usually are closed to prevent contamination of the tympanic cavities, but they can be opened to equalize the air pressure in the tympanic cavity with that of the environment. Opening of the auditory tubes can be precipitated by stretching the walls of the nasopharynx, for example, by yawning or chewing. The **choanae,** or internal nostrils, enter the rostral end of the nasopharynx, but they cannot be seen in this view.

Reflect the epithelium that lines the soft palate and pharynx on the side where the pharynx was cut open (see Table 7-5). You will see a mass of muscles that are oriented obliquely from caudolateral to rostromedial and that insert on the medial surface of the mandibular ramus. The caudal muscle is the portion of the superficial part of the masseter muscle. The more rostral muscle is the **pterygoid muscle,** which originates on the pterygoid process of the sphenoid bone of the skull and inserts on the mandible (see Chapters 4 and 7).

The Larynx, Trachea, and Esophagus

Approach the laryngeal region from the ventral surface of the neck. You will have to remove several muscles, but do not damage larger blood vessels. The walls of the **larynx** are supported by relatively large cartilages (see Fig. 4-19**A**). The body of the **hyoid body,** or **basihyal,** is a transverse rod ventral to the epiglottis. Free it by removing connective tissue, musculature, and epithelium ventral to the epiglottis. Its **greater horns,** or **thyrohyal,** articulate with the rostral processes of the thyroid cartilage of the larynx. In cats, its **lesser horns,** or **ceratohyal,** articulate with the chain of hyoid ossicles; in rabbits, they are connected to the hyoid ligament. Both the hyoid chain of ossicles of cats and the hyoid ligament of rabbits extend dorsally to the mastoid process of the skull (see Figs. 4-19**A**, 4-25, and 5-5). Both will have been cut on the side on which the palatoglossal arch was bisected. Remove enough muscle tissue to expose the hyoid ossicles. The larynx continues caudally as the windpipe, or **trachea,** the walls of which are supported by a series of cartilaginous rings that keep the tracheal lumen open. These **tracheal rings** are incomplete. Their ends do not quite meet dorsally, and they are interconnected by connective tissue and smooth nonvoluntary muscles, which can modify the diameter of the trachea. The esophagus is a collapsed muscular tube lying dorsal to the trachea. Its lumen is pushed open when food is swallowed.

The dark **thyroid gland** lies against the cranial end of the trachea. In cats, it consists of a pair of **thyroid lobes** that are connected across the ventral surface of the trachea by a very narrow band of thyroid tissue called the **thyroid isthmus.** The thyroid isthmus often is inadvertently destroyed. In rabbits, the thyroid isthmus is wide and prominent. Two pairs of **parathyroid glands** are imbedded in the dorsomedial surface of the thyroid, but they cannot be identified without magnification.

Return to the larynx and study it more thoroughly. The very small intrinsic laryngeal muscles are derived evolutionarily from the caudal branchiomeric musculature in fishes and, hence, are innervated by the vagus nerve. They are named according to their attachments on the laryngeal cartilages. Major superficial muscles include, from caudal to rostral, the cricothyroid muscle and the thyrohyoid muscle. Remove these muscles from the outer surface of the larynx to expose the laryngeal cartilages. The large rostral cartilage, which forms much of the ventral and lateral walls of the larynx, is called the **thyroid cartilage.** This cartilage forms the projection known as the Adam's apple in the neck of mature human males. The cartilaginous ring caudal to the thyroid cartilage is the **cricoid cartilage.** It is shaped like a signet ring, because its dorsal portion is greatly expanded and forms most of the dorsal wall of the larynx. Careful dissection will reveal a pair of small triangular cartilages cranial to the dorsal part of the cricoid cartilage. These are the **arytenoid cartilages.** Additional minute cartilages frequently are associated with the arytenoid cartilages, but they are seldom seen. The rostral **epiglottic cartilage** supports the epiglottis.

Cut open the larynx along its middorsal line. The pair of whitish lateral folds that extend from the arytenoid cartilages to the thyroid cartilage are the **vocal cords.** Their tension is controlled by the action of muscles moving the arytenoid cartilages and changing the configuration of the larynx. The **glottis** is the slit between them; normally it is closed, except when air is being inhaled or exhaled. In cats, a pair of accessory folds, also called **false vocal cords,** extend from the arytenoid cartilages to the base of the epiglottis. In rabbits, the paired **epiglottic hamuli,** which are small bumps, lie at the base of the epiglottis. The larynx performs the crucial role of gatekeeper for the respiratory tract by opening or closing the glottis. The glottis needs to be open to allow passage of air, but during deglutition, it needs to be closed to prevent food or liquid from entering the respiratory tract. Sound is generated when the vocal cords clap together after they have been separated by a stream, or puff, of air.

The laryngeal cartilages of mammals have evolved from branchial arches of ancestral fishes and amphibians, but the precise homologies are unclear. The second and third visceral arches evolved into the hyoid skeleton of mammals. The fourth and fifth visceral arches become part of the hyoid apparatus in amphibians, but in most mammals, they form the thyroid cartilage, which is present only in mammals. The sixth and seventh visceral arches form the lateral cartilages of the laryngotracheal chamber of amphibians, but in mammals these arches form the arytenoid and cricoid cartilages. The seventh visceral arch may have evolved into the tracheal rings through multiplication. The epiglottic cartilage may be a new structure in mammals.

The Thorax and Its Contents

The Pleural Cavities

Open the thorax with a pair of strong scissors by making a longitudinal incision about 2 cm to the right side of the midventral line and extending it for the length of the sternum. Spread open the incision and look into the **right pleural cavity.** You will see the dome-shaped muscular **diaphragm** forming the caudal wall of the pleural cavity. Just cranial to the diaphragm, make another cut that extends laterally and dorsally to the back. Follow the line of attachment of the diaphragm to the body wall, but keep on the pleural side. Now feel for the individual ribs along the inner surface of the thoracic wall. With a strong pair of scissors, cut through each rib near its attachment to the vertebrae, but leave the ribs in place. In this way, you can open the right pleural cavity by spreading the thoracic wall laterally. If you simply do this with brute force without first cutting through the ribs, the ribs will break and splinter and may damage the thoracic structures and injure your fingers.

The right pleural cavity and the **right lung** *(pulmo)* are now well exposed. The serosa that lines the walls of the pleural cavity is called the **parietal pleura;** the serosa that covers the surface of the lung is called the **visceral,** or **pulmonary, pleura.** The right lung is divided into four **lobes,** namely, the **cranial, middle, caudal,** and **accessory** lobes (Fig. 10-17). The accessory lobe passes dorsal to a large vein, the **caudal vena cava,** and then projects ventrally into a pocket formed by the **caval fold,** a mesentery that anchors the vein to the diaphragm (Fig. 10-18). By moving the accessory lobe, you can ascertain its presence medial to the caval fold. Another pleural fold, the **pulmonary ligament,** anchors the lobes of the lung to the medial wall of the pleural cavity (Figs. 10-3**C** and 10-18). The **principal bronchus,** one branch of the terminal bifurcation of the trachea, passes through this ligament and bifurcates into smaller **segmental bronchi** and **bronchioles,** forming the **bronchial tree.** The blind endings of these air passages are formed by the thin-walled **pulmonary alveoli,** where gas exchange occurs. Trace some branches of the bronchial tree by removing some lung tissue, but you will not be able to discern the air-filled microscopic alveoli, which are responsible for the spongy consistency of the lung tissue. The **pulmonary arteries** and **veins,** which supply the alveoli with oxygen-depleted blood and return oxygen-rich blood to the heart, respectively, accompany the primary bronchus and divide into smaller blood vessels along the bronchial tree. The primary bronchus and the pulmonary arteries and veins within the pulmonary ligament represent the **root of the lung.**

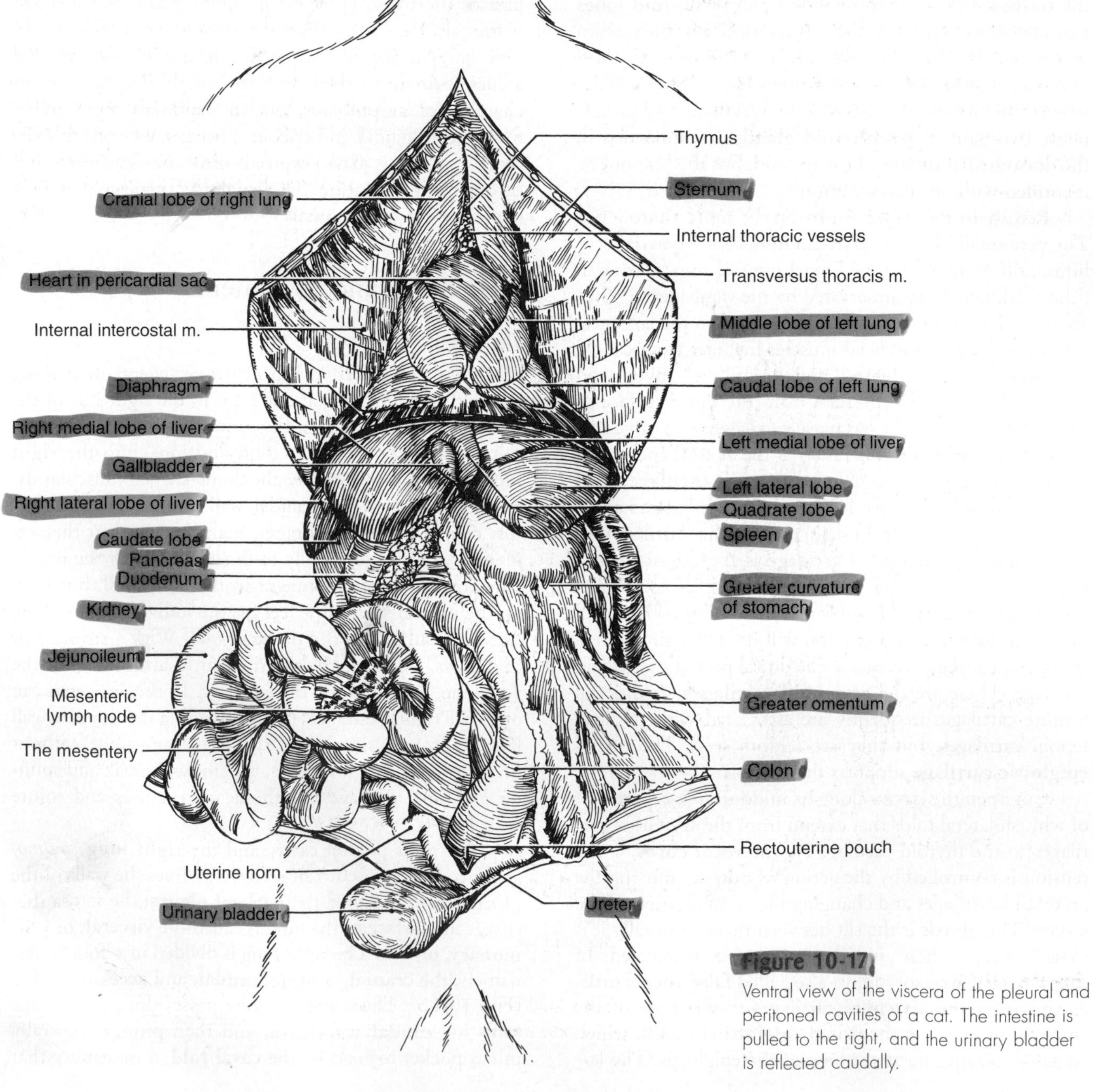

Figure 10-17
Ventral view of the viscera of the pleural and peritoneal cavities of a cat. The intestine is pulled to the right, and the urinary bladder is reflected caudally.

The paired pleural cavities are separated from each other by the **mediastinal septum** along the midline. The mediastinal septum is a mesentery that develops from the embryonic ventral mesentery (Fig. 10-3) and consists of a layer of connective tissue sandwiched between two layers of parietal pleura. The connective tissue–filled space, or potential space, between the two layers of parietal pleura is the **mediastinum.** Part of it contains the pericardial cavity with the heart, which forms the large bulge ventromedial to the lung (Fig. 10-3**C**). The mediastinal septum can be seen caudal to the heart and medial to the accessory lobe of the right lung. The caval fold is an evagination from this portion of the mediastinal septum.

For the time being, leave intact the part of the mediastinal septum that is attached to the ventral thoracic wall. Make a longitudinal incision about 2 cm to the left side of the midventral line and open the left pleural cavity in the same manner in which you opened the right pleural cavity. Examine the left pleural cavity, which will give you a good view of the third muscle layer of the thorax, the transversus thoracis muscle (see Chapter 7). The left lung does not have an accessory lobe. In rabbits, the left middle lobe is not well demarcated from the left cranial lobe. The left lung of mammals is always slightly smaller than the right lung because of the displacement of the heart toward the left side of the body during embryonic development.

Pull the left lung ventrally and examine the region dorsal to it. You can see a large artery, the **aorta,** and the **esophagus** passing through the dorsal portion of the mediastinum. The aorta lies to the left of the vertebral column; the esophagus is situated more ventrally. Let the lungs fall back into place. A pair of white strands, the **phrenic nerves,** can be seen in the central portion of the mediastinum on each side of the pericardial cavity and heart. They pass ventral to the roots of the lungs and run caudad to innervate the diaphragm. The right phrenic nerve follows the caudal vena cava closely, but the left phrenic nerve passes through the caudal part of the mediastinal septum (Fig. 10-18). The fact that the phrenic nerves originate from the ventral rami of the fourth, fifth, and sixth cervical nerves indicates that the diaphragmatic muscles may have evolved from cervical muscles of ances-

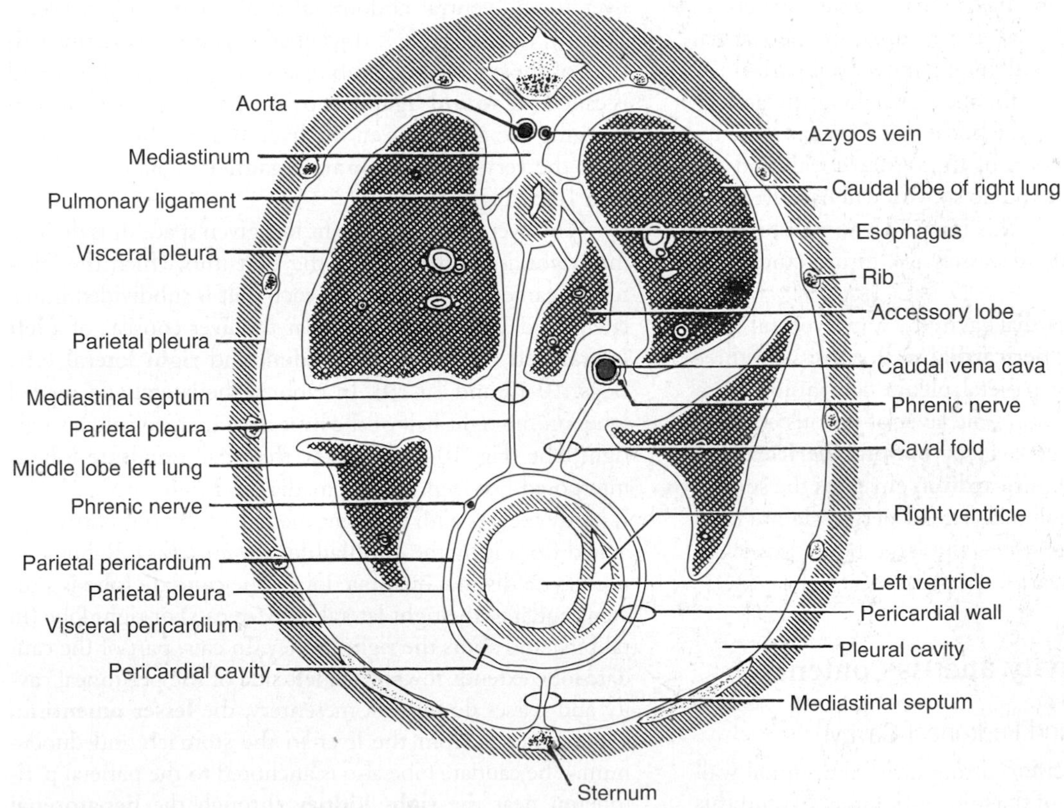

Figure 10-18
Diagrammatic transverse section through the thorax of a cat at the level of the ventricles of the heart showing the serosa and its topographical relationships to the thoracic viscera. The section is viewed from caudal.

tral tetrapods. The portion of the mediastinum ventral and cranial to the heart is occupied by the dark, irregularly lobulated **thymus.** The thymus varies individually in size and is developed best in young individuals (see Chapter 11).

During inspiration, the volume of the thoracic cavity is increased as the diaphragm contracts and pushes the abdominal viscera caudally. Synchronously, the ribs are rotated cranially by the contracting scalenus muscle and the cranial part of the serratus dorsalis muscle. During regular expiration, the volume of the thoracic cavity is decreased as these muscles relax, the viscera and ribs return to their resting position through elastic recoil of their tissues, and the diaphragm reassumes its dome-shaped configuration. Deep and voluntary expiration, as in trained opera singers, requires the contraction of abdominal muscles and the caudal rotation of the ribs by the contracting rectus thoracis muscle and caudal part of the serratus dorsalis muscle. The intercostal muscles may not play a crucial role in the ventilation of the lungs; they probably regulate the proper distance between the ribs while the thorax is expanded and compressed.

The Pericardial Cavity

The **pericardial cavity** and **heart** are the largest structures within the mediastinum. In order to open the pericardial cavity, first cut across the still intact midventral part of the thoracic wall by extending the incisions along the attachment of the diaphragm to the body wall. You can now reflect the midventral portion of the thoracic wall and expose the pericardium. As you do so, you will have to break the mediastinal septum. In this manner, you will preserve the internal thoracic blood vessels for future study (see Chapter 11).

Cut open the pericardial cavity by a midventral incision. Its saclike wall, the **pericardial wall,** consists of three layers: an outer layer of parietal pleura continuing from the mediastinal septum, a middle layer of fibrous connective tissue, and an inner layer of **parietal pericardium** (Fig. 10-3**C**). A **visceral pericardium** envelops the surface of the heart. The parietal and visceral pericardia are continuous with each other where the large blood vessels at the cranial end of the heart leave the pericardial cavity.

The Peritoneal Cavity and Its Contents

The Abdominal Wall and Peritoneal Cavity

Make a longitudinal incision through the abdominal wall slightly to the right side of the midventral line. Extend this longitudinal cut from the diaphragm to the pelvic girdle. Make transverse cuts laterodorsally through the abdominal wall along the attachment of the diaphragm to the body wall. Do this on both sides, thereby isolating the diaphragm as far dorsally as the epaxial musculature of the back. Reflect the flaps of the abdominal wall. Notice that, apart from the skin, you have cut through the external fascia of the trunk, three layers of muscle or their aponeuroses (external oblique, internal oblique, and transversus abdominis muscles), the rectus abdominis muscle near the midventral line, the internal fascia of the trunk, and the parietal peritoneum (see Chapter 7). The portion of the coelom that is opened is the **peritoneal cavity.** Its walls are lined with **parietal peritoneum,** and its viscera are covered with **visceral peritoneum.** Wash out the cavity if necessary.

The Abdominal Viscera and Mesenteries

The concave surface of the dome-shaped diaphragm forms the cranial wall of the peritoneal cavity, and the large **liver** *(hepar)* adjoins it and shapes itself to fit into the dome. Pull the liver and diaphragm slightly apart, just enough to see that the center of the diaphragm is formed by the **central tendon** and that the muscle fibers of the diaphragm insert on it. The vertical **falciform ligament,** which develops from the original ventral mesentery, extends between the diaphragm and the liver and continues from the liver to the ventral abdominal wall on the other. Occasionally, the free edge is thickened because it contains vestiges of the embryonic umbilical vein (see Fig. 11-15) and is called the **round ligament of the liver.** The peritoneum bridging the gap between the liver and diaphragm constitutes the very short **coronary ligament** (Anatomy in Action 10-3, Fig. **E**).

The liver molds itself to fit in a given space. It is divided into right and left halves at the cleft into which the falciform ligament passes. In cats, each half is subdivided into a lateral and a medial lobe so that the liver consists of a **left lateral, left medial, right medial,** and **right lateral lobe** (Figs. 10-17 and 10-19). In rabbits, the lateral and medial lobes on the right half of the liver have merged into a single right lobe (Fig. 10-21). In cats, the small **quadrate lobe** is interposed between the left medial and right medial lobes. It partly merges with the right medial lobe and is partly separated from it by the **gallbladder** *(vesica fellea).* Rabbits do not have a distinct quadrate lobe. The **caudate lobe** is situated caudal to the right lateral lobe (in cats) or right lobe (in rabbits) and abuts the right kidney. In cats, part of the caudate lobe extends toward the left side of the peritoneal cavity and passes dorsal to a mesentery, the **lesser omentum,** which extends from the liver to the stomach and duodenum. The caudate lobe also is anchored to the parietal peritoneum near the **right kidney** through the **hepatorenal ligament.** The **left kidney** lies in a slightly more caudal position on the opposite side of the body.

The major part of the **stomach** *(ventriculus)* is located on the left side of the peritoneal cavity and is more

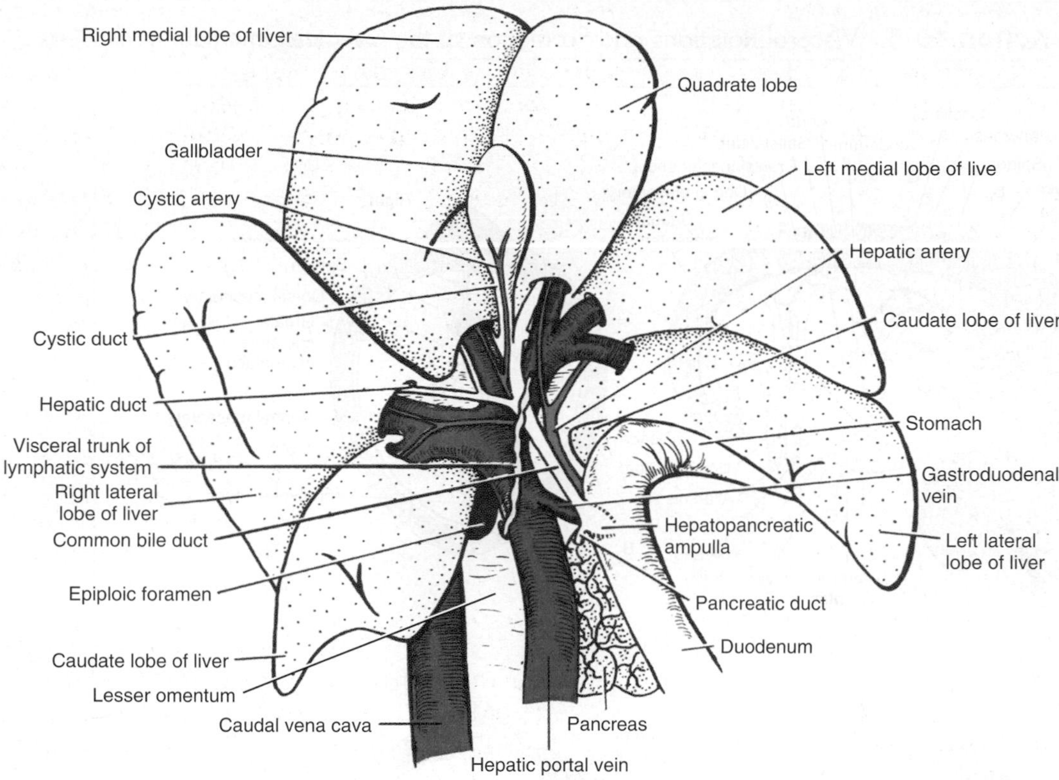

Figure 10-19
Ventral view of the blood and lymphatic vessels and of the ducts of the liver in the lesser omentum of a cat. The right medial, quadrate, and part of the right lateral lobes of the liver are reflected cranially to expose their dorsal surface.

Anatomy in Action 10-3 Visceral Rotations and Formation of the Greater Omentum

The topography of the greater omentum and other mesenteries in the cranial part of the peritoneal cavity is one of the more challenging topics of mammalian anatomy. The best approach to understanding the adult condition is to recapitulate the complexities of embryonic development. At a relatively early embryonic stage, the digestive tube is anchored along the middorsal line of the coelomic cavity by the dorsal mesentery (Figs. **A–C**). The spleen develops within the dorsal mesentery dorsal to the stomach. The duodenal part of the small intestine gives rise to the pancreas, which grows into the dorsal mesentery, and to the liver, which grows into the ventral mesentery.

Subsequently, the stomach embarks on a complex rotational movement about its longitudinal and vertical axes. For didactic purposes, we will consider the rotation about each axis separately, although the rotational movement of the stomach is a composite of both. First, the stomach rotates counterclockwise about its longitudinal axis so that its dorsal side (i.e., its greater curvature) now becomes its left side (Fig. **D**). Simultaneously, the spleen is pulled into the left side of the peritoneal cavity, while the dorsal mesentery anchoring it becomes disproportionately elongated and is now called the greater omentum. The more caudal portion of the dorsal mesentery that anchors the duodenum does not elongate to the same extent as the greater omentum (Fig. **A** and **C**). As a result, a pouch, the omental bursa, is formed as the greater omentum continues to elongate and expand caudad. The second rotation of the stomach proceeds counterclockwise about its vertical axis (as seen from dorsal) so that the greater curvature becomes the caudal side of the stomach and the caudally located pylorus and entrance to the duodenum now move into the right side of the peritoneal cavity (Figs. 10-17, 10-20, and 10-21).

At first, there is a wide communication between the omental bursa and the rest of the peritoneal cavity (Fig. **D**), but subsequent adhesions of the liver to the dorsal part of the diaphragm and adjacent body wall reduce it to a relatively small epiploic foramen (Fig. 10-19).

continued

Anatomy in Action 10-3 Visceral Rotations and Formation of the Greater Omentum *continued*

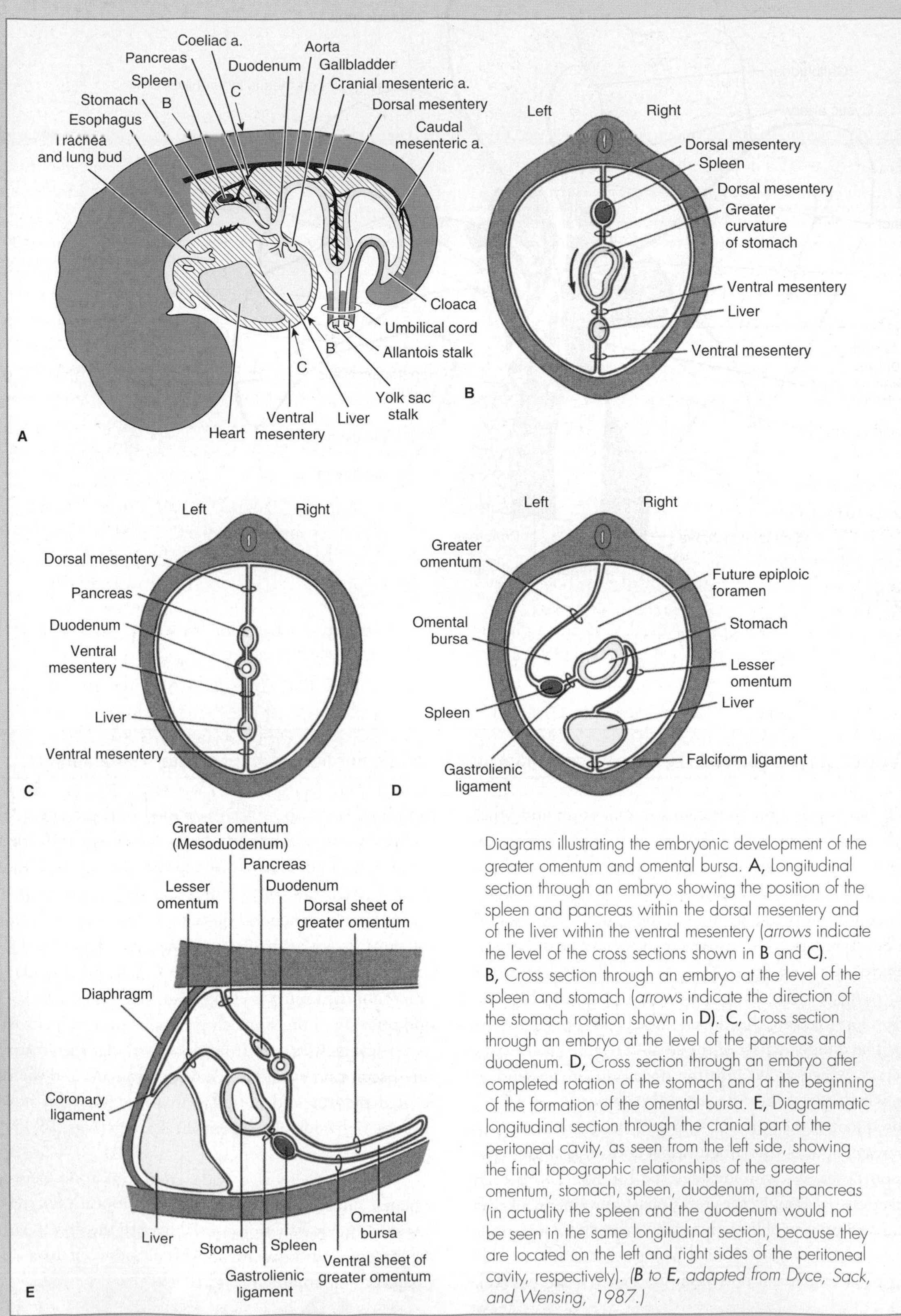

Diagrams illustrating the embryonic development of the greater omentum and omental bursa. **A,** Longitudinal section through an embryo showing the position of the spleen and pancreas within the dorsal mesentery and of the liver within the ventral mesentery (*arrows* indicate the level of the cross sections shown in **B** and **C**). **B,** Cross section through an embryo at the level of the spleen and stomach (*arrows* indicate the direction of the stomach rotation shown in **D**). **C,** Cross section through an embryo at the level of the pancreas and duodenum. **D,** Cross section through an embryo after completed rotation of the stomach and at the beginning of the formation of the omental bursa. **E,** Diagrammatic longitudinal section through the cranial part of the peritoneal cavity, as seen from the left side, showing the final topographic relationships of the greater omentum, stomach, spleen, duodenum, and pancreas (in actuality the spleen and the duodenum would not be seen in the same longitudinal section, because they are located on the left and right sides of the peritoneal cavity, respectively). *(**B** to **E**, adapted from Dyce, Sack, and Wensing, 1987.)*

or less J-shaped as in most vertebrates. Find the place where the esophagus passes through the diaphragm. There is an abrupt change in the diameter of the digestive tract at this point of transition between the esophagus and stomach. Observe the muscle fiber orientation of the diaphragm, and you will see that some muscle fiber bundles form a sling around the entrance to the stomach and function as an external sphincter. The portion of the stomach immediately following the esophagus and lying closest to the heart is the **cardiac region;** the dome-shaped portion extending cranially to the left of the cardiac region is the **fundus;** the main part of the stomach is the **gastric body;** and the narrow caudal recurving portion is the **pyloric region.** These externally defined regions do not correspond exactly to the gastric regions defined by the types of epithelial lining and their gastric glands, which are identified by the same names as the externally defined regions. The caudal end of the stomach is supplied with a thick muscular sphincter, the **pylorus,** which you can identify in a longitudinal section of the stomach. Do not damage any mesenteries. In cats, you will also see longitudinal ridges in the lining of the stomach, the **gastric folds,** which flatten out when the stomach is expanded by food. The long left and caudal margins of the stomach, which used to be the dorsal surface in the embryonic stomach, constitutes the **greater curvature of the stomach.** The shorter, more tightly curved, right and cranial margins, which used to be the ventral side of the embryonic stomach, constitutes the **lesser curvature of the stomach.** The **parietal surface of the stomach** faces cranioventrally and adjoins the liver; the **visceral surface of the stomach** faces dorsocaudally and adjoins the intestine and spleen.

Notice that the lesser omentum, which represents a part of the ventral mesentery, attaches along the lesser curvature of the stomach. The mesentery that attaches along the greater curvature is the **greater omentum,** or **mesogaster,** which is a part of the dorsal mesentery (Anatomy in Action 10-3). The greater omentum of mammals does not extend directly to the middorsal line of the peritoneal cavity, as it does in other vertebrates, but forms the saclike **omental bursa,** which drapes over the intestines and lies directly under the ventral abdominal wall (Figs. 10-17 and Anatomy in Action 10-3, Fig. **E**). In cats, the omental bursa is very large, contains considerable fat in its walls, and extends caudally nearly to the pelvic region. It often is entwined with the intestines and must be untangled carefully. In rabbits, the omental bursa is much smaller. In both species, the greater omentum extends caudad from its attachment along the greater curvature of the stomach as the descending **ventral sheet** (or superficial sheet) and then turns upon itself and extends craniad as the ascending **dorsal sheet** (or deep sheet) of the greater omentum (Anatomy in Action 10-3, Fig. **E**). After incorporating the pancreas, the greater omentum continues dorsad to attach to the dorsal wall of the peritoneal cavity. The spleen (see below) is located within the ventral sheet of the greater omentum on the left side of the stomach (Figs. 10-17 and Anatomy in Action 10-3, Fig. **D** and **E**). The portion of the greater omentum between the stomach and spleen is called the **gastrolienic ligament.** The duodenum, the portion of the small intestine arising from the pylorus, lies within the dorsal sheet of the greater omentum on the right side of the stomach. A small, triangular mesentery, the **gastrocolic ligament,** extends from the portion of the greater omentum that lies dorsal to the spleen to the mesentery (in the strict sense) of the intestine.

The **spleen** *(lien)* is a lymphoid organ that belongs, together with the lymph nodes, to the lymphatic system (see Chapter 11). Make an incision into the spleen and observe the cut surfaces under low magnification. You will discover an intermingling of darker, blood-rich tissue, which stores blood and processes aging red blood cells; and whitish lymphoid tissue, which initiates lymphocyte activities in immune responses. In cats, the very large spleen consists predominantly of dark tissue and is a "storage spleen." In rabbits, the smaller spleen consists predominantly of whitish tissue and is an "immune spleen." In most other mammals, the spleen is of a more intermediary type. In embryos, the spleen produces red blood cells.

The space enclosed between the ventral and dorsal sheets of the greater omentum, although normally collapsed, is called the **lesser peritoneal cavity,** because it communicates with the rest of the peritoneal cavity through the **epiploic foramen** (Figs. 10-19 and Anatomy in Action 10-3, Fig. **D**). Try to grasp one sheet of the greater omentum between the tips of a pair of forceps and cut open the omental bursa. If you do not succeed in separating one sheet from the other, simply tear the omental bursa and verify that it encloses a space. The epiploic foramen lies dorsal to the lesser omentum and between the caudate lobe of the liver and the mesentery of the duodenum. You can pass a probe, or your left little finger, through the epiploic foramen and push it dorsal to the stomach and into the omental bursa.

Carefully dissect the portion of the lesser omentum situated near the caudate lobe of the liver and the epiploic foramen in order to expose the system of bile ducts leading from the liver and gallbladder to the origin of the duodenum. Some lymphatic vessels, which look like chains of small nodules, may have to be mobilized and pushed aside. The **cystic duct** drains the gall bladder and joins several **hepatic ducts** from various parts of the liver in forming the **common bile duct** *(ductus choledochus),* which empties into the duodenum. One particularly prominent hepatic duct originates from the left lobes of the liver and another from the right lateral lobe.

The **small intestine** continues caudad from the stomach, passes through numerous convolutions, and eventually enters the **large intestine** (Anatomy in Action 10-4). Certain features of the intestine and the pancreas differ sufficiently in cats and rabbits to warrant separate treatment (see later), but a few common features can be observed at this time. The small intestine of mammals has differentiated into the proximal **duodenum** and the distal, postduodenal **jejunum** and **ileum.** The duodenum, as will be seen presently, is the first, approximately U-shaped loop of the small intestine. The jejunum is the more cranial portion of the rest of the small intestine, and the ileum is the more

Anatomy in Action 10-4 Packing the Intestine into the Peritoneal Cavity

The numerous convolutions of the very long intestine of mammals are very confusing at first view. It is somewhat easier to learn the different parts of the intestine and their arrangement within the peritoneal cavity if you consider the embryonic development of the digestive tract. As already mentioned, the primitive gut is straight during the very early stages of embryonic development, and it is anchored along the middorsal line of the peritoneal cavity by the dorsal mesentery. As the intestine grows, it elongates at a faster rate than the peritoneal cavity and, therefore, forms a ventral loop (Fig. **A**). Subsequently, the intestine is thrown into additional loops by rotating clockwise (as seen from dorsal) about the part of the mesentery that contains the cranial mesenteric artery (Figs. **B–D**). This coiling process of the intestine results in a compact orderly packing of the very long intestine within the relatively short peritoneal cavity. This process also ensures loops that are as wide as possible to avoid kinking of the intestine and compaction of its contents. Recall the cochlea of the mammalian ear, which is the result of a similar "packing problem" for the greatly elongated lagena (see Chapter 8).

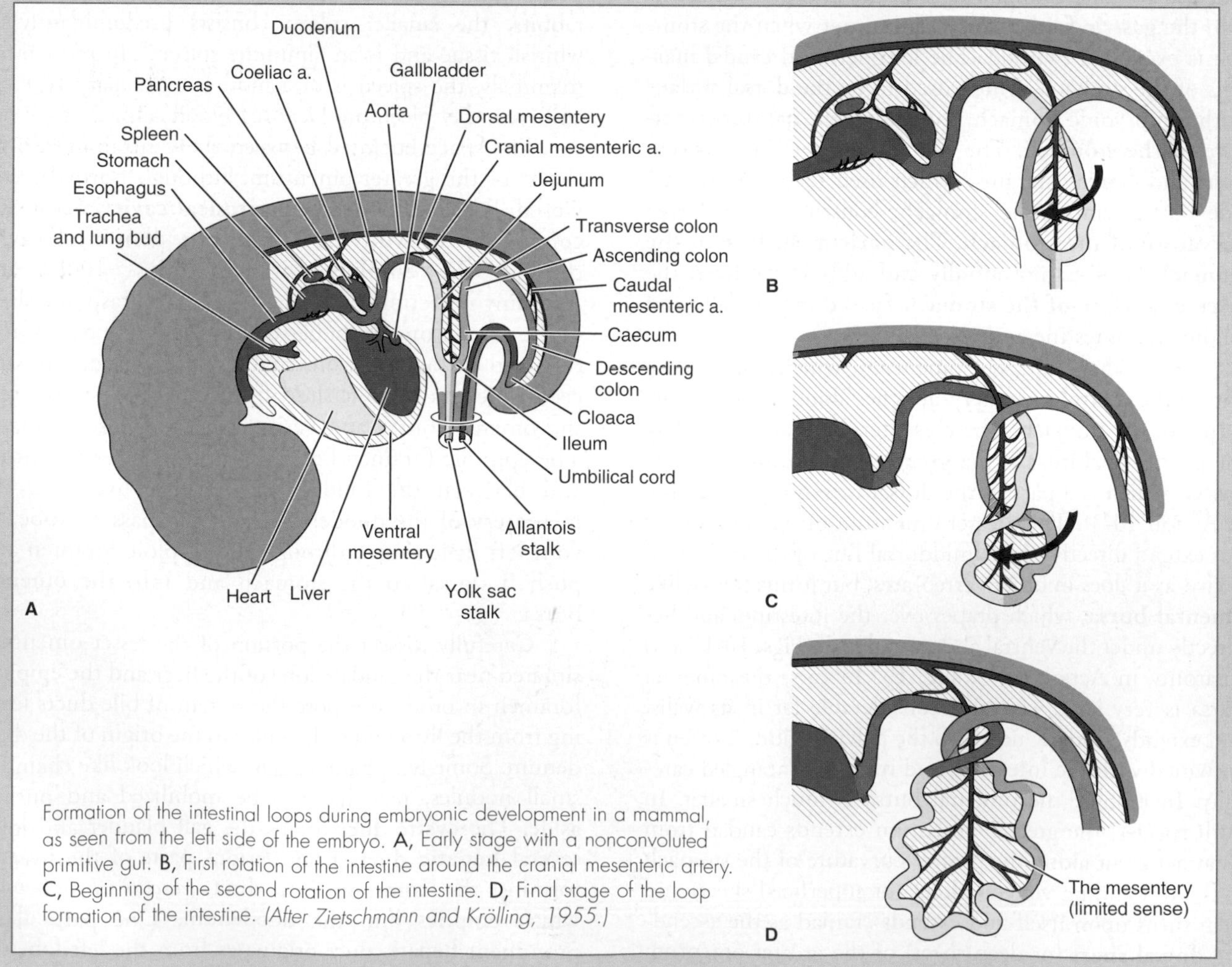

Formation of the intestinal loops during embryonic development in a mammal, as seen from the left side of the embryo. **A**, Early stage with a nonconvoluted primitive gut. **B**, First rotation of the intestine around the cranial mesenteric artery. **C**, Beginning of the second rotation of the intestine. **D**, Final stage of the loop formation of the intestine. *(After Zietschmann and Krölling, 1955.)*

caudal portion in the pelvic region. There is no sharp demarcation between the jejunum and ileum because there is a continuum of structural characteristics; therefore, the distal portion of the small intestine is also called the **jejunoileum.** The small intestine is anchored to the body wall by portions of the dorsal mesentery. The portion anchoring the duodenum is the **mesoduodenum,** and the portion anchoring the jejunum and ileum is **the mesentery** (in the strict sense). Cut open a part of the small intestine and examine its lumen. The lining has a velvety appearance that results from the presence of numerous minute, fingerlike projections called **villi.** These greatly increase the internal surface area. In cats, parasitic roundworms and tapeworms are frequently found within the small intestine.

The large intestine of mammals is much longer than that of other vertebrates. Its diameter is generally greater than that of the small intestine. Most of the large intestine consists of the **colon.** It is anchored to the body wall by the **mesocolon,** which is a portion of the dorsal mesentery. Skip the cranial portion of the colon for now and examine the caudal portion. This caudal portion extends caudad along the dorsal wall of the peritoneal cavity and enters the pelvic canal (Fig. 10-17). In females, this caudal portion of the colon lies dorsal to the Y-shaped **uterus;** in males, it lies dorsal to the pear-shaped **urinary bladder** (see Figs. 12-13, 12-15, and 12-19). Notice that the urinary bladder is anchored to the midventral body wall by the vertical **median vesical ligament,** which is the caudal part of the embryonic ventral mesentery. In addition, the urinary bladder is anchored by a pair of **lateral vesical ligaments** to the sides of the body wall. They usually contain substantial wads of fat. Along their cranial edges, they also contain the **round ligament of the urinary bladder** *(ligamentum teres vesicae),* a remnant of the embryonic umbilical artery (see Fig. 11-15 and Anatomy in Action 11-6). Cut open the colon, clean it out, and notice that its internal surface lacks villi. Also notice the extension of the peritoneal cavity into the pelvic canal.

In males, the portion of the peritoneal cavity that extends caudally between the large intestine and the urinary bladder is the **rectogenital pouch,** and the portion that extends caudally between the urinary bladder and the abdominal wall is the **pubovesical pouch.** In females, the uterus divides the portion of the peritoneal cavity in the pelvic canal into three pouches: the **rectogenital pouch** between the large intestine and the uterus, the shallow **vesicogenital pouch** between the uterus and the urinary bladder, and the **pubovesical pouch** between the urinary bladder and the abdominal wall.

Deep within the pelvic canal, the colon turns into the terminal segment of the large intestine, the **rectum,** which reaches the body surface through the **anus.** You will see the rectal region later when you open the pelvic canal to see the pelvic blood vessels and the urogenital organs (see Chapters 11 and 12).

Specific Structure of Selected Digestive Organs

CAT INTESTINE AND PANCREAS The descending portion of the **duodenum** of cats curves caudad from the pylorus on the right side of the body, then curves toward the left side of the body and ascends toward the stomach (Fig. 10-20**A** and Anatomy in Action 10-4, Fig. **D**). By definition, it turns into the jejunum at its next major

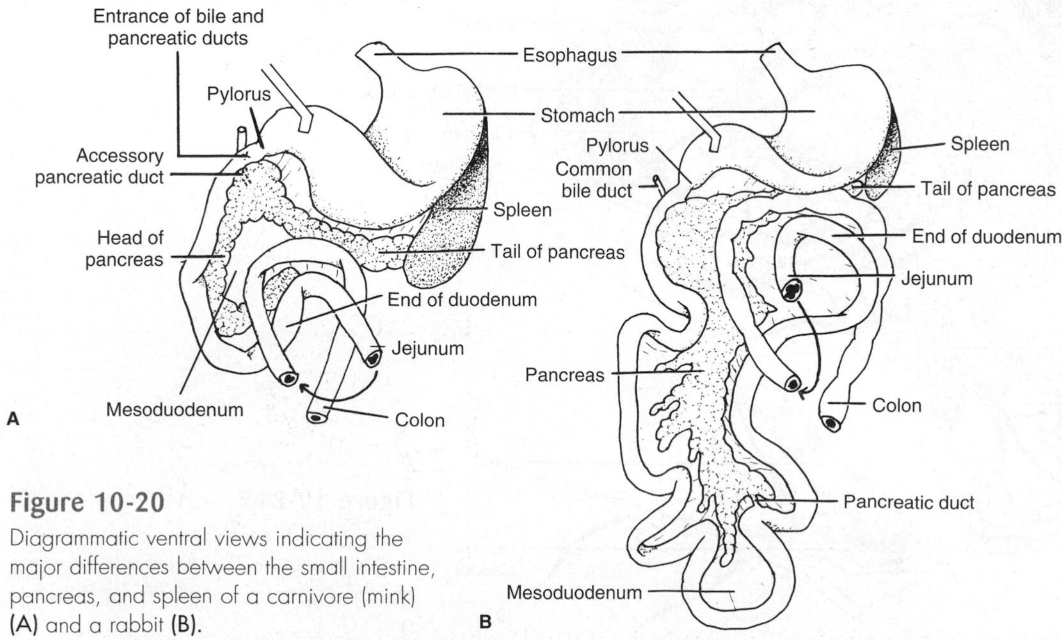

Figure 10-20
Diagrammatic ventral views indicating the major differences between the small intestine, pancreas, and spleen of a carnivore (mink) (**A**) and a rabbit (**B**).

bend. A small, triangular peritoneal fold, the **duodenocolic fold,** connects the caudal end of the duodenum to the mesocolon.

You will recognize the **pancreas** by its lobulated texture. Its **pancreatic head** lies against the descending portion of the duodenum, and its **pancreatic tail** extends transversely across the peritoneal cavity toward the spleen. The pancreatic tail lies within the dorsal sheet of the omental bursa (Anatomy in Action 10-3, Fig. **F**). Two pancreatic ducts are present, because the ducts of both the dorsal and ventral pancreatic primordia are retained. The **main pancreatic duct,** which is the duct derived from the ventral pancreatic primordium, unites with the common bile duct in the **hepatopancreatic ampulla,** a swelling on the wall of the duodenum (Fig. 10-19). You can find both ducts by carefully picking away pancreatic tissue in this region. The **accessory pancreatic duct,** which is derived from the duct of the dorsal pancreatic primordium, enters the duodenum about 1 cm caudal to the main pancreatic duct and on the dorsal side of the duodenum, but it is small and not easily found.

Follow the coils of the jejunum and ileum until you reach the colon (Figs. 10-17). The **caecum,** a blind-ending diverticulum, projects caudally from the beginning of the colon. In humans, the vermiform appendix projects from the tip of the caecum, but cats do not possess an appendix. Cut open the wall of the caecum and colon opposite the entrance of the ileum. Notice how the ileum projects slightly into the lumen of the colon, forming an **ileal papilla,** which prevents the contents of the colon from backing up into the ileum. The colon itself extends craniad on the right side of the body for a short distance as the **ascending colon,** crosses to the left side of the peritoneal cavity as the **transverse colon,** and extends caudad into the pelvic canal as the **descending colon,** finally turning into the rectum and anus. Because of the rotations that occur during embryonic development, the colon loops around the beginning of the jejunum (Fig. 10-20**A** and Anatomy in Action 10-4).

RABBIT INTESTINE AND PANCREAS The **duodenum** of rabbits extends caudally in small coils nearly to the pelvic region (Figs. 10-20**B** and 10-21). It then twists toward the left side of the body and, continuing to coil, ascends nearly to the stomach. In rabbits, the duodenum is approximately 50 cm long, depending on the breed. By definition, the duodenum turns into the jejunum where it curves caudad again.

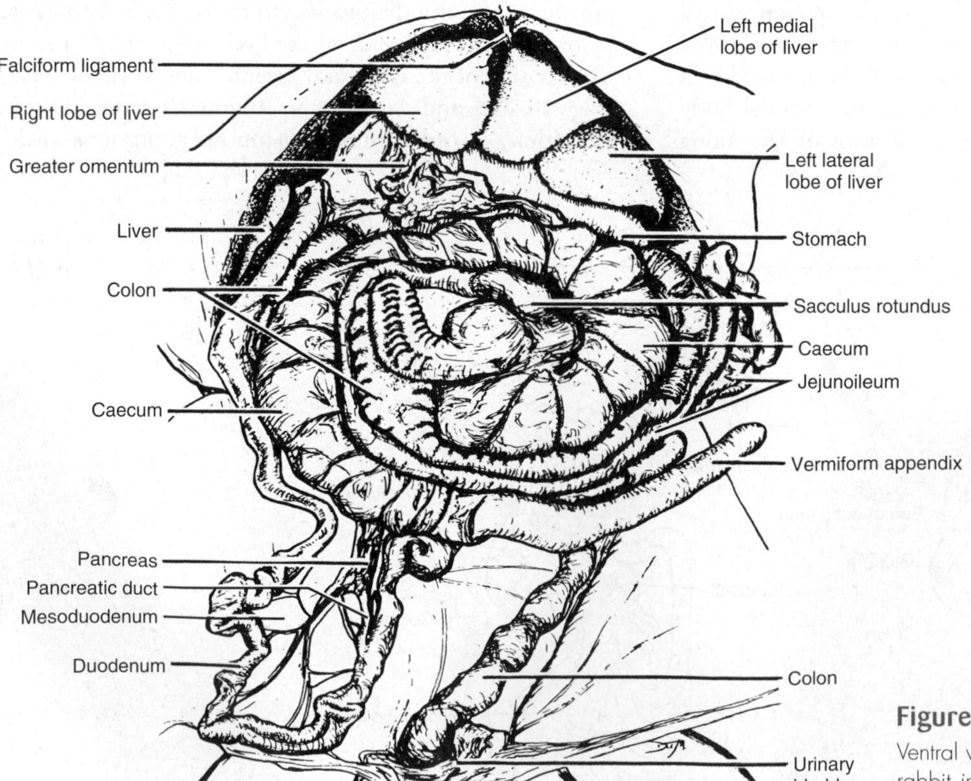

Figure 10-21

Ventral view of the abdominal viscera of a rabbit *in situ,* showing the characteristically enlarged caecum, vermiform appendix, and colon.

Stretch out the mesoduodenum between the descending and ascending limbs of the duodenum. The dark lobulated tissue that lies within the mesoduodenum is the **pancreas.** It is very thin and diffuse. The single **pancreatic duct,** which is derived from the dorsal pancreatic primordium, enters the ascending limb about 5 cm cranial to the most caudal loop of the duodenum.

Follow the coils of the jejunum and ileum until you reach the colon. The caudal end of the ileum is enlarged and forms the **sacculus rotundus,** a round muscular enlargement (Fig. 10-21). The sacculus rotundus may be involved in holding back digesta, while caecal pellets are released from the caecum to travel through the colon to be reingested (Anatomy in Action 10-5). Cut open the colon opposite the entrance of the ileum, clean out its lumen, and find the orifice of the ileum. It is situated on the **ileal papilla,** which prevents the contents of the colon from backing up into the ileum.

Anatomy in Action 10-5 Mammalian Digestion of Cellulose

Metazoan animals, in contrast to bacteria and certain protozoans, cannot produce enzymes that break down cellulose. In order to take advantage of the large potential food source that is represented by plants, herbivorous metazoan animals have entered a symbiotic relationship with cellulase-producing microorganisms by hosting them in their guts. As the microorganisms break down cellulose and metabolize the resulting saccharides, volatile substances are generated. Hence, the portions of the gut housing the microorganisms are enlarged and function as fermentation chambers.

Among mammals, fermentation chambers can be located in either the stomach or the large intestine, such as the caecum or colon. Accordingly, mammals that are specialized on fibrous diets can be classified as **foregut fermenters** (e.g., the ruminant bovids, cervids, and camelids; the leaf-eating sloths, proboscis monkeys, and colobus monkeys; and kangaroos) or **hindgut fermenters** (e.g., horses, elephants, rabbits, guinea pigs, wombats, and koalas).

Foregut fermentation has many advantages. Cellulose is broken down and the resulting saccharides are metabolized further into a variety of metabolites by the symbiotic microorganisms. These bacterial waste products include volatile fatty acids, such as acetic and butyric acids, methane, hydrogen, ethanol, lactate, ammonia, urea; and other substances, such as vitamins. These metabolites, however, are valuable substances for the host organism. The volatile fatty acids are absorbed already in the fermentation chamber and used by the host organism as a source of energy instead of glucose. Ammonia and urea are synthesized into amino acids by some of the symbiotic microorganisms. The amino acids and the microbes themselves represent an important source of protein for the host organism. Hence, most of the energy extracted from the ingested food by the microbes for themselves eventually is returned to the host. Because all these chemical processes have taken place in the stomach, the processed food can be subjected to additional digestive and absorptive processes in the intestine for the most efficient utilization of the ingested food. In addition, urea from the protein metabolism of the host organism can be recycled through the saliva into amino acids in the foregut. However, foregut fermentation has its costs. Although foregut fermenters can utilize foods of poor quality most effectively, the digestive process takes a relatively long time, the intake of food is limited by the size of the fermentation chamber, and relatively large amounts of water are needed by the host organism.

Hindgut fermentation takes place in the caecum or colon of the large intestine, where symbiotic microbes break down cellulose into volatile fatty acids and other microbial metabolic waste products. The volatile fatty acids and vitamins are absorbed by the large intestine of the host organism, but the other microbial metabolites cannot be recovered by the host because they do not pass through the small intestine, where most of the digestive and absorptive processes take place. Nitrogen recycling also is not possible. However, hindgut fermentation has its advantages, as the passage of food through the gut is much faster than in foregut fermenters and the amount of ingested food can be increased significantly if necessary. However, because of this more rapid passage, cellulose is not as completely utilized as in foregut fermenters. The diet of hindgut fermenters seems to be of a usually higher nutritional value, such as leaves, seeds, and roots, of which many nutrients can be digested and absorbed already in the small intestine.

Some hindgut fermenters, such as rabbits, have adopted the strategy of **coprophagy** to counterbalance the drawbacks of hindgut fermentation. They produce two types of feces: regular hard fecal pellets, and soft caecal pellets, which consist of up to 50% bacteria. These caecal pellets are reingested and undergo all the digestive and absorptive processes in the stomach and small intestine so that the host can utilize the microbial metabolites and recycle at least some urea.

As Christine Janis (1976) and R. McNeill Alexander (1993a, 1993b) showed, none of the strategies of cellulose utilization evolved by mammals is intrinsically better than the others, as each strategy is effective under particular environmental conditions and for particular life histories.

Rabbits eat fibrous vegetal food (roots, buds, leaves); hence, they ingest large amounts of cellulose, which is broken down by fermentation in the large intestine (Anatomy in Action 10-5). The large caecum extends in one direction from the sacculus rotundus; the colon, which can be distinguished by its more wrinkled appearance, extends in the other direction. First follow the **caecum.** It is a wide, thin walled, blind ending sac that follows a circular course for about 35 cm and terminates in the thicker-walled **vermiform appendix,** which is about 12 cm long. Cut open the caecum and note that the spiraling line that can be seen on its outer surface represents the point of attachment of a **spiral valve** on the inside. The caecum hosts and nurtures the symbiotic microorganisms that produce cellulase. Cut open the vermiform appendix and notice its thick walls, which comprise large amounts of lymphoid tissue that is active in immune responses (see Chapter 11). The human vermiform appendix, too, is a lymphoid organ and is not simply a functionless vestigial remnant of the ancestral caecum.

Return to the **colon** and follow it. You may have to tear parts of the mesocolon when doing so. The external layer of longitudinal smooth musculature is considerably thickened in two or three places and forms two or three longitudinal muscle bands, the **taeniae coli.** One taenia coli follows the line of attachment of the mesocolon. The wall of the colon between the taeniae coli consists of a very thin layer of longitudinal smooth musculature besides the internal layer of circular smooth musculature. Therefore, it bulges and sacculates into **haustra coli** (singular: haustrum) between the taeniae coli. Farther caudally, the wall of the colon is smoother, bulging only where there are pellet-shaped feces. After a very circuitous course, the colon descends into the pelvic canal to the rectum and anus.

SHEEP FOUR-CHAMBERED STOMACH Cellulose can be broken down by fermentation after it passes through the stomach, as in rabbits, or in the stomach itself (Anatomy in Action 10-5). A stomach of a sheep is relatively easily available and handled, and it is a good example of a stomach of a foregut-fermenting mammal.

In a ruminant, such as a sheep, the stomach is so large that it fills more than half of the peritoneal cavity and consists of four distinct chambers: the rumen, reticulum, omasum, and abomasum (Fig. 10-22). The first three chambers are lined with a stratified squamous epithelium and lack gastric glands, but the abomasum is lined with gastric epithelium and secretes HCl and pepsin like the stomach of nonruminant mammals. All four chambers are derived from the embryonic stomach precursor. The proportional size of the individual chambers is approximately 75% for the rumen, 8% for the reticulum, 4% for the omasum, and 13% for the abomasum.

The anatomy of the four-chambered stomach is understood best on the basis of the physiological processes.

Sheep bite off grass and herbs and swallow them after only perfunctory chewing. From the esophagus, a food bolus always enters the **rumen** and **reticulum** first (Fig. 10-22). The reticulum and rumen form a functional unit; therefore, they often are referred to as the **ruminoreticular compartment.** The rumen is by far the largest chamber and may have a volume of 13 liters in a large sheep. The inner lining is strewn with papillae (Fig. 10-22**A**). The reticulum is much smaller, and its lining is structured into reticular folds. Coordinated contractions of the ruminoreticular compartment throw the fibrous ingesta back and fourth, soaking and softening them. Periodically, portions of the ingesta are regurgitated, masticated, laced with large amounts of saliva, and finally swallowed. The ingesta are segregated within the ruminoreticular compartment by specific weight so that the triturated, smaller, heavier particles sink to the bottom of the compartment, while the larger, less soaked parts tend to float on the surface and are more likely to be regurgitated. While the ingesta are being soaked and broken down mechanically, they also are being broken down chemically by the symbiotic bacteria and protozoans, which thrive in the anaerobic environment of the ruminoreticular compartment. The symbiotic microorganisms not only secrete cellulase to break down cellulose into saccharides, but they also metabolize these further and produce volatile fatty acids, in addition to many other substances. These volatile fatty acids (VFAs) are absorbed by the epithelia of the ruminoreticular compartment and represent the major source of energy for ruminants. Gaseous waste products from the microorganisms' metabolism, such as methane (CH_4), accumulate in the dorsal part of the rumen and are eliminated by regurgitation.

The finely triturated and well-soaked particles eventually make their way into the **omasum.** Contractions of the omasum compact the particles and squeeze out fluid between the deep platelike folds, or **laminae.** Finally, the ingesta reach the **abomasum**, where regular gastric digestion starts and where, in particular, the digestion of protein is initiated, as the symbiotic microorganisms themselves represent a source of protein for the host ruminant. Further digestion and absorption occur in the intestine, as in other mammals.

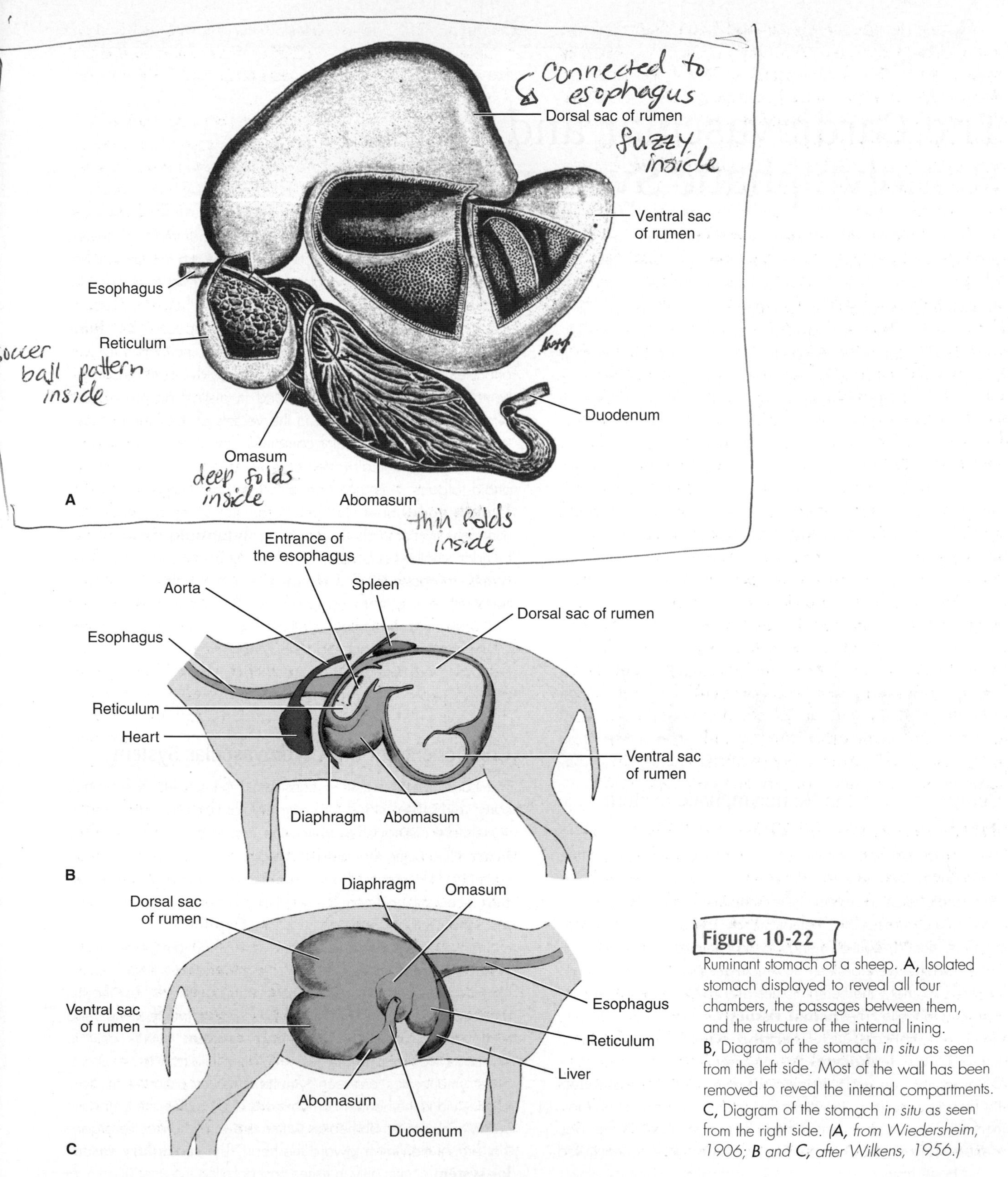

Figure 10-22

Ruminant stomach of a sheep. **A,** Isolated stomach displayed to reveal all four chambers, the passages between them, and the structure of the internal lining. **B,** Diagram of the stomach *in situ* as seen from the left side. Most of the wall has been removed to reveal the internal compartments. **C,** Diagram of the stomach *in situ* as seen from the right side. *(A, from Wiedersheim, 1906; B and C, after Wilkens, 1956.)*

The Cardiovascular and Hemolymphatic Systems

Chapter 11

Continuing with the organ systems that provide for the metabolic needs of the body, we will next consider the **cardiovascular system.** This system is foremost a transportation system as it pumps and distributes extracellular fluid, namely, blood and lymph, to all the parts of the body. Oxygen and nutrients are carried from the respiratory and digestive organs to all tissues and cells; carbon dioxide and other waste products are carried from cells and tissues to organs for metabolic transformation and excretion; hormones are transported from endocrine glands to tissues and cells; and heat is distributed throughout the body.

Through the hemal cells (i.e., blood cells) and lymphocytes that are produced by the **hemolymphatic system,** the cardiovascular system also is involved in immune responses, tissue repair, and homeostasis.

Components of the Hemolymphatic System

The components of the hemolymphatic system are dispersed among various parts of the body and, therefore, are usually noted during the dissection of the nervous, digestive, respiratory, and cardiovascular systems (see Chapters 9 and 10). Much of what is known about the hemolymphatic system is derived from humans; hence, caution needs to be exercised when extrapolating to other vertebrates. Two main subdivisions can be recognized, namely, the hemal tissues and cells, and the lymphoid tissues and cells. The **hemal tissues** are hematopoietic and produce hemal cells, such as **erythrocytes** for the transport of O_2 and CO_2, **leukocytes** for immune reactions (i.e., granulocytes, monocytes, and stem cells of lymphocytes), and **platelets** for blood coagulation in wounds. In nonmammalian vertebrates, hemal tissue appears first in the yolk sac and later in the liver, kidneys, and spleen. In mammals, hemal tissue is concentrated in the bone marrow during the later stages of embryonic development and throughout adulthood. Removal of aging erythrocytes is performed by the liver and spleen. The erythrocytes of nonmammalian vertebrates usually are nucleated and, hence, larger than the enucleated erythrocytes of mammals. However, a certain number of enucleated erythrocytes are found in all vertebrates. Erythrocytes and platelets are restricted to the arterial and venous system, but leukocytes can leave the cardiovascular system and move within the extracellular spaces. The **lymphoid tissues** are sites of proliferation and incubation for lymphocytes after they have been released by the stem cells in the hemal tissues. They are often found in distinct agglomerations and are usually intercalated in the vessels of the lymphatic system. The **spleen** generally combines hemal and lymphoid tissues and is found in all vertebrates. As an exception among lymphoid organs it is not connected to the lymphatic system. **Lymph nodes** are found only in amniotes and are especially numerous in mammals. In mammals, **lymphoid tissue** also is found in smaller tissue concentrations, such as the tonsils and the vermiform appendix. The **thymus** is a lymphoid organ that is especially active in young mammals. The lymphocytes that are incubated and proliferate in the thymus are distinguished as T lymphocytes, as opposed to B lymphocytes, which are incubated and proliferate in the rest of the lymphoid organs.

Components of the Cardiovascular System

The cardiovascular system consists of the vessels that propel, carry, and collect blood and extracellular fluid. Six categories of vessels are distinguished within the cardiovascular system. The **heart** is the pump that propels the blood toward the tissues and contributes to ensuring the return of the blood. The **arterial system** receives blood from the heart and conducts it toward the tissues and capillary beds of the body. **Capillaries** allow the exchange of gases and substances with surrounding tissues. The **venous system** collects blood from the capillary beds and returns it to the heart. In jawless and cartilaginous fishes, the hydrostatic pressure in the veins is very low, and many of the veins form wide sinuses. The **lymphatic system** occurs only in tetrapods. It collects extracellular interstitial fluid from tissues, transports lymphocytes between lymphoid tissues, and returns lymphatic fluid to the heart. In amphibians and reptiles, the lymphatic system includes lymph hearts, which are small contractile organs that propel the lymph toward the heart. The **secondary vascular system** occurs only in fishes and parallels the arterial and venous systems. Its structure and function are not well understood, but it is probably connected to the low-pressure condition of the cardiovascular system of fishes. It will not be discussed further.

The **heart** is the pump that creates the hydrostatic pressure that injects the blood into the arterial system and eventually the capillary beds. This hydrostatic pressure drops continuously as the blood is distributed into smaller and smaller arteries, arterioles, and finally capillaries. Except for erythrocytes and large molecules, fluid and certain substances leave the capillaries and enter the interstitial spaces of the tissues. As the concentration of the large, osmotically active molecules in the capillaries rises, the osmotic pressure of the blood rises above that of the interstitial fluid. In low-pressure cardiovascular systems, such as those of fishes, the interstitial fluid is pulled back into the capillaries by this increased osmotic pressure of the blood. In high-pressure cardiovascular systems, such as those of tetrapods, not all the fluid that leaves the capillaries is returned to them, because the difference between the hydrostatic and osmotic pressures of the blood and interstitial fluid is not as great. This surplus interstitial fluid is picked up by the lymphatic capillaries and returned to the heart by the lymphatic system. In all vertebrates, the hydrostatic pressure is very low after the blood has passed through the capillaries. Blood and lymph in the venous and lymphatic systems also are massaged back to the heart through compression by the surrounding organs and contracting skeletal muscles. Because the veins and lymphatic vessels contain **valves** that prevent backflow, the blood and lymph moves only toward the heart.

Development of the Cardiovascular System

A system of blood vessels is established already at early stages of vertebrate embryonic development so that the metabolic needs of the embryo are met. The basic developmental pattern of the cardiovascular system, which can be observed relatively easily in fishes, has been retained to a large extent by all vertebrates, despite their diverse adult cardiovascular patterns. Hence, the embryonic development of the cardiovascular system of fishes is an excellent model for understanding the diverse adult cardiovascular systems of vertebrates.

The first blood vessels to assume a definite shape in the embryo of any vertebrate are a pair of **vitelline veins** (Fig. 11-1), which are located ventral to the embryonic gut and carry blood and nutrients from the yolk-laden archenteron or from the yolk sac itself, depending on the vertebrate. The anterior portions of these

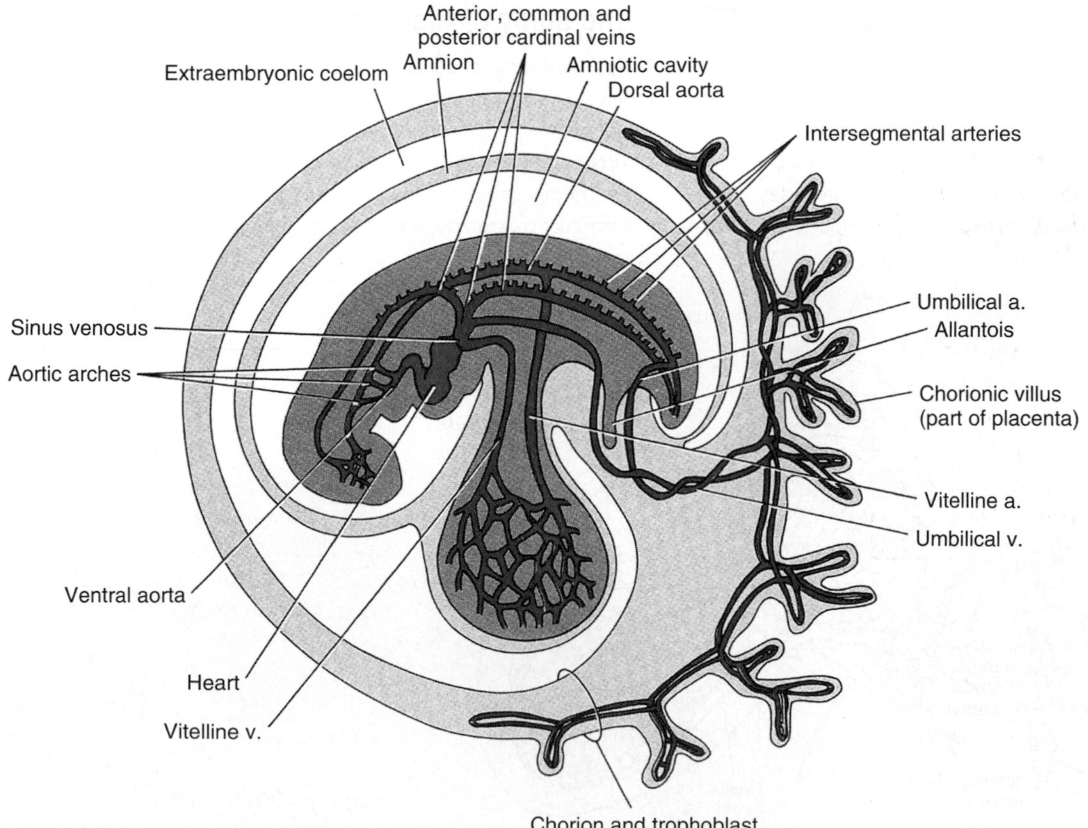

Figure 11-1
Diagram of the embryonic cardiovascular system in a longitudinal section through a 26-day-old human embryo as an example of a mammal. Most blood vessels are paired, but only those on the left side are shown.

veins merge to form the **heart** and the **ventral aorta.** The caudal portions of the vitelline veins are broken up into a capillary network by the growing liver. The portions of the vitelline veins that are situated caudal to the liver develop into the **hepatic portal veins** and their tributaries; the portions situated between the liver and the heart develop into the **hepatic veins.**

Meanwhile, a series of six **aortic arches** branch off from the ventral aorta and extend into the first six branchial arches and interbranchial septa, from where they extend dorsally to converge and form the **paired dorsal aortae.** The paired dorsal aortae merge later into a single **dorsal aorta,** which carries blood caudally to the rest of the embryo and through the **vitelline arteries** to the yolk sac or archenteron. From there, the blood is returned to the heart by the vitelline veins. This completes one circuit.

The early cardiovascular system provides largely a visceral circulation to and from the endomesodermal "inner tube" of the embryo, but a somatic circulation to the ectomesodermal "outer tube" soon appears. Other branches of the dorsal aorta, such as the **umbilical,** or **allantoic, arteries,** carry blood to the body wall and, in amniotes, to the extraembryonic membranes via the allantois (see Chapter 12). Blood from the dorsal region of the embryo returns by way of **anterior** and **posterior cardinal veins.** The anterior and posterior cardinal veins unite cranial to the liver and turn ventrally toward the heart as **common cardinal veins,** or ducts of Cuvier (see Anatomy in Action 11-9, Fig. A). Recall that the common cardinal veins pass through the dorsal part of the transverse septum, which they helped form (see Chapter 10). Blood from the more lateral and ventral portions of the body wall, and from the allantois, returns by a pair of more ventrally situated blood vessels. In anamniotes, these are the **lateral abdominal veins,** and, in amniotes, the **umbilical,** or **allantoic, veins.** These blood vessels enter the base of the common cardinal veins in early embryos, but in later embryos of amniotes, they establish a connection with the hepatic portal system and are drained through the liver (Anatomy in Action 11-6).

As embryonic development proceeds and as sites of nutrition, gas exchange, and excretion grow and change, the configuration of the cardiovascular system changes as well. New channels appear, and some old channels atrophy. Thus, in the course of the development of embryos, especially those of amniotic vertebrates, which have had a complex phylogenetic history, we see a succession of different blood vessels. Furthermore, the arterial, venous, and lymphatic systems exhibit great individual variability. Many of these individual variations can be attributed to the persistence of embryonic channels that normally atrophy and to the failure of certain channels to develop subsequently. Additional variation results from the widening of one channel at the expense of another within particular primordial capillary plexus, which occur in many parts of an embryo (Fig. 11-2). The relative rate of blood flow, as well as hereditary factors, largely determine which channels will develop and which will not.

The Study of Blood Vessels

It is important to keep in mind that arteries and veins are defined by the direction of their blood flow relative to the heart, not by the oxygen content of the blood flowing in them. **Arteries,** which are blood vessels that conduct blood away from the heart, may contain oxygen-rich blood (e.g., the dorsal aorta) or oxygen-depleted blood (e.g., the pulmonary artery in

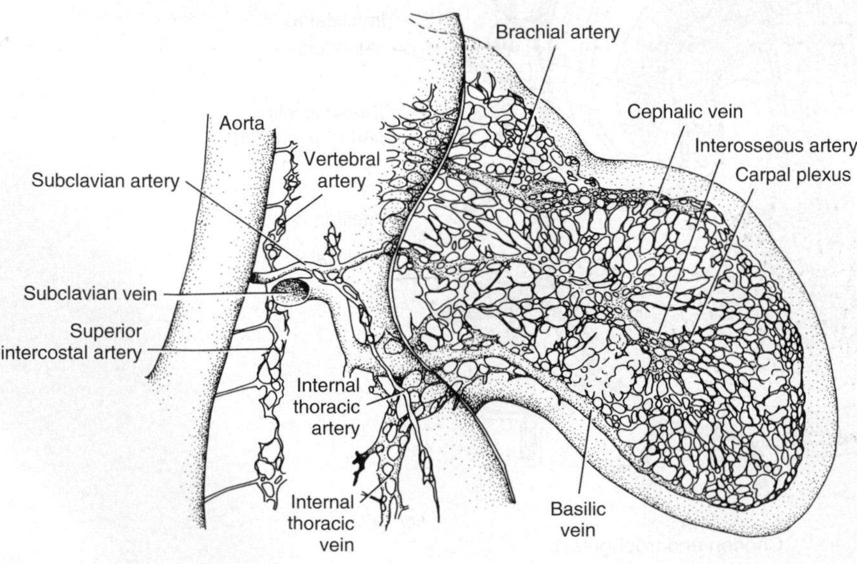

Figure 11-2

Embryonic differentiation of blood vessels at an early stage in the pectoral limb bud of a pig embryo. The definitive blood vessels form through enlargements of certain channels at the expense of others. *(From Woollard, 1922.)*

amniotes or the ventral aorta in anamniotes).[1] **Veins,** which are blood vessels that conduct blood toward the heart, may contain oxygen-depleted blood (e.g., the venae cavae of mammals) or oxygen-rich blood (e.g., the pulmonary veins). **Portal veins** are veins that, after draining one capillary bed, lead to another capillary bed in a different organ (e.g., the hepatic and renal portal veins). Veins leading from a capillary bed directly to the heart are called **systemic veins** (e.g., the caudal vena cava). A **sinus** is an inflated saclike section of a vein (e.g., the anterior cardinal sinus in sharks or the cavernous sinus at the base of the brain in mammals). An **anastomosis** is a blood vessel that is larger than a capillary and connects two blood vessels, such as two arteries, two veins, or an artery and a vein.

Another fact to bear in mind is that arteries and veins tend to follow each other, especially in amniotic vertebrates and in peripheral parts of the body, such as the appendages. For this reason, the arterial and venous systems of a given part of the body often will be described together. Veins usually are more difficult to inject successfully with colored latex, because the injection must be initiated in larger veins, which are near the heart, but valves frequently prevent this retrograde injection mass from reaching the peripheral portions of the veins. If that happens, veins usually can be identified as translucent, fluid-filled tubes lying next to the corresponding artery, because their walls are much thinner than those of arteries. Because of their thinner walls, veins are also usually collapsed and flat. In addition, veins often contain some coagulated blood that was trapped in them because of their low hydrostatic pressure. Arteries can be identified by their thicker, pinkish walls, which comprise a thicker layer of connective tissue and musculature than the walls of veins. Because of their thicker, self-supporting walls, arteries maintain circular cross sections.

While dissecting the blood and lymphatic vessels, you should keep in mind that they vary among different individuals; therefore, they may not be exactly as described here. Because the veins of amniotic vertebrates have had a more complex ontogeny than the arteries and because the venous blood flows more sluggishly and under less hydrostatic pressure, it is to be expected that the venous system will vary more than the arterial system. Odd as it may seem at first, the peripheral blood vessels exhibit fewer variations than the more central and larger blood channels. For example, the left ovarian vein of mammals always drains the ovary, but it may empty into either the left renal vein or the caudal vena cava. If a blood vessel cannot be identified from its point of connection with a major vessel, it usually can be identified if its peripheral distribution can be established.

The blood vessels of vertebrates are exceedingly numerous, because every part of the body needs to be supplied with oxygen and nutrients and relieved from waste products. This chapter emphasizes the major blood vessels in the axis of the body and the vessels branching from or joining them. The blood vessels of the head and appendages and the more distal vessels in the intestinal region are treated in less detail.

In order to learn the cardiovascular system, make your own chart of all the blood vessels mentioned in the dissection instructions for each animal and try to memorize it like a road map. Start with the heart in the center of the chart and proceed by incorporating the arteries and veins that are mentioned in the text and figures. Finally, add capillary beds to connect the arteries and veins where appropriate. See Figure 11-3 for a very simplified example of such a chart, although yours should be much more complex, if complete, and need not be as polished in order to be useful.

[1]Oxygen-rich blood is sometimes also called "arterial" blood; conversely, oxygen-depleted blood is sometimes called "venous" blood. This practice causes confusion, because it leads one to say that the pulmonary artery contains "venous" blood and the pulmonary vein "arterial" blood. The terms "arterial" and "venous" are, therefore, best avoided in this context.

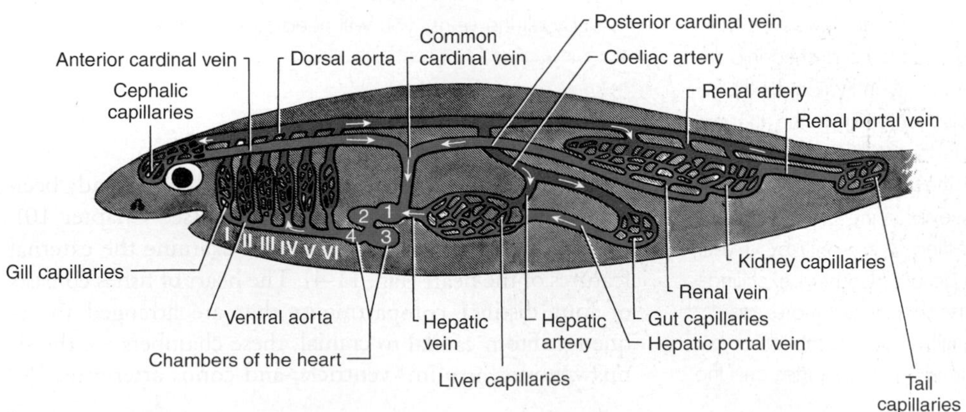

Figure 11-3

Diagrammatic representation of the basic aspects of the cardiovascular system of an ancestral fish. The chambers of the heart are identified with Arabic numerals: *1*, sinus venosus; *2*, atrium; *3*, ventricle; *4*, conus arteriosus. The aortic arches are identified with Roman numerals. *(From Villee, Walker, and Barnes, 1984.)*

FISHES

The basic pattern of the cardiovascular system in chondrichthyan fishes, which do not possess lungs, is shown diagrammatically in Figure 11-3. It is very similar to the pattern described for the early embryos of all vertebrates and may be similar to the pattern that was present in ancestral jawed vertebrates. The heart is located cranial to the pectoral fin girdle and ventral to the gills arches. It receives oxygen-depleted blood from the body. When the ventricle of the heart contracts, the hydrostatic pressure it creates drives the blood through the gill capillaries. After passing through the gills, the blood is distributed to the head and collected by the dorsal aorta. Despite its passage through the gill capillaries and the resulting drop in hydrostatic pressure,[2] the hydrostatic pressure of the blood in the dorsal aorta is sufficiently high for the blood to be distributed to the rest of the body. The blood pressure drops again after passage through the capillaries of the body tissues and again after passage through the capillaries of the hepatic portal system or renal portal system.

The cardiovascular system of fishes is a low-pressure circulatory system. Measurements reported by Satchell (1971) show that the hydrostatic blood pressure in the ventral aorta of the spiny dogfish ranges from 25 to 39 cm H_2O (depending on whether the ventricle is relaxed or contracting). After passing through the gill capillaries, this hydrostatic pressure drops about 23% to about 20 to 29 cm H_2O in the dorsal aorta. There is a further drop in hydrostatic pressure as the blood passes through capillaries in the tissues of the tail, so that the blood pressure in the caudal vein can be as low as 2.4 cm H_2O when the animal is at rest. However, the activity of the caudal muscles probably increases the hydrostatic blood pressure enough to drive the blood through the capillaries in the kidneys (Anatomy in Action 11-1), but the hydrostatic pressure drops again after passage through the renal capillaries. The hydrostatic pressure in the posterior cardinal veins, which drain the kidneys, may be as low as +0.2 to −0.6 cm H_2O so that the blood may need to be pushed into the sinus venosus by external pressure (Anatomy in Action 11-1).

The piscine vascular pattern also determines the thermal physiology of fishes. On its way from the capillary beds of the inner organs and body musculature, the blood picks up heat released by metabolic processes. However, in the gills, where the blood is separated from the surrounding water only by a very thin epithelium, not only gas exchange but also heat exchange is taking place. Actually, heat diffuses ten times more rapidly than oxygen molecules, hence equilibrium is established between the temperature of the blood in the gill capillaries and the temperature of the surrounding respiratory water. In other words, the blood loses heat, and its temperature drops, while the surrounding respiratory water gains the heat released by the blood. Theoretically, the temperature of the water should rise. In reality, however, the rise in temperature of the water is insignificant, because there is so much more respiratory water in the gill pouches than blood in the gill lamellae. The water acts as a heat sink, with the effect that the temperature of the blood drops to that of the surrounding water every time blood passes through the gills, while the surrounding water maintains essentially the same temperature. In this way, a fish is, by necessity, poikilothermic as well as ectothermic.

In certain so-called "warm-blooded" fishes, such as the tuna and mako shark, certain centrally located body muscles maintain a higher temperature than the rest of the body. Francis Carey (1973), who studied such fishes, found that these warmer muscles are capable of working continuously over long periods of time without fatiguing. Fishes with such muscles are pelagic specialists and can cruise for extended times and over vast distances at relatively high velocities. Although tunas and mako sharks are poikilothermic-ectothermic like all fishes, they manage to maintain certain body muscles at a higher temperature by preventing metabolic heat, which is generated by the contracting muscles, from being released to the environment. This is accomplished by *retia mirabilia,* or "wonderful nets," of arterioles and venules that surround the muscles and act as countercurrent heat exchangers, thereby insulating the warm core from the cold periphery of the body. This harnessed metabolic heat also can be used to regulate the temperature of the digestive organs for maximum digestive effectiveness.

The blood vessels of *Squalus* are a good model for the cardiovascular system of most fishes, but it needs to be kept in mind that fishes ancestral to tetrapods possessed lungs as well as gills and, hence, had pulmonary as well as gill circulation. You should study the cardiovascular system on specimens in which at least the arteries and hepatic portal system have been injected. To study all the veins, you will need triply injected specimens.

External Structure of the Heart

The pericardial cavity of the spiny dogfish has already been opened and the heart within observed (see Chapter 10). Return to the pericardial cavity and examine the external features of the heart (Fig. 11-4). The heart of fishes consists of four distinct compartments that are arranged in sequence. From caudal to cranial, these chambers are the sinus venosus, atrium,[3] ventricle, and conus arteriosus. Be-

[2]For a given amount of flowing liquid, the hydrostatic pressure of the liquid is inversely correlated with the size of the cross-sectional area of a tube. The sum of the cross-sectional areas of all the capillaries in a capillary bed is much larger than the cross-sectional area of the artery supplying them.

[3]The terms *atrium* and *auricle* are sometimes used synonymously; at other times, a subtle difference is made between them. We will use *atrium* for the entire chamber (undivided in fishes, divided in mammals) and *auricle* for the ear-shaped part of the atrium of mammals.

Anatomy in Action 11-1 Pumping Blood Toward the Heart in a Low-Pressure System

Because the cardiovascular system in fishes is a low-pressure circulatory system, the hydrostatic pressure alone probably is not sufficient to move the blood from the veins to the heart, especially because fishes possess two portal venous systems (Fig. 11-3), which further reduce the hydrostatic pressure. In addition, the venous system of sharks characteristically comprises distended portions, called sinuses (Fig. 11-5), in which the hydrostatic pressure drops significantly and blood can accumulate. Therefore, the blood in the venous system of sharks needs to be "massaged" back toward the heart. The generally thin-walled veins are compressed by bulging contracting muscles and by the movements of inner organs, while valves inside the veins allow the blood to move only toward the heart.

This mechanism of pumping blood toward the heart has been studied in fishes by Geoffrey Satchell (1971). The "tail pump," which moves oxygen-depleted blood from the caudal vein to the kidneys, is an example of such a mechanism. The caudal vein lies within the hemal canal formed by the hemal arches of the caudal vertebrae (see figure). Therefore, it is not compressed during the waves of muscular contractions that sweep down the tail during swimming (see Anatomy in Action 7-1). However, the contractions of the myomeres compress the intersegmental veins, which drain the epaxial and hypaxial musculature of the tail. Because the intersegmental veins possess multiple valves, the blood in them is automatically squeezed toward the caudal vein. As the intersegmental veins empty, pressure builds up in the caudal vein, because the caudal vein cannot expand within the hemal arch and because the blood cannot escape through the intersegmental veins due to the valves at their entrances to the caudal vein. The flow of the blood craniad toward the kidneys and renal portal veins is ensured by the valves within the caudal vein and by the pulsating caudal artery.

The caudal artery, which runs dorsal along the caudal vein within the hemal arches, is separated from the caudal vein only by a layer of connective tissue (see figure and Fig. 5-2). As it slightly expands with every pulse of blood, it exerts a slight pressure on the caudal vein. The caudal artery supplies the intersegmental arteries with blood when the surrounding myomeres are relaxed. When a subsequent wave of muscular contraction reaches these myomeres, the intersegmental arteries are compressed and the blood is squeezed distally, because it is prevented from flowing back into the caudal artery by a valve.

The venous system probably is subjected to similar pumping mechanisms at other locations. For example, the anterior and common cardinal sinuses run directly underneath the levator palatoquadrati and cucullaris muscles (see Chapter 7 and Fig. 10-8). When these muscles contract and bulge, they probably compress the anterior and common cardinal sinuses against the surface of the hyomandibular and branchial cartilages and the epibranchial musculature, thereby squeezing the blood into the sinus venosus (Anatomy in Action 11-4). Similarly, during undulatory swimming movements of the trunk, each posterior cardinal and gonadal sinus probably is compressed alternatively as the liver is pushed against the wall of the pleuroperitoneal cavity.

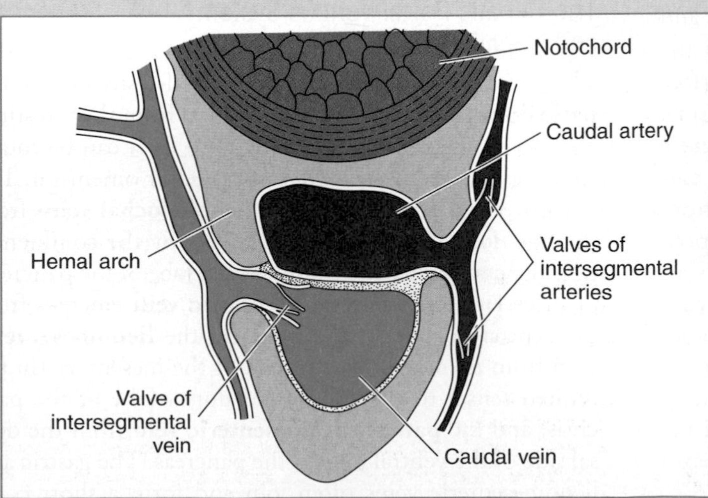

Diagrammatic transverse section through the ventral half of a caudal vertebra of a teleost bony fish *(Clupea)*, showing the location of the caudal blood vessels within the hemal canal. *(After Satchell, 1971.)*

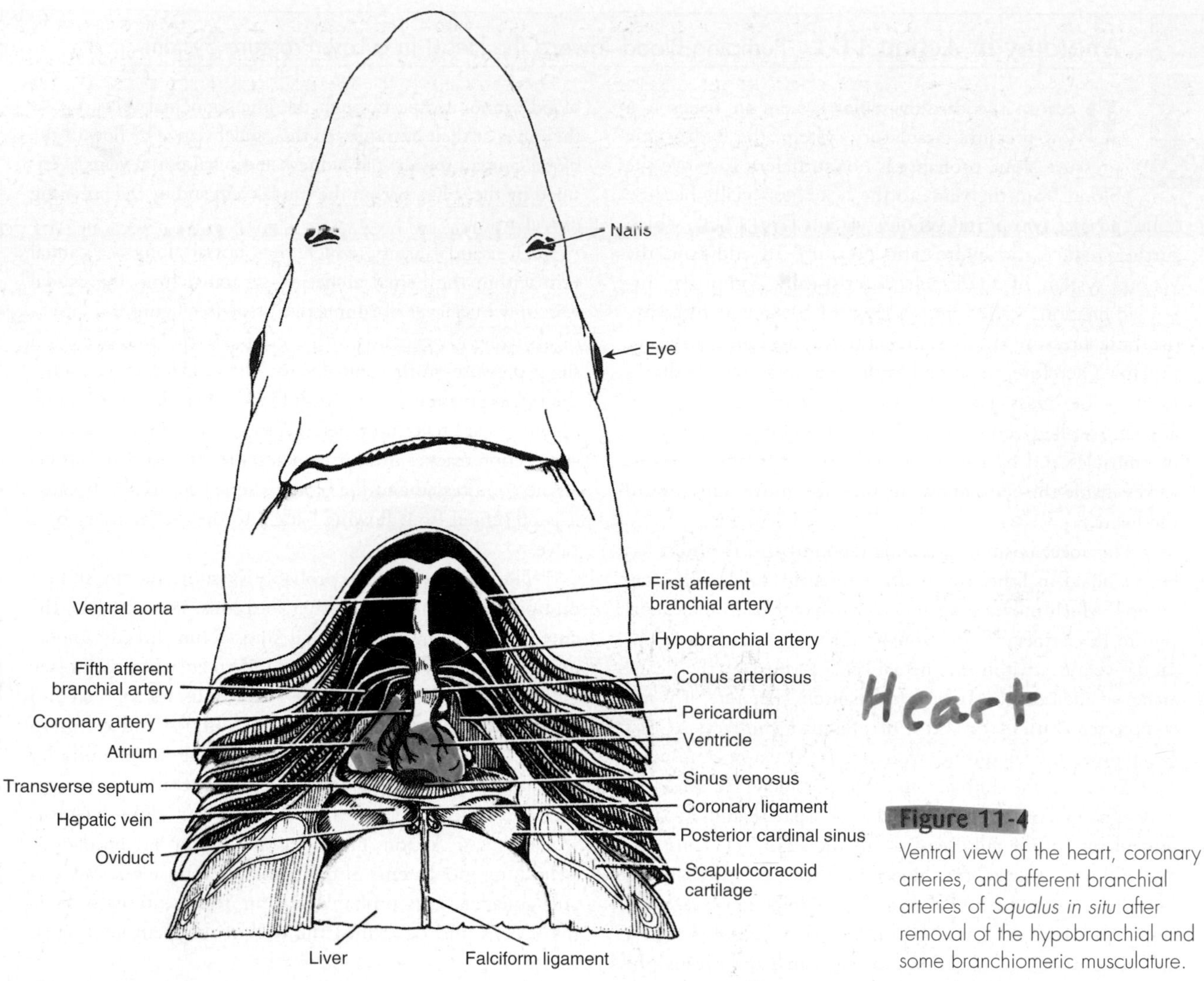

Figure 11-4
Ventral view of the heart, coronary arteries, and afferent branchial arteries of *Squalus in situ* after removal of the hypobranchial and some branchiomeric musculature.

cause the pericardial cavity is shorter than the sequentially arranged heart compartments, the heart assumes an S-shape to fit within the given space (see Figs. 10-2**D,** 11-9, and 11-10). The ventricle and conus arteriosus represent the ventral loop of the "S" and, hence, are seen at the surface when opening the pericardial cavity from the ventral side. The **ventricle** is the thick-walled, muscular, oval structure located in the caudoventral portion of the pericardial cavity; the **conus arteriosus** is the thick tube extending from the rostral end of the ventricle to the rostral end of the pericardial cavity, after which it continues rostrally as the ventral aorta (see later). Lift up the **apex of the heart,** the caudal end of the ventricle; the thin-walled, bilobed chamber dorsal to it is the **atrium.** The **sinus venosus** is the triangular chamber lying caudal to the atrium and extending caudally between its two lobes. The caudal surface of the sinus venosus adheres to the transverse septum and receives the common cardinal and hepatic veins (see later).

The Venous System

The Hepatic Portal Venous System and Hepatic Veins

The principal vein of the hepatic portal system is the **hepatic portal vein,** which collects blood from the intestines and brings it to the liver (see Fig. 10-5). It can be found running next to the bile duct in the lesser omentum. The hepatic portal vein receives small **choledochal veins** from the bile duct, but it is formed mainly by the confluence, near the craniodorsal tip of the dorsal lobe of the pancreas, of three large tributaries. The **gastric vein** emerges from the central portion of the stomach, the **lienomesenteric vein** from the line of attachment of the mesentery (in the limited sense) to the spleen and dorsal lobe of the pancreas, and the **pancreaticomesenteric vein** from the dorsal side of the ventral lobe of the pancreas. The gastric and lienomesenteric veins often join and form a short common trunk before uniting with the pancreaticomesenteric vein, which is the conspicuous vein seen passing just dorsal to the pylorus. These venous tributaries to the hepatic portal vein parallel arteries with corresponding names and drain the abdominal viscera. You will see their peripheral distribution in their catchment areas when you study the arteries.

The hepatic portal vein can be traced into the substance of the liver, where it divides into many branches that lead to the capillary-like **hepatic sinusoids.** The hepatic sinusoids, in turn, are drained by a system of smaller hepatic veins that ultimately are drained by a pair of large **hepatic veins,** which widen into **hepatic sinuses** before emptying into the sinus venosus of the heart (Fig. 11-5). The entrance of the hepatic sinuses into the sinus venosus is guarded by a sphincter that prevents backflow of blood into the liver when the sinus venosus receives blood from the common cardinal veins (Anatomy in Action 11-1). The hepatic veins are systemic veins. Even though they are not injected, you can find at least one of them by scraping away some liver tissue along the lateral edge of the liver about 1 cm caudal to the transverse septum. Make a small incision into the wall of the hepatic vein and pass a probe cranially; you will see it go through the transverse septum

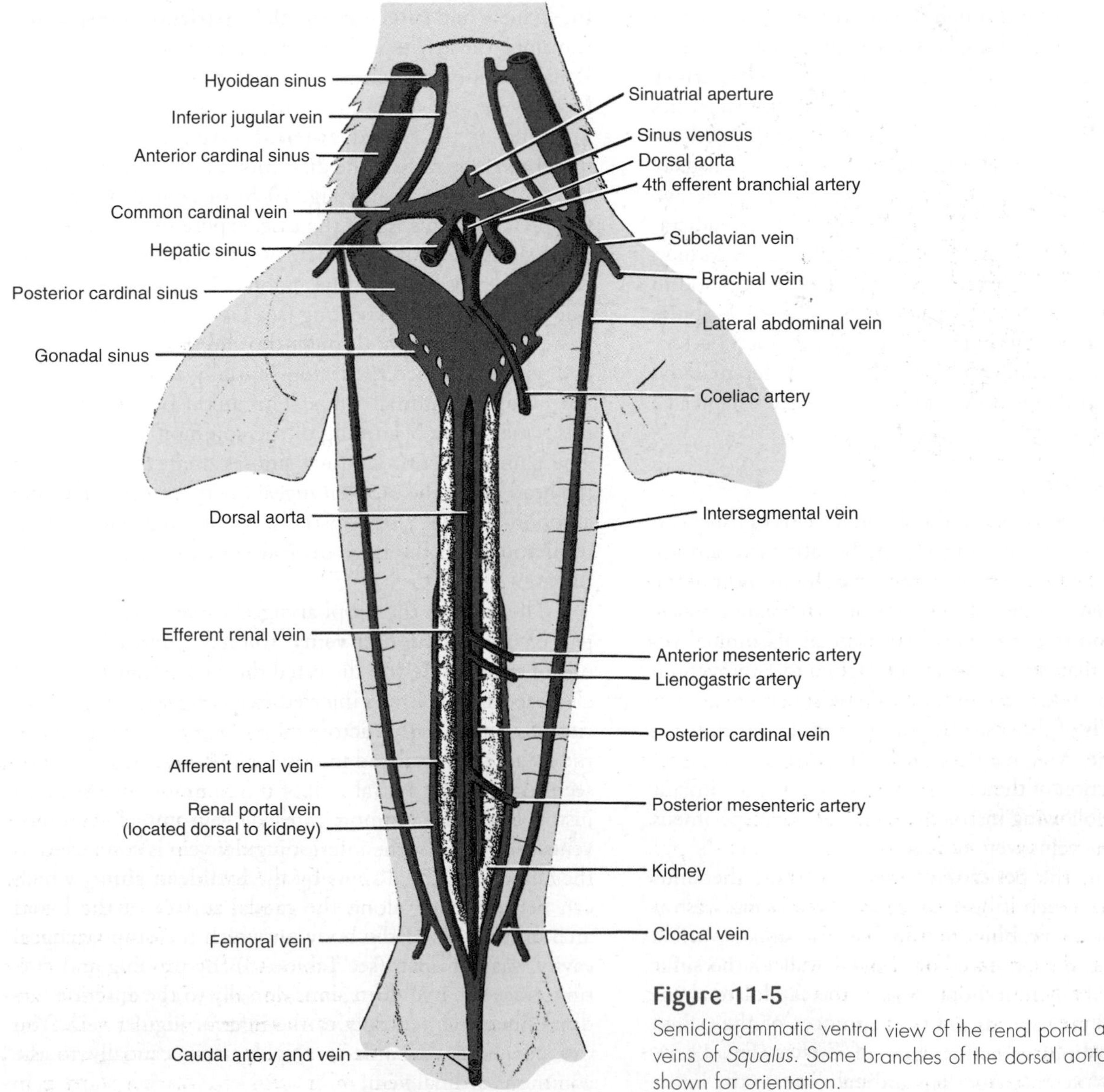

Figure 11-5

Semidiagrammatic ventral view of the renal portal and systemic veins of *Squalus*. Some branches of the dorsal aorta are also shown for orientation.

into the sinus venosus. If you want to, you can trace the hepatic vein caudally well into the liver.

When the portal blood passes through the hepatic sinusoids of the liver, it comes into direct contact with the hepatic cells, because the capillary-like hepatic sinusoids are not lined with endothelium. The liver is the metabolic conversion center of the body, and this direct contact fosters an effective uptake by the liver of substances from the nutrient-rich blood draining the intestine. Excess nutrients in the blood after a meal are stored in the hepatic cells, and deficiencies in the nutrient content of the blood between meals is made up from stores in the hepatic cells. The liver also plays an important role in detoxifying nitrogenous and other metabolic waste.

The Renal Portal System

Return to the section through the tail you made when you studied the vertebral column (see Chapter 5) and look at the cut surface of one of the vertebrae. The **caudal artery** and **vein** will be seen within the hemal arch, the artery lies dorsal to the vein and usually is injected with red latex (Fig. 11-5 and Anatomy in Action 11-1). The vein usually is empty. Cranially, the caudal vein bifurcates into the two **renal portal veins** (Figs. 11-5 and 11-8). They extend cranially, lying dorsal to the medial border of each kidney, but it is impractical to trace them far. They carry blood to the peritubular capillaries associated with the renal tubules by inconspicuous **afferent renal veins** (see Fig. 12-1**A**). These peritubular capillaries are drained by **efferent renal veins** that enter the posterior cardinal veins (see Chapter 12 and Anatomy in Action 11-2).

The Systemic Veins

For purposes of description, the systemic veins of fishes can be arranged into five groups: (1) the hepatic veins already seen; (2) the cardinal venous system that drains most of the trunk and head; (3) the inferior jugular veins that drain the ventral portion of the head; (4) the lateral abdominal venous system that drains the appendages and lateroventral portion of the body wall; and (5) the coronary veins that drain the walls of the heart. If you do not have a triply injected specimen, you may need to refer to Figure 11-5 during the dissection of these veins, because many are difficult to find. The following instructions are based on specimens in which these veins were not injected.

Look into the pericardial cavity and find the sinus venosus again. Open it by a transverse section and wash it out. Confirm, by probing, that the hepatic sinuses, which you have seen earlier, enter the caudal wall of the sinus venosus. The large round openings at the caudolateral angles of the sinus venosus are the entrances of the paired **common cardinal veins,** or ducts of Cuvier (Fig. 11-5). Pass a probe into the common cardinal vein on the intact side of the body (see Table 7-3) and follow it along the lateroventral wall of the esophagus within the transverse septum. It extends into the **posterior cardinal sinus,** which is the large retroperitoneal cavity that lies against the dorsolateral wall of the pleuroperitoneal cavity and curves toward the middorsal line. The posterior cardinal sinuses of opposite sides are interconnected dorsal to the gonads and receive veins from the esophagus as well as the **gonadal veins,** either **ovarian** or **testicular veins**, which drain the **gonadal sinuses** adjacent to the gonads. These veins are difficult to see unless they are injected. Farther caudally, each posterior cardinal sinus is fed by a **posterior cardinal vein,** which runs along the medial border of each kidney. The posterior cardinal sinus is actually just the expanded cranial part of this posterior cardinal vein. Each posterior cardinal vein receives numerous **efferent renal veins** from the kidneys and **intersegmental,** or **parietal, veins,** which run along the myosepta between the myomeres (Fig. 11-8). Renal and intersegmental veins may be difficult to see unless they happen to be filled with coagulated blood.

One of the paired **anterior cardinal sinuses** may have been seen during the dissection of the branchial muscles (see Chapter 7 and Fig. 10-8) or cranial nerves (see Chapter 9). If this is not the case, expose the anterior cardinal sinus in your specimen by carefully separating the cucullaris muscle from the epibranchial muscles on the side you have been dissecting (see Fig. 7-8 and Table 7-3). Pass a probe caudally through the anterior cardinal sinus, and you will notice it curving ventrally and uniting with the posterior cardinal sinus. The union of these two sinuses marks the beginning of the common cardinal vein. The paired anterior cardinal sinuses drain the brain and the head above the oropharyngeal cavity. You can trace the anterior cardinal sinus rostrally by probing and cutting until you reach the large **orbital sinus,** which surrounds the eye.

The floor of the oropharyngeal cavity is drained by a pair of **inferior jugular veins.** You may already have seen one of them while you dissected the coracobranchial muscles, especially if it was injected (see Chapter 7). The inferior jugular vein runs across the branchial arches between the coracobranchial and interbranchial muscles and can be seen entering the rostral wall of the common cardinal vein just before the common cardinal vein enters the sinus venosus. Rostrally, the inferior jugular vein is connected to the anterior cardinal sinus by the **hyoidean sinus,** which can be seen lying along the caudal surface of the hyoid arch on the side of the body on which the oropharyngeal cavity was cut open (see Table 7-3). By probing and cutting, trace the hyoidean sinus dorsally to the anterior cardinal sinus and ventrally to the inferior jugular vein. You can now trace the inferior jugular vein caudally to the common cardinal vein.

Anatomy in Action 11-2 The Renal Portal System of Fishes

The renal portal system is absent in agnathan fishes, embryonic sharks, and mammals, but it is present in adult sharks, bony fishes, amphibians, reptiles, and birds.

In lampreys, the caudal vein leads directly into the two posterior cardinal veins (see Chapter 2). Embryonic sharks at early stages do not yet possess a renal portal venous system, but as embryonic development progresses, a pair of **subcardinal veins** appear ventral to the kidneys and tap cranially into the posterior cardinal veins (Fig. **A**). Meanwhile, a portion of each posterior cardinal vein caudal to this confluence atrophies (dotted line in Fig. **B**). As a result, the caudal portions of the embryonic posterior cardinal veins must pass through the kidneys and become the renal portal veins. Hence, the posterior cardinal veins of adults are composites of the embryonic subcardinal veins and the cranial portions of the embryonic posterior cardinal veins.

The functional significance of the renal portal venous system is not entirely clear, but its presence is correlated with particular vascularization patterns of the nephrons, which are the functional units of the kidney (see Chapter 12 and Fig. 12-1). In all vertebrates, the renal artery brings blood to the glomerulus within the renal corpuscle. Because the renal arteries branch off directly from the aorta, the hydrostatic pressure of the blood in the glomerulus is relatively high so that fluid is filtered into the renal capsule and enters the renal tubule. The renal tubules are surrounded by peritubular capillaries, which may selectively absorb certain substances from the urine through active or passive transport mechanisms, as well as excrete some other substances from the blood into the renal tubule. In fishes, which have a low-pressure cardiovascular system and, therefore, may not be able to filtrate a large volume of blood in the renal corpuscles, the renal portal vein supplies blood directly to the peritubular capillaries. These peritubular capillaries probably are crucial to the overall functioning of the kidneys in relieving the body of metabolic waste products and surplus water.

In bony fishes, the peritubular capillaries are supplied with blood only through the renal portal vein (see Fig. 12-1**B**). This condition enabled many marine teleosts, which must conserve water, to reduce the size of their glomerular capillaries and, hence, the amount of fluid filtered from their blood. Certain marine teleosts have even completely dismantled their glomeruli and rely solely on the peritubular capillaries for excretion of metabolic waste products into the renal tubules.

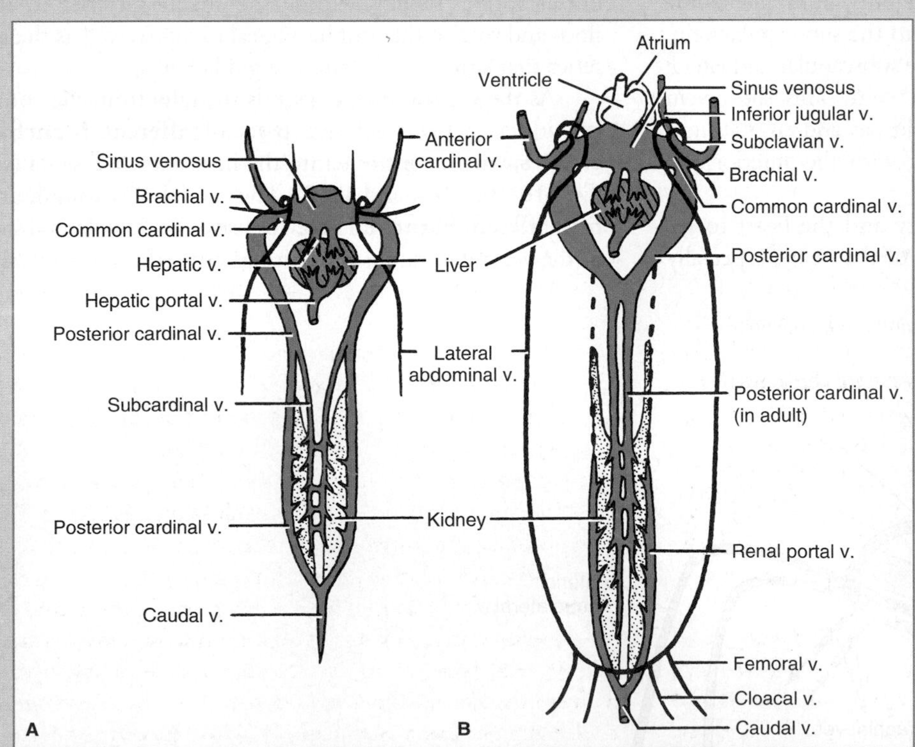

Diagrammatic ventral views of the configuration of the major veins and their development in sharks. **A**, Late embryonic configuration. **B**, Adult configuration. *(A, modified from Hochstetter, 1906; B, modified from Daniel, 1934.)*

The fins and some of the lateroventral portions of the trunk are drained by a pair of **lateral abdominal veins.** These veins are the pair of usually dark longitudinal lines seen on the inside of the body wall beneath the parietal serosa. Examine the caudal end of one of the lateral abdominal veins by probing and cutting it open. After passing dorsal to the pelvic girdle, the vein receives two tributaries. The **cloacal vein** arrives from the lateral wall of the cloaca and the **femoral vein** arrives from the pelvic fin. The cloacal veins of opposite sides of the body are interconnected by an anastomosis. You may not be able to find the two veins in the pelvic region of an uninjected specimen. Cranially, the lateral abdominal vein curves medioventrally just beneath the parietal serosa up to the scapulocoracoid cartilage in the ventral body wall. Here it is joined by the **subscapular vein,** which drains the area around the suprascapular cartilage and follows the caudomedial surface of the scapular process of the scapulocoracoid cartilage (see Figs. 4-4**A,** 4-5**A,** and 11-6). At the level of the joint for the pectoral fin, the subscapular vein receives the **brachial vein.** To find the brachial vein, first mobilize the pectoral fin away from the body musculature by breaking the connecting tissue between them. You will see the brachial vein follow the caudal surface of the metapterygium of the fin. Near the articulation of the basal pterygiophores with the scapulocoracoid (see Chapter 6), it crosses the ventral surface of the metapterygium, pierces the muscular body wall, and empties into the subscapular vein. The vein formed by the union of the subscapular and lateral abdominal veins is the **subclavian vein.** This short vein crosses the dorsal side of the coracoid bar and empties into the common cardinal vein together with the anterior and posterior cardinal sinuses.

Return to the pericardial cavity and the heart to see the **coronary veins** on the surface of the heart, especially the ventricle. They, too, empty into the sinus venosus by a common aperture, which cannot be seen at this stage of the dissection.

The Arterial System

The Branchial Arteries

To dissect the branchial arteries and the ventral and dorsal aortae with their branches, swing open the floor of the oropharyngeal cavity, as you did earlier (see Chapter 10). Remember that the heart in fishes lies cranial to the pectoral girdle (Figs. 11-7 and 11-9). To expose the heart, ventral aorta, and afferent branchial arteries, first remove the epithelium covering the primary tongue and floor of the oropharyngeal cavity, then carefully remove the exposed basibranchial and right hypobranchial cartilages piece by piece. Start by working on the caudal basibranchial cartilage (see Figs. 4-4**B** and 4-5**B**). Notice that it is situated directly above the heart. The afferent branchial arteries are distinct tubes, but they are visually unremarkable and are easily overlooked and damaged because they are not injected. Use Figure 11-7 as a guide to find them. Next, remove the epithelium from the roof of the oropharyngeal cavity to reveal the injected efferent branchial arteries and dorsal aortae. Finally, expose the smaller arteries in the floor and roof of the oropharyngeal cavity, as well as the arteries that supply and drain the gill lamellae.

As the **ventral aorta** extends rostrally from the conus arteriosus, it gives off five pairs of **afferent branchial arteries,** which continue into the interbranchial septa (see Figs. 10-10, 11-4, and 11-7). The caudalmost **fourth** and **fifth afferent branchial arteries** leave the dorsolateral side of the ventral aorta just after it has crossed the rostral wall

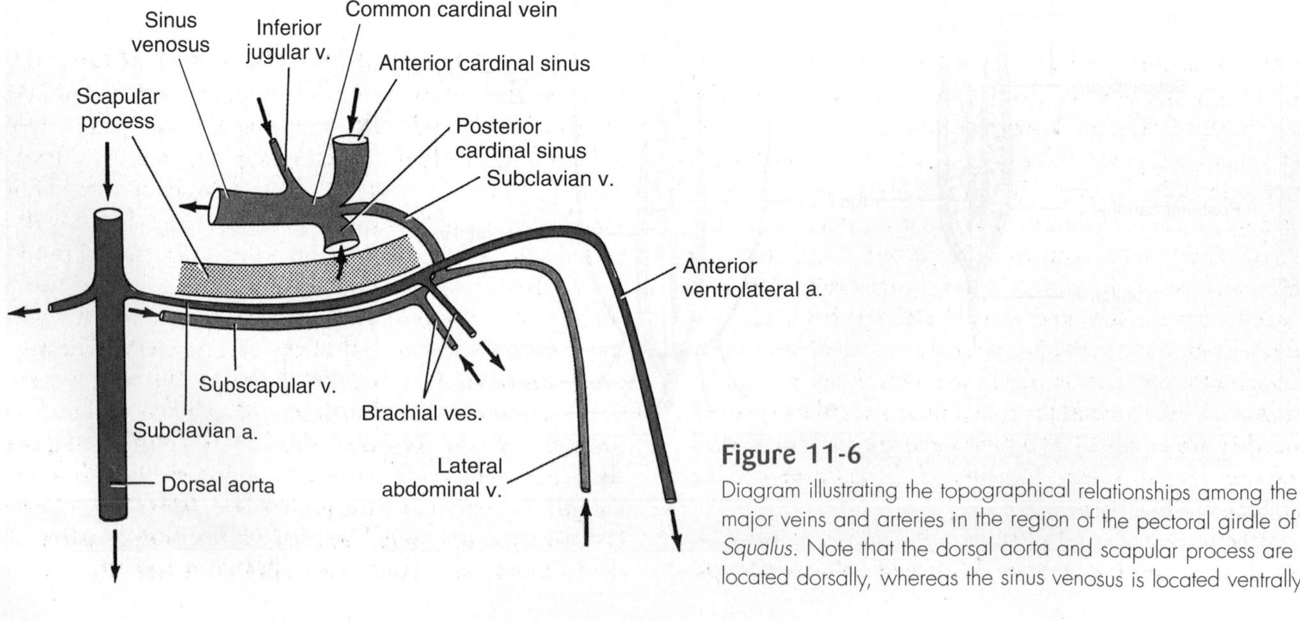

Figure 11-6
Diagram illustrating the topographical relationships among the major veins and arteries in the region of the pectoral girdle of *Squalus*. Note that the dorsal aorta and scapular process are located dorsally, whereas the sinus venosus is located ventrally.

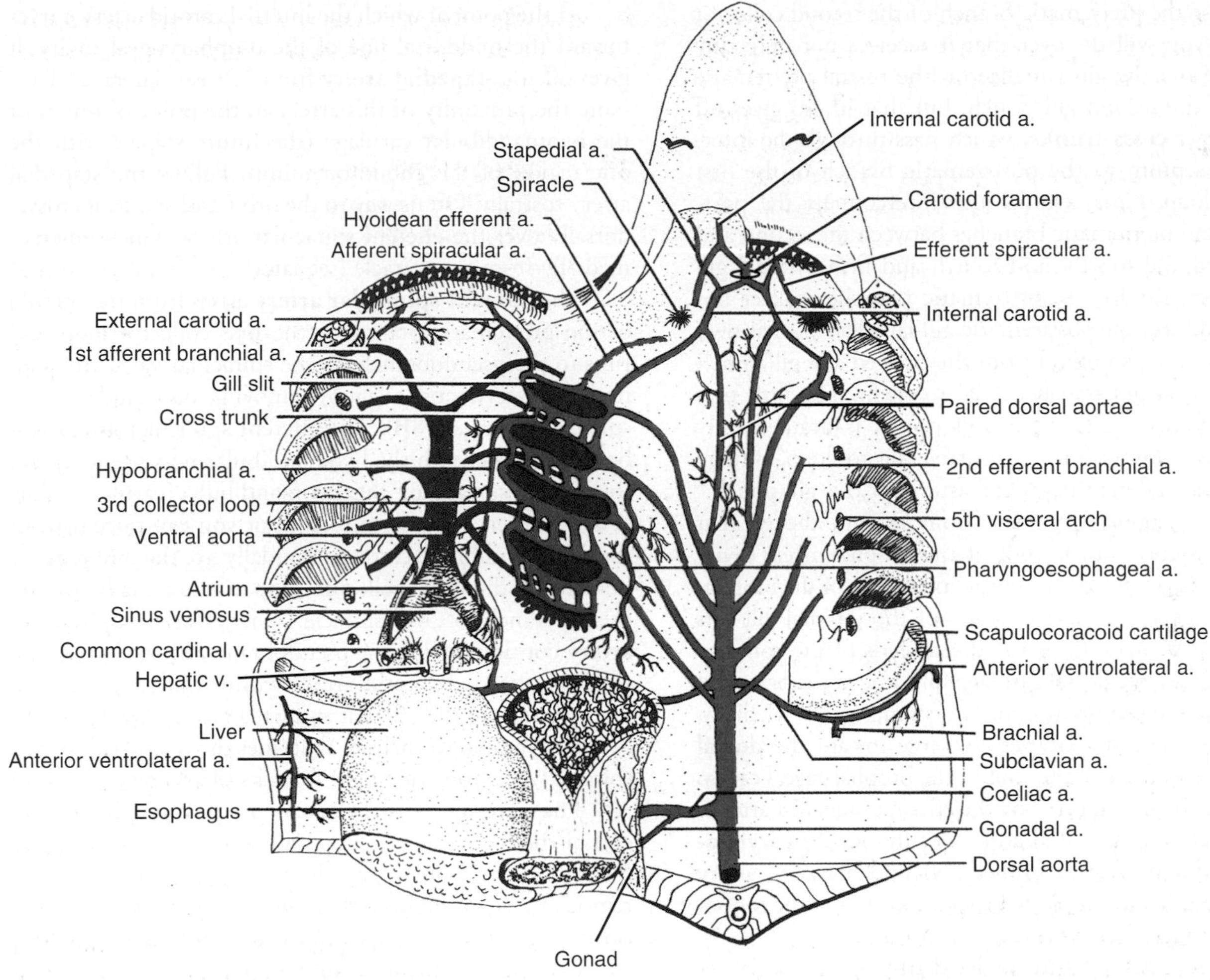

Figure 11-7

Larger arteries of the head of *Squalus* as seen from the oropharyngeal cavity, which has been opened. The afferent branchial arteries are shown in black and are seen in a dorsal view after the tissues of the floor of the oropharyngeal cavity were removed. The efferent branchial arteries are shown in red and are seen in a ventral view.

of the pericardial cavity. They branch off very close together and, occasionally, by a short common trunk. The middle **third afferent branchial arteries** branch off the ventral aorta slightly rostral to the caudal two pairs of afferent branchial arteries. The ventral aorta then continues rostrally for some distance and bifurcates just caudal to the rostral basibranchial cartilage. Trace one of the bifurcations, and you will see that it bifurcates again. The two branches are the **first** and **second afferent branchial arteries.** Trace each of the five afferent branchial arteries far enough into the interbranchial septa on the intact side of your specimen to ascertain which interbranchial septa and gills they are supplying and to observe the numerous small branches that they send into the gill lamellae to saturate the blood with oxygen.

Oxygen-rich blood from the gill lamellae is first collected by a system of four and one-half **collector loops.** You can see a representative collector loop by spreading open the first internal gill slit between the hyoid and first branchial arch and then removing the epithelium (see Fig. 10-10). The **first collector loop** is a vascular ring encircling the first internal gill slit and receiving tiny blood vessels from the gill lamellae. The blood vessel forming the rostral half of the collector loop is the **pretrematic branch** of the **first efferent branchial artery.** It is noticeably smaller than the **posttrematic branch,** the blood vessel that forms the caudal half of the collector loop. The **second, third,** and **fourth collector loops** can be found encircling the second to fourth internal gill slits, and each is drained by the **second, third** and **fourth efferent branchial artery,** respectively. However, at the fifth internal gill slit, only the **fifth pretrematic branch** is present because there are no gill lamellae on the caudal surface of the fifth branchial pouch (see Chapters 4 and 10).

Expose the pretrematic branch of the second collector loop and you will discover that it receives not only tiny branches from the gill lamellae on the rostral pretrematic surface of the second gill pouch, but that it also gives off several larger **cross trunks,** which pass through the interbranchial septum to the posttrematic branch of the first collector loop. Cross trunks also interconnect the posttrematic and pretrematic branches between the second and third, third and fourth, and fourth and fifth internal gill slits. Hence, the larger posttrematic branches collect not only blood from the posttrematic gill lamellae of their own gill pouch, but also blood from the pretrematic gill lamellae of the next gill pouch caudal to them. The fifth pretrematic branch of the fifth gill pouch is drained only through cross trunks into the fourth collector loop, as there is no separate efferent branchial artery for this gill pouch.

Remove connective tissue from around the efferent branchial arteries on the roof of the oropharyngeal cavity (Fig. 11-7) and trace them on the intact side of the oropharyngeal cavity to the dorsal angles of the internal gill slits where they originate from the dorsal ends of the collector loops. To do so, the dorsal cartilages of some of the branchial arches must be removed—can you name them? Medially, the efferent branchial arteries converge toward the **dorsal aorta** in the midline of the roof of the oropharyngeal cavity. Most of the blood collected by the efferent branchial arteries flows caudad via the dorsal aorta, but the small **pharyngoesophageal artery** arises from the second branchial artery and runs caudad to supply the caudal portion of the roof of the oropharyngeal cavity and the esophagus.

Oxygen-rich blood to the head travels in a special set of blood vessels. The **hyoidean efferent artery** arises from the dorsal end of the first collector loop rostral to the first efferent branchial artery and extends rostrally in the rostral portion of the roof of the oropharyngeal cavity. On its medial side, opposite the spiracle, it receives a small blood vessel that arises from the medial end of the first efferent branchial artery. This and the artery on the opposite side represent the rostral portions of the embryonically paired dorsal aorta. They are called simply the **paired dorsal aortae.** The blood vessel resulting from the union of the hyoidean efferent artery with one of the paired dorsal aortae is called the **internal carotid artery.** It also develops embryonically from the rostral portion of the paired dorsal aorta. The internal carotid artery continues rostrally, curves toward the middorsal line, crosses or sometimes unites with its mate of the opposite side, and enters the chondrocranium through the carotid foramen (see Fig. 4-2**B**). Follow the internal carotid arteries by chipping away cartilage from the underside of the chondrocranium. They soon diverge and, at the level of the hypophysis, unite with the arteries on the ventral surface of the brain. The internal carotid arteries are the main arteries supplying the brain with blood.

At the point at which the internal carotid artery curves toward the middorsal line of the oropharyngeal cavity, it gives off the **stapedial artery** from its rostrolateral surface. Note the proximity of this artery to the point of union of the hyomandibular cartilage (the future stapes) with the otic region of the chondrocranium. Follow the stapedial artery rostrally. On its way to the orbit and snout, it crosses dorsally over the efferent spiracular artery, which emerges medially from the spiracle (see later).

The **afferent spiracular artery** arises from the middle of the pretrematic branch of the first collector loop and runs to the pseudobranch on the spiracular valve. You can find this artery most easily by removing skin caudal to the spiracle (see Fig. 7-8**B**). The afferent spiracular artery will be seen just beneath the hyomandibular nerve as it crosses the lateral surface of the hyomandibular cartilage. This artery is often not well injected, but you can trace it from the hyomandibular cartilage caudally to the pretrematic branch of the first collector loop and cranially to the pseudobranch of the spiracle. The **efferent spiracular artery** continues from the pseudobranch medially to unite with the internal carotid artery within the cranial cavity. This portion of the efferent spiracular artery can be found by removing the epithelium that lines the rostral side of the spiracle when approaching it from its pharyngeal entrance.

The **external carotid artery** arises from the ventral end of the first collector loop and runs rostrally to supply the region of the lower jaw. It can be seen most easily by removing the epithelium that covers the gap between the tip of the primary tongue and the mandibular cartilage, where it is revealed as a usually injected blood vessel. Trace it back to its origin on the first collector loop.

Another blood vessel, the **hypobranchial artery,** usually arises from the ventral end of the second collector loop, but it may receive contributions from any of the other collector loops. The hypobranchial artery supplies most of the hypobranchial musculature and then bifurcates at the rostral end of the pericardial cavity. One branch, the **coronary artery,** follows the sides of the conus arteriosus toward the ventral surface of the ventricle (Fig. 11-4). The other branch, the **pericardial artery,** extends caudad and supplies the dorsal and ventral walls of the pericardial cavity. Occasionally, the pericardial arteries form an anastomosis across the dorsal side of the conus arteriosus.

The branchial arteries, despite their modifications to accommodate gills and to supply blood to the brain, have retained the basic configuration of the six aortic arches that are present in fish embryos (see Anatomy in Action 11-3).

The Dorsal Aorta, Its Branches, and Accompanying Veins

The basic branching pattern of the dorsal aorta is shown in Figure 11-8. This pattern is characteristic of all vertebrates. As can be seen, there are three major categories of

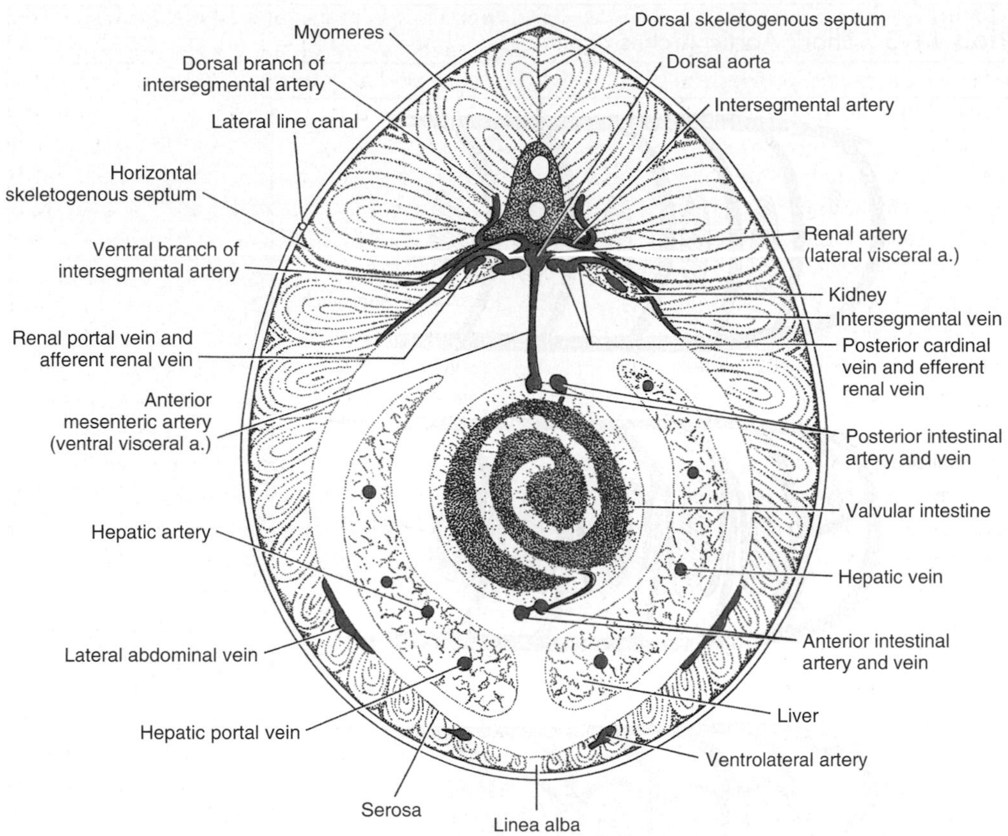

Figure 11-8

Semidiagrammatic cross section through the trunk of *Squalus* showing the topographical relationships and basic branching pattern of the major arteries and veins.

Anatomy in Action 11-3 Shark Aortic Arches

During the course of embryonic development of sharks, six complete aortic arches differentiate from rostral to caudal (Table and Fig. **A**). The dorsal part of the first aortic arch, together with a new connection that develops early on between the first and second aortic arches (Fig. **B**), forms the afferent and efferent spiracular arteries. The dorsal parts of the collector loops, which drain the gill lamellae, develop from the dorsal halves of the second to sixth aortic arches. Branches extend from the ventral parts of the aortic arches and join around the ventral end of each definitive internal gill slit, except the last gill slit (Figs. **B** and **C**). Thus, each collector loop is formed from outgrowths of two successive aortic arches, namely, the pretrematic portion of a collector loop from the caudal branch of a bifurcation of one aortic arch, and the posttrematic portion from the cranial branch of the bifurcation of the next caudal aortic arch. At one stage of development, each collector loop is drained by both parent aortic arches (Fig. **C**). This double drainage persists in the adult only for the first collector loop; the second to fourth collector loops lose their original connection with their cranial parent aortic arch (Fig. **D**). At the same time, cross trunks connecting neighboring collector loops grow through the base of the gills and provide an alternative path for the draining of the gill capillaries. The dorsal part of the second aortic arch develops into the hyoidean efferent artery, whereas the dorsal parts of the third to sixth aortic arches grow into the four efferent branchial arteries.

New arteries extend from these early arteries. The external carotid artery grows out of the ventral end of the first collector loop. The distal portion of the internal carotid artery and the stapedial artery grow out of the rostral end of the dorsal aorta. The hypobranchial artery sprouts from the ventral end of the second collector loop, and the pharyngoesophageal artery grows out of the second efferent branchial artery. The ventral portion of the first aortic arch soon disappears, but the ventral portions of the second to sixth aortic arches form the afferent branchial arteries (Fig. **D**).

continued

Anatomy in Action 11-3 Shark Aortic Arches *continued*

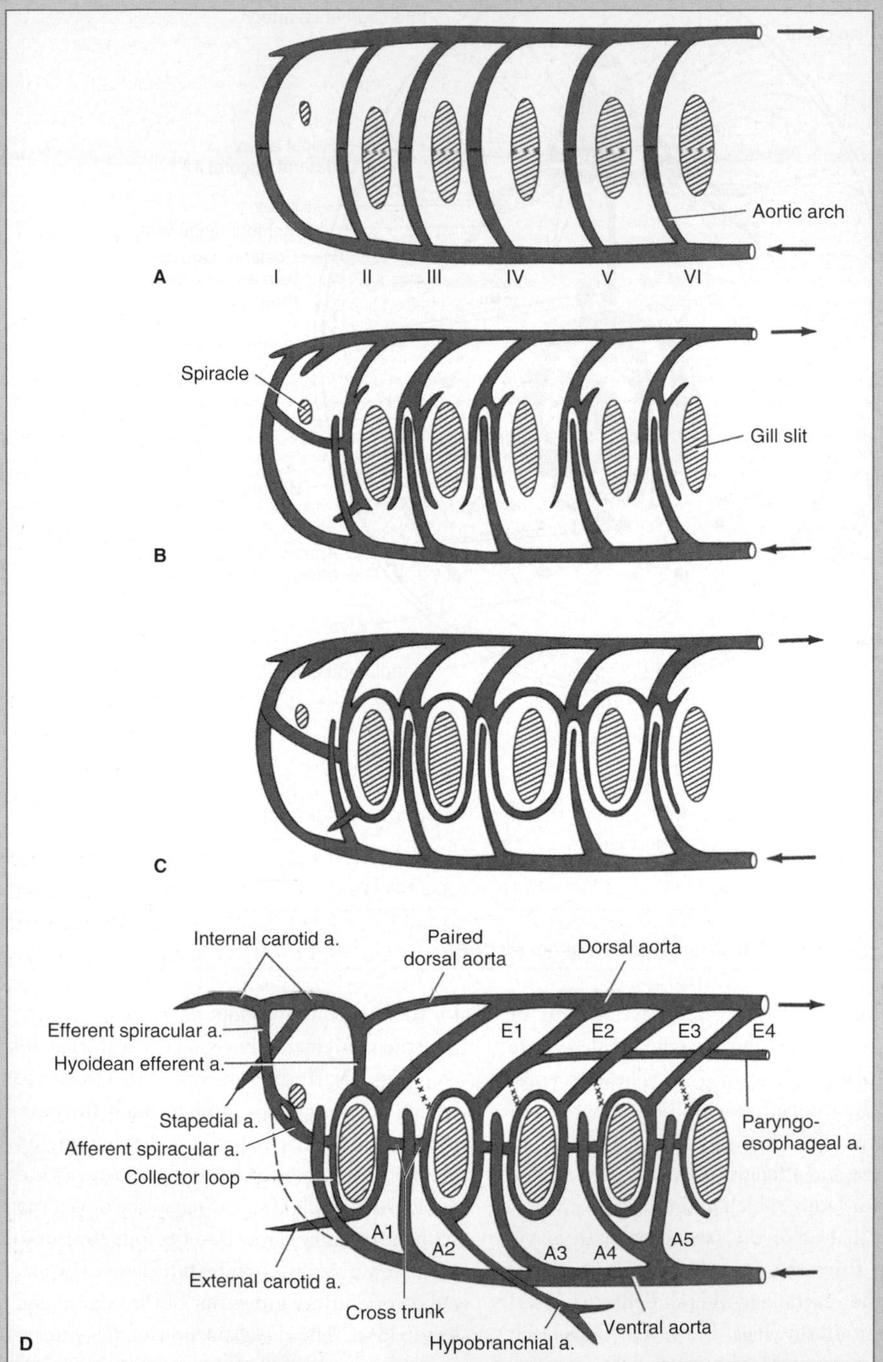

Diagrammatic representation of the aortic arches and their derivatives at different developmental stages of *Squalus* as seen in a lateral view. **A**, Early embryonic configuration. **B**, Intermediate embryonic configuration. **C**, Late embryonic configuration; **D**, Adult configuration. In reality, all arteries, except the caudal portions of the dorsal and ventral aortae, are paired. The afferent branchial arteries branching from the ventral aorta are shown in blue; the rest of the arteries, including the efferent branchial arteries, are shown in red. The first to sixth embryonic aortic arches are identified by Roman numerals. *A1–A5*, first to fifth afferent branchial arteries; *E1–E4*, first to fourth efferent branchial arteries; **xxxx**, dismantled original connection between collector loops.

continued

Embryonic Derivation of the Branchial Arteries of Sharks

Afferent Arteries	Efferent Arteries	Embryonic Origin
	Efferent spiracular artery	Dorsal portion of aortic arch 1
	Afferent spiracular artery	Cross connection between aortic arches 1 and 2
Afferent branchial artery 1	Hyoidean efferent artery	Aortic arch 2
Afferent branchial artery 2	Efferent branchial artery 1	Aortic arch 3
Afferent branchial artery 3	Efferent branchial artery 2	Aortic arch 4
Afferent branchial artery 4	Efferent branchial artery 3	Aortic arch 5
Afferent branchial artery 5	Efferent branchial artery 4	Aortic arch 6
	Interal carotid artery	Rostral extension of rostral portion of one of the paired dorsal aortae
	Stapedial artery	Rostroventral outgrowth from rostral end of one of the paired dorsal aortae
	External carotid artery	Rostral outgrowth from ventral extension of part of aortic arch 2
	Hypobranchial artery	Ventral outgrowth from ventral end of second collector loop (formed by aortic arches 3 and 4)
	Pharyngoesophageal artery	Caudal outgrowth from dorsal part of aortic arch 4

arteries. (1) Paired **intersegmental arteries**[4] branch off between the individual myomeres or body segments and soon bifurcate into a **dorsal branch** to the epaxial region and a **ventral branch** to the hypaxial region. Successive intersegmental arteries may be interconnected by longitudinal anastomoses. The arteries supplying the appendages are simply enlarged ventral branches of intersegmental arteries. (2) Paired **lateral visceral arteries** branch off to supply dorsolateral organs, such as the kidneys, gonads, and adrenal glands. (3) Unpaired median **ventral visceral arteries** develop from the embryonic vitelline arteries and pass through the dorsal mesentery to the viscera.

Trace the dorsal aorta caudad by working, if possible, from the side of the body that has already been cut open. You will have to separate the cranial portion of the esophagus from the dorsal body wall, and some mesenteries will have to be torn. A pair of **subclavian arteries** arise from the aorta, generally between the third and fourth efferent branchial arteries (Fig. 11-7). Follow one subclavian artery as it curves ventrally along the lateral margin of the posterior cardinal sinus. It follows the caudal edge of the scapular process of the scapulocoracoid cartilage and runs parallel to the subscapular vein (Fig. 11-6). It gives off several branches to the adjacent musculature, as well as the **brachial artery.** To see the brachial artery, mobilize the base of the pectoral fin on the intact side of the body wall, if you did not already do so while looking for the brachial vein. You will find the injected brachial artery on the medial surface of the muscular base of the pectoral fin near the brachial vein. Follow the brachial artery cranially and proximally through the body wall until you find its bifurcation from the subclavian artery. After this bifurcation, the subclavian artery continues as the **anterior ventrolateral artery,** which runs along the internal surface of the coracoid bar and parallel to the lateral abdominal vein before it curves sharply caudad to continue within the ventral wall of the body (Fig. 11-8). In well-injected specimens, you may see an artery branching off the anterior ventrolateral artery at that curvature; it supplies the transverse septum.

The subclavian arteries are modified intersegmental arteries. Other intersegmental arteries may be seen at this point, but you should first examine the ventral visceral arteries that branch off the dorsal aorta (Fig. 11-8). These arteries are not accompanied by veins, but their branches are followed by tributaries of the hepatic portal vein, which you have seen earlier. The most cranial visceral artery, the large **coeliac artery** (Figs. 11-5 and 11-7), enters the cranial end of the pleuroperitoneal cavity and extends ventrocaudally along the right side of the stomach to the cranioventral tip of the dorsal lobe of the pancreas (see Fig. 10-5), where it bifurcates. One branch, the **pancreaticomesenteric artery,** follows the pancreaticomesenteric vein dorsal to the pylorus through the ventral lobe of the pancreas and continues onto the ventral side of the intestine as the **anterior intestinal artery.** The **anterior in-**

[4]These blood vessels are also called "segmental" arteries, but the term "intersegmental" is more appropriate because these arteries occupy an intersegmental position.

testinal vein, which drains into the pancreaticomesenteric vein, lies besides the artery. In addition, the pancreaticomesenteric artery sends smaller branches to the pyloric region of the stomach, to the ventral lobe of the pancreas, and into the spiral valve. To be able to follow the entire pancreaticomesenteric artery and vein with their branches, you will have to mobilize the ventral lobe of the pancreas by breaking the gastrosplenic ligament, which anchors it to the pyloric region of the stomach (see Chapter 10). Observe the pancreaticomesenteric artery and vein on the dorsal side of the ventral lobe of the pancreas and follow them by scraping away some of the pancreatic tissue. The other branch of the coeliac artery, the **gastrohepatic artery,** often is extremely short, sometimes only 1 to 2 mm long. It divides into a small **hepatic artery,** which follows the hepatic portal vein to the liver, and a **gastric artery,** which passes to the stomach and gives off a branch to the dorsal lobe of the pancreas. It closely follows the gastric vein and its tributaries.

The next two ventral visceral arteries, which arise closely together or even from a common trunk, will be found in the free caudal edge of the mesogaster (see Chapter 10). These are the lienogastric and anterior mesenteric arteries, which pass through the mesentery to the organs they supply (Fig. 11-5). The **lienogastric artery** supplies the spleen and from there continues through the gastrosplenic ligament to the caudal part of the stomach, where its branches may anastomose with branches of the gastric artery. The lienogastric artery also gives off a branch to the dorsal lobe of the pancreas. The **anterior mesenteric artery** continues as the **posterior intestinal artery** along the dorsal side of the intestine (Fig. 11-8). The last ventral visceral artery is the smaller **posterior mesenteric artery** to the digitiform gland and caudal end of the intestine. These three ventral visceral arteries together supply the area drained by the **lienomesenteric vein** (see Fig. 10-5), whose tributaries are the **lienogastric vein** and the **posterior intestinal vein,** which follow the corresponding arteries.

More caudally, a pair of **iliac arteries** arise from the dorsal aorta at the level of the pelvic fins. The iliac arteries, like the subclavian arteries, are intersegmental arteries (Fig. 11-8). To see them, follow the aorta caudad by separating the medial borders of the kidneys along the midline. Make sure not to destroy the conspicuous white, shiny **caudal ligament,** which lies superficially between the kidneys and runs from the vertebral column to the tail. At the level of the pelvic fins, make an incision through the serosa along the lateral border of one kidney and carefully lift a portion of the kidney away from the body wall so that you can trace the iliac artery to its origin from the aorta. Each iliac artery runs ventrally toward the cloaca. After giving off a **femoral artery,** which passes through the body wall and enters the pelvic fin, the iliac artery continues as the **posterior ventrolateral artery,** which curves craniad in the body wall. The posterior ventrolateral artery joins the anterior ventrolateral artery, which was seen extending caudad from the subclavian artery.

You can see typical **intersegmental arteries** by freeing the lateral edge of a kidney and lifting the kidneys off the body wall. The more conspicuous **ventral branches** of the intersegmental arteries run ventrad along the myosepta accompanied by **intersegmental veins,** which drain into the posterior cardinal veins. The **dorsal branches** of the intersegmental arteries supply the epaxial region of the body (Fig. 11-8). Lateral visceral arteries include a number of renal arteries and a pair of gonadal arteries. The **renal arteries** arise from the dorsal aorta close to, or in common with, the intersegmental arteries and enter the kidneys. You will see them best by spreading the paired kidneys apart as you have done in order to see the iliac arteries. The **gonadal arteries,** either **ovarian** or **testicular arteries,** usually arise from the very base of the coeliac artery or, rarely, from the coeliac artery itself in common with arteries supplying the esophagus. The accompanying gonadal veins are rarely seen and drain into the posterior cardinal veins, which you have seen earlier. In pregnant females, the paired **oviducal arteries** have become large enough to be injected and, thereby, visible. They branch off the aorta less than 2.5 cm caudal to the coeliac and gonadal arteries, follow the oviduct caudally, and divide into multiple branches in the wall of the uterine chamber holding the developing embryos. The vascularization increases as gestation progresses. Occasionally, the oviducal arteries form from a confluence of two branches off the aorta.

Caudal to the iliac arteries, the dorsal aorta enters the tail as the **caudal artery,** which, as you have already seen, lies dorsal to the **caudal vein** within the hemal canal of the vertebrae (see Anatomy in Action 11-1).

The Internal Structure of the Heart

Preceding dissections have explored the general location and structure of the heart and its coronary blood vessels. Place now your shark on its back and open the pericardial cavity by reflecting the flaps of the ventral body wall. Identify the different parts of the heart again. The **sinus venosus** is the thin-walled, transversely triangular chamber that is broadly attached to the cranial side of the transverse septum (Figs. 11-4 and 11-9). It receives blood that is oxygen-depleted through the paired hepatic sinuses and common cardinal veins, which we have already observed. Make a transverse cut through the ventral wall of the sinus venosus by first nicking it with a pair of pointed scissors and then extending the cut laterally from this hole with a pair of scissors. Spread the cut edges of the walls of the sinus venosus and find the slitlike **sinuatrial aperture,** which is located at the cranial end of the sinus venosus and leads into the **atrium** (Fig. 11-10). This aperture is guarded by a pair of lateral connective tissue folds, the **sinuatrial valve,** which prevents the backflow of blood from the atrium into the sinus venosus during contraction of the atrium.

In the present position, the bulbous, thick-walled **ventricle** is the most visible part of the heart. Make a midsagittal cut through the ventral wall of the ventricle with a sharp scalpel. Extend this cut with a pair of scissors rostrally from the caudal end of the ventricle all the way

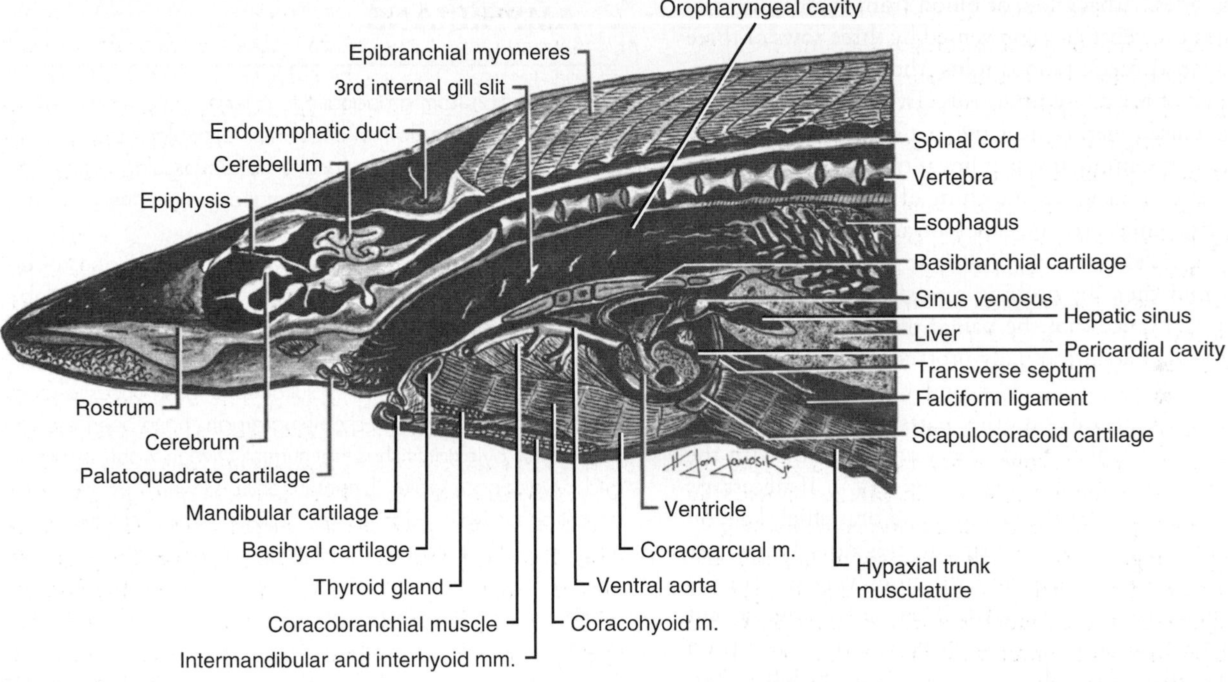

Figure 11-9

Longitudinal section through the head of *Squalus* showing the topographical relationships of the heart to the hypobranchial muscles, basibranchial cartilages, liver, and oropharyngeal cavity. A bristle is shown passing through the sinuatrial aperture.

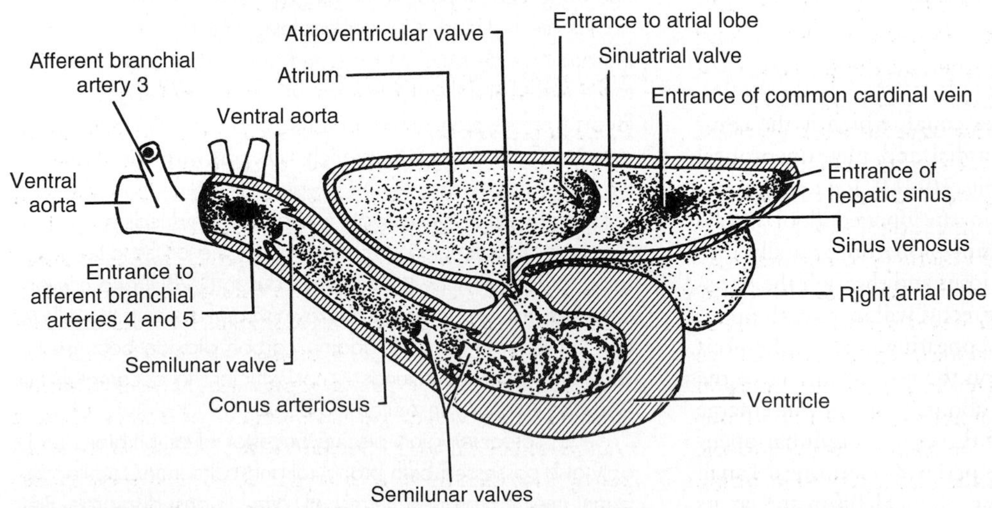

Figure 11-10

Semidiagrammatic longitudinal section through the isolated heart of *Squalus* showing the internal structures and relative thickness of the heart muscle.

through the **conus arteriosus.** The ventricular wall is especially thick and spongy as it is criss-crossed by numerous muscular strands. Notice that the cavity of the ventricle is U shaped because the **atrioventricular orifice** toward the atrium and the exit into the conus arteriosus lie almost next to each other. The conus arteriosus is tubular and has thick walls. Notice the paired coronary arteries along its sides. A backflow of blood from the conus arteriosus into the ventricle is prevented by three rows of three pocket-shaped membranous folds, the **semilunar valves,** on the wall of the conus arteriosus. Two rows of semilunar valves lie close together near the exit of the ventricle; the third circle of semilunar valves lies at the cranial end of the conus arteriosus near the branching of the fourth and fifth afferent branchial arteries. The entrances to the pocket-shaped valves face cranially; if they lie flat against the wall, you can find them by probing caudally along the surface of the wall. Notice that the paired flabby **atrial lobes** of the thin-walled atrium usually protrude ventrally on each side of the ventricle (Figs. 11-4 and 11-10).

To see the atrium and other parts of the heart, return to the opened oropharyngeal cavity and its floor with the exposed ventral aorta and heart (Fig. 11-7). If necessary, remove more of the epithelial cover and branchial skeleton to gain a clear view of the heart. In this dorsal view, the thin-walled flabby atrium and the sinus venosus are the most visible parts of the heart. Make a midsagittal cut through the wall of the atrium and notice that the atrium is an undivided cavity despite its paired atrial lobes. Notice also the muscular strands protruding from the inner surface of the atrial wall and the atrioventricular aperture in the floor of the atrium. This opening leads into the ventricle and is guarded by a pair of folds, the **atrioventricular valve,** which prevents backflow of blood from the ventricle into the atrium when it contracts.

The Pericardioperitoneal Canal

To see the **pericardioperitoneal canal,** which is the communication between the pericardial and pleuroperitoneal cavities mentioned earlier (see Fig. 10-2**D**), look at the dorsal side of the exposed heart in the opened floor of the oropharyngeal cavity. Pass a blunt probe horizontally into the recess dorsal to the sinus venosus and through the transverse septum, which, if done correctly, will meet with no resistance. The probe will be seen appearing in a membranous canal under the visceral serosa on the ventral surface of the esophagus. After a distance of about 3 cm, it will emerge into the pleuroperitoneal cavity through a semilunar opening. This entire passage is the pericardioperitoneal canal. Frequently the pericardioperitoneal canal bifurcates at its caudal end, in which event there would be a pair of openings into the pleuroperitoneal cavity. The pericardioperitoneal canal permits liquid to escape from the pericardial cavity into the pleuroperitoneal cavity (see Anatomy in Action 11-4). Fluid does not move in the other direction in the pericardioperitoneal canal because its delicate walls collapse and act as a valve as soon as it is empty.

AMPHIBIANS

The transition from an aquatic to a terrestrial environment necessitates a transformation of the respiratory system, with fundamental consequences for the cardiovascular system. The most conspicuous of these transformations affect the heart and aortic arches.

As Kjell Johansen (1980) showed, the advantages of being able to exploit the large amounts of oxygen (21%) available in air are especially significant for aquatic vertebrates living in warm, oxygen-poor water. These selective advantages probably were responsible for the evolution of lungs in ancestral bony fishes, and a variety of air-breathing organs have been evolved secondarily by extant fishes that currently live in warm, swampy, and oxygen-poor water. Extant lungfishes, which evolved from ancestral sarcopterygian fishes, represent excellent models to help understand how air breathing could have evolved in aquatic fishes. They can breathe under water as well as in air, and they possess both gills and lungs and, hence, both pulmonary and branchial blood circulations. Even though their heart is only partly divided into right and left sides, radiographic studies have shown that there is, nevertheless, a good separation of oxygen-rich from oxygen-depleted blood within the heart. Oxygen-rich blood is received from the lungs by the left atrium and is directed through the ventricle to the more rostral aortic arches. This blood flows, however, directly to the head and body, because the gills on the more rostral branchial arches are reduced. Oxygen-depleted blood is received from the body by the sinus venosus and right atrium and is directed to the more caudal aortic arches, which supply the gills that are located on the more caudal branchial arches. In addition, the sixth aortic arch sends a branch, the **pulmonary artery,** to the lungs. Depending on the oxygen content of the water and whether the lungfishes estivate in a cocoon in dried-up mud, the blood can be directed preferentially to one or the other respiratory organ, as shown by Kjell Johansen (1980) and coworkers. Even in warm water with a low oxygen content, the gills retain the crucial role of unloading carbon dioxide because the ventilation rate of the lungs is not sufficient to accomplish this (see also Anatomy in Action 10-2).

The sarcopterygian piscine ancestor of amphibians probably also possessed both branchial and pulmonary circulations. Extant metamorphosed terrestrial amphibians dismantle their branchial circulation and depend on their pulmonary circulation supplemented by buccopharyngeal and cutaneous circulations

Anatomy in Action 11-4 The Shark Heart

The heart of the spiny dogfish is located in the pericardial cavity. The pericardial wall, however, does not fit snugly the contours of the heart, and the space between the heart and the pericardial wall is filled with pericardial liquid (see Figs. 10-2**D** and 11-9). The pericardial cavity is encased caudodorsally by the caudal basibranchial cartilage and its ventrally curved spinous process (see Figs. 4-4 and 4-5); caudoventrally by the transverse septum and liver; ventrally by the coracoid bar and coracoarcual muscle; laterally by the third, fourth, and fifth coracobranchial muscles; and cranially by the second coracobranchial muscles (see Figs. 10-8 and 11-9). The parietal pericardium of the floor of the pericardial cavity is heavily reinforced by tough connective tissue, but along the lateral walls, the parietal pericardium is so thin and transparent that the muscle fibers of the coracobranchial muscles can be discerned through it.

When the oropharyngeal cavity is expanding during ventilatory movements (see Anatomy in Action 10-1), the first to fourth branchial arches are unfolded only slightly by the contracting second to fourth coracobranchial muscles. The fifth branchial arch, which comprises the caudal basibranchial cartilage, is unlikely to be unfolded, because it does not support gills and is the site of attachment for the massive esophageal sphincter (see Chapter 10). During the slight compression of the pericardial cavity that occurs during expansion of the oropharyngeal cavity, the liquid surrounding the heart acts as a hydraulic cushion, distributing the pressure evenly over the entire heart. The filled atrium contracts and pushes the blood into the relaxed ventricle, while the sinuatrial valve prevents the blood from flowing back into the sinus venosus. Theoretically, some pericardial liquid could escape through the pericardiopleuroperitoneal canal at this stage, but this is unlikely because the pericardial liquid would be lost from the pericardial cavity after only a few ventilatory cycles with no obvious source to replenish it fast. Furthermore, the pericardiopleuroperitoneal canal is relatively long and, therefore, may serve as an equalizer tube. When the coracobranchial muscles contract, the afferent branchial arteries are compressed because they pass between the bulging muscles. This ensures that blood does not flow to the gills while there is no flow of oxygen-rich water through the gill pouches.

When the oropharyngeal cavity is being compressed and water is pushed from the oropharyngeal cavity through the gill pouches, the coracobranchial muscles relax. The ventricle contracts and drives blood into the conus arteriosus under relatively high hydrostatic pressure. The conus arteriosus expands as it receives this surge of blood and then contracts during ventricular relaxation. In this way, the conus arteriosus functions as a *"windkessel,"* which is an engineering term for a container that equalizes pressure pulses, such as those created by the jets of blood that are ejected with each contraction of the ventricle. The atrium is being filled with blood from the sinus venosus as the pericardial cavity expands. This expansion of the pericardial cavity may create some suction in the atrium and sinus venosus, and the compression of the large cardinal sinuses may support the filling of the venous sinuses (see Anatomy in Action 11-1).

When a large prey item is swallowed, not only is the oropharyngeal cavity expanded more extensively than during ventilation, but the fifth branchial arch is also unfolded (see Figs. 4-4 and 4-5). This unfolding through the contraction of the large fifth coracobranchial muscle compresses the pericardial cavity more than during ventilation. It is in this situation that the hydraulic cushion effect of the pericardial liquid is especially effective in preventing the heart from being crushed. It is unlikely that the pericardial liquid escapes from the pericardial cavity at this point. Because the pericardiopleuroperitoneal canal is membranous and is applied to the ventral surface of the esophagus, it automatically will be stretched and collapsed as the esophagus expands to receive a prey item, thus preventing escape of fluid from the pericardial cavity.

(see Anatomy in Action 10-2). The atrium of the heart of metamorphosed adult amphibians is divided into left and right sides, but the oxygen-rich and oxygen-depleted blood streams are also kept mostly separate within the undivided ventricle. The aortic arches supplying the head and body contain blood with the highest oxygen saturation, whereas the sixth aortic arch, which supplies the lungs and skin, contains blood with the lowest oxygen saturation (Anatomy in Action 11-5D). The lungs tend to be involved primarily in the uptake of oxygen, whereas the skin is primarily involved in unloading carbon dioxide. After a cycle of lung ventilation (see Anatomy in Action 10-2) and when they stay under water, amphibians close their glottis and hold air in their lungs at greater than atmospheric pressure. During prolonged breath holding, oxygen in the lungs is depleted, resistance within the pulmonary arteries increases, and blood is diverted to the skin and the rest of the body. When the lungs are inflated with fresh air during the next cycle of ventilation, the resistance within the pulmonary arteries drops and blood is directed increasingly toward the lungs. The incomplete separation of the heart chambers permits adjustment of the relative blood flow to the lungs and other tissues, depending on the organism's needs during different activities.

In reptiles, the oxygen-depleted and oxygen-rich blood streams in the heart are generally more completely separated than in amphibians. However, the extent to which the ventricles and atria are separated by septa varies greatly among reptiles. Complete separation between the left and right sides of the heart, as in birds and mammals, does not occur in reptiles. Even in alligators and crocodiles, in which interatrial and interventricular septa are developed, shunts between the two sides of the heart can be established. Incomplete separation of the left and right heart chambers allows the blood to be distributed to various parts of the body in a manner appropriate to their levels of activity. For example, during lung ventilation, resistance in the pulmonary blood vessels is low and a greater percentage of the blood passes through the lung. However, as reptiles hold their breath, and the oxygen contents of the air in their lungs is depleted, resistance in the pulmonary blood vessels increases. As a result, more of the blood is shunted away from the lung circulation and recirculated through the body.

In nonmammalian tetrapods, at least the first two aortic arches are dismantled, but the more rostral aortic arches remain and become blood vessels that are not interrupted by gill capillaries. The hydrostatic pressure of their cardiovascular system is higher than that in fishes. It is unclear whether this is a consequence of the absence of capillary beds between the heart and the dorsal aorta or of the need to overcome gravitational forces in the terrestrial environment. For example, the hydrostatic pressure in the dorsal aorta of a frog is 30 mmHg, whereas that of a spiny dogfish is 17 mmHg. This difference is all the more significant because a frog is much smaller than a spiny dogfish.

The venous system of amphibians and early tetrapods is not very different from that of fishes, but significant changes occurred in the lateral abdominal venous system, right hepatic veins, and cardinal veins, of which parts were transformed into a caudal vena cava.

The cardiovascular system of *Necturus* is an adequate model for that of an early tetrapod, except for certain features of its heart and aortic arches. Because it is paedomorphic, *Necturus* covers its oxygen needs by both gill and pulmonary respiration. The cardiovascular system should be studied on doubly or triply injected specimens.

The Heart and Associated Blood Vessels

Find the pericardial cavity, which was opened when you studied the coelomic cavity (see Chapter 10). Open it wide and identify the four chambers of the heart. The large muscular **ventricle** is most conspicuous in this ventral view and occupies the caudoventral portion of the pericardial cavity (Fig. 11-12). Cranial to and to the right of the ventricle, you will see the longitudinally oriented, large **bulbus arteriosus.** It is connected to the ventricle through the **conus arteriosus,** a thinner blood vessel, which emerges from the right side of the craniodorsal end of the ventricle. The conus arteriosus, despite its shape, is a chamber of the heart, but the bulbous arteriosus is a modification of the ventral aorta, even though it is situated within the pericardial cavity. The bulbous arteriosus may be able to expand more and, hence, function as a *windkessel* (Anatomy in Action 11-4). The single **atrium** is visible on the left side of the cranial end of the ventricle and extends toward the right side of the ventricle dorsal to the conus arteriosus. Internally, the atrium is partially divided into a left and a right atrium by an incomplete septum. Lift the caudal end of the ventricle, or **apex of the heart,** and you will see the **sinus venosus** emptying into the right atrium dorsal to the ventricle. Its walls are so thin that they are transparent and can easily be destroyed inadvertently. Its caudal end is fused to the transverse septum.

The sinus venosus receives oxygen-depleted blood from the **caudal vena cava.** The caudal vena cava drains the liver and receives other veins near the cranial end of the liver (see later). To reach the sinus venosus, it runs within the relatively long coronary ligament, which anchors the cranial end of the liver to the transverse septum. The paired **common cardinal veins** enter the caudodorsal corners of the sinus venosus just in front of the hepatic sinuses (Fig. 11-11), but they may be found more easily in connection with the dissection of the cranial systemic veins (see later). The paired **pulmonary veins,** which return oxygen-rich blood from the lungs, pass dorsal to the caudal vena cava and unite to form a single vein that enters the left side of the atrium.

Because the interatrial septum is small and because the two atria have a common opening into the single ventricle, considerable mixing of blood occurs in the heart. This mixing is inconsequential in *Necturus,* because all the blood that leaves the ventricle passes through the gills before it is distributed to the body and lungs (Anatomy in Action 11-5). Thus, at least functionally, the heart of *Necturus* is comparable to that of *Squalus,* and its structure is not remarkable enough to warrant its dissection.

The Venous System

The Hepatic Portal System

Functionally, the hepatic portal system in *Necturus* is the same as in fishes, but the basic branching pattern of its tributaries is representative of that of tetrapods in general. Stretch out the mesentery of the intestine, and you will see a longitudinal blood vessel, the **mesenteric vein,** running cranially and disappearing into the substance of the pancreas. Notice that it receives numerous **intestinal veins** from the intestine. Next, find the tail of the pancreas near the spleen and identify the **lienogastric vein.** It is formed by the confluence of the **lienic vein** from the spleen and

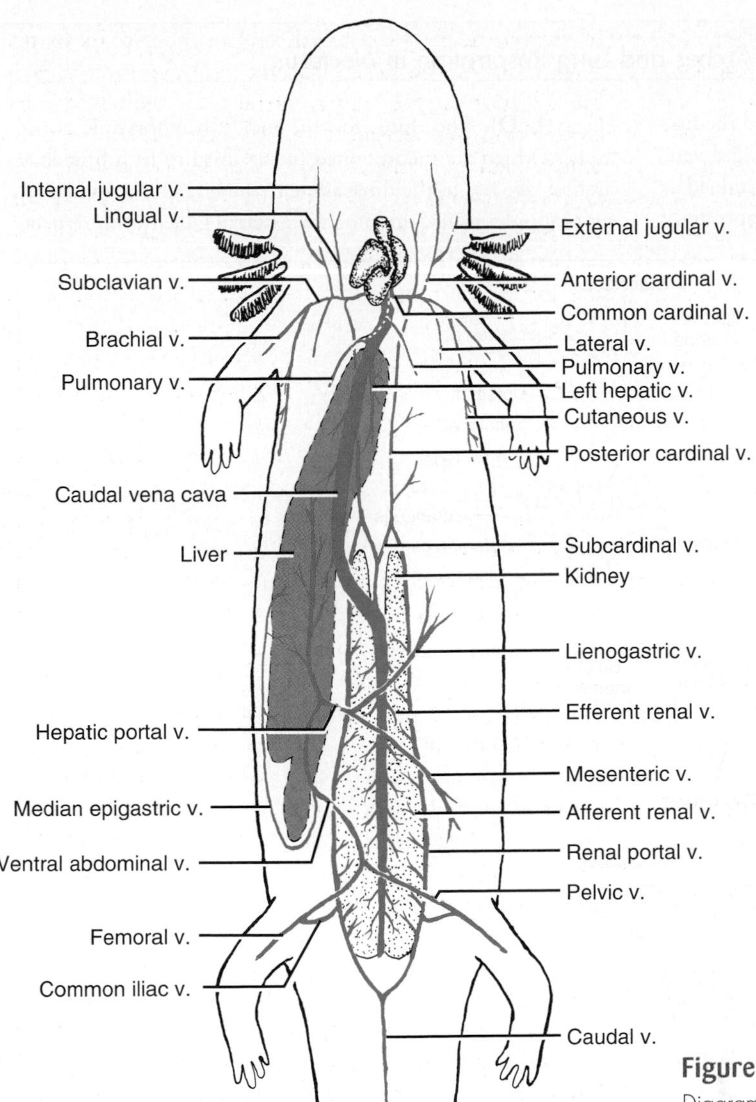

Figure 11-11

Diagrammatic ventral view of the venous system of *Necturus*. Only the proximal portions of the veins are shown.

of several **gastric veins** from the stomach. Carefully scrape away some pancreatic tissue and find the point where the lienogastric and mesenteric veins unite and form the **hepatic portal vein,** which extends cranially to the liver (Fig. 11-11).

The Ventral Abdominal Vein

The median, longitudinal blood vessel that lies within the long falciform ligament is the **median epigastric vein.** It has evolved from the paired lateral abdominal veins of fishes by moving ventrally and merging into a single blood vessel. Cranially, it anastomoses with veins that empty into the caudal vena cava at the cranial and of the liver, but it has established a new main connection to the hepatic portal system through the **ventral abdominal vein.** In this respect, the ventral abdominal vein resembles the umbilical vein in late embryos of amniotes (Fig. 11-1 and Anatomy in Action 11-6). The topographical relationships of the caudal tributaries of the ventral abdominal vein, however, are still very similar to those of the tributaries of the lateral abdominal vein of fishes. The ventral abdominal vein receives several small **vesical veins** from the large urinary bladder (see Chapter 10) and then receives a pair of **pelvic veins.** Each pelvic vein extends caudolaterally and receives on its lateral side, after about 1 cm, a **femoral vein** from the hind limb. At the junction between the pelvic and femoral veins, the **common iliac vein** branches off caudodorsally and connects with the renal portal vein (see below). Hence, blood from the leg may drain either through the ventral abdominal vein, which is

Anatomy in Action 11-5 Aortic Arches and Lung Respiration in *Necturus*

Necturus, like all tetrapods, has dismantled the first two aortic arches, but the paired dorsal and ventral aortae rostral to the third aortic arch are retained as the internal and external carotid arteries, respectively (Figs. **A–D**). The third, fourth, and fifth embryonic aortic arches, which are incorporated in the third to fifth branchial arches, give rise to the three afferent and efferent branchial arteries supplying and draining the external gills. In this respect,

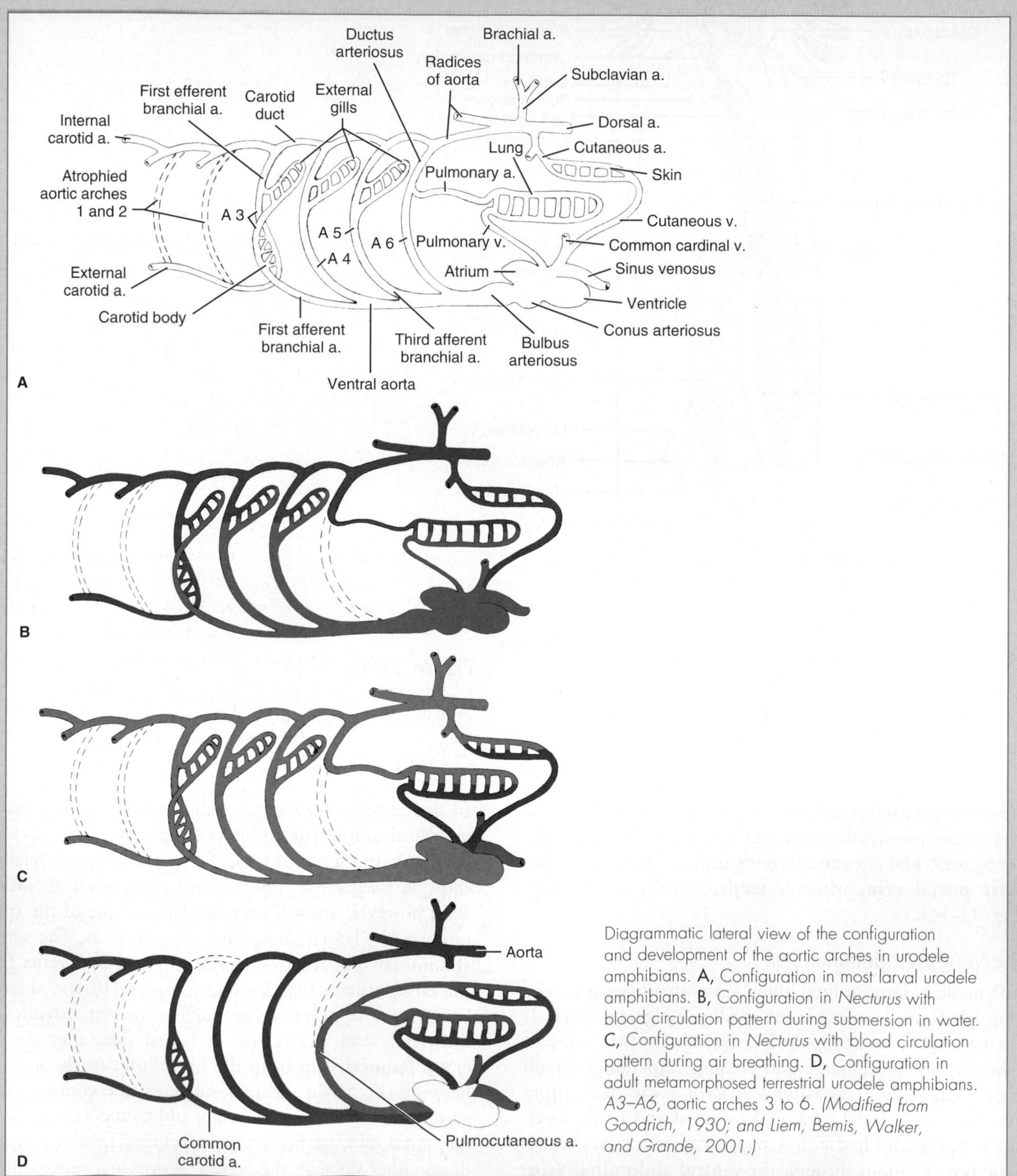

Diagrammatic lateral view of the configuration and development of the aortic arches in urodele amphibians. **A**, Configuration in most larval urodele amphibians. **B**, Configuration in *Necturus* with blood circulation pattern during submersion in water. **C**, Configuration in *Necturus* with blood circulation pattern during air breathing. **D**, Configuration in adult metamorphosed terrestrial urodele amphibians. *A3–A6*, aortic arches 3 to 6. *(Modified from Goodrich, 1930; and Liem, Bemis, Walker, and Grande, 2001.)*

Necturus resembles other larval urodele amphibians (Figs. **A** and **B**). Most other larval urodele amphibians, however, also possess a sixth aortic arch, the ventral part of which leads from the ventral aorta to the pulmonary artery and lungs, whereas the dorsal part continues as the ductus arteriosus to the dorsal aorta (Fig. **A**). *Necturus* is unique, as pointed out by Frank Figge (1930), in having dismantled the ventral portion of the sixth aortic arch (Fig. **B**) so that the pulmonary artery originates from the ductus arteriosus. Because the pulmonary artery is fed blood that originates from the efferent branchial arteries, the degree of blood oxygenation in the lung can be finely tuned to that in the gills. If the oxygen saturation of the water is low and the external gills are unable to take up enough oxygen, blood can be directed to the lung and skin capillaries for additional oxygen uptake and returned directly to the heart, when *Necturus* rises to the surface of the water to breathe air (see Anatomy in Action 10-2 and Fig. **C**). However, if oxygen uptake through the external gills is sufficient or when *Necturus* is submerged, then most of the blood can bypass the lung circulation and be shunted to the body (Fig. **B**). The oxygen-rich blood supplied to the skin can release additional carbon dioxide.

At first view, and as postulated by Frank Figge (1930), it appears that the circulatory pattern of *Necturus* can be functional only if the gills are retained and, therefore, could be a major constraint preventing the larval *Necturus* from metamorphosing into terrestrial adults. However, as we can see from the circulatory pattern of an adult metamorphosed urodele amphibian (Fig. **D**), gills can be dismantled while the aortic arches are retained. Hence, the inability of *Necturus* to metamorphose is unlikely to be causally related to the special arrangement of its aortic arches. *Necturus* is an obligate paedomorphic species. Metamorphosis cannot be induced in *Necturus,* as it can be in facultatively paedomorphic species, by injecting extra amounts of thyroxine, increasing the oxygen tension of the surrounding water, or manipulating other factors. The tissues of *Necturus* appear to have lost their ability to respond to thyroxine.

In urodele amphibians that do metamorphose into terrestrial adults, the carotid duct and ductus arteriosus are dismantled. As a result, the third afferent branchial artery of larval amphibians becomes the common carotid artery of metamorphosed amphibians. The ventral portion of the sixth aortic arch becomes the pulmocutaneous artery, which supplies the lungs and the skin with oxygen-depleted blood.

the ancestral path, or through the common iliac and renal portal veins, which is a novel path in amphibians.

The Renal Portal System

The blood vessel that runs along the lateral margin of each kidney dorsal to the prominent archinephric duct in males, or the oviduct in females, is the **renal portal vein.** As already described, it receives the common iliac vein. Trace the two renal portal veins caudad and find the point where they unite and receive the median **caudal vein** from the tail. The renal portal veins receive blood from the tail, legs, and adjacent body wall and conduct it to the peritubular capillaries of the kidneys, hence their designations as portal veins (see Fig. 12-1**B** and Chapter 12).

The Caudal Systemic Veins

At the very cranial end of the kidneys, the renal portal veins empty into a pair of small **posterior cardinal veins,** which continue cranially on either side of the dorsal aorta. The continuity of the renal portal and posterior cardinal veins is not surprising, because the renal portal veins develop from the caudal portions of the embryonic posterior cardinal veins (Anatomy in Action 11-2). This embryonic condition may have been retained, because *Necturus* does not metamorphose. Notice the **intersegmental veins,** which enter the posterior cardinal veins from the body wall. The posterior cardinal veins diverge at the level of the cranial end of the esophagus and unite with the anterior cardinal veins to form the common cardinal veins.

Return to the kidneys. The blood leaving the peritubular capillaries of the kidneys is collected by numerous microscopic **efferent renal venules.** These venules join the **efferent renal arterioles** draining the glomerular capillaries to form **efferent renal veins,** which are located on the ventral surface of the kidneys (see Fig. 12-1**B**). These efferent renal veins, in turn, join the **gonadal veins,** either **testicular** or **ovarian veins,** from the gonads and empty into the **caudal vena cava** lying in the middorsal line between the kidneys. After anastomosing with the caudal ends of the posterior cardinal veins through the **subcardinal veins,** the caudal vena cava runs cranially through the hepatocavopulmonary ligament (see Fig. 10-12) and enters the substance of the liver dorsally on its right side. Trace the caudal vena cava through the liver. It receives numerous small **hepatic veins** from various parts of the liver and a particularly large **left hepatic vein** at the cranial end of the liver. The caudal vena cava then passes through the coronary ligament and transverse septum to empty into the sinus venosus, as you observed earlier.

The caudal vena cava is a new blood vessel in amphibians. It was formed largely from veins that were present in the piscine ancestors. The portion of the caudal vena cava that is located cranial to the kidneys develops as a caudal extension from the right hepatic vein. Because the right hepatic vein is incorporated into the caudal vena cava, only the left hepatic vein is left at the cranial end of the liver in *Necturus,* instead of the two hepatic veins seen in *Squalus.* In *Necturus,* the caudal extension of the right hepatic vein taps into the embryonic subcardinal veins, as it does in mammalian embryos (Anatomy in Action 11-9, Fig. **B**), and these, particularly the right one, form the caudal segment of the caudal vena cava between the kidneys. In fishes, the embryonic subcardinal veins form the caudal portions of the adult posterior cardinal veins (Anatomy in Action 11-2, Fig. **A**). Although the subcardinal veins of amphibians now contribute to the formation of the caudal vena cava, they retain a connection with the posterior cardinal veins. Blood that has circulated through the kidneys may now drain through either the posterior cardinal veins, which is the ancestral path, or the caudal vena cava, which is the novel path for amphibians. Most of the blood draining the kidneys flows through the larger caudal vena cava, because the posterior cardinal veins are smaller. The advantages of the new venous channel are not clear at this time.

The Cranial Systemic Veins

The cranial systemic veins are difficult to find unless they are filled with blood or unless you are working under magnification. The common cardinal veins may have been seen entering the sinus venosus earlier when you studied the heart and pericardial cavity. Each common cardinal vein receives a number of tributaries, most of which can be found peripherally and then traced to the common cardinal vein (Fig. 11-11). You should study them on the side opposite to that on which you dissected the muscles (see Table 7-4). Do not damage any arteries while dissecting the veins.

Carefully remove the skin from the lateral surface of the brachium and shoulder, and you will see the **brachial vein.** It soon joins the **cutaneous vein** from the skin over the trunk to form the **subclavian vein.** Trace the subclavian vein cranially. It dips into the musculature and enters the **common cardinal vein.** Now remove the skin ventral to the gill slits and separate the hypobranchial muscles from the branchial region (see Chapter 7). The longitudinal vessel lying on the hypobranchial musculature is the **lingual vein.** Trace it caudally, and you will see it entering the common cardinal vein next to, or in common with, the subclavian vein. After locating these two veins, bisect the muscles ventral to their entrance into the common cardinal vein, thereby exposing the common cardinal vein descending to the heart.

Remove the skin from the side of the trunk caudal to the shoulder. The longitudinal vessel lying between the epaxial and hypaxial musculature is the **lateral vein.** Mobilize the scapula by detaching its muscles from the trunk, if you have not already done so while opening the oropharyngeal cavity (see Chapter 10), and trace the lateral vein cranially. It enters the common cardinal vein slightly dorsal to the entrance of the lingual and subclavian veins. Return now to the posterior cardinal vein and trace it cranially. You will see it enter the common cardinal vein caudal to the entrance of the lateral vein. The vein entering the common cardinal vein cranial to the entrances of the lateral and posterior cardinal veins is the **anterior cardinal vein.** Try to trace it cranially as it receives two tributaries. The large **external jugular vein** passes dorsal to the gills, and the smaller **internal jugular vein** drains the roof of the oropharyngeal cavity.

The anterior cardinal vein, together with the internal jugular vein, has evolved from the anterior cardinal vein of fishes. The subclavian and brachial veins have evolved from the cranial portion of the lateral abdominal vein in fishes. The lingual vein has evolved from the inferior jugular vein of fishes. Although the terminology has changed, it is apparent that the major cranial veins of *Necturus* are similar to those of fishes.

The Arterial System

The Aortic Arches and Their Branches

Return to the pericardial cavity and follow the bulbus arteriosus cranially as it leaves the pericardial cavity and turns into the ventral aorta. You may have to remove some of the rectus cervicis muscle (see Chapter 7). Two arteries leave from each side of the cranial end of the bulbus arteriosus. They probably are not injected, and you will need to be careful when you trace them laterally toward the external gills on the intact side of the oropharyngeal cavity (see Table 7-4). They cross the transversi ventrales muscles and then pass deep to the subarcual muscle (see Chapter 7 and Fig. 11-12). The more rostral artery, known as the **first afferent branchial artery,** follows the first branchial arch and supplies the first external gill with oxygen-depleted blood. Cranial to the distal portion of the first afferent branchial artery lies the small, usually well-injected **external carotid artery.** Notice that the external carotid artery supplies the muscles of the floor of the oropharyngeal cavity with oxygen-rich blood through multiple branches. The external carotid artery branches off the first efferent branchial artery (Anatomy in Action 11-5) and, therefore, usually is injected, but it also has tiny anastomoses that connect it to the first afferent branchial artery. One of the paired **thyroid glands** is located in the medial angle

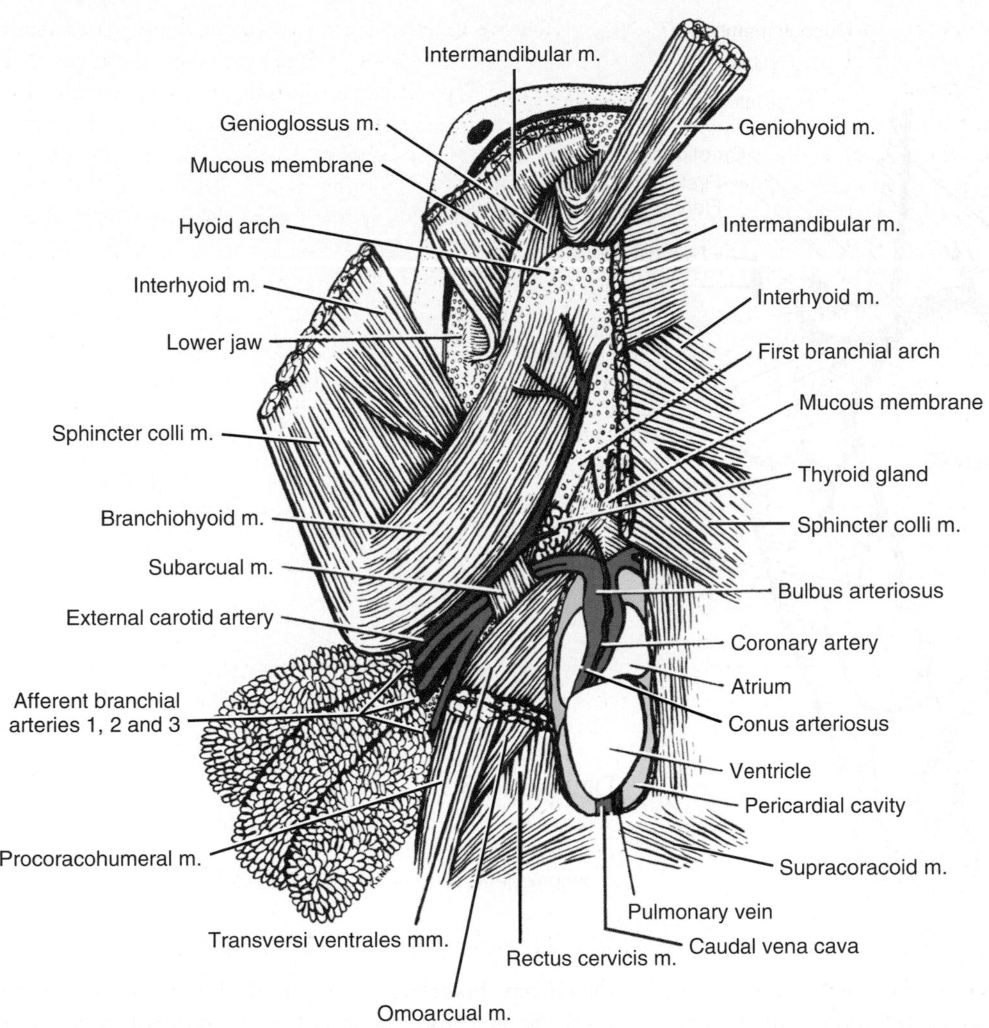

Figure 11-12

Ventral view of the heart and afferent branchial arteries of *Necturus*. Much of the rectus cervicis muscle has been removed.

formed by the first afferent branchial and external carotid arteries. The thyroid gland can be recognized by its texture, because it is composed of many vesicle-like follicles.

The more caudal artery leaving the bulbus arteriosus bifurcates soon after it emerges from the bulbus arteriosus. One branch, the **second afferent branchial artery,** follows the second branchial arch and supplies the second external gill with oxygen-depleted blood. The other branch, the **third afferent branchial artery,** follows the third branchial arch and supplies the third external gill with oxygen-depleted blood.

Swing open the floor of the oropharyngeal cavity (see Chapter 10) and carefully remove the epithelium from the roof of the oropharyngeal cavity. A pair of large arteries, the **radices of the aorta** (singular: radix of the aorta) will be seen converging toward the middorsal line, where they unite to form the **dorsal aorta** (Fig. 11-13). The radices of the aorta receive oxygen-rich blood from the external gills. Carefully trace one radix of the aorta on the intact side of the oropharyngeal cavity toward the internal gill slits (see Table 7-4). Slightly lateral to the vertebral column, the **vertebral artery** branches off and extends cranially, then soon dips into the musculature at the base of the skull to provide the muscles with oxygen-rich blood. Farther laterally, the radix of the aorta curves caudally and receives on its rostral side the short, stout **carotid duct.** The carotid duct is an anastomosis between the radix of the aorta and the **first efferent branchial artery** (Fig. 11-13 and Anatomy in Action 11-5). Trace the first efferent branchial artery laterally as far you can. It drains the capillaries of the first external gill. The **internal carotid artery** branches off the junction between the carotid duct and the first efferent branchial

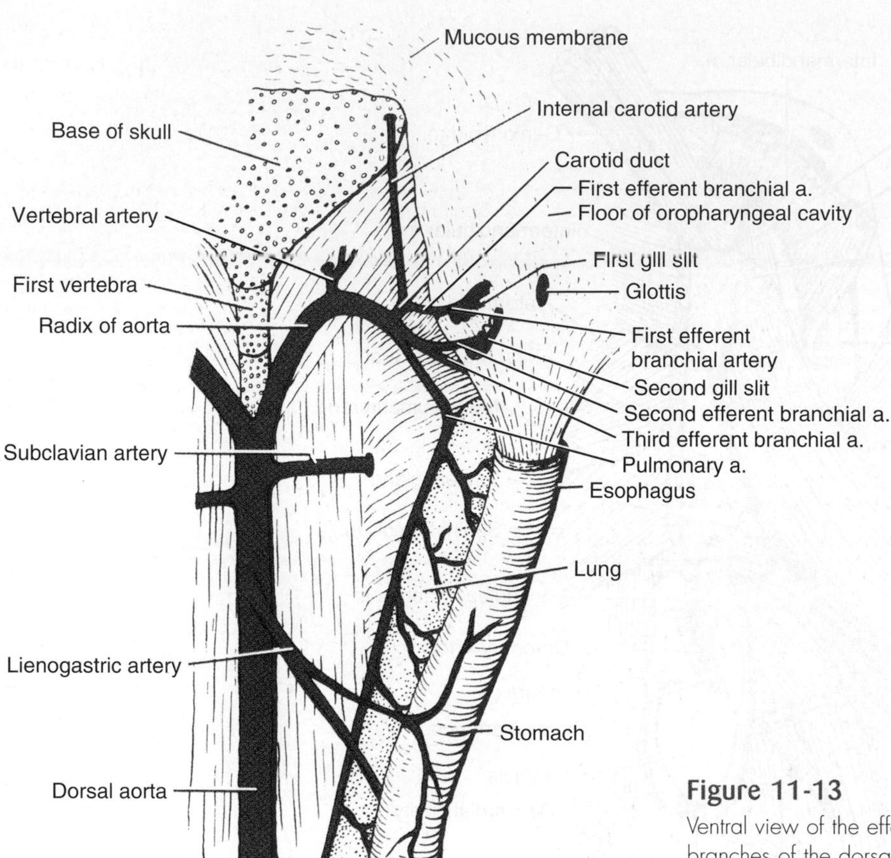

Figure 11-13
Ventral view of the efferent branchial arteries and cranial branches of the dorsal aorta of *Necturus*. The floor of the oropharyngeal cavity is reflected to the left of the specimen.

artery, and runs cranially along the roof of the oropharyngeal cavity. It supplies the facial region and eventually enters the skull to supply the brain with oxygen-rich blood. Return to the radix of the aorta and continue tracing it laterally until it receives the **second** and **third efferent branchial arteries** from the second and third external gills. These arteries bring blood to the radix of the aorta and to the **pulmonary artery,** which branches off on the caudal side of the radix of the aorta (Anatomy in Action 11-5).

The pulmonary artery supplies the lung with blood that is more or less saturated with oxygen, depending on the rate of gas exchange that took place in the capillaries of the external gills (Anatomy in Action 11-5). Trace the pulmonary artery to the lung, but if you have difficulty doing so because the artery is not injected, find the pulmonary artery on the lung and retrace it cranially toward the radix of the aorta. The **pulmonary vein,** which returns oxygen-rich blood to the heart, lies on the opposite side of the lung; we have already noted its entrance into the atrium.

From the present view of the roof of the oropharyngeal cavity, only the points of entrance of the efferent branchial arteries into the radix of the aorta are visible. In order to see the efferent branchial arteries more clearly, you need to remove the skin from the top of the head dorsal to the external gill slits and from the bases of the external gills on the side opposite to that on which you dissected the muscles (see Table 7-4). The three efferent branchial arteries lie near the dorsal edge of the gills. Trace them to the points where you left off their dissection on the roof of the oropharyngeal cavity. You should also find the point where the external carotid artery originates from the first efferent branchial artery.

The Dorsal Aorta and Its Branches

Find the place where the two radices of the aorta form the **dorsal aorta** and follow it caudally. Almost immediately, a pair of **subclavian arteries** branch off laterally (Figs. 11-13 and 11-14). Trace one subclavian artery laterally. At the base of the pectoral limb, it bifurcates into a **brachial artery,** which continues into the arm, and a **cutaneous artery,** which supplies the skin and adjacent muscles. The cutaneous artery can be very large in amphibians, in which cutaneous respiration represents a significant portion of gas exchange.

As you continue to follow the dorsal aorta caudally, you will observe that the **lienogastric artery,** a ventral visceral artery, branches off on its ventral surface. It soon ramifies to supply various parts of the stomach and spleen. The next ventral visceral branch, the **coeliacomesenteric artery,** arises some distance caudad (Fig. 11-14). It runs ventrally to the tail of the pancreas, where it divides into the **lienic artery** to the spleen, the **hepatic artery** to the liver, and the **pancreaticoduodenal artery** to the pancreas and duodenum. You will have to dissect away much of the pancreas to see all these arteries. The remaining ventral visceral arteries consist of a number of **mesenteric arteries** to the intestine and a pair of cloacal arteries to the cloaca (see later). You will have to separate the caudal vena cava from the aorta to see

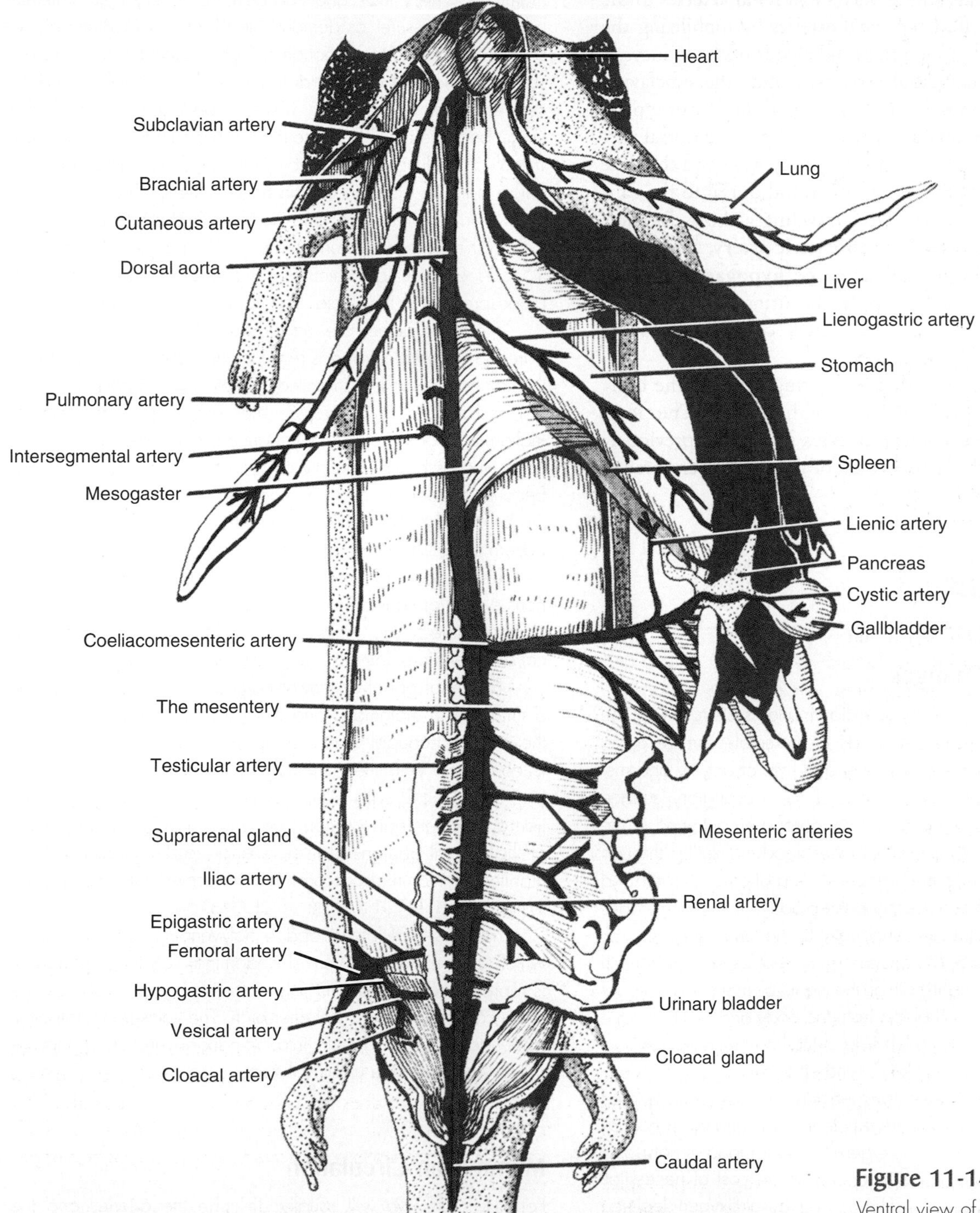

Figure 11-14
Ventral view of the arterial system of *Necturus*.

the point of origin of the mesenteric arteries. Notice that the caudal vena cava lies toward the right of the aorta and the mesenteric arteries. This position reflects the origin of this segment of the vena cava from the right subcardinal vein in the embryo (Anatomy in Action 11-9).

The lateral visceral arteries are represented by a number of paired **gonadal arteries,** either **testicular** or **ovarian arteries,** to the gonads and of very small **renal arteries** to the kidneys. You can find the renal arteries by mobilizing the caudal vena cava between the caudal ends of the kidneys.

Paired intersegmental arteries include the subclavian arteries, which you have already seen; a number of typical **intersegmental arteries,** which arise from the dorsal surface of the aorta and pass into the body wall; and the **iliac arteries.** The iliac arteries can be found dorsal to the caudal ends of the kidneys. After traveling a short distance, each iliac artery gives off the **epigastric artery,** which runs cranially in the body wall, and the **hypogastric artery,** which runs caudally to supply the urinary bladder and cloaca. Subsequently, the iliac artery continues into the hind leg as the **femoral artery.**

Trace the aorta caudad by cutting through the body wall and muscles caudolateral to the cloaca. The aorta gives off the paired **cloacal arteries** referred to previously and then enters the hemal canal of the caudal vertebrae as the **caudal artery.**

MAMMALS

Evolutionary Changes

Most transformations of the cardiovascular system during the evolution from ancestral tetrapods to mammals can be correlated with an increase in the level and rate of activity and metabolism. The lungs have increased their respiratory surface area to such a degree that the blood volume contained in the pulmonary capillaries is equal to the blood circulating through the body. This permits the complete separation of the left and right halves of the heart, because the blood in the ventricle does not need to be distributed unequally to the lung and body, as was necessary in amphibians and most reptiles with their much simpler lungs. The right half of the heart of mammals now receives oxygen-depleted blood from the body and sends it to the lungs for oxygenation; the left side of the heart receives oxygen-rich blood from the lungs and sends it to the body. This complete separation between oxygen-rich and oxygen-depleted sides of the heart allows more efficient gas exchange, because mixing of oxygen-rich and oxygen-depleted blood is not possible. However, this advantage comes at the cost of necessitating a continuous lung ventilation, because oxygen-depleted blood cannot be recirculated for further extraction of oxygen while holding the breath. The complete division of the heart enables the systemic and pulmonary vascular circuits to maintain different hydrostatic blood pressures, which is necessitated by the significantly higher hydrostatic pressures within the mammalian cardiovascular system. For example, the mean hydrostatic blood pressure in the aorta of human beings is 100 mmHg compared to a mean of about 25 mmHg in the dorsal aorta of frogs. The muscular wall of the left ventricle in mammals is very thick and can develop a very high systemic hydrostatic pressure, as demonstrated by the hydrostatic blood pressure in the aorta. Although a high hydrostatic pressure ensures a rapid and efficient distribution of substances to the tissues and cells of the body, it would be inappropriate in the lungs, because it would cause the plasma fluid to filtrate into the alveoli of the lungs and thereby fill the lungs with liquid, a condition that can occur in certain heart patients. The wall of the right ventricle is less thick, and the mean hydrostatic blood pressure in the pulmonary trunk of human beings is only between 15 and 20 mmHg, which is comparable to the hydrostatic blood pressure in the dorsal aorta of nonmammalian tetrapods.

The aortic arches are further reduced in mammals as compared to ancestral tetrapods (Anatomy in Action 11-7), because the fifth aortic arch is dismantled on both sides in addition to the first and second aortic arches. The ventral aorta of nonmammalian tetrapods is completely reduced, because the remaining aortic arches branch directly off the dorsal aorta in mammals. Because the dorsal aorta is the only aorta in mammals, it is simply called the aorta. The branches of the aorta follow the basic pattern established in ancestral vertebrates (Fig. 11-8).

The anterior and common cardinal veins of the cranial venous system of nonmammalian tetrapods evolved into a pair of cranial venae cavae, which is reduced to a single cranial vena cava in certain mammals, such as cats.

The renal portal system of nonmammalian tetrapods was dismantled in mammals, and the caudal vena cava receives the iliac and caudal veins. The loss of the renal portal system is correlated with the evolution of the mammalian nephron (see Fig. 12-1C) with its unique mechanism of urine production and water reabsorption (see Chapter 12). Although most of the cranial portions of the posterior cardinal veins are dismantled in mammals, a part is transformed into the azygos system of veins.

The ancestral lateral abdominal venous system of fishes, which evolved into the ventral abdominal vein in amphibians, is represented in mammals only at the embryonic stage by the umbilical vein, which is lost after birth. The veins from the pelvic appendages empty directly into the caudal vena cava and are not connected to the umbilical vein.

Mammalian Circulation

For mammals, we will usually describe the arteries and the veins together. Because they tend to parallel each other, it is convenient to dissect them at the same time. Veins are often more difficult to find than arteries, because valves frequently prevent the injected latex from reaching the peripheral parts of veins. Identification of the accompanying arteries will help you find the veins. Furthermore, noninjected veins are translucent due to their thin walls and often contain coagulated blood, liquid, and even air bubbles. If you compress veins with a blunt object, you may see air bubbles moving in them, thus revealing the course of the veins.

Before studying the blood vessels on a regional basis, it is helpful to understand the overall circulation pattern of the mammalian cardiovascular system. In an adult mammal (Fig. 11-15), tributaries of the **caudal vena cava** drain the body caudal to the diaphragm, and tributaries of the **cranial vena cava** drain the body cranial to the diaphragm. Both venae cavae return oxygen-depleted blood to the **right atrium** of the heart. From here the blood flows to the **right ventricle,** from where it is

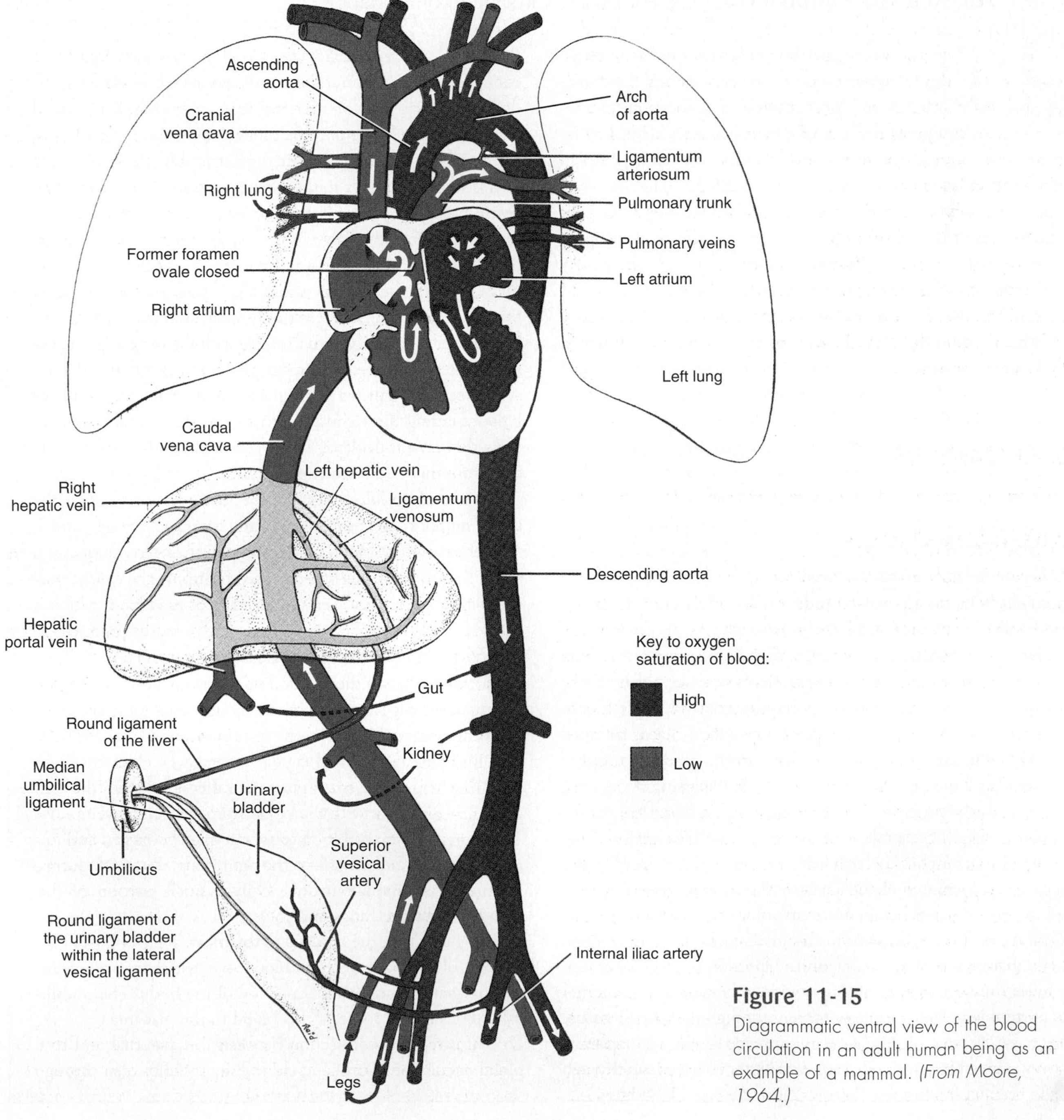

Figure 11-15

Diagrammatic ventral view of the blood circulation in an adult human being as an example of a mammal. *(From Moore, 1964.)*

pumped to the lungs through the **pulmonary trunk** and **pulmonary arteries.** After the blood has been saturated with oxygen in the lungs, it returns to the **left atrium** through the **pulmonary veins.** From here, the blood flows to the **left ventricle,** from where it is pumped into the **aorta.** Branches of the aorta carry oxygen-rich blood to all parts of the body, including the tissues of the lungs. The muscle of the heart itself is supplied with oxygen-rich blood through the separate **coronary arteries,** which leave the base of the aortic arch. The **coronary veins** return oxygen-depleted blood from the heart muscle to the **coronary sinus** and right atrium.

The circulation pattern of a mammalian fetus differs in several aspects from that of an adult mammal. Although we will not dissect a fetal specimen, it is useful and interesting to understand the fetal circulation pattern and the changes that need to take place at birth (Anatomy in Action 11-6). Furthermore, this

Anatomy in Action 11-6 The Embryonic Circulation of Mammals

The circulation pattern of a mammalian embryo can be understood best by considering the drastically different environment and life conditions of a fetus compared to those of a mammal after birth. The fetus is developing in a liquid-filled chamber, which itself lies in a liquid-filled chamber (Figs. 11-1 and 12-25). The buoyancy of the embryo, especially in its early stages with its most delicate tissues, equalizes pressures over the entire developing body, and the double hydraulic cushion system absorbs any mechanical impacts on the womb of the mother.

The metabolic activities and roles of several organs in a fetus are quite different from those in a postnatal mammal. Because nutrients and oxygen are provided by the maternal blood through the umbilical veins and metabolic waste products are eliminated from the fetus through the umbilical arteries, the digestive, excretory, and respiratory organs of fetuses are relatively small and maintain a relatively low metabolic activity. The fetal intestine does not absorb nutrients, and the fetal liver is producing blood cells and is not a metabolic conversion center. Therefore, the veins of the fetal hepatic portal system and the fetal hepatic veins are much smaller than in a postnatal individual, and most of the blood supplied by the umbilical vein is shunted past the liver by the **ductus venosus.** Similarly, the fetal kidneys are not serving as filtering and excretion center. Because tissues of both the digestive and excretory organs require relatively little blood, most of the nutrient- and oxygen-rich blood that is pumped out by the left ventricle is conducted to the head and thoracic region and the umbilical arteries, largely bypassing the viscera. Similarly, because the fetal lungs are collapsed and are not involved in gas exchange for the rest of the body, most of the blood that reaches the right ventricle is largely bypassing the lungs and shunted to the aorta by the **ductus arteriosus.**

Blood that is rich in nutrients and oxygen and low in waste products is transported from the placenta to the fetus through the **umbilical vein** (see figure). A portion of this blood supplies the liver directly through various branches, and another portion joins the hepatic portal vein through the **portal sinus.** However, most of this blood is shunted directly to the caudal vena cava through the **ductus venosus,** which passes through the liver. As a consequence, the blood in the caudal vena cava contains a mixture of nutrient- and oxygen-rich blood from the ductus venosus and of nutrient- and oxygen-depleted blood from the hepatic veins and the caudal half of the body. Most of the blood that is brought to the right atrium by the caudal vena cava is shunted directly to the left atrium through the **foramen ovale,** a valved opening in the interatrial septum. This shunting is possible because a valve in the caudal vena cava directs the blood stream toward the oval foramen. In addition, the blood pressure in the caudal vena cava is high in comparison to that in the left atrium, which receives only small amounts of blood returning from the lungs. After mixing with the oxygen-poor pulmonary blood in the left atrium, the blood flows into the left ventricle, which pumps it into the aorta. A small amount of the blood entering the right atrium through the caudal vena cava joins the oxygen-depleted blood returned by the cranial vena cava from the head to the right atrium.

Most of the blood leaving the left ventricle through the aorta supplies the head, heart, and thoracic region, and a smaller amount of blood is sent through the descending aorta. A small portion of this blood is used to supply the caudal portion of the body. However, the majority of the blood, which is nutrient and oxygen depleted and rich in waste products, is collected by the **umbilical arteries** and returned to the placenta, where carbon dioxide and waste products are released to the maternal blood and oxygen and nutrients are transferred from the maternal blood to the fetal blood (see Chapter 12).

Both the cranial and caudal venae cavae enter the right atrium, where only a small portion of the blood from the caudal vena cava mixes with the blood from the cranial vena cava. Therefore, this mixed blood is not completely oxygen and nutrient depleted and flows into the right ventricle, which pumps it into the pulmonary trunk. Only a small portion of this blood reaches the lungs and most of it is shunted to the descending aorta via the **ductus arteriosus.** Hence, the cranial portion of the body receives blood that is richer in nutrients and oxygen than the caudal portion of the body. The umbilical arteries return some of this blood to the placenta.

At birth, the connection between the maternal and fetal blood circulation is broken, and the supply of oxygen through the umbilical vein is discontinued. The increasing carbon dioxide level in the blood triggers a reflex that induces the newborn to take a breath and inflate the lungs. If this reflex is delayed, human babies are coaxed into crying, because this forces them first to fill their lungs with air. As the lungs inflate, their blood vessels expand and receive a larger proportion of the blood from the pulmonary trunk. As a consequence, more blood, which is now oxygen rich, is returning from the lungs to the left atrium,

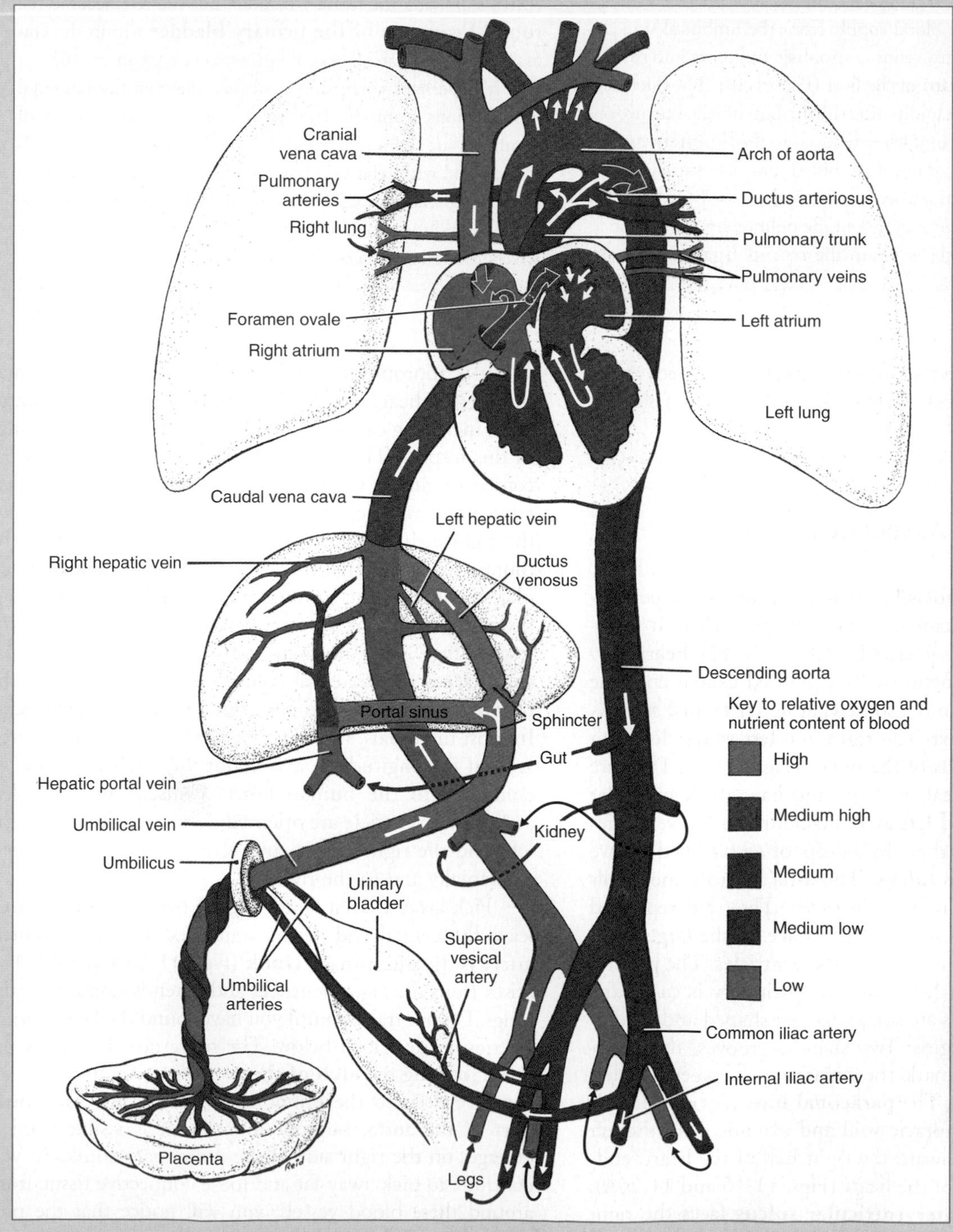

Diagrammatic ventral view of the fetal pattern of blood circulation in a human as an example of the mammalian fetal configuration. *(From Moore, 1976.)*

continued

Anatomy in Action 11-6 The Embryonic Circulation of Mammals *continued*

and the hydrostatic pressure in the left atrium is raised. At the same time, the hydrostatic pressure in the right atrium drops, because the caudal vena cava is no longer supplied with blood from the umbilical vein. With the pressures equalized in both atria, the valve of the oval foramen is closed. With the blood supply from the umbilical vein discontinued, the ductus venosus atrophies and turns into the **ligamentum venosum** of the liver (Fig. 11-19). The ductus arteriosus contracts rapidly after birth, but it takes up to one month for it to be completely reduced to the **ligamentum arteriosum.** Until that moment, blood may flow through it but in the reverse direction, because the hydrostatic blood pressure in the aorta is higher than that in the pulmonary trunk. The fetal umbilical vein shrivels into the **round ligament of the liver,** which may be discernible in the falciform ligament. The proximal portions of the umbilical arteries are retained as the proximal portions of the **internal iliac arteries.** Some of the distal parts of the umbilical arteries persist as small **vesical arteries** supplying the urinary bladder, but the rest become the **round ligaments of the urinary bladder** within the cranial edges of the lateral vesical ligaments (see Chapter 10).

It is perhaps surprising to realize that the fundamental modifications from the fetal to the postnatal circulation of mammals are achieved mostly through rerouting of bloodstreams and with relatively few structural changes. This is one of the well-studied examples in morphology that show how fundamental functional changes can be achieved through relatively modest structural modifications.

will explain the presence of certain structures that may otherwise seem "functionless" within the cardiovascular system of postnatal mammals.

The Heart and Associated Blood Vessels

Carefully make a crosswise incision through the pericardial sac of your specimen so that you can reflect it to reveal the heart and its greater blood vessels. The **heart** *(cor)* is a large compact organ with a pointed caudal end, the **apex of the heart,** and a somewhat flatter cranial surface, the **base of the heart.** The **right** and **left ventricles** form the caudal two thirds of the organ (Fig. 11-16). They are approximately conical in shape and have thick muscular walls. The **right** and **left atria** lie cranial to the ventricles and are set off from them by a deep, often fat-filled groove called the **coronary sulcus.** The atria have thinner walls and are darker than the ventricles. They are separated from each other on the ventral surface by the large arteries that leave the cranial end of the ventricles. The portion of each atrium that lies along the periphery is called the **auricle.** The auricles are somewhat ear shaped and tend to have scalloped margins. Two shallow grooves, the **interventricular sulci,** mark the separation between the left and right ventricles. The **paraconal interventricular sulcus** faces the left thoracic wall and extends from the left auricle diagonally toward the right half of the heart, ending above the apex of the heart (Figs. 11-16 and 11-25**A**). The **subsinuous interventricular sulcus** faces the right thoracic wall and extends from below the coronary sinus to the apex of the heart (Fig. 11-25**C**).

It is appropriate at this point to discuss the orientation of the heart within the thorax because most mammals, including cats and rabbits, differ from human beings in this respect. The names of the parts of the heart were coined originally for the human heart, in which the left atrium and ventricle face the left wall of the thorax, and the right atrium and ventricle face the right wall of the thorax. Consequently, the human paraconal interventricular sulcus faces the sternum and ventral wall of the thorax and is called the anterior interventricular sulcus. Correspondingly, the human subsinuous interventricular sulcus faces the vertebral column and dorsal wall of the thorax and is called the posterior interventricular sulcus. In most mammals, however, the heart is tilted and rotated around its longitudinal axis about 90° toward the left as compared to the human heart. Consequently, the left atrium and ventricle are oriented dorsocaudally and to the left, and the right atrium and ventricle are oriented ventrocranially and to the right.

Pick away the fat from around the large arteries that leave the cranial end of the ventricles. The more ventral artery is the **pulmonary trunk** (Figs. 11-16 and 11-17). It arises from the right ventricle and extends dorsally to the lungs. Do not trace it until you have found the ligamentum arteriosum described below. The more dorsal vessel is the **aorta** forming the **arch of the aorta.** It arises from the left ventricle deep to the pulmonary trunk as the **ascending part** of the aorta, but not much of it is visible until it emerges on the right side of the pulmonary trunk. As you continue to pick away fat and loose connective tissue from around these blood vessels, you will notice that the pulmonary trunk and the **descending part** of the aorta are bound together by a tough band of connective tissue, the

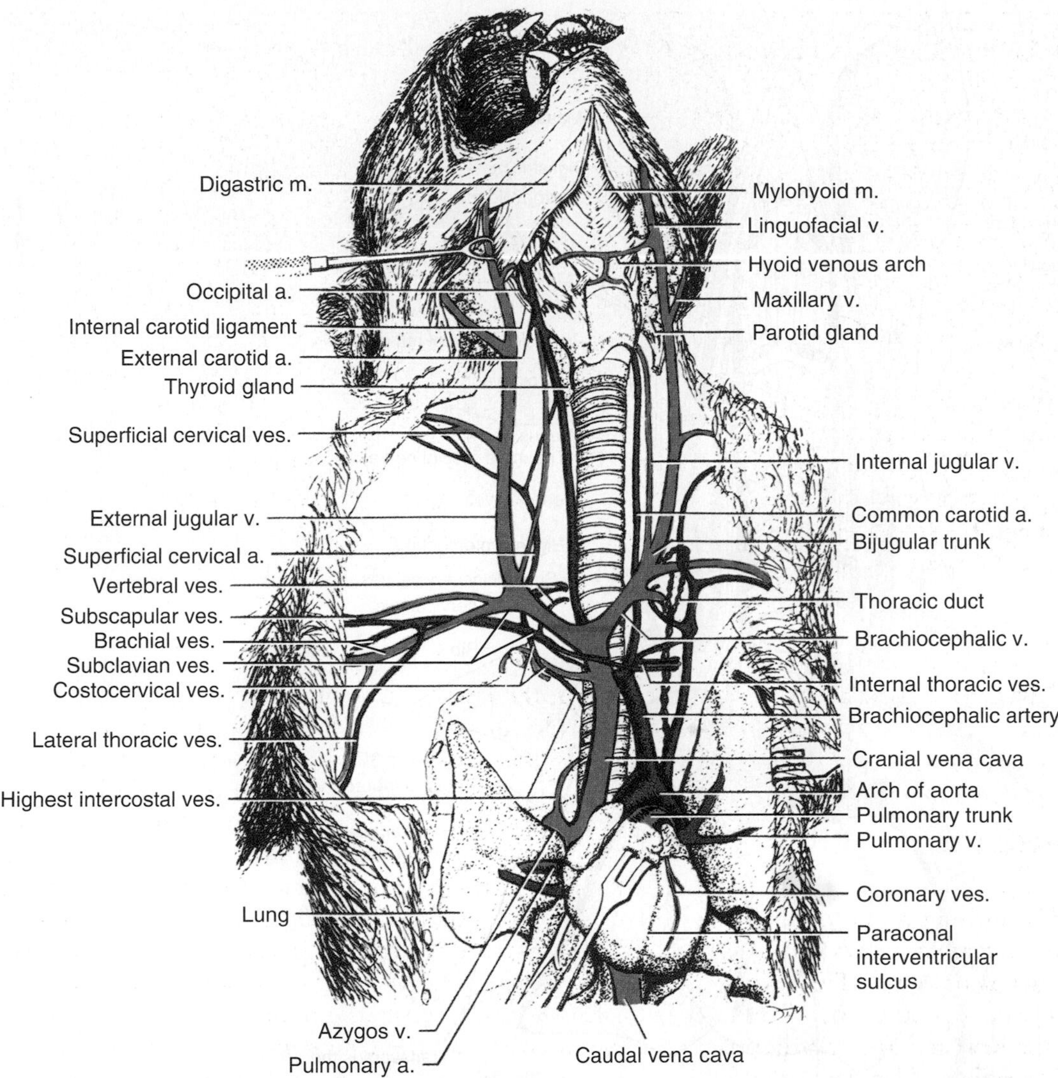

Figure 11-16
Ventrolateral view of the thoracic and cervical blood vessels and larger lymphatic vessels of a cat.

ligamentum arteriosum. Do not destroy it. Just distal to the ligamentum arteriosum, the pulmonary trunk bifurcates into the **pulmonary arteries,** which run to the left and right lungs. You can see this bifurcation better by pushing the pulmonary trunk cranially and removing connective tissue between the pulmonary trunk and the craniodorsal portion of the heart. Two small **coronary arteries** leave the base of the ascending part of the aorta and supply the heart wall. The origin of one coronary artery can be found deep between the pulmonary trunk and left auricle; the origin of the other coronary artery can be found deep between the pulmonary trunk and the right auricle. Distal parts of the **coronary veins** draining the heart wall parallel the arteries. You will have the opportunity to observe the details of the coronary circulation later when you dissect the heart.

Push the heart to the left side of the thorax, and you will see the **caudal vena cava** passing through the diaphragm and entering the right atrium. You will also see a **cranial vena cava** entering this chamber from the right side of the neck. Adult cats normally have only the right cranial vena cava (Fig. 11-25**A** and **B**), but rabbits and sheep also have a left cranial vena cava, which descends the left side of the neck, crosses the dorsal surface of the heart, and enters the right atrium separately (Figs. 11-25**C** and 11-26). Lift the apex of the heart, and you will see the entrance of the venae cavae into the right atrium. In cats, the coronary veins empty into the **coronary sinus** before entering the right atrium. The position of the coronary sinus in cats is comparable to the position of the proximal end of the left cranial vena cava in rabbits. In rabbits, the left

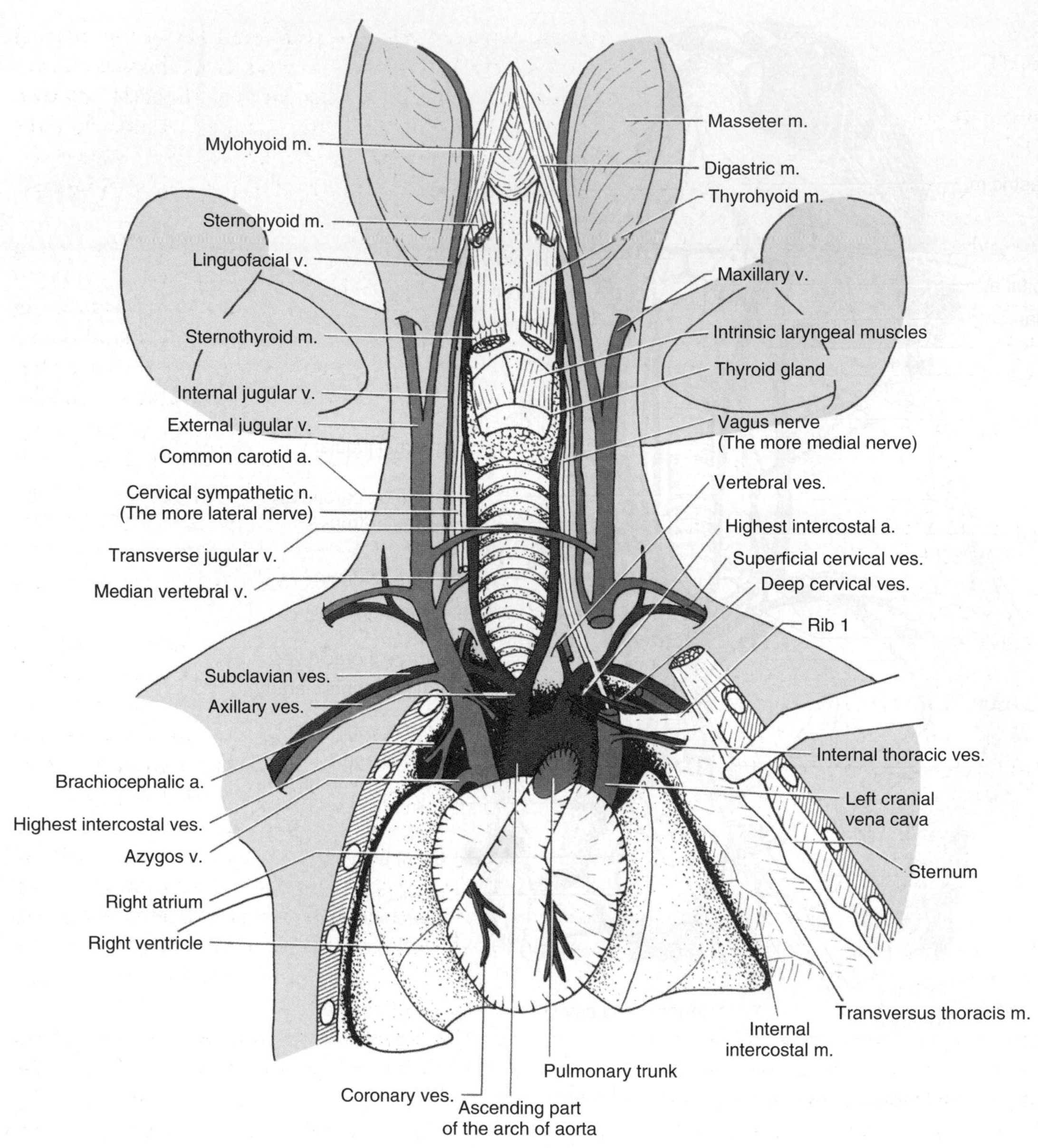

Figure 11-17

Ventral view of the thoracic and cervical blood vessels and nerves of a rabbit. Superficial veins that obscured the view on the subclavian blood vessels were removed.

cranial vena cava enters the right atrium on the dorsal side of the heart almost in common with the caudal vena cava. Carefully pick away connective tissue dorsal to the cranial venae cavae and from the roots of the lungs. You will find the **pulmonary veins** originating from the lungs and entering the left atrium. There are several pulmonary veins, but those of each side generally collect into two main pulmonary veins before entering the left atrium.

You will notice that the mammalian heart contains only two of the primitive four chambers but that these chambers have become completely divided. The ancestral sinus venosus and conus arteriosus are present in the embryo, but become parts of the heart and greater blood vessels in the adult. The sinus venosus is absorbed into the right atrium and forms the part of the atrium that receives the venae cavae. The conus arteriosus (or *bulbus cordis* in the terminology used for mammalian embryology), together with the ventral aorta, divides and forms the bases of the pulmonary trunk and arch of the aorta.

Arteries and Veins Cranial to the Heart

The Blood Vessels of the Thorax, Shoulder, Arm, and Neck

As we have already observed, cats possess a single cranial vena cava, but in rabbits, both the left and right cranial venae cavae are present. Trace them cranially by carefully picking away surrounding portions of the thymus, connective tissue, and fat. The paired **subclavian veins,** which often fail to be injected because of a valve, return from the shoulders and arms just cranial to the first ribs. Each subclavian vein joins a very short **bijugular trunk,** which receives the **external** and **internal jugular veins** from the side of the neck and head (Figs. 11-16 and 11-17). Occasionally, the two jugular veins and the subclavian vein unite at the same point. In cats, the union of the subclavian vein and bijugular trunk forms the **brachiocephalic vein**; the paired brachiocephalic veins then unite to form the single cranial vena cava. In rabbits, the union of the subclavian vein and bijugular trunk forms one of the two cranial venae cavae. The cranial vena cava of cats is comparable to the right cranial vena cava of rabbits. The left cranial vena cava is present in embryonic cats (Anatomy in Action 11-9, Fig. **C**), but it is absorbed when the left brachiocephalic vein develops. On rare occasions, the left cranial vena cava persists in adult cats, together with some other anomalies of the heart.

Examine some of the tributaries of the cranial vena cava and brachiocephalic vein in cats, or the paired cranial venae cavae in rabbits. The most caudal tributary, which enters the dorsal surface of the vena cava, is the **azygos vein,** which receives most of the **intercostal veins** from between the ribs on both sides of the body. Technically speaking, the azygos vein of cats is the right azygos vein. The left azygos vein is not present in adult cats because its embryonic precursor, the left supracardinal vein, atrophies (Anatomy in Action 11-9). In some adult mammals, such as pigs and sheep, the left azygos vein is fully developed and enters the right atrium via the coronary sinus. **Intercostal arteries** will be seen beside the **intercostal veins,** but their origins will be seen later. The **highest intercostal vein,** which drains the cranial intercostal spaces, enters usually the cranial vena cava independently and cranially to the entrance of the azygos vein. In cats, the highest intercostal vein drains occasionally into the deep cervical vein or directly into the vertebral vein (see later), and the right highest intercostal vein drains occasionally into the azygos vein.

The next cranial tributaries of the cranial vena cava are several small veins from the thymus and a larger **internal thoracic vein,** which enters the ventral surface of the cranial vena cava. In cats, it is a single vessel at its entrance, but it receives two branches that drain both sides of the ventral thoracic wall. The distal branches of the internal thoracic vein lie deep to the transversus thoracis muscle and are accompanied by the **internal thoracic arteries,** whose origin will be seen later. The internal thoracic veins and arteries extend into the cranial part of the ventral abdominal wall, where they are called the **cranial epigastric arteries** and **veins.**

Return to the ascending part of the aorta. After giving off the coronary arteries previously described, it curves to the left, forming the arch of the aorta and disappearing dorsally behind the root of the left lung. Two blood vessels arise from the front of the arch of the aorta: the large **brachiocephalic artery** closest to the heart and the smaller **left subclavian artery.** Trace the brachiocephalic artery cranially. It sends off small branches to the thymus and then ramifies into three vessels: Two **common carotid arteries,** which ascend the neck on either side of the trachea, and the **right subclavian artery.** The common carotid arteries continue cranially deep to the cranial vena cava and brachiocephalic veins.

The tributaries of the subclavian vein parallel the branches of the subclavian artery. Because the veins often are not well injected, find the branches of the subclavian artery first, and then look beside them for the accompanying veins. Trace one of the subclavian arteries peripherally. Medial to the first rib, the subclavian artery gives rise to four branches, which are identified best from their peripheral distribution (Figs. 11-16, 11-17, and 11-18): the internal thoracic, vertebral, highest intercostal and deep cervical arteries. The **internal thoracic artery,** which you have identified previously, leaves the ventral surface of the subclavian artery and accompanies the internal thoracic vein to the ventral chest wall.

In cats, the **vertebral artery** arises from the dorsal surface of the subclavian artery opposite the origin of the internal thoracic artery. In rabbits, the vertebral artery arises slightly more proximally. Trace it and the accompanying **vertebral vein** craniomedially. In cats, the right vertebral vein normally enters the dorsal side of the cranial vena cava and the left vertebral vein enters the brachiocephalic vein, but variations from this pattern are common. In rabbits, the vertebral vein normally enters the cranial vena cava; rarely, the left vertebral vein enters a side loop of the brachiocephalic vein. The vertebral blood vessels soon enter the transverse foramina of the cervical vertebrae through which they continue cranially. They eventually enter the cranial cavity and help supply and drain, respectively, the brain (Anatomy in Action 11-8).

A short, caudally pointing **costocervical trunk** arises from the subclavian artery just distal to the origin of the vertebral artery or, rarely, just opposite of the origin of the internal thoracic artery. In cats, it usually divides almost immediately into the highest intercostal and deep cervical

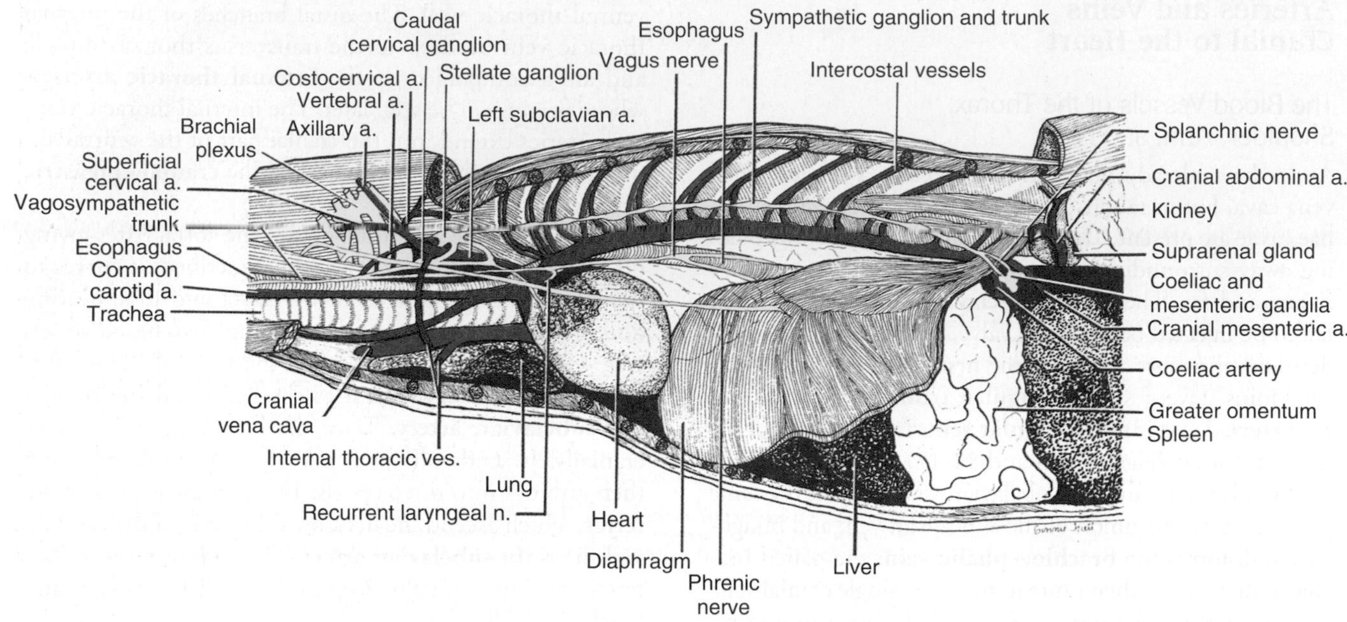

Figure 11-18
Thoracic and cervical blood vessels and nerves of a cat in a lateral view. The left shoulder and arm, the left lung, the thoracic part of the thymus, and most of the veins have been removed. The nerves are drawn slightly thicker than they are in reality.

Anatomy in Action 11-7 Mammalian Aortic Arches

The major arteries supplying the chest, shoulder, arm, and neck develop from the embryonic aortic arches (Fig. **A**). All six aortic arches, which already are present in ancestral vertebrates (Anatomy in Action 11-3 and 11-5), are formed in mammalian embryos and connect the ventral aorta with the dorsal aorta. Note that the ventral aorta bifurcates into the paired ventral aortae rostral to the fifth aortic arches. The dorsal aorta is also paired in its rostral portion, but it merges into a single blood vessel caudal to the heart after giving off the subclavian arteries. The first and second embryonic aortic arches are completely dismantled. The third embryonic aortic arch becomes integrated into the adult internal carotid artery (Fig. **B**), when a segment of the paired dorsal aortae between the embryonic third and fourth aortic arches atrophies. The embryonic paired ventral aortae rostral to the third embryonic aortic arches form the proximal portions of the external carotid arteries. The internal carotid artery in embryos supplies not only the intracranial part of the head but also, through its stapedial branch passing through the stapes, much of the outside of the head. However, in most adult mammals, the external carotid artery taps into the stapedial artery and supplies most, or all, of its peripheral distribution. If this happens, the proximal portion of the stapedial artery atrophies. The paired ventral aortae between the embryonic third and fourth aortic arches become the common carotid arteries in adult mammals.

The fourth embryonic aortic arches continue to connect the paired ventral arches near their bifurcation from the unpaired ventral aorta with the remaining caudal portions of the paired dorsal aortae. They become integrated into the base of the right subclavian artery and the arch of the aorta, respectively, as the roots of the embryonic paired ventral aortae become integrated into the adult brachiocephalic trunk on the right side and the arch of the aorta on the left side (Fig. **B**). The fifth embryonic aortic arches atrophy completely as the embryo continues to develop. On the right side, the segment of the paired dorsal aorta between the branching of the right subclavian artery and the unpaired dorsal aorta atrophies. As a result, the segment of the right half of the paired dorsal aortae becomes part of the adult right subclavian artery. The left subclavian artery in adult mammals leaves the arch of the aorta much closer to the origin of the common carotid arteries than in embryos, because the growth of the arch of the aorta and the adjacent dorsal aorta lags behind that of the other blood vessels during embryogenesis. The segment of the left half of the paired dorsal aortae caudal to the fourth embryonic aortic arch persists as the cranial portion of the descending aorta.

Because the base of the unpaired ventral aorta was divided into two blood vessels in the course of the development of the heart with its four chambers, the sixth embryonic aortic arches originate from the more ventral blood vessel. On the right side, the dorsal portion of the right sixth embryonic aortic arch is dismantled as development progresses, and the ventral portion develops into the right pulmonary artery. On the left side, the ventral portion of the sixth embryonic aortic arch develops into the left pulmonary artery. The dorsal portion persists in the embryo as the ductus arteriosus, which shunts blood from the pulmonary trunk directly to the descending part of the aorta (Anatomy in Action 11-6). After birth, the ductus venosus becomes clogged with proliferating epithelium and turns into the ligamentum arteriosum, which stabilizes the position of the large arteries relative to each other. The base of the embryonic ventral aorta becomes the ascending part of the aorta in adults.

The development of the embryonic aortic arches in birds is very similar to that in mammals, except that the right fourth aortic arch develops into the adult aorta. This fact, in combination with other observations, supports the theory that birds and mammals evolved a four-chambered heart and endothermy-homoiothermy independently from each other.

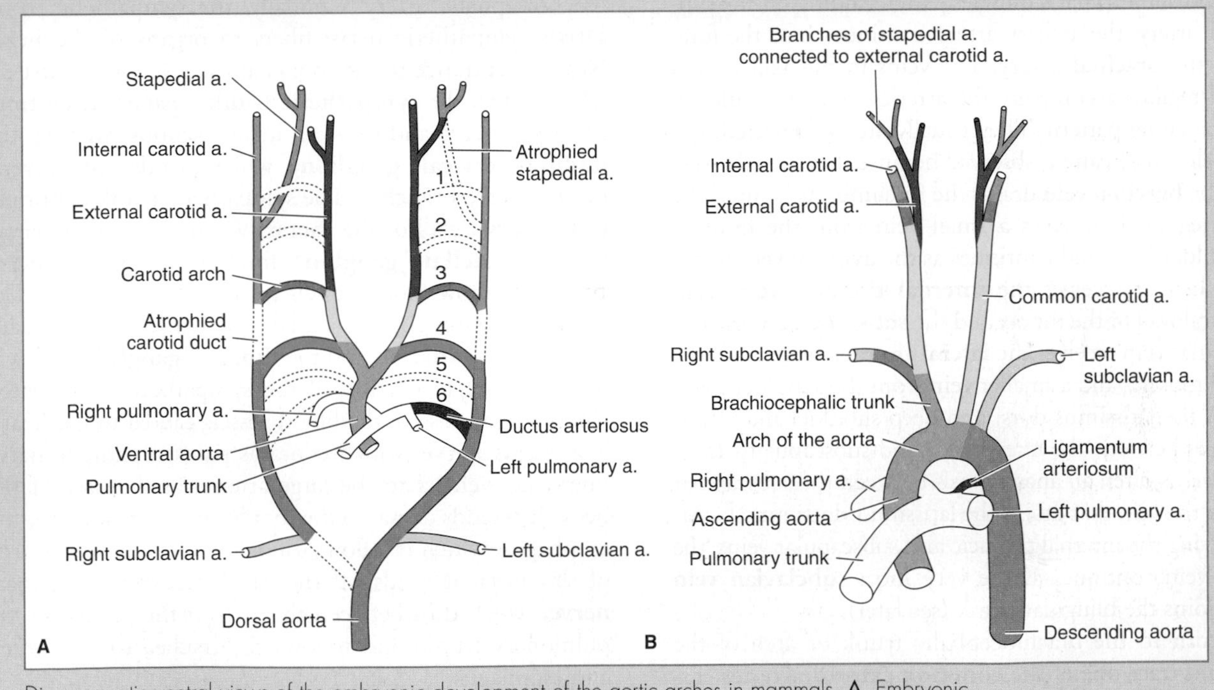

Diagrammatic ventral views of the embryonic development of the aortic arches in mammals. **A**, Embryonic configuration. **B**, Adult configuration in a human. *(Modified from Barry, 1951.)*

arteries; sometimes these two arteries arise independently from each other. In rabbits, the highest intercostal and deep cervical arteries generally arise independently from the subclavian artery. The **highest intercostal artery** extends caudally across the cranial ribs to supply those intercostal spaces drained by the highest intercostal vein, which was identified earlier. The **deep cervical artery** extends dorsally to supply deep muscles of the neck. A major branch of it also passes cranial to the first rib and into the serratus ventralis muscle. The deep cervical artery is accompanied by the **deep cervical vein,** which usually empties into the vertebral vein shortly after the vertebral vein emerges from the transverse foramina of the vertebrae. Occasionally, the deep cervical vein enters the cranial vena cava independently.

The last branch of the subclavian artery is the **superficial cervical artery.** It extends deep to the subclavian vein and follows the **external jugular vein** cranially. Trace both blood vessels. The superficial cervical artery gives off one or more small branches that extend cranially and sometimes reach the thyroid gland, but the main part of the superficial cervical artery continues dorsolaterally to supply muscles on the craniolateral surface of the shoulder. The **superficial cervical vein,** a tributary of the external jugular vein, accompanies the distal part of the superficial cervical artery. One tributary of the superficial cervical vein, the **cephalic vein,** is often a conspicuous vessel along the lateral surface of the brachium.

After giving off the superficial cervical artery and the costocervical trunk, the subclavian artery continues laterally into the armpit *(axilla)* as the **axillary artery** (Fig. 11-16), which sends off the **lateral thoracic** and **subscapular arteries.** The lateral thoracic artery supplies the pectoral muscles. The subscapular artery first sends a smaller branch to the front of the shoulder joint. It then continues and sends a branch to the latissimus dorsi muscle and another branch to the deep shoulder muscles by dipping between the teres major and subscapularis muscles. After sending off the subscapular artery, the axillary artery continues into the forelimb as the **brachial artery.** The veins in the axillary and brachial regions accompany the arteries, but they differ in their branching pattern. They usually are not injected, but nevertheless they can be observed because they are relatively large. The **brachial vein** drains the forelimb. As it enters the armpit region, it receives a small vein from the front of the shoulder joint and continues as the **axillary vein** to the point where it receives the **internal thoracic vein** from the ventral wall of the thorax and the **subscapular vein.** The subscapular vein receives the **lateral thoracic vein** from the pectoral muscles and a smaller vein from the shoulder joint, It drains the latissimus dorsi and deep shoulder muscles as it emerges between the teres major and subscapularis muscles. There is often an anastomosis between the subscapular vein and the vein that drains the latissimus dorsi muscle. After receiving the internal thoracic and subscapular veins, the axillary vein continues as the very short **subclavian vein** until it joins the bijugular trunk (see later).

Return to the brachiocephalic trunk or arch of the aorta, and trace one of the **common carotid arteries** cranially. It passes deep to the brachiocephalic vein or cranial vena cava, and continues cranially. It lies lateral to the trachea and supplies the trachea, thyroid gland, and other cervical structures before it reaches the head. The **internal jugular vein,** which helps drain the inside of the skull, lies lateral to the common carotid artery through most of its course. The internal and external jugular veins usually unite with each other slightly cranial to the subclavian veins to form the **bijugular trunk** previously observed. In rabbits, the **median vertebral vein,** which lies dorsal to the esophagus, enters either jugular vein near the bijugular trunk (Fig. 11-17). Rabbits also have a **transverse jugular vein,** which joins the external jugular veins slightly cranial to the bijugular trunk. Lymphatic vessels also may be seen entering the external jugular veins in this region (Fig. 11-27). They usually are not injected, but they can be recognized by their unique appearance. Lymphatic vessels contain so many valves that they have alternating expanded and constricted segments and resemble strings of beads.

Several nerves may be seen at this time. Find again the **phrenic nerves** innervating the diaphragm in the thorax (see Chapter 10) and trace them cranially through the base of the neck to their origin from the fifth and sixth cervical nerve (Fig. 11-18). The deeper nerve lying between the common carotid artery and the internal jugular vein is the **vasosympathetic trunk,** in which the vagus nerve and sympathetic trunk are bound together by a common sheath of connective tissue throughout most of the neck, although they can easily be dissected apart. The vagus nerve is the larger nerve and passes ventrally to the brachiocephalic artery. Cranially, the sympathetic trunk carries sympathetic nerve fibers to organs of the head. Near the entrance to the thorax at the level of the first rib (Fig. 11-18), the **sympathetic trunk** separates itself from the vagus nerve and bears an inconspicuous swelling, the **middle cervical ganglion,** which sends out several minute nerve branches. The sympathetic trunk continues to the dorsal wall of the thorax, where it bears the large, triangular **stellate ganglion,** from which minute nerve branches follow blood vessels to the heart and lung. The stellate ganglion consists actually of two ganglia, the **caudal cervical ganglion** and the **thoracic ganglion.** We will see the thoracic portion of the sympathetic trunk later, when we will dissect the blood vessels caudal to the heart. The **vagus nerve,** which contains parasympathetic nerve fibers, lies ventral to the large arteries at the base of the neck. It extends dorsal to the root of the lung and onto the esophagus, which it follows caudally. As it crosses the arch of the aorta, it sends off the small **recurrent laryngeal nerve,** which dips between the arch of the aorta and the pulmonary trunk, and follows the trachea to the larynx and esophagus.

The Major Blood Vessels of the Head

On the side on which you dissected the muscles (see Table 7-5), mobilize the external ear so that you can follow the blood vessels at its base. Tributaries of the external jugular vein are superficial to other blood vessels in the head and must be considered first. At about the level of the mandibular gland, the **external jugular vein** is formed by the confluence of the linguofacial and maxillary veins (see Figs. 10-14 and 11-19). Trace the **linguofacial vein** rostrally. It soon receives on its medial side the **hyoid venous arch,** which comes from the opposite side of the body and, in turn, receives two deep unpaired veins, the **unpaired laryngeal vein** from the larynx and

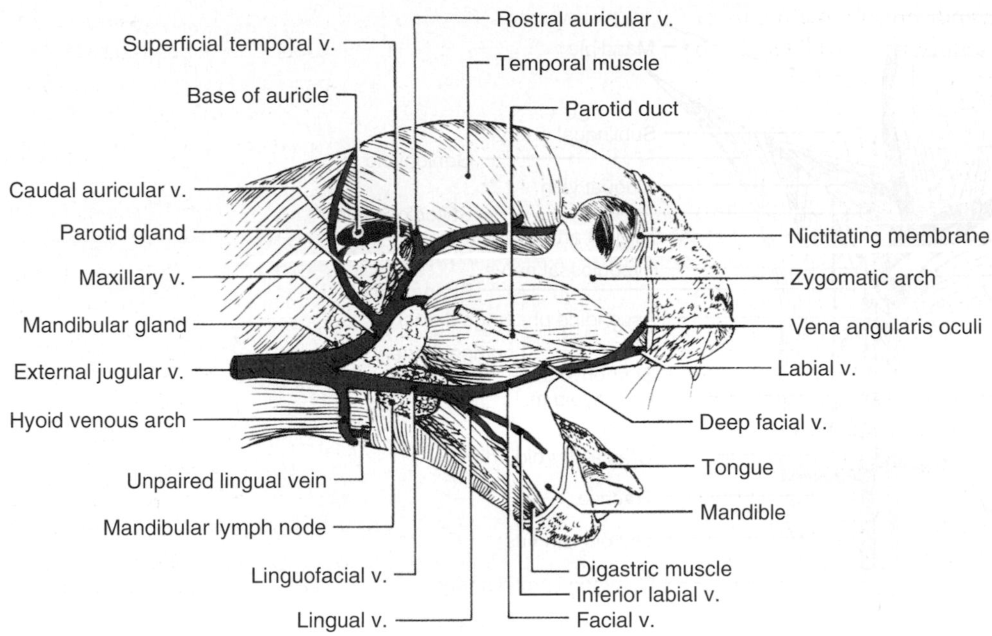

Figure 11-19

Lateral view of the tributaries of the external jugular vein of a cat. Most of the external ear and parotid gland were removed to reveal the underlying veins.

the **unpaired lingual vein.** The latter receives paired tributaries that drain the tongue. Farther rostrally, at the caudoventral edge of the mandible, the linguofacial vein is formed by the confluence of two veins, the lingual vein and the facial vein. The **lingual vein** receives a deep tributary that drains the region around the mandibular ramus and the superficial **inferior labial vein,** which drains the lower lip. The lingual vein also receives a tributary that drains the tongue and the floor of the mouth cavity. The **facial vein** follows the ventral border of the masseter muscle and receives three major tributaries: the **deep facial vein,** which emerges from beneath the masseter muscle and drains venous plexuses in the orbit and the palate (see Anatomy in Action 11-8); the **labial vein** from the upper lip; and the **vena angularis oculi** from the face in front of the eye.

Return to the entrance of the **maxillary vein** into the external jugular vein and trace the maxillary vein dorsally toward the parotid gland and the base of the auricle. It receives a **caudal auricular vein** from behind the ear and the top of the head, and a **superficial temporal vein** from the front of the ear. The superficial temporal vein passes through the parotid gland and receives tributaries from the temporal muscle and from the ear through the **rostral auricular vein.** The maxillary vein drains the pterygoid muscles and the region on the ventral side of the skull. In cats, it also forms the pterygoid and ophthalmic venous plexuses (see Anatomy in Action 11-8), which receive blood from the nasal and orbital regions, as well as most of the drainage from inside the skull, because the internal jugular vein is relatively small (see later). It is impractical to dissect these venous plexuses.

In order to trace the internal jugular vein and common carotid artery rostrally, reflect the mandibular gland and the digastric, stylohyoid, and mylohyoid muscles. At the level of the larynx, the **common carotid artery** gives off a cranial **thyroid artery** and a muscular branch that crosses the longus colli muscle (Fig. 11-20). In rabbits, as in most mammals, the common carotid artery divides into an **internal** and an **external carotid artery** at the level of the cranial end of the larynx. The internal carotid artery, together with the vertebral artery, supplies the brain via the arterial cerebral circle (the circle of Willis; see later) with oxygen-rich blood (Anatomy in Action 11-8, Fig. **B**). In cats, however, such an internal carotid artery is present only during the embryonic stage (Anatomy in Action 11-8, Fig. **D**). As development proceeds, the proximal portion of the internal carotid artery shrivels up and becomes the **internal carotid ligament.** The intracranial portion of the internal carotid artery, however, remains patent and receives blood from branches of the ascending pharyngeal, maxillary, and occipital arteries. The internal carotid ligament arises from the dorsal surface of the common carotid artery slightly caudal to the crossing of the hypoglossal nerve (Fig. 11-20). It then dips deep to the occipital artery (see later), crosses the cranial cervical ganglion of the sympathetic trunk, continues

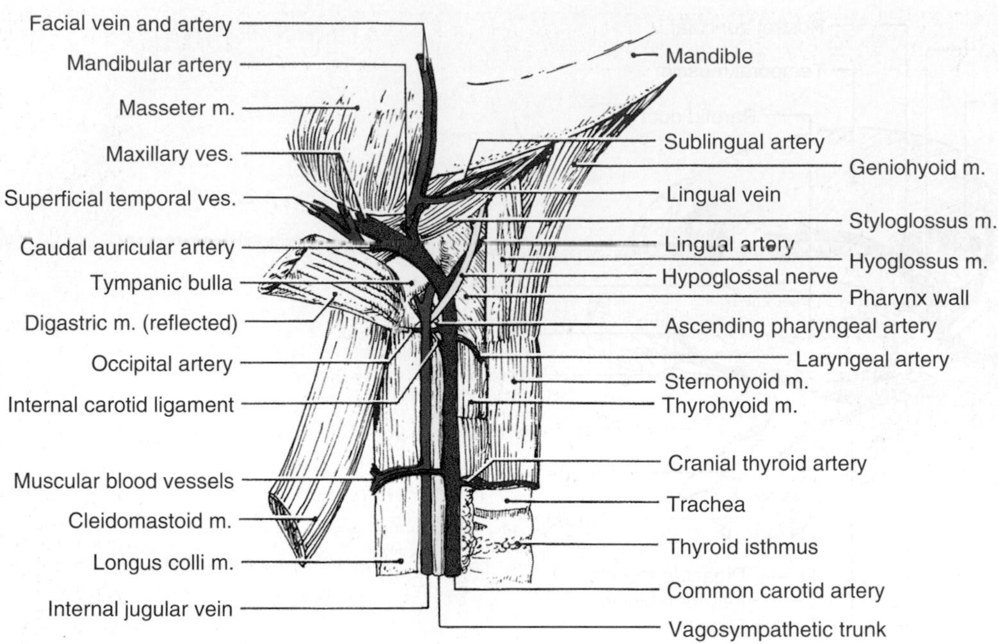

Figure 11-20

Lateroventral view of the tributaries and branches of the internal jugular vein and carotid artery of a cat. The external jugular vein was removed to reveal the deeper blood vessels, but some of its peripheral tributaries are shown.

Anatomy in Action 11-8 The Retia Mirabilia of Cats

In cats, the maxillary artery forms a network of small arteries known as the **maxillary rete mirabile** external to the skull in the region between the optic canal and the foramen ovale of the skull (see Chapter 4 and Figs. **A–D**). This rete mirabile is surrounded by the veins of the **pterygoid venous plexus,** which drains the regions of the nasal cavity, palate, and orbit, as well as the inside of the skull, and empties into the maxillary vein.

Adult cats possess a second intracranial network of small arteries, the **carotid rete mirabile,** which is formed by the intracranial portion of the internal carotid artery, but also receives blood from several other arteries, such as anastomoses with a deep branch of the external carotid artery and the ascending pharyngeal artery (Figs. **B** and **D**). The blood in the distal portion of the internal carotid artery is supplied entirely by the ascending pharyngeal artery, because the proximal portion of the internal carotid artery is ligamentous and impervious. A significant amount of blood is supplied to the carotid rete mirabile by the maxillary rete mirabile through the **anastomotic artery.** Additional blood is supplied to the carotid rete mirabile by the **middle meningeal artery,** which branches off the maxillary artery near the maxillary rete mirabile and enters the cranial cavity through the foramen ovale. In embryonic cats, the carotid rete mirabile is not yet formed, and the brain receives its blood supply mostly from the internal carotid artery, which is still patent, and from anastomotic arteries (Fig. **C**). The carotid rete mirabile in adult cats is situated within a pool of venous blood of the **cavernous sinus,** one of the intracranial venous sinuses that surround the brain and collect the venous drainage from the brain. The cavernous sinus may also receive blood from the pterygoid venous plexus through anastomoses. The cavernous sinus is paired but is connected to the unpaired **basilar sinus** caudal to the hypophysis. From here the blood is drained into the pterygoid venous plexus through the **emissary vein.**

Both the carotid and maxillary retia mirabilia serve to cool the blood before it reaches the brain. The blood within the cavernous sinus is cooler than that within the carotid rete mirabile because it is returning from the periphery of the head, where it has lost heat. Hence, the cooler blood in the cavernous sinus absorbs heat from the blood in the carotid rete mirabile. Similarly, the blood within the pterygoid venous plexus is cooler than that in the maxillary artery, because it receives cooler blood returning from the nasal cavity and, therefore, absorbs heat from the blood in the maxillary artery. The cooling of the blood within the pterygoid venous plexus is a function of the intensity of evaporative cooling in the nasal cavity, which is directly correlated with the rate of breathing

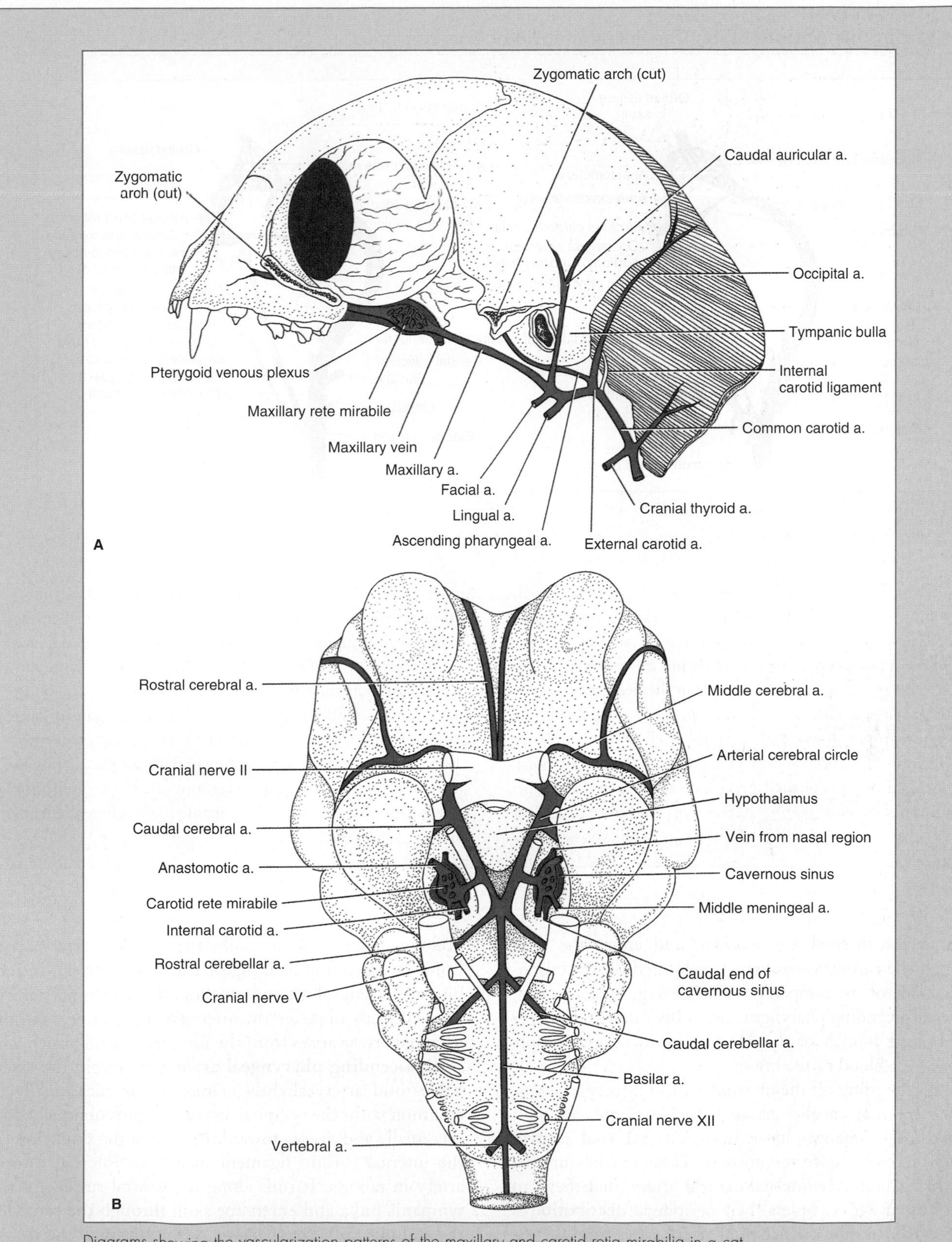

Diagrams showing the vascularization patterns of the maxillary and carotid retia mirabilia in a cat. **A**, Lateral view of the maxillary rete mirabile. **B**, Ventral view of the brain and carotid rete mirabile.

continued

continued

Anatomy in Action 11-8 The Retia Mirabilia of Cats *continued*

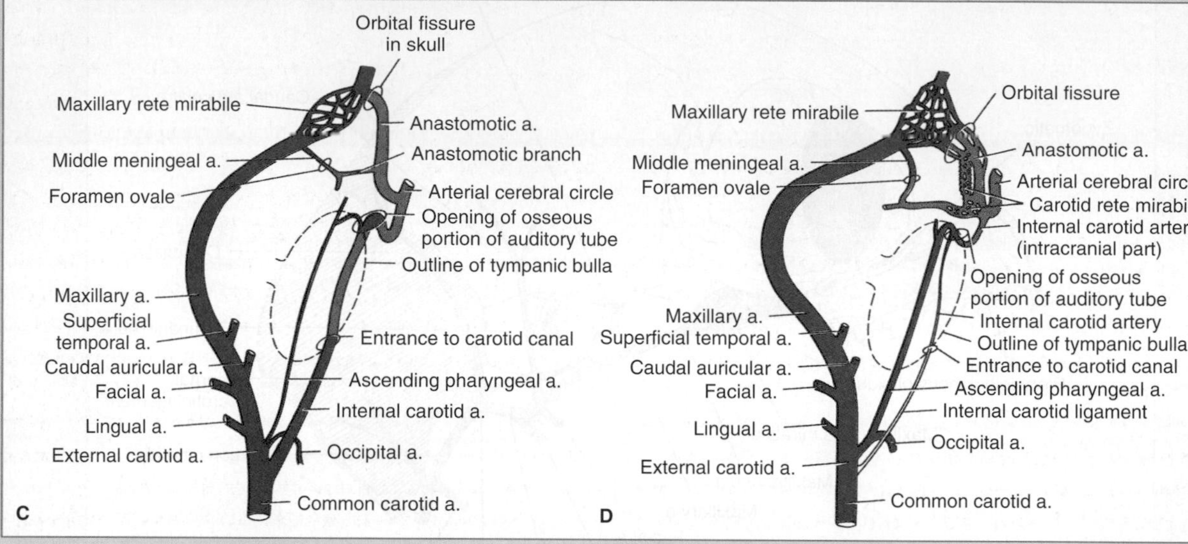

C, Configuration of the maxillary rete mirabile in an embryonic cat. **D**, Configuration of the maxillary and carotid retia mirabilia in an adult cat. *(C and D, after Davis and Story, 1943.)*

and, thus, the degree of physical exertion. In this manner, the cooling of the blood destined for the brain is automatically correlated with the rise of the core temperature of the body and the temperature of the blood within the maxillary artery.

The presence of retia mirabilia in the head of cats allows the body temperature to rise while preventing the brain temperature from rising beyond a critical level. This brain-cooling mechanism has been found in many carnivores (but not in dogs) and in some of their prey, such as sheep and antelopes, but not in rodents, rabbits, horses, pigs, and primates. Mammals that do not possess cranial retia mirabilia possess alternative avenues for cooling the blood destined for their brains during strenuous physical exertions. Dogs and wolves pant by letting their tongue hang out and by producing copious amounts of watery saliva, which evaporates from the large surface area of their tongue. Horses sweat over their entire body, as do humans. Rabbits can lose considerable amounts of heat through the large surface area of their ears. The different types of organs that are used in various mammals for the cooling of their blood during strenuous exertions nicely exemplify the concept of analogy, in which fundamentally different organs serve the same function.

deep to the internal jugular vein, and enters the skull through the carotid canal, which is located on the caudomedial side of the tympanic bulla (see Fig. 4-20). In cats, the small ascending pharyngeal artery (see later) and a larger anastomotic branch of the external carotid artery supply oxygen-rich blood to the brain.

After sending off the internal carotid ligament in cats, or the internal carotid artery in rabbits, the common carotid artery changes its name to the **external carotid artery.** Its first branches arise so close to the internal carotid ligament, or internal carotid artery, that they can be confused with it unless their peripheral distribution is established. In cats, the occipital artery arises from the external carotid artery opposite from and slightly cranial to the **laryngeal artery** (Fig. 11-20). It runs dorsocaudally toward the tympanic bulla to supply neck muscles in the occipital region. Occasionally, the occipital artery arises from the ascending pharyngeal artery (see below). In rabbits, the laryngeal artery branches off the external carotid artery directly opposite the internal carotid artery, and the occipital artery arises from the internal carotid artery. The small **ascending pharyngeal artery** arises from the external carotid artery slightly cranial of, or occasionally in common with, the occipital artery and runs obliquely dorsocranially and deeply toward the base of the skull close to the internal carotid ligament in cats or internal carotid artery in rabbits. It runs along the ventral surface of the tympanic bulla and enters the skull through the canal for the auditory tube (see Chapter 4) and through the rostral portion of the carotid canal (see Fig. 4-20). The ascending pharyngeal artery sends several branches, which usually are not injected, to the pharyngeal constrictor musculature, the soft palate, and some neck muscles. It finally taps into the distal intracranial portion of the internal carotid artery (Anatomy in Action 11-8). The ascending pharyngeal artery is relatively large in cats compared to that in rabbits, because it takes over the function of the proximal portion of the internal carotid artery that has become ligamentous.

After sending off the occipital and ascending pharyngeal arteries, the external carotid artery gives rise to a number of branches that supply various regions of the head external to the skull. These arteries accompany the corresponding veins already observed. The **lingual artery** enters the tongue. The **facial artery** follows the facial vein and the ventral border of the masseter muscle and supplies the jaws and facial structures. In cats, the lingual and facial arteries arise separately from the ventral surface of the external carotid artery, and the **sublingual artery,** which supplies the floor of the mouth cavity, branches off at the base of the facial artery (Fig. 11-20). In rabbits, the facial and lingual arteries have a common origin from the **linguofacial artery.**

Farther distally, the external carotid artery gives off two dorsal branches. The **caudal auricular artery** supplies the region behind the ear. The **superficial temporal artery** supplies the region in front of the ear through the **rostral auricular artery,** and the region on the side of the head through the **transverse facial artery.** In rabbits, the caudal auricular and superficial temporal arteries have a common origin. After sending off the superficial temporal artery, the external carotid artery changes its name to the **maxillary artery** and dips below to the caudal border of the masseter muscle to supply the orbital and palatal regions.

Return to the **internal jugular vein** and trace it rostrally. It receives small tributaries that drain the brain and emerge from the skull through the jugular foramen (see Figs. 4-20 and 4-21), as well as tributaries from muscles at the base of the head. It also anastomoses with the vertebral vein in the upper neck region near the base of the skull. Try to find this anastomosis. To summarize, the brain is drained mainly by the external jugular vein through the maxillary vein, although the relatively small internal jugular and vertebral veins also contribute to the drainage of the brain.

Arteries and Veins Caudal to the Heart

The Blood Vessels of the Dorsal Thoracic and Abdominal Walls

After curving to the dorsal side of the body, the arch of the aorta continues as the **descending part of the aorta.** Trace it caudally. As it passes through the thorax along the left side of the vertebral column, it gives off paired **intercostal arteries** to those intercostal spaces that are not supplied by the highest intercostal arteries (Fig. 11-17), small median ventral visceral branches to the esophagus, and small branches to the bronchi, because the tissue of the lungs, like the wall of the heart, need a separate arterial supply. The **thoracic portion** of the left **sympathetic trunk** can be found at this time by carefully dissecting in the connective tissue near the heads of the ribs dorsal to the aorta. Enlargements along the sympathetic trunk are **sympathetic ganglia,** and the delicate strands passing dorsally are the **white communicating rami** to the spinal nerves (Fig. 9-1). The left **vagus nerve** crosses the ventral surface of the arch of the aorta, passes dorsal to the root of the lung, and extends caudally along the esophagus to the abdominal viscera, which it supplies with parasympathetic nerve fibers. **Phrenic arteries** to the diaphragm may arise from the aorta before the aorta passes through the diaphragm, from the last intercostal arteries, or from arteries caudal to the diaphragm, such as the first lumbar, cranial abdominal, or coeliac arteries (see later).

The **caudal vena cava** was already seen entering the heart. Trace it caudad. As it passes through the diaphragm, it receives several small **phrenic veins** from the diaphragm. It emerges caudally from the liver. Scrape away tissue from the cranial surface of the right medial lobe of the liver in cats, or the right lobe in rabbits, and find the entrances of several large **hepatic veins.** In cats, the major part of the caudal vena cava passes through the right lateral and caudate lobes; you can expose it by scraping away liver tissue. In rabbits, you only need to spread apart the right and caudate lobes to see the caudal vena cava. Hepatic veins, most of which are very small, will be seen entering the caudal vena cava.

Push the abdominal viscera to the right and find the aorta emerging from the diaphragm. Just after emerging, it gives rise to two ventral arteries: first a **coeliac artery** and then a **cranial mesenteric artery** (Figs. 11-21 and 11-22), which supply most of the abdominal viscera. Trace them when you study the vessels of the abdominal viscera (see later). **Coeliac** and **mesenteric ganglia,** which should not be destroyed, lie at the base of the cranial mesenteric artery. They receive one or more **splanchnic nerves** from the sympathetic trunk and send out minute nerve branches that travel along the arteries to the viscera.

Slightly caudal to the cranial mesenteric artery, the aorta lies to the left side of the caudal vena cava. Trace the aorta and caudal vena cava to the pelvic region. Their most conspicuous paired branches are the large **renal arteries** and **veins,** which supply the kidneys. Those of the right side of the body lie slightly cranial to those of the left side, because the right kidney is more cranially situated than the left one. Carefully dissect away fat from around each kidney so that you can lift up its lateral edge and look at the muscles dorsal to it. The blood vessels, which you see supplying the abdominal wall, are the **cranial abdom-**

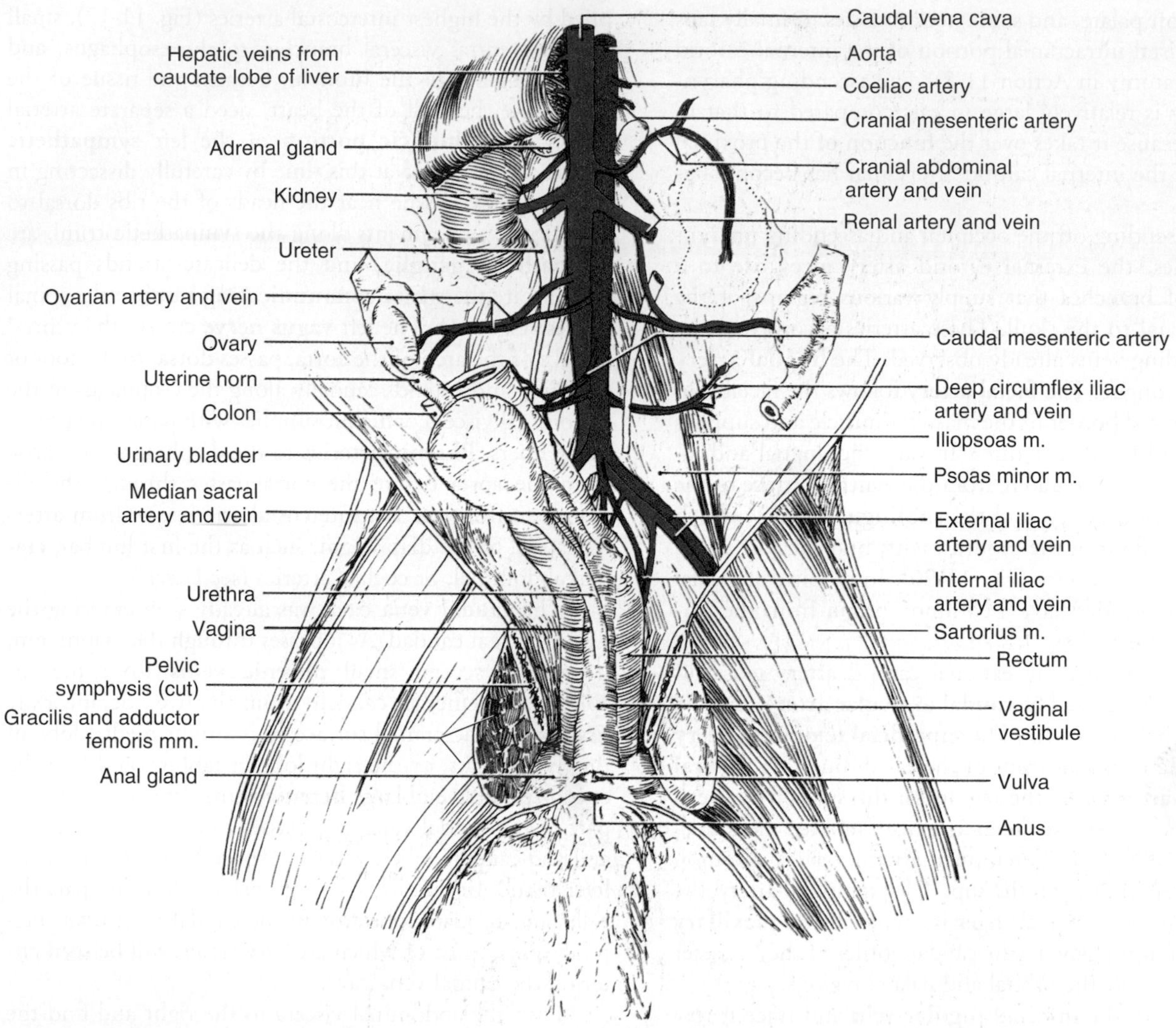

Figure 11-21
Ventral view of the tributaries and branches of the caudal vena cava and the abdominal portion of the aorta in a female cat. The pelvic canal was cut open, and the left kidney and uterine horn were removed to reveal deep blood vessels.

inal artery and **vein.** Trace these blood vessels toward the aorta and vena cava. The cranial abdominal blood vessels usually join the vena cava and aorta just cranial to the renal blood vessels, but they sometimes join the renal blood vessels directly. Before joining these blood vessels, the cranial abdominal blood vessels send branches to supply a small, hard, oval nodule embedded in the fat between the cranial end of the kidney and the aorta and vena cava. This nodule is the **adrenal gland,** or suprarenal gland.

The adrenal gland is an endocrine gland of dual embryonic origin. Its **medullary portion** is derived from postganglionic sympathetic cells of neural crest origin and secretes **epinephrine** (adrenaline), a hormone that reinforces the activity of the sympathetic nervous system and thereby helps the body adjust to stress. Under the influence of the hypophyseal adrenocorticotropic hormone (ACTH), the **cortical portion** of the adrenal gland, which is of mesodermal origin, secretes several types of steroid hormones. **Glucocorticoids,** of which cortisol is the major hormone, promote the conversion of proteins into carbohydrates. **Mineralocorticoids,** of which aldosterone is the major hormone, regulate the potassium and sodium metabolism. **Androgens** act as male sex hormones.

The next paired branches of the aorta are the small **gonadal arteries,** either **testicular** or **ovarian arteries** (see Figs. 12-13, 12-15, 12-18, 12-19, and 12-21). They supply the gonads and are accompanied by the **gonadal veins,** either **testicular** or **ovarian veins,** which drain the

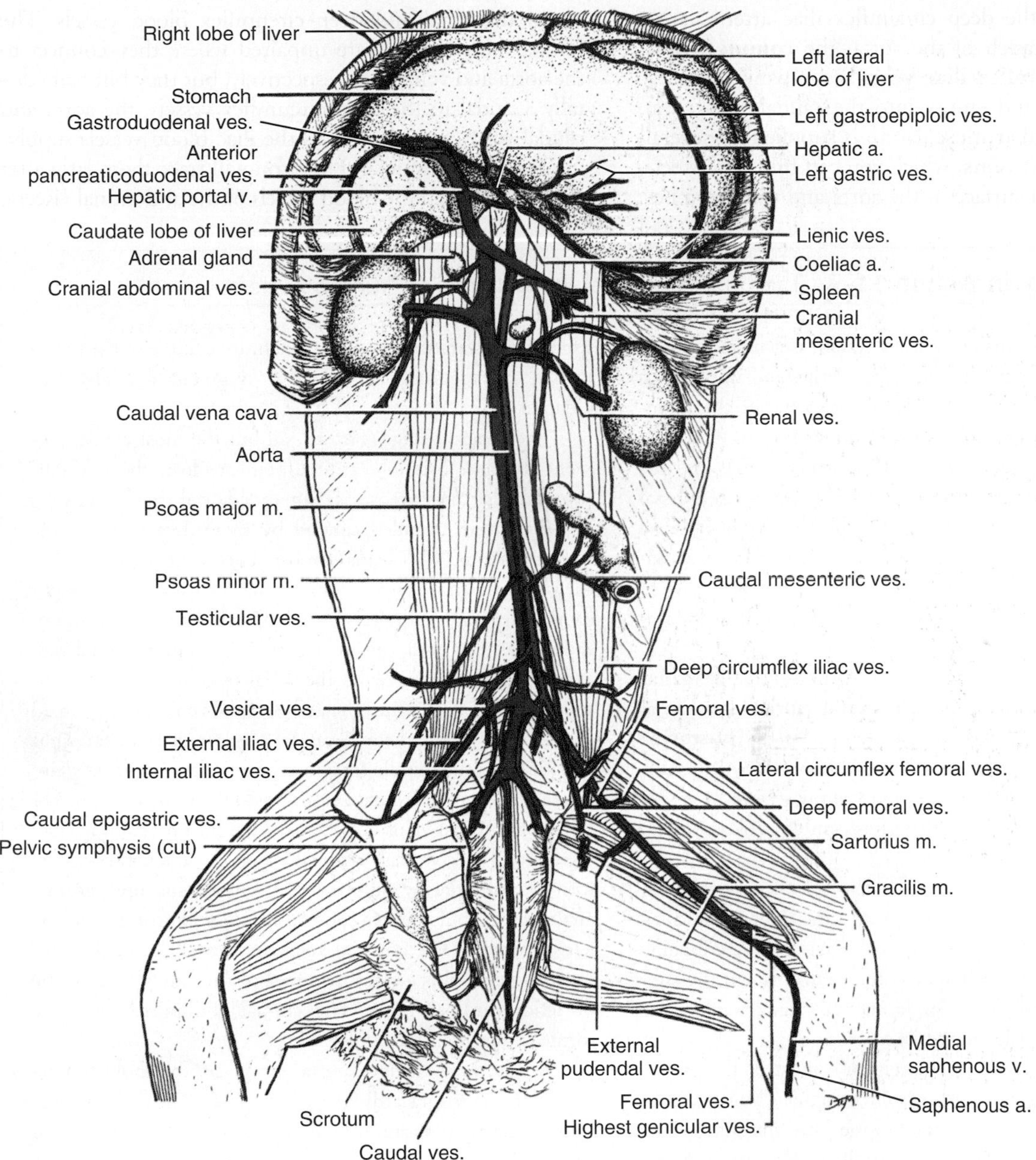

Figure 11-22
Ventral view of the abdominal and pelvic blood vessels of a rabbit. The pelvic canal was cut open.

gonads. The ovaries are small oval bodies that lie near the cranial ends of the Y-shaped uterus. The testes have descended into the scrotum, and, in doing so, each has made an apparent hole, the inguinal canal, through the abdominal wall in the region of the groin. The testicular blood vessels and the sperm duct *(ductus deferens)* can be seen passing through the inguinal canal (see Figs. 12-13 and 12-15). In cats, the right gonadal vein enters the caudal vena cava; the left gonadal vein may do so as well, but it normally enters the left renal vein. In rabbits, both gonadal veins enter the vena cava

The **caudal mesenteric artery** to the colon leaves the ventral surface of the aorta caudal to the gonadal arteries. Again, postpone tracing it until you study the blood vessels of the abdominal viscera. In cats, caudal to caudal mesenteric artery, the aorta gives rise to a pair of **deep circumflex iliac arteries,** which pass laterally to the musculature and body wall that is located ventral to the ilia of the pelvic girdle. In rabbits, the deep circumflex iliac arteries arise from a terminal branch of the aorta, the **common iliac artery. Deep circumflex iliac veins** accompany the deep circumflex arteries and empty into the caudal vena cava. The rest of the lumbar musculature is supplied by several **lumbar arteries** and **veins,** which you can find by dissecting along the dorsal surface of the aorta and vena cava between the renal and deep circumflex blood vessels. The lumbar blood vessels are unpaired where they connect to the aorta and vena cava, respectively, but they bifurcate distally. Caudal to the deep circumflex vessels, the aorta and caudal vena cava give rise to the iliac blood vessels supplying the pelvic region and leg. You will trace them later after you have studied the blood vessels of the abdominal viscera.

Anatomy in Action 11-9 The Development of the Mammalian Venous System

The mammalian venous system is a mosaic of ancestral parts inherited from fishes and amphibians and of new parts that are characteristic of the mammalian condition. The major innovations in mammals include the conversion of parts of the hepatic veins and the transformation of the primitive cardinal venous and renal portal systems typical of fishes and amphibians into the caval and azygos venous systems. The best approach to understanding the evolutionary history of the mammalian venous system is to study the embryonic development of the veins in a mammal as a model.

An early mammal embryo (Fig. **A**) possesses a **cardinal venous system** and an incipient **renal portal system** as some of the blood in the caudal portion of the **posterior cardinal veins** passes through the kidneys to a pair of **subcardinal veins.** At this stage, the venous system of mammalian embryos is similar to that of fish embryos (Anatomy in Action 11-2).

Later in development (Fig. **B**), the right hepatic vein enlarges and extends caudally. It taps into the right subcardinal vein, which develops into the caudal vena cava by enlarging and merging with the left subcardinal vein through the subcardinal anastomosis. This developmental stage resembles the condition in a urodele amphibian.

Still later in development (Fig. **C**), most of the cranial portion of the posterior cardinal vein atrophies, but its caudal portion becomes very large and empties into the subcardinal vein. The evolutionary innovation of the trunk veins in mammals is the subsequent formation of a pair of **supracardinal veins,** which connect cranially and caudally to the remnants of the posterior cardinal veins. The supracardinal veins also connect to the subcardinal veins through a pair of **subsupracardinal anastomoses.** This connection allows the dismantling of most of the caudal portion of the posterior cardinal veins, which are part of the renal portal system in nonmammalian vertebrates (Anatomy in Action 11-2).

During subsequent development (Fig. **D**), the supracardinal veins divide into cranial thoracic portions and caudal lumbar portions. The right subsupracardinal anastomosis and lumbar portion of the supracardinal vein enlarge, but those of the left side do not. Renal veins grow out from the subsupracardinal anastomosis to the definitive kidneys, the mammalian metanephroi, which have migrated cranially (see Chapter 12).

By the adult stage (Fig. **E**), all but the most caudal segments of the posterior cardinal vein are lost, the left subsupracardinal anastomosis is dismantled, and the caudal vena cava has been extended caudad by the enlargement of the right subsupracardinal anastomosis and lumbar portions of the supracardinal veins. In some mammals, only the right caudal supracardinal vein is involved in the transformations of the caudal vena cava, but in cats, the right supracardinal vein expands and merges with the lumbar portion of the left supracardinal vein. Thus, the adult caudal vena cava is formed by the right hepatic vein, a caudal outgrowth from the right hepatic vein, the middle section of the right subcardinal vein, the right subsupracardinal anastomosis, the lumbar portion of the supracardinal veins (especially the right one), and a small segment of the caudal cardinal veins. The renal veins are formed primarily by outgrowths from the subsupracardinal anastomoses, but the left subcardinal vein contributes to the left renal vein. The gonadal veins are formed from the subcardinal veins plus a small segment of the caudal end of the posterior cardinal veins. The cranial abdominal veins are formed from the subcardinal veins.

While these changes are taking place, the thoracic portion of the left supracardinal vein atrophies. The thoracic portion of the right supracardinal vein, however, together with the proximal end of the right posterior cardinal vein, forms the azygos vein. Cats do not have a left highest intercostal vein, but this vein develops in some mammals, such as rabbits, from the stump of the left posterior cardinal vein.

The formation of the cranial vena cava is a simpler process. In some mammals, such as rabbits, the cranial vena cava persists in the condition shown in Figure **C**. The two cranial venae cavae in adult rabbits are formed from the embryonic common cardinal veins together with the proximal portions of the anterior cardinal veins. The more distal portions of the embryonic anterior cardinal veins develop into the internal jugular veins. The external jugular vein is a new outgrowth. In other mammals, such as cats and humans, however, a cross anastomosis develops between the caudal parts of the anterior cardinal veins (Fig. **D**). The right cranial vena cava of cats is formed in the same manner as that in rabbits, but the left cranial vena cava does not develop in cats because the proximal portion of the left anterior cardinal vein atrophies. The cross anastomosis then becomes the left brachiocephalic vein in cats. The portion of the right anterior cardinal vein between the cranial vena cava and the right subclavian vein constitutes the right brachiocephalic vein.

In all mammals, the coronary veins draining the heart enter the embryonic left common cardinal vein. This becomes the base of the left cranial vena cava in rabbits, but it forms a separate coronary sinus in other mammals.

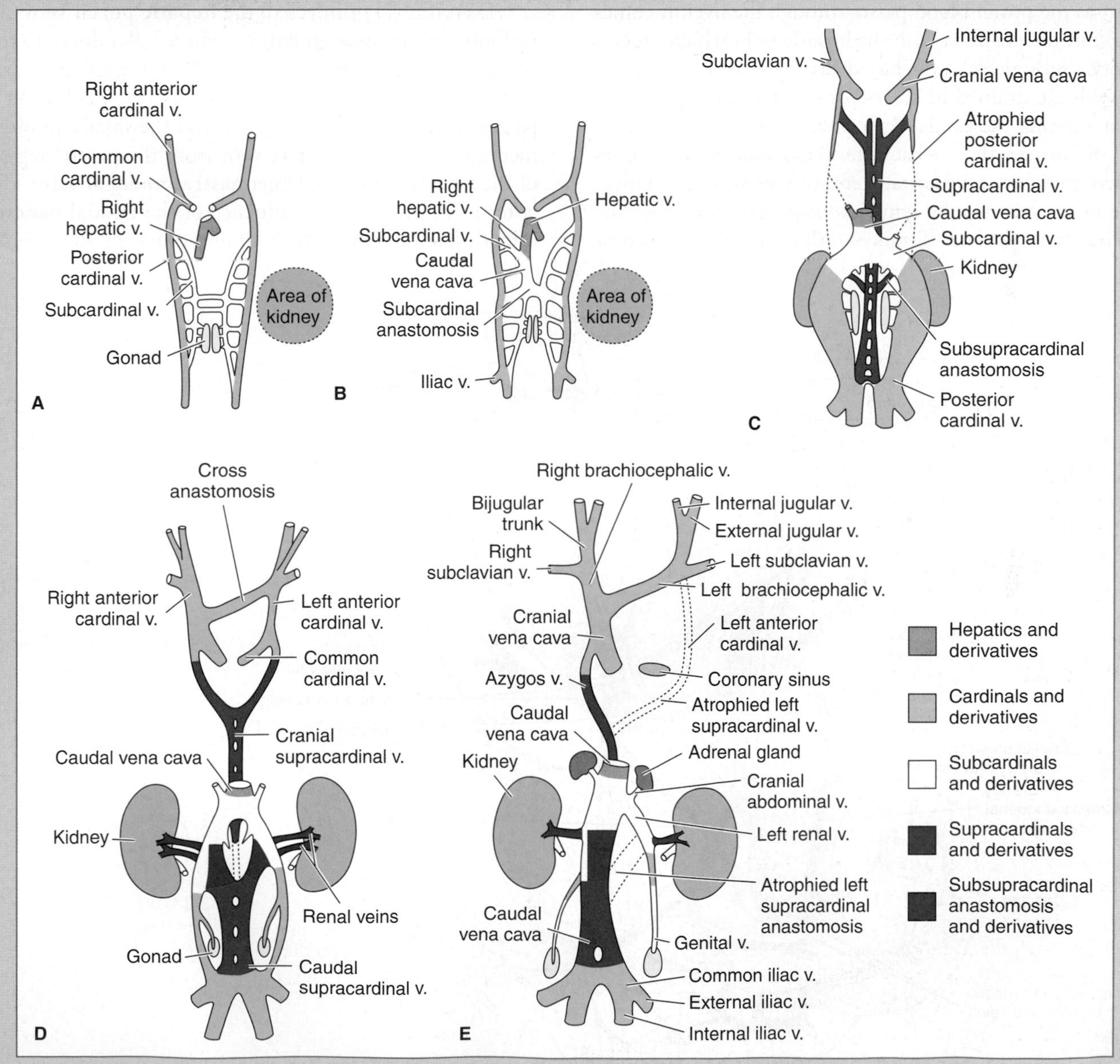

Diagrammatic representation of the embryonic development of the venous system in a cat as a model for the evolutionary transformations of the venous system from the fishlike to the mammalian configuration. The location of the kidneys in A and B is shown only on the right side. A, Early embryonic configuration. B, Intermediate embryonic configuration. C, Later intermediate embryonic configuration. D, Late intermediate embryonic configuration. E, Adult configuration. *(Modified from Huntington and McClure, 1920.)*

The Blood Vessels of the Abdominal Viscera

Blood reaches the abdominal portions of the digestive tract and spleen through vascular branches of the coeliac, cranial mesenteric, and caudal mesenteric arteries. The veins that drain these organs form the **hepatic portal venous system** because they all drain into the hepatic portal vein, which carries blood to capillary-like spaces, called **sinusoids,** in the liver rather than directly to the caudal vena cava. Sinusoids, unlike capillaries and the rest of the circulatory system, are not lined by endothelium. Therefore, as the portal blood passes through the liver, it comes into intimate contact with the hepatic cells, which process various substances in the blood (see Chapter 10). Liver sinusoids are drained by the hepatic veins, which you have seen entering the caudal vena cava.

Return to the coeliac artery and mesenteric arteries where they leave the aorta. Remove surrounding connective tissue, but not the sympathetic ganglia, and trace the coeliac artery a short distance until it divides into two or three branches (Fig. 11-23): the lienic artery to the spleen, the left gastric artery to the lesser curvature of the stomach, and the hepatic artery to the liver, pancreas, duodenum, and part of the stomach. More distal parts of these arteries will be seen with the veins that are described later. You can see the distribution of the **cranial mesenteric artery** to most of the small intestine and adjacent parts of the colon by stretching the mesentery (in the strict sense). The **caudal mesenteric artery** supplies the descending colon and rectum (Figs. 11-21 and 11-22).

Even if it is not injected, the **hepatic portal vein** can be found in the lesser omentum, where it lies dorsal to the bile duct and forms the ventral border of the epiploic foramen (see Fig. 10-19). Trace it caudad (Fig. 11-23). As it passes dorsal to the pylorus, it receives the small and often inconspicuous **right gastric vein** from the pyloric region of the stomach, and the larger **gastroduodenal vein.** The latter is formed by the confluence of the **cranial pancreaticoduodenal vein,** which drains much of the duode-

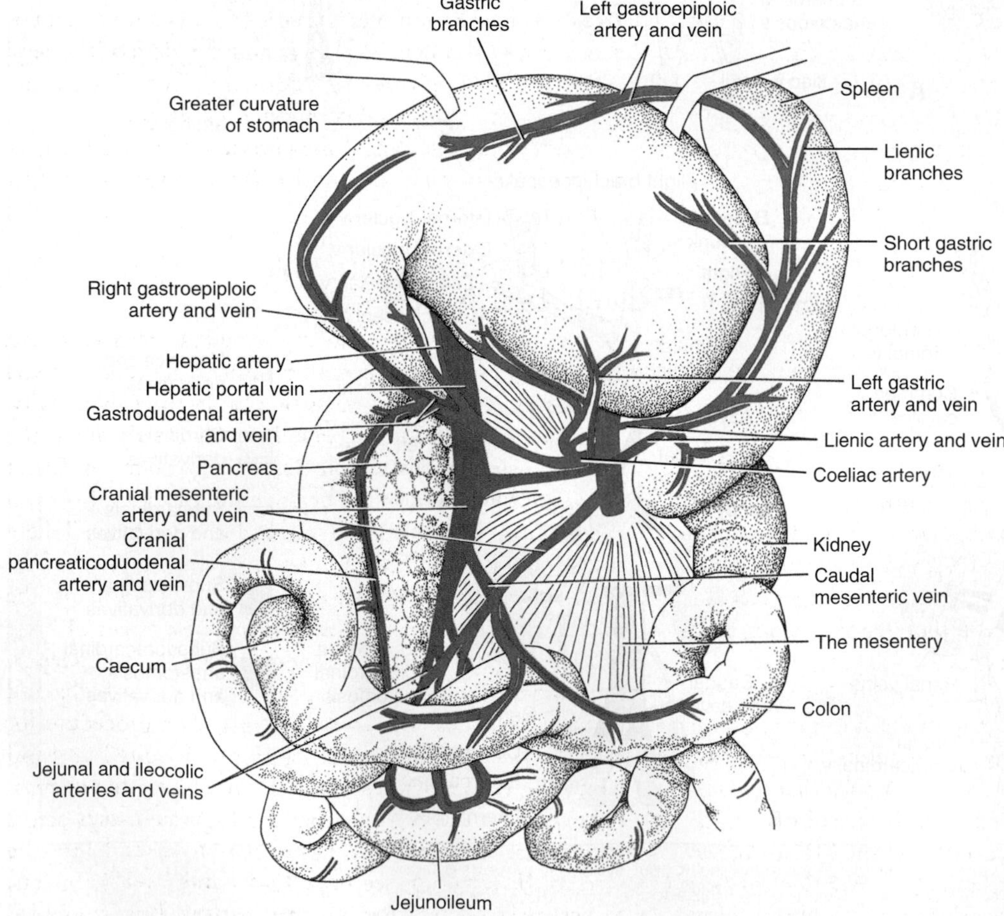

Figure 11-23

Ventral view of the hepatic portal system and accompanying arteries of a cat. The stomach was pulled cranially, and the tail of the pancreas was dissected away.

num and pancreas, and the **right gastroepiploic vein** from the right half of the greater curvature of the stomach and greater omentum.

Return to the coeliac artery and its major branches seen earlier. The first of these branches, the **hepatic artery,** can be seen on the left side of the gastroepiploic foramen and can be followed to the right side of the abdominal cavity. It gives off the large **gastroduodenal artery** before it continues to the liver. The gastroduodenal artery sends off a small but distinct **pyloric artery** to the pyloric region of the stomach and then bifurcates into the **cranial pancreaticoduodenal artery,** which runs between the duodenum and pancreas, supplying both organs, and the **right gastroepiploic artery.** The latter follows the right half of the greater curvature of the stomach and gives off **gastric arterial branches** to the stomach and **epiploic arterial branches** to the greater omentum. The second branch off the coeliac artery, the **left gastric artery,** runs toward the lesser curvature of the stomach near the pylorus. It supplies mainly the parietal side of the stomach and sends a small but distinct **esophageal artery** to the esophagus. Occasionally, the left gastric artery arises from the hepatic artery or from the lienic artery (see later) near its branching point from the coeliac artery. After giving off the left gastric artery, the coeliac artery continues as the **lienic artery.** Its first branch is the **right gastric artery,** which runs toward the lesser curvature of the stomach near the esophagus and supplies mainly the visceral side of the stomach. It usually anastomoses with the left gastric artery. The lienic artery sends off several **lienic arterial branches** to the spleen. Just before dipping into the substance of the spleen, the lienic branches give off **gastroepiploic arterial branches** that cross the gastrosplenic ligament to supply the left half of the greater curvature of the stomach. Occasionally, the lienic artery continues beyond the spleen as the **left gastroepiploic artery** along the left half of the greater curvature of the stomach (Fig. 11-23) and gives off **gastric arterial branches** to the stomach and **epiploic arterial branches** to the greater omentum. If present, the left gastroepiploic artery anastomoses with the right gastroepiploic artery along the greater curvature of the stomach.

Push the stomach cranially and tear through the part of the greater omentum that is going to the spleen and dorsal body wall. Carefully dissect away the tail of the pancreas, which extends toward the spleen, and notice that the hepatic portal vein is formed by the confluence of two tributaries: the **lienic vein,** which enters from the left side of the animal, and the much larger cranial mesenteric vein (see later). Trace the lienic vein by continuing to dissect away pancreatic tissue. The lienic vein receives the **left gastric vein,** which accompanies the left gastric artery and drains the region around the lesser curvature of the stomach. The lienic vein accompanies the lienic artery and drains the spleen through lienic venous branches. It also receives several **gastroepiploic venous branches** from the greater omentum and the left half of the greater curvature of the stomach. Occasionally, the lienic vein receives a **left gastroepiploic vein,** which collects **gastric venous branches** along the left half of the greater curvature of the stomach and **epiploic venous branches** from the greater omentum.

Now trace the **cranial mesenteric vein** and its four main tributaries, whose branching pattern is highly variable. The various veins are best identified by finding first the organs they drain. The **caudal mesenteric vein** drains the rectum and the caudal portion of the colon. The **ileocolic vein** drains the caudal end of the small intestine (i.e., the ileum) and the cranial portion of the colon. The many branches of the **jejunal vein** drain most of the small intestine (i.e., the jejunum). The **caudal pancreaticoduodenal vein** drains the caudal parts of the duodenum and pancreas.

Return to the cranial mesenteric artery. It gives off three main branches: the **caudal pancreaticoduodenal artery,** which anastomoses with the cranial pancreaticoduodenal artery; the **jejunal artery;** and the **ileocolic artery.** These arteries supply those portions of the intestine that are drained by the veins with corresponding names.

Find again the caudal mesenteric artery and the caudal mesenteric vein, and notice the differences in their branching patterns.

The Blood Vessels of the Pelvic Region and Hind Leg

Return to the caudal ends of the caudal vena cava and aorta. The terminal branches of the aorta lie ventral to the tributaries of the vena cava as they enter the pelvic cavity. In order to see the pelvic blood vessels clearly, open the pelvic canal. This is a simple procedure in females. Bisect the ventral ligament of the bladder and push it away from the cranioventral border of the pelvic symphysis. Take a scalpel, cut through the muscles on the ventral face of the pelvic symphysis, and continue to cut right through the midventral pelvic symphysis. You may use a pair of bone scissors to do so, but this is not necessary if you keep in the midventral line. Take a firm grip on the thighs and bend them as far dorsally as you can. The procedure for males is the same, but more caution is required to avoid damaging the reproductive ducts. First locate the cremasteric pouches that extend from the inguinal canals across the ventral surface of the pelvic symphysis and into the skin of the scrotum (see Figs. 12-13 and 12-15). In cats, the cremasteric pouches are very narrow, but in rabbits, they are quite wide. They should be pushed aside before cutting through the pelvic symphysis. Also locate the penis emerging from the caudal end of the pelvic canal;

avoid damaging it. After the pelvic canal has been opened, carefully pick away fat and connective tissue from around the blood vessels, bladder, and rectum. Confine your dissection to one side so as not to damage inadvertently parts of the urogenital system.

The **external iliac artery** extends caudolaterally toward the body wall and hind leg. It is accompanied distally by the **external iliac vein** (Figs. 11-21 and 11-22). The **internal iliac artery** and **vein** enter the pelvic canal. In cats, the iliac arteries arise independently from the aorta, but in rabbits they arise from a **common iliac artery.** In cats, the external and internal iliac veins unite to form a **common iliac vein** before emptying into the caudal vena cava, but in rabbits, they join the vena cava independently.

Trace the external iliac blood vessels from the aorta to the abdominal wall. Before passing through this wall, they give off from their caudomedial surface the **deep femoral artery** and **vein,** which extend deep into the thigh (Fig. 11-24). The **caudal epigastric artery** and **vein** can be seen running on the peritoneal surface of the rectus abdominis muscle. They anastomose cranially with the cranial epigastric artery and vein, respectively, both of which were seen earlier. The caudal epigastric artery usually branches off the deep femoral artery, but it may arise directly from the external iliac artery near the deep femoral artery. The **external pudendal artery** and **vein** can be found in the mass of fat outside the abdominal wall in the groin. They pass through the fat and supply the external genitalia. The external pudendal artery may branch off the caudal epigastric artery or the deep femoral artery. The caudal epigastric and external pudendal veins normally form a short, common **pudendoepigastric trunk** before they join the deep femoral vein. After giving rise to these blood vessels, the external iliac artery and vein perforate the abdominal wall and enter the leg as the **femoral artery** and **vein.** Additional major branches of these blood vessels are shown in Figure 11-24.

Now trace the internal iliac blood vessels. In cats, the internal iliac artery near its origin from the aorta gives rise to the **vesical artery** to the urinary bladder. In rabbits, the

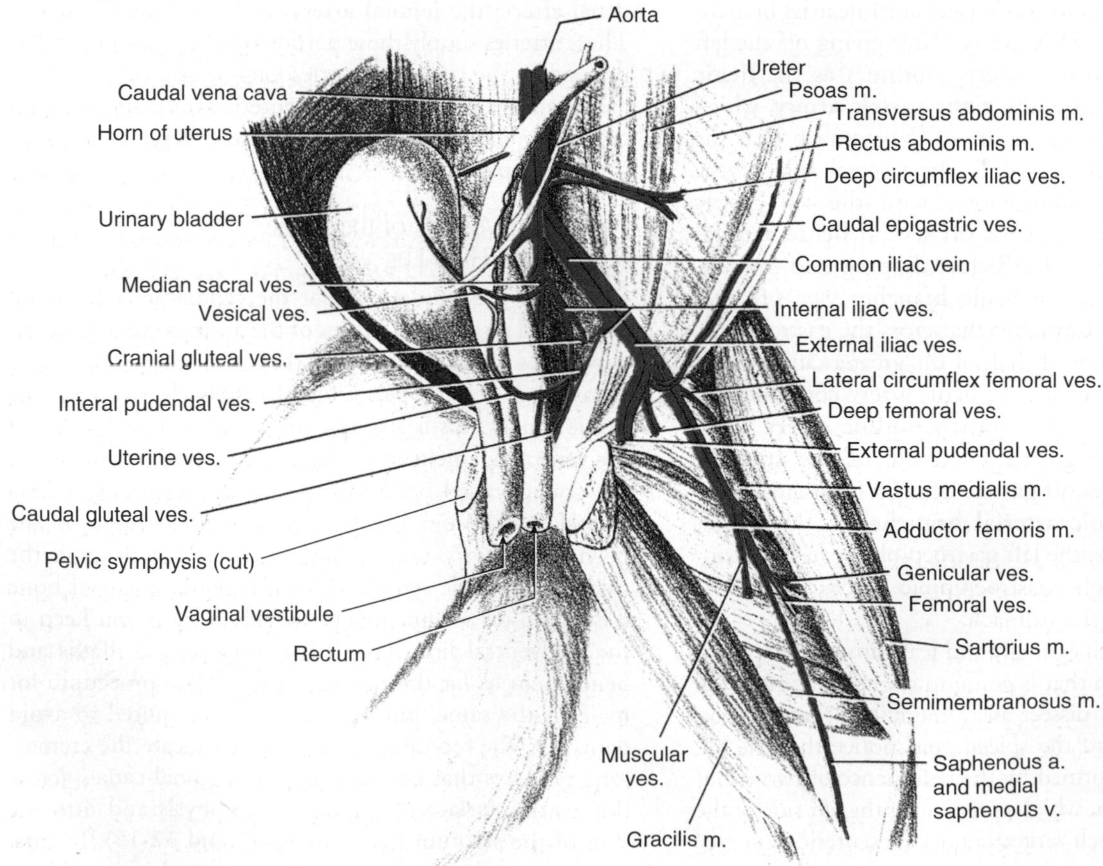

Figure 11-24

Ventral view of the distribution of the left external and internal iliac arteries and veins in a female cat. The pelvic canal was opened, and the pelvic viscera were pushed to the specimen's right side.

vesical artery arises from the external iliac artery. The vesical artery is the remnant of the proximal portion of the large umbilical artery of the embryo, which returns blood from the embryo to the placenta (Anatomy in Action 11-6). Deeper within the pelvic cavity, the internal iliac artery sends off one or two **gluteal arteries** to deep pelvic muscles and the **internal pudendal artery** to the rest of the pelvic viscera. **Gluteal** and **internal pudendal veins** accompany the arteries and drain into the internal iliac vein. In cats, the small **vesical vein** normally joins the internal pudendal vein, but in rabbits, it joins the external iliac vein.

In cats, after the iliac arteries have branched off, the aorta proceeds caudad across the sacrum as the very small **median sacral artery** and then continues as the **caudal artery** into the tail. In rabbits, the median sacral artery branches off the dorsal side of the aorta cranial to the bifurcation of the iliac arteries. The **caudal vein** draining the tail leads to the **median sacral vein,** which usually empties into one of the common iliac veins in cats or one of the internal iliac veins in rabbits.

The Bronchi and Internal Structure of the Heart

Briefly review the external features of the heart seen earlier and find the bifurcation of the trachea into the bronchi referred to earlier. Trace one of the bronchi and part of the respiratory tree into the lung (see Chapter 10). Identify again the chambers of the heart and the great blood vessels entering and leaving it as they appear in a ventral view (Fig. 11-25**A**). Reflect the apex of the heart, carefully clean the dorsal surface of the heart, and identify the chambers and blood vessels in this view (Fig. 11-25**B** and **C**). You can study the internal features of the heart by dissecting either the heart of your own specimen or a separate sheep heart. In the former, the coronary blood vessels may be more easily identified if they are injected; in the latter, the structures are larger and the heart chambers are not clogged with the injection mass. If you are studying a sheep heart, you will have to remove the pericardial sac. Clean and first identify the great blood vessels. They are comparable to those of cats and rabbits, except that the subclavian and common carotid arteries leave the arch of the aorta by a common brachiocephalic trunk. In sheep, the left azygos vein also enters the coronary sinus independently, and there is a conspicuous ligamentum arteriosum.

Open the heart by making the incisions shown in Figure 11-25**A.** To open the right atrium, make an incision that extends from its auricle into the caudal vena cava; to open the left atrium, make an incision that extends from its auricle through one of the pulmonary veins. To open the ventricles, first cut off the apex of the heart in the transverse plane. Cut off enough of the apex of the heart to expose the cavities of both ventricles. Make an incision through the ventral wall of the right ventricle and extend it into the pulmonary artery. This will be a diagonal incision. Open the left ventricle by making an incision through its ventral wall that extends as far cranial as the base of the arch of the aorta. This will be a longitudinal incision. Clean out the chambers of the heart if necessary.

Find the entrance of the **right cranial vena cava** and **caudal vena cava** into the **right atrium.** In cats and sheep, the **coronary sinus** lies along the coronary sulcus on the subsinuous side of the heart. Its entrance into the right atrium is situated just caudal to the entrance of the caudal vena cava. You can explore the extent of the coronary sinus by probing it. In rabbits, there is no coronary sinus, and the left cranial vena cava enters the right atrium directly. Find the entrances of the **pulmonary veins** into the left atrium. In cats, there are six to nine pulmonary veins, but in sheep, there are usually only four. The atria have relatively thin walls, but their auricles form prominent bands of muscles called **pectinate muscles** because they resemble a comb, or pecten, of a chicken.

If time permits, have a look at the coronary blood vessels, which supply and drain the heart muscle. If the blood vessels of your specimen are injected, the coronary blood vessels usually are more easily observed than in an uninjected sheep heart. The coronary blood vessels lie directly under the visceral peritoneum on the external surface of the heart and send and receive branches into and from the heart muscle, respectively. You may have to break the visceral pericardium and remove fat to see some of the blood vessels clearly. The heart is always supplied by two major coronary arteries, but their branching pattern varies among mammalian species and individuals. In most cats and sheep, the **left coronary artery** leaves the very base of the aorta on the paraconal side of the heart and almost immediately branches into a **paraconal interventricular artery** along the paraconal interventricular sulcus and a **left circumflex artery,** which follows the left part of the coronary sulcus and gives off the **subsinuous interventricular artery,** which follows the subsinuous interventricular sulcus. The **right coronary artery** leaves the base of the aorta on the subsinuous side of the heart and follows the right part of the coronary sulcus as the **right circumflex artery.** In some cats (and in human beings), it is the right circumflex artery that gives off the subsinuous interventricular artery and the left circumflex artery that is restricted to the left part of the coronary sulcus. In a few cats, both the left and right circumflex arteries contribute a subsinuous interventricular artery. In all cases, there are anastomoses between the circumflex arteries and the interventricular arteries.

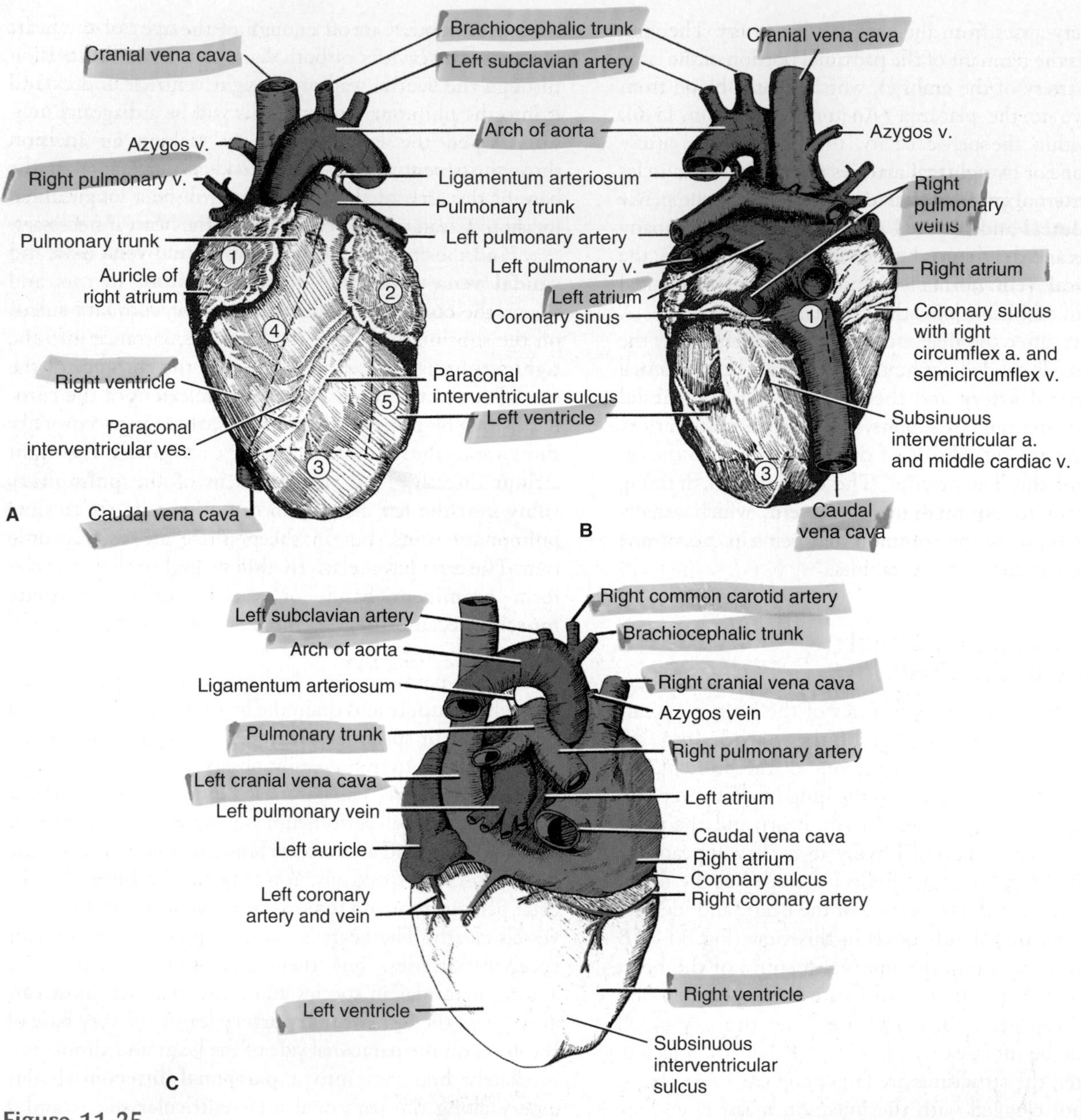

Figure 11-25

Ventral (A) and dorsal (B) views of the heart and great blood vessels of a cat. C, Dorsal view of the heart and great blood vessels of a rabbit. The incisions that need to be made to open the heart are shown in A and B. The sequence of the incisions is indicated by Arabic numerals.

The oxygen-depleted blood from the heart muscle in cats and sheep drains into the right atrium through the coronary sinus, but in rabbits the blood drains into the left cranial vena cava (Fig. 11-25). The **middle cardiac vein** follows the subsinuous interventricular sulcus. The **great cardiac vein** follows the left part of the coronary sulcus and receives the usually paired **paraconal interventricular veins.** The **semicircumflex vein** follows the right half of the coronary sulcus and drains part of the subsinuous side of the heart. In sheep, the semicircumflex vein is replaced by several small **cardiac veins,** which enter the right atrium independently.

The two atria are separated by an **interatrial septum.** Examine the interatrial septum from the right atrium and you will find an oval depression, the **fossa ovalis,** beside the point at which the caudal vena cava enters. Put your

Stretch out a section of the mesentery anchoring the small intestine and hold it up to the light. Very small lymphatic vessels can be seen outlined by little streaks of fat. These lymphatic vessels are called **lacteals,** because the lymph in them is milky white as a result of the fat globules that were absorbed by the small intestine. The lacteals carry the lymph into a conspicuous aggregation of **jejunal lymph nodes,** also called pancreas of Aselli, which are located at the base of the mesentery. The jejunal lymph nodes are drained by one or more larger lymphatic vessels that run along the cranial mesenteric artery to the cisterna chyli. These lymphatic vessels may have been destroyed inadvertently during the earlier dissections. Lymphatic vessels from the stomach, liver, pelvic canal, and hind legs also empty into the cisterna chyli, but they are difficult to see (see Fig. 10-19). The cisterna chyli receives all the lymphatic drainage of the body caudal to the diaphragm and passes it on to the thoracic duct, which then receives the lymphatic drainage of the thorax as it ascends toward the bijugular trunk.

Other lymphatic vessels, which parallel the larger veins, drain the arms, neck, and head. Those of the left side enter the thoracic duct or the left bijugular trunk close to the entrance of the thoracic duct. Those of the right side enter the right bijugular trunk near its union with the subclavian vein, either independently or by a short common trunk, the **right lymphatic duct.**

The Excretory and Reproductive Systems

In the previous chapters, we studied the systems that serve to interact with the environment (the skeletal, muscular, and nervous systems) and to acquire food and transform it into energy that can be used by the body (the digestive, respiratory, and circulatory systems). This last chapter deals with the systems that are involved with the final aspects of the various life processes, namely, the elimination of metabolic waste products (the excretory system) and the generation of new life that transcends the death of an individual (the reproductive system). These two systems are discussed in the same chapter because some of their parts are intimately associated and, especially in the male, are shared by both systems.

In order to understand the intimate topographical and structural relationships of the excretory system with the cardiovascular system on the one hand and with the reproductive system on the other hand, it is necessary to be familiar not only with the structure and function of these systems but also with their embryonic development.

Chapter 12

General Structure and Function of the Excretory System

The main organs of the excretory system are the kidneys.[1] Their functions include the removal of nitrogenous waste products and the maintenance of the proper composition of the body fluids, called **homeostasis,** through differential secretion and reabsorption of water, ions, and various substances.

[1]Remember that the kidneys are not the only organs that can remove metabolic waste products and surplus substances from the body. For example, carbon dioxide, water, ions, and other substances can be excreted also by other organs, such as the gills, lungs, skin, and special excreting glands, such as the digitiform gland of sharks.

Nitrogenous waste products are generated when amino acids are metabolically broken down and their amino groups released. In vertebrates, these amino groups are excreted as ammonia (NH_3), urea, or uric acid. Most teleosts and aquatic amphibians excrete their nitrogenous waste products mainly as **ammonia,** which is highly toxic and soluble in water. It is excreted by the cells of the gill epithelium in teleosts or by the kidneys in aquatic amphibians. Because large quantities of water are needed to dilute and carry away ammonia, terrestrial vertebrates excrete their nitrogenous waste products to a greater or lesser part as urea or uric acid, depending on how much water is available for the production of urine by the kidneys. Chondrichthyan fishes, lungfishes, most terrestrial amphibians, some reptiles, and mammals in general excrete mostly **urea,** which is quite soluble in water. However, it is much less toxic than ammonia and requires ten times less water to be washed out of the body than does ammonia. As an additional advantage, one molecule of urea contains two atoms of nitrogen; thus, it absorbs two amino groups per molecule. Birds and most reptiles excrete mostly **uric acid,** which is less toxic than ammonia and is poorly soluble in water. It precipitates easily and is excreted as a paste, requiring 50 times less water to be removed from the body than does ammonia. Per molecule, uric acid incorporates four atoms of nitrogen.

The structure and physiology of the kidneys vary greatly among vertebrates. The vertebrate kidney that is best understood is the mammalian kidney. In general, the kidneys of other vertebrates are not as well understood and differ considerably in their gross morphology, histology, vascularization, and physiology. However, all vertebrate kidneys consist of the same type of functional unit, the **nephron.** A nephron is composed of two parts: the renal corpuscle and the renal tubule (Fig. 12-1). The **renal corpuscle,** also called Malpighian corpuscle, consists of a capillary tuft, the **glomerulus,** which is surrounded by a double-layered **renal capsule** of thin, flattened epithelial cells, which is also called Bowman's capsule. The inner layer of this capsule is entwined with the capillaries, which it envelopes; and small molecules easily pass through this layer and the capillary walls. The liquid that filters out of the glomerular capillaries into the renal capsule is called the **glomerular filtrate.** Its composition is essentially the same as that of blood, except that it lacks blood cells and large molecules, which do not pass through the walls of the renal capsule and capillaries. The quantity of filtrated liquid is proportional to the size of the glomerulus. The **renal tubule** is a direct extension of the renal capsule and leads to collecting tubules that drain the kidney. Many of the cells of the renal tubule are much larger than those of the renal capsule and are capable of secreting or reabsorbing substances into or from the renal tubule, thereby transforming the glomerular filtrate into the final **urine.** Hence, the nephrons produce a urine that removes a maximum of waste products and surplus substances from the blood and retains substances that are needed by the body, including glucose, water, and various ions.

The blood supply to the nephrons varies among vertebrates. In order to ensure the dual function of filtration on the one hand

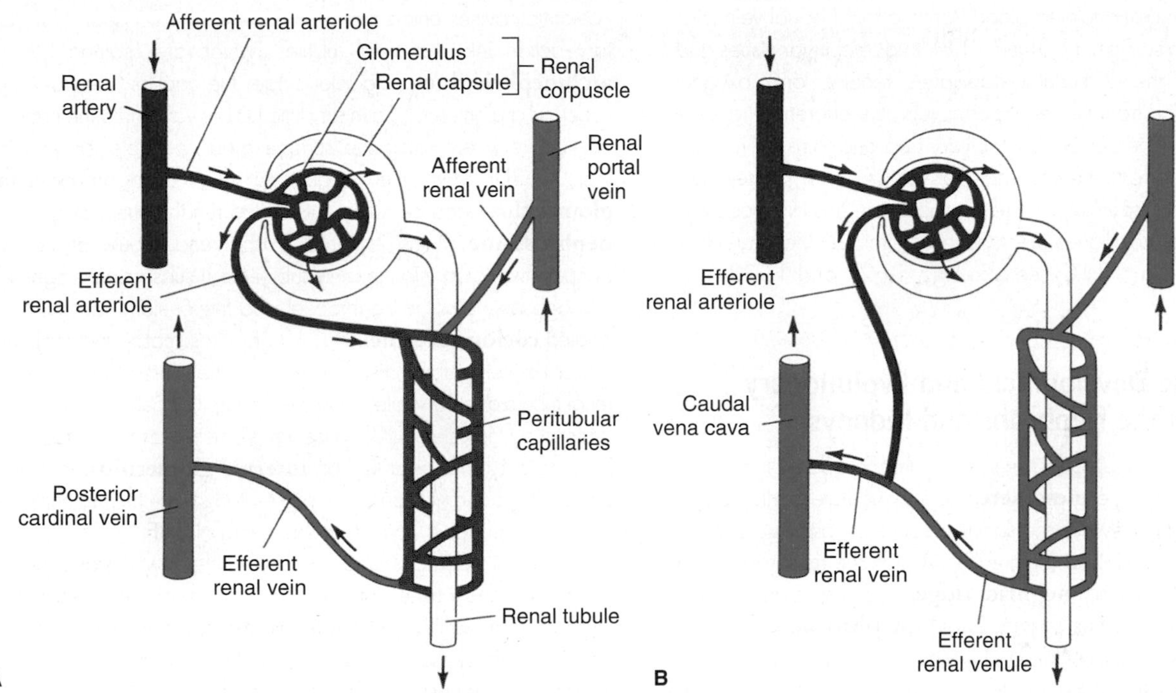

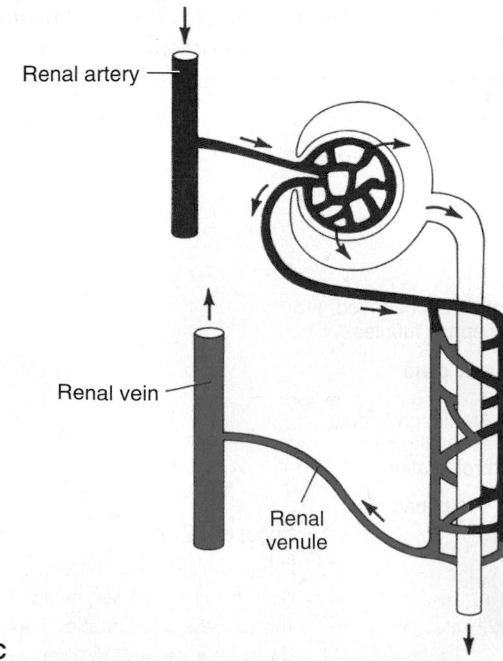

Figure 12-1

Diagrams illustrating the blood supply to a nephron in various vertebrates. Arrows indicate the direction of the blood flow in the blood vessels and the flow of urine in the nephron. A, Chondrichthyan fishes, reptiles, and birds. B, Osteichthyan fishes and amphibians. C, Mammals. Red = oxygen-rich blood; blue = oxygen-depleted blood; purple = mixed blood.

and of secretion and absorption on the other hand, each nephron is provided with two separate capillary networks: the glomerulus to the renal corpuscle and the **peritubular capillaries** to the renal tubule. The glomeruli always receive their blood supply through the renal artery (Fig. 12-1), but the blood supply to the peritubular capillaries varies among vertebrates. In chondrichthyan fishes, reptiles, and birds, the peritubular capillaries receive oxygen-rich blood from the efferent renal arteriole as well as oxygen-depleted blood from the renal portal vein (Fig. 12-1**A**; see also Figs. 11-5 and 11-8). In osteichthyan fishes and amphibians, the peritubular capillaries receive only oxygen-depleted blood from the renal portal vein; the efferent renal arteriole bypasses the peritubular capillary bed and joins the efferent renal venule to form the efferent renal vein (Fig. 12-1**B**; see also Fig. 11-11). In mammals, the renal portal vein has disappeared; the peritubular capillaries are supplied entirely by the efferent renal arteriole (Fig. 12-1**C**; see also Figs. 11-21 and 11-22).

Embryonic Development and Evolutionary History of the Nephrons and Kidneys

The excretory and reproductive systems are mesodermal structures that develop from the **mesomere,** the intermediate portion of the mesoderm lying between the somites (epimeres) and lateral plate (hypomere) (Fig. 12-2 and Fig. 7-5). Most of the mesomere soon differentiates into the **nephric ridge.** Starting cranially, the nephric ridge becomes segmented into **nephrotomes,** but farther caudally the segmentation of the nephric ridge becomes less obvious. The nephrotomes are thin-walled structures containing an extension of the coelom, called the **nephrocoel,** which starts out to be in broad communication with the coelom (Fig. 12-3**A–D**). The unsegmented caudal part of the nephric ridge remains a more massive tissue. Nephrons differentiate sequentially from cranial to caudal. Various types of nephrons are found in different vertebrates at different stages of their development; they are distinguished mainly on the basis of their relationship to the coelom. These various types of nephrons are assumed to represent different stages in a formerly complete embryonic and evolutionary sequence (Fig. 12-3). According to this theory, capillary networks, which are supplied by branches of the aorta and drained by tributaries of the subcardinal veins, at first filter fluid into the nephrocoel and coelom (Fig. 12-3**C**). Surplus fluid that collects in the coelomic cavities enters the nephrocoel and is transported by ciliary action into the renal tubules, which are drained by the **archinephric duct.** In a later stage, the capillary network protrudes as a glomerulus (Latin for "little ball of yarn") into the nephrocoel, thereby enlarging the surface area available for filtration (Fig. 12-3**D**). This type of glomerulus is called an **external glomerulus** because it lies outside the renal tubule. Cilia in the **nephrostome** at the entrance to the renal tubule draw substances and liquid into the renal tubule. In a subsequent stage, the passage between the nephrocoel and the coelom narrows to a ciliated **coelomic funnel** (Fig. 12-3**E**). Presumably, the coelomic funnel ensures that the fluid filtered from the glomerulus enters the renal tubule directly while still allowing surplus coelomic fluid to be drained by the coelom. Because the glomerulus is now separated from the coelom, it is called an **internal glomerulus.** In a final stage, the coelomic funnel is obliterated, and the nephron becomes independent from the coelom (Fig. 12-3**F**).

The first nephrons of an embryo differentiate starting from the cranial end of the nephric ridge, and their distal ends turn caudad, forming the **archinephric duct,** also called the Wolffian duct (Fig. 12-2). As the archinephric duct grows caudally, it induces the formation of nephrons in the more caudal parts of the nephric ridge. The first nephrons that develop at the cranial end of the nephric ridge dorsal to the pericardial cavity in all vertebrate embryos form the **pronephros** (Fig. 12-4**A**). The pronephros functions as a kidney in embryonic and larval hagfishes and lampreys, many osteichthyan fishes, and amphibians,

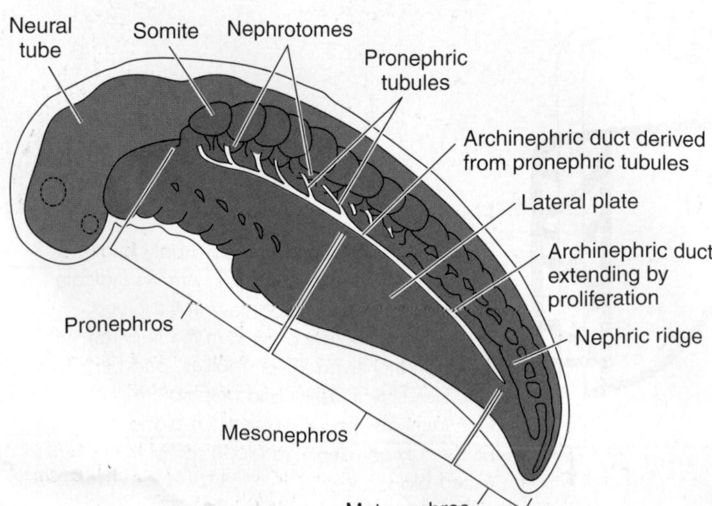

Figure 12-2

Diagram illustrating the embryonic differentiation of the nephric ridge into segmental nephrotomes and nephrons. The process starts cranially and proceeds caudally. *(From Liem, Bemis, Walker, and Grande, 2001; after Hildebrand, 1982.)*

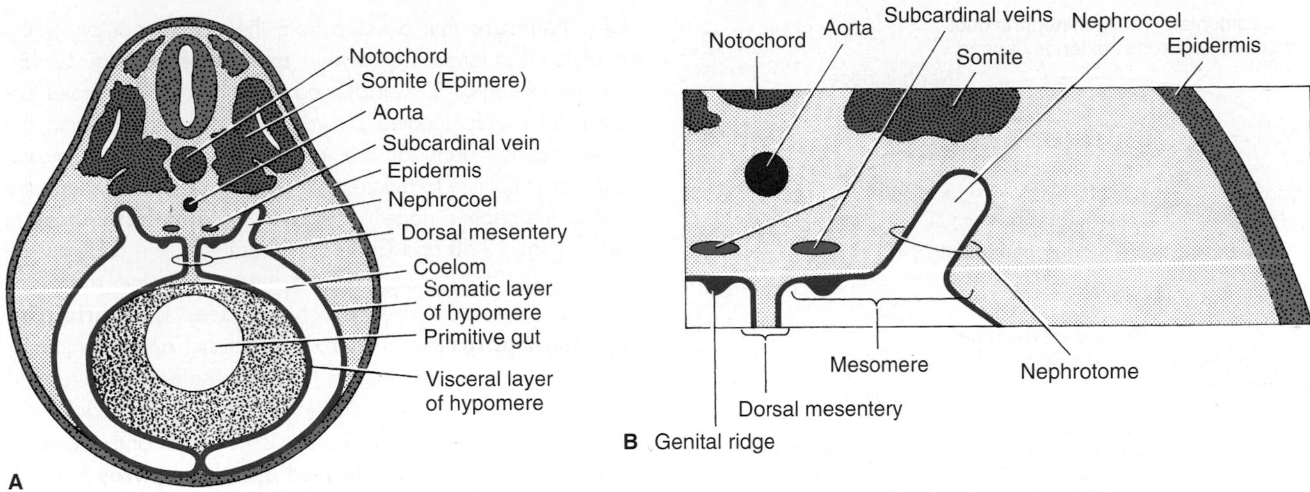

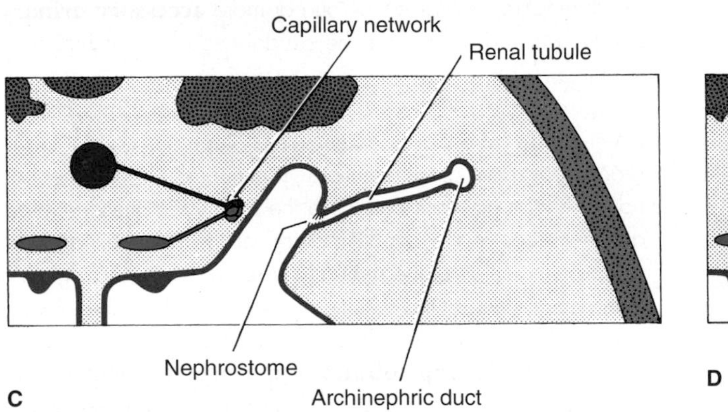

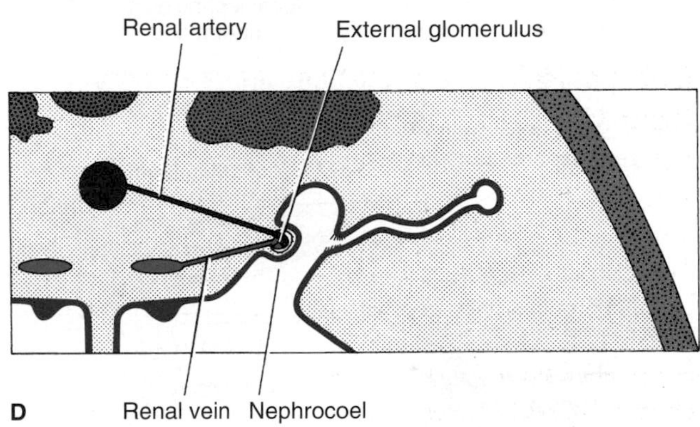

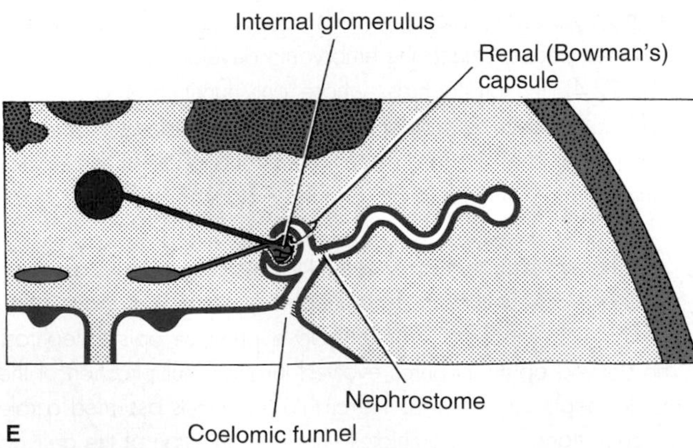

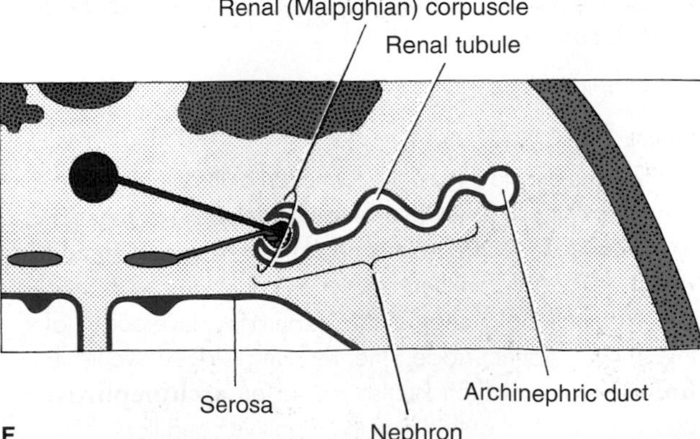

Figure 12-3

Diagrams illustrating the probable sequence of embryonic and evolutionary stages in the development of nephrons. **A**, Cross section of a basal vertebrate embryo showing the location of the nephrocoel, which contributes to the formation of the renal capsule. **B**, Enlargement of the nephric region in **A** showing the nephric ridge, or mesomere, and the adjacent genital ridge. **C** and **D**, Capillary network pushes into the nephrocoel to form a nephron with an external glomerulus as seen in larval lampreys and some larval amphibians. **E**, Further constriction of the nephrocoel forms an internal glomerulus with a coelomic connection as seen in some nephrons of chondrichthyan fishes. **F**, Internal glomerulus with no coelomic connection, which occurs in most vertebrates. *(Adapted from Hildebrand, 1982.)*

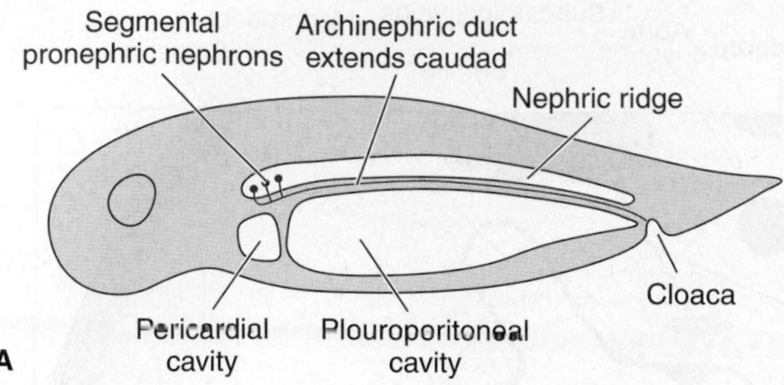

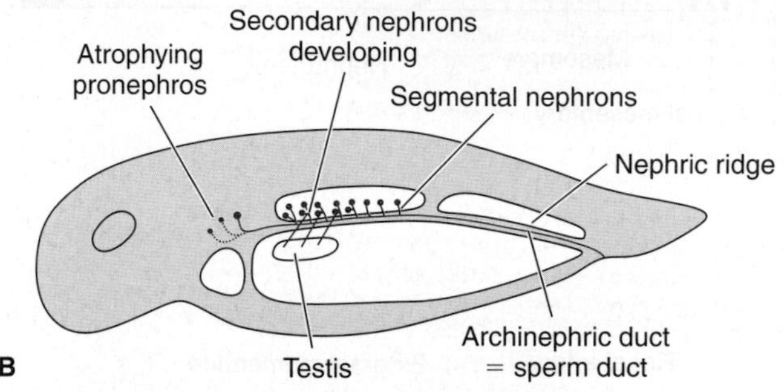

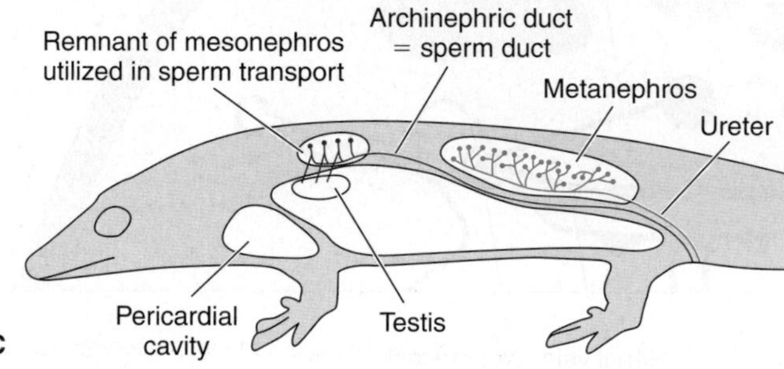

Figure 12-4

Lateral views of the sequence of the developmental stages of the kidneys and its ducts in an amniote. **A,** Pronephros in an early embryo. **B,** Mesonephros in an intermediate stage of an embryo. **C,** Metanephros in a late embryo and adult. *(From Liem, Bemis, Walker, and Grande, 2001.)*

but it serves only to initiate the formation of the archinephric duct in chondrichthyan fishes and amniotes. In adult hagfishes and a few teleosts, a few pronephric nephrons remain as a **head kidney** (Fig. 12-5**B**), but their function is not clear.

If functional, segmental nephrons developed along the entire nephric ridge, the resulting kidney would be a **holonephros,** which is also called an **archinephros,** because it is likely that this was the primitive condition of the kidney in ancestral vertebrates (Fig. 12-5**A**). No extant adult vertebrates have such a kidney, but larval hagfishes and caecilians possess kidneys that are somewhat similar to the holonephros.

The pronephros is a transitory kidney. In a few segments caudal to it, nephrons do not differentiate, thereby creating a gap between the pronephros and the nephrons that are formed later. Nephrons that develop from the middle portion of the nephric ridge form an embryonic **mesonephros** (Fig. 12-4**B**). The mesonephros is well defined in embryos of amniotes because of a gap located between the mesonephros and the more caudal parts of the nephric ridge. In anamniotes, nephrons continue to differentiate without interruption along the rest of the nephric ridge, resulting in an **opisthonephros** in adults (Fig. 12-5**B** and **C**). In adult hagfishes, the nephrons remain segmentally arranged, and the more cranial nephrons possess coelomic funnels. Such a kidney is a **primitive opisthonephros** (Fig. 12-5**B**). In most fishes and amphibians, only the more cranial nephrons remain segmentally arranged and occasionally have coelomic funnels. More caudally, the nephrons multiply, and the kidney is enlarged and excretory. Such a kidney is called a **derived opisthonephros** (Fig. 12-5**C**). The more cranial parts of the kidney and archinephric duct are converted for sperm transport in males. In elasmobranchs and many amphibians, the renal region of the derived opisthonephros is drained by one or more **accessory urinary ducts,** which develop from the caudal end of the archinephric duct (Fig. 12-8).

In amniotes, kidneys differentiate along a somewhat different course from that in anamniotes. Nephrons caudal to the pronephros do not differentiate continuously in the nephric ridge. A gap appears caudal to the embryonic mesonephros, and nephrons in the caudal portion of the nephric ridge develop by induction of the **ureteric bud,** which is an outgrowth of the caudal end of the archinephric duct like the accessory urinary ducts of anamniotes. The ureteric bud branches extensively and forms **collecting tubules,** into which the nephrons tap. The proximal part of the ureteric bud becomes the ureter, which drains the kidney. This type of kidney is found in all adult amniotes and is called a **metanephros.**

In the course of both embryonic development and evolutionary history, renal function proceeds caudally through a succession of kidney stages. During the embryonic development of amniotes (Fig. 12-4), the pronephros, whose only function is to form the archinephric duct, is supplanted by the mesonephros functioning as a kidney (Fig. 12-4). In adults, remnants of the mesonephros participate in sperm transport, and the metanephros functions as a kidney.

The evolutionary history of the kidneys (Fig. 12-5) is assumed to have proceeded from a holonephros, which may have been the kidney of ancestral craniates, to a primitive opisthonephros. The derived opisthonephros evolved through multiplication of the caudal nephrons, whereas the cranial nephrons assumed a role in sperm transport.[2] In amniotes, the renal function of the derived opisthonephros was taken over by the metanephros.

[2]Sometimes the opisthonephros is also called a *mesonephros,* which is a term that should be applied only to an embryonic kidney. The opisthonephros does incorporate the embryonic mesonephros, but also includes the part of the nephric ridge that forms the metanephros of amniotes.

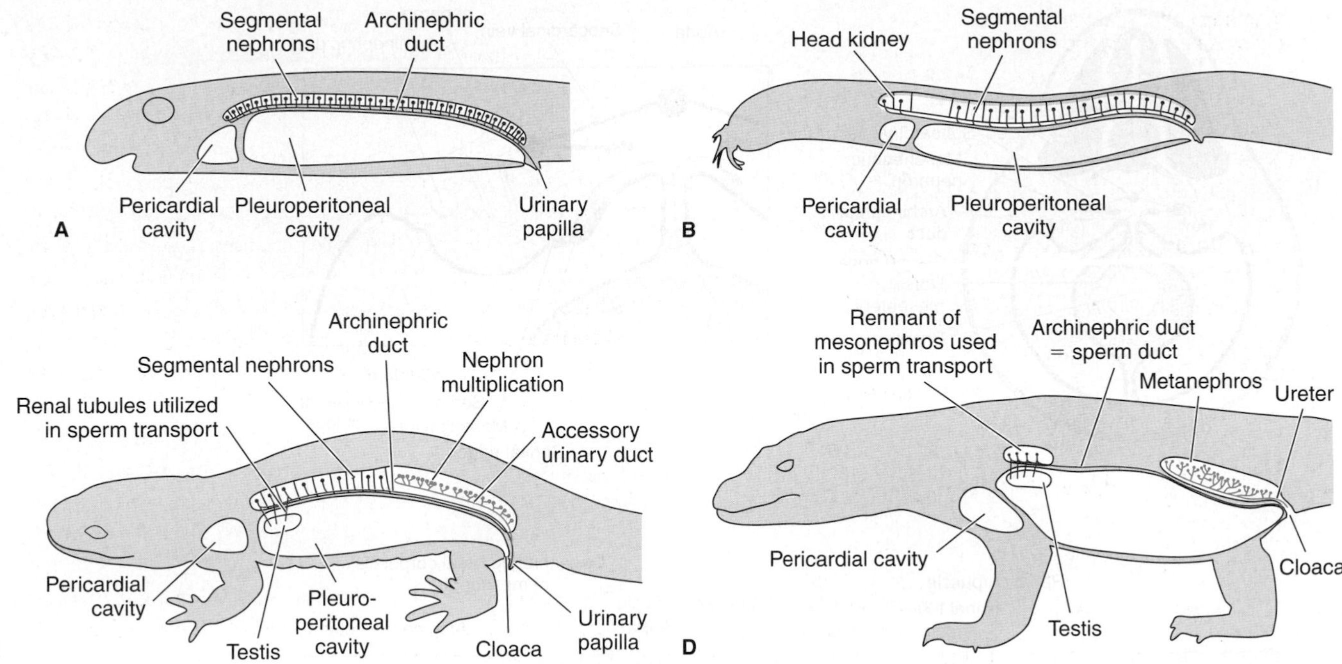

Figure 12-5
Diagrams of the evolutionary stages of the kidney and its ducts in adult craniates. **A,** Hypothetical holonephros of ancestral craniates. **B,** Primitive opisthonephros as seen in a hagfish. **C,** Derived opisthonephros found in most anamniotes. **D,** Metanephros of amniotes. *(From Liem, Bemis, Walker, and Grande, 2001.)*

Sex Determination

The determination of sex in vertebrates is quite complex and variable. In many fishes, turtles, crocodilians, and some lizards, sex is determined by environmental factors. Many teleost fishes can change their phenotypic sex during their lifetime and, thus, are **sequential hermaphrodites;** some teleosts can produce eggs and sperms at the same time and, thus, are **synchronous hermaphrodites.** Temperature often is a critical factor in determining the sex of developing embryos. For example, turtle embryos develop mostly into female adults, if the eggs are situated in the center of a nest where they receive metabolic heat from their surrounding eggs, or are incubated in nests that are exposed to the sun. **Environmentally dependent sex determination** is probably the ancestral condition for amniotes (Graves and Shetty, 2001).

The determination of sex by **sex chromosomes** evolved independently in mammals, some lizards, many snakes, and birds. In mammals, the male is the heterogametic sex with an X and a Y chromosome, and the female is the homogametic sex with two X chromosomes. The sex chromosomes evolved from a pair of autosomes, and the X chromosome has a full complement of various genes. The Y chromosome has lost most of its genes except for a small, tightly linked group of genes that includes the gene SRY, which controls testis development. Hormones produced by the embryonic testis drive the differentiation of an embryo into a male. In the absence of these testicular hormones, differentiation of an embryo results in a female. Both males and females, however, have only a single active X chromosome in each cell, because one X chromosome is inactivated in the cells of females.

In birds and snakes, the female is the heterogametic sex with a Z and a W chromosome, and the male is the homogametic with two Z chromosomes. The Z and W chromosomes also evolved from autosomes, but not from the same autosomes that gave rise to the X and Y chromosomes in mammals. A gene on the W chromosome may have a weak feminizing effect, although the difference in the dosage of gene products of the Z chromosomes may play a critical role in sex determination. No SRY gene has been found. Hormones produced by the embryonic testis drive the differentiation of an embryo into a male.

Embryonic Development and Evolutionary History of the Reproductive System

Irrespective of how the sex of individuals is determined, early vertebrate embryos do not show any morphological sexual differences and are **sexually indifferent.** The gonad appears first as a thickening on the medioventral surface of the mesonephros

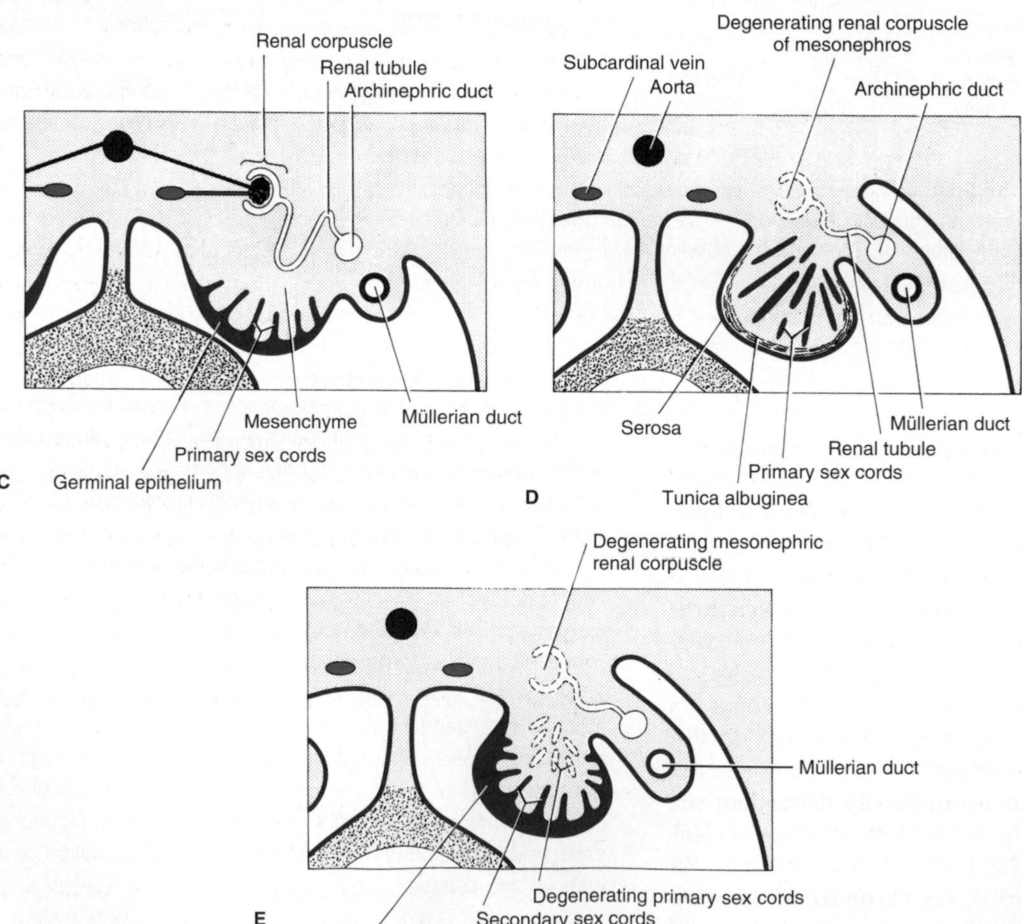

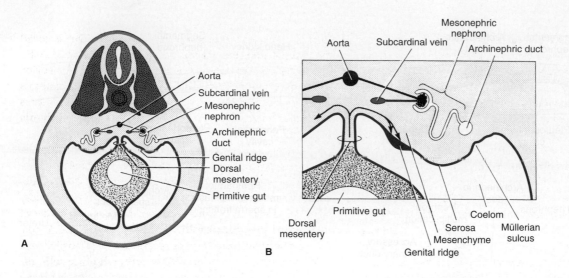

Figure 12-6

Diagrams illustrating the embryonic development of the gonads and genital ducts. **A**, Cross section through an embryo at the level of the mesonephros showing the location of the genital ridges. **B**, Enlarged detail from **A** showing the topographical relationships among the genital ridge, mesonephros, and Müllerian sulcus. Arrows indicate the path of the primordial germ cells. **C**, Formation of the primary sex cords and Müllerian duct in the sexually indifferent stage of the gonad. **D**, Separation of the primary sex cords from the peripheral germinal epithelium in the male gonad. **E**, Formation of the secondary sex cords and degeneration of the primary sex cords in the female gonad. *(A, Adapted from Hildebrand, 1982; B–E, after Browder, Erickson, and Jeffery, 1991.)*

(Fig. 12-6**A** and **B**). This **genital ridge** consists of embryonic connective tissue, called mesenchyme, which forms the **medulla** of the future gonad, and of an overlying segment of thickened coelomic epithelium, called serosa, which forms the **cortex** of the gonad. The **primordial germ cells,** which will give rise to the actual gametes, migrate from the yolk sac via the wall of the primitive gut and the dorsal mesentery into the cortex. The epithelium forming the cortex and containing the primordial germ cells is now called the **germinal epithelium.** As the germinal epithelium proliferates, it sends rodlike extensions, the **primary sex cords,** into the mesenchyme of the medulla (Fig. 12-6**C**). At the same time, the **Müllerian sulcus,** which forms ventrolaterally to the archinephric duct (Fig. 12-6**B**), continues to invaginate and forms the **oviduct,** which often is called the **Müllerian duct** in embryos (Figs. 12-6**C–E** and 12-7**B**). In elasmobranchs, however, the oviduct forms through a splitting of the archinephric duct and opens into the coelom through a modified coelomic funnel of a mesonephric nephron.

If the embryo differentiates into a male (Fig. 12-6**D**), the primary sex cords of the germinal epithelium in the cortex continue to proliferate, penetrate the medulla, and differentiate into the **seminiferous tubules** in amniotes, or the **seminiferous ampullae** in anamniotes. The testis differentiates primarily in the medulla of the sexually indifferent gonad. A secondary proliferation of connective tissue forms a dense **tunica albuginea** just beneath the serosa. During later stages of embryonic development, the testis gradually pulls away from the body wall, to which it remains anchored by a subsidiary mesentery, called the **mesorchium.**

The seminiferous ampullae or tubules contain the stem sperm-forming cells, called the **spermatogonia,** which develop from the primordial germ cells. The sperm cells, which develop from the spermatogonia, are surrounded and nurtured, at least in amniotes, by the **Sertoli cells,** which develop from certain cells in the seminiferous tubules. Some mesenchymal cells between the seminiferous tubules mature into the **interstitial cells of Leydig,** which secrete the male hormone testosterone. In most anamniotes, except teleost fishes, the seminiferous ampullae are connected to certain mesonephric renal tubules by the **rete cords,** or **cords of the urogenital union.** The rete cords become the **efferent ductules.** The mesonephric renal tubules in turn are connected to the archinephric duct. In teleosts, the rete cords form a separate sperm duct, which completely bypasses the opisthonephros. In male amniotes, the sperm passages are comparable to those of nonteleostean anamniotes, but the highly convoluted efferent ductules are formed by mesonephric tubules in the **head of the epididymis.** The convoluted portion of the archinephric duct becomes the **ductus epididymidis,**[3] which is located in the **body** and **tail of the epididymis.** The amniote epididymis is homologous to the cranial part of the opisthonephros and archinephric duct of anamniotes. The rest of the archinephric duct is called the **ductus deferens** (plural: *ductus deferentes*) (Fig. 12-7**B**). The Müllerian duct, which was present in the sexually indifferent stage of the embryo (Figs. 12-6**C** and 12-7**A**), regresses and atrophies. In amniotes, the caudal end of the mesonephros sometimes forms a vestigial adult structure, the **paradidymis** (Fig. 12-7**B**).

In a female embryo, the primary sex cords and traces of the tunica albuginea atrophy. At the same time, the germinal epithelium experiences a second proliferative spurt and extends **secondary sex cords** into the medulla (Fig. 12-6**E**). The secondary sex cords remain close to the cortex and break up into clusters of cells, called **primordial follicles.** Each primordial follicle consists of an **oogonium,** which developed from a primordial germ cell, and is surrounded by **granulosa cells,** or **follicle cells,** derived from epithelial cells of the germinal epithelium. The ovary gradually pulls away from the body wall, to which it remains anchored by a subsidiary mesentery, the **mesovarium.**

In female anamniotes, the anterior opisthonephric renal tubules do not form any direct connection with the ovary and remain a functioning part of the excretory system. In amniotes, the mesonephric renal tubules and the archinephric duct regress and atrophy as the metanephros and the ureter become the definitive excretory organs (Fig. 12-7**C**). Some vestiges of the mesonephros and archinephric duct may remain visible in adult amniote females, such as the **epoophoron** consisting of the cranial mesonephric renal tubules, the **paroophoron** consisting of the caudal mesonephric renal tubules, and **Gartner's duct** consisting of the caudal end of the archinephric duct (Fig. 12-7**C**).

Because the follicles remain near the surface of the ovary and no direct connection is established between the ovary and any duct, the mature oocytes of most vertebrates are released into the coelom after they break through the serosa that envelops the ovary. In lampreys, the oocytes are released directly to the outside through the genital pores (see Chapter 2), but in most vertebrates they are transported toward the entrance of the oviduct, called the **ostium tubae,** and released to the outside through the **oviduct,** which develops from the embryonic Müllerian duct.

Study of the Excretory and Reproductive Systems

We assume that you observed the major parts of the excretory and reproductive systems during previous dissections. In the following exercises, the finer aspects of the systems will be examined and related to the more conspicuous parts. In studying these systems, you should not only dissect your own specimen, but you also should examine the dissection of a specimen of the opposite sex. Because someone else, in turn, will have to examine your specimen, make a particularly careful dissection. If possible, study sexually mature specimens.

[3]Notice the difference between the endings in "epididymis" and "ductus epididymidis" ("the duct of the epididymis").

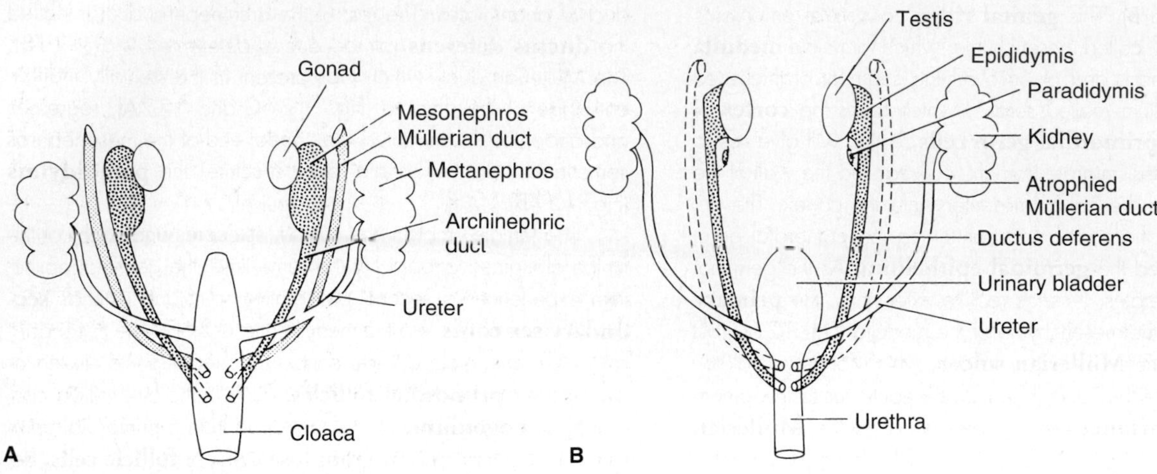

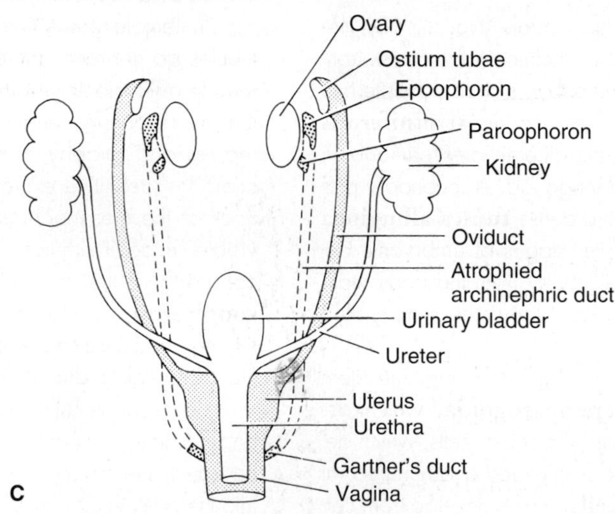

Figure 12-7

Diagrams illustrating the embryonic development of the genital ducts in amniotes. **A**, Sexually indifferent stage. **B**, Male condition. **C**, Female condition. *(After Balinsky, 1981.)*

FISHES

The excretory and reproductive systems of cartilaginous fishes are a good example of an ancestral vertebrate condition in most respects. The kidneys are derived opisthonephroi drained by archinephric ducts, which are supplemented by accessory urinary ducts in males. The gonads are situated far cranial in the pleuroperitoneal cavity. Eggs are discharged into the coelom and through a pair of oviducts; sperm are discharged through the cranial part of the kidneys and a pair of archinephric ducts. A cloaca is present.

The nephrons of cartilaginous fishes have large glomeruli that remove a considerable volume of water from the blood by filtration. This is an advantage in a freshwater, hypotonic environment, which was probably the environment of most of the early vertebrates and may explain the widespread distribution of glomerular nephrons among vertebrates. However, a glomerular nephron poses problems for most marine fishes, which must conserve body water. Because the concentration of osmotically active solutes is greater in sea water than in the body fluids of marine fishes, water tends to diffuse out of the fishes' bodies. Cartilaginous fishes mitigate this problem partly by retaining many nitrogenous molecules, especially urea, in their body fluids, and by having evolved tissues with a greater tolerance for urea. Urea is excreted in the kidneys and gills, but a special transport mechanism returns a considerable amount of urea to the blood (Wood et al., 1995). As a consequence, the osmolarity of the body fluids of elasmobranches is equal to or greater than that of sea water, so that water tends to diffuse into the body. Marine elasmobranch fishes produce a copious and dilute urine. Like other marine vertebrates, cartilaginous fishes also must eliminate excess ions. Divalent ions, such as phosphate and sulfate, are excreted by the nephrons in the kidneys; monovalent ions, such as chloride and sodium, are excreted by certain cells in the gills and by the digitiform gland (see Chapter 10).

Another specialized feature of many cartilaginous fishes, including *Squalus*, is that the females retain their young within their uterus until embryonic development is completed; hence, they are **viviparous.** Hagfishes, lampreys, ancestral sharks, and rays and skates are egg-laying, or **oviparous.** Oviparity probably is the ancestral condition in craniates.

The Kidneys and Their Ducts

The kidneys of the spiny dogfish are a pair of bandlike organs that lie against the dorsal wall of the pleuroperitoneal cavity on either side of the dorsal mesentery (see Chapter 10). Their position is called **retroperitoneal** because they do not project into the pleuroperitoneal cavity (Fig. 12-8). The conspicuous, white **caudal ligament** arises from the ventral side of the vertebral column between the kidneys and passes into the tail. The kidneys of the spiny dogfish are **derived opisthonephroi,** because they extend nearly the entire length of the pleuroperitoneal cavity. Some of the cranial nephrons have a relatively primitive structure with traces of microscopic coelomic funnels (Fig. 12-3**E**). In order to trace one of the kidneys cranially, you may have to cut and reflect the parietal serosa along the lateral border of the kidney. The cranial two thirds of each kidney is narrower and less conspicuous than the caudal one third. In males, this cranial part of the opisthonephros is part of the reproductive system and is drained by the archinephric duct. It is poorly developed in females. Most urine is produced in the caudal third of the opisthonephros, which is drained by the archinephric duct in females and by the accessory urinary ducts in males.

The archinephric duct can easily be seen in mature males; it is a large, highly convoluted tube that lies on the ventral surface of the opisthonephros. It is much smaller

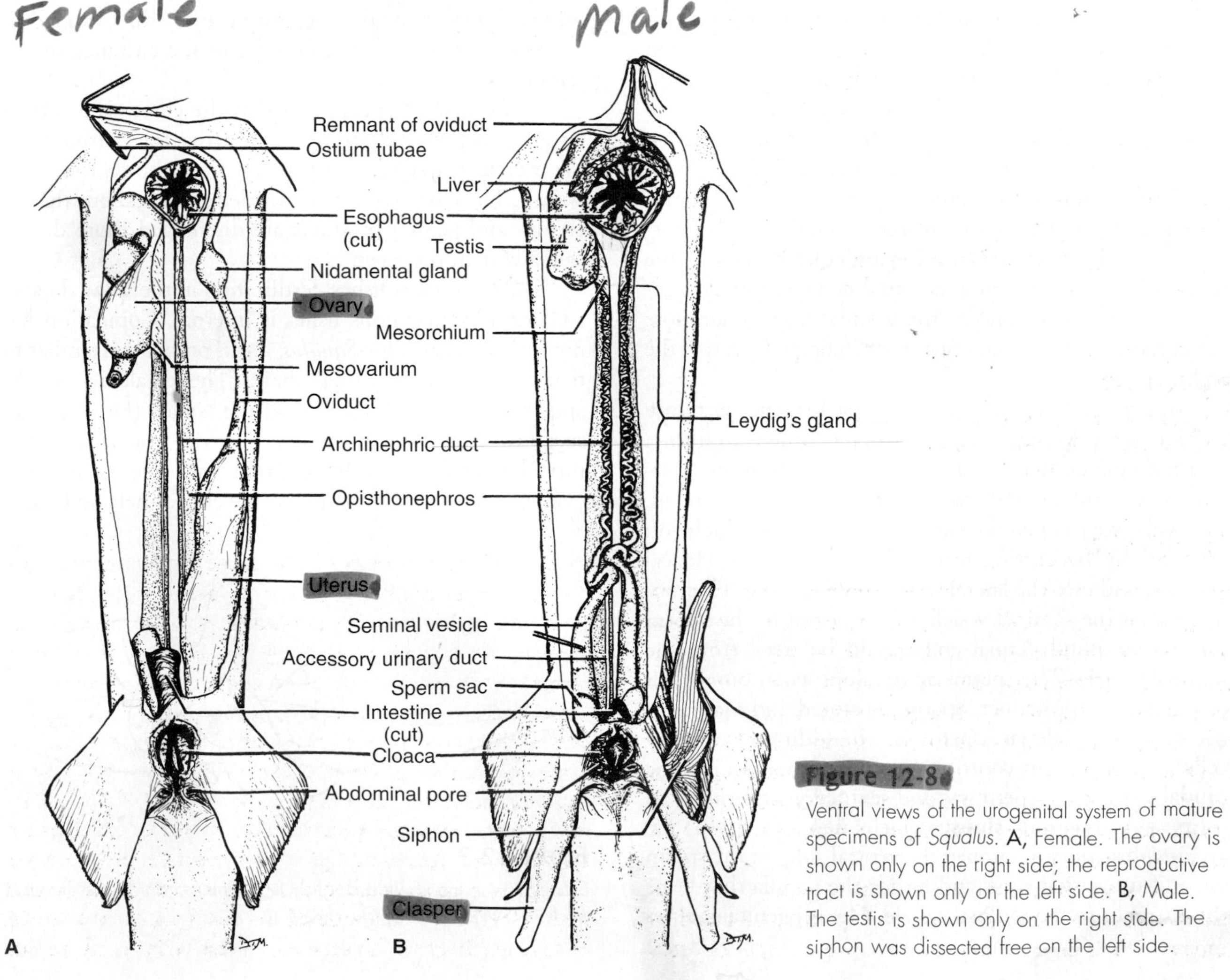

Figure 12-8
Ventral views of the urogenital system of mature specimens of *Squalus*. **A**, Female. The ovary is shown only on the right side; the reproductive tract is shown only on the left side. **B**, Male. The testis is shown only on the right side. The siphon was dissected free on the left side.

and straighter in immature males. The archinephric duct of females resembles that of immature males and cannot be seen until the oviduct is studied.

Further aspects of the urogenital system must be considered separately in each sex.

The Male Urogenital System

Notice that the paired **testes** are located near the cranial end of the pleuroperitoneal cavity adjacent to the cranial end of the kidneys (Fig. 12-8**B**). They are particularly large in specimens caught in January and February during their mating season. Each testis is suspended by a **mesorchium,** through the front of which pass several small, inconspicuous tubules, called the **ductuli efferentes.** The ductuli efferentes carry the sperm from the testis to modified cranial renal tubules. After the sperm pass through these renal tubules, they enter the **archinephric duct,** which lead them to the cloaca. The large size of the archinephric duct in mature males is attributed to its role in sperm transport.

The part of the opisthonephros that receives the ductuli efferentes is homologous to the head of the epididymis of amniotes; the adjacent, highly coiled portion of the archinephric duct is homologous to the ductus epididymidis of amniotes, which comprises the body and tail of the epididymis; and the rest of the archinephric duct is comparable to the ductus deferens of amniotes.

The portion of the opisthonephros between the caudal end of the testis and the enlarged caudal excretory portion of the kidney is known as **Leydig's gland.** Most of the renal tubules in this region are modified to produce a secretion that is analogous to the seminal fluid of amniote vertebrates. This secretion is discharged into the archinephric duct.

As the archinephric duct approaches the excretory portion of the kidney, it straightens and widens to form a **seminal vesicle.** Remove the parietal serosa from this portion of the kidney and trace the seminal vesicle caudad. You will have to open the cloaca on the side on which you are working by cutting through the side of the cloacal aperture and into the lateral wall of the intestine. The caudal end of the seminal vesicle passes dorsal to the **sperm sac,** whose blind cranial end should be freed from the seminal vesicle. The sperm sac develops as an outgrowth of the archinephric duct. Sperm are stored and mature in the seminal vesicle and sperm sac. In addition, secretory cells in these organs contribute to the seminal fluid. The caudal ends of the sperm sac and seminal vesicle merge to form a **urogenital sinus,** which develops from the archinephric duct. Cut open the ventral sides of the sperm sac and urogenital sinus and notice the papilla that bears the entrance to the seminal vesicle. The urogenital sinuses of opposite sides merge caudal to the entrances to the seminal vesicles and extend into the **urogenital papilla** located dorsally in the cloaca. Probe caudally through the urogenital sinus and notice the emergence of the probe through the tip of the urogenital papilla.

Oviducts are present in the sexually indifferent stage of a male embryo. Remnants of them remain in adult males as tubular folds that can be found cranially on either side of the falciform and coronary ligaments. These cranial portions of the oviducts unite in the falciform ligament and have a common entrance, the **ostium tubae,** into the pleuroperitoneal cavity. The ostium tubae can be found along the caudodorsal edge of the falciform ligament. The caudal portions of the oviducts are dismantled.

The renal tubules of the excretory part of the kidney do not enter the seminal vesicle but rather enter an **accessory urinary duct,** which lies against the dorsomedial edge of the seminal vesicle. You can find this duct by freeing the lateral edge of the seminal vesicle and dissecting dorsal to the seminal vesicle. Despite its name, the accessory urinary duct carries virtually all the urine, because the cranial parts of the kidney excrete little, if any, urine in mature males. Trace the accessory urinary duct caudad; it enters the urogenital sinus caudal to the entrance of the seminal vesicle.

You can now understand that the cloaca is a sort of sewer: it receives feces from the digestive tract, urine from the excretory system, and gametes from the reproductive system. The cloaca is not subdivided in males, but the excretory and genital products are discharged more dorsocaudally than the feces.

Unlike in most fishes, fertilization in the spiny dogfish and other cartilaginous fishes is internal. Copulation has not been described for *Squalus,* but it probably is similar to that observed in other small sharks. The female lies quietly, and the male coils itself around the female (Fig. 12-9) in such a way that one of the male's pelvic claspers, which is turned cranially, can be inserted into the cloaca and oviduct of the female. Spurs on the clasper help to hold it

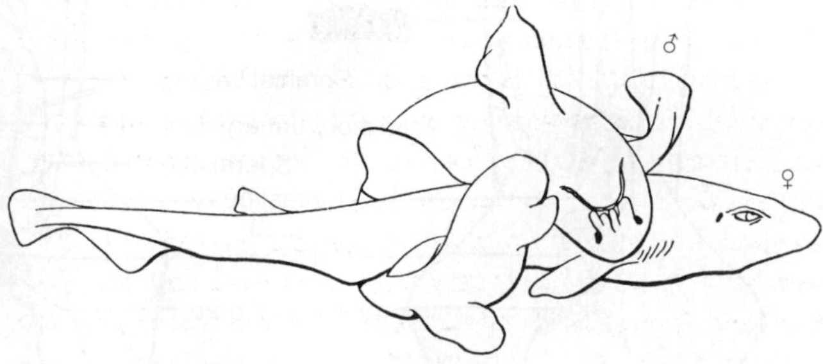

Figure 12-9

Copulation in the spotted dogfish *Scyliorhinus canicularis.* (From Budker, 1971, after Bolau, 1881.)

in place. Sperm pass from the male's cloaca into the groove on the dorsal surface of the clasper. A sac-like, muscular **siphon** is associated with each clasper. If you have not already done so, you can find one siphon by removing the skin from the ventral surface of a pelvic fin. Cut open the siphon and pass a probe through it and into the groove on the clasper. Prior to copulation, sea water is taken into the siphon. When the siphon contracts, sea water and secretions of the siphon are discharged and propel the sperm along the clasper groove and into the cloaca and oviduct of the female. The secretions of the siphon contain 5-hydroxytryptamine, which has been shown *in vitro* to induce contractions of the oviduct. It has been suggested that these contractions help to move sperm through the oviduct to the site of fertilization near the cranial end of the oviduct (for a review of chondrichthyan reproduction, see Wourms, 1977; Shuttleworth, 1988).

The Female Urogenital System

The **ovaries** are a pair of large organs located near the cranial end of the pleuroperitoneal cavity adjacent to the cranial ends of the kidneys (Fig. 12-8**A**). Each is anchored to the body wall by a **mesovarium** and contains eggs within follicles in various stages of maturity. When the eggs are mature, they attain a diameter of nearly 3 cm and contain an enormous amount of yolk. Cut into one of the larger eggs to see the yolk. Each egg is surrounded by a sheath of follicular cells, but this cannot be seen without magnification.

When the eggs are mature, they break through the wall of the surrounding **follicle** and the serosa enveloping the ovary in a process called **ovulation.** They fall into the pleuroperitoneal cavity, and are transported toward the unpaired ostium tubae and into the paired oviducts, probably by the pressure of the surrounding viscera. In mature females, each **oviduct** is a prominent tube that is anchored by a **mesotubarium** to the ventral surface of the kidney. In immature females, the oviducts are small tubes that lie against the kidneys; mesotubaria are lacking. Trace one oviduct cranially. It passes dorsally to the ovary and then curves ventrally and caudally in front of the liver to enter the falciform ligament. The oviducts of opposite sides merge within the falciform ligament and have a common opening, the **ostium tubae,** into the pleuroperitoneal cavity. The ostium tubae is located on the caudodorsal edge of the falciform ligament; you can open the ostium tubae by spreading its lips apart. It is inconspicuous and small in immature females but is very large in mature females. The falciform ligament of mature females is elongated accordingly and tears easily into shreds.

The oviduct is narrow throughout much of its length but widens in two regions. One widening lies dorsal to the ovary. This is the **nidamental gland,** or **shell gland,** where fertilization occurs. In the spiny dogfish, the nidamental gland secretes a thin proteinaceous shell, composed primarily of collagen and known as the "candle," around groups of two or three fertilized eggs. Among elasmobranchs, however, the structure and composition of egg shells vary greatly. Sperm that will fertilize the eggs are stored in the nidamental gland after copulation and may remain viable for several weeks or longer. In mature females, the caudal one third to one half of the oviduct is enlarged as the **uterus.** In pregnant females, the uterus is expanded and its wall is highly vascularized because the embryos develop in it.

Open the cloaca by cutting through the side of the cloacal aperture and into the lateral wall of the intestine. The paired oviducts enter the caudodorsal part of the cloaca just ventral to the **urinary papilla.** The urogenital portion of the cloaca, known as the **urodaeum,** is partially separated by a horizontal fold from the cranioventral fecal portion of the cloaca, called the **coprodaeum.**

Archinephric ducts drain the kidneys of female *Squalus* because they are not associated with the reproductive system as they are in males. Females of some other elasmobranchs, however, have accessory urinary ducts. You can find an archinephric duct by making an incision through the parietal serosa along the lateral border of the caudal portion of the kidney and very carefully reflecting the parietal serosa from the surface of the kidney. The archinephric duct lies on the ventral surface of the kidney directly dorsal to the attachment of the mesotubarium. If you do not see it on the kidney, it probably adheres to the inner surface of the reflected parietal serosa and can be picked off. The archinephric duct of females is much smaller than that of males and is not convoluted. Trace it caudad. The caudal ends of the archinephric ducts of opposite sides widen slightly and merge to form a small **urinary sinus,** which opens through the tip of the urinary papilla. The urinary sinus is too small to be easily dissected.

Reproduction and Embryos

As the eggs, or ova, develop within the follicles in the ovary, some of them begin to increase in size by accumulation of yolk. At the time of ovulation in the early winter of every second year, each ovary contains two or three ova averaging 3 cm in diameter. After ovulation, the follicular cells are converted into a **corpus luteum,** which measures 2 cm in diameter early in pregnancy, but it gradually regresses. The eggs enter the oviduct, which can stretch greatly in a living specimen. They are fertilized in the nidamental gland, which secretes a thin shell enveloping two or three fertilized eggs. The eggs then pass to the uterus, in which they may be seen in specimens in an early stage of pregnancy.

After several months, the shell is reabsorbed, and the embryos develop within the uterus (Anatomy in Action 12-1). The gestation period of *Squalus* lasts 20 to 22 months. Pups that are slightly over one year old, which is a stage often seen in pregnant specimens obtained from biological supply houses, range in length from 12 to 20 cm. Much of the yolk is contained within an **external yolk sac** that protrudes from the ventral side of an embryo. This is a **trilaminar yolk sac** (Fig. 12-12**A**), which contains all three germ layers. The rest of the yolk is carried in an **internal yolk sac,** which can be found within the pleuroperitoneal cavity (Fig. 12-10). Just before birth, pups range in length from 23 to 29 cm. The yolk in the external sac has been consumed by then, but a small reserve of yolk remains in the internal sac. Parturition occurs late in the autumn of the second year.

AMPHIBIANS

The earliest tetrapods, the amphibians, are still anamniotes, and their excretory and reproductive systems have not changed significantly in gross morphology from the condition of these systems in ancestral fishes. The only new feature is a relatively large urinary bladder formed as a ventral outgrowth of the cloaca. In many other respects, the urogenital system in amphibians is even more similar to what is believed to have been the ancestral vertebrate condition than it is in the spiny dogfish. Although most amphibians live in moist habitats on land, they are very much at home in freshwater, which is believed to have been the environment in which the ancestral vertebrates originated. The nephrons in the opisthonephric kidneys of amphibians have large glomeruli and, therefore, eliminate excess water as well as nitrogenous waste products. Amphibians also retain the ancestral vertebrate mode of reproduction. They are

Anatomy in Action 12-1 Reproductive Patterns in Sharks

Many sharks and all rays and skates are **oviparous.** Their eggs, or ova, are heavily laden with yolk, and a well-developed collagenous egg case is secreted around each fertilized egg. The fertilized eggs are released into the surrounding sea water for incubation, and the embryos develop within their protective egg cases.

In sharks with a more derived reproductive mode, the fertilized eggs are enveloped by a much thinner egg case and are retained in the uterus, where development is completed. The young eventually are born as miniature adults. Such sharks are live-bearers, or **viviparous.** This was first observed in the fourth century B.C. by Aristotle (384–322 B.C.), who described that some sharks "reproduce like mammals." Nutritional arrangements for the embryos vary in different viviparous species, ranging from complete dependence for organic nutrients upon the stored yolk, through several types of placental analogues, to the transfer of most materials needed by the embryo from the maternal tissues through a yolk sac placenta. The term *ovoviviparity* has been used to describe cases of live-bearing without the presence of a well-developed placenta, but there is such a continuous range of dependence upon the mother for nutrients that it is hard to determine where ovoviviparity shifts to viviparity. Many authors speak instead of **aplacental** and **placental viviparity.**

Squalus is characterized by aplacental viviparity. Although a *Squalus* embryo derives its organic nutrients from its yolk sac, some exchanges of other substances occur by way of the numerous, vascular **uterine villi** that line the uterus and are applied to the surface of the vascularized yolk sac. Gas exchange and some transfer of water and minerals occur by this route early in development. Later, the uterus appears to be irrigated periodically with sea water. During development, there is a steady decrease in the organic content of the combined yolk sac and embryo, but there is a 78% gain in their combined weight because of the uptake of water and minerals. Early in development, the yolk is digested by endodermal cells lining the yolk sac, and the digested nutrients are absorbed into the vitelline blood circulation. Later, yolk is moved by ciliary action into the embryo's intestine, where it is digested and absorbed.

The uterine villi of *Squalus* are one type of placental analogue. Other placental analogues found in different elasmobranch species include the secretion by the mother of a uterine milk that is rich in organic content. This uterine milk is absorbed by blood vessels in the yolk sac or swallowed by the embryo. In some species, embryos acquire nutrients by eating yolk-rich eggs that are ovulated later. This phenomenon is called **oophagy.** *Mustelus canis* is an example of a small shark with placental viviparity, but even in this species, the embryos get their nutrients from yolk before a union between the wall of the yolk sac and the uterine lining has been established. As in mammals, there is considerable variation in the degree of intimacy of the union between extraembryonic and maternal tissues among placental sharks.

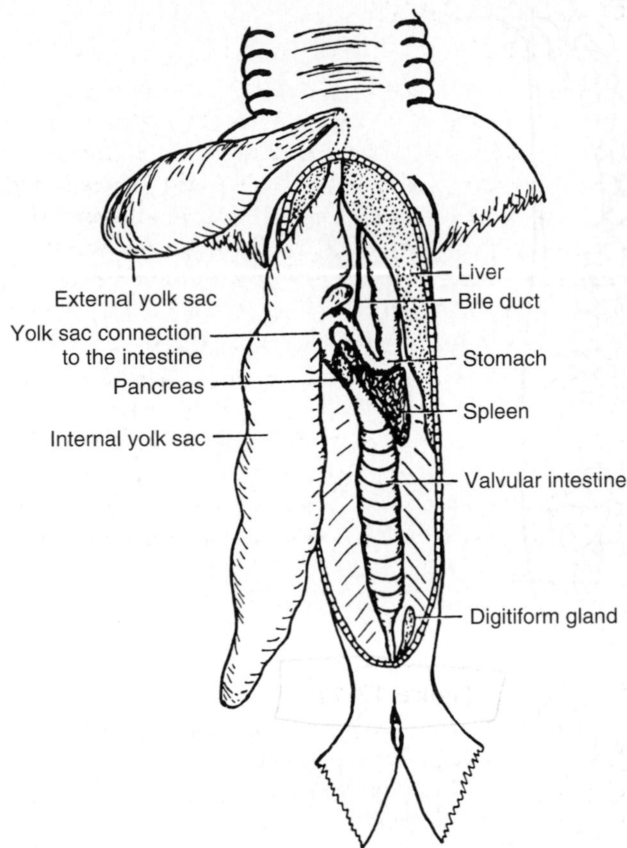

Figure 12-10
External and internal yolk sacs in a 220-mm embryo of *Squalus suckleyi*. The internal yolk sac has been pulled out of the opened body cavity. *(After Hoar and Randall, 1969-1997.)*

oviparous, laying their eggs in water or in very moist areas on land. In most cases, the eggs hatch into free-swimming larvae that later metamorphose into terrestrial adults. Although amphibians have made a start in adapting most of their organ systems to terrestrial conditions, they are generally restricted to moist habitats because of their limited ability to conserve water and reproduce under terrestrial conditions. *Necturus,* although paedomorphic and permanently aquatic, is a good model of this early level of tetrapod evolution.

The Kidneys and Their Ducts

The kidneys of *Necturus* are **derived opisthonephric.** They lie in the caudal half of the pleuroperitoneal cavity on either side of the dorsal mesentery (Fig. 12-11). They are easily seen in males, but you will have to push apart the ovary and oviduct to see a kidney in a female. The kidneys of most vertebrates have a retroperitoneal position, but in *Necturus,* they protrude into the pleuroperitoneal cavity and are nearly completely surrounded by visceral serosa. Also notice that the caudal part of the kidney is much larger than the cranial part. The cranial part of the male kidney functions as part of the reproductive system, whereas that of the female kidney is atrophied. By lifting the lateral edge of the kidney and looking at its dorsolateral surface, you can see the cranial part of the kidney most clearly.

The kidneys are drained in both sexes exclusively by the **archinephric ducts;** accessory urinary ducts are absent in *Necturus.* The archinephric duct of males is a large convoluted tube that extends down the lateral edge of the kidney to the cloaca. The archinephric duct of females is similarly located but is much smaller and is not convoluted. Small **collecting tubules** may be seen entering the archinephric duct from the caudal excretory portion of the kidney.

A **urinary bladder** lies ventral to the large intestine and is connected to the midventral body wall by a mesentery known as the **median ligament of the bladder.** The urinary bladder is not directly connected to the archinephric duct but is a cranioventral evagination of the cloaca. As the urine is discharged from the archinephric ducts into the cloaca, it collects in the urinary bladder, where it is concentrated through selective reabsorption of water by the epithelium lining the urinary bladder. The urinary bladder of amphibians, unlike that of mammals, is not simply a storage organ but an organ with a crucial role in osmoregulation.

The **adrenal glands,** also called suprarenal glands, of amphibians consist of cortical and medullary cells clustered together in irregular patches. In well-injected specimens, you will see a number of small, well-vascularized patches of adrenal tissue along the ventral surface of the kidney of *Necturus.*

The Male Urogenital System

The **testes** are a pair of oval organs located near the cranial ends of the kidneys (Fig. 12-11**A**). Each is suspended by a **mesorchium.** Several inconspicuous **ductuli efferentes** pass through the cranial part of the mesorchium and carry sperm from the testes to the modified renal tubules of the cranial portion of the opisthonephros. From here, the sperm pass to the cloaca through the **archinephric duct,** which serves as a sperm duct as well as an excretory duct, although not simultaneously. Notice that the cranial part of the archinephric duct, which is comparable to the ductus epididymidis of amniotes, is much more convoluted than the rest of the archinephric duct. The small

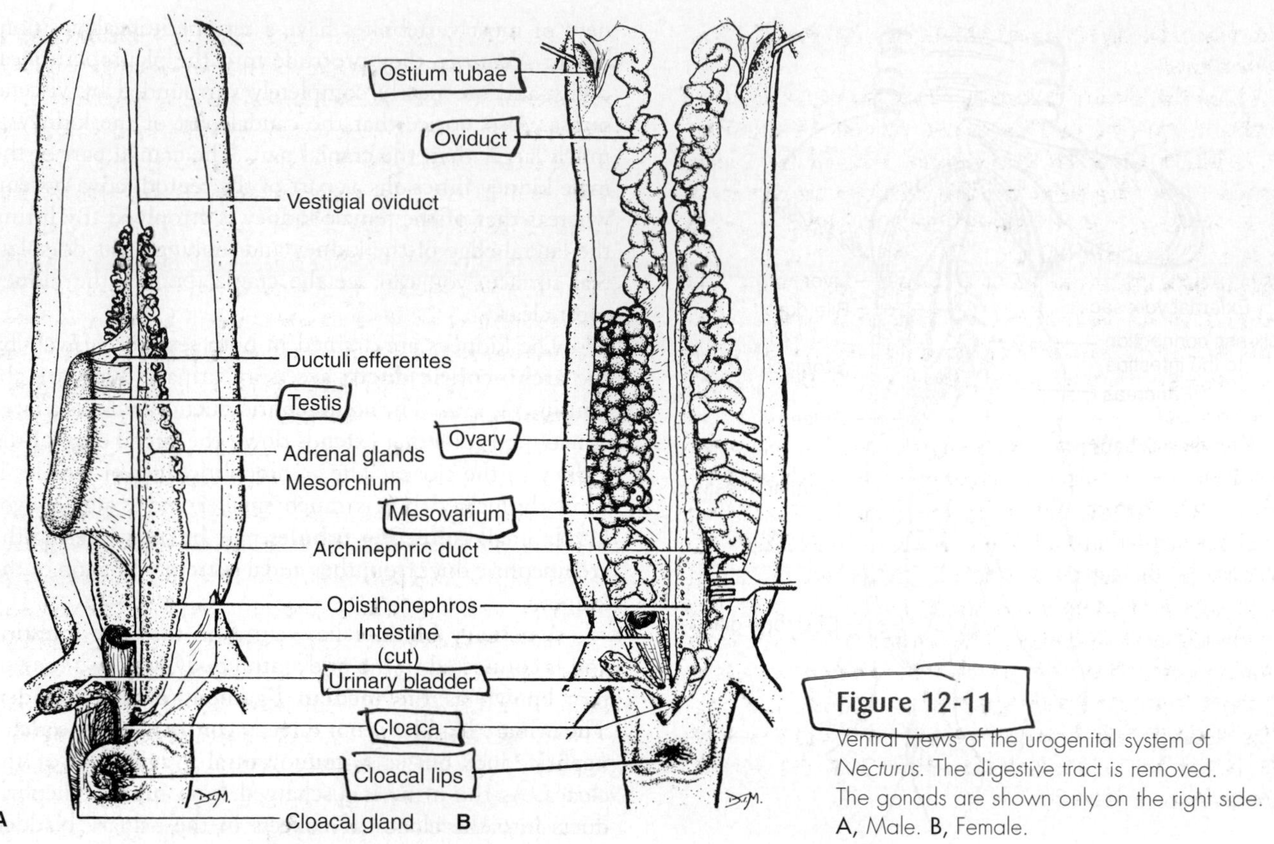

Figure 12-11
Ventral views of the urogenital system of *Necturus*. The digestive tract is removed. The gonads are shown only on the right side. **A**, Male. **B**, Female.

black line located along the edge of the cranial part of the archinephric duct and extending cranially along the side of the posterior cardinal vein is a remnant of the oviduct that has persisted from the sexually indifferent stage of the embryo. A shorter but similar remnant of the oviduct sometimes can be seen along the very caudal end of the archinephric duct.

Cut through the body musculature lateral to the cloaca and find the point at which the archinephric duct passes through the dorsal wall of the cloaca just caudal to the large intestine. Open the cloaca by making an incision that extends from the cranial end of the cloacal aperture into the lateral wall of the intestine. The archinephric ducts enter the craniodorsal wall of the cloaca just caudal to a transverse ridge formed by the entrance of the intestine, but their openings probably will not be seen.

Remove the skin on one side of the ventral surface of the cloaca, if you have not done so yet, and observe the large **cloacal gland,** which consists of many small tubules. The secretions of this gland, together with the secretions of the less conspicuous **pelvic gland** in the dorsal wall of the cloaca, agglutinate the sperm into clumps called **spermatophores.** The spermatophores of salamanders generally are deposited in the water to be picked up by the cloacal lips of a female, but in a few species they are transmitted directly into the cloaca of a female. Finally, notice the numerous papillae on the **cloacal lips,** which are a characteristic feature of males.

The Female Urogenital System

The **ovaries** are a pair of large granular organs located on either side of the dorsal mesentery adjacent to the cranial part of the kidneys (Fig. 12-11**B**). Each ovary is suspended by a **mesovarium** and contains eggs within follicles in various stages of maturity. The **oviducts** are a pair of large convoluted tubes that lie along the lateral side of each kidney and extend cranially nearly to the cranial end of the pleuroperitoneal cavity. Each oviduct terminates cranially in a funnel-shaped opening called the **ostium tubae** and caudally in the cloaca. Cut through the musculature lateral to the cloaca to see that the oviduct opens in the craniodorsal wall of the cloaca just caudal to the entrance of the large intestine. At ovulation, the eggs fall into the pleuroperitoneal cavity, are carried by the action of cilia on the serosa to the ostia tubarum, and descend through the oviducts to the cloaca. The oviducts are not differentiated grossly, but part of their lining is glandular. The glands of the oviduct secrete gelatinous layers around the eggs. These gelatinous layers are semipermeable. They imbibe water when the eggs are laid and form a lightweight protective capsule.

Open the cloaca by cutting from the cranial end of the cloacal aperture into the lateral wall of the intestine. The oviducts enter the craniodorsal wall of the cloaca through a pair of **genital papillae.** If these are not clear, slit an oviduct near its caudal end and pass a probe through it into the cloaca. The caudal end of each **archinephric duct** can be seen leaving the kidney and passing onto the wall of the oviduct, with which it becomes intimately connected. However, the archinephric duct enters the cloaca independently beside the opening of the oviduct.

The **cloacal gland,** which helps to produce the spermatophores in the male, is present in a much reduced state in the female. You can find it by removing the skin on one side of the ventral surface of the cloaca. The **spermatheca** is located in the dorsal wall of the cloaca but cannot be seen grossly. It corresponds to the pelvic gland, which you have already seen in a male. The tubules within the spermatheca store sperm after copulation. In *Necturus,* sperm are stored from the breeding season in the fall until egg-laying in the spring. Finally, notice that the lips of the female cloaca bear smooth folds rather than the papillae characteristic of the male.

MAMMALS

Early in amniote evolution, the evolution of vertebrates diverged into two phylogenetic lines (see Fig. 3-1), a sauropsid line leading through the reptiles to birds, and a synapsid line leading to mammals. Many similar morphological and physiological changes evolved independently in each line as amniotes became adapted to terrestrial life and became more active animals. Birds and mammals became warm-blooded animals with a high level of metabolism. As a consequence of their elevated metabolism, mammals must remove a far larger volume of nitrogenous wastes from their body fluids than other vertebrates, except birds; at the same time, they must conserve body salts and water like all other tetrapods. This is made possible by significant increases in the number of nephrons, in the glomerular filtration rate, and in the amount of tubular reabsorption. Whereas each kidney of a urodele amphibian contains approximately 50 nephrons in its caudal excretory portion, a small mammal, such as a mouse, has approximately 20,000 nephrons per kidney, and a human being has an estimated one to four million nephrons per kidney. Mammals produce an abundant but very concentrated urine. Mammals and, to some extent, birds are the only vertebrates whose kidneys can form a hyperosmotic urine, that is, a urine that is more concentrated than their body fluids.

The bean-shaped kidney of mammals is a **metanephros** drained by a **ureter.** What is left of the embryonic mesonephros and archinephric duct becomes part of the male reproductive, or genital, system. Thus, there is a more complete separation of the excretory and reproductive functions in amniotes than in anamniotes. In most mammals, except monotremes, a division of the cloaca separates the urogenital tract from the digestive tract.

Unlike fishes and amphibians, but like reptiles and birds, mammals do not need to return to an aquatic environment to reproduce. Mating on land requires internal fertilization; therefore, males require a mechanism of direct sperm transfer to the female, usually by a **copulatory organ,** and **accessory genital glands,** which secrete the seminal fluid in which sperm cells are carried. Analogous organs occur only in a few anamniotes, in which internal fertilization has also evolved.

Terrestrial reproduction makes an aquatic larval stage unnecessary. Most non-mammalian amniotes remain oviparous, but their eggs are **cleidoic** (Fig. 12-12). A cleidoic egg is supplied with a large store of yolk, which eventually becomes enclosed in a **yolk sac.** As an egg descends the oviduct, **albumen,** or similar secretions that meet various metabolic needs, and a protective **shell** are deposited around the egg by the oviduct. The embryo itself develops **extraembryonic membranes.** A protective **chorion** and a fluid-filled **amnion,** which provides a local aquatic environment for the embryo, evolve from ectodermal and somatic mesodermal layers that surround the yolk (Fig. 12-12**B**). The yolk sac of amniotes is, therefore, **bilaminar,** with a wall consisting of just visceral mesoderm and endoderm (Fig. 12-12**C** and **D**). Finally, an **allantois** evolves as an outgrowth from the urinary bladder of amphibians. Embryonic blood vessels are carried by the allantois to the chorion, which adheres to the porous shell for gas exchange with the environment. Excretory products, chiefly in the form of inert uric acid, accumulate within the allantois.

Prototherian mammals lay eggs that are similar to those of reptiles, but in therian mammals (marsupial and placental mammals), the eggs (without shell, albumen, and most of the yolk) are retained in the female reproductive tract. In connection with the evolution of viviparity in therian mammals, a uterus and placenta have evolved. The **uterus** is a specialization of the oviduct in which embryos develop. The **placenta,** which is formed by the combination of extraembryonic and maternal uterine tissues, allows the transfer of nutrients and oxygen from the mother to the embryo and the return of waste products from the embryo to the mother. In placental mammals, an immunological reaction of the maternal tissues against the tissues of the embryo, which are partly foreign to the maternal tissues, is prevented by the **trophoblast,** an epithelial layer that surrounds the early embryo, called a **blastocyst.** This trophoblast later becomes part of the chorion and acts as an immunological barrier between the mother and embryo, thereby allowing the typically long gestation periods in placental mammals. Marsupials do not have a trophoblast. Their embryos are born at a very

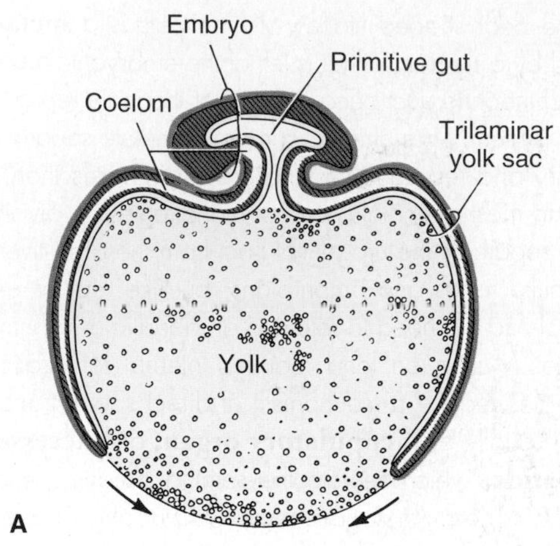

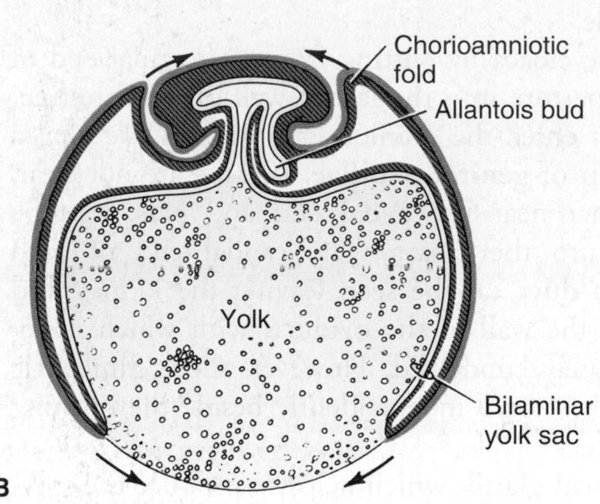

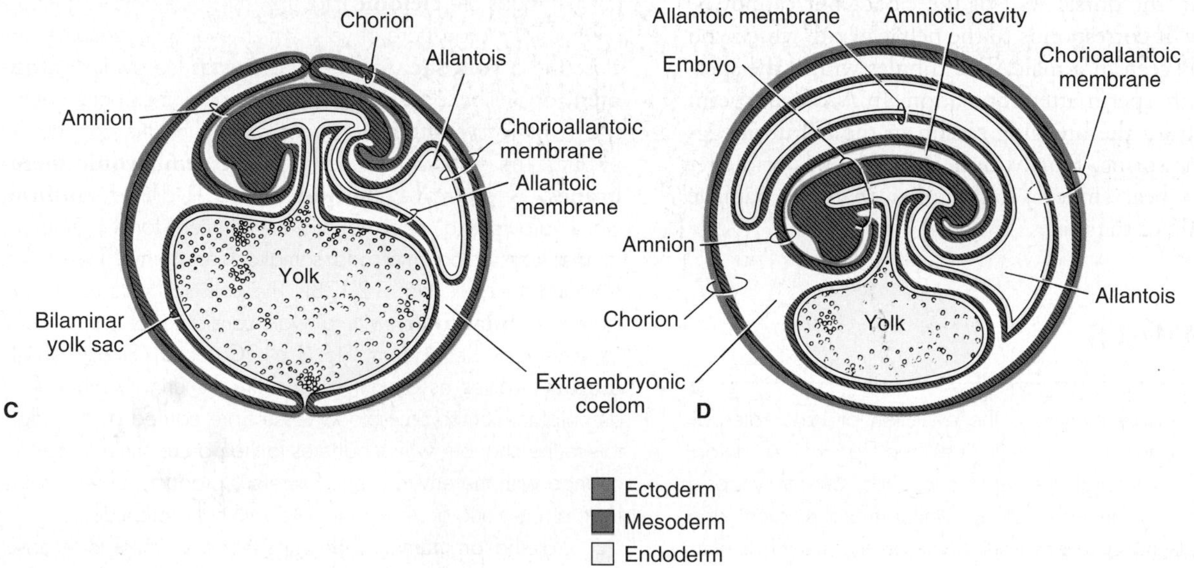

Figure 12-12

Diagrammatic sections through vertebrate embryos and their extraembryonic membranes. The head of the embryo points to the left side of the drawings. A, Developing trilaminar yolk sac of a large-yolked fish embryo. B, Hypothetical derivation of the chorioamniotic folds of an amniote embryo from the superficial layers of the trilaminar yolk sac. C, Extraembryonic membranes of an early amniote embryo. D, Extraembryonic membranes of a later amniote embryo. *(From Villee, Walker, and Barnes, 1984.)*

early stage of development, before the immunological reaction of the maternal tissue has been fully activated, and complete their development in the marsupial pouch of the mother.

THE EXCRETORY SYSTEM

The **kidneys** *(renes)* of mammals, which are **metanephroi,** lie against the dorsal wall of the peritoneal cavity in a retroperitoneal position (see Chapter 10; Figs. 12-13 and 12-15). Each kidney is surrounded by a mass of fat, called the **adipose capsule,** which should be removed. The adipose capsule consists of a special type of fat tissue, called **structural fat.** It serves as a protective shield against mechanical injuries of the kidneys. Unlike the usual fat tissue, it is not used as an energy reserve for the body. Each kidney is closely invested by a **fibrous capsule** (Fig. 12-14**A**). Notice that the mammalian kidney is bean shaped. The indentation on its medial border is called the **hilus** (Fig. 12-14**A**). Carefully remove connective tissue from the hilus, and you will see the **renal artery** and **vein** entering and leaving the kidney.

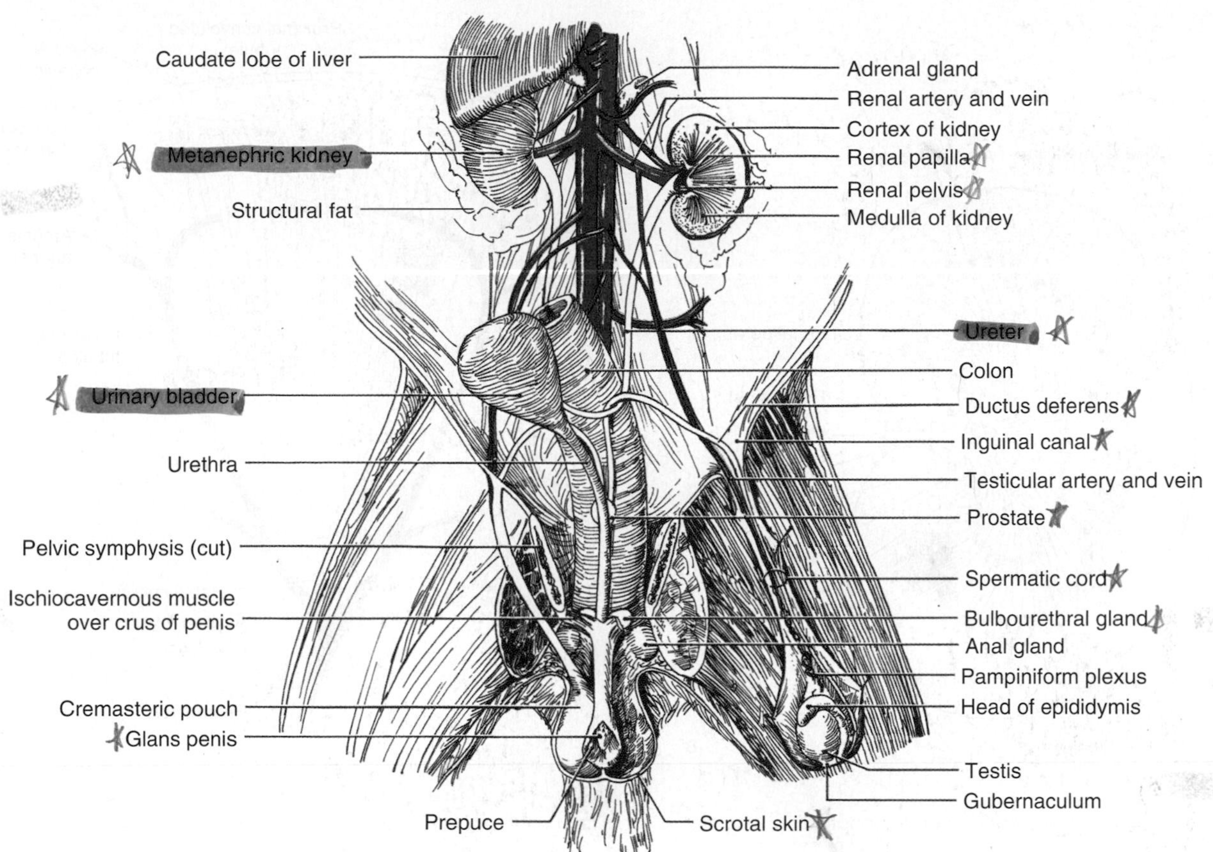

Figure 12-13
Ventral view of the urogenital system of a male cat. The left kidney is sectioned to show its internal structure. The pelvic canal is cut open, and the left cremasteric pouch is dissected, removed from the scrotum, and pinned to the inner surface of the left thigh. The left ischiocavernous muscle is bisected.

Caudal to the renal artery and vein, you will see the **ureter** emerge from the hilus. It drains the kidney.

Trace one of the ureters. It extends caudad and retroperitoneally and then turns ventrally into the lateral vesical ligament (see Chapter 10) to enter the caudal part of the **urinary bladder.** As the ureter enters the lateral vesical ligament, it passes dorsal to the ductus deferens in a male or horn of the uterus in a female (Figs. 12-13, 12-15, 12-18, 12-19, and 12-23). The urinary bladder itself is a pear-shaped organ with a broad, rounded, cranial end, called its **vertex,** and a caudal **body.** Cut open the urinary bladder, and you may be able to see the points of entrance of the ureters in its dorsal wall. Clean away connective tissue from around the urinary bladder, which gradually narrows caudally toward the entrance of the ureters and ends in a tube called the **urethra.** The urethra begins just caudal to the entrances of the ureters and enters the pelvic canal. Its distal parts will be considered with the reproductive organs.

Leave the kidneys in place. With a long-bladed knife, cut one of them longitudinally through the hilus in the frontal plane (Figs. 12-13, 12-14**A,** and 12-15). Study the half that includes the largest portion of the ureter. The hilus expands within the kidney into a chamber called the **renal sinus.** The renal hilus and renal sinus are occupied by the renal blood vessels, the proximal end of the ureter, and fat (Fig. 12-14**A**). In other words, if the blood vessels, nerves, ureter, and renal pelvis were removed from the kidney, the space left would be the renal sinus. Pick away the fat to expose the blood vessels and ureter. The portion of the ureter within the renal sinus is expanded and is called the **renal pelvis.**

Notice in your specimen that the substance of the kidney can be subdivided into two zones. The peripheral and paler **renal cortex** has a finely granular appearance; the central and darker **renal medulla** has a striated appearance (Fig. 12-14**A**). Some stripes extend from the renal medulla into the renal cortex as **medullary rays.** The

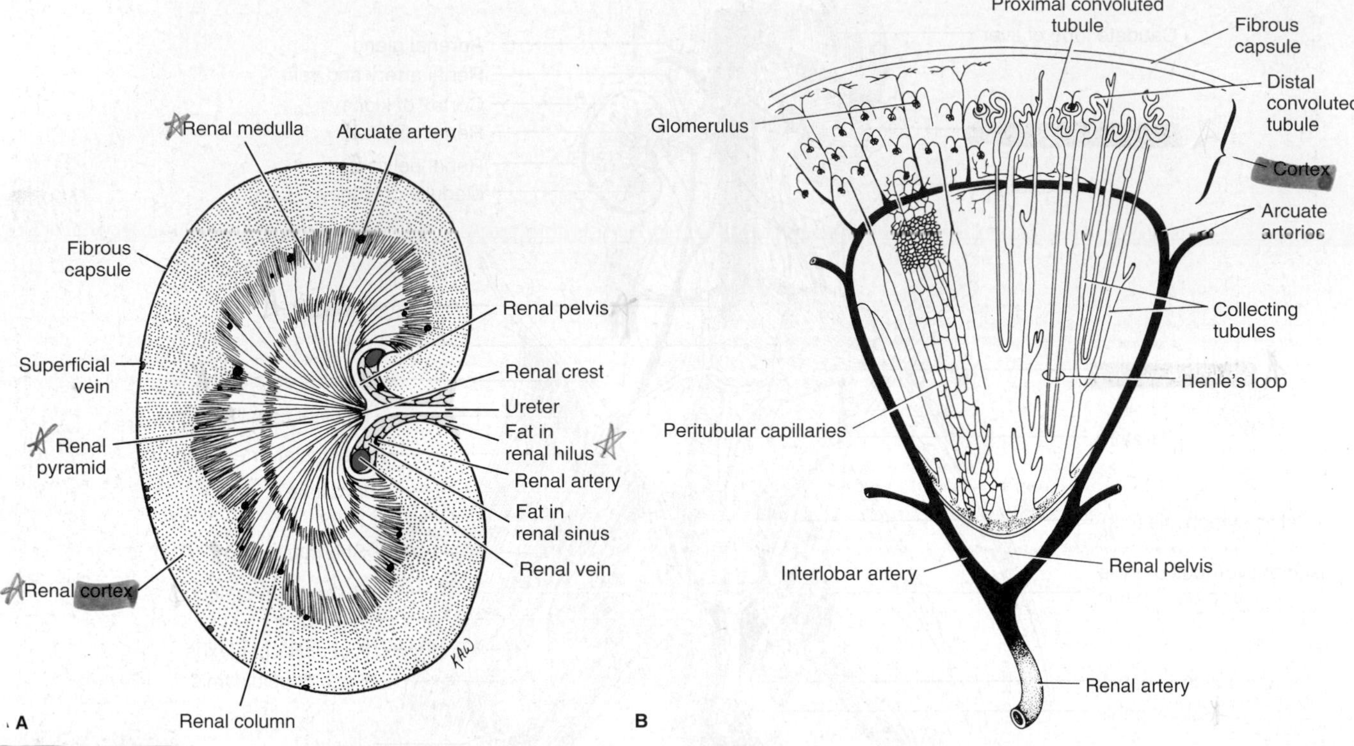

Figure 12-14
Internal structure of the mammalian kidney. **A,** Longitudinal section through a kidney of a cat at the level of the hilus. **B,** Diagram illustrating the arrangement of the blood vessels and nephrons within a kidney. On the left half of the drawing, the nephrons are omitted; on the right half of the drawing, the smaller body vessels are omitted. The veins are completely omitted from the drawing. *(B, Redrawn from Dyce, Sack, and Wensing, 1987.)*

medullary substance in the kidney of cats and rabbits is arranged in such a manner that it forms a single **renal pyramid** (Fig. 12-14). The apex of the renal pyramid is the **renal crest,** which projects into the renal pelvis. Notice that the boundary between the renal cortex and the renal medulla is scalloped because several **renal columns** project to some extent from the renal cortex into the renal medulla. Many mammals have more than one renal pyramid (Anatomy in Action 12-2).

The mammalian kidney can be divided into a renal cortex and a renal medulla, because the nephrons are structurally and functionally quite complex, and their parts are arranged in a particular pattern (Figs. 12-1**C** and 12-14**B**). The **renal corpuscle** and **proximal convoluted tubule** of each nephron lie in the renal cortex. The renal tubule then makes a long **loop of Henle,** which dips into the medulla. The descending limb and part of the ascending limb of this loop have very thin walls. The loop of Henle doubles back to the renal cortex, where it continues as the **distal convoluted tubule.** Several distal convoluted tubules unite to form **collecting tubules** that pass back through the renal medulla and discharge urine into the renal pelvis. This arrangement of the nephrons plays a crucial part in the reabsorption of water and is unique among vertebrate kidneys. Most water is reabsorbed in the loops of Henle and the collecting tubules.

The renal arteries branch into radially arranged **interlobar arteries** (Fig. 12-14**B**). These, in turn, branch into **arcuate arteries,** which run through the substance of the kidney along the interface between the renal cortex and renal medulla. After passing through the glomerulus and peritubular capillaries (Fig. 12-1**C**), the blood is collected by veins that run next to the arteries. The glomeruli sometimes can be seen in well-injected specimens under low magnification.

The Male Reproductive System

The testes of most mammals do not remain in the peritoneal cavity, where they develop embryonically; instead they migrate into paired saclike extensions of the peritoneal cavity, the **processus vaginales** (singular: *processus vaginalis*). The processus vaginales lie within a cutaneous pouch, the **scrotum,** which is located ventral to the anus. In cats, the scrotum is situated caudodorsally to the pelvic symphysis and penis (Figs. 12-13 and 12-18). In rabbits,

Anatomy in Action 12-2 Kidney Architecture in Mammals

Mammalian kidneys vary considerably in their architecture. Cats and rabbits have **unipyramidal kidneys** with a single large pyramid protruding into the renal pelvis (Fig. 12-14**A**). Humans have a **multipyramidal compact kidney** in which the medulla forms 7 to 14 distinct renal pyramids, each of which discharges urine through a conical **renal papilla** into a subdivision of the renal pelvis called a **renal calyx** (Fig. **A**). The entire kidney forms a compact organ. Cattle have a large, **multipyramidal, lobate, compact kidney** (Fig. **B**). Their kidney is subdivided by deep grooves into 25 to 30 lobes, each of which contains one or two renal pyramids and the associated renal cortex, and discharges urine into a single calyx.

Cetaceans, such as whales and porpoises, have **multipyramidal, lobate, loose kidneys;** each lobe is separated, or nearly so, from its neighbors. Cetacean kidneys are much larger than those of cattle, and they have a greater number of lobes, but the number of lobes varies considerably. It is lowest in fish-eating porpoises (about 150 in the black finless porpoise *Neomeris*) and highest in the baleen rorquals (about 3,000). Rorquals ingest large amounts of salts with the krill. Like most marine invertebrates, krill regulates the osmotic pressure of its tissues by adjusting it to the osmotic pressure of the surrounding water; hence, it contains the same salt concentration as sea water. Toothed whales and porpoises ingest less salt through their diet of fish. Teleost fishes, which constitute the overwhelming majority of fishes in the sea in terms of numbers of individuals and species, have a lower salt concentration than sea water. Cetaceans do not have salt-excreting glands, but their kidneys, unlike those of most mammals, can produce a urine with a salt concentration that is higher than that of sea water.

The reasons for the structural variation among mammalian kidneys are not entirely understood. However, the subdivision of the kidney, particularly into distinct lobes, accommodates an increased number of nephrons and makes possible a long medullary region with long loops of Henle and collecting ducts. The topographical configuration is analogous to the increased number of neurons that can be accommodated in a convoluted cerebral cortex (see Chapter 9). Kidneys of desert rodents have more massive and longer pyramids and can reabsorb far more water than the kidneys of related rodents living in habitats with a plentiful water supply.

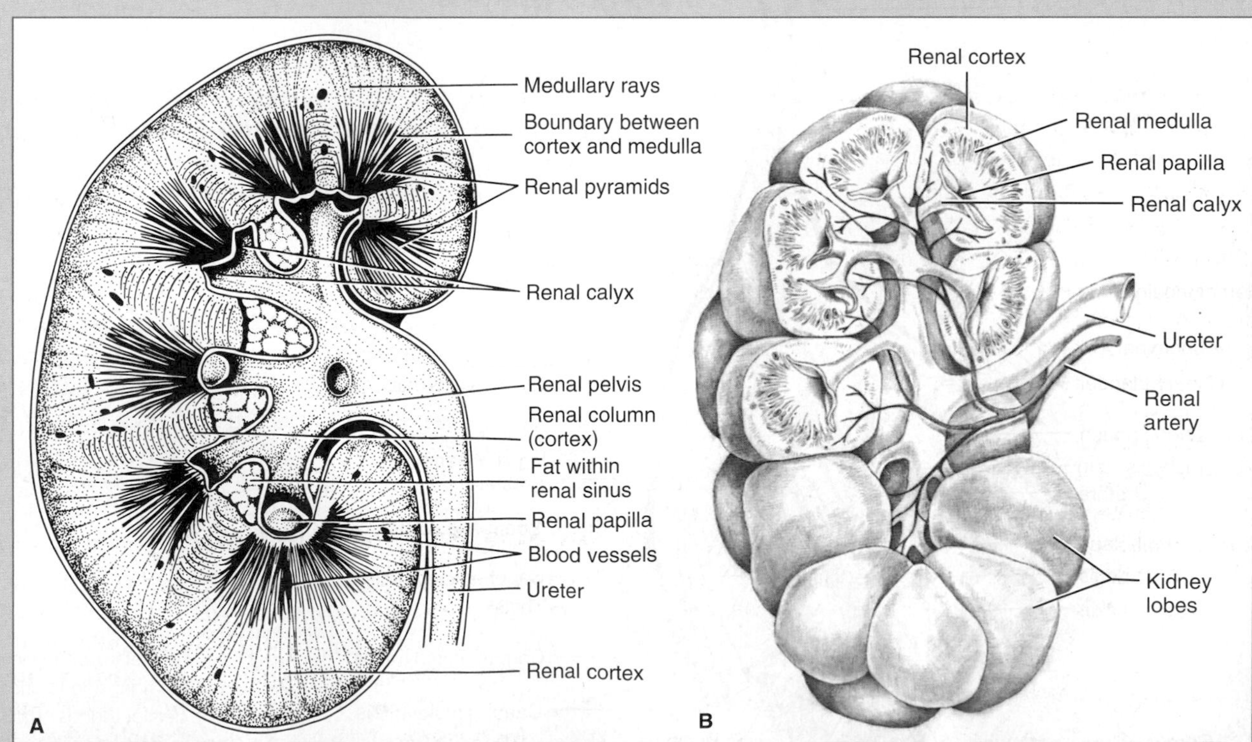

A, Human kidney. B, Kidney from an ox. *(B, After Dyce, Sack, and Wensing, 1987.)*

however, the scrotum lies ventral to the pelvic girdle and penis (Fig. 12-15). This position of the scrotum, incidentally, occurs only in lagomorphs among placental mammals, although a comparable position is seen in marsupials.

Very carefully remove the **scrotal skin** from the scrotum (Figs. 12-16 and 12-17**A**). A dense layer of connective tissue containing some smooth muscle fibers, the **dartos tunic,** is closely associated with the scrotal skin and will be removed with it (Fig. 12-16**A**). A pair of cordlike sacs, the **cremasteric pouches,** will be revealed. They extend cranially from the scrotum, across the ventral surface of the pelvic symphysis, to the abdominal wall. Within the scrotum, they are separated from each other by the **scrotal septum,** an extension of the dartos tunic. The proximal parts of the cremasteric pouches lie just beneath the abdominal skin and should have been seen and preserved when the pelvic canal was opened (see Chapter 11). The testes lie within the distal ends of the cremasteric pouches. In cats, this portion of each pouch is wide but the proximal part is a constricted tube, and the testes remain permanently in the scrotum. In rabbits, however, the entire cremasteric pouch is wide and of nearly uniform diameter, and the testes can be moved in and out of the scrotum.

The wall of the cremasteric pouch is composed of several layers that are continuous with various layers of the abdominal wall (Figs. 12-16**B** and 12-17**A**), but they cannot be distinguished with the naked eye. The outermost layer, the **external spermatic fascia,** is an extension of the external fascia of the trunk, or fascia trunci, which lies directly under the skin (see Chapter 7). The next layer is the **internal spermatic fascia,** which is an extension of the internal fascia of the trunk, or fascia transversalis, which lies directly under the parietal peritoneum. The internal and external spermatic fasciae are separated from each other by loose connective tissue and by the **cremasteric muscle,** which develops as a slip from the internal oblique and transversus abdominis muscles of the abdominal wall. In cats, the cremasteric muscle is atrophied and is replaced by

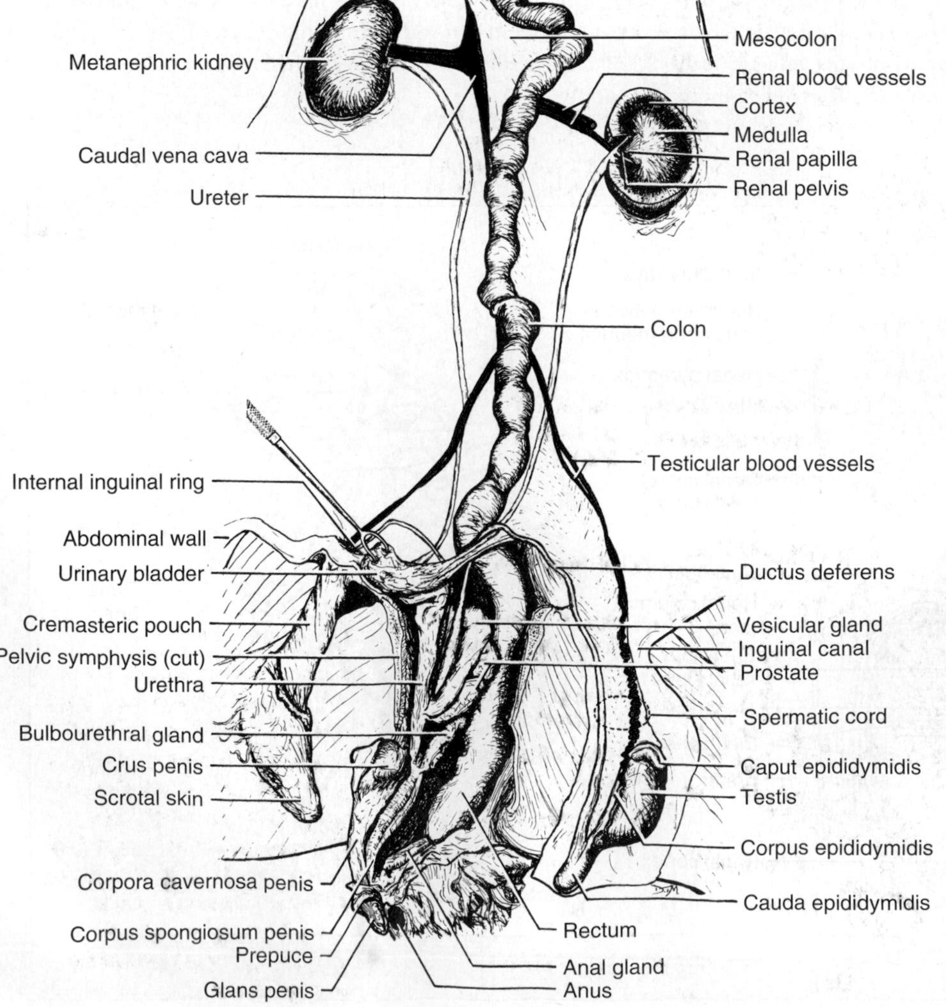

Figure 12-15
Ventral view of the urogenital system of a male rabbit. The left kidney is sectioned to show its internal structure. The pelvic canal was opened, and the urethra and penis are twisted to the right side to show the accessory genital glands. The right cremasteric pouch is shown intact; the left cremasteric pouch was opened.

the **cremasteric fascia.** In rabbits, in contrast, the cremasteric muscle is very large and serves to retract the testis into the peritoneal cavity. The three layers of the wall of the cremasteric pouch are difficult to identify in the small mammals being studied.

Leave the cremasteric pouch intact on one side, but cut open the other one along its ventral surface (Figs. 12-13 and 12-15). Extend the cut from the caudal end of the cremasteric pouch to the muscular abdominal wall, but do not cut through the abdominal wall. Notice that the cremasteric pouch contains a cavity, the **cavity of the processus vaginalis,** or **vaginal cavity** (Figs. 12-16 and 12-17**A**). Probe this cavity through the constricted **vaginal canal** into the peritoneal cavity (Fig. 12-17**B**).

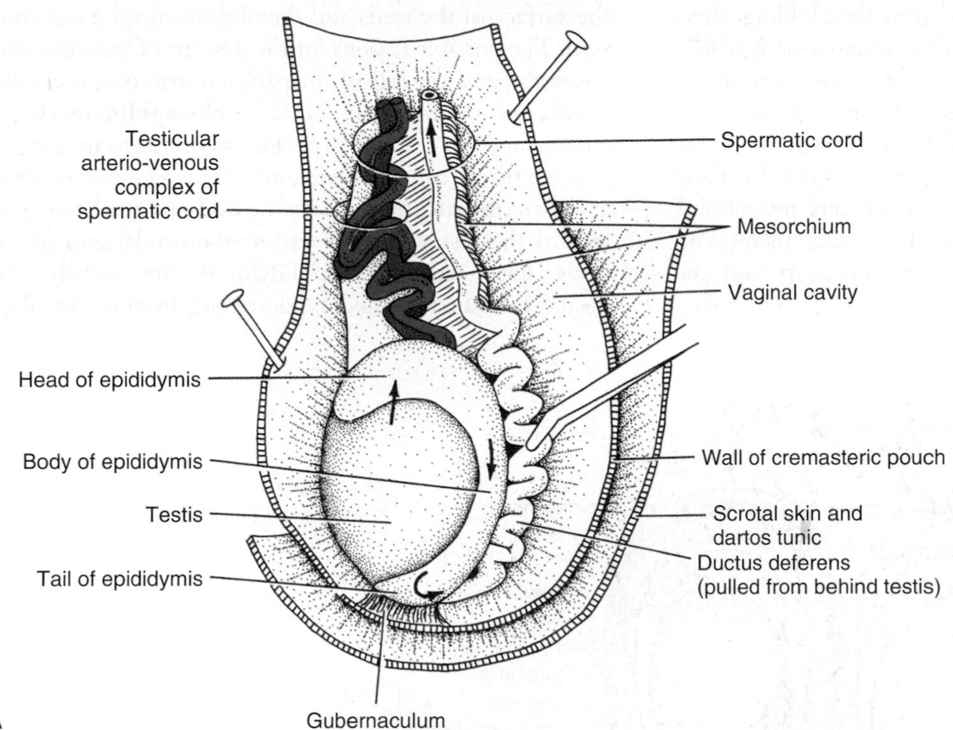

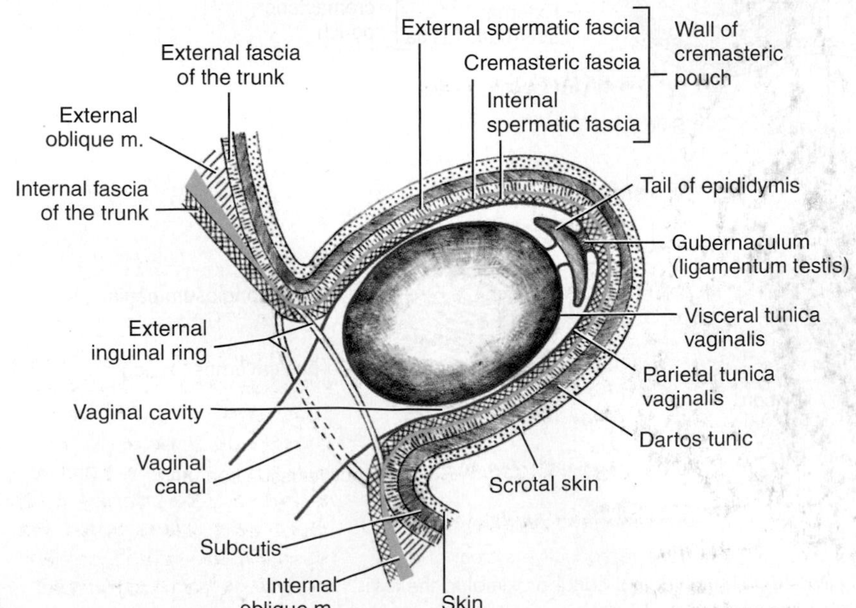

Figure 12-16

The scrotum and its contents in a male cat. **A,** Ventral view of the left testis and epididymis. The cremasteric pouch was cut open to reveal the testis; a piece of scrotal skin and dartos tunic was left attached. **B,** Semidiagrammatic longitudinal section through the scrotum to show the layers of the scrotum and cremasteric pouch and their relationships to the layers of the abdominal wall.

As already mentioned, the processus vaginalis is an outpocketing of the peritoneal cavity into which the testis descends during sexual maturation. Thus, the vaginal cavity is lined with serosa, called the **parietal tunica vaginalis,** which is an extension of the parietal peritoneum. The structures within the vaginal cavity are enveloped by the **visceral tunica vaginalis** and anchored to the dorsal wall of the cremasteric pouch through a mesentery, the **mesorchium** (Fig. 12-17**A** and **B**). The surfaces of the tunica vaginalis are lubricated and provide a gliding interface that allows the testis some limited movement within the vaginal cavity. A very short band of connective tissue, the **gubernaculum** *(Ligamentum testis),* attaches the testis, epididymis, and caudal end of the mesorchium to the distal end of the cremasteric pouch (Fig. 12-16). In many mammals, humans included, but not in cats and rabbits, the vaginal canal atrophies in adults so that there is no communication between the peritoneal cavity and the vaginal cavity.

The contents of the cremasteric pouch can now be examined in more detail. The **testis** is the relatively large, round body in cats (Figs. 12-13 and 12-16) or elongate body in rabbits (Fig. 12-15) lying in the caudal part of the cremasteric pouch. Testicular blood vessels and nerves attach to its cranial end. Notice that the testicular artery coils upon itself before it reaches the testis and is closely invested by a venous network, the **pampiniform plexus,** which receives blood from the venous network covering the surface of the testis and draining it into the testicular vein. The possible reasons for the descent of the testis into a scrotum and the role of the pampiniform plexus are discussed in Anatomy in Action 12-3. The **epididymis** is the band-shaped structure that is closely applied to the surface of the testis. It can be divided into three regions: the **head** at the cranial end of the testis, the **body** on the lateral surface of the testis, and the **tail** at the caudal end of the testis. The head of the epididymis is functionally connected with the testis and has developed from modified re-

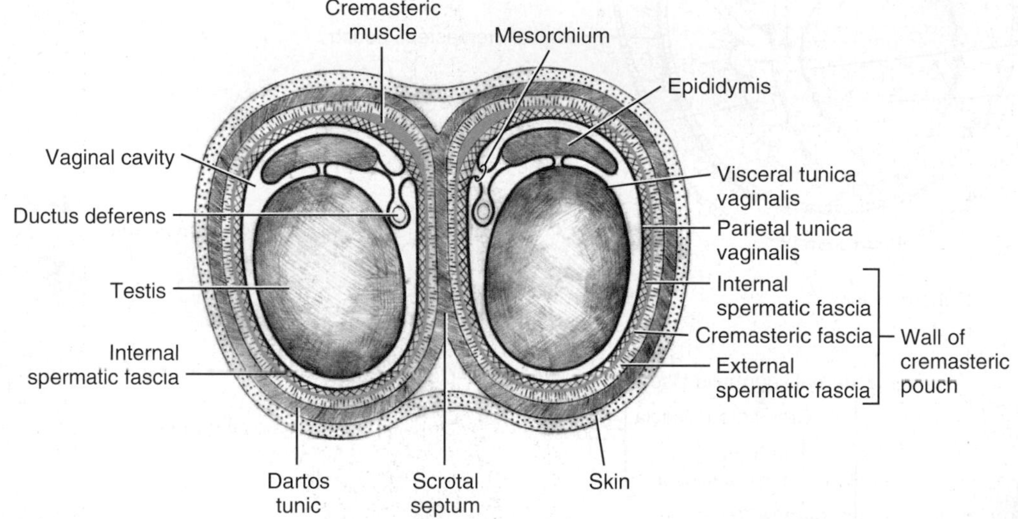

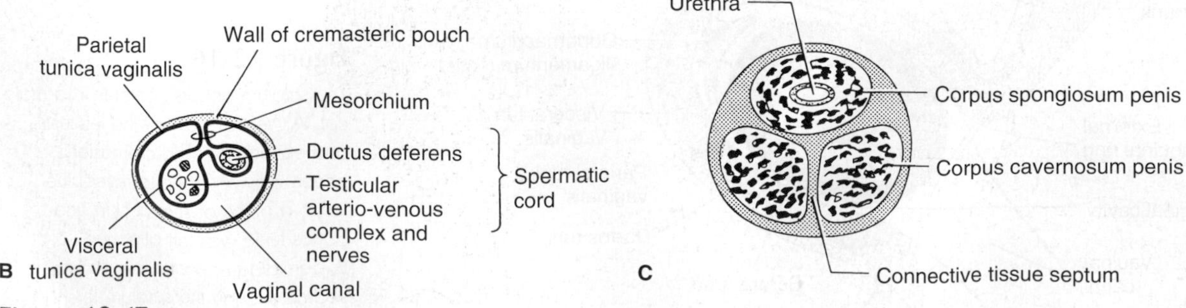

Figure 12-17

Semidiagrammatic, magnified transverse sections through parts of the reproductive system of a male cat. **A**, Section through the scrotum. **B**, Section through the vaginal canal proximal to the testis. **C**, Section through the mid-length of the penis. The skin was removed.

Anatomy in Action 12-3 The Descent of the Testis

As in all vertebrates, the testes of mammals start their development in the cranial part of the peritoneal cavity adjacent to the embryonic mesonephros. However, they do not remain in this position as they do in nonmammalian vertebrates; rather, they undergo a caudal migration or descent. In certain mammals, such as elephants and related groups, including manatees and hyraxes, the descent is only partial. The testes remain in the peritoneal cavity in adults, a position described as **abdominal.** In other mammals, such as insectivores, edentates, tapirs, rhinoceroses, seals, and cetaceans, the testes migrate out of the peritoneal cavity but remain close to the body wall beneath the skin of the groin, a position described as **inguinal.** In the vast majority of mammals, however, the testes migrate into a **scrotum,** where they remain either permanently (e.g., in cats and humans) or only during the reproductive season (e.g., in rabbits). In a phylogenetic analysis of the distribution of these three testis positions, Werdelin and Nilsonne (1999) hypothesized that the scrotal position is the most probable ancestral position for placental and marsupial mammals and that the inguinal and abdominal positions are derived.

The testes of mammals with a scrotum are highly sensitive to elevated temperatures. As clinical and experimental evidence indicates, the final stages of spermatogenesis during which spermatids are transformed into tailed spermatozoa proceed in a normal fashion only at temperatures that are lower than the core body temperature of most mammals. Males of species with a scrotum are sterile if their testes have not descended. If the temperature of the testes is experimentally elevated or if the testes are kept warm for extended times by warm clothing, spermatogenesis is not completed. The pampiniform plexus, which envelops the testis in some species, such as cattle (figure), and the entwined spermatic artery and vein leading to and from the testis in some other species, such as cats (Fig. 12-16), are heat-exchanging devices analogous to the retia mirabilia in the heads of some mammals (see Anatomy in Action 11-8). In both constructions, heat from the blood in the arteries entering the testis is transferred to the cooler blood in the veins returning from the periphery of the body. The release of heat from the scrotum is further facilitated by the sparse hair covering. It also can be regulated by the smooth musculature in the dartos tunic, which crinkles the skin of the scrotum, and by the cremasteric muscle, which can pull the testes closer to the body wall. It is interesting in this context to note, however, that in birds, which have, on average, an even higher temperature than mammals, the testes remain in an abdominal position.

Testes are very sensitive to pressure as Roland Frey (1990) pointed out. Clinical evidence indicates that tight and constricting clothing interferes with normal spermatogenesis. To ensure the production and maturation of sperm, the intratesticular pressure must be kept constant at about 2.3 mmHg, which requires fine control of blood supply and drainage to and from the testis. The pampiniform plexus serves as a reservoir that can be filled when the resistance within the intra-abdominal portion of the testicular vein increases and emptied when the intra-abdominal pressure is low. In addition, the testicular artery acts as a "peripheral arterial pump" because the

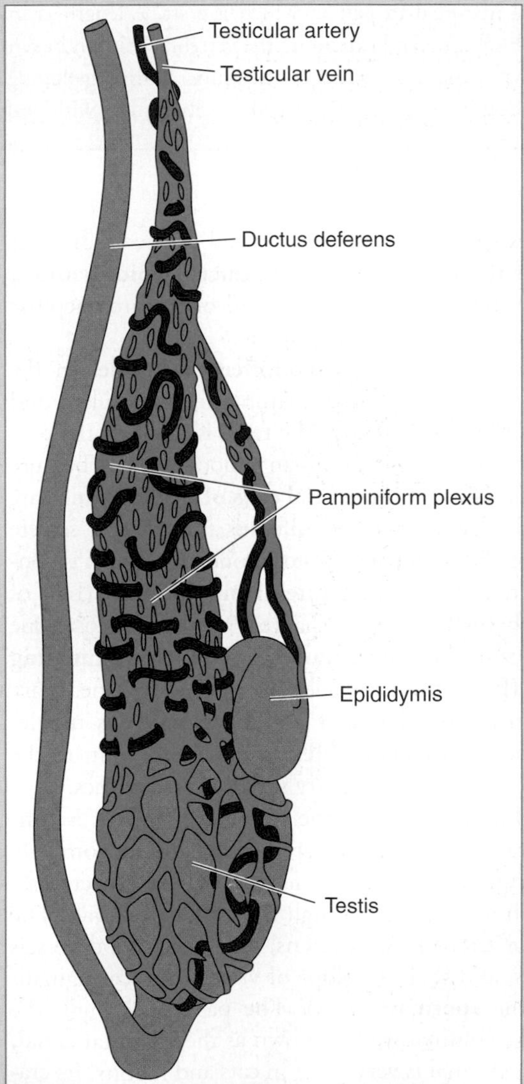

Diagram of the pampiniform plexus and spermatic artery and vein of a bull as an example of a mammal with a scrotum. The connective tissue enveloping the pampiniform plexus is not shown. *(After Frey, 1990.)*

continued

Anatomy in Action 12-3 The Descent of the Testis *continued*

arteriovenous complex of the spermatic cord is enveloped by a nonexpandable layer of connective tissue that keeps the volume of the arteriovenous complex more or less constant. The pulsating pressure of the coiled testicular artery is transmitted to the surrounding veins of the pampiniform plexus and helps to drive the blood back toward the peritoneal cavity. This arterial pump is analogous to the "tail pump" in fishes (see Anatomy in Action 11-1).

Roland Frey (1990) pointed out that the position of the testes in mammals is correlated with locomotory modes. Mammals with scrotal testes are capable of galloping, which is a special mammalian gait in which the stride length is increased by flexing and extending the vertebral column in synchrony with the movements of the limbs. During galloping, intra-abdominal pressure rises to 90 to 150 mmHg, with peak pressures reaching up to 300 mmHg. Such high pressures compress the intra-abdominal thin-walled veins and lymph vessels and temporarily impede the drainage of blood and lymph from the testis via the testicular vein. Primates (except humans) are quadrupedal and strongly flex and extend their vertebral column during galloping on the ground or during jumping or swinging from tree to tree. Humans are bipedal and do not gallop. They do, however, flex and extend their vertebral column during many activities, such as lying down, getting up, and picking up objects from the ground. Pressure within their peritoneal cavity also rises in pulses during speaking, laughing, and coughing. The displacement of the testes out of the peritoneal cavity into a scrotum, in combination with the presence of a pampiniform plexus, isolates the testes from such pressure pulses.

nal tubules (p. 12). The body and tail of the epididymis consists of the highly convoluted **ductus epididymidis,** which develops from the cranial end of the archinephric duct, imbedded in connective tissue.[4]

Sperm develop in the **seminiferous tubules** in the testis, but they complete their maturation, accumulate, and are stored in the epididymis. The testis is not only the production site for sperm. It also is an endocrine gland because it contains clusters of **interstitial cells of Leydig,** which are distributed between the seminiferous tubules and secrete the hormone **testosterone.** Testosterone production is controlled by the gonadotropic **luteinizing hormone** (LH) of the adenohypophysis (see Chapter 9). Both testosterone and luteinizing hormone, as well as **follicle-stimulating hormone** (FSH) from the adenohypophysis, are necessary for the normal development of the seminiferous tubules and for sperm production. Testosterone also controls the development of male secondary sexual characteristics.

The **ductus deferens,** the first part of which is convoluted, leaves the tail of the epididymis and, in company with the testicular vessels and nerves, ascends the cremasteric pouch and passes through the abdominal wall. The complex of the ductus deferens, testicular blood vessels and nerves, and their envelope of visceral tunica vaginalis is called the **spermatic cord.** The passage through the muscular abdominal wall is known as the **inguinal canal.** The inguinal canal is very short in cats and rabbits. Its cranial end, the **internal inguinal ring,** is the entrance into the peritoneal cavity (Fig. 12-18) and its caudal end, the **external inguinal ring,** is the attachment of the external spermatic fascia to the aponeurosis of the external oblique muscle (Fig. 12-16**B**). In humans, the inguinal canal passes diagonally through the abdominal wall and, therefore, is much longer.

Continue to follow the ductus deferens. It passes craniad in the peritoneal cavity for a short distance and then loops over the ureter and extends caudad into the pelvic canal between the urethra and large intestine (Fig. 12-18). The ductus deferentes of opposite sides then converge and soon enter the **urethra.** Much of the urethra in males develops from the urogenital sinus, which forms during subdivision of the cloaca (Anatomy in Action 12-4). The portion of the urethra distal to the union of the ductus deferens carries both sperm and urine. Various accessory genital glands, which secrete the seminal fluid, are associated with the ends of the ductus deferentes and adjacent parts of the urethra. These glands, and the details of the union of the ductus deferentes with the urethra, differ in cats and rabbits. In immature individuals, especially in rabbits, these glands may not be well developed and may be difficult to find.

In cats, the paired ductus deferentes enter the urethra independently, and a small **prostate** surrounds their point of entrance and the adjacent urethra (Figs. 12-13 and 12-18). At the caudal end of the pelvic canal, a pair of **bulbourethral glands,** also called Cowper's glands, enter the urethra, but you will observe these later when you study the penis.

In rabbits, the paired ductus deferentes pass between the urethra and a heart-shaped **vesicular gland** (Fig. 12-15). Carefully separate these structures from one another. The ductus deferentes enter the narrow caudal end of the

[4]See Footnote 3 on page 353.

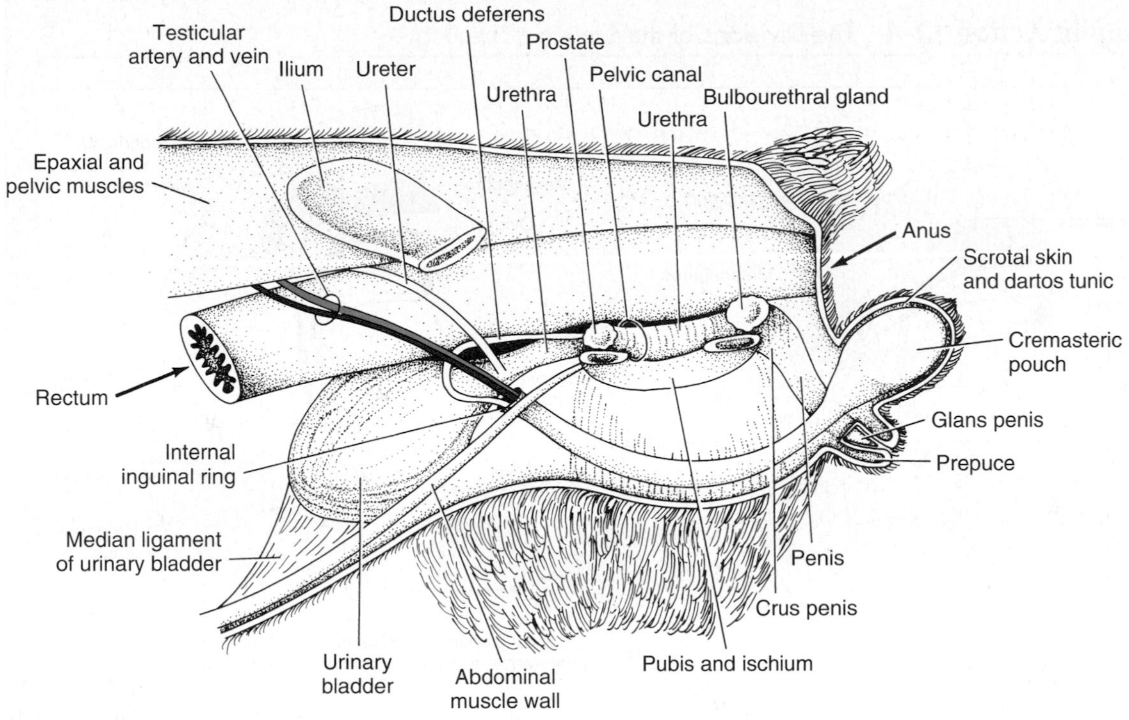

Figure 12-18

Semidiagrammatic lateral view of the topographical relationships of the urogenital system of a male cat. *(Adapted from Dyce, Sack, and Wensing, 1987.)*

Anatomy in Action 12-4 The Divisions of the Cloaca

We have already discussed the evolution of most of the mammalian male and female urogenital tracts from those of their ancestors. However, we deferred a consideration of the cloacal region and its fate in mammals until we were able to study the terminal portions of the urogenital passages.

A cloaca is present in nonmammalian vertebrates and still persists in monotremes and in the embryos of marsupial and placental mammals. However, it becomes subdivided and contributes to the formation of the intestinal and urogenital passages in adult therian mammals. In an early, sexually indifferent eutherian embryo, the **cloaca** consists of a chamber derived from the widening of the caudal end of the hindgut (Fig. **A**). At first this endodermal cloaca is separated from the ectodermal **proctodaeum** by a plate of tissue, the **cloacal membrane,** but this plate soon breaks down, and the proctodaeum then contributes to the formation of the cloaca.

The cloaca receives the intestine dorsally and the allantois ventrally. Even at an early stage (Fig. **A**), the cranial portion of the cloaca is partly divided into the dorsal **coprodaeum,** which receives the intestine, and the ventral **urodaeum,** which receives the allantois, ureters, and archinephric ducts. A small **genital tubercle** is present on the ventral surface of the body cranial to the cloaca. This early embryonic stage of mammals is similar to the cloaca of nonmammalian vertebrates except for the more ventral entrance of the urogenital ducts.

Later in the sexually indifferent period (Fig. **B**), the urodaeum and coprodaeum become completely separated from each other by a fold of tissue, called the **urorectal fold,** and form the urogenital sinus and rectum, respectively. Oviducts now enter the cranial part of the urogenital sinus beside the archinephric ducts, but the attachments of the ureters shift onto the allantois and developing urinary bladder.

In the subsequent differentiation of male mammals (Fig. **D**), the constricted neck of the urinary bladder, which developed from the allantois, and the urogenital sinus form the portion of the urethra that is not included in the penis (segments 1 and 2). The paired archinephric ducts form the ductus deferentes and enter the urethra. Their point of entrance is a landmark that separates the portion of the urethra derived from the allantois from the portion derived from the

continued

Anatomy in Action 12-4 The Divisions of the Cloaca *continued*

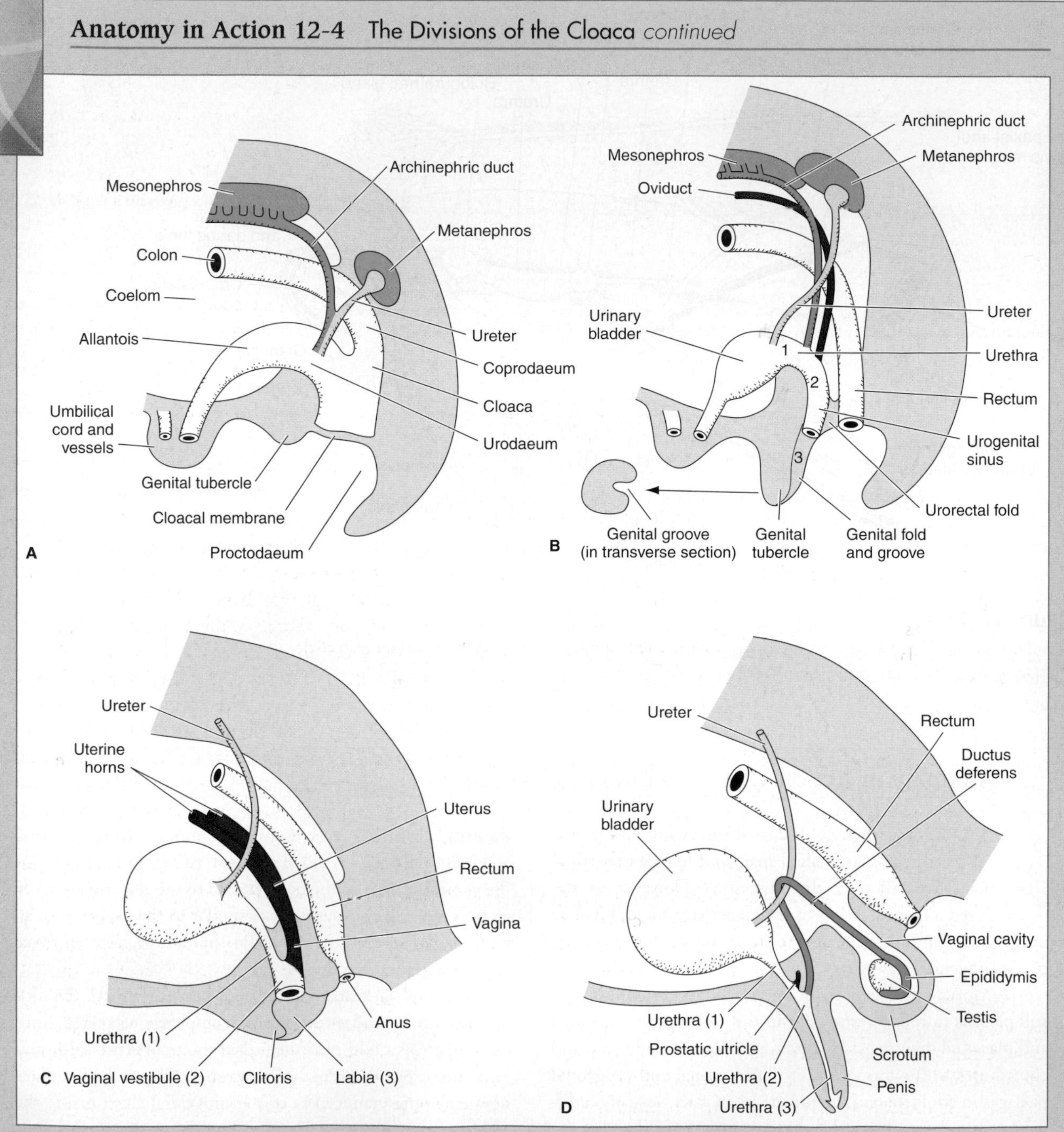

Diagrams in lateral view showing the subdivision of the cloaca that occurs during the embryonic development of a placental mammal. **A**, Early sexually indifferent stage in which the subdivision of the cloaca has just begun. **B**, Later sexually indifferent stage in which the cloaca has become divided into a dorsal rectum and ventral urogenital sinus. **C**, Differentiation of the female condition. **D**, Differentiation of the male condition. *1*, *2*, and *3* indicate comparable regions. Dark green = kidney; pale green = ureter; blue = archinephric duct and derivatives; red = oviduct and derivatives. *(From Liem, Walker, Bemis, and Grande, 2001.)*

Anatomy in Action 12-4 The Divisions of the Cloaca *continued*

urogenital sinus. The oviducts are dismantled in males, although their point of entrance into the urethra may form a small sac, called the **prostatic utricle,** within the prostate. The genital tubercle enlarges to form the penis, and a groove on its ventral surface closes over by the growing together of the **genital folds** to form the penile portion of the urethra (segment 3).

In the differentiation of most female mammals (Fig. **C**) subsequent to the sexually indifferent stage, the constricted neck of the urinary bladder forms the entire urethra. Thus, the female urethra is comparable to only the allantoic segment of the male urethra. The urethra and the two oviducts, whose caudal ends have fused to form the vagina and body of the uterus, enter the urogenital sinus, which becomes the vaginal vestibule (Fig. 12-20). In most female mammals, the vaginal vestibule remains undivided. However, in primates it also becomes divided into the urethra and vagina, which continue nearly to the surface as separate passages. This distal part of the vaginal vestibule forms the shallow vulva. The archinephric ducts are dismantled, and the genital tubercle forms the clitoris. The genital folds of the sexually indifferent stage form the labia minora in mammals that have these labia. The labia majora of females are skin folds comparable to those that form the scrotum in males.

vesicular gland, which in turn enters the urethra. The dorsal wall of the vesicular gland is rather thick and includes the **prostate.** It is possible, by very careful dissection, to free the cranial end of the prostate and turn it caudad. Further dissection will reveal overlapping cranial and caudal lobes of the prostate. Both lobes enter the urethra just caudal to the entrance of the vesicular gland. The **bulbourethral gland,** also called Cowper's gland, enters the dorsal side of the urethra caudal to the prostate, but will be seen later, when you dissect the penis.

The **penis** of mammals, like the claspers of sharks, is an organ that allows internal fertilization. (Recall that *Necturus* does not have a penislike organ.) The base of the penis is anchored to the pelvic ischia (see later and Fig. 12-18) and encloses the part of the urethra that projects caudally beyond the pelvic canal (Figs. 12-13 and 12-15). The free end of the penis, the **glans penis,** lies in a pocket of skin called the **prepuce** *(preputium).* Cut open the prepuce to see the glans penis and the opening of the urethra. In cats, a number of small spines are strewn over the glans penis. The rest of the penis is a firm cylindrical structure, which you should expose by removing the skin and surrounding loose connective tissue. Make a cross section of this portion of the penis and examine it under low magnification. The urethra lies along the dorsal surface of the penis, when the organ is flaccid, and is imbedded in a column of spongy erectile tissue called the **corpus spongiosum penis** (Fig. 12-17**C**). A pair of columns of spongy erectile tissue that are separated by a connective tissue septum, which often is indistinct, lie along the ventral surface of the penis and are surrounded by a ring of dense connective tissue. These paired columns are the **corpora cavernosa penis.** The glans penis is actually a caplike fold of the corpus spongiosum penis that covers the distal ends of the corpora cavernosa penis. The **crura penis** (singular: *crus penis*) are the diverging proximal ends of the corpora cavernosa penis. The crura penis are anchored to the ischia (Fig. 12-18), and each crus penis is covered by muscular tissue, the **ischiocavernous muscle** (Figs. 12-13 and 12-15). If the crura penis were not torn when the pelvic canal was opened, you may have to bisect one now to see the paired (cat) or bilobed (rabbit) bulbourethral gland clearly. During erection of the penis, the spaces within the spongy erectile tissue of the corpus spongiosum and corpora cavernosa become engorged with blood. Make a cross section through the glans penis and look for a small bone, called the **os penis** or baculum, which lies along the urethra and helps to stiffen this part of the penis. (The baculum of the walrus is much larger in accordance with its larger body size and was prized by whalers as a club.) Most mammals, including humans, lack an os penis; their penis is a purely hydrostatic skeleton that stiffens through the increased pressure of the blood accumulating within the erectile tissue.

Before leaving the urogenital system, dissect beneath the skin on either side of the rectum near the anus and find a pair of **anal glands** (Figs. 12-13, 12-15, and 12-19). These glands are spherical in cats but elongated in rabbits. They produce an odoriferous secretion that contains pheromones and is discharged through the anus. The pheromones are thought to have a role in synchronizing the reproductive behavior between individuals of opposite sexes.

The Female Reproductive System

The **ovaries** are a pair of small oval bodies (Fig. 12-19). In adult females, they lie slightly caudal to the kidneys, because they have undergone a partial descent and because the metanephroi have shifted cranially during embryonic development. The small size of the mammalian ovaries is correlated with the intrauterine development of the embryos. Fewer **eggs,** or ova, are produced, and they contain

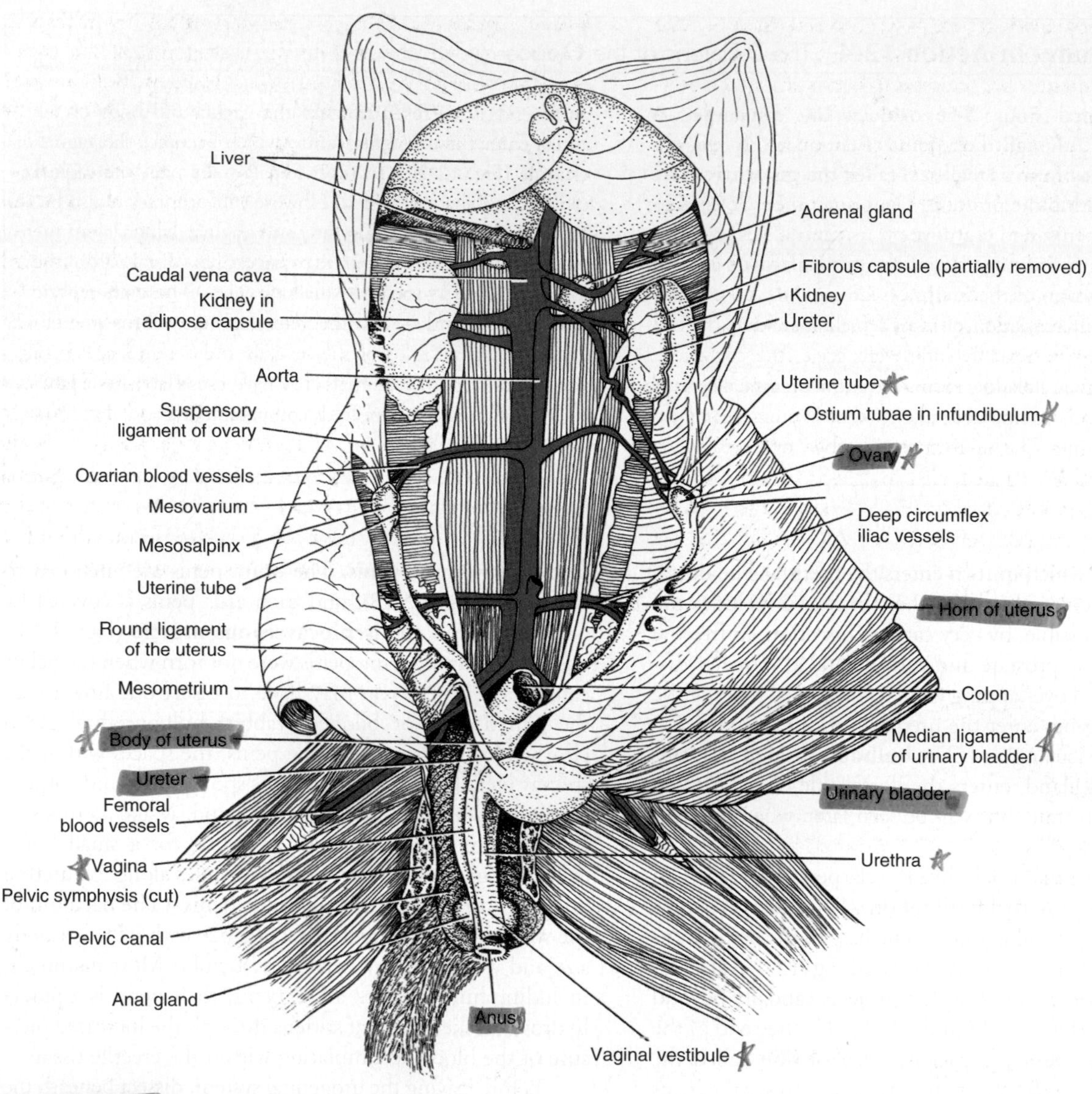

Figure 12-19
Ventral view of the urogenital system of a female cat. The pelvic canal was cut open.

very little yolk. The eggs are microscopic, but you may see small vesicles, the **Graafian follicles** *(folliculi vesiculosi),* protruding from the surface of the ovary. Each follicle contains one egg surrounded by liquid. The ovaries of cats and rabbits protrude into the body cavity and are anchored to the body wall by mesenteries, the **mesovaria.** At **ovulation,** the eggs are discharged from the follicles and break through the serosa enveloping the ovary. After ovulation, the follicles remain imbedded within the ovary and are transformed into the hormone-producing **corpora lutea** (singular: *corpus luteum*), which also may be seen protruding from the surface of the ovary, especially in pregnant specimens. The corpora lutea secrete female hormones that maintain pregnancy.

The eggs of mammals normally do not fall into the peritoneal cavity at ovulation, unlike the eggs of sharks and amphibians. In many mammals, the cranial ends of the oviducts form funnellike expansions, **infundibula,** with fringed rims. These fringes, called **fimbriae,** sweep over the surface of the ovaries and ensure that the eggs are collected into the infundibula. In some mammals, such as cats, the hoodlike infundibula almost completely envelop the ovary; in other mammals, such as minks, the infundibula form ovarian bursae that completely enclose the ovaries.

Paired, simple oviducts are present in early mammalian embryos, but the oviducts differentiate into several regions during development, and their caudal ends merge in varying degrees (Figs. 12-7, 12-19, and 12-20). Thus, the reproductive tract of adult females is more or less Y shaped. The cranial part of each wing of the Y forms a narrow, convoluted **uterine tube,** also called Fallopian tube, lying lateral to the ovary. Notice that a uterine tube curves over the front of the ovary and forms the infundibulum with its fimbriated lips. Spread open the lips of the infundibulum, and you will see the coelomic opening of the uterine tube, the **ostium tubae,** which receives the eggs from the ovary.

The proximal part of each wing of the Y lies caudal to the ovary and forms a much wider tube: the **horn of the uterus** in cats or the **uterus** in rabbits (Figs. 12-19 and 12-20**A** and **B**). In cats, this section is straight, but in rabbits it is somewhat convoluted. It is very large in pregnant specimens because the embryos develop within it.

The ovary and reproductive tract of females are anchored to the body wall by a mesentery known as the **broad ligament.** Often a great deal of fat lies within it. The portion of the broad ligament attaching to the uterus is the **mesometrium;** the portion attaching to the uterine tube is the **mesosalpinx;** and the portion attaching to the ovary is the **mesovarium** (Figs. 12-19 and 12-21). In addition, each ovary is anchored by the **suspensory ligament of the ovary** to the body wall and by the **ovarian ligament** to the uterine horn or uterus (Fig. 12-21). Pull the uterine horn of a cat or the uterus of a rabbit toward the midline, thereby stretching the mesometrium. The mesenteric fold that extends diagonally across the mesometrium from a point near the cranial end of the uterine horn or uterus to the body wall is the **round ligament of the uterus** *(ligamentum teres uteri)* (Fig. 12-19). Notice that the round ligament of the uterus attaches to the body wall at a point comparable to the location of the inguinal canal in the male. The round ligament of the uterus is the female counterpart of the male gubernaculum, which is a connective tissue strand that plays an important role in the descent of the testis.

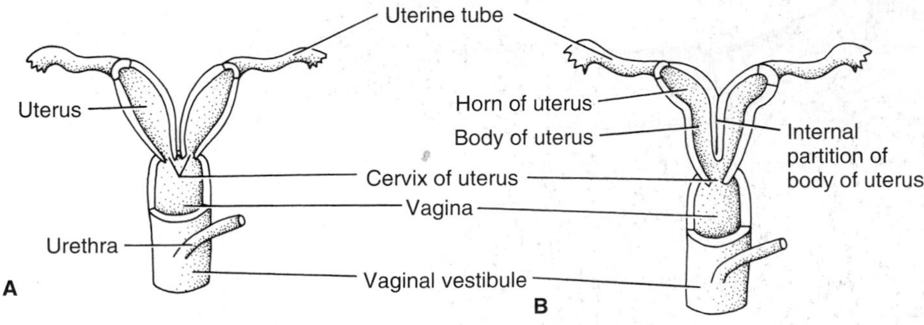

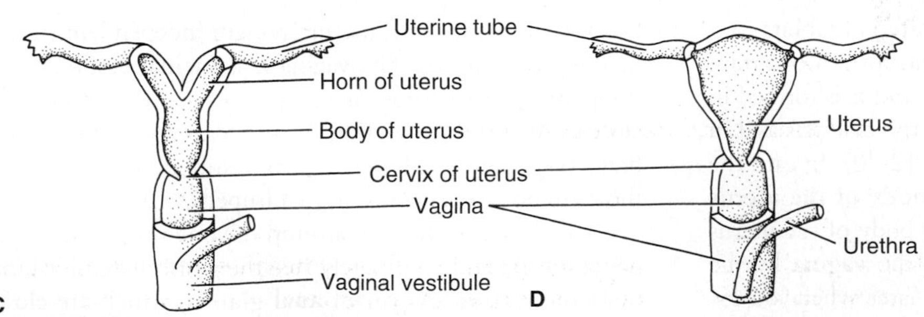

Figure 12-20
Diagrams showing, in ventral views, the progressive fusion of the caudal ends of the paired oviducts in placental mammals. The uterus and part of the vagina are cut open. **A,** Duplex uterus type, in which the caudal ends of the oviducts have united to form a vagina, but the uteri have remained distinct, each with its own cervix. This uterus type is found in rodents and lagomorphs. **B,** Bipartite uterus type, in which the caudal ends of the oviducts have merged to form a median body of the uterus, from which paired horns of the uterus extend. An internal partition is present in the cranial part of the body of the uterus. This uterus type is found in carnivores. **C,** Bicornuate uterus type, in which the body of the uterus is not subdivided by a cranial partition; paired horns of the uterus extend from the body of the uterus. This uterus type is found in ungulates. **D,** Simplex uterus type, in which the paired uteri have completely merged to form a large median body of the uterus, into which the paired uterine tubes open. The vaginal vestibule is subdivided, so that the vagina and urethra open closer to the body surface. This uterus type is found in primates. *(Modified from Wiedersheim, 1906.)*

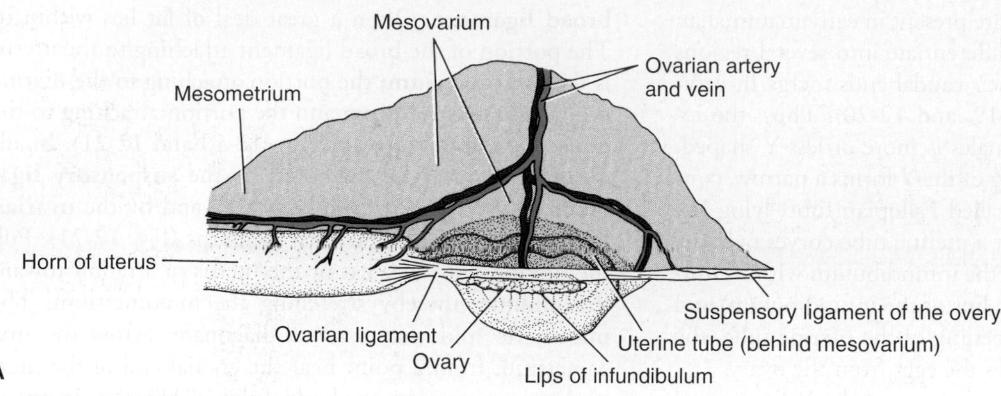

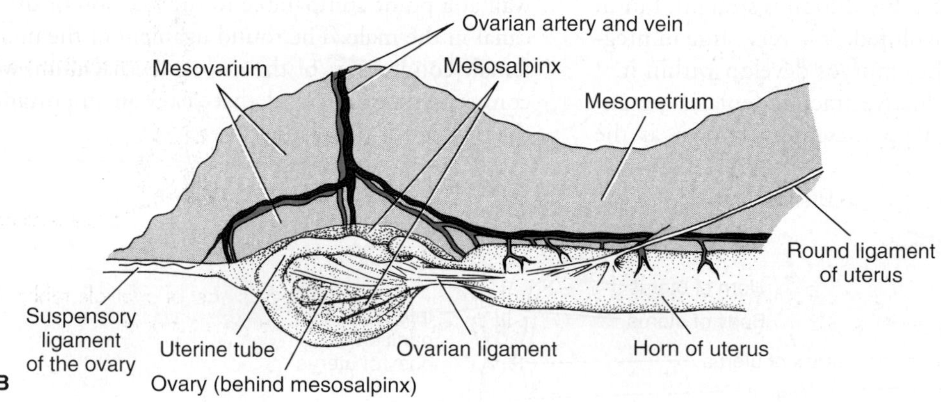

Figure 12-21

Left ovary, uterine tube, uterine horn, and associated mesenteries and ligaments in a female cat. **A**, Median view. **B**, Lateral view. *(From Merkt, 1948.)*

The paired uterine horns of cats or uteri of rabbits converge cranial toward the pelvic canal and form a common median tube. This is the stem of the Y, and it is formed in part by the fusion of the caudal ends of the oviducts and in part by a subdivision of the cloaca (Fig. 12-20). In cats, the cranial part of this median tube is the **body of the uterus** (Figs. 12-19, 12-20**B**, and 12-22). The body of the uterus is relatively short. It is separated from the **vagina** by the **cervix of the uterus,** which you will see later, when you cut open the genital tract. In rabbits, the two uteri enter the vagina independently; each uterus has its own cervix (Figs. 12-20**A** and 12-23). The vagina continues caudally through the pelvic canal, lying between the urethra and large intestine. Carefully separate these structures from one another and find the point where the vagina and urethra unite. The common passage from here to the body surface is the **vaginal vestibule,** also called urogenital canal. It is a relatively long passage in quadrupeds. The comparable area in humans is known as the **vulva,** but the vulva is very shallow because the vagina and urethra remain independent nearly to the body surface. The vagina and vaginal vestibule develop from the ventral, or urogenital part, of the divided cloaca (Anatomy in Action 12-4). Skin folds, the **labia,** flank the opening of the vaginal vestibule, or vulva, but these are not conspicuous in quadrupeds.

Cut through the skin around the opening of the vaginal vestibule and completely free the vestibule and vagina from the rectum. A pair of **anal glands,** which are elongate in rabbits and round in cats, can be found by dissecting beneath the skin on the lateral surface of the vestibule in rabbits (Fig. 12-23) or of the rectum near the anus in cats (Fig. 12-19). Their functions are the same as in the male (p. 373).

Now open the median portion of the genital tract by making a longitudinal incision through its dorsal wall, which extends from the vestibule to the horns of the uterus in cats or the uteri in rabbits. Veer away from the middorsal line toward one of the uterine horns as you open the

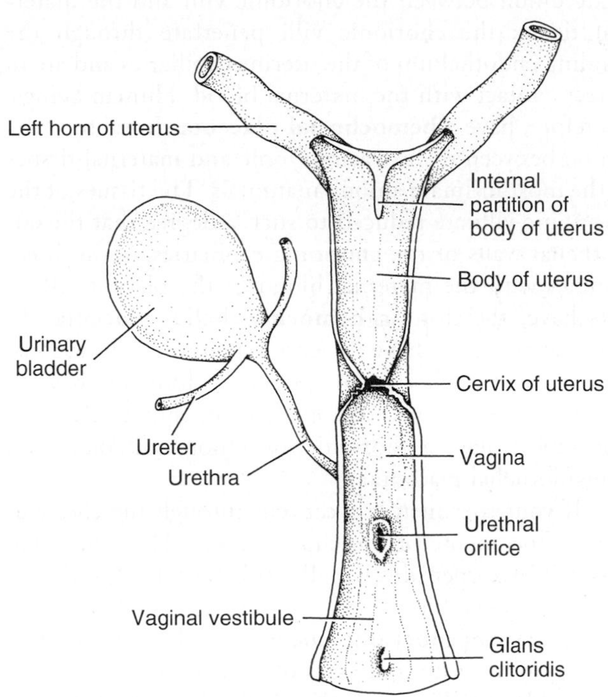

Figure 12-22

Dorsal view of the reproductive organs of a female cat. The uterus, vagina, and vaginal vestibule were cut open.

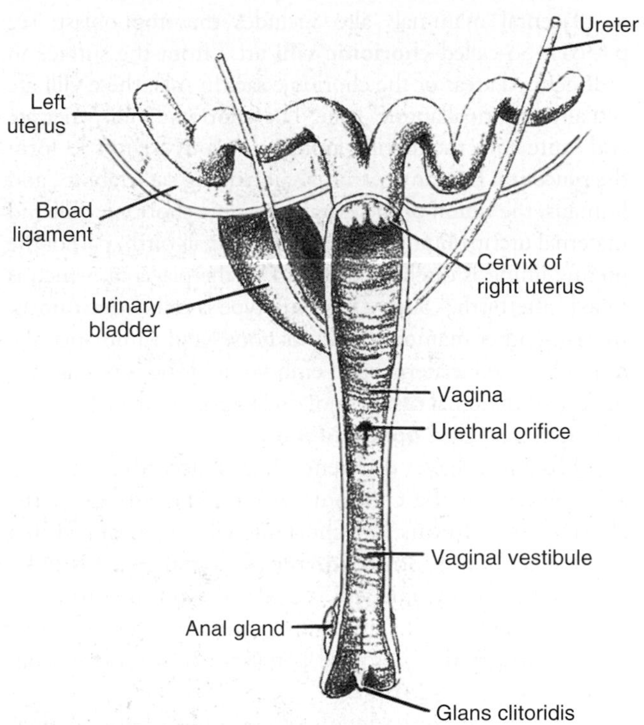

Figure 12-23

Dorsal view of the reproductive tract of a female rabbit. The vagina and vaginal vestibule were cut open.

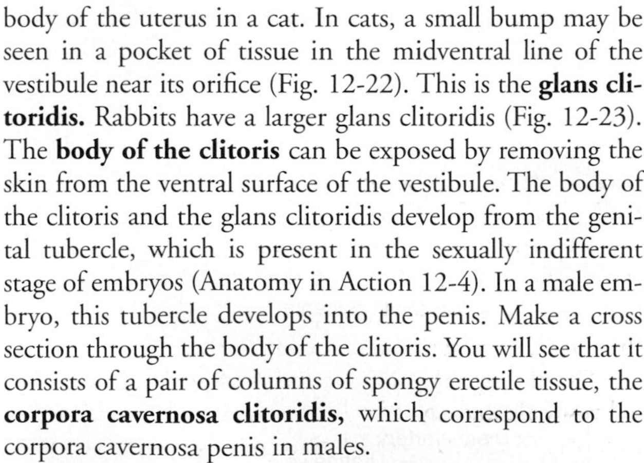

body of the uterus in a cat. In cats, a small bump may be seen in a pocket of tissue in the midventral line of the vestibule near its orifice (Fig. 12-22). This is the **glans clitoridis.** Rabbits have a larger glans clitoridis (Fig. 12-23). The **body of the clitoris** can be exposed by removing the skin from the ventral surface of the vestibule. The body of the clitoris and the glans clitoridis develop from the genital tubercle, which is present in the sexually indifferent stage of embryos (Anatomy in Action 12-4). In a male embryo, this tubercle develops into the penis. Make a cross section through the body of the clitoris. You will see that it consists of a pair of columns of spongy erectile tissue, the **corpora cavernosa clitoridis,** which correspond to the corpora cavernosa penis in males.

More cranially in the vaginal vestibule, you will see the entrance of the urethra, also called the **urethral orifice.** The genital passage cranial to the urethral orifice is the vagina, and it continues forward to the sphincterlike neck, or **cervix, of the uterus** (Figs. 12-20, 12-22, and 12-23). In cats, the single cervix lies about halfway between the urethral orifice and the horns of the uterus. It appears as a pair of folds that constrict the lumen of the reproductive tract. The body of the uterus lies between the cervix and the horns of the uterus. Notice that the cranial part of the body of the uterus is subdivided into right and left sides by a vertical partition. The uterus of cats is therefore **bipartite** (Fig. 12-20**B** and 12-22). In rabbits, the vagina extends cranially to the paired uteri, and each uterus has a sphincterlike **cervix** that projects into the vagina. This type of uterus is called a **duplex** uterus (Fig. 12-20**A** and 12-23).

Reproduction and Embryos

Eutherian mammals are viviparous: the embryos develop within the uterus. If your specimen is pregnant, cut open one horn of the uterus and examine the embryos within. Various extraembryonic membranes that are characteristic of amniotes envelop the embryos (see Figs. 11-1, 12-12**C** and **D,** and 12-25). The whole complex of embryo and extraembryonic membranes is called the **chorionic sac** (Figs. 12-24**A** and **B**), even though the outermost extraembryonic membrane in many mammals, including cats, is formed by a fusion of the **chorion** and **allantoic membrane** to form a **chorioallantoic membrane** (Figs. 12-12**D** and 12-25). In humans and rabbits, however, the outermost extraembryonic membrane is formed by the chorion alone, because the allantois remains small and does not fill the extraembryonic coelom (see Figs. 11-1 and 12-12**D**). (Recall that the chorion in placental mammals also includes the trophoblast; see p. 361). So-called **chorionic villi** arise from the surface in well-defined areas of the chorionic sac. In cats, these villi are actually "chorioallantoic" villi. The chorionic villi penetrate and unite with the uterine lining to various degrees to form the **placenta.** In many mammals, including cats, rabbits, and humans, the union of the extraembryonic chorionic villi and maternal uterine lining is so intimate that, at birth, part of the uterine lining is discharged as part of the placenta, which is called "afterbirth." Such a placenta type is called **deciduous.** In many other mammals, such as horses and ruminants, the union between maternal and embryonic tissue is not as intimate, and maternal tissue is not discharged at birth. This type of placenta is called **nondeciduous.**

Placentas have different shapes according to the arrangement of the chorionic villi on the surface of the chorionic sac. In cats, the chorionic villi are arranged in a beltlike band and form a **zonary placenta** (Fig. 12-24**A**). This band of chorionic villi is easily seen on the surface of the chorionic sac. In rabbits and humans, the chorionic villi are arranged in a round patch and form a **discoidal placenta** (Fig. 12-24**B**).

The microscopic details of the union between maternal and extraembryonic tissue in the placenta differ among mammals. In cats, the epithelium lining the uterine wall is dismantled. The chorionic villi penetrate the substance of the uterine lining and come into direct contact with the endothelial walls of the maternal uterine capillaries. Cats, therefore, have an **endotheliochorial placenta.** In humans, there is an even more intimate union between the chorionic villi and the maternal tissue: the chorionic villi penetrate through the eroding endothelium of the uterine capillaries and are in direct contact with the maternal blood. Human beings, therefore, have a **hemochorial placenta.** In rabbits, the union between the extraembryonic and maternal tissues is the most intimate among mammals. The tissues of the chorionic villi are reduced to such a degree that the endothelial walls of the chorionic capillaries are in direct contact with the maternal blood in the placenta. Rabbits have, therefore, a **hemoendothelial placenta.** To summarize, cats have a chorioallantoic, deciduous, zonary, endotheliochorial placenta; humans have a chorionic, deciduous, discoidal, hemochorial placenta; and rabbits have a chorionic, deciduous, discoidal, hemoendothelial placenta.

If your specimen is a cat, cut through the chorioallantoic membrane of the chorionic sac. The cavity that you will have opened is the **allantois** (Fig. 12-25). If your specimen is a rabbit, you will cut through the chorion and open the **extraembryonic coelom;** recall that the morphology of the chorionic sacs of humans and rabbits are very similar (see Fig. 11-1). You will then see that the embryo is enclosed by yet another extraembryonic membrane, the **amnion** (see Figs. 11-1, 12-12**D,** and 12-25). In most mammals, including cats, but not humans and rabbits, the original amnion is fused to the allantoic membrane and forms, strictly speaking, an "amnioallantoic membrane" (Fig. 12-25). Cut open the amnion to reveal the embryo within the **amniotic cavity.** Note that al-

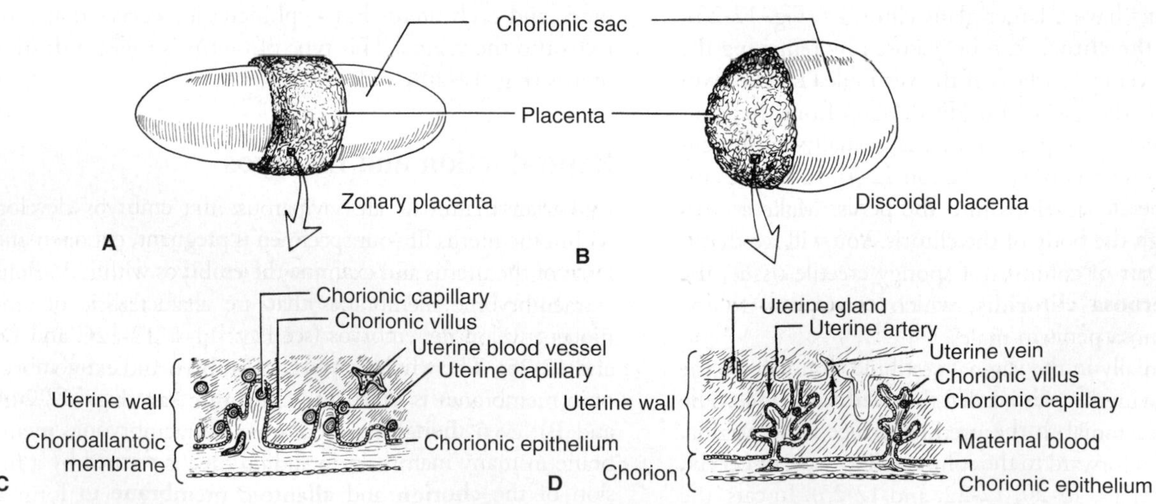

Figure 12-24

Examples of the mammalian placenta. **A,** Chorionic sac of a cat embryo with zonary placenta. **B,** Chorionic sac of a human embryo with discoidal placenta. **C,** Diagrammatic microscopic section through the placenta of a cat. **D,** Diagrammatic microscopic section through the placenta of a human. Arrows indicate the direction of blood flow.

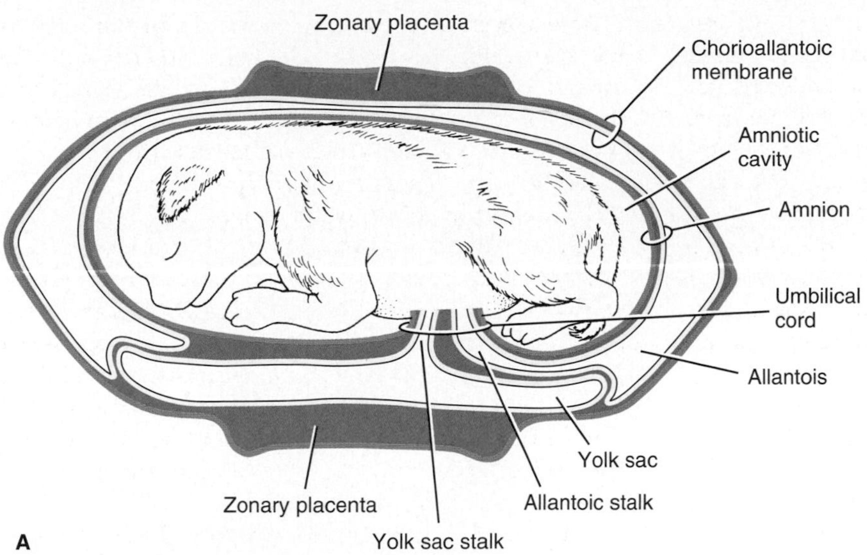

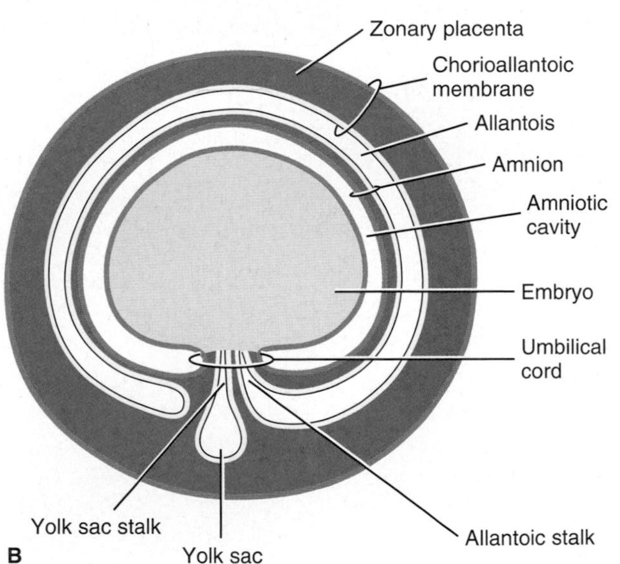

Figure 12-25
Semidiagrammatic sections through the chorionic sac and extraembryonic membranes of a cat embryo. The colored lines indicate the germ layer origin of the various layers of the extraembryonic membranes. Blue: Ectoderm. Red: Mesoderm. Yellow: Endoderm. **A,** Longitudinal section. **B,** Transverse section at the level of the umbilical cord. *(After Dyce, Sack, and Wensing, 1987.)*

though all mammalian embryos are surrounded by two fluid-filled chambers, the components of the extraembryonic membranes enclosing these chambers vary among mammals. In cats, these chambers are the allantois, which has completely filled the original extraembryonic coelom, and the amniotic cavity; the two extraembryonic membranes that enclose these chambers are the chorioallantoic membrane and the amnion. In rabbits and humans, the two chambers are the extraembryonic coelom and the amniotic cavity; the two extraembryonic membranes that enclose these chambers are the chorion and the amnion. The cord of tissue that extends from the belly of the embryo to the placenta is the **umbilical cord.** It contains the **allantoic stalk,** the umbilical blood vessels, and the vestige of the **yolk sac stalk.** The umbilical cord is surrounded by the amnion.

APPENDIX A Word Roots

When studying anatomy, you will encounter many unfamiliar terms for anatomical structures and organisms. You should become familiar with them because effective communication requires their use. Understanding the etymology of these terms will help you to fix their meaning and spelling in your mind. About 400 prefixes, suffixes, and combining forms in Greek or Latin form word roots that make up most of the medical vocabulary. A list of the more important word roots, and the Greek (Gr.) and Latin (L.) words from which they are derived, is presented below. Each entry is exemplified by anatomical terms, names of major groups of organisms, or common words in which the word roots are used. By periodically consulting this list, you will soon be able to see how the same word roots are used in different combinations and be able to infer the meanings of unfamiliar words. In this list we also include some other widely used terms that are the same in Latin or Greek and English, and some of the more common eponyms, or terms derived from a person's name. The definitions of terms can be found through the index.

In the following list, usually only the nominative singular is given for a noun, but the genitive (gen.), nominative plural (pl.), or diminutive (dim.) is included if it is different enough and if it is the basis for an example cited. For a verb only the infinitive is given, unless another form, such as the past participle (p.p.), is needed to recognize the word root. This list of word roots is far from complete. If this list whets your appetite, you should consult standard dictionaries, medical dictionaries [e.g., Dorland (2000), Stedman (2000)], or references, such as Jaeger (1978), Nybakken (1985), Allaby (1999), or Brown (2000), for further study.

a— [Gr. *a-* or *an-*, prefix meaning without, not]: acelous, Agnatha, amorphous

ab— [L. *ab-*, prefix meaning away from]: abducens nerve, abduction

abdom— [L. *abdomen,* abdomen, probably from *abdo,* to conceal]: abdomen, abdominal

acanth— [Gr. *akantha,* spine, thorn]: Acanthodii

acetabulum [L. *acetabulum,* vinegar cup, from *acetum,* vinegar]: acetabulum, acetic acid

acinus [L. *acinus,* berry, grape]: pulmonary acinus

acro— [Gr. *akros,* topmost, extreme]: acromiodeltoid, acromion, Acropolis, acronym

acti— [Gr. *aktis,* gen., *aktinos,* ray]: Actinopterygii

ad—, af— [L. *ad-* (may be changed to *af-* before certain words), prefix meaning motion toward]: adduction, afferent

adeno— [Gr. *aden,* gland]: adenohypophysis

adip— [L. *adeps,* gen., *adipis,* fat]: adipose tissue

af— See *ad—*.

al— [L. *ala,* wing]: alar foramen, aliform, alisphenoid bone

all— [Gr. *allos,* different, other]: Allotheria, allele, allometry

allant— [Gr. *allas,* gen., *allantos,* sausage]: allantoic artery, allantois

alveolus [L. *alveolus,* small pit]: alveolus

amelo— [Old Fr. *amel,* enamel]: amelogenins

amn— [Gr. *amnion,* fetal membrane]: amnion, anamniote

amphi— [Gr. *amphi,* on both sides of, double]: amphibian, amphicoelous, amphioxus

ampulla [L. *ampulla,* a small jug]: ampulla of Lorenzini, seminiferous ampullae

an— See *a—*.

anastomosis [Gr. *anastomosis,* an opening, a connection between two seas]: subsupracardinal anastomosis

anc— [Gr. *ankon,* elbow, bend of the arm]: anconeus muscle

ant— [L. *ante-*, prefix meaning in front of, previous to]: antebrachium, anterior, antorbital process

anus [L. *anus,* anus]: anus

aort— [Gr. *aorte,* aorta, from *aeirein,* to raise up]: aortic arch

apo— [Gr. *apo-*, prefix meaning away from, far]: apophysis, synapomorphy

apsid— [Gr. *apsis,* gen., *apsidos,* a loop]: anapsid skull, synapsid

aqu— [L. *aqua,* water]: aqueous humor, aqueduct of Sylvius

arachn— [Gr. *arachne,* spider]: arachnid, arachnoid membrane, arachnophobia

arbor [L. *arbor,* tree]: arbor vitae, arboretum, arboreal

arch— [Gr. *arche-*, prefix meaning primitive]: archenteron, archinephric, archipallium

arcu— [L. *arcuare,* to bend like a bow]: arcualia, coracoarcual muscle, arcuate arteries

arthro— [Gr. *arthron,* joint]: arthrology, arthropod, arthritis

arter— [L. *arteria,* artery]: artery, arterio-venous

aryten— [Gr. *arytaina,* ladle]: arytenoid cartilage
Aselli [Gasparo Aselli, 1581–1626, Italian anatomist]: pancreas of Aselli
atlas [Gr. mythology *Atlas,* a demigod condemned to support the earth upon his shoulders]: atlas
atri— [L. *atrium,* entrance room]: atrium, atrioventricular
aur [L. *auris,* dim., *auricula,* ear]: auricle
auto— [Gr. *auto-,* prefix meaning self]: automatic, autostylic
avi— [L. *avis,* pl., *aves,* bird]: Aves, aviary, aviation
axon [Gr. *axon,* axle]: axon
axill— [L. *axilla,* armpit]: axillary artery
bas— [Gr. *basis,* base, bottom]: basal nuclei, basapophysis
bi—, *bin*— [L. *bi-, bin-,* prefix meaning two]: biceps brachii muscle, binocular, biped
bi—, *bio*— [L. *bios,* life]: amphibian, biology
blast— [Gr. *blastos,* germ, or a precursor]: blastocoel, lymphoblast, osteoblast
Bowman [Sir William Bowman, nineteenth-century English anatomist]: Bowman's capsule
brachi— [Gr. *brachion,* upper arm]: antebrachium, brachialis muscle, brachium pontis
branchi— [Gr. *branchion,* gill]: branchial, branchiomere, Elasmobranchii
bronch— [Gr. *bronchos*]: bronchus, bronchitis
bulla [Gr. *bulla,* a bubble, seal]: tympanic bulla
bucc— [L. *bucca,* cheek]: buccal glands
bursa [Gr. *bursa,* hide, wineskin]: ovarian bursa, omental bursa
caec— [L. *caecus,* blind, ending blind]: caecum, ileocaecal valve
calc— [L. *calx,* gen., *calcis,* lime, heel]: calcaneus, calcify, calcium
call— [L. *callum,* hard skin]: callous, corpus callosum
calyx— [Gr. *kalyx,* cup]: renal calyx
can— [L. *canis,* dog]: canine teeth
capill— [L. *capillus,* hair]: capillary
capit— [L. *caput,* gen., *capitis,* dim., *capitulum,* head]: capitulum, caput, rhomboideus capitis muscle
carapace [French, *carapace,* a shell]: carapace
cardi— [Gr. *kardia,* heart]: cardiac, cardiologist
cardinal [L. *cardinalis,* principal, red]: cardinal vein
carn— [L. *caro,* gen., *carnis,* flesh]: carnal, carnivore, trabeculae carneae
carotid [Gr. *karotides,* large neck arteries, from *karoun,* to stupefy]: carotid artery, carotid rete
carp— [Gr. *karpos,* wrist]: carpal bone, carpus
caud— [L. *cauda,* tail]: caudal, caudofemoral muscle
cav— [L. *cavus,* hollow]: cavity, vena cava, cave
cel— See *coel*—.
ceph— [Gr. *kephale,* head]: cephalic, cephalochordate, Holocephali
—*ceps* [L. *capere,* p.p., *captus,* suffix —*ceps,* to take, to be at the top, head]: biceps muscle, triceps muscle
cerat— [Gr. *keras,* geno, *keratos,* horn]: ceratobranchial cartilage, ceratotrichia, keratin
cerc— [Gr. *kerkos,* tail]: heterocercal fin
cerebr— [L. *cerebrum,* dim., *cerebellum,* brain]: cerebrum, cerebellum, cerebral
cervic— [L. *cervix,* gen., *cervicis,* neck]: cervical, cleidocervicalis muscle, uterine cervix, rectus cervicis muscle
chiasma [Gr. *chiasma,* cross]: optic chiasm
choan— [Gr. *choane,* funnel]: choana
chol— [Gr. *chole,* bile]: ductus choledochus, choledochal
chondr— [Gr. *chondros,* cartilage]: Chondrichthyes, chondrify, chondrocranium
chord— [L. *chorda,* string]: chordae tendineae, Chordata, notochord
chorio— [Gr. *chorion,* skin, afterbirth]: chorion, choroid layer of eyeball, choroid plexus, chorionic villi
choro— See *chorio*—.
chromato— [Gr. *chroma,* color]: chromatophore
cilium [L. *cilium,* pl., *cilia,* eyelash]: cilia on a cell
ciner— [L. *cinereus,* the color of ashes, from *cinis,* ashes]: incinerate, tuber cinereum)
cingul— [L. *cingulum,* girdle]: cingulate cortex
circ— [L. *circum,* about, around]: circular, circumvallate papilla
cisterna [L. *cisterna,* reservoir, cistern]: cisterna chyli
clast— [Gr. *klastos,* broken]: osteoclast
clav— [L. *clavis,* dim., *clavicula,* key, the clavicle]: clavicle, subclavian artery
cleid— [Gr. *kleis,* gen., *kleidos,* key, the clavicle]: cleidoic egg, cleidomastoid muscle
cleithr— [Gr. *kleithron,* a key, bar]: cleithrum
clitor— [Gr. *kleitoris,* gen., *kleitoridos,* small hill]: clitoris, glans clitoridis
cloaca [L. *cloaca,* sewer]: cloaca
cnem— [Gr. *kneme,* shin or tibia]: gastrocnemius muscle
coccyx [Gr. *kokkyx,* gen., *coccygis,* cuckoo, bone in the shape of cuckoo's bill]: coccyx, coccygeus muscle
coch— [Gr. *kokhlos,* land snail]: cochlea, vestibulocochlear nerve
coel— [Gr. *koilos,* hollow]: coeliac artery, coelom, neurocoel, nephrocoel
coll— [L. *collum,* neck]: collar, longus colli muscle
coll— [L. *collis,* dim., *colliculus,* hill]: superior colliculus
colon [Gr. *kolon,* colon]: colon, mesocolon
commissura [L. *commissura,* point of union of two things]: anterior commissure of brain
conch— [Gr. *konkhe,* sea shell]: concha, conchology
condyl— [Gr. *kondylos,* articulatory prominence at a knuckle joint]: epicondyle, occipital condyle
conju— [L. *conjungere,* p.p., *conjunctus,* to join together]: conjugal, brachium conjunctivum, conjunctiva of eye

contra —[L. *contra,* opposite]: contralateral, contradiction
conus [L. *conus,* cone]: conus arteriosus
copro— [Gr. *kopros,* dung]: coprodaeum, coprophagy
coraco— [Gr. *korax,* gen., *korakos,* a raven or crow]: coracobrachial muscle, coracoid (resembles a crow's beak)
corn— [L. *cornu,* horn]: cornea, cornify, hyoid cornua
coron— [L. *corona,* crown, or something curved]: coronary arteries, coronation, coronoid process
corpus [L. *corpus,* pl., *corpora,* dim., *corpusculum,* body]: corpora cavernosa, corpse, corpus callosum, renal corpuscle
cort— [L. *cortex,* gen., *corticis,* bark]: cerebral cortex, corticospinal tract, renal cortex
cost— [L. *costa,* rib]: costal cartilage, intercostal space
Cowper [William Cowper, seventeenth-century English anatomist]: Cowper's gland
cox— [L. *coxa,* hip bone]: coxal, articularis coxae muscle
cran— [Gr. *kranion,* skull]: chondrocranium, cranial
cre— [Gr. *kreas,* flesh]: cremasteric muscle, to cremate, pancreas
cribr— [L. *cribrum,* sieve]: cribriform plate
cric— [Gr. *krikos,* ring]: cricoid cartilage
cru— [L. *crus,* pl., *crura,* lower leg shank]: abductor cruris muscle, crural fascia, crus penis
cucull— [L. *cucullus,* cap, hood]: cucullaris muscle
cun— [L. *cuneus,* wedge]: cuneiform bone, funiculus cuneatus
cupula [L. *cupa,* dim., *cupula,* small tub]: cupula
cut— [L. *cutis,* skin]: cutaneous, cuticle, cutis
Cuvier [Baron Georges Cuvier, eighteenth-century French scientist]: duct of Cuvier
cyclo— [Gr. *kyklos,* circle]: cyclic, cyclostome
cyst— [Gr. *kystis,* bladder]: cyst, cystic duct
cyt— [Gr. *kytos,* hollow vessel, cell]: lymphocyte, osteocyte
—*daeum* [Gr. *hodaios,* on the way]: proctodaeum, stomodaeum, urodaeum
decid— [L. *decidere,* to fall off]: deciduous placenta
delta [Gr. *delta,* fourth letter of the Greek alphabet,]: deltoid muscle
dendr— [Gr. *dendron,* tree]: dendrite of a nerve cell
dens, dent— [L. *dens,* gen., *dentis,* tooth]: dens, dentary bone, dentine, dentist
derm— [Gr. *derma,* skin, leather]: dermal bone, dermis, ostracoderm, mesoderm
—*deum* See —*daeum.*
di— [Gr. *di-,* prefix meaning two]: diapsid reptile, digastric muscle
dia— [Gr. *dia-,* prefix meaning across, between]: di[a]encephalon, diapophysis, diaphragm
didym— [Gr. *didymos,* gen., *didymidis,* testicle]: ductus epididymidis, epididymis
digit— [L. *digitus,* finger]: digit, digitigrade, extensor digitorum muscle
dont— See *odont*—.
dors— [L. *dorsum,* back]: dorsal, longissimus dorsi muscle
duct [L. *ductus,* a leading, from *ducere,* to lead]: abduction, duct, oviduct
duoden— [New L. *duodenum,* the first part of the intestine, which is about 12 fingerbreadths long (a contraction of *intestinum duodenum digitorum,* from *duodeni,* 12 each)]: duodenal, duodenum
dur— [L. *durus,* hard]: durable, dura mater
ecto— [Gr. *extos,* outside]: ectoderm, ectoparasites
ef— See *ex*—.
elasm— [Gr. *elasmos,* a thin plate]: Elasmobranchii
encephal— [Gr. *enkephalos,* brain, from *en* + *kephale*]: diencephalon, encephalitis, mesencephalon
endo—, *ento*— [Gr. *endon, entos,* within]: endoderm, endolymph, endostyle, endotherm, endotympanic bone
enter— [Gr. *enteron,* intestine]: archenteron, mesentery
ento— See endo—.
epi— [Gr. *epi-,* prefix meaning upon, above]: epicondyle, epidermis, epididymis, epithalamus
epiplo— [Gr. *epiploon,* the greater omentum]: epiploic foramen, gastroepiploic artery
esophagus [Gr. *oisophagos,* gullet]: esophagus, esophageal
ethm— [Gr. *ethmos,* sieve]: ethmoid bone
eu— [Gr. *eu-,* prefix meaning good, true]: Eutheria
Eustachio [Bartolommeo Eustachio, sixteenth-century Italian anatomist]: Eustachian tube
ex— [L. *ex-, ef-,* prefix meaning out of or away from]: ductuli efferentes, efferent neuron, extrinsic muscle
faci— [L. *facies,* form, face]: facial nerve
falx, falc— [L. *falx,* gen., *falcis,* sickle]: falciform ligament, falx cerebri
Fallopio [Gabriele Fallopio, sixteenth-century Italian anatomist]: Fallopian tube
fascia [L. *fascia,* band, bandage]: fascia
fasci— [L. *fascis,* dim., *fasciculus,* bundle]: muscle fascicle
fauces [L. *fauces,* gen., *faucium,* throat]: isthmus faucium
femur [L. *femur,* thigh, femur]: femur, femoral artery
fenestra [L. *fenestra,* window]: fenestra cochleae, temporal fenestra
fer— [L. *ferre,* to carry]: afferent neuron, deferent duct, efferent artery, seminiferous tubules
fibula [L. *fibula,* buckle, pin]: fibula
fid— See *fiss*—.
fil— [L. *filum,* dim., *filamentum,* thread]: filamentous, filiform papillae
fimbria [L. *fimbria,* pl., *fimbriae,* fringe]: fimbria of hippocampus, fimbria of infundibulum
fiss— [L. *findere,* past, *fidi,* p.p., *fissus,* to split]: fissure, multifidus muscle
flex— [L. *flectere,* p.p., *flexus,* to bend]: circumflex iliac artery, flexion
flocc— [L. *floccus,* dim., *flocculus,* tuft of wool]: flocculonodular lobe

foli— [L. *folium,* leaf]: cerebellar folia, foliage, foliate papillae
follicle [L. *folliculus,* small sac]: hair follicle, ovarian follicles
foramen [L. *foramen,* pl., *foramina,* an opening]: foramen magnum, superficial ophthalmic foramina
form— [L. *forma,* shape, rule]: cribriform plate, digitiform gland, falciform ligament, formula, lentiform nucleus
forn— [L. *fornix,* a vault or arch, a brothel]: fornicate, hippocampal fornix
foss— [L. *fossa,* a ditch, from *fodere,* p.p., *fossus,* to dig]: fossil, fossorial, mandibular fossa
fovea [L. *fovea,* small pit]: fovea in retina
fron— [L. *frons,* gen., *frontis,* forehead]: frontal bone
funic— [L. *funis,* dim., *funiculus,* cord, rope]: funiculi of spinal cord, funicular
fundus [L. *fundus,* bottom]: gastric fundus
fung— [L. *fungus,* mushroom]: fungiform papilla, fungus
gamet— [Gr. *gametes,* husband, *gamete,* wife]: gamete, gametogenesis
ganglion [Gr. *ganglion,* a swelling, tumor]: ganglion
Gartner [Hermann Gartner, eighteenth- to nineteenth-century Danish anatomist]: Gartner's duct
gastr— [Gr. *gaster,* stomach]: digastric muscle, gastric glands, gastrocnemius muscle
gem— [L. *geminus,* or *gemellus,* twin, double]: gemellus cranialis muscle, trigeminal nerve
gen— [L. *genus,* race or kind, from *genere,* to beget]: genesis, genital, genus
gen— [L. *genu,* knee]: geniculate bodies, genu of corpus callosum
genio— [Gr. *geneion,* chin]: genioglossus, geniohyoid
glans [L. *glans,* dim., *glandula,* acorn]: gland, glans penis
glen— [Gr. *glene,* cavity or socket]: glenoid cavity
glia [Gr. *glia,* glue]: neuroglia
gloss—, *glott*— [Gr. *glossa,* or *glotta,* tongue]: genioglossus muscle, glottis, hypoglossal nerve
glut— [Gr. *gloutos,* buttock]: gluteus muscle
gnath— [Gr. *gnathos,* jaw]: Agnatha, gnathostome
gon— [Gr. *gonos,* seed]: gonad
Graaf [Reijnier de Graaf, seventeenth-century Dutch anatomist]: Graafian follicle
gracil— [L. *gracilis,* slender]: fasciculus gracilis, gracile, gracilis muscle
gubernaculum [L. *gubernaculum,* a rudder, from *gubernare,* to steer]: gubernaculum of testis, gubernatorial
gular [L. *gula,* throat]: gular
gustat— [L. *gusto,* to taste, p.p., *gustatus,* tasted]: gustatory
gyr— [Gr. *gyros,* circle, round]: cerebral gyri, gyrate
habenula [L. *haben,* dim., *habenula,* a strap]: habenula of epithalamus
haem— See *hem*—.
hallu— [L. *hallux,* gen., *hallucis,* big toe]: flexor hallucis longus muscle
ham— [L. *hamus,* dim., *hamulus,* a hook]: hamate bone, hamulus of pterygoid bone
Harder [Johann Harder, seventeenth-century Swiss anatomist]: Harderian gland
Hatschek [Bernard Hatschek, 1854–1941, Austrian anatomist]: Hatschek's pouch
hem— [Gr. *haima,* blood]: hemal arch, hemoglobin, hemorrhage
hemi— [Gr. *hemi,* prefix meaning half]: hemibranch, hemichordate
Henle [Friedrich Henle, nineteenth-century German anatomist]: Henle's loop
hepa— [Gr. *hepar,* gen., *hepatos,* liver]: hepatic duct, hepatic portal vein, hepatogastric ligament
hetero— [Gr. *heteros,* other, different]: heterocercal, heterogeneity
hippo— [Gr. *hippos,* horse]: hippocampus
histo— [Gr. *histion,* web, tissue]: histology
holo— [Gr. *holos,* whole, entire]: Holocephali, holonephros, Holostei
homo— [Gr. *homos,* alike, the same]: homodont, homology
hormo— [Gr. *hormao,* pr.p. *hormono,* to rouse or set in motion]: hormone
humerus [L. *humerus,* humerus]: humerus
humor [L. *humor,* liquid; specifically, one of the body fluids thought in ancient times to affect disposition]: aqueous humor, humor
hyalo— [Gr. *hyalos,* glass]: hyaline cartilage
hyo— [Gr. *hyoeides,* in the form of the letter upsilon (U-shaped)]: hyoglossus muscle, hyoid bone, hyomandibula
hyper— [Gr. *hyper-,* prefix meaning above, over]: hyperactive, hypertrophy
hypo— [Gr. *hypo-,* prefix meaning under, beneath]: hypaxial, hypoglossal nerve, hypophysis, hypothalamus
ichthy— [Gr. *ichthys,* gen., *ichthyos,* fish]: Chondrichthyes, Osteichthyes, ichthyology
ile—, *ili*— [L. *ileum* or *ilium,* the groin]: ileum, iliac, ilium, ileal papilla
incis— [L. *incidio,* to cut into, p.p., *incisus,* cut]: incision, incisor teeth
incus [L. *incus,* an anvil]: incus of middle ear
ineum [Gr. *inan,* to excrete]: perineum
infra— [L. *infra-,* prefix meaning below]: infraorbital foramen, infraspinatus muscle
infundibulum [L. *infundibulum,* a funnel]: infundibulum of the uterine tube
inguin— [L. *inguen,* groin, *inguinalis,* pertaining to the groin]: inguinal canal
integument [L. *integumentum,* covering]: integument
inter— [L. *inter-,* prefix meaning between]: intercalary plate, intercostal muscle, internuncial neuron, interparietal bone

intra— [L. *intra-,* prefix meaning within]: intracellular

ipsi— [L. *ipse,* same]: ipsilateral

iri— [Gr. *iris,* gen., *iridos,* rainbow]: iridescent, iris of the eye

ischi— [Gr. *ischion,* hip]: ischial, ischium

iso— [Gr. *isos,* equal]: isocortex, isometric contraction

isthm— [Gr. *isthmos,* narrow passage, constriction]: isthmus faucium, thyroid isthmus, isthmus of the pancreas

Jacobson [Ludwig L. Jacobson, nineteenth-century Danish surgeon and anatomist]: Jacobson's organ

jejun— [L. *jejunus,* fasting, empty]: jejunum (so called because it usually is found to be empty in dissection)

jug— [L. *jugum,* dim., *jugulum,* collarbone]: jugular vein

kerat— [Gr. *keras,* gen., *keratos,* horn]: keratin

kine— [Gr. *kineo,* to move]: cinema (moving pictures), kinetic skull

labi— [L. *labium,* pl., *labia,* lip]: labial cartilages, labia majora

labyrinth [Gr. mythology *Labyrinthos,* the maze in which the Minotaur was confined]: membranous labyrinth of the ear

lachr—, *lacrim*— [L. *lacrima,* tear]: lacrimal bone, nasolacrimal duct

lacuna [L. *lacuna,* a small cavity or hollow]: lacunae in cartilage or bone

lagen— [Gr. *lagenos,* flask]: lagena

lam— [L. *lamina,* dim., *lamella,* layer or thin plate]: lamellar bone, lamina terminalis, laminate

Langerhans [Paul Langerhans, nineteenth-century German anatomist]: islets of Langerhans

laryn— [Gr. *larynx,* gen., *laryingis,* gullet]: laryngeal artery, larynx

lat— [L. *latus,* gen., *lateris,* side, broad]: fascia lata, lateral, latissimus dorsi muscle

lemniscus [L. *lemniscus,* ribbon]: lateral lemniscus

len— [L. *lens,* gen., *lentis,* a lentil]: lens, lentiform nucleus

Leydig [Franz von Leydig, nineteenth-century German anatomist]: interstitial cells of Leydig

lien— [L. *lien,* spleen]: gastrolienic ligament, lienogastric artery

ligament [L. *ligamentum,* band]: ligament

limb— [L. *limbus,* edge, border]: limbic lobe

lingu— [L. *lingua,* dim., *lingula,* tongue]: lingualis muscle, lingual nerve, sublingual gland

lip— [Gr. *lipos,* fat]: lipids, liposuction

liss— [Gr. *lissos,* smooth]: Lissamphibia

lith— [Gr. *lithos,* stone]: lithosphere, otolith

—*logy* [Gr. *logos,* study]: biology, embryology

lumb— [L. *lumbus,* loin]: lumbar artery, thoracolumbar fascia

lun— [L. *luna,* moon]: lunate bone, lunatic, semilunar valve

lute— [Gr. *luteus,* yellow]: corpus luteum

lymph— [L. *lympha,* water]: endolymph, lymphocyte

macula [L. *macula,* pl., *maculae,* spot]: maculae in inner ear

magn— [L. *magnus,* great]: foramen magnum, magnificent

mall— [L. *malleus,* hammer]: mallet, malleus

Malpighi [Marcello Malpighi, seventeenth-century Italian histologist]: Malpighian corpuscle

mam— [L. *mamilla,* nipple]: mamillary body

mamma— [L. *mamma,* breast]: mammal, mammography

man— [L. *manus,* hand]: manual, manubrium, manus, manipulate

mandibul— [L. *mandibula,* lower jaw]: mandible, mandibular gland

marsup— [L. *marsupium,* purse, pouch]: Marsupialia, marsupium

masseter [Gr. *masseter,* a chewer]:masseter muscle

mast— [Gr. *mastos,* breast]: mastoid process, neuromast

mater [L. *mater,* mother; *maternus,* motherly]: dura mater, maternal

maxilla [L. *maxilla,* upper jaw]: maxilla

meatus [L. *meatus,* passage]: acoustic meatus

Meckel [Johann F. Meckel, eighteenth-century German anatomist]: Meckel's cartilage

mediastin— [L. *mediastinus,* median, from *medius,* middle]: mediastinum

medulla [L. *medulla,* marrow]: medulla oblongata, renal medulla

melan— [Gr. *melas,* gen., *melanos,* black]: melancholy, melanin, melanocyte

meninges [Gr. *meninx,* pl., *meninges,* membrane]: meninges covering the brain and spinal cord

ment— [L. *mentum,* chin]: mental foramen

mer— [Gr. *meros,* part]: branchiomeric, myomere

mes— [Gr. *mesos,* middle]: mesencephalon, mesentery, mesoderm, mesonephros

met—, *meta*— [Gr. *meta,* beside, after]: metacarpal, metacromion, metanephros, Metatheria

metr— [Gr. *metra,* uterus]: endometrium, mesometrium

mol— [L. *mola,* millstone]: molar teeth

mon— [Gr. *monas,* single]: monarch, Monotremata, monocular

Monro [Alexander Monro, eighteenth-century Scottish anatomist]: foramen of Monro

morph— [Gr. *morphe,* shape]: morphology, amorphous

muc— [L. *mucosus,* slimy]: mucus, mucous membrane

Müller— [Johannes P. Müller, nineteenth-century German anatomist and physiologist]: Müllerian duct

mult— [L. *multus,* many]: multifidus muscle, multiply

myel— [Gr. *myelos,* marrow, spinal cord]: myelencephalon, myelin

myl— [Gr. *myle,* millstone, molar]: mylohyoid muscle

myo— [Gr. *mys,* gen., *myos,* muscle]: myology, myomere, myotome

naris [L. *naris,* pl., *nares,* external nostril]: naris
nas— [L. *nasus,* nose]: nasal bone
nav— [L. *navis,* dim., *navicula,* ship]: navicular bone, navy
neo— [Gr. *neos,* new, young]: neo-Darwinism, neopallium
nephr— [Gr. *nephros,* kidney]: mesonephros, nephron, protonephridia, nephritis
neuro— [Gr. *neuron,* tendon, nerve]: neurocoel, neuro mast, neuron
nict— [L. *nictare,* to wink]: nictitating membrane
nidament— [L. *nidamentum,* nesting material]: nidamental gland
noct— [L. *nox,* gen., *noctis,* night]: nocturnal
noto— [Gr. *notos,* back]: notochord
nucleus [L. *nucleus,* kernel of a nut]: nucleated, nucleus, nucleus pulposus
nuch— [L. *nucha,* nape of neck]: nuchal crest
obturat— [L. *obturare,* p.p., *obturatus,* to close by stopping up]: obturator foramen, obturator internus muscle
occipit— [L. *occiput,* gen., *occipitis,* the back of the head]: occipital bone
octav— [L. *octavus,* eighth]: octavolateralis system, octave
ocul— [L. *oculus,* eye]: bulbus oculi, orbicularis oculi muscle, oculomotor nerve
odont— [Gr. *odous,* gen., *odontos,* tooth]: thecodont, odontoblast
—*oid* [New L. *-oid,* from Gr. *o* + *eidos,* form, suffix that indicates resemblance to]: arachnoid, diploid, pterygoid, sphenoid, xiphoid
olecranon [Gr. *olekranon,* elbow tip]: olecranon
olfac— [L. *olfacere,* p.p., *olfactus,* to smell]: olfactory nerve, olfaction
om— [Gr. *omos,* shoulder]: metacromion, omohyoid muscle, omotransversarius muscle
oment— [L. *omentum,* membrane]: greater omentum, omental bursa
oo— [Gr. *oon,* egg]: epoophoron, oogenesis
ophthalm— [Gr. *ophthalmos,* eye]: ophthalmic nerve, ophthalmologist
operculum [L. *operculum,* lid, cover]: operculum
opisth— [Gr. *opisthe,* behind, at the rear]: opisthonephros, opisthotic
opt— [Gr. *optikos,* pertaining to sight]: optic nerve
or— [L. *os,* gen., *oris,* mouth]: oral cavity, oropharyngeal cavity, orbicularis oris muscle
orb— [L. *orbis,* dim., *orbiculus,* circle]: orbicularis oris muscle, orbit
orchi— [Gr. *orchis,* testicle]: mesorchium, orchid (from shape of root)
os, oss— [L. *os,* gen., *ossis,* dim., *ossiculum,* bone]: ossicle, ossify, os penis
oste— [Gr. *osteon,* bone]: Osteichthyes, osteoblast, Teleostei
osti— [L. *ostium,* a door or mouthlike opening]: ostium of oviduct
ostrac— [Gr. *ostrakon,* shell, pot-sherd for voting]: ostracoderm, ostracized
ot— [Gr. *otikos,* pertaining to the ear]: otic capsule, otolith, parotid gland
ov— [L. *ovum,* egg]: ovary, oviduct, oviparous
paed— [Gr. *pais,* gen., *paidos,* child]: paedomorphic, pediatrics
palae—, *pale*— [Gr. *palaios,* ancient]: paleontology, paleopallium
palat— [L. *palatum,* roof of the mouth]: palate, palatine bone, tensor veli palatini muscle
palli— [L. *pallium,* a Roman cloak]: neopallium
palpebr— [L. *palpebra,* eyelid]: levator palpebrae superioris muscle
pampin— [L. *pampinus,* tendril]: pampiniform plexus
pan— [Gr. *pan,* all]: Pan-American, pancreas
Papez [James W. Papez, nineteenth- to twentieth-century American anatomist]: Papez circuit in brain
par— [L. *pareo,* p.p., *partus,* to bring forth]: parturition, viviparous
par—, *para*— [Gr. *para,* beside]: paradidymis, parapophysis, parasympathetic nerve, parotid gland
pariet— [L. *paries,* gen., *parietis,* wall]: parietal, peritoneum
pat— [L. *patina,* dim., *patella,* dish or plate]: patella of knee
path— [Gr. *pathetikos,* sensitive, liable to suffer, from *pathos,* suffering]: pathology, sympathetic nervous system
pect— [L. *pectus,* gen., *pectoris,* chest]: pectoral girdle, pectoral muscle
pectin— [L. *pecten,* gen., *pectinis,* comb]: pectinate muscles of heart, pectineus muscle
ped—, See *paed*—.
ped—, See *pes*.
pes, ped— [L. *pes,* gen., *pedis,* dim., *pedunculus,* foot]: bipedal, cerebral peduncle, pes, quadruped, pedicle
pellucid— [L. *pellucidus,* clear, transparent]: septum pellucidum, zona pellucida
pelv— [L. *pelvis,* basin]: pelvic girdle, renal pelvis
peri— [Gr. *peri-,* prefix meaning around]: periosteum, peritonaeum, perineum
perine— [Gr. *perineon,* region between anus and genitals]: perineum
peron— [Gr. *perone,* pin, fibula]: peroneal nerve, peroneus muscle
petr— [Gr. *petros,* rock]: Peter, petrify, petrous portion of temporal bone
phago— [Gr. *phago,* to eat]: phagocytosis, coprophagy
phalang— [G. *phalanx,* gen., *phalangos,* a line of soldiers]: phalanges
phall— [Gr. *phallos,* penis]: phallic, phallus
pharyn— [Gr. *pharynx,* gen., *pharyngos,* pharynx]: glossopharyngeal nerve, pharynx, oropharyngeal cavity
phor— [Gr. *phoros,* from *pherein,* to bear]: epoophoron, pterygiophore, spermatophore
phragm— [Gr. *phragmos,* fence, partition]: diaphragm

phren— [Gr. *phren,* diaphragm, mind]: phrenic nerve, phrenology
phys— [Gr. *physis,* a growth]: apophysis, epiphysis, hypophysis, symphysis
pia [L. *pia,* tender]: pia mater
pin— [L. *pineus,* pertaining to a pine tree]: pineal gland
pir— [L. *pirum,* pear]: piriform lobe, piriformis muscle
pis— [L. *pisum,* pea]: pisiform bone
plac— [Gr. *plax,* gen., *plakos,* flat plate]: placoderms, placoid scale
placenta [L. *placenta,* small, flat cake]: placenta
plant— [L. *planta,* sole of the foot]: plantaris muscle, plantigrade
plast— [contraction of Gr. *emplassein,* p.p., *emplastros,* to daub on]: epiplastron, plaster, plastron
platy— [Gr. *platys,* flat or broad]: duck-billed platypus, platysma muscle
pleur— [Gr. *pleura,* side, rib]: metapleural fold, pleural cavity, pleurapophysis
plexus [L. *plexus,* an interweaving, network]: brachial plexus
pod— [Gr. *pous,* gen., *podos,* foot]: podocyte, tetrapod
poie— [Gr. *poieo,* to make]: hemopoietic
poll— [L. *pollex,* gen., *pollicis,* thumb]: abductor pollicis muscle
pon— [L. *pons,* gen., *pontis,* bridge]: brachium pontis, pons
poplit— [L. *poples,* gen., *poplitis,* knee joint]: popliteal fossa, popliteus muscle
port— [L. *porta,* gate]: portal vein
post— [L. *post-,* prefix meaning after, behind]: posterior, postorbital process
pre— [L. *prae-,* prefix meaning before, in front]: precaval vein, premaxillary bone
prepu— [L. *praeputium,* the foreskin]: prepuce
prim— [L. *primus,* first]: primates, primitive
pro— [Gr. *pro-,* prefix meaning before, in front]: pronephros, prosencephalon, protraction
proct— [Gr. *proktos,* the anus]: proctodaeum
pron— [L. *pronus,* bending, leaning forward]: pronator teres muscle, prone, pronation
prostat— [Gr. *prostates,* one who stands before]: prostate (stands before the bladder)
proto— [Gr. *protos,* first]: protoplasm, Prototheria
pseudo— [Gr. *pseudos,* false]: pseudobranch
psoa— [Gr. *psoa,* the loin]: psoas major muscle
ptery— [Gr. *pteryx,* gen., *pterygos,* wing, fin]: Archaeopteryx, Actinopterygii, metapterygium, pterygoid process
pub— [L. *pubes,* young adult]: puberty, pubic hair, pubis
pudend— [L. *pudendum,* external genitals, from *pudere,* to be ashamed]: pudendal artery, pudendum
pulmo— [L. *pulmo,* gen., *pulmonis,* lung]: pulmonary artery
pulvin— [L. *pulvinus,* cushion]: pulvinar
Purkinje [Johannes von Purkinje, nineteenth-century Bohemian (Czech) anatomist]: Purkinje cells
—*pus* [Gr. *pous,* gen., *podos,* foot]: platypus
putamen [L. *putamen,* a pod]: putamen
pylor— [Gr. *pyloros,* gate keeper]: pyloric region of stomach, pylorus
quadr— [L. *quadrus,* fourfold]: quadriceps femoris muscle, quadruped
quadrat— [L. *quadratus,* squared]: quadrate bone, quadratus lumborum muscle
radi— [L. *radius,* ray, spoke]: corona radiata, radium, radius
radix [L. *radix,* pl., *radices,* root]: radix of aorta
ram— [L. *ramus,* branch]: ramify, ramus
raphe [Gr. *raphe,* seam, suture]: raphe (in muscles)
re— [L. *re-,* prefix indication indicates backward]: retractor muscle, reaction
rect— [L. *rectus,* straight, *rectum,* straight intestine]: rectangular, rectum, rectus abdominis muscle
ren— [L. *ren,* kidney]: renal artery, renal corpuscle
rept— [L. *reptare,* p.p., *reptum,* to creep]: reptile
rest— [L. *restis,* rope]: restiform body
ret— [L. *rete,* dim., *reticulum,* net]: rete mirabile, reticular formation, retina
retinaculum [L. *retinaculum,* band, holdfast]: extensor retinaculum
retro— [L. *retro,* behind]: retroperitoneal
rhin— [Gr. *rhis,* gen., *rhinos,* nose]: rhinal fissure, rhinarium, rhinencephalon, rhinoceros
rhomb— [Gr. *rhombus,* parallelogram with oblique angles and unequal adjacent sides]: rhombencephalon, rhomboideus muscle
rostr— [L. *rostrum,* beak, ship's prow]: rostral, rostrum
ruga [L. *ruga,* gen., *rugae,* fold]: gastric rugae
rum— [L. *rumino,* to chew the cud, from *rumen,* throat]: rumen, ruminate
sac— [L. *saccus,* dim., *sacculus,* bag]: sacculus, sac
sacr— [L. *sacrum,* sacred]: sacral nerve, sacrum
sagitt— [Gr. *sagitta,* arrow]: sagittal plane, Sagittarius
salpinx [Gr. *salpinx,* trumpet]: mesosalpinx
saphen— [Gr. *saphene,* manifest, clearly visible]: saphenous vein
sarco— [Gr. *sarx,* gen., *sarkos,* flesh]: sarcophagus, Sarcopterygii
sartor— [L. *sartor,* tailor]: sartorius muscle
scala [L. *scala,* ladder]: scala vestibuli
scale— [Gr. *skalenos,* a triangle with three unequal sides]: scalenus muscle
scaph— [Gr. *skaphe,* bowl, boat]: scaphoid bone
scapul— [L. *scapula,* shoulder blade]: scapula, scapulocoracoid
Schlemm [Friedrich Schlemm, nineteenth-century German anatomist]: canal of Schlemm

sclera— [Gr. *skleros,* hard]: sclera of eyeball, sclerous, sclerotome
scrotum [L. *scrotum,* pouch, scrotum]: scrotum
scut— [L. *scutum,* shield]: scute
seb— [L. *sebum,* grease, wax]: sebaceous gland, sebum
sella [L. *sella,* saddle, seat]: sella turcica
sem— [L. *semen,* gen., *seminis,* seed]: semen, seminal vesicle, seminiferous tubules
semi— [L. *semi-,* prefix meaning partly, half]: semicircular canal, semispinalis muscle, semitendinosus muscle
septum [L. *septum,* a partition]: transverse septum
serr— [L. *serra,* saw]: serrated, serratus ventralis muscle
Sertoli [Enrico Sertoli, nineteenth-century Italian histologist]: Sertoli cells
sesam— [Gr. *sesame,* seed of the sesame plant]: sesamoid bone
sinus [L. *sinus,* cavity]: sinus venosus, sphenoidal sinus
sole— [L. *solea,* sandal]: sole of the foot, soleus muscle
somat— [Gr. *soma,* gen., *somatos,* body]: somatic, somatopleure
sperm— [Gr. *sperma,* gen., *spermatos,* seed, semen]: sperm, spermatic fascia, spermatophore
sphen— [Gr. *sphen,* wedge]: sphenoid bone
sphinct— [Gr. *sphinkter,* that which binds tightly]: sphincter muscle
spin— [L. *spina,* thorn or spine]: erector spinae muscle, spinalis muscle, supraspinatus muscle
spirac— [L. *spiraculum,* pore, air hole]: spiracle
splanchn— [Gr. *splanchnon,* viscus, inner organ]: splanchnic nerve
splen— [Gr. *splen,* spleen]: spleen, splenic
spleni— [Gr. *splenion,* bandage]: splenial bone, splenius muscle
squam— [L. *squama,* scale]: squamate, squamosal bone, squamous epithelium
stap— [L. *stapes,* stirrup]: stapedius muscle, stapes bone
stell— [Gr. *stella,* star]: stellate
stern— [Gr. *sternon,* breast, chest]: sternebrae, sternomastoid muscle, sternum
stom— [Gr. *stoma,* gen., *stomatos,* mouth]: cyclostome, stomochord, stomodaeum, gnathostome
stomach— [Gr. *stomachos,* stomach]: stomach
stratum [L. *stratum,* layer]: stratum corneum
stri— [L. *striare,* p.p., *striatus,* to make furrows or stripes]: corpus striatum, striated
styl— [Gr. *stylos,* pillar, stalk]: endostyle, hyostylic, styloid process, stylomastoid foramen
sub— [L. *sub-,* prefix meaning beneath, below]: subscapular fossa, subscapular muscle
sulcus [L. *sulcus,* furrow]: sulcus of brain
super— [L. *super,* above, beyond]: superficial
supra— [L. *supra,* on the upper side]: suprarenal gland, supraspinatus muscle, suprascapular process
sur— [Old French *sur-,* prefix, from L. *super,* above, beyond]: surangular bone, surcharge
sura [L. *sura,* gen., *surae,* calf of the lower leg]: triceps surae muscle
sutur— [L. *sutura,* seam, suture]: suture
Sylvius [Jacques Dubois Sylvius, fifteenth- to sixteenth-century French anatomist]: aqueduct of Sylvius
sym— See *syn*—.
syn— [Gr. *syn-, sym-,* prefix meaning together or with]: symbiosis, sympathy, symphysis, synapsid skull, synchondrosis, synotic tectum, synapomorphy
syst— [Gr. *systema,* a composite whole]: organ system, systemic veins
taenia [L. *taenia,* ribbon, band]: taenia coli
talus [L. *talus,* ankle]: talus bone
tape— [Gr. *tapes,* dim., *tapetion,* carpet]: tapestry, tapetum lucidum
tars— [Gr. *tarsos,* flat surface, sole of foot]: metatarsal bone, tarsus
tectum [L. *tectum,* roof]: optic tectum
tegmentum [L. *tegmentum,* covering]: tegmentum of mesencephalon
tel— [Gr. *tele,* far off, distant]: telencephalon, teleost, television, telephone
tela— [L. *tela,* weblike membrane]: tela choroidea
tempor— [L. *tempus,* gen., *temporis,* time]: temple of the head, temporal bone, temporal muscle
ten— [L., *tendere,* p.p., *tentus,* to stretch]: tendon, tensor tympani muscle, tent, tentorium
—teny [Gr. *teinein,* to extend]: neoteny
teres [L. *teres,* round, smooth]: teres major muscle, ligamentum teres
test— [L. *testis,* a witness (originally an adult male)]: testify, testicular, testis
tetra— [Gr. *tetra,* four]: tetrapods
thalam— [Gr. *thalamos,* inner chamber]: epithalamus, thalamus
thec— [L. *theca,* sheath]: theca, thecodont
thel— [Gr. *thelys,* tender, delicate]: endothelium, epithelium
ther— [Gr. *therion,* wild beast]: Eutheria, Prototheria
thora— [Gr. *thorax,* gen., *thorakos,* chest]: thoracic, thorax, thoracolumbar fascia
thym— [Gr. *thymos,* the thymus (akin to *thymon,* the herb thyme or a lump resembling a bunch of thyme)]: thymine, thymus
thyr— [Gr. *thyreos,* an oblong shield]: thyrohyal muscle, thyroid cartilage
tibia [L. *tibia,* shin bone]: tibia
tom— [Gr. *tomia,* a cutting or segment]: dermatome, myotome
ton— [Gr. *tonos,* something stretched]: peritoneum, tone, tonus

trab— [L. *trabs,* dim., *trabecula,* beam]: basitrabecular processes, trabeculae carneae

trach— [L. *trachea,* windpipe, from *trachys,* rough]: trachea

tract— [L. *trahere,* p.p., *tractus,* to pull or draw out]: corticospinal tract, protractor muscle, retractor muscle

trans— [L. *trans,* across, beyond]: transcend, transverse section, transversus thoracis muscle

trapez— [Gr. *trapeza,* small table of characteristic shape]: trapezius muscle, trapezoid bone

trem— [Gr. *trema,* hole]: Monotremata, pretrematic nerve branch

tri— [L. *tri,* three]: triceps muscle, tricuspid valve, trigeminal nerve

triquetr— [L. *triquetrus,* triangular]: triquetrum bone

trochanter [Gr. *trochanter,* a runner]: greater and lesser trochanters of the femur

trochlea [L. *trochlea,* pulley]: trochlea of humerus, trochlear nerve

tuber [L. *tuber,* dim., *tuberculum,* bump, lump]: tuber cinereum, tuberculum of a rib

tunic— [L. *tunica,* garment]: tunicates, vascular tunic

turb— [L. *turbo,* gen., *turbinis,* a spinning thing]: turbinate bone, turbine

tympan— [L. *tympanum,* drum]: tympanic membrane

ulna [L. *ulna,* elbow, lower arm]: ulna

unci— [L. *uncinus,* hook]: uncinate process

ungu— [L. *unguis,* nail, claw]: ungulate, unguligrade

ur— [Gr. *oura,* tail]: Anura, Urochordata, Urodela

ur— [L. *urina,* urine]: urea, ureter, urethra, urinary, urogenital

uterus [L. *uterus,* the womb]: uterus

utriculus [L. *utriculus,* leather bag or bottle]: utriculus of inner ear

vag— [L. *vagus,* wandering, undecided]: vague, vagus nerve, vagabond

vagina [L. *vagina,* sheath]: vagina, processus vaginalis

vall— [L. *vallare,* to surround with a rampart]: vallate papilla

vast— [L. *vastus,* large area, immense]: vast, vastus lateralis muscle

velum [L. *velum,* gen., *veli,* veil, covering]: rostral medullary velum, tensor veli palatini muscle

ven— [L. *vena,* vein]: vena cava, venous

ventr— [L. *venter,* dim., *ventriculus,* belly, womb]: ventral, ventricle

vermis [L. *vermis,* worm]: vermis of the cerebellum, vermiculite

vertebra [L. *vertebra,* joint, vertebra]: vertebra, vertebrates

vesic— [L. *vesica,* dim., *vesicula,* bladder]: vesica fellea, vesicular gland, lateral vesical ligament

vestibul— [L. *vestibulum,* entrance chamber]: vestibule, vestibulocochlear nerve

vibr— [L. *vibrare,* to agitate]: vibrate, vibrissa

vill— [L. *villus,* pl., *villi,* shaggy hair]: villus, villi

visc— [L. *viscus,* pl., *viscera,* entrails]: visceral, viscous

vita [L. *vita,* life]: arbor vitae, vital

vitell— [L. *vitellus,* yolk]: vitelline veins

vitr— [L. *vitrum,* glass]: vitreous body of eye, vitrify

vivi— [L. *vivus,* alive]: viviparous

vomer [L. *vomer,* plowshare (cutting blade of a plow)]: vomer bone

vor— [L. *vorare,* to devour]: carnivore, herbivore, voracious

vulva [L. *vulva,* covering]: vulva

Willis [Thomas Willis, seventeenth-century English physician]: circle of Willis

Wolff [Kaspar Friedrich Wolff, eighteenth-century German embryologist]: Wolffian duct

xiph— [Gr. *xiphos,* sword]: xiphihumeralis muscle, xiphisternum

zoo— [Gr. *zoon,* animal]: zoology

zyg— [Gr. *zygon,* yolk, union, pair]: azygos vein, zygapophysis, zygomatic arch, zygote

APPENDIX B
A Note on the Handling of Specimens

Specimens for a comparative anatomy course are available through biological supply companies. Students and instructors who wish to prepare their own specimens should consult Hildebrand (1969).

There are various well-established techniques that help maintain specimens in good condition for the duration of a semester-long comparative anatomy course, but here we describe only some techniques that have proved highly successful in our own courses (see also Blaney and Johnson, 1989; Wineski and English, 1989).

Specimens must be treated with toxic and noxious substances, such as formalin, alcohol, and phenol, in order to prevent microorganisms from attacking them. Unfortunately, these preserving agents are volatile and quickly fill the air of a laboratory. The resulting bad air bothers most students greatly and may prevent many from even considering a teaching and research career in the anatomical sciences. Although soaking the specimens in running cold water removes the smell, it also makes the specimens vulnerable to decay.

An excellent alternative technique that avoids virtually all problems of the older techniques involves the use of a 1% phenoxyethanol solution as a wetting agent for the preserved specimens. Phenoxyethanol (2-phenoxyethanol, practical grade; Eastman Kodak Co., Rochester, NY, USA) is a nontoxic liquid that has a faint but pleasant scent. It often is used as a preserving agent in cosmetic products and externally applied medications. Because it is used as a 1% solution, it is relatively inexpensive. It is nonflammable, effectively prevents the growth of microorganisms, and softens and rehydrates the tissues of specimens. European museum collections have used phenoxyethanol as a storage solution for wet specimens for several decades.

Phenoxyethanol is slightly lipophilic and sinks as liquid globules to the bottom if it is mixed with water in a concentration that exceeds 2%. A 1% solution is best prepared by mixing exactly one part of phenoxyethanol with 99 parts of very hot tap water. Use of a magnetic stirrer for about 30 minutes ensures a homogeneous solution.

Every student should be supplied with a plastic dispensing bottle filled with the 1% phenoxyethanol solution. During the dissection, the specimen should be repeatedly and liberally sprayed or doused with the solution. Between laboratory sessions, the specimens should be wrapped in rags or cheesecloth soaked with the phenoxyethanol solution and then stored in tightly closed plastic bags or in storage bins that are filled with the phenoxyethanol solution.

The condition of specimens with excessively dry, hardened, or brittle tissues can be greatly improved by immersing them in a 1% phenoxyethanol solution to which some fabric softener has been added. Any fabric softener commonly used for laundry can be used, preferably in the proportion of one part of fabric softener to 99 parts of the 1% phenoxyethanol solution. The specimens can be left in this conditioning solution for the duration of a semester or more. The occasional precipitation of the fabric softener into whitish clouds does not diminish the effectiveness of the solution.

APPENDIX C References

The references given below include those cited in the text, those of particular value for laboratory studies in comparative anatomy, and certain key references on the functional significance and evolutionary interrelationships of the various organs. More inclusive bibliographies can be found in many of the works cited below.

Students who are interested in current research in comparative or functional anatomy may want to look up a few journals in their library. *Scientific American, Discover,* and *The American Scientist* are journals in magazine format. They publish articles on a variety of scientific subjects, including anatomical topics, which are current, readable, and easily understood even by beginning students. *The American Zoologist* publishes excellent review articles on a variety of zoological subjects, including anatomy. They are accessible to advanced undergraduate students. Original research articles in comparative and functional anatomy are published in several scientific journals, such as *Acta Anatomica, The American Journal of Anatomy, The Anatomical Record, The Biological Journal of the Linnean Society, The Zoological Journal of the Linnean Society, The Journal of Anatomy, The Journal of Experimental Biology, The Journal of Experimental Zoology, The Journal of Morphology, The Journal of Zoology, Proceedings of the Royal Society of London Series B, Zoologischer Anzeiger, Zoology,* and *Zoomorphology.*

Abell, N. B.: "A comparative study of the variations of the postrenal vena cava of the cat and rat and a description of two new variations." *Denison University Journal of the Science Laboratory,* vol. 40, pp. 87–117, 1947.

Aleev, Y. G.: *Function and Gross Morphology in Fish.* Washington, DC, Smithsonian Institution, 1969.

Alexander, R. M.: *Animal Mechanics.* 2nd ed. Oxford, Blackwell Scientific Publications, 1983.

Alexander, R. M.: *The Chordates.* 2nd ed. Cambridge, Cambridge University Press, 1981.

Alexander, R. M.: "The energetics of coprophagy: A theoretical analysis." *Journal of Zoology (London),* vol. 230, pp. 629–637, 1993a.

Alexander, R. M.: "The relative merits of foregut and hindgut fermentation." *Journal of Zoology (London),* vol. 231, pp. 391–401, 1993b.

Allaby, M.: *A Dictionary of Zoology.* 2nd ed. New York, Oxford University Press, 1999.

Allin, E. P.: "Evolution of the mammalian inner ear." *Journal of Morphology,* vol. 147, pp. 403–438, 1975.

Arey, L. B.: *Developmental Anatomy.* 7th ed. revised. Philadelphia, W. B. Saunders Company, 1974.

Baker, M. A.: "A brain-cooling system in mammals." *Scientific American,* vol. 240 (May), pp. 130–139, 1979.

Balinsky, B. I., assisted by Fabian, B. C.: *An Introduction to Embryology.* 5th ed. Philadelphia, Saunders College Publishing, 1981.

Banks, W. J.: *Applied Veterinary Histology.* 2nd ed. Baltimore, Williams & Wilkins, 1986.

Barone, R., Pavaux, C., Blin, P. C., and Cuq, P.: *Atlas d'Anatomie du Lapin.* Paris, Masson et Cie, 1973.

Barrington, E. J. W.: "The supposed pancreatic organs of *Petromyzon fluviatilis* and *Myxine glutinosa.*" *Quarterly Journal of Microscopical Science,* vol. 85, pp. 391–417, 1945.

Barrington, E. J. W.: *The Biology of Hemichordata and Protochordata.* San Francisco, W. H. Freeman and Company, 1965.

Barry, A.: "The aortic arch derivatives in the human adult." *Anatomical Record,* vol. 111, pp. 221–238, 1951.

Baumel, J. J., King, A. S., Breazile, J. E., Evans, H. E., and Vanden Berge, J. C. (eds.): *Handbook of Avian Anatomy: Nomina Anatomica Avium.* 2nd ed. Cambridge, Massachusetts, Publications of the Nuttall Ornithological Club, No. 23, 1993.

Bels, V. L., Chardon, M., and Vandewalle, P.: "Biomechanics of Feeding in Vertebrates." In *Advances in Comparative and Environmental Physiology.* Vol. 18. Berlin, Springer-Verlag, 1994.

Bels, V. L., Gasc, J.-P., and Casinos, A.: *Vertebrate Biomechanics and Evolution.* Oxford, BIOS Scientific Publishers Ltd., 2003.

Bemis, W. E., and Grande, L.: "Early development of the actinopterygian head: I. External development and staging in the paddlefish, *Polydon spatulata.*" *Journal of Morphology,* vol. 213, pp. 47–83, 1992.

Benton, M. J.: *Vertebrate Paleontology.* 2nd ed. London, Chapman and Hall, 1997.

Berson, D. M., Dunn, F. E., and Tako, M.: "Phototransduction by retinal ganglion cells that set the circadian clock." *Science,* vol. 295, pp. 1070–1073, 2002.

Blackburn, D. G.: "Evolutionary origins of the mammary gland." *Mammal Reviews,* vol. 21, pp. 81–96, 1991.

Blaney, S. P. A., and Johnson, B.: "Technique for reconstituting fixed cadaveric tissue." *Anatomical Record,* vol. 224, pp. 550–551, 1989.

Bolau, H.: "Ueber die Paarung und Fortpflanzung der Scyllium-Arten." *Zeitschrift für die Wissenschaftliche Zoologie,* vol. 32, pp 321–325, 1881.

Bolis, L., Keynes, R. D., and Maddrell, S. H. P. (eds.): *Comparative Physiology of Sensory Systems.* New York, Cambridge University Press, 1984.

Bolk, L., Göppert, E., Kallius, E., and Lubosch, W. (eds.): *Handbuch der vergleichenden Anatomie der Wirbeltiere.* Six volumes. Berlin and Vienna, Urban und Schwarzenberg, 1931–1938. Reprinted in 1967 by A. Asher and Co., Amsterdam.

Bone, Q., Marshall, N. B., and Blaxter, J. H. S.: *Biology of Fishes.* 2nd ed. London, Stanley Thornes Publishers, 1995.

Boord, R. L., and Campbell, C. B. G.: "Structural and functional organization of the lateral line system of sharks." *American Zoologist,* vol. 17, pp. 431–441, 1977.

Brainerd, E. L.: "Mechanics of lung ventilation in a larval salamander *Ambystoma trigrinum.*" *Journal of Experimental Biology,* vol. 201, pp. 2891–2901, 1998.

Brainerd, E. L., Ditelberg, J. S., and Bramble, D. M.: "Lung ventilation of salamanders and the evolution of vertebrate air-breathing mechanisms." *Biological Journal of the Linnean Society,* vol. 48, pp. 163–183, 1993.

Bramble, D. M.: "Origin of the mammalian feeding complex, models and mechanism." *Paleobiology,* vol. 4, pp. 271–301, 1978.

Browder, L. W., Erickson, C. A., and Jefferey, W. R.: *Developmental Biology.* 3rd ed. Philadelphia, Saunders College Publishing, 1991.

Brown, R. W.: *Composition of Scientific Words.* Washington, DC, Smithsonian Institution, 2000.

Budker, P., and Whitehead, P. J.: *The Life of Sharks.* New York, Columbia University Press, 1971.

Burke, A. C.: "Development of the turtle carapace: Implications for the evolution of a novel Bauplan." *Journal of Morphology,* vol. 199, pp. 363–378, 1989.

Burke, A. C., and Alberch, P.: "The development and homology of the chelonian carpus and tarsus." *Journal of Morphology,* vol. 186, pp. 119–131, 1985.

Butler, A. B.: "Chordate evolution and the origin of craniates." *Anatomical Record,* vol. 261, pp. 111–125, 2000.

Butler, A. B, and Hodos, W.: *Comparative Vertebrate Neuroanatomy: Evolution and Adaptation.* New York, Wiley-Liss, 1996.

Cahn, P. H. (ed.): *Lateral Line Detectors.* Bloomington, Indiana University Press, 1967.

Capdevila, J., and Belmonte, J. C. I.: "Perspectives on the origin of tetrapod limbs." *Journal of Experimental Zoology,* vol. 288, pp. 287–303, 2000.

Capranica, R. R.: "Morphology and physiology of the auditory system." In Llianas, R., and Precht, W. (eds.): *Frog Neurobiology.* Berlin, Springer-Verlag, 1976.

Carey, F. G.: "Fishes with warm bodies." *Scientific American,* vol. 228 (February), pp. 36–44, 1973.

Carey, F. G., and Gibson, Q. H.: "Heat and oxygen exchange in the rete mirabile of the bluefin tuna, *Thunnus thynnus.*" *Comparative Biochemistry and Physiology,* vol. 74A, pp. 333–342, 1983.

Carrier, D. R.: "Action of the hypaxial muscles during walking and swimming in the salamander *Dicamptodon ensatus.*" *Journal of Experimental Biology,* vol. 180, pp. 75–83, 1993.

Carroll, R. L.: *Vertebrate Paleontology and Evolution.* New York, W. H. Freeman & Co., 1988.

Cave, A. J. E.: "The morphology of mammalian cervical pleurapophyses." *Journal of Zoology (London),* vol. 177, pp. 377–393, 1975.

Chase, S. W.: "The mesonephros and urogenital ducts of *Necturus maculosus* Rafinesque." *Journal of Morphology,* vol. 37, pp. 457–532, 1923.

Clark, J. A.: *Gaining Ground: The Origin and Evolution of Tetrapods.* Bloomington, Indiana University Press, 2002.

Coates, M. I., and Clark, J. A.: "Polydactyly in the earliest known tetrapod limbs." *Nature,* vol. 347, pp. 66–69, 1990.

Corwin, J. T.: "Audition in elasmobranchs." In Tavolga, W. N., Popper, A. N., and Fay, R. R. (eds.): *Hearing and Sound Communication in Fishes.* New York, Springer-Verlag, 1981.

Corliss, C. E.: *Patten's Human Embryology.* New York, McGraw-Hill, 1976.

Crompton, A. W., and Parker, P.: "Evolution of the mammalian masticatory apparatus." *American Scientist,* vol. 66 (March-April), pp. 192–201, 1978.

Crouch, J. E.: *Text-Atlas of Cat Anatomy.* Philadelphia, Lea & Febiger, 1969.

Daniel, J. F.: *The Elasmobranch Fishes.* 3rd ed. Berkeley, University of California Press, 1934.

Davis, D. D.: "The giant panda, a morphological study of evolutionary mechanisms." *Fieldiana, Zoological Memoirs,* vol. 3, pp. 1–340, 1964.

Davis, D. D., and Story, H. E.: "The carotid circulation in the domestic cat." *Zoological Series Field Museum of Natural History,* vol. 28, pp. 1–47, 1943.

Davis, R. E., and Northcutt, R. G. (eds.): *Fish Neurobiology. Vol. 2. Higher Brain Areas and Functions.* Ann Arbor, University of Michigan Press, 1983.

Dean, B.: *Fishes, Living and Fossil.* New York, Macmillan, 1895.

De Beer, G. R.: *The Vertebrate Skull.* Oxford, Clarendon Press, 1937. Reprinted in 1985 by Chicago University Press, Chicago.

De Long, K. T.: "Quantitative analysis of blood circulation through the frog heart." *Science,* vol. 138, pp. 693–694, 1962.

Demski, L. S.: "Electrical stimulation of the shark brain." *American Zoologist,* vol. 17, pp. 487–500, 1977.

Demski, L. S.: "Evolution of LHRH system and terminal nerve in the vertebrate reproductive brain." *American Zoologist,* vol. 24, pp. 809–839, 1984.

Demski, L. S., Fields, R. D., Bullock, T. H., Schreibmann, M. P., and Margolis-Nunn, H.: "The terminal nerve of sharks and rays." *Annals of the New York Academy of Sciences,* vol. 519, pp. 15–32, 1987.

Dorit, R. L., Walker, W. F., and Barnes, R. D.: *Zoology.* Philadelphia, Saunders College Publishing, 1991.

Dorland, W. A. N.: *Dorland's Illustrated Medical Dictionary.* 29th ed. Philadelphia, W. B. Saunders Company, 2000.

Duellman, W. E., and Trueb, L.: *Biology of Amphibians.* Baltimore, Maryland, Johns Hopkins University Press, 1994.

Dutta, H. M., and Datta Munshi, J. S. (eds.): *Vertebrate Functional Morphology: Horizon of Research in the 21st Century.* Enfield, New Hampshire, Science Publishers Inc., 2001.

Dyce, K. M., Sack, W. O., and Wensing, C. J. G.: *Textbook of Veterinary Anatomy.* Philadelphia, W. B. Saunders Company, 1987.

Dyce, K. M. Sack, W. O., and Wensing, C. J. G.: *Textbook of Veterinary Anatomy.* 3rd ed. Philadelphia: W. B. Saunders, 2002.

Eckert, R., Randall, D., and Augustine, G.: *Animal Physiology—Mechanisms and Adaptations.* 3rd ed. New York, W. H. Freeman & Co., 1988.

Edgeworth, F. H.: *The Cranial Muscles of Vertebrates.* London, Cambridge University Press, 1935.

Evans, D. H.: *The Physiology of Fishes.* 2nd ed. Boca Raton, Florida, CRC Press, 1998.

Evans, H. E.: *Miller's Anatomy of the Dog.* 3rd ed. Philadelphia, W. B. Saunders Company, 1993.

Evans, H. E., and deLahunta, A.: *Miller's Anatomy of the Dog.* 5th ed. Philadelphia, W. B. Saunders Company, 2000.

Fawcett, D. W. (ed.): *Bloom and Fawcett, A Textbook of Histology.* 12th ed. Philadelphia, Lippincott, Williams & Wilkins, 1994.

Federative Committee on Anatomical Terminology: *Terminologia Anatomica—International Anatomical Terminology.* Stuttgart, Thieme, 1998.

Ferry, L. A., and Lauder, G. V.: "Heterocercal tail function in leopard sharks: A three-dimensional kinematic analysis of two models." *Journal of Experimental Biology,* vol. 199, pp. 2253–2268, 1996.

Figge, F. H.: "A morphological explanation for the failure of *Necturus* to metamorphose." *Journal of Experimental Zoology,* vol. 56, pp. 241–265, 1930.

Flood, P. P.: "Fine structure of the notochord of *Amphioxus.*" In Barrington, E. J. W., and Jefferies, R. P. S. (eds.): *Protochordates.* Symposia of the Zoological Society of London, no. 36. London, Academic Press, 1975.

Francis, E. B.: *The Anatomy of the Salamander.* London, Oxford University Press, 1943.

Franz, V.: "Vergleichende Anatomie des Wirbeltierauges." In Bolk, L., Göppert, E., Kallius, E., and Lubosch, W. (eds.): *Handbuch der vergleichenden Anatomie der Wirbeltiere.* Vol. 2, part 2. Berlin and Vienna, Urban und Schwarzenberg, 1934. Reprinted in 1967 by A. Asher and Co., Amsterdam.

Frewein, J., and Vollmerhaus, B.: *Anatomie von Hund und Katze.* Berlin, Blackwell Wissenschafts-Verlag, 1994.

Frey, R.: "Zur Ursache des Hodenabstiegs (Descensus testiculorum) bei Säugetieren." *Zeitschrift für Zoologische Systematik und Evolutions-Forschung,* vol. 29, pp. 40–65, 1990.

Frick, H., Kummer, B., and Putz, R.: *Wolf-Heidegger's Atlas of Human Anatomy.* 4th ed. Karger, Basel, 1990.

Gans, C.: *Biomechanics: An Approach to Vertebrate Biology.* Philadelphia, J. B. Lippincott Company, 1974. Reprinted in 1980 by University of Michigan Press, Ann Arbor.

Gans, C., et al. (eds.): *Biology of the Reptilia.* Multiple volumes. London, Academic Press, 1969–1998.

Gans, C., and Parsons, T. S.: *A Photographic Atlas of Shark Anatomy.* Chicago, University of Chicago Press, 1981.

German, R. Z., and Crompton, W. W.: "Ontogeny of feeding in mammals." In Schwenk, K. (ed.): *Feeding in Tetrapod Vertebrates: Form, Function, and Phylogeny.* San Diego, California, Academic Press, 2000.

Getty, R. (ed.): *Sisson and Grossman's Anatomy of the Domestic Animals.* 5th ed. Philadelphia, W. B. Saunders Company, 1975.

Gilbert, S. F.: *Developmental Biology.* 6th ed. Sunderland, Massachusetts, Sinauer Associates, Inc., 2000.

Gilbert, S. G.: *Pictorial Anatomy of the Dogfish.* Seattle, Washington University Press, 1973.

Gilbert, S. G.: *Pictorial Anatomy of the Necturus.* Seattle, University of Washington Press, 1973.

Gilbert, S. G.: *Pictorial Anatomy of the Cat.* 2nd ed. Seattle, University of Washington Press, 1975.

Gonyea, W., and Ashworth, R.: "The form and function of retractile claws in the Felidae and other representative carnivores." *Journal of Morphology,* vol. 145, pp. 229–238, 1975.

Goodrich, E. S.: "On the development of the segments of the head of *Scyllium.*" *Quarterly Journal of Microscopical Science,* vol. 63, pp. 1–30, 1918.

Goodrich, E. S.: *Studies on the Structure and Development of Vertebrates.* London, Macmillan and Co., 1930. Reprinted in 1986 by University of Chicago Press, Chicago.

Grassé, P. P. (ed.): *Traité de Zoologie.* Vols. 11–17 deal with protochordates and vertebrates. Paris, Masson et Cie, 1948–1973.

Grande, L., and Bemis, W. E.: *A Comprehensive Phylogenetic Study of Amiid Fishes (Amiidae) Based on Skeletal Anatomy. An Empirical Search for Interconnected Patterns of Natural History.* Society of Vertebrate Paleontology, Memoir 4, Supplement to *Journal of Vertebrate Paleontology,* vol. 18, 1998.

Graves, J. A. M, and Shetty, S.: "Sex from W to Z: Evolution of sex chromosomes and sex determining genes." *Journal of Experimental Zoology,* vol. 290, pp. 449–462, 2001.

Gray, J.: *Animal Locomotion.* London, Weidenfeld and Nicolson, 1968.

Gregory, W. K.: *Evolution Emerging.* Two volumes. New York, Macmillan, 1951. Reprinted in 1974 by Arno Press, New York.

Griffiths, M.: *The Biology of Monotremes.* New York, Academic Press, 1978.

Gruber, S. H., and Cohen, J. L.: "Visual system of the elasmobranchs: State of the art 1960–1975." In Hodgson, E. S., and Mathewson, R. F. (eds.): *Sensory Biology of Sharks, Skates, and Rays.* Arlington, Virginia, Department of the Navy, Office of Naval Research, 1978.

Gudo, M., and Homberger, D. G.: "The functional morphology of the pectoral fin girdle of the spiny dogfish *(Squalus acanthias):* Implications for the evolutionary history of the pectoral girdle of vertebrates." *Senckenbergiana lethaea,* vol. 82, pp. 241–252, 2002.

Gutmann, W. F.: "Relationships between invertebrate phyla based on functional-mechanical analysis of the hydrostatic skeleton." *American Zoologist,* vol. 21, pp. 63–81, 1981.

Gutmann, W. F.: "The hydraulic principle." *American Zoologist,* vol. 28, pp. 257–266, 1988.

Hamlett, W. C.: *Sharks, Skates, and Rays: The Biology of the Elasmobranch Fishes.* Baltimore and London, Johns Hopkins University Press, 1999.

Hanken, J., and Hall, B. K. (eds.): *The Skull.* Three volumes. Chicago, University of Chicago Press, 1993.

Harris, J. P.: "*Necturus* papers: The skeleton of the arm. The pelvic musculature. The muscles of the forearm. The levator anguli scapulae. The musculus depressor mandibulae. Natural history." *Field and Laboratory,* vols. 20, 21, 22, 25, and 27, 1952–1959.

Hatter, S., Liao, H.-W., Takao, M., Berson, D. M., and Yau, K.-W.: "Melanopsin containing retinal ganglion cells: Architecture, projections, and intrinsic photosensitivity." *Science,* vol. 295, pp. 1065–1070, 2002.

Heath, G. W.: "The siphon sacs of the smooth dogfish and spiny dogfish." *Anatomical Record,* vol. 125, p. 562, 1956.

Herold, R. C., Graver, H. T., and Christner, P.: "Immunohistochemical localization of amelogenins in enameloid of lower vertebrate teeth." *Science,* vol. 207, pp. 1357–1358, 1980.

Herrick, C. J.: *The Brain of the Tiger Salamander, Ambystoma tigrinum.* Chicago, University of Chicago Press, 1948.

Herring, S. W.: "The ontogeny of mammalian mastication." *American Zoologist,* vol. 25, pp. 339–349, 1985.

Hetherington, T. E., Jaslow, A. P., and Lombard, R. E.: "Comparative morphology of the amphibian opercularis system. I. General design features and functional interpretation." *Journal of Morphology,* vol. 190, pp. 42–61, 1986.

Hildebrand, M. *Anatomical Preparations.* Berkeley, University of California Press, 1969.

Hildebrand, M.: *Analysis of Vertebrate Structure.* 4th ed. New York, John Wiley and Sons, 1982.

Hildebrand, M., and Goslow, G.: *Analysis of Vertebrate Structure.* 5th ed. New York, John Wiley and Sons, 2001.

Hildebrand, M., Bramble, D. M., Liem, K. F., and Wake, D. B. (eds.): *Functional Vertebrate Morphology.* Cambridge, Massachusetts, Belknap Press of Harvard University, 1985.

Hoar, W. S., and Randall, D. J.: *Fish Physiology.* Multiple volumes. New York, Academic Press, 1969–1997.

Hochstetter, F.: "Die Entwicklung des Blutgefässystems." In Hertwig, O. (ed.): *Handbuch der vergleichenden und experimentellen Entwickelungslehre der Wirbeltiere.* Vol. 3, part 2. Jena, Fischer, 1906.

Hodgson, E. S., and Mathewson, R. F. (eds.): *Sensory Biology of Sharks, Skates, and Rays.* Arlington, Virginia, Department of the Navy, Office of Naval Research, 1978.

Holland, N. D., and Holland, L. Z.: "Amphioxus and the utility of molecular genetics data for hypothesizing body part homologies between distantly related animals." *American Zoologist,* vol. 39, pp. 630–640, 1999.

Homberger, D. G.: "The case of the cockatoo bill, horse hoof, rhinoceros horn, whale baleen, and turkey beard: The integument as a model system to explore the concepts of homology and non-homology." In Dutta, H. M., and Datta Munshi, J. S. (eds.): *Vertebrate Functional Morphology: Horizon of Research in the 21st Century.* Enfield, New Hampshire, Science Publishers Inc., 2001.

Homberger, D. G. "The avian linguo-buccal system: Multiple functions in nutrition and vocalization." In Adams, N., and Slotow, R. (eds): *Proceedings of the 22nd International Ornithological Congress.* Durban, South Africa, University of Natal, 1999.

Hughes, G. M.: *Comparative Physiology of Vertebrate Respiration.* Cambridge, Harvard University Press, 1963.

Hughes, G. M.: "The relationship between cardiac and respiratory rhythms in the dogfish, *Scyliorhinus canicula* L." *Journal of Experimental Biology,* vol. 57, pp. 415–434, 1972.

Hughes, G. M., and Hills, B. A.: "Oxygen tension distribution in water and blood at the secondary lamella of the dogfish gill." *Journal of Experimental Biology,* vol. 55, pp. 399–408, 1971.

Huntington, G. S., and McClure, C. F. W.: "The development of the veins in the domestic cat." *Anatomical Record,* vol. 20, pp. 1–31, 1920.

Hyman, L. H.: *Comparative Vertebrate Anatomy.* 2nd ed. Chicago, University of Chicago Press, 1942.

International Committee on Veterinary Gross Anatomical Nomenclature: *Nomina Anatomica Veterinaria.* 3rd ed. *Nomina Histologica.* 2nd ed. Ithaca, NY, International Committee on Veterinary Gross Anatomical Nomenclature, 1983.

Jaeger, E.: *A Source-Book of Biological Names and Terms.* 3rd ed. Springfield, Charles C. Thomas, 1978.

Janis, C.: "The evolutionary strategy of the Equidae and the origins of the rumen and cecal digestion." *Evolution,* vol. 30, pp. 757–774, 1976.

Jarvik, E.: *Basic Structure and Evolution of Vertebrates.* Vol. 1. London, Academic Press, 1980.

Jayne, J.: *Mammalian Anatomy. Part I. The Skeleton of the Cat.* Philadelphia, J. B. Lippincott Company, 1898.

Jenkins, F. A.: "The evolution and development of the dens of the mammalian axis." *Anatomical Record,* vol. 164, pp. 173–184, 1969.

Jenkins, F. A.: "Limb posture and locomotion in the Virginia opossum, *Didelphis marsupialis,* and other non-cursorial mammals." *Journal of Zoology (London),* vol. 165, pp. 303–315, 1971.

Jenkins, F. A.: "The movement of the shoulder in claviculate and aclaviculate mammals." *Journal of Morphology,* vol. 144, pp. 71–84, 1974.

Johansen, K.: "Air-breathing fishes." *Scientific American,* vol. 219 (Oct.), pp. 102–111, 1968.

Johnson, J. I., Sudheimer, K. D., Davis, K. K., and Winn, B. M.: *Atlas of the Sheep Brain.* East Lansing, Michigan State University, Neuroscience Program. Available online at *http://www.msu.edu/user/brains/sheepatlas/*

Jollie, M.: *Chordate Morphology.* Huntington, NY, Robert E. Krieger Publishing Company, 1962.

Kardong, K. V.: *Vertebrates: Comparative Anatomy, Function, Evolution.* 3rd ed. Boston, McGraw-Hill, 2002.

Kent, G. C., and Carr, R. K.: *Comparative Anatomy of the Vertebrates.* 9th ed. Boston, McGraw-Hill, 2001.

Kimmel, C. B., Miller, C. T., and Keynes, R. J.: "Neural crest patterning and the evolution of the jaw." *Journal of Anatomy,* vol. 199, pp. 105–119, 2001.

Kinsbury, B. F.: "On the brain of *Necturus maculatus.*" *Journal of Comparative Neurology,* vol. 5, pp. 138–205, 1895.

Knospe, C.: "Periods and stages of the prenatal development of the domestic cat." *Anatomia, Histologia, Embryologia,* vol. 31, pp. 37–51, 2002.

König, H. E.: *Anatomie der Katze mit Hinweisen für die tierärztliche Praxis.* Stuttgart, Gustav Fischer Verlag, 1992.

Langenbach, G. E. J., and Eijden T. M. G. J.: "Mammalian feeding motor patterns." *American Zoologist,* vol. 41, pp. 1338–1351, 2001.

Lauder, G. V., and Reilly, S. M.: "Metamorphosis of the feeding mechanism in tiger salamanders *(Ambystoma tigrinum):* The ontogeny of cranial muscle mass." *Journal of Zoology (London),* vol. 222, pp. 59–74, 1990.

Lauder, G. V., and Shaffer, H. B.: "Ontogeny of functional design in tiger salamanders *(Ambystoma tigrinum):* Are motor patterns conserved during major morphological transformations?" *Journal of Morphology,* vol. 197, pp. 249–268, 1988.

Le Douarin, N.: *The Neural Crest.* New York, Cambridge University Press, 1982.

Li, W., Scott, A. P., Siefkes, M. J., Yan, H., Liu, Q., Yun, S.-S., and Gage, D. A.: "Bile acid secreted by male sea lamprey that acts as a sex pheromone." *Science,* vol. 296, pp. 138–141, 2002.

Liem, K. L., and Woods, L. P.: "A probable homologue of the clavicle in the holostean fish *Amia calva.*" *Journal of Zoology (London),* vol. 170, pp. 521–532, 1973.

Liem, K. F., Bemis, W. E., Walker, W. F. Jr., and Grande, L.: *Functional Anatomy of the Vertebrates.* 3rd ed. Fort Worth, Texas, Harcourt College Publishers, 2001.

Lindvall, M., Edvinsson, L., and Owman, C.: "Sympathetic nervous control of cerebrospinal fluid production from the choroid plexus." *Science,* vol. 201, pp. 176–178, 1978.

Lombard, R. E., and Bolt, J.: "Evolution of the tetrapod ear: An analysis and reinterpretation." *Biological Journal of the Linnean Society,* vol. 11, pp. 19–76, 1979.

Lombard, R. E., and Straughan, I. R.: "Functional aspects of anuran middle ear structures." *Journal of Experimental Biology,* vol. 61, pp. 1–23, 1974.

Maisey, J. G.: "An evaluation of jaw suspension in sharks." *American Museum Novitates,* no. 2706, pp. 1–7, 1980.

Mallatt, J.: "Ventilation and the origin of jawed vertebrates: A new mouth." *Journal of the Linnean Society (London),* vol. 117, pp. 329–404, 1996.

Marinelli, W., and Strenger, A.: *Vergleichende Anatomie und Morphologie der Wirbeltiere. I. Lieferung: Lampetra fluviatilis L. III. Lieferung: Squalus acanthias L.* Vienna, Franz Deuticke Verlag, 1954 and 1959.

McClanahan, L. L., Ruibal, R., and Shoemaker, V. H.: Frogs in the deserts. *Scientific American,* vol. 270 (March), pp. 82–88, 1994.

Meier, S., and Tam, P. L.: "Metameric pattern development in the embryonic axis of the mouse. I. Differentiation of the cranial segments." *Differentiation,* vol. 21, pp. 95–108, 1982.

Merkt, H.: "Die Bursa ovarica der Katze. Mit einer vergleichenden Betrachtung der Bursa ovarica des Hundes, Schweines, Rindes und Pferdes sowie des Menschen. Ph. D. dissertation. Hannover, Germany, University of Hannover, 1948.

Miller, M. E., Christensen, G. C., and Evans, H. E.: *Anatomy of the Dog.* Philadelphia, W. B. Saunders Company, 1964.

Miller, W. S.: "The vascular system of *Necturus maculatus.*" *University of Wisconsin Science Series,* vol. 2, pp. 211–226, 1900.

Mivart, St. G.: *The Cat. An Introduction to the Study of Backboned Animals, Especially Mammals.* London, John Murray, 1881.

Moore, J. A. (ed.): *Physiology of the Amphibia.* New York, Academic Press, 1964.

Motta, P. K., and Wilga, C. D.: "Advances in the study of feeding mechanisms, mechanics, and behavior in sharks." Environmental Biology of Fishes, vol. 60, pp. 131–156, 2001.

Moyle, P. B., and Cech, J. J.: *Fishes: An Introduction to Ichthyology.* 4th ed. Englewood Cliffs, New Jersey, Prentice-Hall, 1999.

Nelson, J. S.: *Fishes of the World.* 3rd ed. New York, John Wiley and Sons, Inc., 1994.

Nickel, R., Schummer, A., and Seiferle, E.: *The Anatomy of the Domestic Animals.* Volumes 1, 2, and 3. New York, Springer-Verlag, 1986, 1979, 1981.

Nishi, S.: "Muskeln des Rumpfes." In Bolk, L., Göppert, E., Kallius, E., and Lubosch, W. (eds.): *Handbuch der vergleichenden Anatomie der Wirbeltiere.* Vol. 5. Berlin, Urban und Schwarzenberg, 1938. Reprinted in 1967 by A. Asher and Co., Amsterdam.

Noble, G. K.: *The Biology of the Amphibia.* New York, Dover Publications, 1954.

Noden, D. M.: "The embryonic origins of avian cephalic and cervical muscles and associated connective tissues." *American Journal of Anatomy,* vol. 168, pp. 257–267, 1983.

Noden, D. M.: "Craniofacial development: New views on old problems." *Anatomical Record,* vol. 208, pp. 1–13, 1984.

Noden, D. M.: "Vertebrate craniofacial development: The relation between ontogenetic process and morphological outcome." *Brain, Behavior and Evolution,* vol. 38, pp. 190–225, 1991.

Norris, H. W., and Hughes, S. P.: "The cranial, occipital, and anterior spinal nerves of the dogfish, *Squalus acanthias.*" *Journal of Comparative Neurology,* vol. 31, pp. 293–402, 1920.

Northcutt, R. G.: "Elasmobranch central nervous system organization and its possible evolutionary significance." *American Zoologist,* vol. 17, pp. 411–429, 1977.

Northcutt, R. G.: "Ontogeny and phylogeny: A reexamination of some conceptual relationships and some application." *Brain, Behavior and Evolution,* vol. 36, pp. 116–140, 1990.

Northcutt, R. G., and Davis, R. E.: *Fish Neurobiology. Vol. 1. Brain Stem and Sense Organs.* Ann Arbor, University of Michigan Press, 1983.

Northcutt, R. G., Kenneth, L. W., and Barber, R. P.: *Atlas of the Sheep Brain.* 2nd ed. Champaign, Illinois, Stiles Publishing Company, 1966.

Nursall, J. R.: "Swimming and the origin of paired appendages." *American Zoologist,* vol. 2, pp. 127–141, 1962.

Nybakken, O. E.: *Greek and Latin in Scientific Terminology.* Ames, Iowa State University Press, 1985.

O'Donoghue, C. H., and Abbot, E. B.: "The blood vascular system of the spiny dogfish, *Squalus acanthias* Linné, and *Squalus sucklii* Gill." *Transactions of the Royal Society of Edinburgh,* vol. 55, pp. 823–894, 1928.

Oguri, M.: "Rectal glands of marine and fresh water sharks, comparative histology." *Science,* vol. 144, pp. 1151–1152, 1964.

Pang, P. K. T., Griffith, R. W., and Atz, J. W.: "Osmoregulation in elasmobranchs." *American Zoologist,* vol. 17, pp. 365–377, 1977.

Parker, T. J., and Haswell, W. A.: *Textbook of Zoology.* London, Macmillan, 1972.

Pivorunas, A.: "The feeding mechanism of baleen whales." *American Scientist,* vol. 67 (July–August), pp. 432–440, 1979.

Pough, F. H., Heiser, J. B., and McFarland, W. N.: *Vertebrate Life.* 4th ed. Upper Saddle River, New Jersey, Prentice Hall, 1996.

Prosser, C. L. (ed.): *Neural and Integrative Animal Physiology, Comparative Animal Physiology.* 4th ed. New York, John Wiley and Sons, Inc., 1991.

Prosser, C. L. (ed.): *Environmental and Metabolic Animal Physiology, Comparative Animal Physiology.* 4th ed. New York, John Wiley and Sons, Inc., 1991.

Randall, D., Burggren, W., and French, K.: *Eckert Animal Physiology: Mechanisms and Adaptations.* 5th ed. New York, W. H. Freeman, 2002.

Ranson, S. W.: *The Anatomy of the Nervous System.* 10th ed. Revised by S. L. Clark. Philadelphia, W. B. Saunders Company, 1959.

Rasmussen, A. T.: *The Principal Nervous Pathways.* 4th ed. New York, The Macmillan Company, 1952.

Reighard, J. E., and Jennings, H. S.: *Anatomy of the Cat.* 3rd ed. Revised by R. Elliot. New York, Henry Holt and Company, 1935.

Robson, P., Wright, G. M., Sitarz, E., Maiti, A., Rawat, M., Youson, J. H., and Keeley, F. W.: "Characterization of lamprin, an unusual matrix protein from lamprey cartilage." *Journal of Biochemistry,* vol. 268, pp. 1440–1447, 1993.

Romer, A. S.: *Osteology of Reptiles.* Chicago, University of Chicago Press, 1956.

Romer, A. S., and Parsons, T. S.: *The Vertebrate Body.* 6th ed. Philadelphia, Saunders College Publishing, 1986.

Ruppert, E. E.: "Structure, ultrastructure and function of the neural gland complex of *Ascidia interrupta* (Chordata, Ascidiacea): Clarification of hypotheses regarding the evolution of the vertebrate anterior pituitary." *Acta Zoologica (Stockholm),* vol. 71, pp. 135–149, 1990.

Ruppert, E. E.: "Evolutionary origin of the vertebrate nephron." *American Zoologist,* vol. 34, pp. 542–553, 1994.

Ruppert, E. E., and Smith, P. R.: "The functional organization of filtration nephridia." *Biological Reviews,* vol. 63, pp. 231–258, 1988.

Sadler, T. W.: *Langman's Medical Embryology.* 8th ed. Baltimore, Williams & Wilkins, 2000.

Sanderson, S. L., and Wassersug, R.: "Suspension-feeding in vertebrates." *Scientific American,* vol. 262 (March), pp. 96–101, 1990.

Satchell, G. H.: *Circulation in Fishes.* Cambridge, Cambridge University Press, 1971.

Satchell, G. H.: *Physiology and Form of Fish Circulation.* Cambridge, Cambridge University Press, 1991.

Schmidt-Nielsen, K.: *Animal Physiology: Adaptation and Environment.* 5th ed. Cambridge, Cambridge University Press, 1997.

Schmidt-Nielsen, K., Bolis, L., and Taylor, C. R. (eds): *Comparative Physiology: Primitive Mammals.* Cambridge, Cambridge University Press, 1980.

Schwenk, K.(ed.): *Feeding: Form, Function, and Evolution in Tetrapod Vertebrates.* San Diego, California, Academic Press, 2000.

Shaffer, H. B. and Lauder, G. V.: " The ontogeny of functional design: Metamorphosis of feeding behavior in the tiger salamander *(Ambystoma tigrinum).*" *Journal of Zoology,* vol. 216, pp. 437–454, 1988.

Shuttleworth, T. J. (ed.): *Physiology of Elasmobranch Fishes.* Berlin, Springer-Verlag, 1988.

Simons, R. S., Bennett, W. O., and Brainerd, E. "Mechanics of lung ventilation in a post-metamorphic salamander, *Ambystoma tigrinum.*" *Journal of Experimental Biology,* vol. 203, pp. 1081–1092, 2000.

Sivak, J. G.: "Elasmobranch visual optics." *Journal of Experimental Zoology Supplement,* vol. 5, pp. 13–21, 1991.

Slijper, E. J.: "Comparative biological-anatomical investigations on the vertebral column and spinal musculature of mammals." *Koninklijke Nederlandse Academie Van Wetenschappen (Tweede Sectie),* vol. 42, pp. 1–128, 1946.

Smeets, W. J. A. J., and Nieuwenhuys, R.: "Topological analysis of the brain stem of the sharks *Squalus acanthias* and *Scyliorhinus canicula.*" *Journal of Comparative Neurology,* vol. 165, pp. 333–368, 1976.

Smith, K. K., and Kier, W. M.: "Trunks, tongues, and tentacles: Moving with skeletons of muscle." *American Scientist,* vol. 77, pp. 29–35, 1989.

Sordino, P., van de Hoeven, F., and Duboule, D. F.: "Hox gene expression in teleost fins and the origin of digits." *Nature,* vol. 375, pp. 678–681, 1995.

Starck, D.: *Embryologie: Ein Lehrbuch auf allgemein biologischer Grundlage.* 3rd ed. Stuttgart, Georg Thieme Verlag, 1975.

Starck, D.: *Vergleichende Anatomie der Wirbeltiere auf evolutionsbiologischer Grundlage.* Volumes 1, 2, and 3. Berlin, Springer-Verlag, 1978, 1979, 1982.

Stedman, T. L.: *Stedman's Medical Dictionary.* 27th ed. Philadelphia, Lippincott, Williams & Wilkins, 2000.

Strother, G. K.: *Physics with Applications in Life Sciences.* Boston, Houghton Mifflin Company, 1977.

Terminologia Anatomica. See: Federative Committee on Anatomical Terminology.

Thomason, J. J., and Russell, A. P.: "Mechanical factors in the evolution of the mammalian secondary palate: A theoretical analysis." *Journal of Morphology,* vol. 189, pp. 199–213, 1986.

Thomson, K. S.: "The shape of a shark's tail." *American Scientist,* vol. 78, pp. 499–501, 1990.

Thomson, K. S., and Simanek, D. E.: "Body form and locomotion in sharks." *American Zoologist,* vol. 17, pp. 342–354, 1977.

Ullian, E. M., Sapperstein, S. K., Christopherson, K. S., and Barres, B. A.: "Control of synapse number by glia." *Science,* vol. 291, pp. 657–661, 2001.

Villee, C. A., Walker, W. F., and Barnes, R. D.: *General Zoology.* 6th ed. Philadelphia, Saunders College Publishing, 1984.

Villee, C. A., Solomon, E. P., Martin, C. E., Martin, D. W., Berg, L. R., and Davis, P. W.: *Biology.* 2nd ed. Philadelphia, Saunders College Publishing, 1989.

Wagner, G. P., and Chiu, C. -H.: "The tetrapod limb: A hypothesis on its origin." *Journal of Experimental Zoology,* vol. 291, pp. 226–240, 2001.

Wainwright, S. A., Vosburgh, F., and Hebrank, J. H.: "Shark skin: Function in locomotion." *Science,* vol. 202, pp. 747–749, 1978.

Wake, M. H. (ed.): *Hyman's Comparative Vertebrate Anatomy.* 3rd ed. Chicago, University of Chicago Press, 1979.

Walker, W. F.: *Dissection of the Frog.* 2nd ed. San Francisco, W. H. Freeman and Company, 1981.

Walker, W. F., and Homberger, D. G.: *A Study of the Cat with Reference to Human Beings.* 5th ed. Philadelphia, Saunders College Publishing, 1993.

Walker, W. F., and Homberger, D. G.: *Anatomy and Dissection of the Rat.* 3rd ed. New York, W. H. Freeman and Company, 1997.

Walker, W. F., and Homberger, D. G.: *Anatomy and Dissection of the Fetal Pig.* 5th ed. New York, W. H. Freeman and Company, 1998.

Walls, G. L.: *The Vertebrate Eye and Its Adaptive Radiation.* Bloomfield Hills, Michigan, Cranbrook Institute of Science, Bulletin No. 19, 1942.

Webb, P. W., and Weihs, D. (eds.): *Fish Biomechanics.* New York, Praeger Publishers, 1983.

Werdelin, L., and Nilsonne, A.: "The evolution of the scrotum and testicular descent in mammals: A phylogenetic view." *Journal of Theoretical Biology,* vol. 196, pp. 61–72, 1999.

Wheeler, R. A., Baldwin, A. E., Reid, R. S., Quinn, I. J., and Cannon, J. T.: *Dissection of the Sheep Brain.* Scranton, Pennsylvania, University of Scranton, Behavioral Neurosciences Laboratory, 1996. Available online at *http://academic.uofs.edu/department/psych/sheep/pge2.html*

Wiedersheim, R.: *Vergleichende Anatomie der Wirbeltiere.* 6th enlarged edition. Jena, Gustav Fischer, 1906.

Wilder, H. H.: "The skeletal system of *Necturus maculatus* Rafinesque." *Memoirs of the Boston Society of Natural History,* vol. 5, pp. 387–439, 1903.

Wilder, H. H.: "The appendicular muscles of *Necturus maculosus.*" *Zoologische Jahrbücher,* suppl. 15, pt. 2, pp. 383–424, 1912.

Wilga, C. D.: "A functional analysis of jaw suspension in elasmobranchs." *Biological Journal of the Linnean Society,* vol. 75, pp. 483–502, 2002.

Wilga, C. D., and Lauder, G. V.: "Function of the heterocercal tail in sharks: Quantitative wake dynamics during steady horizontal swimming and vertical maneuvering." *Journal of Experimental Biology,* vol. 205, pp. 2365–2374, 2002.

Wilga, C. D., and Motta, P. J.: "Conservation and variation in the feeding mechanisms of the spiny dogfish, *Squalus acanthias.*" *Journal of Experimental Biology,* vol. 201, pp. 1345–1358, 1998.

Wilga, C. D., Wainwright, P. C., & Motta, P. J.: "Evolution of jaw depression mechanics in aquatic vertebrates: Insights from Chondrichthyes." *Biological Journal of the Linnean Society,* vol. 71, pp. 165–185, 2000.

Wilga, C. D., Hueter, R. E., Wainwright, P. C., and Motta, P. J.: "Evolution of upper jaw protrusion mechanisms in elasmobranches." *American Zoologist,* vol. 41, pp. 1248–1257, 2001.

Wilkens, H.: "Zur Topographie der Verdauungsorgane des Schafes unter besonderer Berücksichtigung von Funktionszuständen." *Zentralblatt für Veterinärmedizin,* vol, 3, pp. 803–816, 1956.

Williams, P. L., Bannister, L. M., Berry, M. M., Collins, P., Dyson, M., Dussek, J. E., and Ferguson, M. W. J. (eds.): *Gray's Anatomy.* 38th ed. New York, Churchill Livingstone, 1995.

Williamson, R. M., and Roberts, B. L.: "Sensory and motor interactions during movement in the spinal dogfish." *Proceedings of the Royal Society of London, Series B,* vol. 227, pp. 103–119, 1986.

Wilson, J. M., and Laurent, P.: "Fish gill morphology: Inside out." *Journal of Experimental Zoology,* vol. 293, pp. 192–213, 2002.

Wineski, L. E., and English, A. W.: "Phenoxyethanol as a nontoxic preservative in the dissection laboratory." *Acta Anatomica,* vol. 136, pp. 155–158, 1989.

Withers, P. C.: *Comparative Animal Physiology.* Fort Worth, Texas, Saunders College Publishing, 1992.

Wood, C. M., Part, P., and Wright, P. A.: "Ammonia and urea metabolism in relation to gill function and acid-base balance in a marine elasmobranch, the spiny dogfish, *Squalus acanthias.*" *Journal of Experimental Biology,* vol. 198, pp. 1545–1558, 1995.

Woollard, H. L.: "The development of the principal arterial system in the fore-limb of the pig." *Carnegie Institute Contributions to Embryology,* vol. 14, pp. 139–154, 1922.

Wourms, J. P.: "Reproduction and development in chondrichthyan fishes." *American Zoologist,* vol. 17, pp. 379–410, 1977.

Yoshikawa, T.: *Atlas of the Brains of Domestic Animals.* University Park, Pennsylvania State University Press, 1968.

Young, J. Z.: "The autonomic nervous system of selachians." *Quarterly Journal of Microscopical Science,* vol. 75, pp. 571–624, 1933.

Young, J. Z.: *The Life of Mammals.* Oxford, Clarendon Press, 1963.

Young, J. Z.: *The Life of Vertebrates.* 3rd ed. London, Oxford University Press, 1981.

Young, J. Z., and Hobbs, M. J.: *The Life of Mammals.* 2nd ed. New York and Oxford, Oxford University Press, 1975.

Zangerl, R.: "Chondrichthyes: I. Paleozoic Elasmobranchs." In Schultz, H. P. (ed.): *Handbook of Paleontology.* Vol. 3. Stuttgart, Gustav Fisher, 1981.

Zietzschmann, O., and Krölling, O.: *Lehrbuch der Entwicklungsgeschichte der Haustiere.* Berlin and Hamburg, Paul Parey, 1955.

INDEX

Terms are indexed under their defining noun, e.g., artery, bone, canal, cartilage, cavity, duct, fascia, fenestra, foramen, fossa, ganglion, gland, ligament, muscle, nerve, process, vein, etc., as well as by name. These lists of arteries and other major organs may also be useful for review. Organ systems are indexed by name or by major part (e.g., skull), and under mammals, *Necturus, Squalus,* etc. Most figures are indexed under the name of the animal illustrated. References to figures are indicated by (f); references to tables by (t).

O

S